Agribusiness

FUNDAMENTALS AND APPLICATIONS

2nd Edition

FFA

Agribusiness

FUNDAMENTALS AND APPLICATIONS
2nd Edition

Dr. Cliff Ricketts

Dr. Kristina Ricketts

DELMAR
CENGAGE Learning

Australia • Brazil • Japan • Korea • Mexico • Singapore • Spain • United Kingdom • United States

Agribusiness: Fundamentals and Applications, 2nd Edition

Cliff Ricketts and Kristina Ricketts

Vice President, Career and Professional Editorial: **Dave Garza**

Director of Learning Solutions: **Matthew Kane**

Acquisitions Editor: **David Rosenbaum**

Managing Editor: **Marah Bellegarde**

Product Manager: **Christina Gifford**

Editorial Assistant: **Scott Royael**

Vice President, Career and Professional Marketing: **Jennifer McAvey**

Marketing Director: **Deborah Yarnell**

Marketing Coordinator: **Jonathan Sheehan**

Production Director: **Carolyn Miller**

Production Manager: **Andrew Crouth**

Content Project Manager: **Kathryn B. Kucharek**

Art Director: **Dave Arsenault**

Technology Project Manager: **Mary Colleen Liburdi**

Production Technology Analyst: **Tom Stover**

Library of Congress Control Number: 2007942523

ISBN-13: 978-1-4180-3231-9

ISBN-10: 1-4180-3231-X

Delmar
5 Maxwell Drive
Clifton Park, NY 12065-2919
USA

Cengage Learning products are represented in Canada by Nelson Education, Ltd.

For your lifelong learning solutions, visit **delmar.cengage.com**

Visit our corporate website at **cengage.com**

2 3 4 5 6 7 12 11 10 09

This book is dedicated to my family: my parents, Hall and Louise Ricketts, who were production agriculturalists during the Great Depression and overcame great odds to leave a rich heritage for their family and the agribusiness community; my wife, Nancy; and my three children, John, Mitzi, and Paul, who are all employed in the world of agribusiness.

DR. CLIFF RICKETTS

This book is also dedicated to my family, who have been a great source of love and inspiration.

DR. KRISTINA RICKETTS

Contents

Preface

Agriculture has changed. For years, agriculture was a career and a way of life. Nearly everyone lived on farms, and farmers were self-sufficient. Very few manufactured supplies and materials were available. Today, however, the agricultural industry is a technology-oriented industry that includes production, agriscience, and agribusiness. Agriculture is a big business, and agribusiness itself is massive. This book discusses the kinds, sizes, fundamentals, and applications of this phase of the agricultural industry.

Most people who work in the agricultural industry do not work on farms and ranches, but rather are employed in the feed, seed, farm machinery, fertilizer, and chemical supply businesses, which make up the input sector. There are also many agribusiness workers in marketing firms that move food and fiber from production agriculturalists to consumers; these businesses form the output sector. This book also explains different agribusiness areas, discusses their economic importance to the agricultural industry, and explains basic principles of agricultural economics, marketing, and finance.

Introduction to Agribusiness is divided into four major sections. Unit 1 includes a comparison of agriculture and agribusiness, which sets the stage for the book. The history of agriculture is briefly discussed, including the evolution of agribusiness and today's technology. Agribusiness is explained in terms of where it fits into the "big picture" of the agricultural industry and its relationship with production agriculture and agriscience. The size, importance, and impact of agribusiness on the economy are discussed.

Unit 2 includes eight chapters on starting and running an agribusiness. They provide the information necessary for students to understand what is involved in owning and operating an agribusiness. Chapters include information on planning and organizing an agribusiness, types of agribusinesses, financing an agribusiness, personal financial management, recordkeeping and accounting, managing human resources, farm management, and production economics.

Unit 3 explains the agribusiness inputs that go to production agriculturalists. The input sector of agribusiness includes four major areas (a chapter is devoted to each): agribusiness supplies (feed, seed, fertilizer, and chemicals), farm machinery, and equipment; economic activity and analysis; agricultural policy and governmental agribusiness services, such as the USDA and state departments of agriculture; and private agribusiness services, such as farm and commodity organizations.

Unit 4 explains the agribusinesses that are involved with agricultural commodities and products once they have left the farm (the agribusiness output sector). Chapters are included on basic principles of marketing, commodity (futures) marketing, international agriculture marketing, and the agrimarketing channels through which products move from the farm to the consumer.

Throughout, the basic economic principles involved in agribusiness are emphasized and explained. Information includes the economic principles that a national association on economic education says should be taught.

Pedagogical Features

Each chapter begins with a list of performance objectives, which relate to the main topics of each chapter. The objectives are followed by a list of "Terms to Know," to enable students to better understand the chapter content. The definitions of all the terms are found in the glossary, and in most cases within the unit. Each chapter ends with a conclusion and a summary. Also at the end of each chapter are review questions, which encourage students to read the whole chapter, and fill in the blank and matching questions, which are designed to reinforce students' understanding of important ideas.

Activities are the last feature of each chapter. These are application exercises that enable students to apply newly learned concepts and practices, and also to draw on their own experiences and share those with others.

About the Authors

Dr. Cliff Ricketts came through the ranks of agricultural education and the Future Farmers of America; he is a product of that program. He has achieved every degree offered by the FFA, including the Greenhand, Chapter, State, and American FFA (Farmer) degrees; Honorary Chapter; Honorary State; and Honorary American FFA; the last of which was received the same year by former President George H.W. Bush and Lee Iacocca. As a high school agricultural education instructor, his students won the State FFA Parliamentary Procedure Contest, State FFA Creed Speaking Contest (four times), and were awarded runner-up in Public Speaking and Extemporaneous Speaking. Ricketts was also a recipient of the National Vocational Agriculture Teachers Association's National Outstanding Teacher Award while a high school teacher.

Ricketts received his B.S. and M.S. degrees from the University of Tennessee and his Ph.D. from Ohio State University. While teaching at Middle Tennessee State University (MTSU), he received both the MTSU Foundation's Outstanding Teacher Award (twice) and its Outstanding Public Service Award. He was also a finalist for its Outstanding Research Award.

Ricketts is still involved in production agriculture with a beef cattle operation on his family farm 20 miles east of Nashville, Tennessee. His children, John, Mitzi, and Paul, are the fourth generation on the family farm. Cliff Ricketts is also involved with alternative fuels research and has run engines off ethanol from corn, methane from cow manure, soybean oil, and hydrogen from water. He and his students from Middle Tennessee State University presently hold the world's land speed record for a hydrogen-fueled vehicle.

Dr. Kristina Ricketts received her B.S. and M.S. degrees from the University of Nebraska and her Ph.D. from the University of Florida. She has taught at Pennsylvania State University and is presently teaching at the University of Kentucky. Dr. Kristina Ricketts was actively involved with the college livestock judging teams, and one of her specialties is agricultural leadership. She is married to Paul Ricketts and has a daughter, Ava.

Acknowledgments

The authors are appreciative of several people who had input into this book as a resource for course content or as a consultant.

Beth Cripps, Hendersonville High School

Keith Gill, Lincoln County High School

Becky Thomas, Middle Tennessee State University

Carol Gardner, Middle Tennessee State University

Dawn Mosley, Middle Tennessee State University

Glenn Ross, McEwen High School

Michael Barry, University of Tennessee Extension Service

Kim Webb, Tennessee Farmers Cooperative

Janet Kelly, Middle Tennessee State University

Jamie Mundy, Powell Valley High School

Chaney Mosley, Middle Tennessee State University

Debbie Weston, Lochinvar Corporation

Mitzi Ricketts, Beech High School

Julie Shew, Middle Tennessee State University

The authors and Delmar Cengage Learning also wish to express their appreciation to the following reviewers:

Colleen Geurink
Marathon High School
Marathon, WI

Rebecca Carter
Essex High School
Tappahannock, VA

Glenn Alexander
Grand Ridge High School
Grand Ridge, FL

Stacie D. Rhonemus
Spencerville, OH

Mike Cambell
Imperial High School
Imperial, CA

Lastly, we would like to express our gratitude to Brooke Graves for her hard work and dedication in bringing this product to completion.

UNIT 1

Introduction to Agribusiness

Chapter 1
Agriculture and Agribusiness

Chapter 2
Agricultural Economics and the American Economy

Chapter 3
The Size and Importance of Agribusiness

Chapter 4
Emerging Agribusiness Technologies

Chapter 1

Agriculture and Agribusiness

Objectives

After completing this chapter, the student should be able to:

- Explain agribusiness.
- Describe the "big picture" of agribusiness.
- Explain how agribusiness affects us daily.
- Discuss farming and agriculture before agribusiness.
- Discuss the beginning of agribusiness in America.
- Describe the historical development of farm machinery and equipment.
- Describe the Steam Era.
- Discuss the historical development of the internal combustion engine.
- Discuss the historical development of farm tractors.
- Discuss the success of American agribusiness.

Terms to Know

A.D.
agribusiness
agriscience
agronomic
B.C.
biotechnology
compression
Cooperative Extension Service
cotton gin
crawler-type
crop rotation
cultivate
cylinder
Dairy Herd Improvement Association (DHIA)
domesticate
draft animals
drought
exports
fallow
Federal Land Banks
futures
hog cholera
hybrid
hydraulic lift
hydrostatic transmission
implements
indigo
input
irrigate
mechanical power
organic fertilizer

continued

output	resources	tertiary
pesticides	selective breeding	threshing machine
pneumatic	sickle	torque amplification
power take-off (PTO)	Soil Conservation Service (SCS)	tricycle-type
production agriculturalists	surveying	turbochargers
raw material	terminal	turpentine
reaper		vaccines

Introduction

When most people think of agriculture, they picture farmers producing animals and crops. Agriculture is often thought of as "cows, sows, and plows" or "weeds, seeds, and feeds." But agriculture has changed. For years agriculture was farming, as previously stated. Nearly everyone lived on farms, and farmers were self-sufficient. Very few manufactured supplies and materials were available. Today, however, agriculture is a technology-oriented industry that includes production, **agriscience**, and agribusiness.

Most people who work in the agricultural industry do not work on farms and ranches, but rather are employed in the feed, seed, farm machinery, fertilizer, chemical supply, and food-processing businesses. There are also many agribusiness workers in finance, distribution, and marketing firms who provide service to **production agriculturalists**. Agriculture is big business.

What Is Agribusiness?

There are many different definitions of **agribusiness**. According to *Merriam-Webster's Collegiate Dictionary*, agribusiness is "an industry engaged in the producing operations of a farm, the manufacture and distribution of farm equipment and supplies, and the processing, storage, and distribution of farm commodities."[1] As this definition implies, some people interpret the word narrowly to mean only very large businesses within the agricultural industry. However, John Davis and Ray Goldberg, in their early research on agribusiness, defined it as all operations involved in the manufacture and distribution of farm supplies; production operations on the farm; and the storage, processing, and distribution of the resulting commodities and items.[2] A similar definition of agribusiness describes it as any profit-motivated enterprise that involves providing agricultural supplies and/or the processing, marketing, transporting, and distributing of agricultural materials and consumer products.[3] Ewell Roy defined agribusiness as the "coordinating science of supplying agricultural production inputs and subsequently producing, processing, and distributing food and fiber."[4]

The Missouri Department of Agriculture provided an expanded definition of agribusiness, which it considers an all-encompassing term that comprises many activities. Farmers and ranchers who produce food, fiber, and other raw materials lay the foundation for a complex series of actions thereafter: following production, processors, handlers, transportation agents, wholesalers, and retailers perform various activities on the raw products. Some of these middlemen take farm-produced raw materials and change them into more useable forms through processing; others move them around the country so that wholesalers and retailers can sell finished products to consumers.[5]

1. *Merriam-Webster's Collegiate Dictionary*, 11th ed. (Springfield, MA: Merriam-Webster, Inc., 2003).
2. John H. Davis and Ray A. Goldberg, *A Concept of Agribusiness* (Boston: Harvard Business School, Research Division, 1957).
3. Joseph J. Timka and Robert J. Birkenholz, *Introduction to Agribusiness Unit* (Columbia, MS: Instructional Material Laboratory, 1984).
4. Ewell P. Roy, *Exploring Agribusiness* (Danville, IL: Interstate Printers & Publishers, 1980), p. 1.
5. Missouri Department of Agriculture, http://www.mda.mo.gov (accessed 2003; publication now discontinued).

An inclusive definition of agribusiness was provided by the Australian Department of Agriculture, Fisheries and Forestry. The DAFF called the agribusiness sector a "chain" of industries directly or indirectly involved in the production, transformation, or provision of food, fiber, and chemical and pharmaceutical substrates. "Links" in the agribusiness chain include:

- primary production of raw materials ("commodities"), such as unprocessed food, fiber, and substrates
- **tertiary** transformation of commodities into value-added products where the value is derived from the process of transformation
- supply of inputs to the primary and tertiary sectors
- wholesale and retail provision of processed or unprocessed foods, fibers, and related products to consumers
- provision of educational, financial, and technical services to all sectors

Therefore, agribusiness encompasses all activities "from the paddock to the consumer" that contribute to the eventual production, processing (value-addition), distribution, and retailing of food, fiber, and products based on food or fiber.[6]

A more concise definition calls agribusiness "a generic term that refers to the various businesses involved in food and fiber production, including farming, seed supply, agrichemicals, farm machinery, wholesale and distribution, processing, marketing, and retail sales."[7] All these definitions agree that agribusiness includes all the activities that take place in the production, manufacturing, distribution, and wholesale and retail sales of agricultural commodities. Modern agribusiness is a dynamic and growing industrial complex that provides Americans with the highest-quality, lowest-cost food supply in the world.

Is Farming an Agribusiness?

Some authorities exclude "farming" or the "production" of food and fiber from the definition of agribusiness.[8] If we include production agriculture, agriscience, and agribusiness under the umbrella of the agricultural industry, we must include production agriculture (farming) within the context of this book. We would be mistaken to think that farming is not a business. A production agriculturalist must make decisions, develop plans, and solve problems, all of which require business-related skills. A typical farmer or production agriculturist manages interest, taxes, repair and replacement of equipment, fertilizers, wages, fuel, electricity, and many other items.[9] As you can see, production agriculturalists must be financial and business managers or they will fail. Production agriculture is indeed a business, but in the context of this book farming is addressed directly only in Chapter 11 (Farm Management) and Chapter 12 (Production Economics).

Is Agribusiness the Same as Agricultural Economics?

Agricultural economics refers to the monetary and physical factors that affect the profitability of the agribusiness. According to the American Agricultural Economics Association, "Agricultural economics is the study of the economic forces that affect the food and fiber industry. Specific areas of study in agricultural economics include:

- community and rural development
- food safety and nutrition
- international trade
- natural resource and environmental economics
- production economics
- risk and uncertainty
- consumer behavior and household economics
- analysis of markets and competition
- agribusiness economics and management

Agricultural economists can be found at every level of business, government, and education around the globe.[10]

6. Australian Department of Agriculture, Fisheries and Forestry, http://www.daff.gov.au (accessed 2004; no longer available on this site).
7. http://www.agribusiness.asn.au/index.php?h=biotechnology &content=biotechnology_biotechresources.php# (accessed August 2, 2007).
8. Roy, *Exploring Agribusiness*, p. 1.
9. Alfred H. Krebs and Michael E. Newman, *Agriscience in Our Lives*, 6th ed. (Danville, IL: Interstate Publishers, 1994), p. 13.
10. American Agricultural Economics Association, http://www.aaea.org/info/whatis (accessed August 10, 2007).

Figure 1–1 This Tractor Supply Company (TSC) store is an example of an agribusiness input supply company. Many agribusiness inputs are sold to production agriculturalists. (Courtesy of Tractor Supply Company)

Figure 1–2 Trucks are delivering milk to a milk-producing plant. Both the trucking and the milk-producing plant are examples of agricultural output. (Courtesy of Cliff Ricketts)

The Big Picture of Agribusiness

Agribusiness companies provide **input** supplies to the production agriculturalist (farmer), as shown in Figure 1–1. The production agriculturalist produces food and fiber (cotton, wool, etc.), and the **output** is taken by agribusiness companies that process, market, and distribute the agricultural products, as shown in Figure 1–2.

Many services are needed in agriculture, such as transportation, storage, refrigeration, credit, finance, and insurance. Agribusiness manufacturers furnish production agriculturalists with the supplies and equipment needed to produce, store, and transport their crops. Government agencies inspect and grade agricultural products for quality and safety.[11] Hundreds of agribusiness trade organizations, commodity organizations, committees, and conferences educate, promote, advertise, coordinate, and lobby for their agricultural products. Science, research, engineering, and education help improve agribusiness. Millions of people are employed in agribusiness throughout the world, and people throughout the world also depend on agribusiness for their food, clothing, and shelter. Refer to Figure 1–3 for the "Big Picture of Agribusiness."

11. Marcella Smith, Jean M. Underwood, and Mark Bultmann, *Careers in Agribusiness and Industry*, 4th ed. (Danville, IL: Interstate Publishers, 1991), p. xiv.

Agribusiness Affects Us Daily

Consider one of the best-selling fast-food items, the cheeseburger. To get an idea of how agribusiness affects our daily lives, imagine what is involved in assembling a cheeseburger with all the trimmings (refer to Figure 1–4).[12]

Many items had to be brought together to create the cheeseburger, such as land, labor, seeds, fertilizer, chemicals, machinery, credit, transportation, and marketing, among other **resources**. In addition, agribusiness plants had to be built to process the food products that go into the cheeseburger from their **raw material** state. Many engineering and technological systems were used to obtain the needed quality and standardization. When the various ingredients that go into making the cheeseburger left the processing plants, they were transported by air, ground, or water, requiring airplanes and runways, trucks and roads, railroads and tracks, barges and waterways, and communications and employees. As these food products arrived in your community, wholesalers and distributors needed receiving **terminals**, depots, or warehouses to store the products for distribution in smaller quantities to retail food stores and restaurants.[13] As you can see, agribusiness is an essential part of our lives and crucial to the economy of the United States.

12. Jasper S. Lee, *Working in Agricultural Industry* (New York: Gregg Division, McGraw-Hill, 1978), p. 15.
13. Roy, *Exploring Agribusiness*, p. 2.

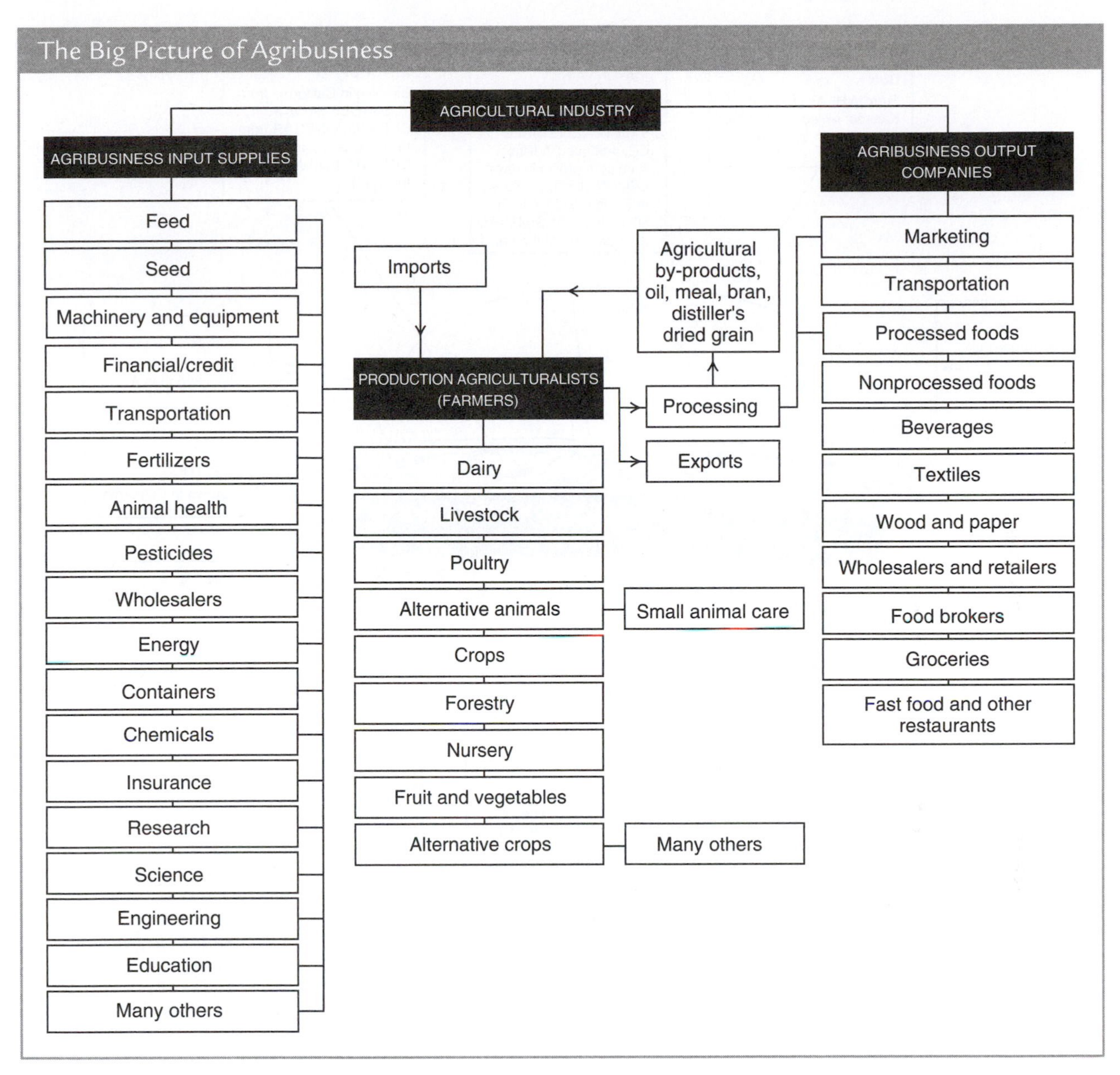

Figure 1–3 The agricultural industry is massive. This flowchart illustrates where agribusiness fits into the "big picture" of the agricultural industry.

Farming and Agriculture before Agribusiness

People have searched for ways to feed themselves since prehistoric times. If people did not eat one day, they would hardly have enough energy to find food the next day. Thus, nearly all their waking time was spent searching for food by hunting or gathering nuts and other naturally grown foods. As farming developed, it gave people time to develop other things, such as language, art, and religion. Today, if you want to eat, you need only go to the nearest fast-food restaurant to get a cheeseburger or another favorite food.

Life before Agriculture

Several thousand years ago, early humans roamed great distances to gather plants and hunt animals for food. Nutrition and health were so poor that people seldom lived past 25 years of age. They hunted large game, but they were not very successful. They had more luck gathering vegetables

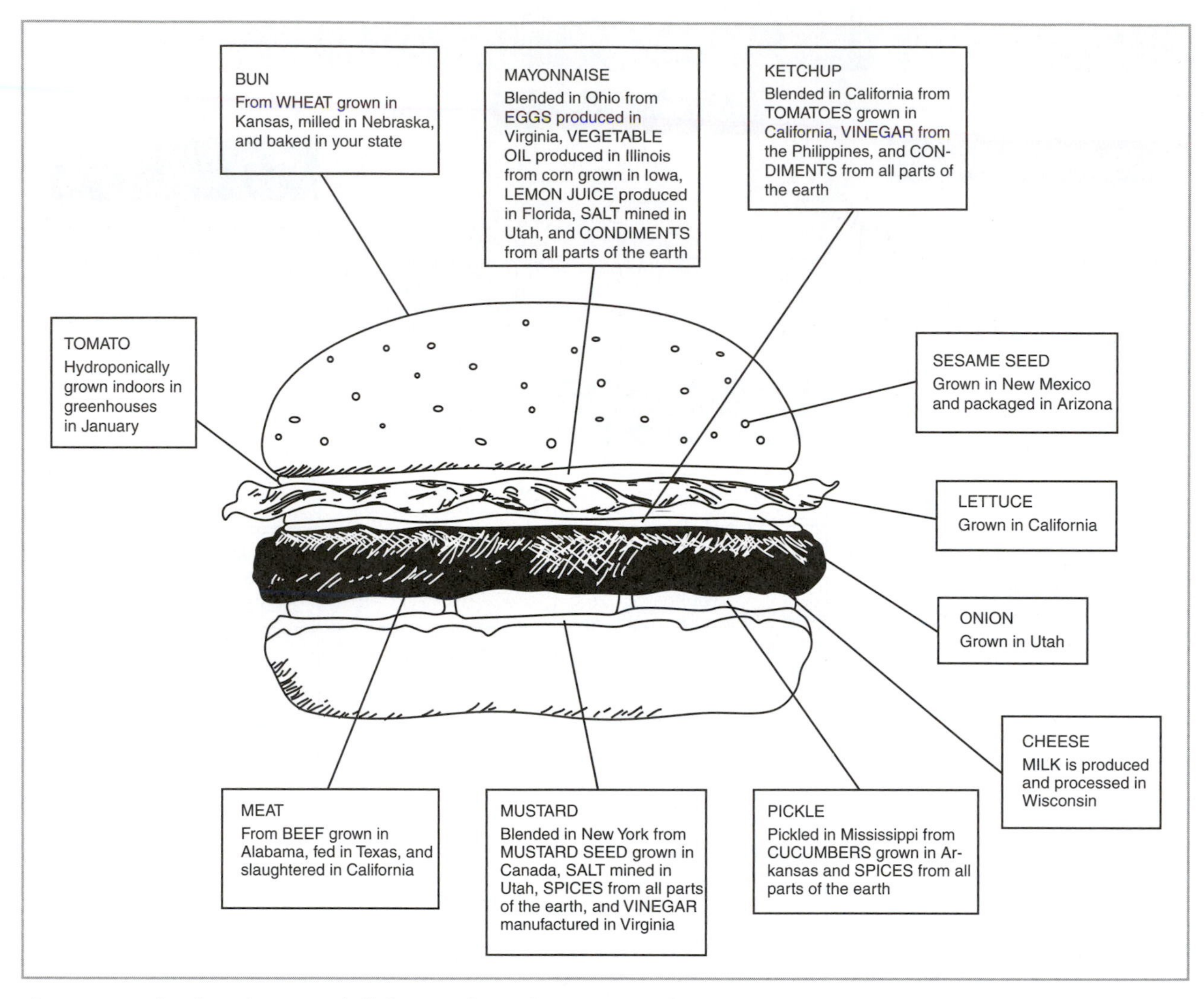

Figure 1–4 The cheeseburger and all the ingredients that go into it illustrate the vastness, diversity, and closeness to our everyday lives of agriculture.

and insects. Insects were a favorite. Early humans considered them so tasty that some species were almost exterminated. Common wild vegetables included turnips, onions, radishes, squash, cabbage, and mushrooms. However, two important developments took place before humans discovered agriculture. People learned to fish, and they learned to use fire for cooking. Civilization, however, evolved very slowly until people learned to farm.

Early Agricultural Development

To many people, the idea of farming seems simple. All one does is plant some seed, wait a while, and then harvest the food. But prehistoric humans had a problem with growing food. There was no one around to teach them that seed would grow into plants. Eventually, however, the puzzle to farming was solved and people began to raise crops and **domesticate** animals.

Farming changed the way people lived. They no longer had to roam great distances in search of food. Instead, they could settle in one spot and build a home. They also had time to perfect methods of farming. Simple sticks evolved into wooden hoes and plows. Some clever person invented the **sickle** to make the harvesting of grain and grasses easier.[14]

14. *Introduction to World Agricultural Science and Technology* (8379B-2) (College Station, TX: Instructional Materials Service, 1989), p. 1.

The Bronze Age

During the Bronze Age (around 3,000 **B.C.**), wooden **implements** were made sharper and more durable by using metal. This allowed people to **cultivate** larger areas of land faster. Agriculture spread throughout the world and became a way of life for most people. Some of the developments during the Bronze Age included:

- Bronze tools and plows made for easier and faster farming.
- The Nile River was used by Egyptians to **irrigate** crops.
- The wheel was discovered, making the transport of crops possible.
- World population rose from 3 million before the invention of agriculture to nearly 100 million.[15]

The Iron Age

The year 1000 B.C. was the dawn of the Iron Age. The use of iron gave people the ability to produce even more crops. When people could not use all the crops themselves, they had to do something with them, and trade among people developed as a result. Other developments of the Iron Age were:

- Iron hand tools and plows were created, some of which are similar to those used today.
- Money was developed because of the need to trade excess crops.
- Leaving land **fallow** gave the soil a chance to rebuild and store moisture.[16]

The Middle Ages (A.D. 400–1500)

The Middle Ages ranged from **A.D.** 400 to A.D. 1500. The fall of the Roman Empire slowed the growth of agriculture. Farming was still a way of life, but only a few important developments occurred. These included **crop rotation**, a new harness for plowing, and **selective breeding** of livestock.

Farmers in the Middle Ages began to understand the importance of conserving soil moisture and nutrients. They accomplished this by dividing their fields and leaving parts bare during certain years. This innovation led to the development of fences to mark separate fields.

15. Ibid.
16. Ibid.

Before the Middle Ages, oxen were the main **draft animals**. A new harness was developed so that farmers could use horses to work their fields faster. Also, during the late Middle Ages, livestock producers began to select certain animals for breeding based on desirable characteristics. Before the development of this practice, there were few different breeds of domestic animals. Using selective breeding, livestock producers developed many of the breeds we now have.

Near the end of the Middle Ages, Columbus discovered America, and people began to travel to the "New World." Until this time, most agricultural developments had come from Europe and Asia. Now there was an opportunity to develop American agriculture.[17]

The Evolution of Farming and Agribusiness in America

As the first European settlers became familiar with America, they encountered Native Americans who hunted, fished, and raised crops, several of which are still important today, including corn, beans, squashes, pumpkins, tomatoes, tobacco, and cotton. Both sweet and white potatoes were brought to North America by South American Indians around the same time the Pilgrims landed in New England.[18]

Early American agriculture was a trial-and-error process for the Pilgrims, who found initial attempts to feed their families difficult. American Indians had developed crop and tillage practices different from those of European farmers, and served as mentors to the early settlers. In fact, the first agriculture educators in American were the Native Americans, and their students were the Pilgrims. American Indians in New England advised settlers to bury fish under each hill of corn. Although they would not have known that the decaying fish gave off nitrogen, they did know that it helped the corn to grow.[19]

Although early American agriculture provided much of what the settlers needed, they did not gain economically from their newfound ability to farm

17. Ibid.
18. Randall D. Little, *Economics: Applications to Agriculture and Agribusiness,* 4th ed. (Danville, IL: Interstate Publishers, 1997), p. 9.
19. Ibid.

in the New World. However, in 1613, Virginia colonist John Rolfe shipped tobacco to England, which proved to be a large market for the new crop. Much of early America was built on economic growth from the export of crops such as tobacco, rice, and **indigo**.[20] Later, colonists shipped furs, timber, and grain to England and southern Europe, while grain, flour, and dried fish were sent to the West Indies. As time passed and American agriculture began to flourish, more markets opened, marking the beginning of agribusiness in America.[21]

Many developments in agriculture during the 17th and 18th centuries eventually led to the way we farm today. Some of the other important developments included:

- The practice of putting dead fish into the ground along with corn seed led to the development of **organic fertilizer**.
- Rice, the world's most popular grain, was first grown in the United States.
- George Washington created one of the first experimental farms.
- Thomas Jefferson experimented with seeds and livestock, invented farm implements, and was active in establishing a local agricultural society.[22]

The Era after the American Revolution

After the American Revolution, more people moved to the United States, spreading out and using more land. People went west and developed new ways to produce food. Some new developments were:

- **Surveying** of land was used to separate property.
- The **cotton gin** was invented by Eli Whitney in 1793.
- Edward Jenner discovered **vaccines** to prevent diseases.
- The first one-piece, cast-iron plow was invented in 1819 by Jethro Wood.
- Interchangeable parts were developed so that people could fix their equipment.[23]

The Agricultural and Industrial Revolution

During the 1840s and 1850s, the Industrial Revolution created much change and spurred the growth of production agriculture and agribusiness. During these important years of rapid evolution, many technological advances made a huge impact on the world and in the agriculture industry, including the steam engine, railroads, the sewing machine, and the powered loom for weaving cloth, among many others.

A by-product of the Industrial Revolution was large-scale movement from farms to the cities. The new and productive machinery required much manpower, leaving many farms vacant as the men worked in factories. This allowed those still farming for a living to increase production for those who no longer did so.[24]

However, the transfer of labor from the farm to the factory left the remaining farmers with few workers. To increase production, many of these farmers turned to the newly invented machinery to decrease the time it took to accomplish a task and increase the output of each task. In doing so, farms began to commercialize, as they became much better equipped to feed the ever-increasing number of people who no longer fed themselves. Tilling, planting, and harvesting were just a few areas where new-age farmers discarded animal power in favor of machinery. As the demand rose, the industrial advancements reflected the need for more output on the farm. This important era in agriculture marked the shift from animal power to **mechanical power**.[25]

During the late 19th century, it became evident that the Industrial Revolution had had a great impact on the way farmers would operate. The switch from animals to machines began with the development of the steam engine in England. It was first used for pumping water out of coal mines, and later to power sawmills, textile mills, and many other industrial machines. Power animals like horses, mules, and oxen were used less, as machines proved to shorten human labor hours and increase production. In fact, before 1830, it took nearly 56 hours of human labor to produce just one acre of wheat. Today, according to the USDA, with the modern farm machinery that has roots in the Industrial Revolution

20. Little, *Economics*, p. 10.
21. Ibid.
22. Willard W. Cochrane, *The Development of American Agriculture: A Historical Analysis*, 2d ed. (Minneapolis: University of Minnesota Press, 1993).
23. Ibid.
24. Lee, *Working in Agricultural Industry*, p. 12.
25. Ibid., pp. 12–13.

Figure 1–5 Agriculture advanced another step with the design of a better one-piece, steel plow by John Deere. (Courtesy of Deere & Company)

and the switch to mechanical power, less than two hours are required to produce that same acre.[26]

The Agricultural and Industrial Revolution also brought us the following developments:

- Henry Ford developed the automobile.
- Crop rotation was promoted by Charles Townsend.
- Advances in livestock breeding were achieved by Robert Bakewell.
- The first workable seed drill was invented by Jethro Tull.
- Cyrus McCormick invented a mechanical **reaper** to reduce hand labor in the harvest of grain.
- A stationary grain **threshing machine** was developed to separate grain from waste.
- John Deere designed a better one-piece, steel plow (refer to Figure 1–5).
- Barbed wire was invented to keep livestock away from cropland.
- The first gasoline-powered tractor was built in 1892.[27]
- Seed and plant genetics were developed by Gregor Mendel.

By the end of the 1800s, farming had become the world's most important industry. Many things had been done to make agriculture a better, more rewarding way of life.

First Half of the Twentieth Century

Around the year 1900, things began to improve for the American farmer. New machines made farm work much easier, and better transportation was developed, which meant that farmers were able to market their products to many more people. Farm prices were high, and farmers were making a good living. The extra money made allowed people to pay for the research and development of new ways to farm. Some of these new developments were:

- The U.S. government established the Bureau of Forestry.
- A vaccine was developed for **hog cholera**.
- The Panama Canal opened for shipping.
- The Smith-Hughes Act established vocational agriculture in high schools.
- The U.S. government established the **Cooperative Extension Service**.
- **Federal Land Banks** were established to give credit to farmers.
- **Hybrid** plant seed was developed for better-quality, higher-producing crops.[28]
- Many agricultural scientists, such as George Washington Carver, were developing new products (refer to Career Option).

Bad times occurred after World War I, when farmers could not sell as much food overseas. There were not as many soldiers in foreign countries, and thus there was not as much need for American-grown food overseas. Farm prices began to drop, and many farmers were forced out of business. A terrible **drought** in the midwestern states also caused farmers to go broke. Those days were called the "Dust Bowl" days. Many farmers had bought and used tractors to plow and plant many more acres of land. This added to the problem of the Dust Bowl. Large areas of land were plowed, leaving bare surfaces. When no rain came, the soil blew away.

As farmers made less and less money, the whole country began to lose money. In 1929, the stock market "crashed" and the United States entered the Great Depression. Something had to be done to pull America out of this hole. Many agricultural developments helped get the United States back on track. Some of these developments were:

- The U.S. government began to pay farmers for using soil conservation practices.
- The **Soil Conservation Service (SCS)** was established in 1935 to keep the Dust Bowl from reoccurring.
- The Future Farmers of America (FFA) organization was started in 1928.

26. Ibid., p. 50.
27. *Introduction to World Agricultural Science and Technology*, p. 17.
28. Ibid., p. 18.

Career Options

Agricultural Scientist—George Washington Carver

There are thousands of agricultural scientists working with the U.S. Department of Agriculture (USDA), universities, and agribusiness companies throughout the world. An early agricultural scientist who worked at the beginning of the 20th century was George Washington Carver. This great African American is known for his work with peanuts, sweet potatoes, and cotton.

Carver was a science professor at Tuskegee Normal and Industrial Institute (now called Tuskegee Institute), started by Booker T. Washington. He was one of the first scientists to teach crop rotation. He found that cotton grew much better after growing sweet potatoes one year in the same field, and growing cowpeas the next. The third year, cotton was grown again. Carver planted peanuts ("goobers") to control the boll weevils. He also extracted the oils, sugars, and starches from peanuts to make products such as oil, rubbing oil, milk, cheese, and margarine. From cotton, he was able to make paper, rugs, insulating boards, and many other products. He found an amazing 118 uses for the sweet potato.

George Washington Carver at work in his laboratory conducting research on one of his experiments. (Courtesy of George Washington Carver Museum)

- Higher crop yields resulted from better management practices.
- The U.S. government began to pay for more research and education in agriculture.
- Antibiotics were first used to treat animals.[28]
- The **Dairy Herd Improvement Association (DHIA)** was organized to monitor dairy herds.
- Groves of trees were planted along the edges of farmsteads to prevent wind erosion.

Latter Part of the Twentieth Century

The United States came out of the Depression, after which World War II (1941–1945) caused an increase in farm prices. This trend got people started on developing more advanced farming methods. Many advances were made in the areas of production, marketing, and agricultural mechanics.

- **Artificial insemination** was more widely used in the livestock industry.
- Use of new technology increased productivity.
- Farmers began to use electric fences.
- Disc plows came into widespread use.
- Chemical fertilizers and **pesticides** were widely used.
- **Futures** trading was used to control risk.
- Computers became popular as agricultural management tools.[29]

In recent years, **agronomic** and animal production systems have been improved and refined. Plant varieties have now been developed for specific climate zones. Per-acre plant populations have increased, and specific fertilization rates are now easier than ever to determine efficiently. All these trends illustrate that as a need was determined, adjustments were made and agriculture productivity increased further.

In the 1980s, many developments allowed livestock production to improve and even surpass crop production advancements. Improved

29. Ibid.

livestock-handling facilities, increased feed efficiency, better and more uniform records to monitor individual animal performance, and improved disease prevention and control all allowed the livestock farmer to flourish.[30]

Biotechnology is now a vital part of agriculture as well, with advances in gene splicing, cloning, and gene mapping. All these advances allow farmers and scientists to work together to create the most desirable, profitable product possible.

The establishment of agribusiness enterprises began in the 1970s. These agribusinesses performed functions that farmers had previously been required to do. Before, the farmer would handle all operations required to get the product from the farm to the table. However, with agribusiness operations, farmers could spend all their efforts on the pure production of a product, and allow companies to haul, process, package, and market. Agribusiness enterprises also supply the equipment and machinery required for these jobs. Agribusinesses now include:

- farm machinery dealerships
- commodity (futures) brokers
- artificial breeding services
- research consulting firms
- agricultural (ag) chemical companies
- veterinary supply companies
- livestock supply companies
- animal feed companies
- biotechnology firms
- export companies

Figure 1-6 illustrates the evolution of farming, agriculture, and agribusiness

Examples of Agribusinesses
Farm Machinery Dealerships
Commodity (Futures) Brokers
Artificial Breeding Services
Research Consulting Firms
Ag Chemical Companies
Veterinary Supply Companies
Livestock Supply Companies
Animal Feed Companies
Biotechnology Firms
Export Companies

Figure 1-6 The business evolution of farming, agriculture, and agribusiness.

Historical Development of Modern Farm Equipment

The development of modern farm equipment began before tractors came on the scene. One of the first agricultural machines that had a significant impact on farming was the cotton gin, invented in 1784 by Eli Whitney. Three years later the cast-iron plow was patented by Jethro Wood. This plow worked very well in eastern soils but not the hard soils of the Midwest. In 1837 John Deere, founder of John Deere Tractor Company, made the first successful steel plow from a saw blade, and by 1846 he was building 1,000 steel plows per year. The steel did not wear out as fast as cast iron, and the soil did not stick to the new plow as it did with the old one, so farmers were very happy with it.

Methods of production agriculture changed very little until the 1850s. Practices such as sowing, tilling, and harvesting were still performed by human muscle power, and techniques were passed from father to son and mother to daughter. As late as 1800, it is estimated that almost 90 percent of the U.S. population still lived on farms. These Americans were agriculturists so that they could feed, clothe, and shelter their families. They were almost totally self-sufficient, as they produced nearly everything they needed, including clothing, tools, feed, and farm equipment. Even those who farmed only to feed their families had a difficult time producing enough to stay comfortable. Understandably, there was little to no surplus production left to exchange.[31]

Beginning of Change

The 19th century brought the beginnings of change. The century began with the Louisiana Purchase in 1807 and the opening of the farmland west of the Allegheny Mountains.

30. Ibid.

31. James G. Beierlein and Michael W. Woolverton, *Agribusiness Marketing: The Management Perspective* (Englewood Cliffs, NJ: Prentice-Hall, 1991), p. 106.

The first cotton planter was patented in 1825, and a corn planter was developed three years later. In 1831, Cyrus McCormick, one of the founders of International Harvester Company, developed a successful grain reaper. Although this was not the first reaper invented, it was the first to achieve wide acceptance, and by 1860, 20,000 reapers were being sold each year. The reaper not only cut labor time and cost, it also reduced the risk of weather damage by more than 50 percent by reducing harvest time. In 1885, more than 250,000 harvesting machines were produced and sold. The earlier farm machines were built for horsepower rather than tractors, but they adjusted very quickly when the source of power changed.

Between 1850 and 1880, the amount of land used for farming increased 82 percent, from 294 million acres to 536 million acres, and the number of farms rose 167 percent, to 4 million.[32] The advent of the Industrial Age near the middle of the century brought about shortages of farm workers as people flocked to northern industrial cities. The Civil War brought increased interest in the development of labor-saving tillage and harvesting tools by simultaneously raising farm commodity prices and reducing the supply of farm laborers.

From Manpower to Horsepower

In contrast to the early 19th century, only 50 percent of Americans lived on farms. In addition, production agriculturalists were not only self-sufficient, but had raised their productivity to allow for extra income to be used to purchase horse-drawn farm equipment and other tools. By the end of the 1800s, the transition from manpower to horsepower had been mostly completed.[33]

The Steam Era

Although steam power had its major impact on the industrial sector of the economy, it also played a major role on the farm between 1850 and 1900, which is generally called the "Steam Era." It is estimated that more than 70,000 steam engines were produced for farm use. The first steam engines were stationary, and farm jobs had to be brought to the engine. The second phase produced a portable steam engine, which was taken to the fields for specific jobs such as powering wheat threshing machines or sawmills. The third type of farm steam engines, called "traction engines," were much more useful as a source of farm power. They pulled themselves as well as other pieces of equipment. The first steam traction engine for farm use was made in 1869 by J.I. Case Company.[34]

The steam traction engine ushered in a new era in agriculture by providing an alternate mobile source of power on the farm. However, steam engines had numerous problems, and it was obvious that they were not the ideal source of power that farmers needed. The engines were extremely heavy, bulky, dangerous, expensive, and adaptable to very few chores.

Internal Combustion Engine

The development of the internal combustion engine occurred in stages similar to that of the steam engine. First, small, stationary, one-**cylinder** engines were made for small jobs around the home and on the farm. Then, larger two-cylinder motors were mounted on wheels or sleds and taken to the fields, shops, or wherever they were needed. Finally, two-cylinder engines were mounted on frames with wheels and a transmission so they could pull themselves as well as some type of equipment.

Fuels Used

A wide variety of fuels were used in early internal combustion engines. Some of the major types were gun powder, **turpentine**, coal dust, and kerosene, which was commonly called coal oil. Even though early tractors were called gasoline tractors, the major source of fuel was kerosene. Most early tractors were made with a small tank for gasoline and a large one for kerosene. The farmers started the engine with gasoline and then switched to kerosene because it was cheaper and more efficient to use.

The Beginning of Internal Combustion Engines

The exact date when internal combustion engines were first made is not clear. However, some engines have been traced to the latter part of the 17th century.

32. Ibid.
33. Beierlein and Woolverton, *Agribusiness Marketing*, p. 107
34. Farm and Industrial Equipment Institute, *Men, Machines, and Land* (Chicago: Author, 1974), pp. 19–21.

The first practical power unit was developed in 1876 by two German inventors, N. A. Otto and Eugene Langen. This engine was very popular and became known as the Otto engine.

In 1899 there were more than 100 firms making internal combustion engines in the United States, not counting automobile engines, and by 1911 more than 500 companies were in operation.[35] Small engines continued to be a popular source of power through the 1940s. Currently, these small engines are experiencing a comeback through restoration by private collectors. However, they are used for fun, not for work.

Figure 1–7 These loggers still need a mule to drag these cedars from the woods so that modern-day equipment can to complete the loading process. (Courtesy of Cliff Ricketts)

Farm Tractors

The record is not clear as to when the first tractor was made or who made it. However, historian R. B. Gray reports that the Charter Gas Engine Company built six gasoline tractors in 1889 and that in 1890 George Taylor applied for a patent on a walking-type motor plow.[36]

The First Gasoline-Powered Tractor

In 1892 John Froehlich built what is sometimes called the first successful gasoline-powered tractor. The Froehlich tractor was the forerunner of the Waterloo Boy and the modern John Deere line of tractors. The Case Threshing Machine Company and the Dissinger and Bros. Company also built tractors in 1892, but they were experimental in nature and were not put to practical use until several years later. Several other companies began experimenting with gasoline tractors in the late 1890s, but the switch from horsepower and steam power to tractor occurred primarily during the first 30 years of the 20th century. The term *tractor* was first coined in 1906 by a salesman for the Hart-Parr Tractor Company (a predecessor of Oliver, White Farm Equipment Company and today's AGCO). Previously, they were called "gasoline traction engines." Although tractors have been around for nearly 100 years, mules are still needed for certain jobs. Refer to Figure 1–7.

Effect of World War I on Tractor Production

Tractor production expanded rapidly in the early 1900s. In 1910, 15 tractor companies sold 4,000 tractors. The onset of World War I marked another turning point in the development of agriculture and caused a rapid increase in tractor production, and in 1920, 166 tractor companies sold more than 200,000 tractors. Spurred by higher incomes from feeding war-ravaged Europe, U.S. production agriculturalists began the process of replacing their horse-drawn equipment with gasoline-powered tractors and the larger tillage implements that those tractors could pull.

Effect of the Depression Years on Tractor Production

After 1921, the number of tractor companies decreased about as fast as it had increased. This was due to the Depression of the 1920s. In 1921, 186 companies sold only 68,000 tractors, and by 1925 only 58 companies had survived, although the number of tractors sold increased. By 1935, 20 companies had sold more than 1 million tractors, with 90 percent of sales coming from 9 major companies: International Harvester, John Deere, J. I. Case, Massey-Harris, Oliver, Minneapolis Moline, Allis Chalmers, Cleveland Tractor Company, and Caterpillar Tractor Company.[37]

Henry Ford and the Tractor

The first gasoline-powered tractors had the same problems as the steam engines: they were expensive, big, bulky, hard to drive, and very limited in

35. R. B. Gray, *The Agricultural Tractor, 1855–1950* (St. Joseph, MI: American Society of Agricultural Engineers, 1975), pt. 1, p. 14.

36. Ibid., pt. 1, p. 15

37. Ibid., pt. 2, p. 29.

their application. Tractor companies soon began experimenting with smaller tractors that were more suitable for small farms and less expensive. One of the most successful small tractors was the "Fordson" tractor made by Henry Ford. It was also the first mass-produced tractor on the market. The Fordson accounted for about 70 percent of the total tractor market by 1925. Refer to Figure 1–8.

Power Take-Off Units, Tricycle-Type Tractors, and Rubber-Tired Tractors

In 1918, International Harvester announced a **power take-off (PTO)** unit, which allowed the operator to control mounted and drawn equipment with the engine of the tractor. The **tricycle-type** tractor, introduced by International Harvester in 1924, was very popular for cultivating as well as plowing. In 1932, Allis Chalmers, in cooperation with Firestone Rubber Company, introduced a **pneumatic** rubber-tired tractor, which completed the basic design of a light, versatile tractor that could handle most farm jobs. This essentially finalized the transition from horses and mules to tractors with internal combustion engines. Animals were no longer needed as a major source of power, although many smaller farmers continued to use horses and mules through the 1950s.

Effect of the Shift from Animal Power to Tractor Power

The shift of animal power to tractor power affected American farming in two major ways.

Decreased Demand for Animal Feed. A large portion of the land that had been used to produce animal feed was shifted to the production of food. Horses, for example, would be a large drain on food production if they still were a major source of power on the farm, as they are estimated to consume about 25 percent of grain acreage.[38] Mechanical power benefited production agriculture not only by increasing output and productivity significantly, but also by increasing the amount of food available for human consumption and lessening the amount needed by work animals. Refer to Figure 1–9.

Figure 1–8 The Fordson tractor had a major impact on American agriculture as production agriculturists switched from horse to tractor power. (Courtesy of New Holland North America, Inc.)

Reduced Labor Time and Cost. In 1936, the Iowa State University Experiment Station reported that production agriculturalists with rubber-tired, two-plow tractors were producing 100 acres of corn with 51 days of fieldwork. The same operation with horses required 141 days.[39]

Advent of Various Fuels

Machine power continued to change and improve. In 1931, Caterpillar Tractor Company developed a diesel-powered, **crawler-type** farm tractor. The crawler-type tractor did not fit most farm needs, but the diesel engine had a major impact a few years later. The gasoline engine itself has been improved through the development of high-**compression** engines. Also, in 1941 liquefied petroleum (LP) gas tractors were introduced by the Minneapolis Moline Company. This made it possible for farmers to use clean-burning, low-cost butane and propane fuels, especially in areas near these energy sources.

Modern Tractor Accessories

Today, **hydraulic lifts**, **torque amplification**, **hydrostatic transmission**, power steering, **turbochargers**, heated and air-conditioned cabs, and many other features provide an efficient and comfortable power unit for modern production agriculturalists. The production agriculturalist of today, operating a 100-horsepower tractor, can do the work of more than 1,000 workers who are without machine or animal power. It is no wonder that the average American farmer produced enough for more than 131 people. Refer to Figure 1–10.

Increased Size and Four-Wheel Drive

During the decades of the 1960s and 1970s, major changes included the shift to diesel as the major fuel,

38. Beierlein and Woolverton, *Agribusiness Marketing*, p. 107.
39. Farm and Industrial Equipment Institute, *Men, Machines, and Land*.

Figure 1–9 If horses were still the major source of power on the farm, they would consume the output of approximately 25 percent of grain acreage. (Courtesy of Massey-Ferguson Operations)

Figure 1–10 A production agriculturalist today operating a 100-horsepower tractor can do the work of more than 1,000 workers without machine or animal power. (Courtesy of New Holland North America, Inc.)

an increase in horsepower, and a shift to four-wheel-drive power. Currently, more than 80 percent of farm tractors use diesel, and most major tractor companies offer tractors with a horsepower rating of 200 or more. The major change in the 1970s was the shift to four-wheel drive. The major advantages of four-wheel drive include the ability to use more power efficiently, better traction and flotation with less soil compaction, and increased safety. Four-wheel drive is now standard on extremely large models and optional on medium and small models.

Success of American Agribusiness

Due to the effectiveness and efficiency of agribusiness, one American farmer can now supply enough food and fiber for more than 150 people. Also, Americans spend less of their income on food than any other people in the world: only 9 percent of total personal disposable income.[40]

40. Hiram M. Drache, *History of U.S. Agriculture and Its Relevance to Today* (Danville, IL: Interstate Publishers, 1996), vii, ix.

American agriculture and agribusiness have truly become a world wonder. No nation has better fed itself and still had the ability to contribute heavily to world food supplies. U.S. agribusiness supplies each citizen with close to 1,500 pounds of food annually while still producing **exports** of vast amounts of grains, vegetable oils and fats, cotton, tobacco, and many other products. In fact, production agriculture became so efficient that the federal government had to establish restrictions of various kinds to ensure that production was controlled and prices were maintained. It is thought by many, however, that if the restrictions were lifted, American agriculture would be able to meet the world's needs for food, clothing, and shelter.[41]

Because of the success and power of American business, it has an enormous responsibility to promote world peace and security. At the least, it is a major factor. The existence of underfed, poorly housed, and badly clothed people in different parts of the world represents a threat to global peace and security.[42]

Conclusion

Agribusiness was born in America when John Rolfe shipped the first tobacco to New England and farmers began buying newly invented production machinery and other products from suppliers. Farmers became able to purchase things they had formerly had to produce themselves on the farm. Horseshoers, barrel makers, harness makers, and other skilled craftspeople contributed to the first supplies, marking the decentralization of one area of agriculture. More separation came as farmers began to rely on butchers and millers to prepare meat and grain. As the farmer became more focused on production and less on preparation, distribution and other required steps in production agriculture became more and more specialized.[43]

The changes in agricultural techniques that came with the introduction of machinery stimulated the growth of agribusiness, which, in turn, brought about more changes in the agricultural industry. Functions that once were performed on farms and ranches now are performed by workers in agribusiness.

41. Roy, *Exploring Agribusiness*, p. 7.
42. Ibid.
43. Lee, *Working in Agricultural Industry*, p. 13.

Summary

Today, agriculture is a technology-oriented industry that includes production, agriscience, and agribusiness. Most people who work in the agricultural industry do not work on farms and ranches, but rather are employed in the feed, seed, farm machinery, fertilizer, chemical supply, food-processing, and related businesses.

There are many different definitions of agribusiness. Some definitions include, and some exclude, production agriculture. Within the context of this book, agribusiness means the manufacture and distribution of farm supplies to the production agriculturalist and the storage, processing, marketing, transporting, and distributing of agricultural materials and consumer products that were produced by production agriculturalists.

Within the big picture of agribusiness are agribusiness companies that provide supplies to production agriculturalists. The production agriculturalists produce food and fiber, and the output is taken by agribusiness companies, which process, market, and distribute the agricultural products. Many other support services, such as research, education, and finance, are also involved. Agribusiness surrounds us in our daily lives. Consider all the agribusiness functions involved in getting a cheeseburger to you, and you will fully appreciate the magnitude of agribusiness.

People have searched for ways to feed themselves since prehistoric times. Some spent nearly all their waking time searching for foods by hunting or gathering nuts and other naturally growing foods. Eventually, the puzzle to farming was solved and people began to raise crops and domesticate animals.

The American colonists found conditions difficult. They faced starvation until they adopted the crops and tillage practices of the American Indians. American agribusiness was started when John Rolfe shipped tobacco to England. Later, many other agricultural exports were shipped, which provided the capital necessary for growth and development in American agriculture.

Many agricultural inventions and practices were developed in America, all of which led up to the agribusiness of today; these include inorganic fertilizer; rice, which was first grown in America; the cotton gin; vaccines; the cast-iron plow; the drill mechanical reaper, the threshing machine; barbed wire; crop rotation; the gasoline-powered tractor; hybrid seed; artificial insemination; electric fences; disc plows; futures trading; and many more.

This power shift began in the latter half of the 19th century and the first half of the 20th century. The shift from human and animal power to mechanical power originated in the Industrial Revolution, which began in the 18th century in England with the invention of the steam engine.

The exact date when internal combustion engines were first made is not clear. However, some engines have been traced to the latter part of the 17th century. In 1899, more than 100 firms were making internal combustion engines in the United States, not counting automobile engines, and by 1914 more than 500 companies were in operation.

In 1892 John Froelich built what is sometimes called the first successful gasoline tractor. The term *tractor* was first coined in 1906 by a salesman for the Hart-Parr Tractor Company. Tractor production expanded rapidly in the early 1900s. In 1910, 15 tractor companies sold 4,000 tractors. The onset of World War I caused a rapid increase in tractor production, and in 1920, 166 tractor companies sold more than 200,000 tractors.

In 1932 Allis Chalmers, in cooperation with Firestone Rubber Company, introduced a pneumatic rubber-tired tractor, which completed the basic design of a light, versatile tractor that could handle most jobs. In 1931, Caterpillar Tractor Company developed a diesel-powered, crawler-type farm tractor. Today, hydraulic lifts, torque amplification, hydrostatic transmission, power steering, turbochargers, heated and air-conditioned cabs, and four-wheel drives provide an efficient and comfortable power unit for modern-day production agriculturalists.

Because of the development of production agriculture, agriscience, and agribusiness, U.S. agriculture represents one of the greatest achievements in the history of humankind's struggle for food, clothing, and shelter. Indeed, American agriculture has become the marvel of the world.

End of Chapter Activities

REVIEW QUESTIONS

1. Define the Terms to Know.
2. List 20 businesses or services that are needed to get a cheeseburger ready for consumption.
3. What two important developments took place before humans discovered agriculture?
4. Name five developments during the Bronze Age.
5. Name four developments during the Iron Age.
6. Name four developments during the Middle Ages.
7. What eight crops did American Indians develop that are still of great importance?
8. What were nine crops that were exported by colonists in the New England area?
9. List four agricultural developments during the 17th and 18th centuries in America.
10. List five agricultural developments during the era following the American Revolution.
11. Name 10 developments during the Agricultural and Industrial Revolution.
12. What were nine developments during the first half of the 20th century?
13. List seven developments during the latter part of the 20th century.
14. Describe the three types of steam engines used in the Steam Era.

15. List five reasons why steam engines were not the ideal source of power for farmers.
16. Briefly discuss the three phases of use in the development of the internal combustion engine.
17. Name four fuels used in early internal combustion engines.
18. What effect did World War I have on tractor production?
19. What effect did the Depression years have on tractor production?
20. What were the nine major tractor companies in 1985?
21. What were the two effects of shifting from animal power to tractor power?
22. What were the three major changes in tractors in the 1960s and 1970s?
23. What are the three major advantages of four-wheel-drive tractors?

FILL IN THE BLANK

1. Agriculture, to many, is cows, sows, and ___________, or weeds, seeds, and ___________.
2. Several thousand years ago, prehistoric people's nutrition and health were so bad that they seldom lived past ___________ years of age.
3. Early humans considered ___________ so tasty that some species were almost exterminated.
4. The fall of the ___________ ___________ slowed the growth of agriculture.
5. The first agricultural education teachers in America were the ___________.
6. In 1613, John Rolfe shipped ___________ to England, and the Virginia colonists discovered a crop with a large, untapped market.
7. A consequence of the Industrial Revolution was the movement of people away from the ___________ to the ________________.
8. A time of terrible drought in the midwestern states, which caused farmers to go broke, was called the _____________ _____________ days.
9. One American farmer can supply enough food and fiber for more than ___________ people.
10. Before 1830, it took nearly ___________ hours of human labor to produce one acre of wheat.
11. Today, with modern farm machinery and equipment, fewer than ___________ hours are needed to produce one acre of wheat.
12. In 1800, approximately ___________ percent of all the people in the United States lived on farms.
13. In 1900, approximately ___________ percent of Americans lived on farms.
14. Even though early tractors were called gasoline tractors, the major source of fuel was ___________.
15. The ________ tractor accounted for about 70 percent of the total tractor market by 1925.
16. If horses were still the major source of power on the farm, they would consume the output of approximately ___________ percent of grain acreage.
17. In 1936, production agriculturalists with rubber-tired, two-plow tractors were producing 100 acres of corn with _________ days of fieldwork, whereas the same operation with horses required _________ days.
18. Today, the production agriculturalist operating a 100-horsepower tractor can do the work of more than _________ workers who do not have machine or animal power.
19. Currently, more than ___________ percent of tractors use diesel.
20. Americans spend just ___________ percent of their personal disposable income on food.

MATCHING

a. A.D. 400–1500
b. Output
c. 1,000 B.C.
d. Input
e. 3,000 B.C.
f. Dust Bowl days
g. Great Depression
h. 1935
i. John Deere
j. 1928
k. 1892
l. Eli Whitney
m. Cyrus McCormick
n. Allis Chalmers
o. Minneapolis Moline Company
p. Fordson
q. J. I. Case Company
r. John Froelich
s. Caterpillar Tractor Company
t. John Deere
u. International Harvester
v. Oliver

_____ 1. supplies sold to production agriculturalist
_____ 2. processing, marketing, and distribution of agricultural products
_____ 3. Bronze Age
_____ 4. Iron Age
_____ 5. Middle Ages
_____ 6. result of the stock market crash in 1929
_____ 7. SCS was established
_____ 8. FFA organization was established
_____ 9. first gasoline-powered tractor was built
_____ 10. invented the cotton gin in 1784
_____ 11. made the first successful steel plow from a saw blade
_____ 12. developed a successful grain reaper in 1831
_____ 13. developed the first steam tractor engine for farm use in 1869
_____ 14. in 1892, built what is sometimes called the first successful gasoline tractor
_____ 15. first mass-produced tractor on the market
_____ 16. first tractor with power take-off (PTO)
_____ 17. first pneumatic rubber-tired tractor
_____ 18. in 1931, developed a diesel-powered, crawler-type farm tractor
_____ 19. introduced liquefied petroleum (LP) gas tractors in 1941
_____ 20. the term *tractor* was coined by a salesman in 1906 for a company that evolved into this company

ACTIVITIES

1. Compile a list of the agribusinesses in your community.
2. Select one of the agribusinesses that you selected and show where it fits into "The Big Picture of Agribusiness." For example, is it an input or output agribusiness company?
3. Determine the agricultural products contained in a pizza. Then, identify and list the types of agribusiness companies involved in getting the pizza ready for your consumption.
4. Select one raw agricultural product and one processed agricultural product and list all the agribusinesses and agriservices (example, extension service) involved in getting it ready for your purchase. Present this report to your class.
5. Select an agricultural product (machinery, equipment, fertilizer, medicine, etc.) and complete a two- or three-page report on its development. Present this report to your class.
6. Select a particular farm machinery and equipment product. Go to the library and use other resources to prepare a two- to three-page paper on the historical development of the product. Share this with the class by giving an oral report.
7. Interview an older relative, friend, or farmer and get his or her impression of the changes in production agriculture over the years with the advances in farm machinery and equipment. Share these findings with your class.
8. Are four-wheel-drive tractors a luxury or a convenience? Defend your answer.
9. Go to an Internet search engine and type in: "Agribusiness, food Industry, and forest industry associations on the Internet." Select one and report to the class.

Chapter 2

Agricultural Economics and the American Economy

Objectives

After completing this chapter, the student should be able to:

- Define economics.
- Explain three major components of economics.
- Discuss three basic economic questions.
- Explain six types of economic systems.
- Discuss economics from a historical perspective.
- Discuss the role of government versus individuals in the economic system.
- Describe the characteristics of the American economy.
- Differentiate between macroeconomics and microeconomics.
- Differentiate between positive and normative economics.
- Explain agricultural economics.

Terms to Know

aggregate
agricultural economics
allocation
applied science
basic science
capital
capitalism
classical economic theory
communism
cost-price squeeze
economics
fascism
free markets
goods
initiative
Keynesian
laissez-faire
macroeconomics
microeconomics
needs
normative economics
Payment in Kind (PIK)
positive economics
profit
scarce
scarcity
services
socialism
subsectors
subsidies
synthesis
wants

Introduction

Today's world is complex and continually changing. The successful agribusiness manager must possess a basic understanding of economic principles to react to these changes. To understand **agricultural economics**, the agribusiness manager must first understand basic economic principles.

Definitions of Economics

There are many definitions of **economics**. Consider each of the following definitions, look for key words and phrases, and then form your own definition:

- Economics is the study of allocation of scarce resources among competing alternatives.[1]
- Economics is the study of how individuals and countries decide how to use scarce resources to fulfill their wants.[2]
- Economics is the "study of how society allocates scarce resources and goods."[3]
- Economics is "the science of allocating scarce resources (land, labor, capital, and managements) among different and competing choices and utilizing them to best satisfy human wants."[4]
- Economics "is a study of how to get the most satisfaction for a given amount of money or to spend the least money for a given need or want."[5]
- Economics is the study of how scarce resources are transformed into goods and services to satisfy our most pressing wants, and how these goods and services are distributed.[6]
- Economics is "the study of the decisions involved in producing, distributing, and consuming goods and services."[7]
- Economics "is a social science that studies how consumers, producers, and societies choose among the alternative uses of scarce resources in the process of producing, exchanging, and consuming goods and services."[8]
- Economics is "concerned with overcoming the effects of scarcity by improving the efficiency with which scarce resources are allocated among their many competing uses, so as to best satisfy human wants."[9]

Three Major Components of Economics

Three key words or phrases can be drawn from each of these definitions: **scarcity**, types of resources, and wants and needs.

Scarcity

Scarcity is the economic term for a situation in which there are not enough resources available to satisfy people's needs or wants. Economics is the study of society's **allocation** of **scarce** resources. These resources are considered scarce because of a society's tendency to demand more resources than are available.

Although most resources are scarce, some are not, such as the air that we breathe (except in places where smog or other air pollutants are severe). A resource that is not scarce is called a free resource or good. However, economics is mainly concerned with scarce resources and goods. Scarcity is what motivates the study of how society allocates resources.[10]

Shortage versus Scarcity. Shortage and scarcity are not the same. Scarcity always exists because it relates to an unlimited or unsatisfied want, whereas shortages are always temporary. Shortages often exist after natural disasters destroy goods

---

1. *Advanced Agribusiness Management and Marketing* (8735-B) (College Station, TX: Instructional Materials Service, 1990), p. 33.
2. Roger LeRoy Miller, *Economics Today and Tomorrow* (New York: Glencoe/McGraw-Hill, 1995), p. 8.
3. John Duffy, *Economics* (Lincoln, NE: Cliffs Notes, 1993), p. 1.
4. Randall D. Little, *Economics: Applications to Agriculture and Agribusiness,* 4th ed. (Danville, IL: Interstate Publishers, 1997), p. 3.
5. Ibid.
6. Fred M. Gottheil, *Principles of Economics* (Cincinnati, OH: South-Western, 1996), p. 6.
7. Sanford D. Gordon and Alan D. Stafford, *Applying Economic Principles* (New York: Glencoe/McGraw-Hill, 1994), p. 5.
8. John B. Penson, Rulon D. Pope, and Michael L. Cook, *Introduction to Agricultural Economics* (Englewood Cliffs, NJ: Prentice-Hall, 1986), p. 7.
9. Gail L. Cramer, Clarence W. Jensen, and Douglas D. Southgate, Jr., *Agricultural Economics and Agribusiness,* 8th ed. (New York: John Wiley & Sons, 2001), p. 5.
10. Duffy, *Economics,* p. 1.

Figure 2–1 Land is a valuable resource. Cities, manufacturing, highways, and other factors associated with population growth have resulted in the permanent loss of much fertile farmland ideal for agricultural use. (Courtesy of USDA)

and property.[11] Temporary shortages of products such as gasoline may be caused when imports are dramatically decreased for any reason.

Types of Resources

Resources are the inputs that society uses to produce outputs. Traditionally, economists have classified resources as natural resources (land), human resources (labor), manufactured resources (capital), and entrepreneurship (management).

Natural Resources (Land). Land and the mineral deposits in it constitute a huge natural resource in the agricultural industry. The population growth in this country has caused an increasing demand for land for commercial and residential uses. This has resulted in the permanent shift of land use (some of it formerly fertile farmland) away from agricultural use. Water is another natural resource that is limited in many areas.[12] Refer to Figure 2–1.

Human Resources (Labor). The services provided by laborers and managers for the production of goods and services are human resources and are also considered scarce. Agribusinesses may not be able to obtain all the labor services they need at a wage they desire to pay.[13]

Manufactured Resources (Capital). All the property people use to make other goods and services is **capital**. These resources take the form of machines, equipment, and structures.

11. Miller, *Economics Today and Tomorrow,* p. 9.
12. Penson, Pope, and Cook, *Introduction to Agricultural Economics,* pp. 4, 5.
13. Ibid., p. 4

Figure 2–2 Without proper management, agribusinesses will fail. It is not how much you make, but how you spend. Excellent management can lead to a profit. (Courtesy of Farm Credit Services)

Entrepreneurship (Management). Entrepreneurship refers to the ability of individuals to start new businesses and to introduce new products and techniques. It involves **initiative** and individual willingness to take risks and make a profit. Without proper management or entrepreneurship, agribusinesses would cease operating efficiently. Refer to Figure 2–2.

Today the four resources just discussed are called the *factors of production*. They are used to produce goods and services. **Goods** are the items that people buy. **Services** are the activities done for others for a fee.

Wants and Needs

Needs include things that are really crucial to daily living. Basic needs include enough food, clothing, and shelter to survive. Most of us would also consider a good education and adequate health care to be needs. Of course, there are other needs depending on your situation.

Wants are things that are not crucial to daily living. A basic tractor is a need to a farmer, but a tractor with a cab, air conditioner, and radio may be a want. The difference between needs and wants is not always clear. The tractor with an air-conditioned cab may be a need if the operator has severe allergies. Refer to Figure 2–3.

Economists use the term *insatiable* (unlimited, unsatisfied) wants. This means that human wants cannot be satisfied no matter how much or how many goods we have. You have probably heard individuals say, "If I had a million dollars I could buy everything I want." The fact is that most

Figure 2–3 Whether this big tractor is a need or a want depends on the situation. (Courtesy of Keith Mason)

people are not satisfied with whatever level of consumption they have. The more they get, the more they want. The people who have thousands of dollars want millions, and the people who have millions want billions. People who have small houses want large houses, and those who have small cars want large cars. As soon as we get what we thought would satisfy us, we suddenly discover a need (really a want) for something else. This human trait, in conjunction with scarcity of resources, creates economic problems. The efforts to solve these problems are the basis of the discipline of economics.

Three Basic Economic Questions

Because of the relationship between scarce resources and unlimited/unsatisfied wants, all societies have to answer some basic economic questions. These questions entail trying to decide what to sell, how to sell it, and who should receive the benefits. Therefore, all economic systems must solve the following three questions:

- What goods should be produced, and how much of each?
- How should these goods be produced?
- Who should get what and how much?

What Goods, and How Much, to Produce

This question is answered every time people buy goods. In a market economy, consumers determine what products will be made when they buy things. For example, when sales are made, agribusiness retailers usually order additional products to replace those that were sold. Manufacturers set their production schedules according to these orders.[14] The consumer also tells the manufacturer how much to produce.

How to Produce Goods

This question is answered by agribusiness producers or manufacturers according to what will yield the greatest profits. Agribusiness firms are in business to make money. One way to improve profits is to find the least expensive means of production. Therefore, the least expensive way should be the most efficient way. Consumers' choices guide the producer or manufacturer on what to produce, but the producer or manufacturer determines how to produce it.[15]

Who Should Get What?

This question refers to who will receive the benefits of the goods. The question is answered by determining who has the greatest needs, wants, and ability to pay.

Combining the Economic Ingredients

To simplify the economic questions, assume that all the goods and services in our society are represented by a pie.

What Type of Pie to Produce? We have to decide what type of pie to produce—chocolate, strawberry, or lemon—or if more than one, how many of each. The type and amount of resources, as well as the current level of technology, limits the size and the variety of pies available to us.

What Combination of Ingredients to Use? Once we decide the type and number of pie or pies to produce, we have to decide what combination of ingredients to use: how much labor versus capital, and so on. The ideal combination depends on the availability of each and the management skills of the pie makers. Some countries emphasize more labor; others emphasize more capital.

How to Divide the Pie? The most controversial economic question that societies have to deal with is how to divide the pie. Should everyone get an equal share? Should those who are more

14. Gordon and Stafford, *Applying Economic Principles,* p. 41.
15. Ibid.

productive get a larger share? How much pie, if any, should people get, who do not, for reasons beyond their control, make any contribution to producing the pie? Once a decision is made about how much goes to each individual or group, we must determine which economic system will best accomplish this goal, and what role governments should play in the economy.

Economic Systems

Each society answers the three basic questions (what, how, and for whom) according to its view of how best to satisfy the needs and wants of its people. The values and goals that a society sets for itself determine the kind of economic system it will have. Economists have identified six types of economic systems: traditional, **capitalism**, **fascism**, **socialism**, command (**communism**), and mixed. These terms are often used to designate political systems as well as economic systems. The major distinction between these classifications is the degree of control by private individuals versus the group represented by government.

Traditional System

A pure traditional economic system answers the three basic questions according to tradition. In a traditional system, things are done "the way they have always been done." Economic decisions are based on customs, religious beliefs, and ways of doing things that have been passed from generation to generation. Today, traditional economic systems exist in very limited parts of Asia, Africa, the Middle East, and Latin America.[16]

Capitalism

In capitalism, individuals have free reign over their time and resources, and can determine exactly how to use those assets, with few legal controls by the government. It is a self-regulating system that excludes the government from economic decisions. What capitalism depends on is the will and desires of those involved in the system. Market forces determine prices, assign resources, and distribute income. Market prices indicate the value of resources and economic goods.[17]

Major Components of Capitalist Economic System

- There is private ownership of property and resources.
- The free market determines prices.
- There is freedom of occupational choice.
- There is minimal involvement from government.
- There is freedom for economic gain through profit or wages.
- Competition is a driving force.
- Efficiency is rewarded and inefficiency is penalized.

Figure 2–4 The United States tends to be a capitalistic economic system, even though government guides production in many industries.

Ownership is private and the factors of production are controlled by individuals, allowing them to make decisions based on the potential for **profit**. Any change in profit is directly related to the individual's choice to increase or decrease production. Therefore, any profits earned or losses incurred result from right or wrong business decisions.[18]

Individuals and firms depend on free markets to make production, exchange, and compensation decisions. Competition is the driving force in every economic activity and decision. Refer to Figure 2–4.

Socialism

In sharp contrast to capitalism, the basis of socialism as an economic system is public ownership of all productive resources. Instead of little to no involvement in the system (as with capitalism), the government or "state" directs all decisions regarding the utilization of resources (both human and nonhuman) by the various sectors of the community. Society as a whole, through the state, equally owns all industries and controls all property for the mutual benefit of all people. Individual economic incentives are limited, as competition would not encourage equality. Decisions are therefore made on a centralized basis by government

16. Miller, *Economics Today and Tomorrow,* p. 32.
17. *Advanced Agribusiness Management and Marketing,* p. 45.
18. Ibid., p. 45.

Components of a Socialist Economic System

- The state owns all public utilities and large-scale industries.
- Property is owned by society for the benefit of society.
- Prices are determined by the government.
- Occupations are assigned by the government.
- The government makes all economic decisions.
- There are no economic incentives.
- There is no economic competition.
- There is inefficient production.

Figure 2–5 The basis of socialism as an economic system is that the government directs all decisionmaking.

planners. There is only one major employer and one owner: the government.[19]

Because all resources are considered the property of everyone, the state has control over all economic effort, as well as all consumer and producer processes. The lack of free prices and **free markets** eliminates economic competition. The state is the only entity that can initiate new business activity.[20] Refer to Figure 2–5.

Fascism

Fascism is an economic system in which productive property, though owned by individuals, is used to produce goods that reflect government or state preferences. Private individuals have a high degree of economic power if they support the government in power. If they do not, they have little or no control.

Fascism suppresses opposition, censors criticism, and denies some freedoms to individuals. Although property is privately owned and businesses control production, the government controls labor, employers, and consumers.[21]

Communism

Communism is a totalitarian system of government in which a single, authoritarian political party or body controls government-owned means of production. In other words, the government has total control of economic matters and private individuals have none. Production is organized by an economic plan, usually centralized. Everyone contributes according to ability and distribution is made according to need.[22]

Mixed Economic Systems

With the exception of the traditional economic system, it is doubtful whether any country fits any of these classifications exactly. The real question is which classification most closely resembles the actual system of each country. No society has been willing to grant all economic decisionmaking to individuals, as characterized by capitalism. Conversely, no society has been willing to give all power to the government, as characterized by communism. Each country has to decide how much power to give to each. Obviously, there is not one ideal ratio for all countries or even for any one country, because all societies are constantly changing the ratio of each. During some periods, more government control is emphasized, while in other periods less government and more individualism is emphasized.

American Economy Is Mixed. Most people refer to the United States as capitalistic. However, the U.S. government guides production in many industries. Government regulations apply to nearly all sectors of the U.S. economy. Government **subsidies** and grants act as incentives to increase or reduce production of goods and services.[23]

American Agriculture and the Mixed Economic System. Supply of and demand for specific agricultural products are directly influenced by the government's intervention in agriculture, with supply-control and demand-expansion programs. One example of such government involvement in agriculture is the **Payment in Kind (PIK)** program. Government intervention is not limited to the agriculture industry. Loan guarantees to businesses like Lockheed and Chrysler are also types of government intervention in the private sector.[24]

19. Ibid., p. 46.
20. Ibid.
21. Little, *Economics,* p. 27.
22. Gottheil, *Principles of Economics,* p. 862.
23. *Advanced Agribusiness Management and Marketing,* p. 47.
24. Penson, Pope, and Cook, *Introduction to Agricultural Economics,* p. 9; *Advanced Agribusiness Management and Marketing,* p. 47.

Economics—A Historical Perspective

Normally, an academic discipline evolves over a long period of time, and it is difficult to pinpoint a place or time when one begins. This is not the case with economics. Of course, economic problems have been around since resources became limited relative to the number of people. However, economic historians generally agree that modern economics as a discipline began in 1776 with the publication of a book by Adam Smith entitled, *An Inquiry into the Causes of the Wealth of Nations*. This book is commonly referred to as *The Wealth of Nations* and is generally considered to be one of the literary classics of that period.

The Father of Economics

England was the major center of intellectual activity during the 18th century and much current economic policy is based on economic ideas presented during that period. Economic historians contend that all the ideas presented by Adam Smith had been introduced before by other economists. However, his contribution was one of **synthesis**. He organized the ideas of predecessors into a major body of thought relative to economic problems of his time. As a result, Adam Smith is commonly referred to as the Father or Founder of Economics.

In *The Wealth of Nations,* Adam Smith was responding to the economic environment of his period. In the latter part of the 18th century, the government in England was highly involved in economic matters. Under that system, a small percentage of the population was wealthy, but the masses were extremely poor with no hope for improvement. Most economists of that period were discussing economic problems and proposing solutions.

Adam Smith's Proposed Economic System

Adam Smith, like most economists of that time, was looking for new directions in economic policy to deal with the severe economic conditions he saw around him. He proposed a system completely opposite to the system operating in England and many other countries at that time. Several names have been coined for the system he proposed. The major ones include pure capitalism, free enterprise, and **laissez-faire**. All of these terms are used interchangeably, and they mean basically the same thing. The basic meaning is that individuals can control the economy without any government interference.

The Influence of David Ricardo and Thomas Malthus

David Ricardo and Thomas Malthus were also prominent English economists during the 18th century, and each made a major contribution to economic thought in a special area. Ricardo was concerned primarily with the role of land as a resource and its effect on the total economic picture. Malthus was interested primarily in the effect of increasing population relative to limited food production.

The Role of Government Versus the Role of Individuals

A major question in all economic systems is, "What is the role of government versus the role of individuals?" Adam Smith said, "The government should provide a police force for internal and external protection and nothing else. Economic matters should be the sole responsibility of private individuals." He said, "Let each individual seek his or her own personal gain," and that society would be guided by an "invisible hand" in such a way that as a whole it would be better off than it would with government intervention. In other words, let each individual operate independently rather than having some government agency decide what should and should not be done.

Economic Classification According to the Role of Government in Economic Decisionmaking

Economic theory is often classified according to the role of government in economic decisionmaking. These two classifications are **classical economic theory** and **Keynesian** economics.

Classical Economic Theory. Classical economic theory contends that an economic system is self-sufficient in itself and any outside interference by government does more harm than good. Adam Smith and his followers are called classical economists. They advocated little or no government intervention in economic matters. This type of economic thought, in various degrees, was dominant in the United States until the 1930s. However, the Depression raised serious questions concerning the effectiveness of this system, and numerous economists, as well as political leaders, began to look for something different.

Keynesian Economics. The demise of the concept of classical economics began with the publication of *The General Theory of Employment Interest and Money* by John Maynard Keynes in 1936. Keynes, an Englishman, said that economic systems are not always self-sufficient and sometimes need outside help. Of course, the only help available outside the private sector is the public sector or government. He did maintain that public intervention should only be temporary. Once the system was working properly, the private sector should take over again.

Keynesian economics, which requires the government to do whatever is necessary to accomplish economic goals, began to play a major role in the United States with President Franklin D. Roosevelt's New Deal program. Ironically, the temporary role of government in economics evolved into permanent government intervention; the major question today is not whether the government should be involved in economic matters, but rather how much it should be involved. The economic policy in the United States has definitely shifted from classical economics to Keynesian economics.

Competition

One might ask, "What is the major characteristic of the capitalist economic system that would balance it out and keep a small group from getting control of a majority of the wealth?" The answer is a high degree of competition. Adam Smith envisioned a society of large numbers of small businesses. If there were excess profits in one industry, more firms would be developed, increasing the output of that product, and the price would be reduced to a level that would provide normal or acceptable profits. If an industry did not provide an acceptable profit, the least efficient firms would shift to other industries that offered potential for profit, thus keeping the economy balanced.

Individual Economic Freedom

The major characteristic of *pure capitalism* is individual economic freedom. However, in pure capitalism, a high degree of individual economic freedom requires a high degree of responsibility from each individual: if an individual cannot make a decent living, it is his or her fault, not that of some government agency. Those individuals who cannot get a job or cannot feed and clothe their families have no government agency to blame. The only sources of economic help for individuals under capitalism are the extended family and civic or religious groups. Pure capitalism is a very impersonal system, which provides a high degree of freedom but turns a deaf ear to individuals who cannot seem to get their fair share of the pie.

Political Party Philosophy

Adam Smith's concepts of pure capitalism had a major impact on many countries around the world, especially the United States. The economic system in the United States is generally considered the major example of capitalism, but it is mixed with small portions of other economic systems, as discussed earlier. Americans are also mixed on their philosophy as to the role of government versus individuals. The Republican Party philosophy tends to support *individual* economic control, whereas the Democratic Party philosophy tends to promote *economic assistance* from the government.

Characteristics of the American Economy

A common term for the U.S. economic system is the *free enterprise system*. It is the freedom of private businesses to organize and operate for profit in a competitive environment. Government interference is necessary only for regulation to protect the public interest and to keep the national economy in balance.[25]

The American economy has six major characteristics:

- little or no government control
- freedom of enterprise
- freedom of choice
- the right to own private property
- profit incentive
- competition[26]

25. *Advanced Agribusiness Management and Marketing,* p. 47.
26. Miller, *Economics Today and Tomorrow,* p. 37.

Figure 2–6 Although many individuals are opposed to government intervention, we take for granted a safe food supply. Most food products are inspected by the USDA. (Courtesy of USDA)

These characteristics are interrelated and, to varying degrees, all are present in the American economy. The six characteristics could be referred to as *free enterprise with some regulation*. Notice the influence of Adam Smith in these six characteristics, which are discussed in the following subsections.

The Role of Government

Although the role of government was discussed earlier, a few more comments will help put it in perspective. Capitalism, as currently practiced in the United States, can be defined as "an economic system in which private individuals own the factors of production and decide how to use them within the limits of the law."[27]

The Founders of the United States originally limited the role of government to national defense and keeping the peace. More recently, however, the role of the government has significantly increased, especially in the regulation of business and provision of public services. The work of federal agencies now includes regulating the quality of various foods and drugs, monitoring the nation's money and banking system, ensuring that workplaces are free of hazardous conditions, and protecting the environment. Refer to Figure 2–6.

Figure 2–7 The United States is not a pure capitalist economy because we have laws to control factors of production. Can you imagine these children working eight hours daily? Some countries continue to operate sweatshops. (Courtesy of USDA)

Freedom of Enterprise

Our economic system is also called the free enterprise system. This term emphasizes that individuals are free to own and make decisions about the factors of production.[28] Still, there are limits on free enterprise. In most states, teenagers must be 16 years of age before they can work, and then laws are set to limit how many hours they can work. Minimum wages are also set by law to protect workers. Refer to Figure 2–7.

Freedom of Choice

Part of freedom of choice is the freedom to fail. Freedom of choice means that buyers make the decisions about what should be produced. The success or failure of a good or service in the marketplace depends on individuals freely choosing what they want. Nevertheless, the government sets laws to protect buyers. Laws set safety standards for such things as labels or tags, appliances, and automobiles. Laws also regulate public utilities.

Private Property

Private property is simply what is owned by individuals or groups rather than by the federal, state, or local government. You are free to buy whatever you can afford, whether it is land, an agribusiness, a home, or a car. You can also control how, when, and by whom your property is used. If you own an

---

27. Ibid.

28. Ibid.

Figure 2–8 The freedom to own our land, manage it, and conserve it is a great characteristic of the American economy. (Courtesy of USDA)

agribusiness, you can keep any profit the business earns.[29] Refer to Figure 2–8.

Profit Incentive

Whenever a person spends time, knowledge, money, and other capital resources in an agribusiness, that investment is made with the idea of making a profit. The desire to make a profit is called the *profit incentive*. The goal of profit is what prompts people to produce things that others want to buy.[30]

Competition

Producers or sellers of goods hope to win more business by offering lower prices or better quality—this rivalry is competition in the free market. The skill in competition is the balance between keeping prices low enough to attract customers and high enough to still turn a profit. This forces businesses to keep the cost of production as low as possible. Competitors are successful because they are able to produce products at a price the average consumer can afford.[31]

Macroeconomics Versus Microeconomics

As disciplines evolve over time, the body of knowledge accumulates and eventually the discipline is broken into divisions and specialties. Economics is no exception. The total body of economic knowledge today is too extensive for a person to be a general economist. The most common division of economics is into **macroeconomics** and **microeconomics**.

Macroeconomics

The prefix *macro* means "large," indicating that macroeconomics is the study of the economy on a large scale or nationally. Macroeconomics looks at the **aggregate** (total) performances of all the markets in the national economy and is concerned with the choices made by large **subsectors** of the economy. These subsectors include the household sector, which includes all consumers; the business sector, which includes all companies; and the government sector, which includes all government agencies.[32] Seven key economic concepts are related to macroeconomics: gross domestic product, aggregate (total) supply, aggregate (total) demand, unemployment, inflation and deflation, monetary policy, and fiscal policy. A short explanation of each follows.

Gross Domestic Product. The gross domestic product (GDP) of a country is defined as the market value of all final goods and services produced within that country in a given period of time.

Aggregate Supply. Aggregate supply is the total supply of goods and services produced by a national economy during a specific time period.

Aggregate Demand. Aggregate demand is the total demand for goods and services in a national economy during a specific time period.

29. Ibid.
30. Miller, *Economics Today and Tomorrow,* p. 39.
31. Ibid., p. 40.
32. Duffy, *Economics,* p. 1.

Unemployment Rate. The unemployment rate is the number of unemployed workers divided by the total civilian labor force, which includes all those willing and able to work for pay—both unemployed and employed.

Inflation and Deflation. Inflation is a rise in the general level of prices, as measured against some baseline of purchasing power. Deflation is a decrease in the general price level, over a period of time.

Monetary Policy. Monetary policy is the government process of managing the money supply to achieve specific goals, such as constraining inflation, maintaining an exchange rate, and achieving full employment or economic growth.

Fiscal Policy. Fiscal policy determines changes in taxes, government spending on goods and services, and transfer payments that are intended to affect overall (aggregate) demand in the economy.[33]

Microeconomics

The prefix *micro* means "small," indicating that microeconomics studies the economy on a small scale. Microeconomics considers the individual markets that make up the economy and is concerned with the decisions made by small economic units such as individuals, single firms, and individual government agencies such as the USDA and state governments.[34]

Six key economic concepts are related to microeconomics: markets and prices, supply and demand, competition and market structure, income distribution, business failures, and the role of government. A short explanation of each follows.

Markets and Prices. Markets are institutions that enable buyers and sellers to exchange goods and services. The amount of money people pay in exchange for a good or service is the price of that good or service.

Supply and Demand. Supply is the different quantities of a resource, good, or service that will be offered to consumers at various prices during a certain amount of time. Demand is economic want backed up by purchasing power, expressing different amounts of a product buyers are willing and able to buy at possible prices.

Competition and Market Structure. Competition is determined by the number of buyers and sellers in a particular market. Market structures are determined by the success or failure of competition in particular markets.

Income Distribution. There are two classifications of income distribution. The first is functional distribution, where the division of an economy's total income goes into wages and salaries, rent, interest, and profit. Income distribution also can be classified by personal distribution of income, which groups different populations by the number of people receiving various amounts of income.

Business Failures. Business failures occur when there is a lack of competition, no reliable information, resource immobility, and not enough public demand for goods.

Role of Government. The government's role includes establishing a basic social order of law in which a market economy can function.[35]

Positive Versus Normative Economics

The media often contain statements about economic issues. These statements can be classified into "what is" and "what should be." These two main categories are called positive and normative economics, respectively.[36]

Positive Economics

Positive economics consists of objective statements dealing with matters of fact and questions

33. Miller, *Economics Today and Tomorrow,* p. 3.
34. Duffy, *Economics,* p. 2.
35. Miller, *Economics Today and Tomorrow,* pp. 2–3.
36. http://www.tutor2u.net/economics/content/topics/introduction/positive_and_normative_economics.htm (accessed August 13, 2007).

about how things actually are. Positive statements do not contain obvious value judgments or emotional content. They may suggest an economic relationship that can be tested against evidence or by research as being factual or valid.

Positive economics can be described as "what is, what was, and what probably will be" economics: statements based on economic theory rather than emotion or social philosophy. Often these statements express a hypothesis that can be analyzed and evaluated, such as the following examples:

- Higher interest rates will cause a rise in the exchange rate and an increase in the demand for imports.
- Lower taxes may stimulate an increase in the active labor supply.
- A nationwide minimum wage will probably cause a contraction in the demand for low-skilled labor.[37]

Cost-Benefit Analysis. "That person is hungry." What is the cost of feeding that person? What benefit accrues to you or society if that person is not fed? What is the cost of not feeding that person? What is the benefit of not feeding that person? Positive economics tries to answer such questions objectively, by doing what is called cost-benefit analysis.

Normative Economics

Normative economics consists of subjective statements based on opinion only, often without a basis in fact or theory. They are value-based, emotional statements that focus on "what ought to be." Consider the following examples:

- A national minimum wage is undesirable because it does not help the poor and causes higher unemployment and inflation.
- The national minimum wage should be increased as a method of reducing poverty.
- Protectionism is the only good way to improve the living standards of workers whose jobs are threatened by outsourcing and imports.[38]

Political Campaign Statements. It is important to be able to distinguish between positive and normative statements, particularly during arguments or debates. Most economists tend to take a positive approach. During political election years, we hear a great deal of normative economics from the candidates for office. Every four years, a new person comes along claiming to have the answer to all of the country's economic problems. "We should raise taxes." "We should lower taxes." "The rich do not pay enough taxes." These are all normative economics statements.[39]

Interrelationship between Positive and Normative Economics

Most statements are not easily categorized as purely positive or purely normative. However, economists have found the positive-normative distinction useful because it helps people with very different views to communicate. People with different values, beliefs, and cultures may have very different ideas about what is desirable. When they disagree, they can try to learn whether their disagreement arises from differing normative views or differing positive views. If their disagreement is normative, they know that their disagreement lies outside the scope of economics, so economic theory and evidence will not help them reach agreement. However, if their disagreement is positive, further discussion and study may draw them closer to agreement or compromise.

Few economists limit themselves to positive statements because this constrains what they can say about government policy issues. Both positive and normative views are included in a policy statement. One must decide what goals are desirable (the normative part), and choose a way of reaching those goals (the positive part). Often people propose courses of action that will never achieve the intended results. If economists confine themselves to evaluating whether proposed actions will bring about the intended results, they do only positive analysis. Nevertheless, although economists can lend special authority to positive issues, they can still be wrong.[40]

37. Ibid.
38. Ibid.
39. Robert Schenk, CyberEconomics: *An Analysis of Unintended Consequences*. http://www.ingrimayne.com/econ/Introduction/Normativ.html (accessed August 13, 2007).
40. Ibid.

Agricultural Economics

Agricultural economics may be defined in several ways, but basically it refers to *the application of economic concepts to agricultural problems*. Therefore, as stated earlier, to be a productive agricultural economist, one must first understand basic economic principles. Economics is sometimes called a **basic science**, whereas agricultural economics is an **applied science**. The following definition is a good example of merging general economics with agriculture. Agricultural economics is an applied social science dealing with how humans choose to use technical knowledge and scarce productive resources such as land, labor, capital, and management to produce food and fiber and to distribute it for consumption to various members of society over time. Refer to the Career Options for a further explanation of a career as an agricultural economist.

Agricultural economics, like most disciplines, did not begin in a given year with a specific publication, as economics did. Instead, it evolved from specific farm problems in the United States around the beginning of the 20th century. It evolved as a special study of agricultural problems, not as a specialty within the discipline of economics. Refer to Figure 2–9 for an example of a key agricultural economic concept.

Early Agricultural Economics Centered around Farm Production

Agricultural economics began as farm economics, with an emphasis on farm production and management problems. During the early 1900s, farmers were being pressured by the increasing cost of resources, on one hand, and lower prices received from marketing firms, on the other, which is commonly known as the **cost-price squeeze**.

Career Option

Agricultural Economist

Agricultural economics could be described as the application of economic concepts to problems of agriculture. Obviously, the job of economists is to alert us when the factors within the economy are beginning to get unbalanced. Therefore, to be a productive agricultural economist, one must first understand basic economic principles, including the relationship of the economy to federal, state, and local governments. Agricultural economics deals with how humans choose to use their knowledge and scarce productive resources, such as land, labor, capital, and management, to produce and market food and fiber.

The field of agricultural economics includes a wide array of content knowledge. It is such a broad area that it is very difficult for an individual to absorb all the economic information, so one usually specializes in one or more phases of the discipline. Some of the specialties include production economics (farm management), agriculture finance, rural development, price analysis, agriculture policy, international development, and marketing. If one chooses the marketing phase, he or she usually concentrates on a specific product or group of products, such as grain marketing, livestock marketing, cotton marketing, or fruit and vegetable marketing. Agricultural economists tend to have become economists working with agriculturalists rather than agriculturalists working with economists.

Beyond production, the key to making a profit in the agricultural industry is having a keen knowledge of agricultural economics. These agricultural economists are evaluating an array of options for their client's cattle-breeding program. (Courtesy of USDA)

Example of Key Agricultural Economic Concept

Utilization of Existing Resources: A Key Concept in Agricultural Economics

A very simple economic principle is that of the utilization of existing resources. Whatever business venture you plan, look around and determine what resources are already available for your use. Remember, the four factors of production discussed in an earlier chapter (land, labor, capital, and management). These resources should be considered in any business decision.

Let us suppose that you are trying to decide whether to go into the dairy farming business or the flower and plant production business. One of your uncles has a 200-acre farm that he wants to keep in the family, and another uncle has just retired and has three 30- by 80-foot greenhouses equipped with heat, water, and tables. Both relatives have agreed that your only cost for using the facilities would be to keep them maintained, neat, and orderly, plus pay the property taxes. The taxes are approximately $2,000 yearly.

Which would you choose? Both are rare opportunities. Considering that dairying and flower and plant production are equally attractive, the greenhouse facilities would be the best utilization of available resources because most of the resources are in place. Except for consumable supplies, with the greenhouse selection, you are ready for production. With the dairy, 200 acres in land is available, but over $200,000 in capital would still be needed to build the dairy barn and silos and to purchase tractors, equipment, and cows. Assuming an interest rate of 8 percent on $200,000 of borrowed capital, $16,000 interest would have to be paid the first year alone. These costs can be avoided by selecting the greenhouse option.

Although the utilization of resources is a simple concept, it is often overlooked. Even for those who have already started a business, it is wise to make production decisions based on resources that already exist whether it be family labor, existing land, or other utilized tractors and equipment. Dr. Warren Gill, University of Tennessee Beef Specialist, jokingly presents the "UDE" ("use daddy's equipment") principle of agricultural economics to explain the concept of utilization of resources. You should utilize whatever resources are available and affordable to maximize profit.

Figure 2–9 Many agribusinesses fail because they spend unnecessary money by not utilizing the resources already at their disposal.

In an effort to solve these problems, specialists from a wide variety of established disciplines were recruited; as a result, a new discipline evolved. The major emphasis of agricultural economics is still farm production economics. However, the discipline has expanded considerably, and today it encompasses the total agricultural industry.

Specialty Areas in Agricultural Economics

Almost a century of study of agricultural economics has yielded a vast body of knowledge, and new knowledge is being added constantly. As a result, it is very difficult for an individual to absorb the total amount of economic information, so we have to specialize in one or more phases of the discipline.

Some of the specialties include production economics (farm management), agricultural finance, rural development, price analysis, agricultural policy, international development, and marketing. Those students interested in marketing usually concentrate on a specific product or group of products, such as grain marketing, livestock marketing, cotton marketing, or fruit and vegetable marketing.

Conclusion

Agricultural economics has many subspecialties, including crop, livestock, land, resource, and water economics. Some of the many areas in which agricultural economists work include marketing, finance, business organization, policy, statistics, and resource development.[41]

41. Little, *Economics,* p. 6.

Agricultural economics is also related to consumer economics. The production agriculturalist's home and family are integral parts of the business, as the family competes for funds with the agribusiness itself and is a integral component of its processes.[42]

Agricultural economists increasingly act as advisors or consultants to other social sciences and disciplines because of their knowledge of, and concern for, the efficient use of resources. Agricultural economists have become economists working with agriculture rather than agriculturalists working with economics.[43]

Summary

Today's world is complex and continually changing. The successful agribusiness manager must possess a basic understanding of economic principles to react to these changes. To understand agricultural economics, the agribusiness manager must first understand basic economic principles.

There are many definitions of economics. Three key words or phrases can be gotten from each of these definitions: scarcity, types of resources, and wants and needs.

Because of the relationship between scarce resources and unlimited/unsatisfied wants, all societies have to answer some basic economic questions. These questions entail trying to decide what to sell, how to sell it, and who should receive the benefits.

Each society answers these basic questions according to its view of how best to satisfy the needs and wants of its people. The values and goals that a society sets for itself determine the kind of economic system it will have. Economists have identified six types of economic systems: traditional, capitalism, fascism, socialism, command (communism), and mixed economic systems.

Economic historians generally agree that modern economics as a discipline began in 1776 with the publication of Adam Smith's *An Inquiry into the Causes of the Wealth of Nations*, which is considered one of the literary classics of that period. Because of the influence of this book, Adam Smith is commonly referred to as the Father or Founder of Economics.

A major question in all economic systems is, "What is the role of government versus individuals?" Therefore, economics is often classified as to the role of government in economic decisionmaking. These two classifications are classical economic theory and Keynesian economics. Classical economic theory contends that an economic system is self-sufficient and that any outside interference by government does more harm than good. Keynesian economics contends that economic systems are not always self-sufficient and sometimes they need outside help.

A common term for the U.S. economic system is *free enterprise system*. This guarantees the freedom of private businesses to organize and operate for profit in a competitive environment. Government interference is necessary only for regulation to protect the public interest and keep the national economy in balance. The American economy has six major characteristics: little or no government control, freedom of enterprise, freedom of choice, private property, profit incentive, and competition.

The two most common divisions of economics are macroeconomics and microeconomics. Macroeconomics is concerned with the study of the economy on a large scale or nationally. Microeconomics is concerned with the study of the economy on a small scale. Small-scale economic units include individual consumers, firms, farms, or agribusinesses.

Positive economics are objective statements dealing with matters of fact or questions about how things actually are. Positive statements are made without obvious value judgments or emotion. Normative economics are subjective statements based on opinion, often without consideration of fact or theory. They are value-based, emotional statements that focus on "what should be."

Agricultural economics is the application of economic concepts to agricultural problems. To be a productive economist, you must first understand basic economic principles. Agricultural economists have become economists working with agriculture rather than agriculturalists working with economics.

42. Ibid.
43. Little, *Economics,* p. 7.

End of Chapter Activities

REVIEW QUESTIONS

1. Define the Terms to Know.
2. Explain the difference between shortage and scarcity.
3. Name and briefly discuss four types of resources that are classified by economics and that affect the economy.
4. Explain the difference between a need and a want.
5. Name and briefly explain the three basic economic questions.
6. Compare or match the three basic economic questions with three analogies of goods and services in our society as represented by making a pie.
7. List seven major components of a capitalistic economic system.
8. List eight components of a socialistic economic system.
9. Briefly describe Adam Smith's economic beliefs.
10. List the three major characteristics of an American economy.
11. Name three large subsectors of the economy that macroeconomists study.
12. Name and briefly discuss seven economic concepts related to macroeconomics.
13. Name and briefly discuss six economic concepts related to microeconomics.
14. Briefly describe how agricultural economics began.
15. Name seven specialty areas of agricultural economics.

FILL IN THE BLANK

1. Economics is concerned with overcoming the effects of _________ by improving the efficiency with which scarce _________ are allocated among many competing uses, so as to best satisfy human _____ _____________.
2. _________ is the study of how society allocates scarce resources.
3. A resource that is not scarce is called a _________ resource or good.
4. The book __________________ was written by Adam Smith and is considered the beginning of economics as we know it today.
5. A common term for the U.S. economic system is _________ enterprise system.
6. Part of freedom of choice is the freedom to _________.
7. The goal of _________ is what moves people to produce things that others want to buy.
8. _________ is the rivalry among producers or sellers of similar goods to win more business by offering the lowest price or better quality.
9. The prefix *macro* means "large," indicating that __________ is concerned with the study of the economy on a large scale or nationally.
10. The prefix *micro* means "small," indicating that _________ is concerned with the study of the economy on a small scale.
11. ___________ economics are objective statements dealing with matters of fact.
12. ___________ economics are subjective statements based on opinion.

13. Economics is sometimes called a basic science, whereas agricultural economics is called an _________ science.
14. Agricultural economists have become economists working with agriculture rather than agriculturalists working with _________.

MATCHING

a. fascism
b. Adam Smith
c. David Ricardo
d. capitalism
e. Thomas Malthus
f. traditional economic system
g. Democratic Party
h. mixed economic system
i. Republican Party
j. John Maynard Keynes
k. communism
l. socialism

_____ 1. "the way things have always been done"
_____ 2. the basis of this economic system is public ownership of all productive resources
_____ 3. this economist was interested primarily in the effect of increasing population relative to limited food production
_____ 4. the government has total control of economic matters and private individuals have none
_____ 5. Father of Economics
_____ 6. an economic system that possesses the characteristics of both capitalism and socialism
_____ 7. an economic system in which individuals own the land but produce goods that reflect government preferences
_____ 8. economist who believed that economic systems are not always self-sufficient and sometimes need outside help
_____ 9. political party that tends to favor individual economic control
_____ 10. political party that tends to favor government assistance
_____ 11. economist who was concerned primarily with the role of land as a resource and its effect on the total economic picture
_____ 12. economic system in which individuals own resources and have the right to use their time and resources however they choose

ACTIVITIES

1. Read the nine definitions of economics. Select the definition that makes the most sense to you and write a paragraph explaining why.
2. List five wants and five needs that you have. Share these with the class. Do your classmates agree with you as to those that are wants and those that are needs?
3. Write a 100-word essay on why entrepreneurship or management is considered a factor of production.

4. Pretend that you wish to start an agribusiness. Select the type of agribusiness and answer the three basic economic questions.
5. Select and defend the economic system that you favor. Write a short essay and present it to your class. Give an explanation of features that appeal to you, along with any weaknesses.
6. Select an economist such as Adam Smith, David Ricardo, Thomas Malthus, John Maynard Keynes, or anyone else. Using the library, Internet, or other sources, write a one- to two-page report on who this person was and his or her economic philosophy or contribution to the field of economics.
7. Select and defend your position as to whether you favor individual economic control or governmental assistance with the economy. Your teacher may have you debate this or participate in a panel discussion. Either way, write a one-page essay defending your position.
8. Study the six major characteristics of the American economy covered in this chapter. Select the one that you could most easily give up and the one that you would most want to hold onto. Write a one-page essay defending your choice and share this with the class.
9. There are seven key economic concepts related to macroeconomics. Refer to these in this chapter. Select one concept and prepare a one- to three-page report further elaborating on your selection.
10. Do the exercise in question 9 with one of the key microeconomic concepts.
11. Write a paragraph (or more if needed) explaining why economics is a basic science whereas agricultural economics is an applied science.
12. Select a product that you use every day. Identify as many factors as you can that were used in making the product. Classify each factor as either a natural resource, human resource (labor), manufactured resource (capital), or management (entrepreneurship).

Chapter 3

The Size and Importance of Agribusiness

Objectives

After completing this chapter, the student should be able to:

- Discuss the size and importance of production agriculture.
- Analyze the efficiency of production agriculture.
- Discuss the impact of U.S. agriculture on the global economy.
- Discuss the agribusiness economy.
- Describe the agriservice sector of the agricultural industry.
- Explain the importance of agribusiness and foreign trade.
- Describe the relationship between agribusiness and energy.
- Describe the relationship between agribusiness and the environment.

Terms to Know

agribusiness input sector
agribusiness output sector
agriservices sector
distillation
E-85
ecologists
economists
enterprises
environmentalists
ethanol
exports
fermentation
gasohol
gross domestic product (GDP)
humus
imports
industrialization
outstanding loans
private agriservices
production efficiency
public agriservices
value-added

Introduction

In Chapter 1, we discussed the fact that the agricultural industry is changing. We learned that nearly everyone was a farmer in early America. Today, however, the effectiveness and efficiency of the production agriculturalist is so great that less than 3 percent of the American population is actively involved in the production of food or fiber.

Because farming (production agriculture) and agribusiness are often presented as the same thing, distorted pictures of the agricultural industry are reinforced. For example, because the number of farms has been decreasing for a number of years, many people have concluded that the agricultural industry is declining. Mistakenly, high school counselors often advise students that opportunities in agricultural occupations are declining because, they note, the number of farmers is decreasing. However, they do not take into consideration the fact that agricultural exports and imports and agricultural inputs and outputs are at all-time highs.

Today's modern agricultural industry of production agriculturalists, agrisciences, and agribusinesses encompasses a massive and complex group of organizations and the people who run them. This chapter discusses the size and importance of various areas of agribusiness that make up the agricultural industry. However, first let us look at production agriculture, which makes all agribusiness possible.

Size and Importance of Production Agriculture

There could be no agribusiness without production agriculture. It all starts with the land. Goods from the earth come from the land, from farms and the hard work of production agriculturalists.[1] Day by day, the role of American production agriculturalists grows in importance, as more and more people become dependent on them.

1. Bureau of Economic Analysis, United States Department of Commerce (Washington, DC: U.S. Government Printing Office, 1999).

Land

Because land is the major resource in modern farming, we are especially interested in how much is available and how it is being used. The United States has about 2.3 billion acres. About 21 percent of the land is used for crops, 25 percent for livestock, and 30 percent for producing forest products. The remaining 24 percent is used for nonagricultural purposes.

Some people contend that because of increasing nonagricultural uses of land, we are running out of good farmland. The facts do not support this conclusion. Even though the number of acres of farmland is decreasing, the economic land base is increasing. The productivity of land is increasing faster than the acres are decreasing.

Farm numbers have declined since the 1920s, but at different rates. Currently, about 2.1 million farms are producing the total food supply for Americans and numerous foreign countries. Although farm numbers are decreasing, the average size of farms is increasing. Currently, the average American farm is 469 acres. This means that most farms that go out of business are bought by other farmers. **Economists** predict that the trend toward fewer farms will continue, but the decrease will occur in small- and medium-sized farms. Larger farms are expected to increase in number and volume of sales. Presently, about 176,990 farms contain 1,000 acres or more; 161,550 farms contain 500 to 1,000 acres; more than 1,611,090 farms have 10 to 500 acres; and approximately 179,350 farms have less than 10 acres.[2]

Products

To see the impact of production agriculture on the U.S. economy, we need to reference the yardstick that measures the value of goods and services that America produces in a year, the **gross domestic product (GDP)**. The agricultural industry accounts for 17 percent of the GDP and provides

2. National Agricultural Statistics Service, U.S. Department of Agriculture, *2002 Census of Agriculture*, vol. 1, Geographic Area Series Part 51 (Washington, DC: U.S. Government Printing Office, 2004); http://www.nass.usda.gov/Census_of_Agriculture/Publications/2002.

2002 Census of Agriculture
Value and Percentage of Agricultural Products Sold

Item	Farms	Sales ($1,000)	Rank by Sales	Percenct of Total Sales
Total sales (see text)	2,128,982	200,646,355	(X)	100.0
Cattle and calves	851,971	45,115,184	1	22.5
Grains, oilseeds, dry beans, and dry peas	485,124	39,957,698	2	19.9
Poultry and eggs	83,381	23,972,333	3	11.9
Milk and other dairy products from cows	78,963	20,281,166	4	10.1
Nursery, greenhouse, floriculture, and sod (see text)	56,070	14,686,390	5	7.3
Fruits, tree nuts, amd berries	107,707	13,770,603	6	6.9
Vegetables, melons, potatoes, and sweet potatoes	59,044	12,785,898	7	6.4
Hogs and pigs	82,028	12,400,977	8	6.2
Other crops and hay (see text)	359,262	7,929,618	9	4.0
Cotton and cottonseed	24,721	4,005,366	10	2.0
Tobacco	56,879	1,616,533	11	0.8
Horses, ponies, mules, burros, and donkeys	128,045	1,328,733	12	0.7
Aquaculture	6,653	1,132,524	13	0.6
Other animals and other animal products (see text)	29,391	721,738	14	0.4
Sheep, goats, and their products	96,249	541,745	15	0.3
Cut Christmas trees and short-rotation woody crops	14,744	399,848	16	0.2

Figure 3–1 These sales, rankings, and percentages represent the value of agricultural products sold by production agriculturalists. (Courtesy of USDA, *2002 Census of Agriculture*)

more than 20 percent of all the jobs in the country. Two percent of the GDP comes from firms or people that sell goods and services to production agriculturalists. However, 13 percent of the GDP comes from related industries. These related industries include ice cream makers, textile mills, flour mills, tanneries, breakfast food makers, and a host of others (mentioned in Chapter 1). These related industries purchase food and fiber from production agriculturalists and then process and package it so they will have a **value-added** product to sell to consumers.

In 2002, America's GDP was more than $8.5 trillion, and the agricultural industry accounted for 17 percent, which comes to an impressive $936 billion.[3] To place the $936 billion in perspective, it would be enough money to feed the entire U.S. population for almost five years.

The products produced by production agriculturalists are very diverse. The relative percentages of the various farm products, from production agriculturalist estimates, are shown in Figure 3–1. There are many kinds of production agriculturalists in America. Where they are located and what

3. Economic Research Service, U.S. Department of Agriculture, available at http://www.ers.usda.gov/briefing/baseline/present2003.htm.

they raise depend on conditions such as the type of soil, topography, climate, rainfall, and markets.

The animal **enterprises** with the greatest value of production for the United States are beef cattle and calves, dairy cattle, hogs, and poultry. The leading plant enterprises in America, in value and production, are corn for grain, soybeans, wheat, and cotton. Among the fruits, the leaders are grapes, oranges, and apples. Almonds are the leader in the nut category. Refer to Figure 3-2 for a list of production agriculturalist enterprises.

Percent of Food Dollar to the Production Agriculturalist

The production agriculturalist's share of each dollar spent for food by the consumer is about

Production Agricultural Enterprises			
Crops			
Barley	Oats	Sunflowers	Nursery Crops
Beans	Peanuts	Tobacco	Others
Corn	Popcorn	Wheat	
Cotton	Rice	Timber (lumber)	
Hay	Sorghum	Timber (paper)	
Maple Syrup	Soybeans	Christmas Trees	
Mint Oil	Sugarcane	Floriculture	
Vegetables			
Asparagus	Cabbage		
Beans, green/lima	Carrots		
Beans, snap	White Potatoes		
Beets	Sweet Potatoes		
Broccoli	Pumpkins		
	Lettuce		
	Others		
Animals			
Broilers (meat)	Sheep and Lambs	Ostriches	
Cattle and Calves	Wool	Emus	
Chicken (eggs)	Turkeys	Other Fowl	
Hogs and Pigs	Bees for Honey	Others	
Dairy (milk)	Horses		
Fruits and Nuts			
Apples	Grapefruit	Peaches	Strawberries
Apricots	Grapes	Pears	Tangerines
Blueberries	Lemons	Pecans	Tangelos
Cantaloupes	Limes	Persimmons	Tomatoes
Cherries	Oranges	Pomegranates	Watermelons
Cranberries	Papayas	Plums and Prunes	Other

Figure 3-2 Many agricultural products are produced by production agriculturalists. Those listed here represent a majority of them.

America's Share of World Food Production			
Product	**Share**	**Product**	**Share**
Soybeans	64%	Grapefruit	56%
Corn	46%	Sorghum	31%
Oilseeds	42%	Eggs	17%
Poultry	24%	Oranges	25%
Beef	23%	Green Peas	23%
Pork	13%	Wheat	17%
Cotton	17%	Milk	15%

Figure 3–3 American agriculture is a showcase for the world. These percentages represent the impact of American agriculture on the world food supply. (Courtesy of USDA)

30 cents. This means that approximately 70 cents of every dollar the consumer spends for food products go to pay the cost of transportation, processing, other marketing services, and profits after the products leave the hands of the production agriculturalist. Currently, farmers take in about $150 billion a year from the sale of farm products, plus another $45 billion from off-farm sources.[4]

Efficiency of Production Agriculture

One of the most outstanding characteristics of production agriculture in America has been the tremendous increase in **production efficiency**. The efficiency of American agriculture is second to none. With less than 0.3 percent of the world's production agriculturalists, the United States produces a major percentage of the world's total food supply. Refer to Figure 3–3 to see America's share of the world's food production.

There are many different ways to measure farm efficiency, and the degree of efficiency depends on the type of measurement used. One of the most common indicators used is the number of persons supplied by the average farmworker. Presently, the average farmworker supplies more than 150 persons with food and fiber, compared with only 20 in 1955 and 15 in 1945. Annual consumption of food in the United States is 1,365 pounds per person. Production agriculturalists produce an average of about 108,000 pounds, which is almost 54 tons of food! This productivity frees others to pursue careers in science, government, the arts, medicine, computers, and many other areas.[5]

American farmers are recognized throughout the world for their ability to continually increase output while inputs remain relatively constant. U.S. farm output has increased more than 60 percent since 1950, while total inputs into farming have remained the same. In the past 20 years, agricultural productivity has increased more than three times faster than industrial productivity per hour worked. Today, an hour of farm labor produces 16 times as much food and fiber as it did 60 years ago. One production agriculturalist creates six agribusiness jobs for people who produce the things production agriculturalists need and who process, transport, and merchandise the things farmers produce.[6]

In 1946, bread cost 10 cents a loaf and steak was 50 cents a pound. These prices may sound like a bargain, but the pay for an hour's work was a little over a dollar before taxes and other deductions. Although retail food prices are more than four times higher than a half century ago, the average worker's paycheck is 11 times higher. When comparing costs over time, effort required is a better way to visualize the cost to consumers. An hour of work today is the same as it was at any time in history, but a dollar today is not the same as it was in 1930. Refer to Figure 3–4.

Today, Americans spend less than 9 percent of family income on food. In some countries, as much as 60 to 70 percent of income goes for food. Americans spent 16.2 percent of income on food in 1973 and 18.7 percent in 1963.[7]

4. Ibid.

5. Ibid.
6. Ibid.
7. Ibid.

Minutes of Work Equal to the Price of Selected Food Items[1]

Item	Amount	1930	1950	1970	1980
			Minutes		
Round Steak	1 lb.	48.4	43.8	28.8	29.4
Potatoes	10 lb.	40.9	21.6	19.9	20.2
Bacon	1 lb.	48.3	29.8	21.0	15.5
Eggs	1 doz.	50.6	28.3	13.6	8.9
Bread (2 loaves)	3 lb.	29.3	20.1	16.1	16.2
Butter	1 lb.	52.7	34.1	19.2	20.0
Milk	1 gal.	64.1	38.6	29.2	22.3
Coffee	1 lb.	44.9	37.2	20.2	33.3
Sugar	5 lb.	34.7	22.7	14.4	22.8
Rice	5 lb.	54.0	39.3	21.2	27.2
All the above	1 ea.	467.9	315.5	203.6	215.8

Source: USDA.

[1]Price of food item relative to manufacturing wage rate after taxes and employee social security contributions. Through the Internet or other sources, determine what the current prices would be.

Figure 3–4 This figure represents the efficiency and effectiveness of American agriculture. Notice the eggs: it took 50.6 minutes to buy a dozen eggs in 1930, but only 8.9 minutes in 1980. What do you suspect the figure is today? (Courtesy of USDA)

Consider the following agriculture facts:

- More than 90 percent of the food items consumed in the United States is produced partly or completely by American farmers.
- Per capita income in the United States increased 48 percent from 1970 to 1997. The amount of money spent on food increased only 23 percent; the increase was caused mostly by a rise in the amount spent on dining out.
- It takes the average American 40 days to earn enough money to buy food for a year. It takes 131 days for the average American to earn enough to pay all taxes (federal, state, and local) for one year.
- Since 1978, the share of the American food dollar spent on processing and marketing has increased 180 percent.[8]

8. http://www.alfafarmers.org/ag_facts/food_dollar.phtml (accessed August 13, 2007).

Why Is Efficiency So Important?

Increased efficiency of production agriculture releases manpower for other work and for increasing **industrialization**. Greater industrialization leads to a healthier economy, the result of which has been gradual elevation of the standard of living of all people.

U.S. Agriculture and the Global Economy

Agriculture is one of the world's largest industries. On a worldwide basis, more people are in some way involved in agriculture than in all other occupations combined. The United States produces more food than any other nation in the world. Export revenues account for 25 to 30 percent of U.S. farm cash receipts and are a key factor in determining gains in net farm income. According to the 2002 survey from the USDA's National Agricultural Statistics Service, in addition to providing an abundant food supply for domestic markets, crops from nearly one-third of U.S. farm acreage are exported to overseas customers.[9]

Issues

Agriculture in the United States is now a large part of a complex global food system. Although production agriculture is a solid foundation, other factors, such as the international business climate, worldwide consumer needs, and the application of technology, increasingly will affect the food system in the future.

Despite the fact that the increase in agricultural productivity has slowed over the past 40 years, production continues to grow to meet rising world demand. An increase in the number of cultivated acres is not likely; rather, rising yields will create this growth. Biotechnology, plant-breeding innovations, and better pest resistance are largely responsible for the higher yields.

Agricultural economic experts expect U.S. exports of value-added commodities to increase significantly from about 2010 to 2015, but also foresee raw commodity exports gaining as livestock production takes place nearer to the end consumer.[10]

Magnitude

Fifty years ago, the United States was the largest agricultural exporter in the world, with about $3 billion in sales per year. Of its 10 best customers, 6 were developed countries in Western Europe; 2 others (Japan and Canada) also were developed countries. The only developing countries that were major markets were India, a food aid recipient, and Cuba (before the Castro regime took power). Today, U.S. agricultural exports exceed $50 billion annually. Six of its 10 best customers are developing countries, and three-fourths of U.S. agricultural exports go to Asia and the Americas.[11]

9. National Agricultural Statistics Service, U.S. Department of Agriculture, *2002 Census of Agriculture.*
10. Mace Thornton, "The Food System of the Future," American Farm Bureau Association *FB Focus*, available at http://www.fb.org/index.php?fuseaction=newsroom.focusfocus&year=1998&file=fo1012.html (accessed August 13, 2007).
11. David W. Raisbeck, "The Role of Agriculture in the Global Economy," May 18, 2003, http://www.cargill.com/news/media/030518raisbeck.htm (accessed August 13, 2007).

Events Shaping the Global Agricultural Market

Three events have transformed the global agricultural market. The first was the formation of the European Community and its Common Agricultural Policy (CAP). The CAP's farm supports changed the EC from a 20-million-ton-per-year net grain importer in the 1960s to a 20-million-ton-per-year net grain exporter by the 1980s.

The second notable event was the collapse of many of the socialist economies, particularly that of the former Soviet Union. At one time, the USSR and the People's Republic of China were importing 40 to 50 million tons of grain annually; now, Russia and China are net grain exporters.

The third event was developing countries' takeover of the 80-plus million tons of grain imports no longer needed in traditional markets. Developing areas took up this slack, and more, as they hugely expanded their commercial grain imports. Although total world grain trade has not grown significantly in the past few decades, the grain trade patterns have changed dramatically. It appears that the future of world grain trade depends on increasing food demand in the developing world.[12]

World Trade

World trade is the common thread in these changes. The General Agreement on Tariffs and Trade (GATT) world trade agreement established baseline trade positions and protectionism levels for each signatory country. However, GATT has not resulted in noticeable reductions in protectionism; what is effectively a variable trade levy has kept Europe in essentially the same status it had before GATT. The agreement did, nevertheless, set forth principles to build on during future talks. As world trade negotiations increasingly focus on the hotly disputed issue of how products are grown, regional trade agreements such as the North American Free Trade Agreement seem better positioned to enhance trade.[13]

12. Ibid.
13. Thornton, "The Food System of the Future."

The Agribusiness Economy

Agriculture is the largest industry in the United States. As mentioned earlier, agribusiness accounts for approximately 17 percent of the nation's total economic output. It is also the country's largest single employer, providing more than one out of every five jobs. Production agriculturalists sell approximately $123 billion in farm products, but by the consumer purchases those products, the value-added brings the amount to more than $546 billion.[14] This means that the dollar increase from farmer to consumer is nearly fourfold. The importance of agribusiness in America is reflected throughout the world because agribusiness is the single most important contributor to the world's economy.[15] Agribusiness worldwide represents approximately one-fourth of the total world economic production and provides employment for nearly half the population on earth.[16]

The Agribusiness Input Sector

One of the most neglected areas of study in agriculture concerns the supplier of inputs for farm products, the **agribusiness input sector**. Agriculturists have assumed that this group constituted a section of general business and is not a part of agriculture. However, input suppliers play a major role in the production of food and fiber, and the sector is currently recognized as a major phase of agribusiness. In 2002, more than $200 billion was spent on agribusiness inputs by American production agriculturalists.[17]

Agricultural input provides production agriculturalists with the feed, seed, fertilizer, credit, machinery, fuel, chemicals, and various other things that they need to operate. Improvements in the quality of agricultural inputs have been the major reason for the outstanding efficiency of production agriculturalists. Examples of some larger firms within the agribusiness input sector include John Deere (farm equipment), Ralston Purina (feed), Pioneer Seed, International Mineral and Chemical (fertilizer), and Monsanto (chemicals).

Farm equipment, including tractors, other motor vehicles, and machinery, cost production agriculturalists approximately $8.8 billion each year and requires 140,000 employees to produce. Industry requires 40,000 workers to produce the 7 million tons of steel needed for farm equipment, trucks, cars, fencing, and building materials.[18]

Fuel, lubricants, and maintenance for machinery and motor vehicles used by production agriculturalists cost $10.1 billion each year. Production agriculture requires more petroleum than any other single industry.

Other yearly expenditures for agricultural inputs include:

- $3.9 billion for seed
- $18.4 billion for feed
- $10.5 billion for livestock
- $8.8 billion for fertilizer and lime
- $10.3 billion for hired labor
- $21.7 billion for depreciation and other consumption of farm capital[19]

In view of the trend toward specialized farming, the input sector is expected to become an even more important factor in food production. Production agriculturalists spent $6 billion more in 2002 than in 1997, and this trend is expected to continue because of the demand for more inputs by farmers as they attempt to supply the world's need for more food. Figure 3–5 shows the major agribusiness expenditures as a percentage of the total.

The Agribusiness Output Sector

The **agribusiness output sector** includes all agribusinesses and individuals that handle agricultural products from the farm to the final consumer. This includes agribusinesses involved in buying, transporting, storing, warehousing, grading, sorting, processing, assembling, packing, selling,

14. National Agricultural Statistics Service, *2002 Census of Agriculture.*
15. N. Omri Rawlins, *Introduction to Agribusiness,* 3d ed. (Murfreesboro, TN: Middle Tennessee State University, 1998), p. 16.
16. Ibid.
17. National Agricultural Statistics Service, *2002 Census of Agriculture.*
18. Rawlins, *Introduction to Agribusiness*, p. 80.
19. National Agricultural Statistics Service, *2002 Census of Agriculture.*

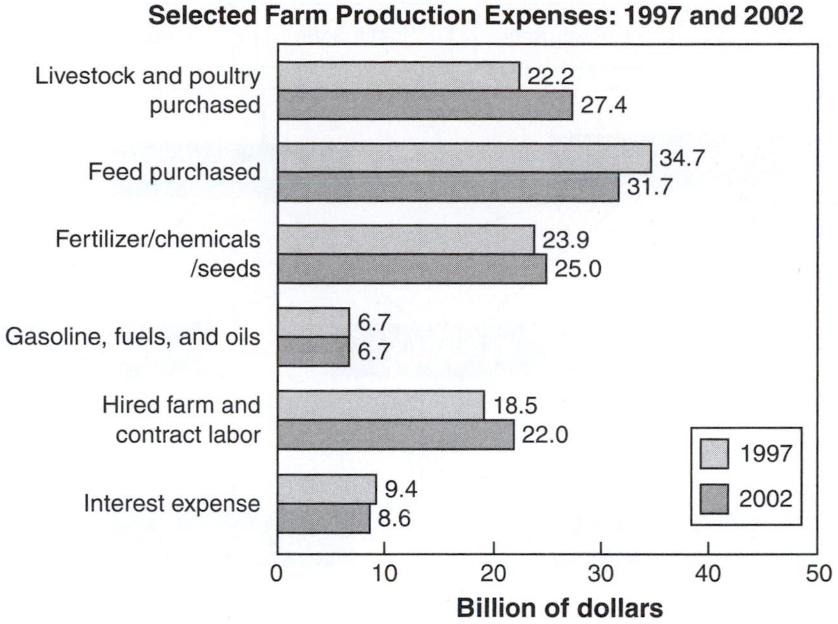

Selected Farm Production Expenses: 1997 and 2002

	1997	2002
Livestock and poultry purchased	22.2	27.4
Feed purchased	34.7	31.7
Fertilizer/chemicals /seeds	23.9	25.0
Gasoline, fuels, and oils	6.7	6.7
Hired farm and contract labor	18.5	22.0
Interest expense	9.4	8.6

0 10 20 30 40 50

Billion of dollars

Figure 3–5 Many agribusiness inputs are needed by production agriculturalists. This graph shows the percentage of total expenses of each of the agribusiness inputs. (Courtesy of USDA) Source: http://www.nass.usda.gov/Census_of_Agriculture/Publications/2002/Quick_Facts/selectfarm.asp.

merchandising, insuring, regulating, inspecting, communicating, advertising, and financing. Restaurants, fast-food chains, and grocery stores are also a part of the agribusiness output sector. Figure 3–6 illustrates a few of the various agribusinesses in the output sector.

Processing and Manufacturing Sector. Approximately 22,000 companies are active in the commodity processing and food manufacturing sector. They add more than $202 billion of value to the items they handle. The 100 largest firms account for about half of the output of this sector.

The agribusiness output sector needs approximately 20 million workers to handle the output of the nation's farms. The following are examples of where these workers are employed:

Meat-processing industry	350,000
Dairy-processing industry	175,000
Baking industry	225,000
Canned, cured, and frozen foods industry	240,000
Cotton/textile industry	175,000[20]

The USDA estimated that more than 600,000 businesses are involved in marketing food products.[21] Nearly $425 billon was spent by consumers for food products alone within the agribusiness output sector. This represents about 9 percent of consumers' disposable personal income. Some of the better known agribusiness firms include Kellogg's, Pillsbury, General Mills, Campbell's Soup, Beatrice Foods, Carnation, H. J. Heinz, Hershey Foods, Archer Daniels Midland, IBP Meat Packers, Mid-America Dairymen, and Sunkist. Figure 3–7 shows sales and rank of the top agribusiness corporations (both input and output) in America.

Agriservices

The **agriservices sector** of the agriculture industry is concerned with researching new and better ways to produce and market food and to protect food producers and consumers, and with providing special, customized services to all the other phases of agriculture. Public agencies have played a dominant role in the agriservices area, but private

20. Ibid.

21. Rawlins, *Introduction to Agribusiness*, p. 29.

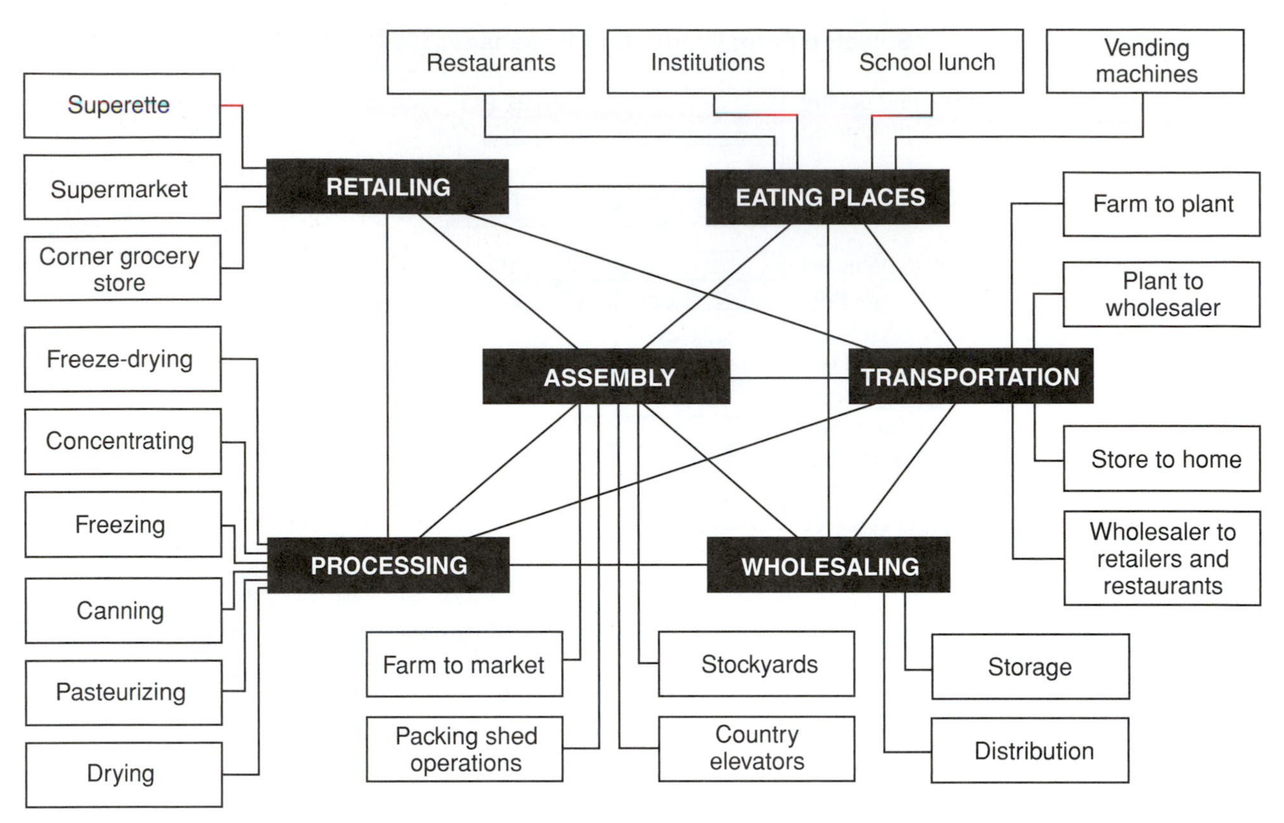

Figure 3–6 The agribusiness output sector involves many agencies and functions after the agricultural products leave the farm. (Courtesy of USDA)

agencies are rapidly increasing their offerings of farm services.

Public Agriservices

The **public agriservices** group provides special services at the federal, state, and local levels. The major areas of emphasis include research, education, communication, and regulation. The USDA alone has more than 100,000 employees, 12,000 of whom work in Washington, D.C. Their work is often closely aligned with that of state and county agriculture workers. Close to 17,500 employees work in the state and federal extension services, and 24,000 more are associated with the state agriculture experiment stations. About 25,000 county committee members administer the agriculture programs established by Congress.[22] There are approximately 12,000 teachers of agricultural education. Other public agriservices include the Food and Drug Administration; the Department of the Interior, which includes the National Park Service; and a small portion of the Department of Commerce. Public agriservices are discussed further in Chapter 15.

Private Agriservices

According to the USDA, production agriculturalists paid more than $1 billion yearly for **private agriservices** such as veterinary care, feed grinding and mixing, machine harvesting, contract labor, and spraying. More than 30,000 firms provide agricultural services, some of which are nonfarm. This group employs more than 100,000 workers and has a payroll of about $600 million. This does not include about 30,000 nonpaid owners and family members and other part-time workers. Specifically, three major areas of private agriservices serve the agricultural industry: financial services, trade associations, and agricultural cooperatives.

22. Ibid., pp. 28, 115.

Largest Agribusiness Corporations in America

Agribusiness Company	Rank	Sales ($ Millions)	Agribusiness Company	Rank	Sales ($ Millions)
Wal-Mart Stores[1]	4	93,627.00	Weyerhaeuser	104	11,787.70
Philip Morris	10	53,139.00	Deere	124	10,290.50
State Farm Group	12	40,809.90	Pfizer	127	10,021.40
E. I. Du Pont De Nemours	13	37,607.00	McDonald's	132	9,794.50
Procter & Gamble	17	33,434.00	Publix Super Markets	136	9,470.70
Pepsico	21	30,421.00	Monsanto	146	8,962.00
Conagra	26	24,108.90	CPC International	154	8,431.50
Kroger	27	23,937.80	General Mills	156	8,393.60
Dow Chemical	36	20,957.00	H. J. Heinz	162	8,086.80
International Paper	60	19,797.00	Eli Lilly	171	7,535.40
American Stores	45	18,308.90	Campbell Soup	177	7,278.00
Coca-Cola	48	18,018.00	Farmland Industries	178	7,256.90
Sara Lee	50	17,719.00	Ralston Purina	180	7,210.30
Fleming	52	17,501.60	Kellogg	187	7,003.70
Supervalu	57	16,563.80	Champion International	188	6,972.00
Safeway	58	16,397.50	James River Corp. of Virginia	194	6,799.50
Caterpillar	63	16,071.00	Coca-Cola Enterprises	196	6,773.00
RJR Nabisco Holdings	64	16,008.00	Quaker Oats	206	6,365.20
Georgia-Pacific	75	14,292.00	American Brands	221	5,904.90
Kimberly-Clark	78	13,788.60	W. R. Grace	228	5,784.20
Archer Daniels Midland	92	12,671.90	Aramark	235	5,600.60
IBP	94	12,667.60	Tyson Foods	239	5,511.20
Albertson's	96	12,585.00	Mead	256	5,179.40
Anheuser-Busch	97	12,325.50	Case	261	5,105.00
Winn-Dixie Stores	103	11,787.80	Vons	263	5,070.70
Boise Cascade	265	5,057.70	Circle K	348	3,565.60
Union Camp	300	4,211.70	Food4Less Holdings	353	3,494.00
Supermarkets Genl. Holdings	303	4,182.10	Smith's Food & Drug Centers	399	3,083.71
Dole Food	304	4,152.80	Hormel Foods	409	3,046.20
Stop & Shop	308	4,116.10	Louisiana-Pacific	449	2,843.20
Chiquita Brands International	312	4,026.60	Sonoco Products	462	2,706.20
Giant Food	318	3,695.60	Dean Foods	472	2,630.20
Hershey Foods	336	3,690.70			

Source: *Fortune*, April 29, 1996.

[1]Note: Wal-Mart is included because of the grocery division. However, these revenues represent all Wal-Mart sales.

Figure 3–7 Few people outside the agricultural industry are aware of the magnitude of the agribusiness input and output sectors. (Courtesy of *Fortune* magazine)

Financial Services. Financial services are a vital part of the agricultural industry. Lending money to all three sectors of the agricultural industry (input, production, and output) is big business. In 1995, **outstanding loans** for farm real estate alone amounted to nearly $80 billion, in addition to other outstanding loans and non-real estate debt, which came to nearly another $151 billion.[23] The funds that production agriculturalists need to buy land, equipment, livestock, machinery, seed, fertilizer, and the other products they use in daily farming operations are provided by commercial banks, the Farm Credit System, the Farm Service agency, individual businesses and cooperatives, and insurance companies. Many of these agencies also provide financial services to agribusiness input suppliers and companies that buy, transport, process, and market agricultural products. These agribusinesses need credit and capital for their day-to-day operations, buildings, and equipment, just as production agriculturalists do.

Trade Associations. Trade associations, as well as dairy and livestock associations, are vital to the agricultural industry. Every agribusiness and production agriculture enterprise has a trade association, society, or institute, which are supported by members who are active in a particular enterprise. Thousands of trade associations operate in America. They serve agribusiness and promote certain agricultural products, and they have become an essential part of agribusiness relationships.

Trade associations perform many and varied services for their members, especially in the areas of public relations, promotion, publicity, communications, sales training, auditing and record-keeping, transportation, research, legislative lobbying, and legislative and marketing information. All these activities are more effectively carried out by groups than by individuals. Members of associations pay dues to receive benefits, and producers may pay a check-off from sales of their products.[24]

Agricultural Cooperatives. Agricultural cooperatives provide services essential to the agriculture industry and serve a variety of purposes. More than 4,200 cooperatives market agricultural products and furnish the agriculture industry with production supplies and services. The annual business volume of cooperatives is more than $83 billion. More than one-fourth of all agricultural products are marketed by cooperatives, and they provide one-quarter of all production supplies for farmers.[25] They also furnish electricity to the agriculture industry. Livestock producers are better able to market their animals and improve dairy products through dairy herd improvement associations and the assistance of cooperatives.[26] (Cooperatives are discussed further in Chapter 15.)

Agribusiness and Foreign Trade

The United States has long played a major role in world agricultural trade and is rapidly increasing its role relative to most other countries. Currently, the United States is the leading participant in international trade of agricultural products. The five major farm commodities sold in world markets in 1994 were feed grains and feed grain products, soybeans and soybean products, wheat and wheat products, live animals, meat and meat products, and vegetables, for a total value of more than $50 billion.[27] In comparison, the total value of all agricultural **exports** was only $6 billion in 1966. Almost every country in the world purchases some agricultural products from the United States.

In addition to being a major world exporter of agricultural products, the United States is also a major importer of agricultural products. Over a 10-year period, annual farm imports increased from $4.45 billion to $10.50 billion, a 136 percent increase. Although the largest quantities are imported from Latin America, Asia, and Europe, U.S. **imports** come from every continent except Antarctica. Despite the fact that import amounts

23. U.S. Department of Agriculture, *Agriculture Fact Book 1997* (Washington, DC: U.S. Government Printing Office, 1997); available at http://www.usda.gov/news/pubs/fbook97.
24. Marcella Smith, Jean M. Underwood, and Mark Bultmann, *Careers in Agribusiness and Industry*, 4th ed. (Danville, IL: Interstate Publishers, 1991), p. 287.
25. Rawlins, *Introduction to Agribusiness*, p. 150.
26. Ibid.
27. National Agricultural Statistics Service, *2002 Census of Agriculture.*

are larger than ever, the U.S. trade surplus (dollars exported minus dollars imported) was $18.9 billion in the mid-1990s.[28]

In addition to providing markets for American farmers, foreign exports have additional advantages. Between 25,000 and 30,000 jobs are created in the United States for each billion dollars of agricultural exports. In 1994, between 1.2 and 1.4 million full-time jobs could be attributed to our agricultural exports. Exports account for approximately one-third of U.S. agricultural production. In many states, as much as one-half of farm income comes from agricultural exports.[29]

Agribusiness and Energy

Agriculture as a User of Energy

The agricultural industry is both a producer and a user of energy. The agricultural industry as a whole consumes 10 to 20 percent of the nation's energy. About one-third of the energy is used by production agriculturalists, and the remaining two-thirds is used by agribusiness.

Energy Produced from Renewable Agricultural Products

Agriculture is becoming a larger producer of energy. Four major areas include direct burning, ethanol production, biodiesel, and methane gas production.

Direct Burning. An estimated 5 million Americans rely on wood-burning stoves for all their home heating needs. This resurgence of the wood-burning stove occurred because of the rising costs of home heating fuels. Direct burning of wood and wood wastes constitutes a considerable energy source.

Ethanol Production. **Fermentation** and **distillation** of grains into drinking alcohol have been done for thousands of years. In **ethanol** production, sugars and/or high-starch grains feed yeast cells in an oxygen-free environment. Corn liquor, or "moonshine" as it is sometimes called, is highly combustible when it is 80 percent pure (160 proof) or higher. When blended with gasoline, **E-85** (formerly called **gasohol**) is the result. Brazil pioneered the development of gasohol. Large-scale fermentation plants used local sugarcane to produce this form of alcohol. The use of gasohol increased rapidly in the early 1980s.

E-85 is the way ethanol is presently marketed for highway vehicles. The blend is 85 percent ethanol and 15 percent gasoline. Not all vehicles can run off E-85. Check the Internet for vehicles that can use E-85 by using such terms as "E-85," "ethanol approved vehicles," or "ethanol fueled vehicles" in a good search engine.

As of 2006, more than 4 billion gallons of ethanol are being produced annually, using about 10 percent of the U.S. corn crop. Farmers are often owners or investors in ethanol plants, the construction and operation of which create local jobs, income, and tax revenue as well as using locally produced corn. Both rising oil prices and the Renewable Fuels Standard enacted in 2005 virtually ensure that demand for ethanol will only increase.[30] Refer to Figure 3–8.

Biodiesel. Worldwide, biodiesel is the most used renewable agricultural alternative energy source. Sometimes referred to as soy diesel, this fuel is made from soybeans and is used either as a pure soybean oil or mixed with diesel fuel. Diesel engines can use this fuel without changes as long as the soybean oil and diesel fuel are in at least a 50/50 blend. However, 80 percent diesel and 20 percent soybean oil is the mixture most commonly used. Refer to Figure 3–9.

Methane Gas Production. Methane, a natural gas, accounts for nearly two-thirds of biogas. Methane is also produced as a product of sewage sludge and landfill decomposition. Availability of

28. Ibid.
29. USDA, *Agriculture Fact Book* 1997.

30. John M. Urbanchuk, "Contribution of the Ethanol Industry to the Economy of the United States," February 21, 2006, available at http://www.ethanol.org/pdf/contentmgmt/Ethanol_Economic_Contribution_Feb_06.pdf; *Issue Brief: Economic Impacts of Ethanol* (Spring 2006), available at http://www.ethanol.org/pdf/contentmgmt/Economic_Impacts_of_Ethanol_Production.pdf; American Coalition for Ethanol, "Ethanol 101," available at http://www.ethanol.org/index.php?id=34&parentid=8.

Figure 3–8 This vehicle has been driven more than 26,000 miles on pure ethanol. It was converted to ethanol by agribusiness and agriscience students at Middle Tennessee State University. After 25,000 miles, it raced and won in its division at the Bonneville Salt Flats. (Courtesy of Cliff Ricketts)

Figure 3–9 This Corvette, "The Senator," is powered by an Allis Chalmers four-cylinder turbodiesel engine that is fueled by soybean oil. Built by agriculture students of Dr. Cliff Ricketts, professor of agricultural education at Middle Tennessee State University, "The Senator" holds several speed records for a soybean-oil-fueled vehicle. (Courtesy of Cliff Ricketts)

concentrated quantities of manure, such as in feedlots or dairy lots, is the primary factor in production of pure methane. The strong advantage of using methane is that use of the manure for energy production eliminates the need for its immediate disposal. The residue material is nutrient-rich **humus**, which serves as a valuable soil conditioner.

Sun and Water

Can vehicles actually run off water? Yes, they can. By linking various technologies, researchers at Middle Tennessee State University developed a process to run vehicles with sun and water as the only power sources.

A 10-kilowatt solar unit is used to produce electricity from the sun. The electrical energy is stored in the local power company's grid; in essence, the electricity is banked. When an electric truck is charged, the electricity used is measured and taken from the bank.

In a vehicle that runs off water, a solid polymer electrolysis unit separates the hydrogen from water. The electrolysis unit is powered by stored (banked) electricity. The hydrogen produced is stored and used to run an internal combustion engine that powers the vehicle.

Agribusiness Energy as a Petroleum Alternative. Dr. Cliff Ricketts and his students at Middle Tennessee State University have converted engines to run off ethanol, biodiesel oil (soybean oil), and hydrogen. These agriculture students have entered and won races in their alternative fuel divisions in Utah, California, North Carolina, and Florida. Certainly, agribusiness energy is a viable alternative to petroleum and fossil fuels.

Agribusiness and the Environment

The environment is becoming more important to the business of agriculture. Unfortunately, our resources and environment took a back seat to production and profit in American industries for most of the 20th century. Concerned citizens and scientists began to speak out about the various types of pollution when they realized that the quality of air and water around them was deteriorating. In the early days of the environmental movement, these people were known as **ecologists**. At first they were considered radicals and alarmists, and very few people paid any attention to their warnings. However, in the late 1960s, an increasing number of Americans joined the effort to reduce pollution of all types.

Rachel Carson's *Silent Spring*, published in 1962, was the first major publication calling attention to environmental issues. The Sierra Club is an organizational pioneer that called attention to

Case Study 1

MTSU's Dr. Cliff Ricketts' Hybrid Hydrogen Truck

Dr. Cliff Ricketts, professor of agricultural education at Middle Tennessee State University (MTSU), has a second job. For the past 25 years, Dr. Ricketts has been researching and building alternative fuel vehicles. "This is my passion," he says. "I'm able to be creative here, since we don't have an engineering department."

When Dr. Ricketts first started his research, he was building ethanol engines. In 1982, he gave six presentations on corn-derived fuel at the World's Fair in Knoxville. From there, Dr. Ricketts moved on to methane engines and is now working on a vehicle powered by the sun and hydrogen from water, which he plans to drive across Tennessee this spring.

The vehicle's source of solar power comes from photovoltaic panels installed by Big Frog Mountain next to MTSU's Agriculture Education building. The panels store power in the grid line of Murfreesboro Electric Service, which is under the umbrella of TVA. Ricketts says he already has 10,000 kilowatts stored in the grid bank. The solar unit provides enough "fuel" for a 70-mile trip, which Ricketts points out is really enough for 85 percent of U.S. commuters. However, he wants to prove the ability to drive long distances, which is where hydrogen power comes in. The Nissan truck Dr. Ricketts is using has a hybrid engine that uses hydrogen. The solar unit powers a hydrogen electrolysis unit, which purifies water to prepare it for separation. Once the hydrogen is separated from the oxygen, it goes into a metal hydride tank and solidifies. It then goes through another process of converting to gaseous hydrogen in order to power the truck. The hydrogen allows the truck to travel an additional 70 miles, but Ricketts says with the proper equipment and a lighter vehicle, he can produce and store enough power to travel 300 miles.

Select students in the agriculture department at MTSU work with Dr. Ricketts, performing the majority of the mechanical work and installation involved. "It takes a special student to do this," he says. He points out that few people have the ability to work on a project from concept to production, but Dr. Ricketts has worked with students whose talents have enabled his research to get to the next level. "If I could hire about four of my former students, I would, and I'd put them up against anyone in the world," he boasts.

A lack of funding for things like manpower and equipment is all that's stopping Dr. Ricketts from his cross-state drive. Tractor Supply Company, headquartered in Nashville, provides Ricketts' program with $9,500 in annual funding, which MTSU matches. "I can buy supplies and do maintenance off the $19,000," Dr. Ricketts says, "but I can't buy any equipment with that." The university's technology access fee provides a lot of the funding that goes toward equipment. With the right amount of funding, Dr. Ricketts is sure he can prove hydrogen's pertinence to the automobile industry. He says, "If a professor can pull this off, surely the auto companies can. We put people on the moon, but we have not yet shown the public how to run engines off sun and water."

In the future, Dr. Ricketts would like to purchase a Porsche 914 to build a hybrid that will run at 50 to 60 miles per hour using the sun and hydrogen.[31]

environmental problems. The club's membership tripled between 1965 and 1970 and since that time has increased yearly by about 10,000 members, mostly young people. Other major environmental groups include the Audubon Society, the National Wildlife Federation, Friends of the Earth, the Environmental Defense Fund, and the League of Conservation Voters.

Pollution control became a major priority for the nation. Federal, state, and local legislators began to develop and pass laws to prevent and

31. Shanta McGahey, "MTSU's Dr. Cliff Ricketts' Hybrid Hydrogen Truck," *ENVIROLink Handbook Southeast*, April 21, 2005; available at http://www.envirolinkhandbook.com/www/docs/174.123.

Career Options

Environmental Manager, Conservationist

Environmental managers and conservationists are involved in the management, protection, and preservation of the environment and natural resources. Natural resources include all the things that help support life, such as sunlight, water, soil, and minerals. Plants and animals are also natural resources. Without the work of environmental managers and conservationists, most of the earth's resources would be wasted, degraded, or destroyed.

Environmental managers may specialize in wildlife, air and water quality, fire control, automotive emissions, and forestry emissions. Conservationists may specialize in soil, forest, grazing lands, minerals, energy, and water conservation. Many people with an interest in the outdoors pursue a career in forestry; wildlife ecology; or water, soil, air, or energy conservation. Refer to Figure 3–10 for a list of several environmental agriculture and conservationist job opportunities.

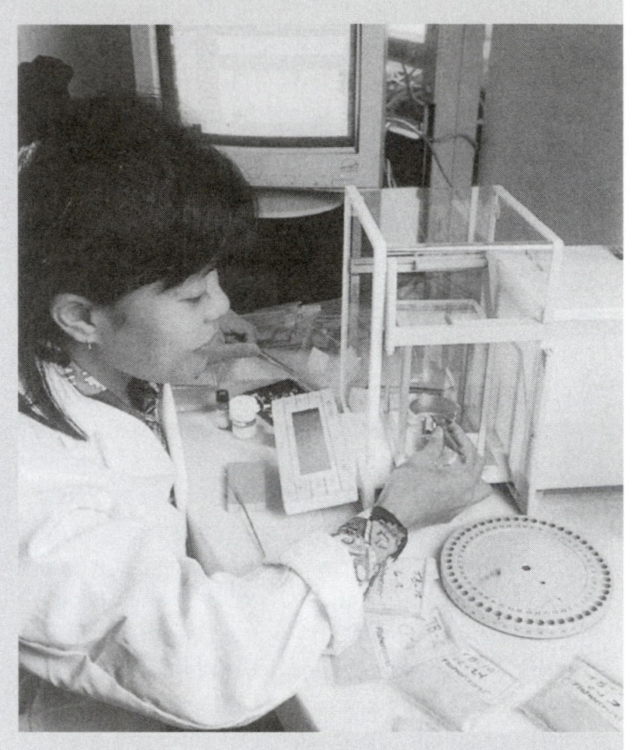

Soil scientist Marife Corre prepares to analyze soil samples from a riparian buffer. The carbon and nitrogen status of riparian zone soils indicates their potential to remove nitrate from shallow groundwater and to improve water quality. (Courtesy of USDA)

control pollution. The National Environmental Policy Act of 1969 created the Council on Environmental Quality (CEQ), which initiated the requirement of environmental impact studies for controversial environmental activities. The Office of Environmental Quality was established in 1970, under the Environmental Quality Improvement Act, to support the work of the CEQ.

The Environmental Protection Agency (EPA) and the National Oceanic and Atmospheric Administration (NOAA) were established in 1970. The Federal Water Pollution Control Act Amendments of 1972 set up the National Commission on Water Quality and established water discharge requirements. Much other legislation has been approved to develop and enforce air, pesticide, noise, drinking water, occupational safety, and health requirements.

Many jobs have developed that have environmental agriculture implications. High schools have courses and colleges have environmental agriculture majors and produce graduates ready to enter the workforce. Research and development efforts are creating jobs with ecology as the focal point. Agricultural careers within environmental areas include soil, water, air, and energy sources. Refer to Figure 3–10 for a list of job opportunities for **environmentalists** in each of these four areas.

Environmentalists in Agribusiness

As soil is certainly something agriculturalists should protect, those involved in the use of the soil for growing crops can be called environmentalists. Most fruit, vegetable, and grain producers have been environmentalists for hundreds, or perhaps thousands, of years. To control soil erosion and protect

Environmental Agricultural Job Opportunities

Job Opportunities Related to Soil Resources

- Agricultural Economist
- Agronomist
- Biologist
- Botanist
- Chemical Sales Representative
- Chemist
- Conservation Officer
- Conservation Scientist
- County Agricultural Extension Agent
- Crop Research Scientist
- Disease and Insect Control Technician
- Drainage Design Coordinator
- Ecologist
- Education Consultant
- Entomologist
- Environmental Control Scientist
- Fertility Expert
- Field Auditor
- Forester
- Geologist
- Horticultural Garden Curator
- Horticulture Extension Agent
- Land Planner
- Landscape Architect
- Landscape Contractor
- Landscape Drafter
- Lobbyist
- Management Specialist
- Mediator
- Miner
- Mineralogist
- Nursery Gardener
- Nursery Manager
- Plant Breeder
- Plant Propagator
- Range Conservationist
- Seed Specialist
- Soil Conservation Field Inspector
- Soil Engineer
- Soil Scientist
- Soil Science Technician
- Soil Tester
- Urban Planner
- Wastes Engineer
- Wastes Technician

Job Opportunities Related to Water Resources

- Agriscience Teacher
- Aquaculturist
- Aquatic Facility Manager
- Attorney
- County Agricultural Extension Agent
- Dairy Operator
- Drainage Design Coordinator
- Ecologist
- Environmental Control Technician
- Financial Institution Officer
- Fish Culturist
- Fish Hatchery Manager
- Food Processing Manager
- Game Warden
- Gravity Flow Irrigator
- Hydroelectric Facility Supervisor
- Hydroponics Nursery Manager
- Ichthyologist
- Information Scientist
- Irrigation Engineer
- Lawn Sprinkler Installer
- Lobbyist
- Marine Biologist
- Marine Life Hatchery Manager
- Park Naturalist
- Park Ranger
- Plumber
- Public Health Microbiologist
- Public Works Engineer
- Rice Producer
- SCUBA Diver
- Shellfish Hatchery Superintendent
- Slaughtering and Processing Manager
- Soil and Water Conservation Agent
- Sprinkling System Irrigator
- Technical Publications Writer
- Water Purification Chemist
- Water Quality Board Member
- Water Quality Control Engineer
- Water Quality Tester
- Water Treatment Plant Engineer
- Waterworks Supervisor

Job Opportunities Related to Air Resources

- Agricultural Engineer
- Allergist
- Automobile Designer
- Automobile Mechanic
- Diesel Engine Mechanic
- Environmental Research Technician
- Environmental Scientist
- Furniture Manufacturer
- Government Agency Employee
- Industrial Chemist
- Market Research Analyst
- Meteorologist
- Planning Engineer
- Plant Ecologist
- Plant Pathologist
- Public Health Technician
- Research Technician
- Tobacco Market Analyst
- Tractor Mechanic
- Vehicle Inspection Station Worker
- Weather Observer

Job Opportunities Related to Energy Resources

- Agricultural Economist
- Building Inspector
- Computer Specialist
- Construction Worker
- Crew Boss
- Equipment Repairer
- Geologist
- Line Repairer
- Loan Officer
- Materials Scientist
- Mechanical Engineer Technician
- Methods Study Analyst
- Research Director
- Research Engineer
- Safety Engineer
- Safety Inspector
- Sales Representative
- Service Technician
- Surveyor
- Utilities Supervisor
- Welder

Figure 3–10 It is evident from this list that many jobs are available in the area of environmental agriculture. (Courtesy of USDA)

soil is to show concern for the environment. The first agricultural environmentalists were the ancient peoples of Rome, India, and Peru—they valued soil enough to build anti-erosion terraces.[32]

Natural Resources Conservation Service

The first field research on soil and water conservation in the United States was conducted by M. F. Miller and F. L. Daley at the University of Missouri.[33] This study was started in 1917 and the first results were published in 1923. This provided the foundation for the soil conservation movement. After several name changes, the soil conservation agency is now called the National Resource Conservation Service, under the USDA. Its mission is to "provide leadership and administer programs to help landowners and land users to conserve, improve, and sustain our natural resources and the environment, while enabling the United States to continue as the world's preeminent producer of food and fiber."[34]

Some traditional soil and water conservation practices that protect the environment include:

- minimum tillage
- no-till practices
- strip cropping
- contour farming
- surface drainage
- subsurface drainage
- farm ponds
- grassed waterways
- irrigation systems
- field windbreaks
- brush management pastures and ranges
- rotation grazing
- pasture and hay land seedings
- range reseeding
- improved tree harvesting
- tree planting
- windbreak renovation[35]

32. Frederick R. Troeh, J. Arthur Hobbs, and Roy L. Donahue, *Soil and Water Conservation for Productivity and Environmental Protection* (Englewood Cliffs, NJ: Prentice-Hall, 1980), p. 11.
33. Ibid., p. 629.
34. USDA, *Agriculture Fact Book 1997*, http://www.usda.gov/news/pubs/fbook97/10natres.pdf.
35. Troeh, Hobbs, and Donahue, *Soil and Water Conservation*.

Many workers are needed to design, construct, and manage these conservation practices. Refer to the Career Options about environmental managers and conservationists.

Summary

Today's modern agricultural industry of production agriculturalists, agriscience, and agribusinesses encompasses a huge, complex group of organizations and the people who run them.

There could be no agribusiness without production agriculture and land. The United States has about 2.3 billion acres. About 21 percent of the land is used for crops, 25 percent for livestock, and 30 percent for producing forest products. The remaining 24 percent is used for nonagricultural purposes. The agricultural industry accounts for 17 percent of the GDP and provides more than 20 percent of all the jobs in the country. The production agriculturalist's share of each dollar spent for food by the consumer is about 25 cents.

One of the most outstanding characteristics of production agriculture in America has been a tremendous increase in production efficiency. The average farmworker supplies more than 150 persons with food and fiber. Annual consumption of food in the United States is 1,365 pounds per person. One production agriculturalist produces an average of 108,000 pounds, which is about 54 tons of food per year. Americans spend about 9 percent of their personal disposable income on food.

Agriculture is one of the world's largest industries. On a worldwide basis, more people are in some way involved in agriculture than in all other occupations combined. The United States produces more food than any other nation in the world and is the world's largest exporter of agricultural products. Export revenues account for 25 to 30 percent of U.S. farm cash receipts, and crops from nearly one-third of U.S. farm acreages are exported to overseas customers. Foreign trade is a big part of agribusiness. Almost every country of the world purchases some agricultural products from the United States. Approximately one-third of U.S. agricultural production is exported. The top five exported agricultural commodities sold for $46.7 billion. In 1994, between 1.2 and 1.4 million full-time jobs could be attributed to U.S. agricultural exports.

Agribusiness is the largest industry in the United States. Products from production agriculturalists valued at $123 billion increase in value to $546 billion once they are purchased by the consumer, a fourfold increase. In 1996, more than $180 billion was spent on agribusiness inputs by production agriculturalists. Nearly $425 billion was spent by consumers for food products alone within the agribusiness output sector. The USDA estimates that more than 600,000 businesses are involved in the agribusiness output sector.

The agriservice sector of the agricultural industry includes both public and private services, which research new and better ways to produce and market food and fiber, disseminate new technology, develop and enforce laws to protect food producers and consumers, and provide special services to all other phases of agriculture within the public agriservice sector. The USDA alone employs more than 100,000 workers. Within the private agriservice sector, there are more than 30,000 firms, which employ more than 100,000 people and have a payroll of about $600 million. The three major areas of private agriservices available to the agricultural industry are financial services, trade associations, and agricultural cooperatives.

The agricultural industry is both a producer and user of energy. The agricultural industry consumes 10 to 20 percent of the nation's energy. Agriculture is becoming a greater producer of energy, through direct burning and production of ethanol, biodiesel, and methane gas.

The environment is becoming more important to the business of agriculture. For much of the 20th century, our resources and environment took a back seat to production and profit in American industries. Today, however, many jobs have developed that have environmental agriculture implications. High schools have courses and colleges have environmental agriculture majors and produce graduates ready to enter the workforce.

End of Chapter Activities

REVIEW QUESTIONS

1. Define the Terms to Know.
2. What percent of the gross domestic product is attributed to the agricultural industry?
3. List five crop, five vegetable, five animal, and five fruit and nut enterprises.
4. Why is production agriculture efficiency so important?
5. What is the largest industry in the United States?
6. What types of agribusinesses are included in the agricultural input sector?
7. Name five agribusiness input firms.
8. List the yearly national expenditures for each of the following agricultural inputs: (a) seed, (b) feed, (c) livestock, (d) fertilizer and lime, and (e) hired labor.
9. List 18 types of agribusiness outputs.
10. Name 10 agribusiness output firms.
11. How many workers are employed in each of the following? (a) meat-processing industry, (b) dairy-processing industry, (c) baking industry, (d) canned, cured, and frozen food industry, (e) cotton/textile industry, and (f) agricultural exports.
12. Explain the role of the agriservice sector.
13. What are the three major areas of private agriservices?
14. List 10 things that trade associations do.
15. What are the four major energy sources produced from renewable agricultural products?
16. List the names of six environmental groups.
17. What are four areas of concern to environmentalists?

FILL IN THE BLANK

1. Only ________ percent of the American population is actively involved in the production of food and fiber.
2. There could be no agribusiness without ____________ ____________.
3. The average American farm consists of _________ acres.
4. The production agriculturalist's share of each dollar spent for food by the consumer is about ______.
5. America produces ______ percent of the world's soybeans, _____ percent of the world's corn, ______ percent of the world's poultry, and ______ percent of the world's beef.
6. Annual consumption of food in the United States is ________ pounds per person.
7. One production agriculturalist produces an average of about _____ pounds.
8. One production agriculturalist creates ______ agribusiness jobs.
9. Agribusiness worldwide represents approximately ______ of the total world economic production and provides employment for nearly _______ of the population on earth.
10. Cooperatives market more than _______ of all agricultural products.
11. Between 25,000 and 30,000 jobs are created in the United States for $________ in agricultural exports.
12. Exports account for approximately _________ of U.S. agricultural production.

MATCHING

a. 30%
b. 5,000
c. 24%
d. $300 billion
e. 2.3 billion acres
f. 600,000
g. 3 million acres
h. 21%
i. 100,000
j. $80 billion
k. 25%
l. 2.1 million

_____ 1. land in United States
_____ 2. percent land in United States used for crops
_____ 3. percent land in United States used for livestock
_____ 4. percent land in United States used for forests
_____ 5. percent land in United States used for nonagricultural purposes
_____ 6. number of farms producing the total food supply for Americans and numerous foreign countries
_____ 7. production agriculturist share of each dollar spent by consumers on food
_____ 8. number of businesses involved in the agribusiness output sector
_____ 9. approximate number of USDA employees
_____ 10. outstanding loans on farm real estate alone
_____ 11. number of cooperatives marketing agricultural products

ACTIVITIES

1. The author uses the terms *agricultural industry, production agriculture, agriscience*, and *agribusiness*. Draw a picture illustrating the difference and relationship of these terms. Use whatever words are needed to explain your picture.
2. Prepare a brochure titled, "Reasons for Taking Agricultural Education Courses." Include material from this chapter and other places illustrating the size and importance of agriculture. Ask the school guidance counselor to select the best brochure from your class.
3. Identify five agribusiness input suppliers and five agribusiness output firms in your community.
4. Identify three public agriservices and three private agriservices in your community or county.
5. Identify three trade associations in your community. Briefly describe the purpose of each.
6. Name and describe two cooperatives in your community.
7. Prepare a short, two- to three-page essay on an agricultural alternative energy resource.
8. Copy an article from a magazine or newspaper on an environmental agriculture issue. Share the article with your class by making a presentation about it.
9. With the aid of the Internet or a USDA source, determine the minutes of work equal to the prices of selected food items (as shown in Figure 3-4) from the latest available data.

Chapter 4

Emerging Agribusiness Technologies

Objectives

After completing this chapter, the student should be able to:

- Discuss the value of global positioning.
- Explain the importance and use of genetic engineering.
- Explain the importance and use of cloning.
- Describe the procedure of determining gender selection.
- Discuss four types of production hormones.
- Give examples of animal research that results in advances in human medicine.
- Describe domestic animals of the future.
- Describe computerized and electronic animal management technologies.
- Explain emerging mechanical technologies in the plant industry.
- Explain the importance and use of tissue culture.
- Discuss selected emerging plant technologies.
- Discuss the importance and use of integrated pest management.
- Explain the use of hydroponics and aquaculture in extended space travel.
- Describe eight other emerging agribusiness technologies.

Terms to Know

augmentation
bovine somatotropin (BST)
callus
cloning
deoxyribonucleic acid (DNA)
embryo splitting
embryo transfer
flaming
gender selection
gene mapping
gene splicing
genetic engineering
hormones
host
hybrid
hydroponics
implant

integrated pest management (IPM)
pathogens
pessary
pheromone
physiologically
porcine somatotropin (PST)
precision farming
restriction enzyme
solarization
tissue culture
transgenic
xenotransplantation

Introduction

Agricultural technologies are applied sciences. However, once they have been developed, it is the agribusiness input or agribusiness output sector that sells or markets the emerging technologies. These technologies are used to improve agricultural production and to improve methods of processing, transporting, and distributing agricultural goods.

Agricultural technology has expanded at a rapid rate since the beginning of recorded history, but technology development has a "snowball effect." Many of America's older production agriculturalists started their careers behind teams of horses or mules. Today, production agriculturalists operate air-conditioned tractors, which permit them to do more work in a few hours than the previous generation used to complete in a week.

The agricultural industry is moving at such a rapid rate that it is difficult, if not impossible, to label any process or procedure "new" or "futuristic." However, it is the goal of this chapter to give a brief overview of outstanding technological developments in the agricultural industry, as viewed by any generation. For example, we cannot forget the value of hybrids, artificial insemination, and the round haybaler. Although these technologies are not new, the agricultural industry made significant advances as a result of their use.

Max Lennon, former president of Clemson University, made a most appropriate evaluation when he stated: "In the area of technology alone, there will be more new developments in the next 15 years than in all of history to this point. Simply put, the world is changing, and whether the agricultural industry will benefit from these changes, or suffer because of them, depends on the industry itself."

Global Positioning

The Global Positioning System (GPS) is a technology that uses satellites to view specific fields and crops. The GPS satellite array uses its vantage points in space to provide a computerized picture that will aid field crop management and increase yield per acre.[1]

Purpose of Global Positioning and Field Management

In days past, before agriculture was industrialized, production agriculturalists were familiar with their fields as a whole, and successful production agriculturalists adapted farming techniques to the variations of their fields. Now, crop producers are concerned not only with managing large fields, but also with field conditions, which can vary from one square yard to the next. A single crop producer's field may contain three or four different soil types, with varying fertility, drainage, organic content, and nitrogen levels. Traditionally, agricultural production methods required that fields be treated uniformly despite these within-field variances in soil type, weed and disease incidence, and topography. Often this led to over- or underapplications of inputs such as fertilizer, seed, water, and pesticides. Overuse often results in wasted resources and pollution. By mapping fields using the GPS, production agriculturalists can vary their seeding, irrigation, fertilization, and cultivation practices to accommodate each section of a field and thus increase yield.[2] Refer to Figure 4–1.

Origin of the Global Positioning System

The GPS was originally designed for the military defense systems of the U.S. Department of Defense. The system consists of 24 satellites that circle the globe every 12 hours, orbiting 10,900 miles above

1. "Satellite System Can Control Tractors," *USA Today,* 126, no. 2631 (December 8, 1997): 11; available at http://www.findarticles.com/p/articles/mi_m1272/is_n2631_v126/ai_20077711.
2. "Applications and Products of GPS in Agriculture," available at http://www.age.uiuc.edu/classes/age221/project/1996/team3/index.html.

Figure 4–1 In Missouri, corn is harvested by a combine that is linked to the satellite-based Global Positioning System. Precise yield and location data will be correlated with soil samples taken earlier at sites throughout the field. This information will help growers plan optimal fertilizer rates for the next crop. (Courtesy of USDA)

the earth. Infrared camera technology in the satellites can distinguish healthy crops from sick ones, locate problem spots, and distinguish high-weed and other less productive areas.[3] The GPS system consists of four components: a GPS receiver, a crop yield monitor, a digital soil fertility map, and variable rate application technology (VRT). In many cases, crop producers use an inexpensive handheld GPS receiver, or attach the GPS system to equipment that will automatically vary application rates. The application rate of seed, fertilizer, pesticides, and water is varied by sensors mounted on a truck or tractor that allow computer-controlled nozzles to know the location in the field; here, soils and crop needs are met by specific applications of inputs.[4]

Global Positioning and the Environment

By mapping a field using GPS, crop producers obtain information that will help them make decisions that will increase productivity by better management of all areas of the field. Not only will this reduce the cost of production, but also, many argue, this type of **precision farming** will improve farm productivity while reducing the need for fertilizer, herbicides, and pesticides. Site-specific information could thus help decrease the impact of adverse environmental agricultural practices. Additionally important for the environment, water use and management are enhanced by the GPS system.[5]

Precision Farming

Precision farming may become another of the wide range of tools used to help crop producers increase productivity while maintaining the environmental integrity of the land. Precision farming combines crop planning with specific calculations for tillage, planting, chemical applications, harvesting, and postharvest processing of the crop.[6]

The five major objectives of comprehensive precision farming are:

- increasing production efficiency
- improving product quality
- using chemicals more efficiently
- conserving energy
- protecting soil and groundwater

Precision farming requires the use of digital, geographically referenced data in cropping operations. Precision farming is principally associated with soil fertility and soil management.[7] Simply put, the farm field and crop are referenced as a computer model of each specific section of a crop producer's field. Then, each section is individually assessed for increased production, decreased cost, and environmentally sound techniques.

Genetic Engineering

Genetic engineering can be used to speed up the process of selectively breeding livestock. It involves inserting a specific gene into the chromosome of an organism in order to develop a desired trait in

3. "Hitting the Spot," *Economist (U.S.)* 338, no. 7950 (January 27, 1996): 74.
4. Stephen A. Wolf and Spencer D. Wood, "Precision Farming: Environmental Legitimization, Commodification of Information, and Industrial Coordination," Rural Sociology 62, no. 2 (summer 1997): 180–206.
5. Michael D. Weiss, "Precision Farming and Spatial Economic Analysis: Research Challenges and Opportunities" (paper presented at the annual meeting of the American Agricultural Economics Association, San Antonio, Texas, July 23–28, 1996).
6. Jess Lowenberg-DeBoer, "Precision Farming and the New Information Technology: Implications for Farm Management, Policy, and Research: Discussion (paper presented at the annual meeting of the American Agricultural Economics Association, San Antonio, Texas, July 23–28, 1996).
7. Gary T. Roberson, "Precision Agriculture: A Comprehensive Approach" (North Carolina State University, Biological and Agricultural Engineering Department); available at http://www.bae.ncsu.edu/programs/extension/agmachine/precision.

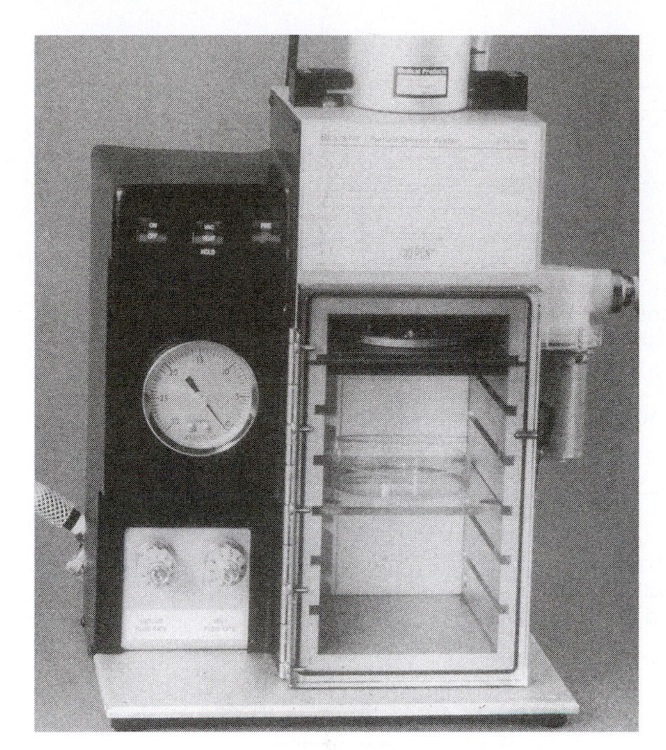

Figure 4–2 Using a device known as a "gene gun" (biolistic particle delivery system), a gene can be propelled into growing tissue by putting material containing the gene on a projectile and shooting it at the target cells. The projectile, which is similar to a BB, is stopped, but the genetic material continues to move and penetrates the target cells. (Courtesy of DuPont Co. Biotechnology Systems)

the recipient organism. See Figure 4–2. The gene placed in the target chromosome comes from another organism. Refer to Figure 4–3.[8]

Advantages of Using Genetic Engineering

Transforming a Single Gene. With traditional and classical breeding methods, a plant scientist has to cross and backcross several generations of a plant to eliminate undesirable traits or characteristics. With genetic engineering, the plant scientist can transfer a single gene for the desired trait into the plant's genetic material without changing the rest of its genetic makeup. Thus, a breeder working with genetically engineered plants can easily predict the characteristics of the modified plant, need not hunt for a rare desirable plant, and need not spend many years backcrossing, as no undesirable genes have been added. The result is faster production of new varieties.

8. L. DeVere Burton, *Agriscience and Technology*, 2d ed. (Albany, NY: Delmar Publishers, 1998), p. 30.

Transfer to a Totally Unrelated Plant. Classical breeding is also limited to plants that are cross-fertile. With genetic engineering it is possible to transfer genes containing desirable traits from one plant to another, totally unrelated plant. The genetic engineer can take a gene from any organism and add that gene to the chromosome of another organism. The recipient of the gene need not be related to the donor. Scientists are now adding bacterial genes to plants, plant genes to bacteria, and animal genes to plants.

Making Hybrids from Plants That Cannot Be Cross-Pollinated. Another well-known breeding technique is hybridization. One can create **hybrids** by pollinating plants back to themselves and planting the seed. This inbreeding concentrates the undesirable characteristics. There is more chance that recessive ("bad") genes will pair up and express themselves, which will then affect the plant.

You save seed from better-quality plants. You inbreed still further and discard still more plants with undesirable characteristics. Usually, inbred plants lose vigor, but when you cross these inbred lines, you get a boost in vigor and better quality than you started with. You have created a hybrid.

Through the use of biotechnology, we can now make hybrids from plants that cannot be crossed using pollen. For example, tomatoes can be crossed with potatoes. Further, in a year or less, we can produce drought resistance or other qualities that would take 15 years using classical breeding techniques.

Producing Disease-Resistant Plants. Genetic engineering makes it possible to produce plants that are resistant to certain diseases. Refer to Figure 4–4. An example is a tobacco plant that is resistant to crown gall disease.

Producing Plants That Are Toxic to Insects But Not Humans. It is also possible to concentrate gene messages that tell a leaf to produce materials that are toxic to insects, but avoid having the edible fruit of the plant produce that same toxin.

Producing Crops That Are Tolerant of Herbicides. Plants are being engineered to be tolerant of certain herbicides so that environmentally safer chemicals can be used to selectively kill weeds and not the cultivated crop plants. This should enhance competition among agribusiness chemical companies; the winners will be those that make

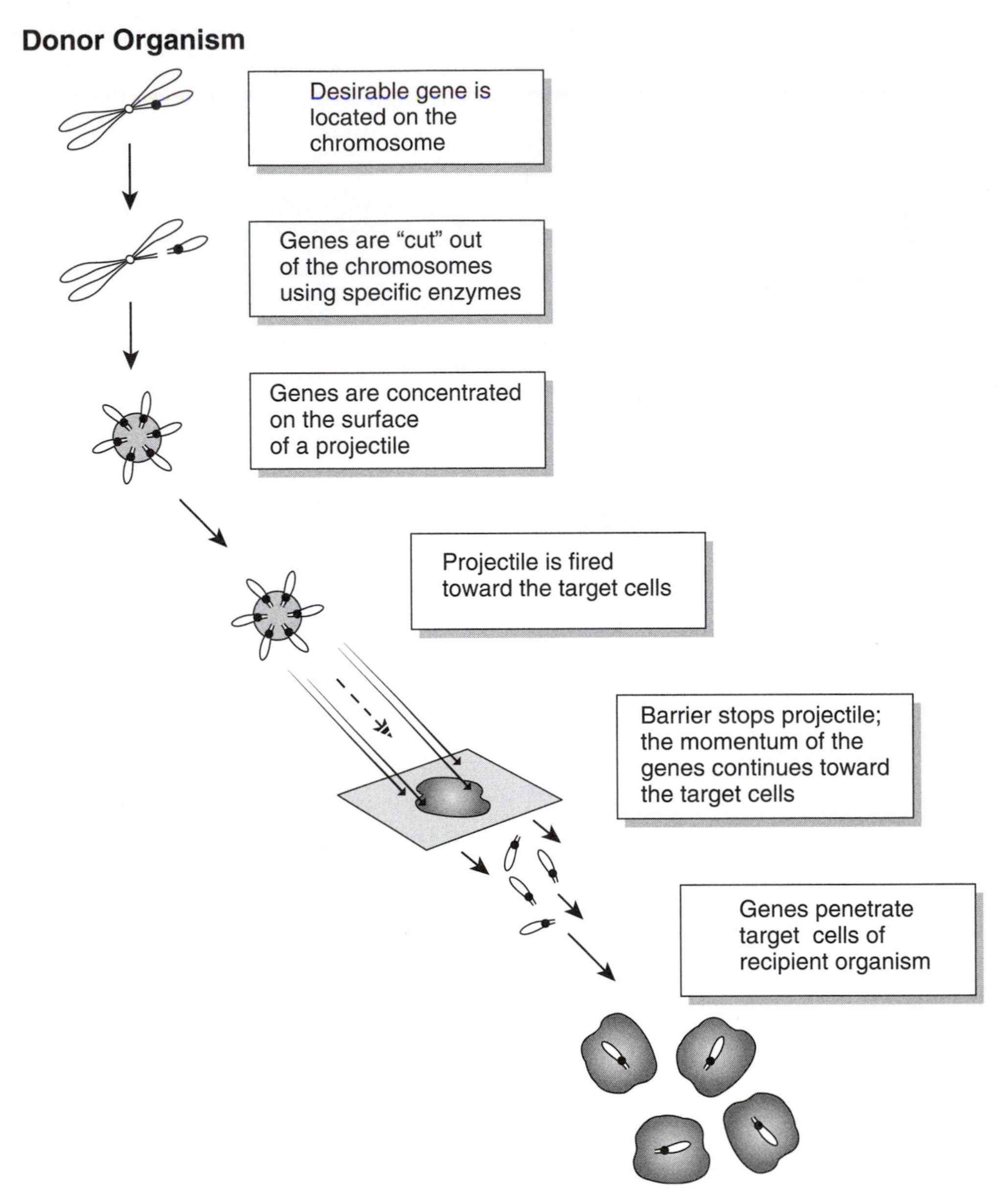

Figure 4–3 Desirable genes may be transferred from one organism to another.

the herbicides that are most desirable to crop producers, other consumers, and regulatory agencies.[9] Refer to Figure 4–5 for examples of plants that have been genetically engineered.

Gene Splicing

Important research in biotechnology is being conducted to locate and record where particular genes are found on a chromosome. This type of research is called **gene mapping**. See Figure 4–6. To change the gene code of a chromosome, the exact location of the gene must be known. Although there are millions of genes on the many different chromosomes in plants and animals, only a few genes have actually been located and mapped.[10]

Gene mapping requires the scientist to isolate the chromosome that contains the desired gene and find exactly where on the chromosome the gene is located. Once the location of the gene is known, the scientist must find a **restriction enzyme**. A restriction enzyme allows a particular gene to be cut out of a chromosome[11] much like a person uses scissors to

9. Cliff Ricketts, *Science 1A (Agriscience)* (Nashville, TN: Division of Vocational and Technical Education, 1994).

10. Burton, *Agriscience and Technology*, p. 33.

11. Ibid., p. 33.

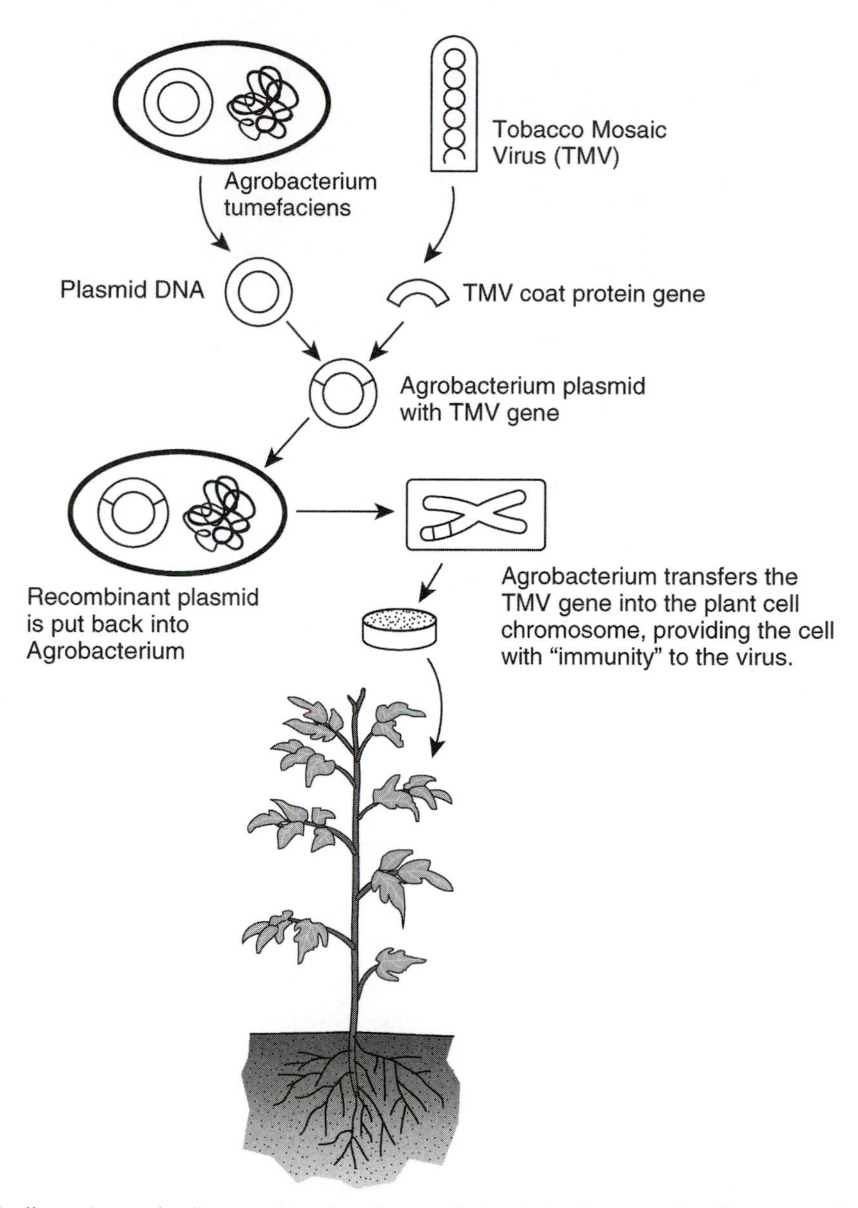

Figure 4–4 Genetically engineered cells are cultured and grown into whole plants; each cell contains the tobacco mosaic virus (TMV) gene. When the plants are reinfected with the virus, they do not contract the disease.

clip an article from a newspaper. After carefully choosing and preparing the appropriate enzyme, the scientist can remove a very specific gene from its location on the chromosome and replace it with a different gene. This technique is called **gene splicing**. Refer to Figure 4–7. Gene splicing is useful because scientists can develop a desired trait within or between species more quickly than by waiting for the trait to develop naturally through gene mutations. Figure 4–8 shows how gene splicing works.[12]

Selective breeding is still a valuable process, but thanks to science, it can now be used to reproduce an animal that carries a desired genetic trait. Gene splicing enables scientists to make genetic changes more rapidly and in a more orderly manner than selective breeding alone could produce.[13]

Cloning

Cloning is the process of creating genetically identical organisms. In this process, the nucleus of an unfertilized ovum from one organism is replaced with the nucleus of a cell from the organism to be duplicated. The result is a new organism

12. Ibid., p. 34.

13. Ibid., p. 37.

Examples of Plants That Have Been Genetically Engineered

- Corn plants with a higher photosynthesis rate, in order to manufacture food more efficiently.
- Salt-tolerant barley and tomato plants.
- Strawberry plants that produce fruit all summer long, sometimes well into October.
- Black cherry trees that produce higher-quality fruit.
- Flower varieties that produce colors that were previously nonexistent for those particular varieties. A brick-red petunia flower is an example.
- Walnut trees that resist the codling moth and blackline disease. (These pests can kill walnut trees or severely affect production.)
- More perfect potatoes that chip, french fry, or bake better, resist local or regional potato diseases, and mature at the right time for states of different latitudes.

Figure 4–5 These are just a few examples of plants that have been genetically engineered.

with the exact genetic makeup of an existing one. Every cell in an organism has a full set of genes. As new tissues are formed, only part of the genetic information is used; the genes that perform other functions in the cell are turned off.

Cloning Plants

Cloning in plants occurs when a single plant cell is divided to create a completely new plant. For this to happen, the cell must be temporarily altered so that it does not have the structure or function of any specific tissue. If the cells come from leaves, hormones must later be added to the culture medium so that the genes in the cell begin to function as before. The cells resulting from this type of treatment are called **callus**.

Nutrients and hormones are then added to the callus. As a result, the genes in the cell being cultured begin to switch on and off in proper order, and the growth of roots, stems, leaves, and other plant organs begins.[14] This procedure is frequently referred to as **tissue culture**. (Tissue culture is discussed later in this chapter.)

The cloning process is used by plant breeders to accelerate the creation and reproduction of new varieties of plants. By combining cloning and gene splicing, even more useful plant varieties can be produced.

Embryo Splitting

Mammals have been effectively cloned through a process called **embryo splitting**. In this type of cloning, a developing embryo is divided using a type of microscopic surgery. The best results occur when an embryo in the 16- to 32-cell stage of development is divided into two or more independent embryos. This process is expensive, but when the embryos are used to impregnate receptor females, the offspring produced are genetically identical. This practice is most common in the dairy and beef cattle industries.[15]

Embryo Transfer

The practice of implanting embryos in surrogate animals is called **embryo transfer**. Thousands of calves have been born using this procedure. When embryos are transferred without being split, the pregnancy rate approaches 65 percent. Refer to Figure 4–9. However, when the embryos are split before being transferred, scientists have been able to produce nearly one calf for each healthy embryo harvested. Only about one-half of the split embryos transferred result in a pregnancy; however, because splitting produces twice as many embryos for transfer, about 35 percent more calves are born overall. If superior females are used in this procedure, genetic progress is enhanced even more.[16]

Advantages of Embryo Transfers

Cloning will make it possible to create multiple, identical copies of a champion bull, prize steer, or any other domestic animal. Also, USDA Agricultural Research Service scientists at Beltsville, Maryland, are a step closer to breeding cattle with natural immunity against diseases that have plagued ranchers for decades and are estimated to cost them $9 billion a year. Using split embryos implanted in surrogate mother cows, the scientists have produced five animals with identical immunity genes—possibly the foundation for a whole herd of genetically matched cattle, which could offer an unprecedented opportunity to study the immune system.

14. Ibid.

15. Burton, *Agriscience and Technology*, p. 38.
16. Ibid., p. 38.

Gene Mapping of DNA

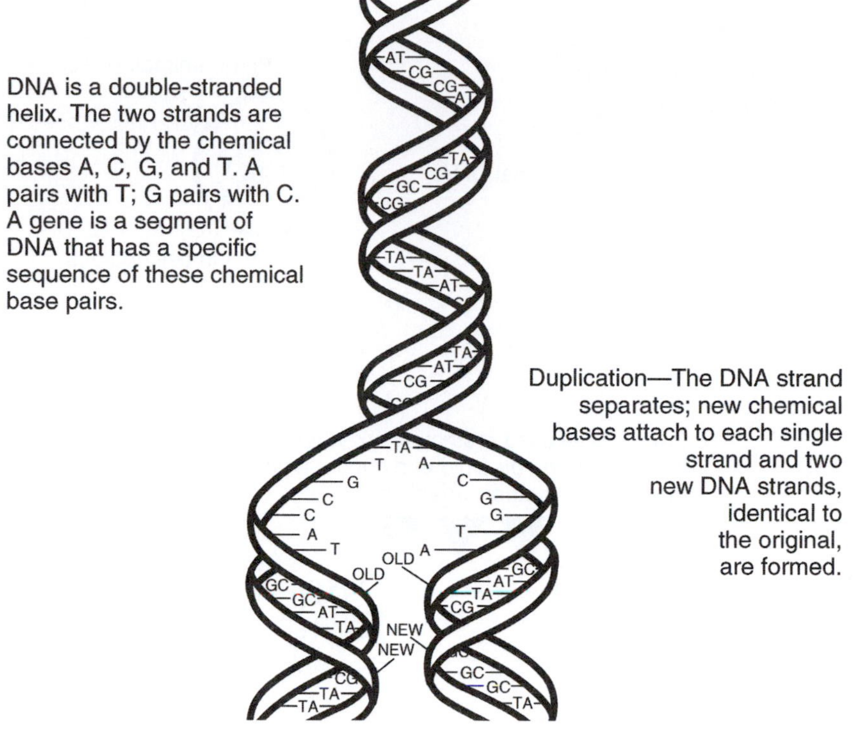

Protein synthesis—Special enzymes copy a single DNA strand to make a single messenger RNA (mRNA) strand in which U replaces T as a base.

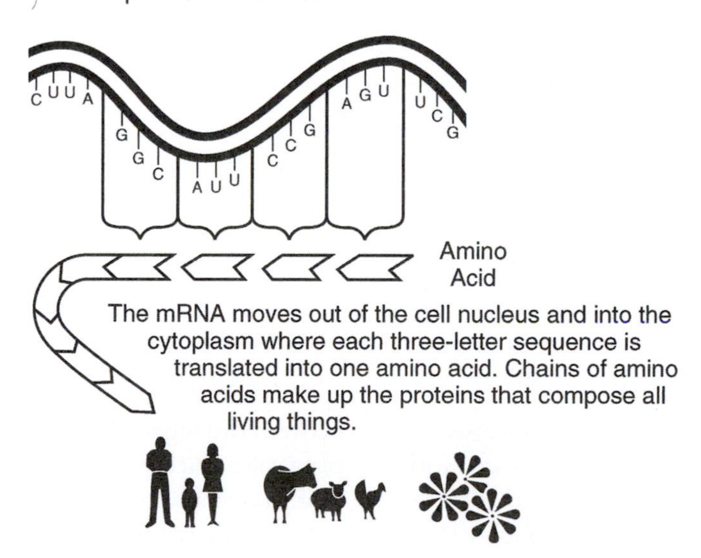

Figure 4–6 Before gene splicing, a gene must be mapped.

Gender Selection

Gender selection is the ability to control the sex of offspring at the time the parents mate. Animal scientists have attempted for many years to control the sex of animal offspring.

Sperm-Sexing Procedure

The only measurable difference between X (female) and Y (male) sperm is their **deoxyribonucleic acid (DNA)** content. The X chromosome is larger and contains slightly more DNA than the Y chromosome.

Gene Splicing

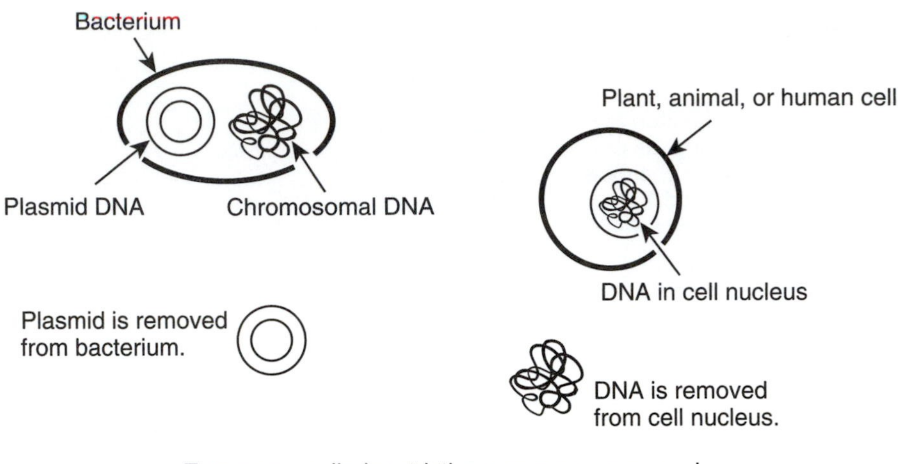

Enzymes—called restriction enzymes—are used to cut open the plasmid and cut out a gene from the DNA of another organism.

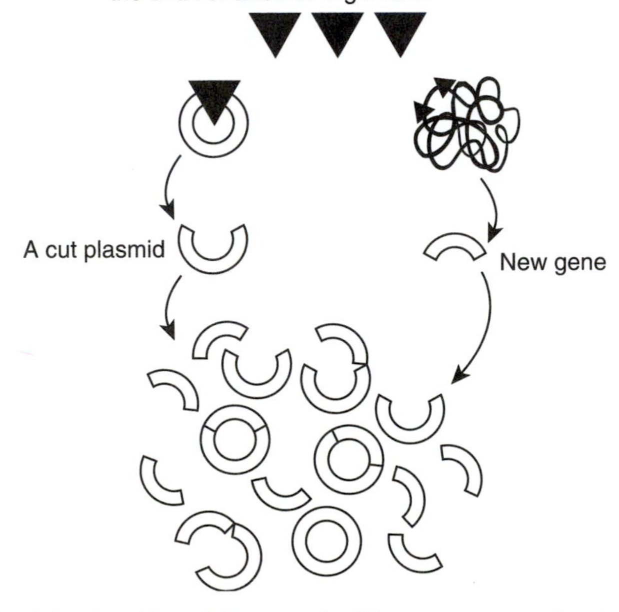

Figure 4–7 The cut ends of the plasmids and the cut ends of the new genes are chemically "sticky," so they will attach to each other–recombine–to form a new loop containing the inserted gene. This technique is called gene splicing of recombinant DNA.

An instrument called a *flow cytometer* can identify X and Y sperm cells after they have been treated with fluorescent dye, which stains the DNA so that it lights up when the sperm are passed through a laser beam. The amount of fluorescent light given off is measured by a computer. Because the X chromosome is larger and contains more DNA, female sperm give off more light. From this analysis, the ratio of X to Y sperm in a semen sample can be determined. Also, the X and Y sperm can be separated after identification. Refer to Figure 4–10.

The sperm cells can be encased in a droplet of liquid, given a positive or negative charge, and passed between two high-voltage steel deflection plates. Each plate will attract sperm with the opposite charge, separating them into collection tables. At present, the separated sperm cannot fertilize eggs because their tails have been removed, but researchers are seeking to use a variant of this process to separate intact sperm.[17]

Advantages of Sperm Sexing

With sperm sexing, dairy farmers could produce female calves for herd replacement, thus reducing

17. Ibid., p. 75.

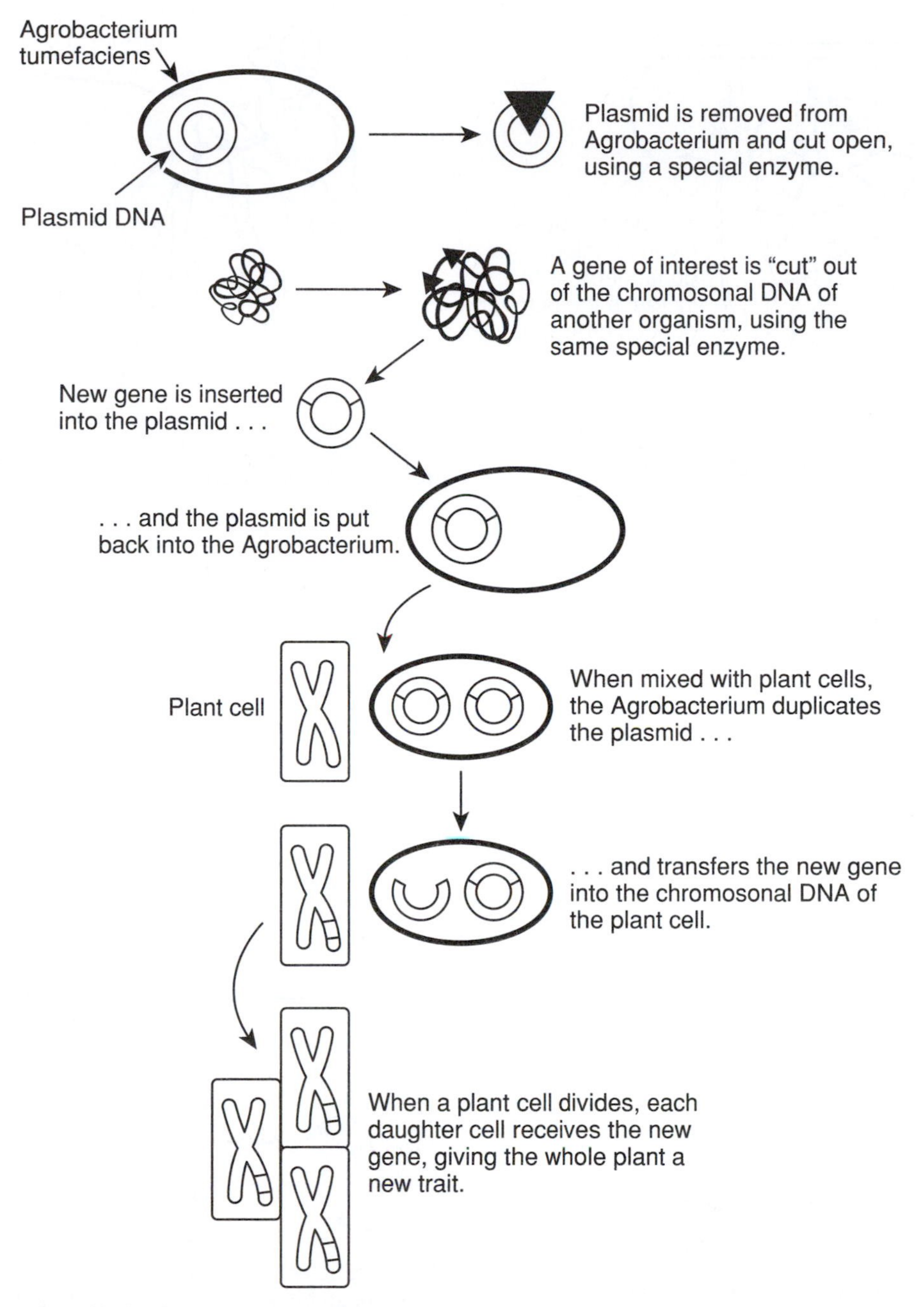

Figure 4–8 Examples of how plants are genetically engineered.

the time and expense of raising bull calves for slaughter. Beef producers want mainly male cattle because they gain weight faster than females and usually command a higher price at slaughter.

Sorting Embryos According to Sex

Today it is possible to select the gender of livestock before birth. Embryos can be removed from females and sorted by sex. In fact, one high-quality embryo of either gender can be split into as many as four identical embryos. The gender-selection process is completed when an embryo of the desired sex is implanted in a female and grows until birth.

Modifying Gender in Fish by Ultraviolet Light

The fish industry has also become interested in gender selection. Scientists have found that the process of meiosis can be modified through the use of ultraviolet light. After being exposed to such light, the sperm

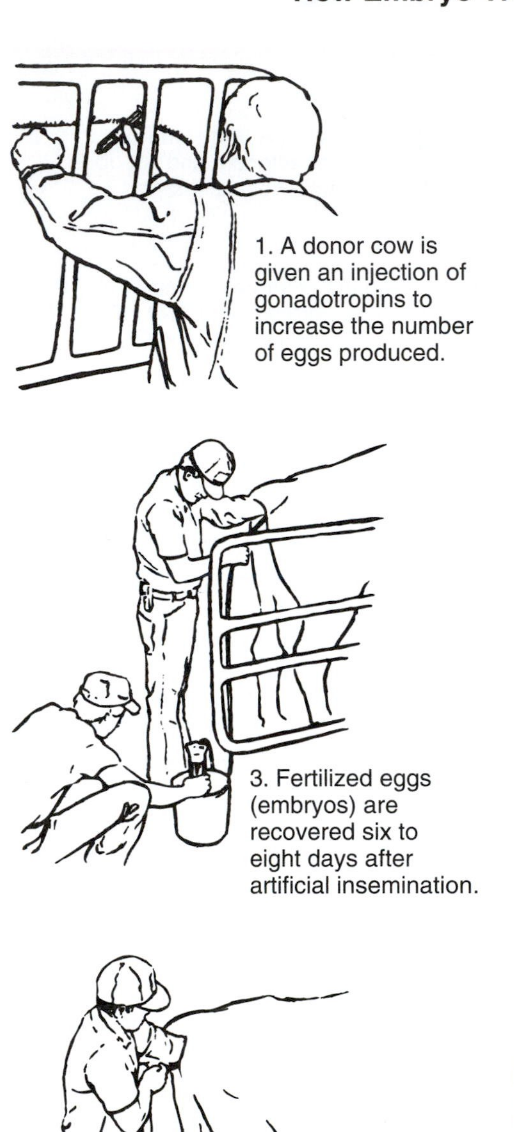

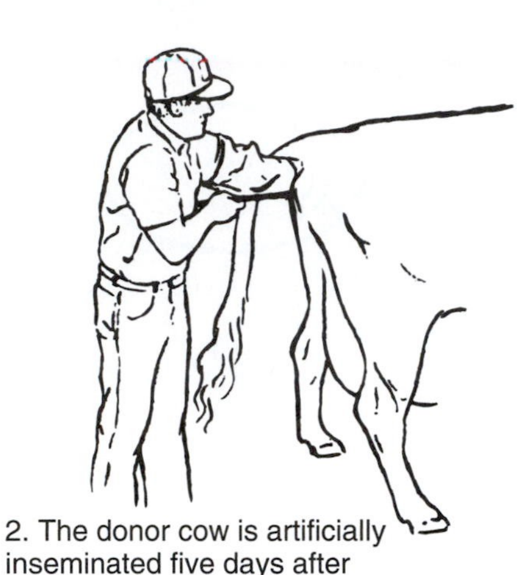

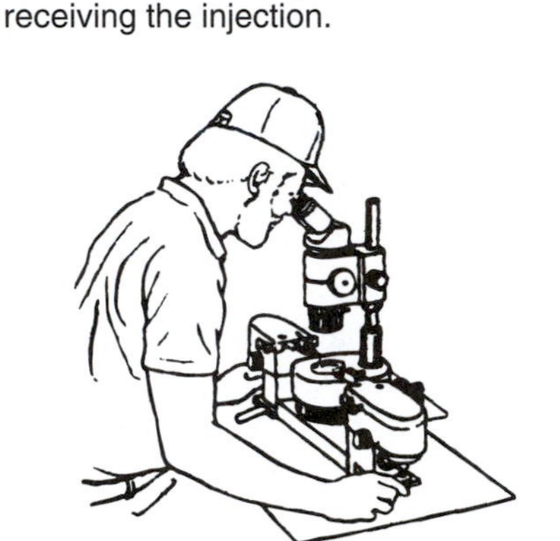

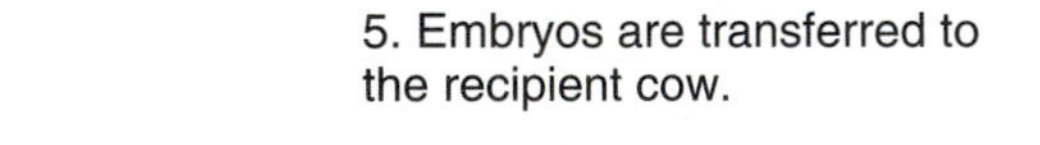

Figure 4–9 Explanation of the embryo transfer procedure.

from male fish become unable to combine with chromosomes in the eggs of the female. The eggs are still fertilized and still develop, but they hatch only female fish. Male hormones are then used to treat the female fish thus hatched so that they take on the sexual characteristics of males. However, the sperm from the treated fish contain only X chromosomes. Female fish eggs also contain only X chromosomes. Hence, all the eggs that are fertilized will yield female fish. Refer to Figure 4–11. In some species of fish, such as trout, the female is more desirable in the marketplace because it produces a higher-quality fillet that can be sold at a premium price.[18]

18. Ibid., p. 76.

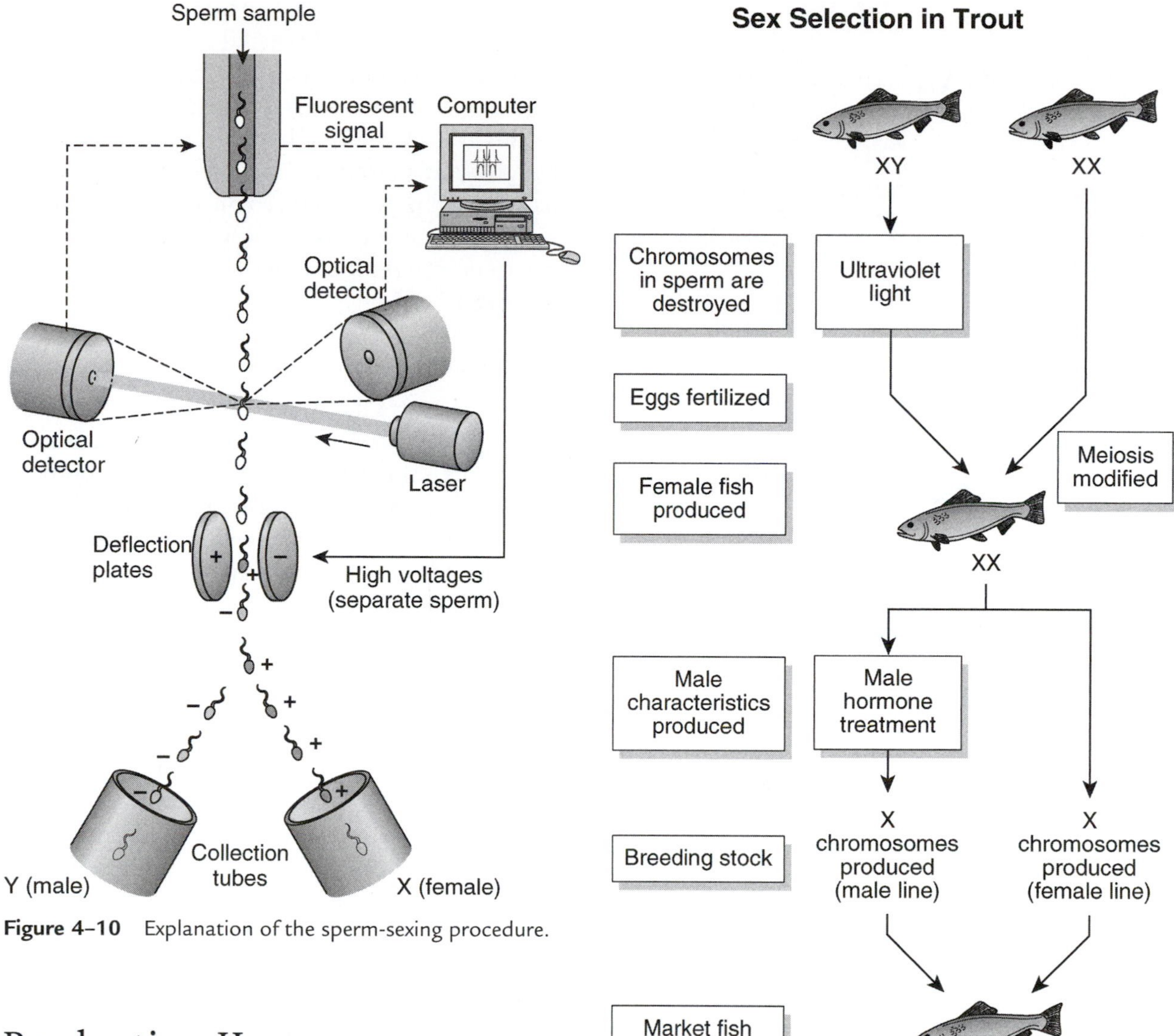

Figure 4–10 Explanation of the sperm-sexing procedure.

Figure 4–11 Gender in fish can be modified by ultraviolet light.

Production Hormones

Hormones are substances that are formed in the organs of the body and dispersed throughout other tissues by body fluids. As the tissues absorb the hormones, specific body functions are triggered or enabled. Today scientists can actually produce and modify certain hormones. These are called *synthetic hormones* because they are not formed naturally.

Animal Growth Hormones

Scientists have discovered that they can accelerate growth in meat animals by treating them with certain hormones. There are a number of ways to accomplish this. Perhaps the simplest is to add small amounts of a desired hormone to the feed of animals. Scientists have also found that growth hormones can be injected directly into the animals' tissues for fast absorption, and that they can be formed into small pellets and injected into an animal's ear just beneath the skin for more gradual absorption.[19]

Bovine Somatotropin

The U.S. Food and Drug Administration (FDA) and the USDA have approved the use of a growth hormone that increases milk production in dairy cows by as much as 30 to 40 percent. The hormone, called **bovine somatotropin (BST)**, is a protein that occurs naturally in dairy animals and

19. Ibid., pp. 52–53.

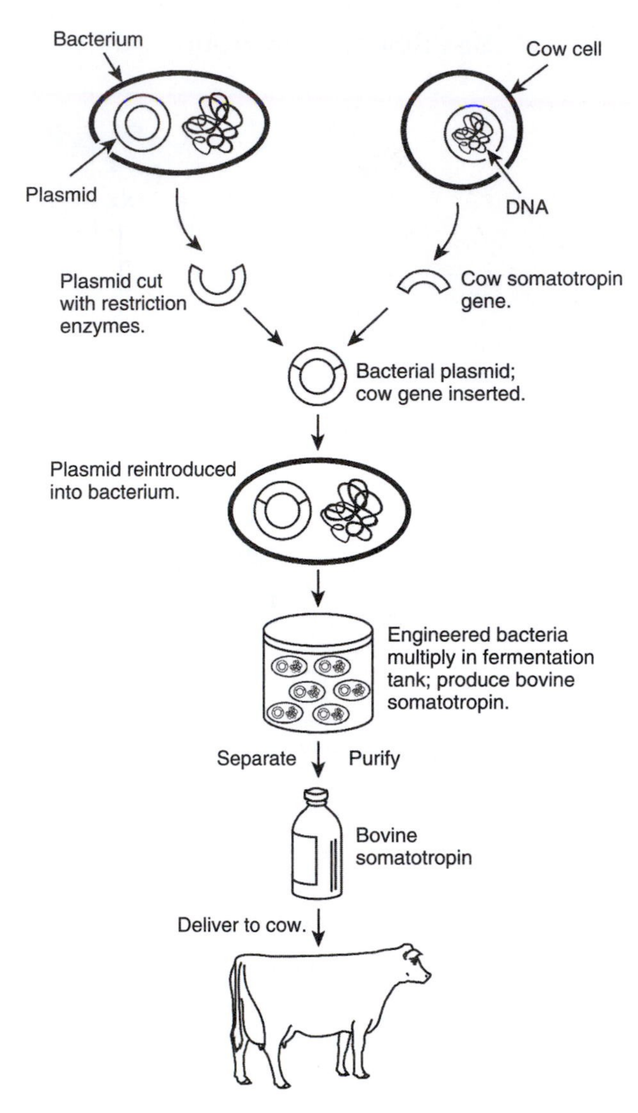

Figure 4–12 Explanation of how bovine somatotropin is produced.

stimulates milk production. Now, BST can be artificially produced in a laboratory through a fermentation process. When it is injected into dairy cows, it enables them to more efficiently turn feed into milk. There is no noticeable difference between the milk of treated and untreated cows.[20] Refer to Figure 4–12.

Porcine Somatotropin. The pork industry may soon have available **porcine somatotropin (PST)** hormone to enhance muscle growth in young pigs. This growth-enhancing hormone has been shown, in experimentation at Auburn University, to increase slaughter weight and litter size by an average of one pig per litter.

Research conducted at the U.S. Experiment Station in Beltsville, Maryland, using PST has been encouraging. A strain of hogs has been developed that grows at a faster rate and produces less fat than normal pigs. Genetic engineering has made this possible by increasing the usual amount of growth hormones found in the pigs through the use of PST.[21]

Fish Growth Hormones

The fish industry has also been working to genetically engineer growth hormones. In Canada, fish injected with these growth hormones have been found to grow 50 percent faster than untreated fish. In the United States, growth hormone genes have been taken from trout and injected into carp eggs. The fish that hatch, called **transgenic** fish, can grow as much as 20 percent faster than the offspring of untreated carp.

Heat Synchronization

By changing the hormone balance in any type of farm animal, the estrus cycle can be modified. This can be accomplished in one of several ways. A carefully measured amount of hormones may be injected directly into the animal, or a hormone **implant** may be placed just beneath the animal's skin. Scientists have also found that hormone material can be placed in the female reproductive tract in a sponge **pessary**. The hormones from the implants and pessaries are applied several days before breeding so that the animals' bodies have time to absorb them. Pessaries must be removed before breeding can take place.

Controlling the estrus in farm animals has the distinct advantage of allowing a large number of animals to be bred in a relatively short time. This is useful, for example, when artificially inseminating cattle that are scattered for miles on open ranges in the western United States. By controlling when the cows come into heat, the normal three-week breeding period for the herd can be reduced to a period of only a few days.[22]

20. Ibid., p. 53.

21. "Injections Increase Muscle," *Hogs Today*, June 1997, p. 40.
22. Burton, *Agriscience and Technology*, pp. 73–74.

Animal Research and Human Medicine

Animals have been used for scientific investigation in veterinary and human medicine for generations. There is a close relationship between the health of humans and the health of animals. Illnesses that are devastating to humans and animals are studied using animal research to determine the cause, transmission, mechanism of infection, prevention, treatment, and control of the organisms that cause these diseases. The result is a longer, healthier, and more productive life for both animals and humans.

Swine Research for Human Health

More than almost any other species, the anatomy of the pig most closely resembles that of the human. **Physiologically**, swine have organs similar to the human heart, circulatory system, and digestive tract, as well as similar enzyme production and teeth.

Swine and humans often suffer from similar diseases, such as tuberculosis, influenza, peptic ulcers, brucellosis, and cardiovascular disease. Pigs have similar nutritional requirements and digest and process their food in much the same manner as humans. For these reasons, swine have become subjects for medical research that may aid in reducing human death and disease. For decades, the swine have produced insulin for enzyme replacement use by diabetics.[23] Refer to Figure 4–13 to see the wide range of medical products that come from swine.

Medical Products from Swine

Adrenal glands—produce steroids, cortisone, and epinephrine
Intestines—produce heparin, an anticoagulant
Overies—produce estrogen and progesterone
Stomach—produces pepsin, an intrinsic factor that aids in vitamin assimilation
Liver—produces desiccated liver to aid in iron metabolism
Pancreas—produces insulin, glucagon, and a host of digestive enzymes
Heart and heart valves—hog heart valves, specially preserved and treated, are surgically implanted in humans to replace heart valves weakened by disease, injury, or congenital malformation. Since the first operation in 1971, tens of thousands of hog heart valves have been successfully implanted in human recipients of all ages.
Pituitary gland—yields antidiuretic hormone (TSH), which regulates metabolism.
Skin—for skin grafts in severe burn cases.

Figure 4–13 Examples of several medical products from swine.

Transplantation Technology

Animal organs may soon be routinely used for transplantation in critically ill humans. According to statistics, 300 patients die each year awaiting an organ suitable for transplantation, and another 100,000 more die without qualifying for the transplantation list. Specialists agree that thousands of lives might be saved if organs for transplantation were readily available.[24]

It is believed that pigs raised without known diseases would make excellent sources of transplantable organs. The technology now exists to control the molecular basis of the immune-system rejection that previously prevented the use of animal organs for transplantation to critically ill humans.

Xenotransplantation. **Xenotransplantation** is the transplantation of organs and tissues between species. Swine may be genetically altered to produce proteins that are sufficiently similar to human proteins that they will withstand the assault of the rejection process of the human immune system. Also, it is now possible to create drugs that will counteract the rejection process. In cases where patients must have a human donor organ, the animal organ may serve as a stopgap measure, allowing the patient to buy time while waiting for an acceptable organ from a human donor. This has been studied especially with pig livers, which may be genetically altered to enhance their already close anatomical and physiological similarities to human livers.

23. "In Praise of Research Animals," in *The Science of Animals That Serve Humanity: Animal Research in Retrospect and Prospect* (New York: McGraw-Hill, 1985), 730–741.
24. Lawrence M. Fisher, "Down on the Farm, a Donor: Genetically Altered Pigs Bred for Organ Transplants," *New York Times,* January 5, 1996, p. C1.

Pharmaceutical and Other Medical Products Made from Animal Products

Help for Heart Patients. Scientists are working on ways to perfect the use of genetically engineered cows that will act as living factories, producing in their milk such products as TPA (a biotech product given to heart attack victims). The blood-clotting factor used to treat hemophilia is an example of a rare protein that may someday be extracted from milk.

Help for Burn Victims. Doctors are cultivating cells from burn victims' healthy areas of skin in layers of cow protein. This new technique, which shows great promise, allows grafts to anchor to the patients' bodies within nine days, compared to the usual four weeks to six months or a year associated with conventional techniques.[25]

Help for AIDS Patients. Cows may hold the key to a measure of relief for some AIDS patients. Because of their limited immunity, certain AIDS patients become infected with an internal parasite that also infects animals. The parasite—found in contaminated food or water—causes severe diarrhea in humans, but it does not cause AIDS. USDA Agricultural Research Service scientists have obtained an immunoregulatory chemical from the lymph nodes of cows infected with the parasite. When medical researchers from New York University gave the chemical to AIDS patients, 75 percent stopped having diarrhea without experiencing unwanted side effects. (The treatment has no effect on the primary AIDS infection itself.)

Production of Hepatitis B Vaccine. In Japan, scientists have developed a method that uses silkworms to produce a superior hepatitis B vaccine.

Help for Leprosy Patients. Approximately 4,000 people in the United States are affected every year by the disease known as leprosy (Hansen's disease). Leprosy is rarely fatal, but it does produce skin and nerve lesions, and when left untreated it can cripple or severely disfigure. Scientists had trouble testing new drugs that might rid humans of leprosy because no other animal was known to get the disease. However, a biologist in Florida doing reproductive and other studies on armadillos noticed a similarity between these animals and humans. Leprosy attacks the cooler areas of the body, such as the nose and ears. Armadillos have a normal body temperature (between 92° and 95°F) lower than a human's normal body temperature. Armadillos are therefore susceptible to leprosy and can be used as laboratory test animals in finding a cure for this dreaded disease.

Using Leeches to Increase Blood Circulation. Leeches are making a comeback in medical science. However, they have become harder to find in the wild, which has spurred ingenious animal producers to start leech farms to answer the need for quantities of leeches. When burn patients start to reject skin transplants, doctors apply leeches to the affected area to increase the blood flow and thus speed healing. Leeches also produce an anticlotting agent called hirudin. This substance is used to treat hemorrhoids, rheumatism, thrombosis, and contusions. Another enzyme produced by leeches, hyaluronidase, is used to speed the perfusion of injected drugs or anesthetics in the body.

Domestic Animals of the Future

Dairy Cattle

The big news in dairy cattle for the 21st century will be milk output—herds yielding as much as 40,000 pounds of milk per animal per year, or approximately 15 gallons of milk per day. Through nationwide genetic improvement efforts, cows will be capable of producing 70 pounds of milk more per year than the cows of today, because they will be descendants of animals that produce more milk. Compare this with an annual gain of 10 to 20 pounds prior to innovations in genetic evaluation and improvement starting in the early 1960s.

Beef Cattle

In the future, the speed at which beef cattle grow to market size will change drastically. Cattle will grow 50 percent faster because producers will know more about nutritional requirements and will feed their cattle precisely to those requirements, to more fully use their genetic ability for meat production.

---

25. Ricketts, *Science 1A (Agriscience),* p. 20–14.

Cloning of cattle, as discussed earlier, is certain to be part of the 21st-century scene, as will shipping whole "herds" of cattle into and out of the country in the form of frozen fertilized embryos. Animal geneticists predict that in the United States, producers will likely see new breeds, selected genetically to match the climate. They are conducting experiments now to see how vigorous these crosses are, and their preliminary results are very encouraging.

Agricultural scientists also expect an increase in twinning of calves. A project is under way in which scientists are selecting intensely for twinning, using daughters of cows with outstanding twinning capabilities and crossing them with bulls with unusual twinning capabilities in their daughters. Presently, these researchers have had 13 percent twins in their fall calves and 8 to 9 percent in their spring calves, compared to the norm of only 1 to 2 percent.[26]

Swine

The pigs of tomorrow will be bigger, but not fatter. With somatotropin, producers can increase the yield of muscle tissue 20 percent and simultaneously reduce the amount of body fat up to 70 percent. Currently producers slaughter pigs at 220 to 240 pounds. The carcass is about 70 percent of the pig's live weight. Of that, the fat-to-lean ratio is now two-to-one. A ratio of one-to-one is within producers' grasp.

The ability to control fat levels will change the market weight of hogs. Producers slaughter them when they do mainly because beyond that weight, the hogs become too fat. If producers can control the amount of fat, there is no reason to stop at 240 pounds. They could go as high as 300 pounds. The key is body composition.

Poultry

In 1950, it took 84 days to produce a 4-pound broiler, and the feed conversion rate was 3.25 pounds of feed for every pound of meat produced. Today it takes 42 days, and the conversion rate is 1.9. In the 21st century, farmers should be able to produce a 4-pound broiler in 25 days. They will make these gains the same way they reduced production time from 84 to 42 days: by making the system better via genetics, disease control, nutritional control, and environmental control. However, these processes will have to be synchronized.

Also, it has been predicted that breeders will probably make a dwarf breeder hen; there are some of these already. That will have no effect on the size of the chick, but it will save on feed for the hens.

Sheep

Some scientists are predicting that average lamb slaughter weight at 5 months of age will rise to 170 pounds by 2040, compared with 120 pounds today and 80 pounds 50 years ago. Producers have increased their lambs' weight almost a pound a year, and there is no reason, some scientists think, to believe that the trend will not continue.

Catfish

Fish farmers are hoping that a hormone called dihydrotestosterone (DHT) will do the trick in producing all male catfish. Their aim is to make the most of the finding that male catfish grow faster than females, because even a small advantage could mean big profits to producers in one of the nation's fastest growing agricultural endeavors. (Earlier in the chapter, we learned that female trout were preferred.)

Bees

In the future, honey producers will use specific bee species to pollinate particular plants. An example is that when *Habropoda laboriosa*, commonly known as the southeastern blueberry bee, is used, rather than the common European honeybee, the blueberry honey yields increase. This is because blueberry bees have longer tongues and can pollinate rabbiteye blueberries. Honeybees' tongues are not long enough to do the job on this type of blueberry plant. Also, pollination by blueberry bees of rabbiteye blueberries produces more and better-quality berries earlier in the season.

Transgenic Animals

As discussed earlier, gene transfer is a method whereby scientists transfer a desirable gene from one species to another. Some scientists are predicting

26. Ibid., pp. 20–17.

that transgenic animals will find a place in the human food supply within the next decade. Consider the following examples:

- Scientists have introduced cattle genes into pigs in an attempt to produce pork with more meat and less fat.
- Biologists at the University of California at Davis have produced a "geep" by fusing a sheep embryo and a goat embryo in a test tube, and then implanting the fused embryo into a goat's womb for development of the new animal.
- Some scientists are replacing traditional or classical breeding methods with gene transfer to develop new breeds of swine. Nonnative swine species are being imported into the United States to act as donors of genetic material in crossings with U.S. breeds of swine.
- A rare, wild, ox-like animal called the gaur was crossed with dairy cattle, and the resulting animals proved to be superior in meat production. Scientists in the future may use genes from exotic species of animals to improve domestic species, whether or not they are related genetically.[27]

Computerized and Electronic Animal Management Technologies

Agricultural Computer Use

As in most areas of modern life, the computer has become an important tool for efficient and time-saving progress in agriculture. Computer software applications record milk productivity in dairy herds and compare milk output to feed consumption. Other agricultural software packages offer farm tax information, electronic tax filing, and specific bookkeeping programs. Still other software applications help gardeners and landscaping professionals design and prepare low-maintenance and energy-saving beautification projects for rural, urban, and suburban sites.

Computers aid the livestock industry by allowing producers to calculate balanced animal diets consisting of effective combinations of feeds that will provide all necessary nutrients at the least cost. Many computer programs evaluate production records to aid in the selection of superior animals for breeding. Refer to Figure 4–14. The agriculture student and scientist use the computer as a storage medium for their many records and scientific reports, which can be called up instantly for review of the scientific literature in any given research area.

Figure 4–14 These agricultural biologists are analyzing data to confirm the DNA sequence for receptor genes. (Courtesy of USDA)

Electronic Animal Management Systems

Robotic Milking System. In the livestock industry, the use of electronic devices has considerably reduced the amount of manual labor needed to complete chores. Milking cows is one example. Very few cows today are milked by hand, as good electronic milking systems have been developed. Thanks to ongoing research, this and other routine and repetitive farm tasks may be completed in the future by specially designed robots.

In one research facility, an electronic milking system is being developed that can wash, milk, and feed a cow. As the cow voluntarily enters the milking area, a robotic arm connects the machine to the cow. When the cow has been milked, the robotic arm automatically removes the milking machine. Daily computer records indicate the amount of milk produced and the amount of food consumed by each cow.[28]

27. Ibid., pp. 20–65.

28. Burton, *Agriscience and Technology*, p. 55.

Electronic Sensors. Electronic sensors are being used to control the environment when livestock are in confined areas. Such sensors relay information to central systems that ventilate the area, apply insecticides, or mist the animals with water when the weather is hot.[29]

Robotic Sheep Shearing. Robotic arms may soon be able to shear sheep, and thus eliminate one more tedious task. Human laborers will place the sheep in a restraining device. At that point, the robot will measure the sheep, program the shearing tool accordingly, and remove the wool.[30]

Birthing Sensors. Electronic sensors implanted in an animal's body cavity can help producers detect when she is ready to give birth. This helps the producer to prevent losses due to difficulties during parturition. The same sensors have proven to be more than 90 percent accurate in detecting when cows are in heat.

Live Animal and Carcass Evaluation Devices

X-Ray Scanners. X-ray scanners can provide an excellent way to evaluate meat quality while the animal is still alive. The technique used, called computerized tomography, produces monitor-screen images of segments right through the animal's body. This technique offers a complete picture of conformation without the need to slaughter the animal.

Measuring Percentages of Red Tissues. A commercial company has developed a machine that can measure percentages of red tissues as an animal walks through a scanning chamber. The machine does this by sensing the conductivity of electricity. Red tissues conduct electricity 20 times better than white tissues. The higher the conductivity reading that is recorded, the greater the lean meat percentage. The machine can measure 240 hogs per hour at better than 98 percent accuracy.

Ultrasound Imaging. Animal scientists are borrowing from human medicine in their search for harmless ways to see inside livestock. Ultrasound imaging (sonography) permits researchers to measure body fat and lean tissue in growing animals.

29. Ibid.
30. Ibid.

Emerging Mechanical Technologies in the Plant Industry

Many technological developments are occurring in the plant industry. Space does not permit an explanation of each, but short introductory comments are included here on several mechanical technological developments.[31]

- A device has been developed that helps farmers in the early detection of insect pests. When early detection is used, the farmer can use various methods to prevent costly and damaging insect infestations. The system is called *sodar* (for "sound detection and ranging") and works along with a second system of infrared light detectors. The device looks like a rocket-shaped weather vane and is mounted on a pole. A rudder orients the device into the wind and releases pheromone downwind. Male insects pick up the scent and follow it to the echo receiver. Each echo causes one radio signal to be sent to a receiver and computer located away from the field.

A second pest detection system is a basic cone trap baited with a pheromone that has been modified for counting. The pheromone attracts a male, and as he realizes that this is a false alarm, he flies up into a wire mesh cone. As the insect flies up through the top of the cone, he trips a triggering infrared light beam, which again passes a radio signal to the computer receiver. The devices can detect and count the moths of several major pests.

- A computerized apple has been developed that helps shippers pinpoint problem areas where major fruit damage is occurring. Apple losses after harvest, caused by damage from handling, shipping, and storing, total millions of dollars annually. Because fresh apples bring

31. Ricketts, *Science 1A (Agriscience),* pp. 44–48.

higher prices than processed apples, the growers want to minimize damage. The artificial apple automatically records the bumps that real apples receive as they are moved from orchards to retail stores.

- Someday in the future, crop dusters may be flying planes by remote control. The planes will have a wingspan of only 8 feet, and a four-horsepower chainsaw engine, but they will be able to fly at much lower speeds than conventional crop-dusting aircraft. The lower airspeeds enable crop dusters to apply whatever chemical or natural biological control is being used at more efficient rates. This method of crop dusting will, in many cases, also be less expensive to operate.
- A high-tech microphone can tell grain operators when insects are most active inside stored grain. The microphone is part of a durable acoustic system that can detect the feeding sounds—amplified up to 75,000 times—of the lesser grain borer, rice weevil, and Angournois grain moth. This information can cut costs for crop producers and grain operators by telling them, with no need for grain samples, when insecticidal fumigants will do the most good.
- An underground plow originally designed to pulverize hard, thin planes of soil in dryland wheat-growing regions may become a boon to conservation tillage farmers. The inventor is patenting a sweep plow that he modified by attaching four steel shanks underneath to break up the soil. A sweep is a flat, V-shaped blade with a "wingspan" of up to 5½ feet. In operation, the blade is pulled horizontally through the soil, point first, about three to four inches below the surface.
- A robot that does monotonous, backbreaking chores could one day help to automate the labor-intensive work of raising, shipping, and transplanting vegetable and tree seedlings in fields, nurseries, and greenhouses. Researchers have designed and filed for a patent on the first component of a robotic transplanting system that could work at least four times faster than human-dependent systems, with each row in the machine processing and planting 180 to 240 seedlings per minute, compared to a typical 40 that can be done by hand.
- Melons that look and smell ripe at the supermarket are often picked too soon to be sweet. To solve this problem, Agricultural Research Service (ARS) engineers have developed a device that uses light rays to measure the sweetness of melons such as honeydew, watermelon, and cantaloupe. The breadbox-size device can monitor sweetness in melons by measuring the amount of near-infrared light the fruit absorbs. The more infrared light absorbed, the sweeter the fruit.
- *Laser* is an acronym for "light amplification by stimulated emission of radiation." Lasers contain a medium that, when stimulated by energy such as electricity, produces a light that is amplified by a reflection process between mirrored surfaces. The portion of the amplified light that is allowed to come out of the laser produces a thin, straight, extremely bright beam. This characteristic makes lasers suitable for many measurement applications.

In agriculture, lasers are used to alter the land to improve irrigation, as well as in building levees and terraces. Fields can also be leveled with consistently high accuracy. Irrigation systems can be built that improve crop yield and reduce water usage at the same time. Drainage materials can be installed at a more economical rate. With proper use, lasers can be a valuable tool in soil, water, and energy conservation. They can also be a tool for increasing food production.

- Scientists are researching weather modification and analysis. The United States and other countries have been working on perfecting techniques for seeding clouds. Some progress has been made: in tests in Florida, a 50 percent increase in rainfall was obtained.
- To get more accurate information on long-term weather patterns, scientists use remote sensing technology. By analyzing atmospheric variations, they are better able to predict general weather patterns.

Tissue Culture

In a relatively new procedure, tissue cultures from plants are being used to rapidly produce many identical new plants from one parent plant. In this procedure, plant scientists take tissues from the buds, leaves, or shoots of a parent plant. They sterilize the tissue and place it on a growing material such as a sterile agar gel. Growth-regulating hormones are applied, and then the culture is sealed to prevent any type of contamination. Soon, tiny sprouts appear from the culture. These are removed with tweezers and placed in a new growing medium that will promote root growth.

Career Options

Aquaculture Research, Genetic Engineering, Hydroponics, Plant Breeding, Plant Propagation, and Tissue Culture

Aquaculture enterprises are big business in many nations. Catfish farming is one of the fastest-growing food production enterprises in the United States. The rapid growth of aquaculture has spurred research and development activities. These, in turn, have stimulated career opportunities in animal science, nutrition, genetics, physiology, aquaculture construction, facility maintenance, pollution control, fish management, harvesting, marketing, and other areas.

Genetic engineering cuts across many fields of endeavor. Procedures for the genetic modification of organisms have been developing for more than a decade. Biologists, microbiologists, plant breeders, and animal physiologists are some examples of specialists who might use genetic engineering in their work. Work settings include the field, laboratory, classroom, and commercial operations.

The practice of **hydroponics** is not new, but hydroponics for commercial production has captured the world's imagination. The increased popularity of hydroponics operations has resulted in new career opportunities, especially in urban areas.

Plant breeders' objectives might include making plants faster growing, disease resistant, frost tolerant, more beautiful, or better flavored, depending on the particular uses.

Tissue culture, a procedure developed in biotechnology, permits the production of thousands of new plants identical to one, superior plant. The procedure is relatively cheap and easy and is used extensively to reproduce ornamental plants. Many jobs are available in the area of plant reproduction.

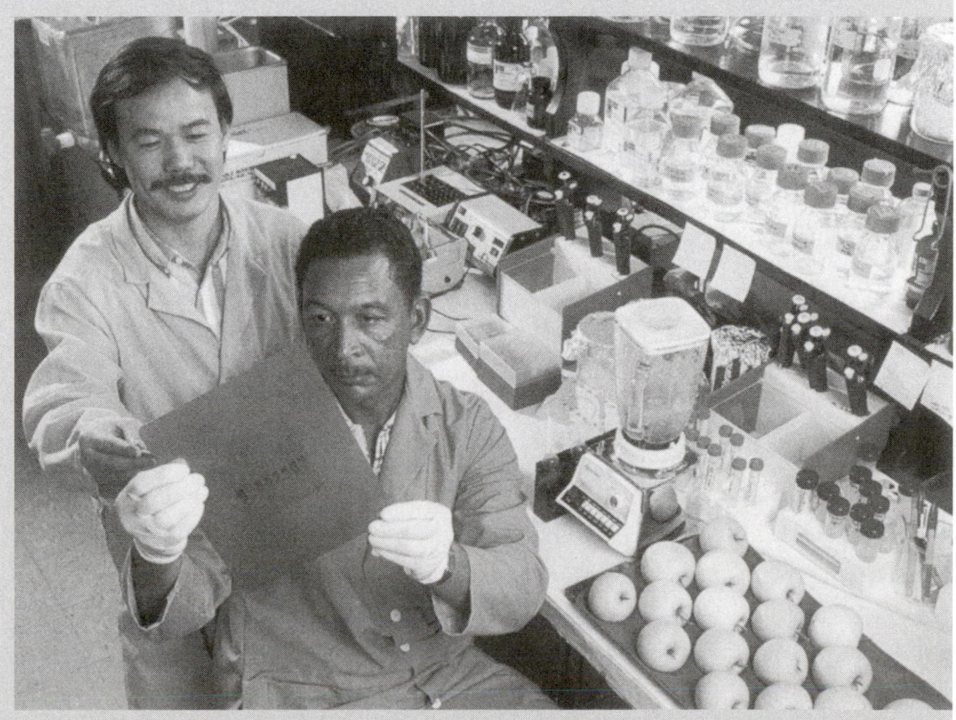

These agricultural scientists (a plant pathologist and a biological technician) are checking apples that have been genetically engineered to produce larger amounts of rot-inhibiting protein. (Courtesy of USDA)

The key to using plant tissue cultures is to maintain a sterile, contamination-free environment. When done properly, propagation through this method has proven to be one of the most valuable plant technologies developed in recent years. It makes possible the exact duplication of large numbers of valuable plants.[32]

Special Skills Required

An understanding of general laboratory procedures is needed, as is an understanding of general cleanliness. Anyone can be trained to perform the cutting, prepare the media, and clean the lab. However, someone in the operation must be familiar with horticulture and plant physiology for the laboratory to be successful and profitable.

Advantages and Disadvantages of Tissue Culture

Advantages. The major advantage of tissue culture is that one leaf-tip cutting can produce 4,096 plants in a year. In the nursery industry, many hard-to-propagate, woody ornamental plants are produced economically through tissue culture. Another advantage is that each plant produced is a clone, an exact replica of the parent plant. This assures that the plants produced and sold will be true to species and culture. During the early phases of tissue culture, the only facilities required are enough room for jars of media and growing plants. Only after the plants have been transplanted is a greenhouse for storage needed. The plants can be transplanted into containers with media or into a hydroponic system.

Disadvantages. The largest disadvantage of tissue culture is the chance of introducing a contaminant into the laboratory, which could spread and kill all the plants in the lab. This would result in great monetary loss as well as a huge loss of time and labor.

Emerging Plant Technologies

Frost-Protected Plants

Freezing temperatures may radically affect the productivity of certain fruit and vegetable plant species. The amount of damage plants sustain depends on how long temperatures remain below freezing, how far the temperatures are below freezing, and the type of plant in question. With some plants, critical damage occurs whenever freezing occurs during the growing season. With others, damage occurs only during the flowering stage or when the fruit is immature.[33]

Biotechnology has enabled researchers to develop a special frost-free spray, that, when applied to plants, prevents them from freezing. Some bacteria help ice to form on plants at temperatures just above the freezing point.

Plant Growth Regulators

Plant growth hormones cause plants to produce and activate certain genes. They also help regulate the growth of roots, stems, and leaves, and control the development of seeds and fruit.[34] The plant growth process can now be duplicated and controlled scientifically through the use of plant growth regulators. These regulators were developed from certain chemical compounds found naturally in plants. They are used by nursery owners and flower growers to help them better produce and market plants for the appropriate season. For example, there is no demand for Easter lilies after Easter. Moreover, regulating the size of plants also has marketing advantages.

Salt-Tolerant Plants

Through genetic engineering, salt-tolerant genes have been successfully transferred into various types of forage crops. The significance of this scientific advance is that ocean water, which is available in abundance, but has a high salt content, may be used to water plants and possibly even irrigate deserts. In fact, good crop yields of salt-tolerant plants have been produced using the ocean as the only source of water.[35]

Integrated Pest Management

The modern concept of **integrated pest management (IPM)** uses many methods to control pests, rather than depending on a single approach. Integrated pest management uses an array of cultural, mechanical, biological, and chemical

32. Burton, *Agriscience and Technology*, p. 114.

33. Ibid., pp. 84–85.

34. Ibid., p. 92.

35. Ibid., p. 94.

methods to keep pest damage below economic levels. All this is done while keeping the balance of the ecosystem and environmental protection in mind. It involves a single, unified program, the goal of which is to reduce a pest population to an acceptable level and then maintain it at that level.[36] Furthermore, integrated pest management is an ecosystem-based approach to controlling insect problems. It takes into account the effects that a particular form of insect control might have on the other living things found in the ecosystem.

Rationale for IPM

The use of IPM is being promoted heavily for two major reasons. First, consumers increasingly oppose the use of chemical insecticides on food, animals, and crops; second, many insects are becoming genetically immune to the effects of certain insecticides. As a result, there is an emphasis on finding alternatives to chemical insecticides. An effective alternative to insect control other than complete reliance on insecticides is needed.[37]

Cultural Control of Pests

Cultural practices make the environment less favorable for the survival, growth, and reproduction of the pest. Some cultural methods that are used include frequent cultivation, adjustment of dates for planting and harvest, crop rotation, appropriate choice of good seed and proven varieties, water management, and **solarization.**

Mechanical Control of Pests

Mechanical methods use machines and equipment to remove or destroy pests outright. To be effective, action must be immediate. The advantage of using mechanical methods is that there is no chemical residue. Mechanical methods (depending on whether a field or a greenhouse is being used) include:

- use of barriers such as screen and netting
- artificially controlled temperatures rises or drops
- direct mechanical destruction, such as the use of shredders, rollers, plows, and soil pulverizers
- sterilization of soil
- **flaming**

36. Ricketts, *Science 1A (Agriscience),* pp. 25–19.
37. Burton, *Agriscience and Technology*, p. 89.

Biological Control of Pests

In biological control, we depend on the action of parasites, predators, and **pathogens** on a **host**. This results in reducing the pest population. Although biological control does not effect an immediate reduction in pest numbers, over a long period of time it is more effective and economical than chemicals on certain plants. Traditional biological control includes:

- introduction of species
- conservation of parasites and predators
- **augmentation** of parasites, predators, and pathogens

Example of Biological Control. Interference with the natural life cycle of insects is one means that has been shown to be effective. Sterile males of certain insect populations are introduced, which alter the reproductive success of the entire population. For example, the screwworm fly, a livestock pest, has been controlled by the introduction of large numbers of sterile male insects. These males, which have been made sterile through irradiation, are released into the natural environment to mate with fertile females, which then lay infertile eggs.

Pheromones (naturally occurring organic compounds that act as hormones) may be introduced into the environment to alter insect behavior and reproduction.

Genetic insect resistance. Insect resistance is natural in some plants. In some species, the plant juices act as a natural insecticide; other species emit odors that repel insects.

Production of natural insecticides. For plants that are not naturally resistant to insects, genetic engineering has developed an encouraging solution. Scientists are now able to transfer genes for insect resistance from donor plants into plants that are not naturally insect resistant. This technique is preferable to using chemicals to control insects because the only insects killed are those that try to eat the plant. Insects that simply pollinate or that are natural enemies of the harmful insects are not affected.[38]

Pesticide Control of Pests

Pesticides (chemicals) often have a primary role in pest management. The potential dangers of pesticide chemicals to humans, food products, animals, and

38. Ibid., p. 90.

the environment make them the least desirable method of controlling agricultural pests. However, pesticides are often the only effective method of control. Pesticides provide a barrier between the plant and the attacking pest. The major benefits derived from the use of chemical pesticides include effectiveness, ease and speed of pest control, and reasonable cost compared to other alternatives. Some valuable crops are so susceptible to devastation by insects or pathogens that without chemical control they would simply disappear.

Hydroponics and Aquaculture in Extended Space Travel

For extended space travel involving several people, a way to produce food in space is being researched. Scientists at the Kennedy Space Center are conducting a project called the Closed Ecological Life Support System (CELSS). This system will allow plants to be grown in space, hydroponically, in an environmentally controlled chamber called a biomass production center.

Aquaculture in Space

Scientists are also researching the possibility of raising fish in space as a source of protein. Presently, the National Aeronautics and Space Administration (NASA) is experimenting with tilapia. Research is being conducted to see if the fish can be raised using hydroponics, in a solution along with nutrients produced from residue. Refer to Figure 4–15.

Utilization of Plant Residue (Stalks) in Space. By using microbiological digestion, plant residue (stalks) is broken down into a "soupy" liquid. At present, 75 percent of the cellulose can be broken down into sugar. At this stage there exist other possibilities. First, the digested sugar solution could be fermented

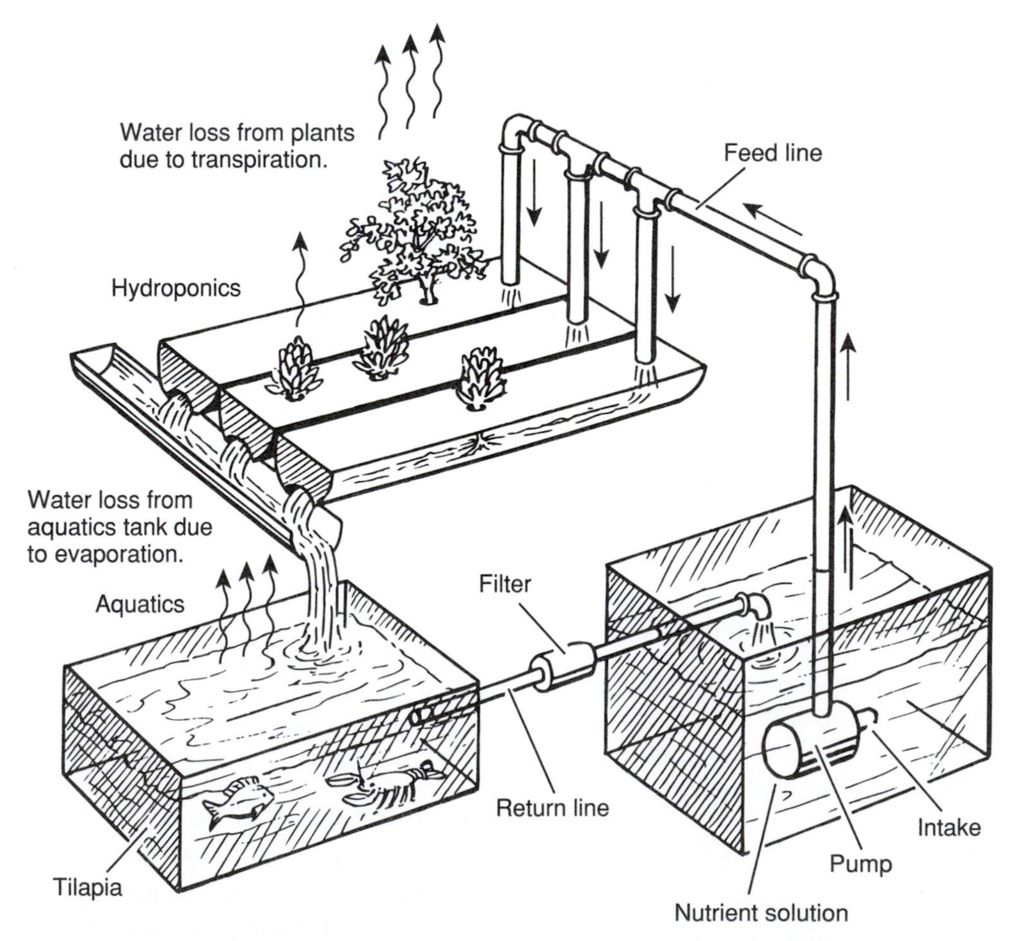

Figure 4–15 Hydroponics and aquatics in a closed system where the nutrient solution is used to feed plants, and the runoff is used to feed aquatic animals. All water and excretions are recycled to keep the system going and basically self-sustaining. The temperature and light are synthetic and controlled. (Note: The original supply of water will come from the engine fuel cell, where hydrogen and oxygen produce electricity, with the by-product being water.)

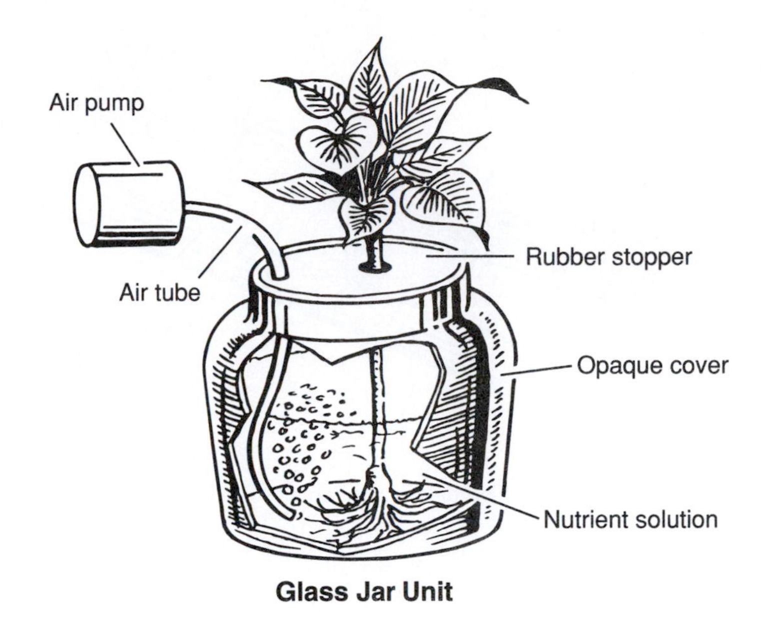

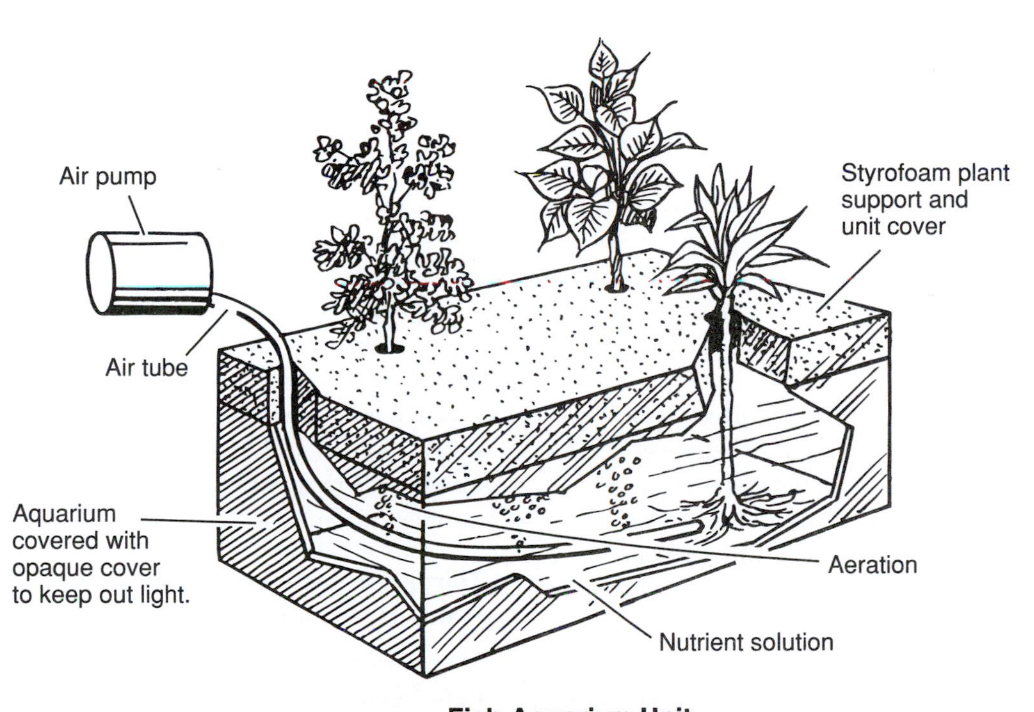

Figure 4–16 Two simple examples of hydroponics tank culture systems.

and ethanol could be produced. Second, the solution could be consumed by fish such as grass carp.[39]

Hydroponic Plant Growth

The types of hydroponic systems are classified according to the support medium available for the plants. A brief discussion of each follows.

39. Ricketts, *Science 1A (Agriscience),* pp. 27-16.

Tank Culture. The first type is called tank (or water) culture. In this type of system, the root hangs in a nutrient solution while the plant is supported on wire, line, wood, or some other material. Greenhouse tomatoes and "floatbed" tobacco plants are raised this way. Refer to Figure 4–16.

Modified Drip System. The modified drip system is an open system in which the nutrients are not recycled. Plants are grown in a fine-textured substance.

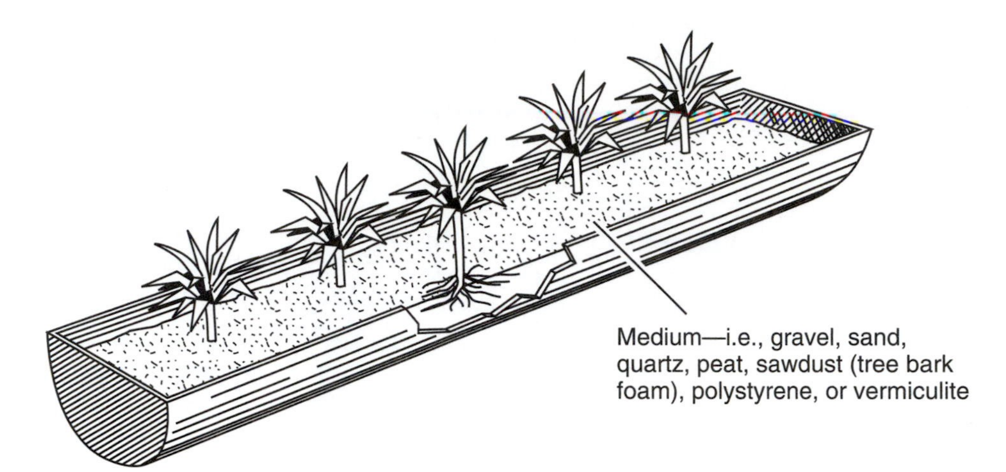

Figure 4–17 With the hydroponics modified drip system, an intermittent-mist watering system provides the essential nutrients for plant growth.

Refer to Figure 4–17. The roots of the plant support it by anchoring themselves in an inert medium such as sand, vermiculite, perlite, or peat moss. The nutrient solution is held within the particles of the medium. Today, this system is widely used in the nursery industry for propagation beds. This type of system generally has an intermittent-mist watering system, which works in conjunction with the nutrients to produce a healthy, rooted cutting.

Flood System. The flood system often has an irrigation system installed beneath the gravel that holds the plants in place. This system pumps nutrients and water into a reservoir beneath the plant roots. So far, this arrangement has been the easiest system to sterilize between crops. Refer to Figure 4–18.

Airoponics. An airoponics system simply sprays a hydroponic mist on the plants and root system, which are both suspended in air. Disney's EPCOT Center demonstrates such a system at "The Land Pavilion." Refer to Figure 4–19.

Differences When Using Hydroponic Systems in Space and on Earth

Unlike hydroponic systems on earth, in space farming plants will be unable to absorb water or nutrients in the form of water droplets. Capillary action in microgravity will cover the surface with a film of water (nutrients). Refer to Figure 4–20. Liquids will have to be supplied as a thin film to keep them from floating away.

To make this happen, agricultural scientists Steven Britz and Todd Peterson, at the USDA Research Station in Beltsville, Maryland, are using two plastic pipes, fitted one inside the other. The inner tube is wettable and porous. It serves as the source from which the roots draw nutrients and as a support surface for the roots.[40]

By increasing the suction in the pipes against which the plant must extract moisture or nutrients through a porous membrane (the inner tube), the researchers will be able to subject plants to a measurable range of water stress while delivering sample nutrients.[41]

Thirteen Essential Elements for Growing Plants Hydroponically

Macroelements. Essential elements in predetermined quantities are needed by all plants and must be provided by the grower. Often these required elements are called *macroelements* because they are needed in larger quantities than other trace elements. Plants cannot produce essential elements by themselves. The essential nutritional elements required by a plant are:

- Nitrogen (N)—The element demanded most by a plant. Nitrogen accumulates in the tissues of

40. Ibid., 27–6.
41. Ibid., 27–13.

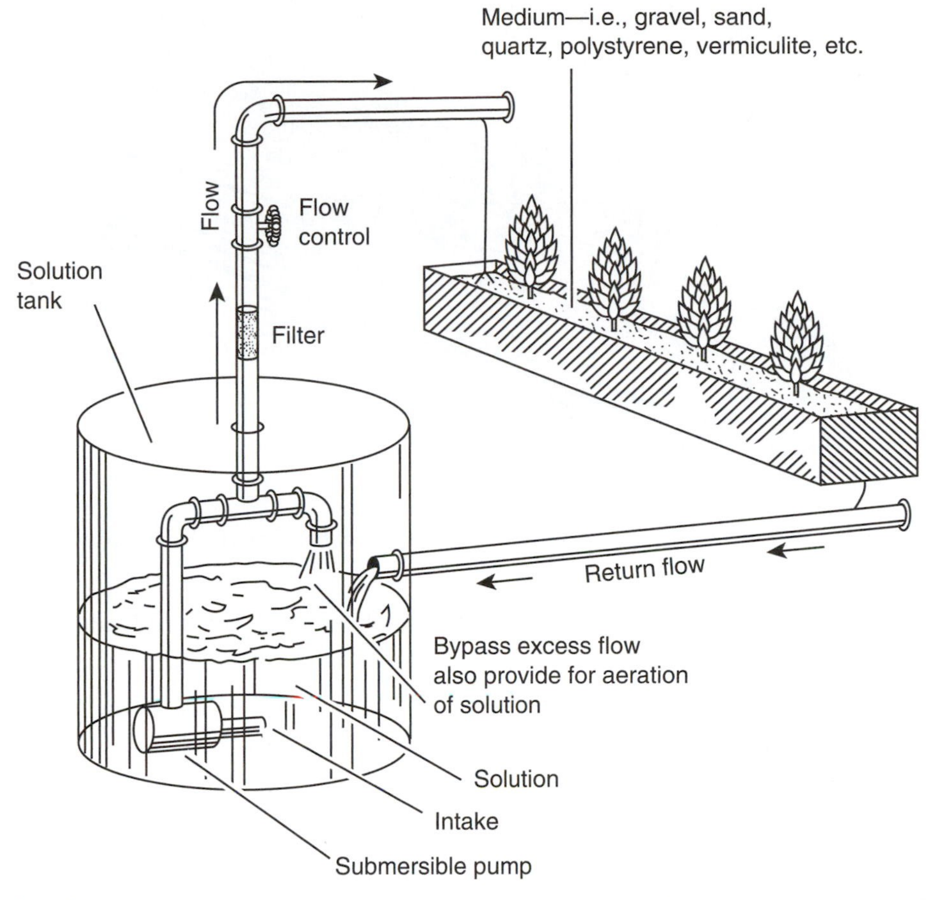

Figure 4–18 With the hydroponics flood system, the water is recirculated after providing the nutrients to the plant and returning to a holding tank.

young plants. Nitrogen can readily move from one area of a plant to another, from the area of older growth to younger leaves.

- Phosphorus (P)—Most often found in great concentration in the fruiting tissue of a plant, rather than the vegetative tissue. Phosphorus can move throughout a plant, but it generally gathers in actively growing tissue.
- Potassium (K)—Cell plasma and leaf tips are the site of potassium storage.
- Magnesium (Mg)—The element moves from one plant tissue to another with ease.
- Sulfur (S)—This element is not mobile in a plant and is required only in small amounts.
- Calcium (Ca)—Calcium is not a transferable element. It does not move from an area of older tissue to younger tissue. Saline water usually contains sufficient calcium for hydroponic growing.

Figure 4–19 Besides growing plants hydroponically, in Disney's Epcot Center, genetic engineering is conducted in a biotechnology laboratory. (Courtesy of USDA)

Microelements (Trace Elements). Seven micronutrients are required to produce hydroponically grown plants. Because plants can absorb some nutrients through leaf tissue, a topical spray is often used to

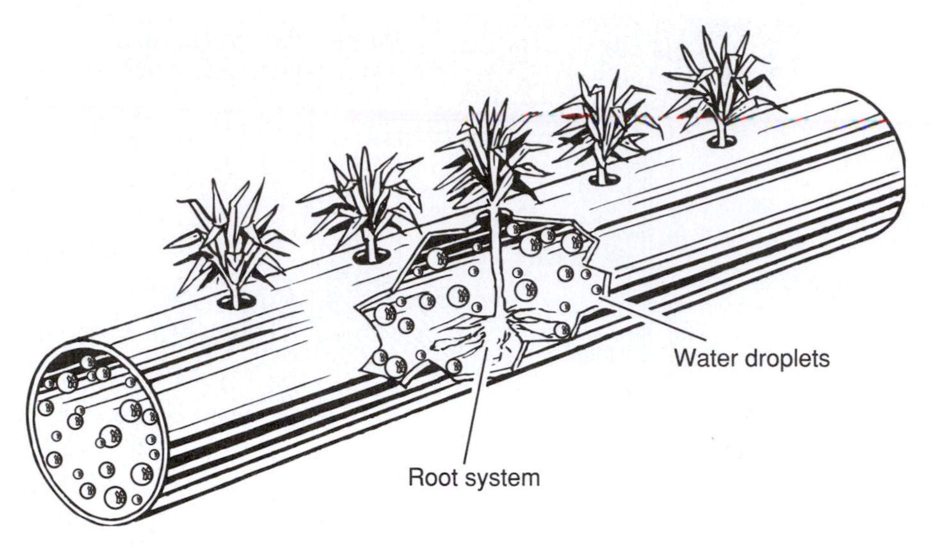

Figure 4–20 In space, because of the lack of substantial gravitational pull, plants will have to be grown in suction pipes to keep the hydroponics solution from floating away.

apply microelements to a crop. The micronutrients needed by hydroponic crops are:

- Iron–Used to prevent chlorosis in the tissue of the plant.
- Boron–Supplied to a plant as boric acid or borax.
- Manganese–Helps keep tissue healthy.
- Zinc–Works with light quantity and helps leaf tissue grow.
- Copper–Helps head off chlorosis.
- Chlorine–About 10–20 parts per million (ppm) is needed by plants for optimum growth.
- Molybdenum–Needed only in minute quantities; sometimes is identified as an impurity in the nutrient solution.

Figure 4–21 shows the ingredients that would provide enough nutrients from 1,000 gallons of water. Several agribusiness companies sell a premixed solution that can be added to the appropriate amount of water.

Hydroponic Mixture for 1,000 Gallons of Water
Potassium Nitrate: 5 pounds, 13 ounces (2.6 kilograms)
Ammonium Sulfate: 1 pounds (0.5 kilograms)
Magnesium Sulfate: 4.5 pounds (2 kilograms)
Monocalcium Phosphate: 2.5 pounds (1.1 kilograms)
Calcium Phosphate: 5 pounds (2.3 kilograms)

Figure 4–21 These are the ingredients necessary to provide enough nutrients for 1,000 gallons of water. Several agribusiness companies sell such a mixture.

Keeping the Hydroponic Solution Fully Active

Once the solution is in the tank, 1 gallon (3.8 liters) of water containing 1 ounce (28 grams) of manganous sulfate and 3 to 5 drops of concentrated sulfuric acid should be added to each 1,000 gallons (3,800 liters) of solution once a month. Four ounces (113 grams) of ferrous sulfate in 1 gallon (3.8 liters) of water should be added once a week. Oxygen in the water solution must be replaced as fast as it is taken up by the roots. Therefore, air must be continuously pumped or mixed into the solution.

Technologies Used for Hydroponic Growing

Scientists have developed a computer program that can automatically control the application and mix of the hydroponic solution. Through the use of probes and sensors, the essential elements are added as needed to get the optimum mixture.
A Zybetron sterilizer is used to kill algae as the water circulates. Metal halide lighting is used on a revolving track. In this way, fewer lamps can be used to achieve the same growth rate. Use of a simple kerosene lamp substantially increases growth rate, because of the additional carbon dioxide produced. Also, research is continuing on the use of fiber optics as a means of transporting light and heat. To create conditions similar to space (except for gravity and oxygen), scientists are conducting this research in a cave-type atmosphere.[42]

42. Ibid., 27–17.

NASA Discoveries from the CELSS Project

The crops being tested in the CELSS project are wheat, rice, soybeans, white potatoes, tomatoes, green beans, cowpeas, sugar beets, lettuce, and sweet potatoes. The NASA researchers have made the following discoveries:

- Twenty square meters of plant surface area will produce all the gases needed to keep one person alive in space indefinitely. The plants will use enough carbon dioxide (CO_2) and generate enough oxygen (O_2).
- High-pressure sodium lamps are more efficient than fluorescent lamps for growing plants in space, but they do not provide a full spectrum for reproduction.
- To better utilize space, only one meter is allowed between the light and plants. Therefore, dwarf varieties are selected that will not get as tall.
- Wheat is grown in plastic trays with one-quarter square meter surface area for every 400 plants. This will provide enough food for one person per day.
- When a 1/8-inch-thin film hydroponic solution is used, 1 liter supplied continuously will grow 25 to 30 plants.
- It takes 65 to 85 days to grow a crop of wheat in space.[43]

Adoption of NASA Research

Because of the research at NASA, and the creativity of high school agricultural education instructors, demonstration projects combining hydroponics and aquaculture are being conducted at several high school and agricultural education programs. A program at Mountain City, Tennessee, under the direction of Harvey Burniston, Kenneth McQueen, and Thomas Boyd, is combining a greenhouse, hydroponics, and aquaculture heated by geothermal energy. Their innovative work has attracted visitors from throughout the United States and foreign countries.

Hydroponics Projects

The horticulture program at Johnson County Vocational School in Mountain City, Tennessee, teaches students how to grow bedding plants, hanging baskets, chrysanthemums, and other potted plants, and how to use hydroponics. The focus is on three economical crops: tomatoes, Bibb lettuce, and European cucumbers.

Tomatoes are grown using bag culture. Both rockwool and perlite bags are used to show potential vegetable producers two different types of medium. Students grow tomatoes in the aquacenter and in the traditional hydroponics greenhouse. Tomatoes are sold to local restaurants. Students grow a variety called "Trust," a tomato plant that will produce for a year or more.

Salina Bibb lettuce is grown using the hydroponic nutrient film technique (NFT) system, in which nutrient water is continually recycled. Seeds are started in rockwool cubes and transferred to NFT channels when the plants are two to three weeks old. All traditional crops of bedding plants and hanging baskets at the aquacenter are watered with water from the fish tanks, using drip irrigation and sprinklers. The system is designed as a symbiotic relationship between the plants and the fish. Lettuce is grown at the aquacenter using exclusively recycled fish water. A major portion of the lettuce is used by the school system's food service program each year.

European seedless cucumbers are also grown hydroponically at the aquacenter using perlite bag culture. Three to four ounces of nutrient solution are injected into each plant once an hour during daylight hours, very similar to the tomato production techniques. Cucumber crops last 90 to 120 days, resulting in 3 to 4 crops a year.

Geothermal Heating Systems. The Alternative Farming Center (Aquacenter) utilizes an innovative "geothermal" heat-pump system to both heat and cool the structure and water. Refer to Figure 4–22.

Figure 4–22 The agricultural education program at Mountain City, Tennessee, combines hydroponics and aquaculture in a geothermally heated and cooled greenhouse. (Courtesy of Harvey Burniston)

43. Ibid., 27–16.

This system is designed on geothermal or ground-source heating and cooling principles, based on the fact that, at a depth of 5 feet or more, the earth maintains a constant temperature of 53–54°F in this area. This principle is used to equalize the temperature of the system by pumping water through a continuous pipe in the ground called the ground loop. In the system used at the aquacenter, a total of 2 miles of continuous pipe is run through 15 vertical wells at a depth of 250 to 300 feet. The water is pumped through the pipe, where it equalizes to 54°, then returned to the heat pumps.

Aquaculture Projects

Aquaculture is a growing agricultural industry in the United States. The Johnson County agricultural education program in Mountain City, Tennessee, is studying the feasibility of aquaculture as an alternative crop for the region and has designed a system to raise fish as well as hydroponic plants, traditional bedding plants, and hanging baskets.

The aquaculture system, which is designed for fish, consists of four tanks that are 72 by 8 by 5 feet and one tank that is 50 by 6 by 5 feet. The system holds 90,000 gallons of water, which is recirculated through bead filters for removal of solids and some bacterial exchange nine times per day.

Biotowers are used to help degasify the water. The water is pulled across the tanks from the bottom and sent through filters by pumps; gravity then causes it to flow back to the raceways at six intervals along the sides. The system can handle anywhere from 200 to 400 pounds of feed per day.

Air is added to the water through air stones placed 24 to 32 inches deep in the water at 1-foot intervals throughout the tanks. Liquid oxygen is added through saturators after the water has circulated through the heat exchanger. The water, which is heated by geothermal heat pumps, is sent through a heat exchanger and returned to the raceways by 1-inch pipes located in the bottom of the tanks.

Currently, tilapia and koi are being studied. They seem to work well together, as one is a food fish and the other an ornamental fish. Tilapia are also top feeders, whereas koi are bottom feeders. Tilapia have been cultivated for years all over the world. The fish were first believed to have been grown and eaten by the Egyptian pharaohs more than 4,500 years ago. Presently, there is a strong market for tilapia. Koi have also been studied as an alternative type of fish. With a growing market for indoor and outdoor water gardens, a market for these colorful fish is also growing.

Other Emerging Agribusiness Technologies

Many other agribusiness technologies are arriving daily. A computer search using one of the many available search engines can help you discover these: use keywords such as "new agriculture technologies," "farming innovations," and the like. Eight of these more recent developments are discussed here.

Light Detection and Ranging Systems

Light detection and ranging (LIDAR) systems (also called laser range finders), can be used to create topographic maps. The Forestry Service uses LIDAR systems attached to planes to create maps not only of ground elevation, but also of tree elevation and low-level vegetation. The technology provides three-dimensional views of land and even the ocean. Researchers have found that they can also use the data collected to calculate tree diameter and levels of sunlight throughout a forest.

Feed Formulation Software

Computer programs are now available for balancing diets for dairy cows, beef cattle, swine, poultry, or any other species. Feed formulation software allows farmers to calculate the least-cost feed formulation. This type of software also lets farmers adjust inputs to create a custom blend of feed to provide a balanced diet for livestock. These custom blends and products can be manufactured for sale.

Biosensors

Biosensors could be used in a variety of agricultural industries, but are currently being used primarily in poultry processing to detect pathogens. These sensors use optical wave-guides to measure microorganisms. The use of biosensors is quicker and less expensive than the traditional assay method for measuring microorganisms. This same technology could easily be adapted to detect bacteria or contaminants in food or water. The biosensor is extremely sensitive, able to detect pathogens at levels of 500 cells per milliliter.[44]

Ranch Management Software

The Ranch Vision interactive software is used to manage grazing livestock herds. The software can be used to test alternative scenarios, do cost/benefit analyses, and perform other functions. Ranch Vision can analyze up to 18 different cattle, sheep, and/or goat enterprises.[45]

Auction Tools and Technologies

AgriAuctions is a cattle-auction tool developed to record, verify, and handle all financial transactions for animals auctioned in local sales arenas. It can handle an unlimited number of customers, cattle, and transactions. The program has advanced search tolls and can generate various reports.[46]

Another useful technology for the beef cattle industry is the video auction. Sellers of feeder calves can sell large lots of calves to buyers hundreds of miles away. Buyers simply watch a live video of the calves rather than having to be present at the site; the highest bidder purchases the calves, as at a traditional auction.

Radio-Frequency Identification Tags

A radio-frequency identification (RFID) tag is a small object that can be attached to or implanted into a product, animal, or person. RFID tags contain silicon chips and antennas that enable them to receive and respond to radio-frequency queries from an RFID transceiver. Passive tags require no internal power source; active tags do need a power source.

These identification tags can store information useful in the livestock industry, such as locations an animal has visited and any diseases or vaccinations the animal has had. This type of identification system is becoming more important because of issues regarding bovine spongiform encephalopathy (BSE) and avian influenza.

Heat-Map Systems

Heat-map systems such as FARMAX are used to find cold or hot spots in buildings. Some can produce color-contour temperature maps. This technology is useful in the greenhouse, nursery, and poultry industries. The ability to locate hot or cold areas in a chicken house or greenhouse allows the owner or manager to repair any problems with the construction or heating/cooling system. Other similar technologies can be found at the Agricultural Research Service Website.[47]

Biofuels

Because of the rapidly increasing price of gasoline, biofuels are being used more and more frequently. Ethanol made from corn has been popular for several years. Recently, ethanol made from cellular material is gaining momentum. Switchgrass is favored because of its potential yield, drought resistance, and ability to grow for 10 to 15 years in the same field before reseeding is necessary. For cellular ethanol, wood products are also popular. Studies estimate that Americans could fuel half our cars just with the stuff we waste and throw away: grass clippings, garbage, wood waste, corn stover, old newspapers—anything with carbon in it. Ethanol is used in spark-ignited engines. See Figure 4–23 for a picture of a Corvette that runs on ethanol and biodiesel.

Biodiesel fuel also requires alcohol, which could be produced from hydrolysis and fermentation of plant cellulose. However, Michael Briggs, from the University of New Hampshire Physics Department, advocates using widespread biodiesel made from algae, envisioning it as a possible substitute for all petroleum-based transportation fuels. Algae were first investigated as a means to sequester carbon in emissions from power plants. However, the high oil content and fast growth rates of some species prompted a shift in focus to algae that could be grown specifically for biodiesel production, using shallow saltwater pools in desert areas or enclosed photobioreactors.

---

44. Jane M. Sanders, "Biosensor That Detects Pathogens in Poultry and Other Foods Being Tested in Metro Atlanta Processing Plant," *Research Horizons,* http://gtresearchnews.gatech.edu/reshor/rh-win00/rnoteW00.html#motel (accessed August 13, 2007).
45. http://www.ranchvision.com/index.html (accessed August 13, 2007).
46. http://www.ag2networks.com (accessed August 13, 2007).
47. http://www.ars.usda.gov/business/patents.htm (accessed August 13, 2007).

Figure 4–23 A Corvette that has been modified to run on ethanol/biodiesel. (Courtesy of Cliff Ricketts)

The United States would need 140.8 billion gallons of biodiesel to replace all transportation fuels. Algae farms to produce that amount would require between 9.5 and 29 million acres (much less than the 450 million acres now used for crop farming, and the more than 500 million acres used as grazing land). Use of such algae farms would also negate the need to displace food or animal-feed cropland with biofuel crops. Furthermore, such farms could be located in various areas around the country, and built to use waste streams as food sources and yield high-nitrogen, high-phosphorus fertilizer as a by-product.[48]

Conclusion

Although most of the emerging technologies discussed in this chapter are applied agriscience technologies, there must be a way to get these technologies to the consumer. Research completed that is not then applied is research lost. This is where the agribusiness input and output sectors, which sell or market these technologies, enter the picture. Many of the companies that market these emerging technologies are discussed in later chapters.

Summary

Agricultural technologies are applied sciences. However, once they have been developed, the agribusiness input or output sector must market these emerging technologies. The agricultural industry is developing and adapting new processes and procedures at an unprecedented rate.

The Global Positioning System uses satellites to view specific fields and crops. The view from space provides a computerized picture that can aid field crop management and increase yield per acre. Because of GPS, precision farming is now a reality. In precision farming, a computer model of each specific section of a crop producer's field is created. Each section is then individually assessed for increased production, decreased cost, and environmentally sound techniques.

---

48. Michael Briggs, "Widescale Biodiesel Production from Algae," August 2004; available at http://www.unh.edu/p2/biodiesel/article-alge.html (accessed August 13, 2007).

Genetic engineering bypasses and augments selective breeding and natural selection by inserting a desired gene directly into the chromosomes of a living organism. The target cells receive the new genes that were obtained from another organism. With gene splicing, a scientist can cut and trim a chromosome using enzymes in much the same way that a person uses scissors to clip an article from a newspaper.

Cloning is a process through which genetically identical individuals are produced. By replacing the nucleus of an unfertilized ovum with the nucleus from a cell of the organism to be duplicated, individuals of identical genetic makeup can be produced. Embryo splitting and embryo transfer are parts of the cloning process. Cloning may be accomplished in plants by stimulating cell division of a single plant cell to produce a complete plant.

Gender selection involves the ability to control the offspring's sex at the time of mating. Flow cytometers can identify X (female) and Y (male) sperm cells after they have been treated with fluorescent dye. After identification, the X and Y sperm can be separated after being given a positive or negative charge, passed between two high-voltage deflection plates, and separated into collection tubes. Also, with cattle, embryos can be collected from donor females, selected by sex, and implanted into recipient cows. With fish, ultraviolet light can determine the sex.

Hormones are substances that are formed in the organs of the body. Some hormones are formed into tiny pellets and injected under the skin of an animal's ear. This increases the rate and efficiency of growth in meat animals. Production of milk can be increased by treating dairy cows with a growth hormone, bovine somatotropin. The pork industry is experimenting with porcine somatotropin to enhance the muscle growth of young pigs. Heat synchronization is another result of hormone injection.

Animals have been used for scientific investigation in veterinary and human medicine for generations. There is a close relationship between the health of humans and the health of animals. Illnesses that are devastating to humans and animals are studied via animal research to determine the cause, transmission, mechanism of infection, prevention, treatment, and control of the organisms that cause these diseases. Many pharmaceutical and other medical products have animal origins.

Domestic animal husbandry of the future will be much more efficient. Dairy cows will produce as much as 40,000 pounds of milk per year. Beef cattle will grow 50 percent faster. Swine will be bigger, but not fatter. Poultry broilers will reach 4 pounds in 25 days. Transgenic animals will become more common.

As in most areas of modern life, the computer has become an important tool for efficient and time-saving operations in agriculture. Computer software applications record milk productivity in dairy herds; offer farm tax information, electronic filing, and specific bookkeeping programs; help gardeners and landscaping professionals create designs; and aid the livestock industry by calculating balanced animal diets. Electronic animal management systems include robotic milking systems, electronic sensors, robotic sheep shearing, birthing sensors, and ultrasound imaging.

Many mechanical technologies are being introduced in the plant industry. Some examples include a device for early detection of insect pests, a computerized apple to pinpoint shipping damage, remote-controlled crop dusters, an underground plow, lasers to alter the land, and many others.

Plant propagation using tissue culture is a relatively new tool for plant scientists. Plants are grown from tissue, such as buds, leaf parts, or terminal shoots, obtained from a parent plant. This technique allows rapid propagation of many new plants that are genetically identical to the parent stock.

Many technologies are emerging for use with plants. Biotechnology has enabled researchers to develop a special frost-free spray that can be applied to plants to prevent them from freezing. Plant growth regulators regulate the growth of roots, stems, and leaves, which helps nursery owners and flower growers market their crops at the appropriate season. Also, scientists are researching plants that are tolerant to ocean water.

Integrated pest management uses an array of cultural, mechanical, biological, and chemical methods to keep pest damage at acceptable levels while keeping the balance of the ecosystem in mind. IPM is a single, unified program that aims to reduce a pest population to an acceptable level and then maintain it at that level.

For extended space travel, scientists are researching hydroponics and aquaculture. The hydroponic medium, consisting of water and the essential nutrients for plant growth, will be used to grow wheat, rice, soybeans, and various vegetables. Also, scientists are researching the possibility of raising fish in space in the same hydroponic solution. Further, scientists want to know if the fish can consume crop residue, such as stalks or stems from wheat plants.

Many other agribusiness technologies are arriving daily. These include, but are not limited to, light detection and ranging systems, feed formulation software, biosensors, ranch and auction management software, remote and Internet video auctions, radio-frequency identification tags, heat-map systems, and biofuels.

Although most of the technologies discussed in this chapter are applied agriscience technologies, someone must get these technologies to the consumer. This is where agribusiness enters the picture.

End of Chapter Activities

REVIEW QUESTIONS

1. Define the Terms to Know.
2. Explain the purpose of global positioning and field management.
3. What are four components of the Global Positioning System?
4. What are the five major objectives of comprehensive precision farming?
5. List five advantages of genetic engineering.
6. Briefly explain the gene-splicing process.
7. Briefly explain the cloning process.
8. Briefly explain embryo splitting.
9. Briefly explain embryo transfer.
10. List two advantages of embryo transfers.
11. Briefly explain the sperm-sexing procedure.
12. List two advantages of sperm sexing.
13. List six reasons why illnesses that are devastating to both humans and animals are studied using animal research.
14. Give 12 examples of how swine are physiologically similar to humans.
15. List six pharmaceutical and other medical products made from animal products.
16. Name one characteristic or technological development in the future of each of the following domestic animals: (a) dairy cattle, (b) beef cattle, (c) swine, (d) poultry, (e) sheep, (f) catfish, (g) bees.
17. List four examples of research resulting in transgenic animals.
18. List eight examples of computer use in the agricultural industry.
19. Briefly explain four electronic animal management systems.
20. Briefly explain three technological live-animal and carcass evaluation devices.
21. Briefly explain nine emerging mechanical technologies in the plant industry.
22. Briefly explain the basic procedure for tissue culture.
23. List two advantages of tissue culture.
24. Briefly explain three emerging plant technologies.
25. What two major factors have accelerated the adoption of integrated pest management?
26. List five examples of mechanical control of pests.

27. List three traditional biological controls.
28. Briefly discuss two examples of biological controls.
29. List three major benefits derived from the use of chemical pesticides.
30. Explain how fish could possibly be raised in space.
31. Name and briefly discuss four types of hydroponic systems.
32. What is the main difference between using hydroponic systems in space and on earth?
33. List six essential elements (macroelements) in a hydroponic solution and give one purpose of each.
34. List seven trace elements (microelements) required to produce hydroponically grown plants.
35. List five technologies used in hydroponic growing of plants.
36. List six discoveries made by NASA from the CELSS project.
37. List and briefly describe eight more emerging agribusiness technologies.

FILL IN THE BLANK

1. Although agricultural technologies are applied science, it is __________ that will sell or market these technologies.
2. ________ ________ ________ is a technology that uses satellites to view specific fields and crops.
3. Scientists have discovered that they can modify gender of fish by _________________.
4. More than almost any other species, the anatomy of the _________ most closely resembles that of the human.
5. ________ ________ may soon be routinely placed for transplantation in critically ill humans.
6. The largest disadvantage of tissue culture is the chance of introducing a ________ into the laboratory.
7. Integrated pest management uses an array of _________, _________, _________, and _________ methods to keep pest damage below economic levels.
8. _________ is a NASA project researching a system that will allow plants to be grown in space.
9. Following the research at NASA and elsewhere, demonstration projects combining hydroponics and aquaculture are being conducted in several high school _________ ________ programs.
10. _________ and _________ are two agricultural practices being researched to be used in a combined effort in space.

MATCHING

a. porcine somatotropin
b. ultrasound imaging
c. micronutrients
d. heat synchronization
e. pesticide control of pests
f. animal growth hormones
g. mechanical control of pests
h. biological control of pests
i. cultural control of pests
j. bovine somatotropin
k. macroelements

_____ 1. treatments that increase the rate and efficiency of growth in meat animals

_____ 2. increases the production of milk by as much as 20 to 40 percent

_____ 3. enhances muscle growth in young pigs

_____ 4. modifies the estrus cycle of farm animals by changing their hormone balance

_____ 5. permits researchers to measure body fat and lean tissue in growing animals

_____ 6. practices that make the environment less favorable for the survival, growth, and reproduction of a pest

_____ 7. uses machines and equipment to remove or destroy pests outright

_____ 8. depends on the action of parasites, predators, or pathogens on a host

_____ 9. chemical used on plants

_____ 10. essential nutrients

_____ 11. trace elements

ACTIVITIES

1. Research periodicals (magazines, newspapers, Internet, professional journals, etc.) to determine current economic trends concerning the effects of biotechnology. This could include concentration on the use of BST and/or any other product. State your opinion as to whether you think the biotechnology product should be used, taking into consideration the current status of the economy. Support your opinion with facts, and cite all sources used. Present your findings to the class.
2. Biotechnology has given production agriculturalists new crops that are herbicide resistant. The crop producer can then apply herbicides that normally kill a broad range of plants without killing the herbicide-resistant crop. The benefit is that production agriculturalists can now grow these crops in areas where it was previously impossible because of high weed infestations. However, some people are predicting that herbicide-resistant crops will backfire on the crop producers. It has been proposed that the natural evolutionary process will produce weeds that are resistant to the herbicides. This could then cause the crop producer to need even more herbicide.

   New data is being published constantly that could support either side of this debate. Research biotechnology using articles in periodical publications or the Internet, as well as library sources such as magazines and newspapers. After analyzing the data you collected, decide how you think the possible problems can be solved. Give evidence and bibliographic information supporting your opinion.
3. Genetically engineered organisms are being released into the environment to fight agricultural problems such as frost damage on strawberries, insect and weed pests, and plant diseases. Organisms can mutate and exhibit new traits, which may be beneficial or detrimental. Some people believe that engineered organisms could mutate after being released into the environment and thereby become a new agricultural or environmental pest. Research the theory of the causes of mutations (biology textbooks covering genetics are a good source). If you were a lawyer protesting the release of engineered organisms into the environment, could you cite cases where problems were caused because of this practice? Research this information using library sources such as magazines and newspapers. List bibliographic information to support your case.
4. Select a prominent biotechnology scientist and study the individual's education, work, and contribution to society. Examples include: Barbara McClintock and Paul Mangeldorf (corn), Norman E. Borlaug (wheat), and Steven F. Lindow (frost-free bacteria). Use library sources, and list all bibliographic information for the sources used. Write a report on the biotechnology scientist of your choice and present it to the class.

5. Write a two-page report on a current article concerning a hydroponic system. Present your report to the class.
6. Design a plan of an ideal tissue culture laboratory. Consider using "off-the-shelf" items or salvage parts. Present your findings to the class.
7. Design a hydroponic demonstration operation. With the help of three to five classmates, construct the hydroponic demonstration unit. Consider using the following materials: tank, nutrients, water pump and/or aerator, sheet of Styrofoam, litmus paper, plants.
8. Space within this chapter did not permit the inclusion of all the emerging technologies. Research and present a report to the class on a new or emerging technology not discussed in this chapter.
9. Approximately 14 different categories of technologies were presented in this chapter. Identify an agribusiness in your community that exists because of one of these technologies. Write a brief report on this business and present it to the class.
10. Select and write a brief report on new technology in either the agribusiness input sector or the agribusiness output sector. Present your report to the class.

UNIT 2

Starting and Running an Agribusiness

Chapter 5

Planning and Organizing an Agribusiness

Objectives

After completing this chapter, the student should be able to:

- Explain the importance of small businesses.
- Examine whether entrepreneurship is for you.
- Describe the challenges of entrepreneurship.
- Describe why agribusinesses fail.
- Analyze a potential agribusiness venture.
- Prepare a business plan.
- Explain the five major areas of agribusiness management.
- Establish goals for an agribusiness.
- Discuss the importance of problem solving and decisionmaking.

Terms to Know

business cycles
business plan
business survey
capital
capital intensive
capitalism
collateral
diligence
entrepreneur
financial institutions
financial resources
free enterprise
investors
niches
rapport
small business
Small Business Administration (SBA)
undercapitalization

Introduction

Part of the American dream is that everyone who wants to should be able to own their own business and be their own boss. It is still possible for a person to start with very little **capital** and build a successful business, but it is a very difficult process and is becoming more difficult each year. Even though most agribusiness students will never own their own businesses, it is important to understand the basic processes involved in business planning and organizing.

If you pursue a career in agribusiness, you will first need to choose whether you would like to be self-employed or work for someone else. In making this decision, it is important to think carefully about your personal needs, preferences, and desires. Consider your abilities, your training and experience, the lifestyle you desire, and your financial situation. Use these factors to set goals that will help you move along your chosen career path. Realize also that your decision to work for yourself or someone else depends on the value you place upon security, self-expression, others' perceptions of you, and your ability to build **rapport** with other people.[1] The student shown in Figure 5–1 is assessing whether to start an agribusiness or work for someone else.

Figure 5–1 Starting an agribusiness involves much planning and decisionmaking. This student is accessing information for deciding whether to start a business or work for someone else. (Courtesy of Cliff Ricketts)

Importance of Small Businesses

In business and economic terminology, most agribusinesses are small businesses. There is no single definition of **small business**, but the U.S. **Small Business Administration (SBA)** defines it as a business that is independently operated, is not dominant in its field, and meets certain size standards in terms of number of employees and annual receipts.

Small Businesses Are Not Really Small

When it comes to jobs, small businesses are not small in the number of people they employ. Ninety percent of the nation's new jobs in the private sector are in small businesses. Two-thirds of all new jobs are in companies with fewer than 25 employees.[2] Thus, there is a very good chance that you will either work in a small business someday or start one. The following are some interesting statistics illustrating the importance of small businesses[3]:

- There are about 25.4 million full- and part-time home-based businesses in the United States.[4]
- Of all nonfarm businesses in the United States, almost 97 percent are considered small by SBA standards.
- Small businesses account for more than 40 percent of gross domestic product (GDP).

1. William H. Hamilton, Donald F. Connelly, and D. Howard Doster, *Agribusiness: An Entrepreneurial Approach* (Albany, NY: Delmar Publishers, 1992).

2. "Where the New Jobs Are," *BusinessWeek,* March 20, 1995, p. 24.
3. William G. Nickels, James M. McHugh, and Susan M. McHugh, *Understanding Business,* 5th ed. (Chicago: Irwin/McGraw-Hill, 1999), pp. 170–171.
4. Cynthia E. Griffin, "Going Home," *Entrepreneur,* March 1995, pp. 120–125.

- The total number of employees who work in small business is greater than the populations of Australia and Canada combined.
- The first jobs of about 80 percent of all Americans are in small business.
- The number of women who own small businesses has increased elevenfold since 1960, and studies predict that women will own half of the small businesses by 2000.
- The number of minority-owned businesses increased more than 64 percent between 1962 and 1987.[5]
- Small companies produce 90 percent of new jobs.[6]

As you can see, small businesses are really a big part of the U.S. economy. The reason why small businesses, including small agribusinesses, will always have a market can be explained by an analogy of the differences between rocks (big businesses) and sand (small businesses). If you fill a jar with sand, there are no spaces between the grains of sand. However, if you fill it with big rocks, there are many empty spaces between them. That is how it is in business. Big business cannot serve all the needs of every consumer. Therefore, there is plenty of room for small businesses to make a profit filling those **niches**.[7]

Entrepreneur in Agribusiness—Is It for Me?

The American economy is based on the **free enterprise** system, or **capitalism**. This simply means that we have the right to own our own business and to make a profit. It also means that we can lose money, perhaps even all the money that was invested. Those who start their own businesses take big risks. Therefore, entrepreneurship may not be for everyone.

5. Wilma Randle, "Sowing Seeds," *Chicago Tribune,* October 27, 1993, sec. 7, pp. 3, 14; Cynthia Todd, "Black Businesswomen Make Their Own Success," *St. Louis Post-Dispatch,* April 4, 1994, pp. 1A, 6A.
6. Gene Koetz, "Small Business Is Putting Some Snap in the Job Market," *BusinessWeek,* April 25, 1994, p. 26.
7. Nickels, McHugh, and McHugh, *Understanding Business,* pp. 171, 398.

General Characteristics of Entrepreneurs

Entrepreneurs are people who have the initial vision, **diligence**, and persistence to follow through. An **entrepreneur** is a person who accepts all the risks pertaining to forming and operating a small business. This also entails performing all business functions associated with a product or service, fulfilling social responsibilities, and meeting legal requirements.[8] There are many reasons why people are willing to take the risks of starting a business:

- Entrepreneurs work for themselves, are independent, and make their own business decisions.
- Whatever income they earn above their financial obligations is theirs to keep.
- They can test their own theories and ideas on how to run a business.
- They set their own working hours.
- They themselves set prices, determine production levels, and control inventory according to the market.
- They determine the product or service offered and control its quality as well as the overall reputation of the business.
- They solve the problems.
- They perform all the human resource functions, such as hiring, training, and firing.
- They set company policy.[9]

Not everyone satisfies the qualifications to be an entrepreneur. It takes a lot of hard work, and there may be little job security. Many small businesses fail each year for a number of reasons, as discussed later in this chapter. You should thoroughly investigate all the aspects of starting your own business. Make sure it is what you really want and can actually handle. Refer to other characteristics of the entrepreneur in the Career Option section.

Personal Characteristics of Entrepreneurs

Every person is different and unique. This includes entrepreneurs. However, entrepreneurs generally have common qualities that set them apart. The

8. *Entrepreneurship in Agriculture* (8747-A) (College Station, TX: Instructional Materials Service, 1988), p. 1.
9. Ibid., pp. 1–2.

skills needed by entrepreneurs vary widely, depending on the type and nature of the business. One person may possess technical knowledge about an agribusiness but lack leadership and management skills. A good guideline is that an entrepreneur should have the following qualities, as well as knowledge of or expertise in the particular agribusiness that he or she is pursuing. The characteristics listed here are found most often in successful entrepreneurs.

Independent. Entrepreneurs believe that they can do the job better than anyone. They prefer to control their own destiny.

Self-Confident. People who believe in themselves and what they can do have a definite edge in self-employment.

Energetic. Entrepreneurs usually are sick less often than other people. They enjoy good health and use it to improve their business.

Organized. These people are able to organize their work according to their own unique system.

Visionary. Entrepreneurs can keep tabs on the whole business and see how the different parts of the business fit together. They are in command of the entire business operation.

Persistent. They are able to keep the business moving forward even during tough times. If they do encounter a roadblock, they will find another way.

Optimistic. No matter what the situation, the outlook is always good to the entrepreneur. Such people are natural optimists.

Committed. They accomplish what is necessary to carry out their ideas.

Problem Solver. Entrepreneurs willingly take on new challenges. They capitalize on opportunities to make use of their time, talents, and ideas.

Self-Nurturing. They have little concern for what others think about them. Entrepreneurs have a conviction that they are on the right track regardless of what others may think.

Risk Taker. They are willing to give up things (such as the security of a steady job) to achieve their personal and business goals. They are not gamblers; rather, they take risks based on their confidence in their own abilities.[10]

Action-Oriented. Great business ideas are not enough. The truly important thing is a strong desire to turn dreams into reality.[11]

Has a Sense of Urgency. Entrepreneurs are restless unless they are working, and they may often seem to want challenges if everything is going too smoothly.

Flexible. If one option falls through, they will find another way to do the job or finance the change.

Emotionally Stable. Entrepreneurs do not give in to emotional highs and lows.[12]

Challenges of Entrepreneurship

Being successful is not easy. You do not simply start a business and enjoy the profits. Three major challenges must be addressed by entrepreneurs.

Total Responsibility

A beginning entrepreneur is in charge of everything. The success or failure of the agribusiness depends on just one person: the owner. Entrepreneurs must manage workers, manufacturing, and shipping. They have to find customers, sell the product, and be certain that orders are met. No matter what size the business is, the owner always has total responsibility. Refer to Figure 5-2.

Long, Irregular Hours

Being your own boss requires much work. People who start their own businesses work more hours than those who work for someone else. It is not uncommon for an entrepreneur to work more than 60 hours per week. Weekends are often spent working in the business as well.

Financial Risks

The most serious disadvantage of a small business is the need for money. Obviously, it takes money

10. Ibid.
11. Nickels, McHugh, and McHugh, *Understanding Business,* p. 165.
12. Hamilton, Connelly, and Doster, *Agribusiness: An Entrepreneurial Approach.*

Career Option

Entrepreneur

The entrepreneur is the person who organizes a business or trade or improves an idea. Entrepreneurship is the process of planning and organizing a small business venture. It is the entrepreneur who visualizes the business strategy and is willing to take the risk of getting the business started.

Entrepreneurial opportunities cut across the food and nonfood spectrum of agribusiness. This includes the agribusiness input sector and the agribusiness output sector. Entrepreneurial opportunities exist in every sector in which goods are bought, sold, or produced.

Many high school and college students become well established as entrepreneurs before they finish their education. Some popular agribusiness entrepreneurships include lawn services, lumber businesses, greenhouse or nursery operations, machinery repair, agricultural supplies, home and garden centers, florist shops, and retail flower sales.

Entrepreneurs may sell services such as animal care, crop spraying, or recreational fishing privileges. Planning is very important before starting an agribusiness, because many new businesses end in failure. Careful planning increases the chances of success.

Preparation for careers in agribusiness management includes both formal education and on-the-job training. High school agribusiness and agriscience programs provide excellent training for business. Such programs should include classroom and laboratory training, supervised agriscience experience, and leadership development. Advanced agribusiness programs at technical schools, colleges, and universities offer appropriate training in economics, finance, and management.

The entrepreneur is constantly seeking ways to provide better products, to improve profits, and to meet the needs of consumers. (Courtesy of USDA)

to get a business going. While the business is getting started, the entrepreneur has to pay bills and wages, which will likely exceed income. The owner will probably try to borrow money, but lending institutions are often reluctant to lend to a new business because of the high risk.

The chances of a business reaching its fourth birthday are only about 50/50. In the wholesale business or manufacturing, the odds are that about 40 percent will fail, and more than 50 percent of retail trade businesses will fail during the first four-year period.[13]

13. Ibid, p. 8.

Figure 5–2 Although this student is attending a national FFA convention, during a break she is attending to her agribusiness back home, for which she, as the owner, has total responsibility.

Why Agribusinesses Fail

Agribusinesses fail for many reasons. Usually failure is due to factors in the areas of management, labor, or **financial resources**. Many people go into business or expand without adequate planning, and without analyzing the added costs associated with the additional returns or the potential risk. Too many agribusinesses have poor recordkeeping and inadequate management information systems, which means that the owner is unable to control costs, identify and correct problems, and recognize profit opportunities. Agribusiness failures usually display one, several, or many of the following characteristics within the categories of management, labor, or financial resources.

Management

People do not plan to fail, they only fail to plan. If income and expenses have been projected in the planning process and a profit is not foreseen, there will probably be no profit. Therefore, stop the plans for your business. However, even if the projections show a profit, many factors can still cause a business to fail. The point is that if the business does not work "on paper," it will not work in the real world; even if the business works on paper, there is still no guarantee that it will succeed. Avoid the following pitfalls as you plan and manage your business:

- jumping in without first testing the waters on a small scale
- buying too much on credit
- underestimating how much time it will take to build a market
- going into business with little or no experience and without first learning something about it
- attempting to do too much business with too little capital
- not allowing for setbacks and unexpected expenses
- not understanding **business cycles**[14]
- failing to develop an effective marketing program
- spending too much time and/or money on nonproductive and nonprofitable activities
- making unnecessary capital investments to minimize income taxes
- underpricing or overpricing goods or services
- extending credit too freely
- extending credit too rapidly[15]

Labor

The strength of a business is in its people. Hiring undependable and unqualified employees can quickly destroy your business. Failure can also result if you attempt to support too many people from the business. For example, bringing your children into the business or using it to help them get started can result in business failure. Too much debt, compounded by too much pride or a sense of obligation, prevents many parents from making the necessary adjustments until it is too late.[16] As mentioned previously, the entrepreneur has to work hard. Some owners mistake the freedom of having their own business for the liberty to work only when or as much as they wish.

14. The Service Corps of Retired Executives (SCORE), and Judith Gross, "Autopsy of a Business," *Home Office Computing*, October 1993, pp. 52–60.
15. Nickels, McHugh, and McHugh, *Understanding Business*, p. 172; *Advanced Agribusiness Management and Marketing* (8735-B) (College Station, TX: Instructional Materials Service, 1990), pp. 2–3.
16. *Advanced Agribusiness Management and Marketing*, pp. 2–3.

Financial Resources

Management of financial resources is not merely an exercise to be carried out at the beginning and end of every year. It is an ongoing activity, which involves comparing plans to actual performance and taking appropriate action. Avoid the following as you manage your financial resources:

- starting with too little capital
- starting with too much capital and being careless about how it is used
- borrowing money without planning how and when to pay it back
- failing to keep complete, accurate records, so that you drift into trouble without realizing it
- extending habits of personal extravagance into the business
- forgetting about taxes, insurance, and other costs of doing business[17]
- depending too much on **collateral**
- structuring loans improperly and/or improperly matching loan repayment period with loan repayment ability
- failing to control living expenses and withdrawing more from the business than the business actually earns[18]

Failure is always a possibility. However, if we become afraid to fail, we will not even try. Plan, study, and research every possible angle—then give it your best. Refer to Figure 5-3.

Don't Be Afraid to Fail

You've failed many times, although you may not remember. You fell down the first time you tried to walk. You almost drowned the first time you tried to swim, didn't you? Did you hit the ball the first time you swung a bat? Heavy hitters, the ones who hit the most home runs, also strike out a lot. R. H. Macy failed seven times before his store in New York caught on. English novelist John Creasey got 753 rejection slips before he published 564 books. Babe Ruth struck out 1,330 times, but he also hit 714 home runs. Don't worry about failure. Worry about the chances you miss when you don't even try.

Figure 5-3 When it comes to selling, or, trying to achieve productive goals in life in general, it is better to have tried and failed than never to have tried at all.

Undercapitalization

Undercapitalization is an important factor in small business failure. Many small businesses do not have adequate startup capital to survive the initial one- to two-year period of business establishment. This is analogous to building a car without the wheels. It may run, but it will not move forward.

Analyzing Your Agribusiness Venture

Most people would like to become wealthy by starting a business that proves to be successful. However, it is one thing to dream; it is another to make that dream become reality. Careful planning from the beginning is one way to ensure the success of your agribusiness. Planning includes recognizing a need, considering startup factors, applying business fundamentals, and conducting a business survey.

Recognizing a Need

A business succeeds only when it fills an economic need, and it is the responsibility of the prospective business organizer to determine what types of businesses are needed. A free enterprise system does not, however, guarantee anyone a successful business just because the business is of a desirable type. It is up to the individual to select the right type of business and to make it work. Thousands of firms go bankrupt each year because they provided the wrong type of product or service or were unable to make a potentially good business work. Many new businesses succeed because they provide a product or service that others need and for which they are willing to pay.

Agribusiness Startup Factors to Consider

Anyone trying to analyze whether to start an agribusiness should consider the following factors.[19]

17. Nickels, McHugh, and McHugh, *Understanding Business,* p. 172; SCORE and Gross, "Autopsy of a Business."
18. *Advanced Agribusiness Management,* pp. 2-3.
19. Alfred H. Krebs and Michael E. Newman, *Agriscience in Our Lives,* 6th ed. (Danville, IL: Interstate Publishers, 1994), pp. 656-659.

What Financial Resources Are Needed to Get Started? When entering any field of agribusiness, the entrepreneur should determine all startup costs. This is especially important if the entrepreneur intends to borrow money to get started.

What Labor Does the Agribusiness Need? Determining how much hired labor is necessary to conduct business is a crucial decision for the entrepreneur. Consideration should be given to the amount of capital currently available, as well as how much potential income the business may generate.

What Management Requirements Exist? Although managing an agribusiness requires technical knowledge of agriculture and business operations, the entrepreneur must also deal with day-to-day issues. It is crucial to understand recordkeeping, labor regulations, local business laws, markets, sales taxes, credit, pricing methods, inventory management, human relations, and general business procedures.

Does a Market Exist for the Product or Service? To make a profit, an agribusiness must be marketable. That is, people must want to purchase the item or service offered. The entrepreneur must accurately assess the market potential of the agribusiness being considered.

Where Should the Agribusiness Be Located? A prime location for an agribusiness improves its chance for success. Being located in a high-traffic area will increase the possibility that customers will notice the agribusiness.

Should You Buy an Agribusiness or Start a New One? Instead of starting a completely new agribusiness, the entrepreneur may choose to purchase one that is already established. Consider these advantages and disadvantages of buying an existing agribusiness:

Advantages

- allows a quicker start
- provides ready-made customers
- eliminates some competition
- reduces the cost of getting established
- has a base of financial information for estimating costs and profits[20]

20. Ibid., p. 659.

Disadvantages

- more capital resources needed at the very beginning
- no time to learn while the business is developing
- possibility of misjudging and buying a loser
- problem of having to either accept the location or move the business
- loss of the safety that comes from expanding and growing into a business as resources permit
- risk of missing some critical and costly aspect of the business when the sales deal is finalized
- cost of legal assistance needed for making the purchase
- cost of going out of business if the business fails[21]

Applying Business Fundamentals

Entrepreneurs must learn some business basics if they are to be successful in beginning and growing an agribusiness. Regardless of the size of the new agribusiness, the following fundamentals will prove useful when getting started:

- Keep the size of the business consistent with the capital resources available and with progress in developing management capability.
- Select a business that is labor intensive rather than **capital intensive.**
- Devote adequate time to the development of management capability.
- Keep all employees fully employed, either in the business or elsewhere.
- Develop a good plan of operations and follow it.
- Study and improve the business continually.
- Maintain an adequate cash operating fund so that you will not have to borrow at high interest rates.
- Maintain an inventory level that allows quick response to customer requests.
- Use as much of the profits as possible to expand the business.
- Use the physical facilities of your home to the fullest extent possible.
- Set the prices charged for products and services at a level that will result in a reasonable profit while still being lower than or equal to the prices charged by your competition.

21. Ibid.

- Buy good-quality materials and necessary supplies at the lowest price possible.
- Treat customers and potential customers courteously and fairly.
- Treat employees fairly.[22]

Conducting a Business Survey

Some individuals go into business based on a "feeling" or "intuition" that the business will be successful. A business founded on this basis may, with good luck, be successful. However, the risk is extremely high, and most businesspeople prefer a more solid foundation on which to start. The procedures to follow in investigating the economic potential of a new business vary with the specific type of business under consideration. However, certain basic information on income, expenses, and market potential should be collected and analyzed regardless of the type of business. Regardless of the type of **business survey** you choose to do, some basic questions should be answered:

- What resources are needed?
- Are these resources available?
- What are the costs of these resources?
- What level of management is required?
- Does the prospective owner have the experience necessary to operate the business?

Furthermore, the economic feasibility of marketing the products or services should be analyzed. Some of the questions that should be answered include:

- Are there one or more markets for the products or services?
- At what prices can the products or services be sold?
- Are the prices sufficient to cover the costs and give the seller a satisfactory profit?
- How reliable are the potential markets?
- What is the potential for future growth?

The answers to these and other relevant questions should be collected and analyzed to determine the market potential for the products or services.

Preparing a Business Plan

It's amazing how many people are eager to start an agribusiness yet have only a vague idea of what they want to do. Eventually, they come up with an idea for their agribusiness and begin discussing the idea with friends. It is at this stage that the entrepreneur needs a business plan.

A **business plan** is a written description of a new business venture that describes all aspects of the proposed agribusiness. It helps you focus on exactly what you want to do, how you will do it, and what you expect to accomplish. The business plan is essential for receiving help in starting the business from potential **investors** and **financial institutions**. Although there is no standard business-plan format, there are many similarities.[23] Refer to Figure 5-4 for one example.

The business plan for each venture is unique. Writing a business plan helps you further analyze the agribusiness that you want to start. It helps you be realistic, honest, detailed, and objective about your plans. A business plan forces you to set your goals and objectives. Furthermore, a business plan will help convince financial institutions that they

Business Plan Outline

- Introduction:
- Company Description:
- Product and/or Service:
- Management Plan:
- Marketing Plan:
- Legal Plan:
- Location Analysis:
- Business Regulations:
- Capital Required:
- Financial Plan:
- Financing Arrangements:
- Competition:
- Operating Plan:
- Appendix:

Figure 5-4 People do not plan to fail, they only fail to plan. Proper planning, including the development of a business plan outline, is essential for the success of an agribusiness.

22. Krebs and Newman, *Agriscience in Our Lives*, pp. 659–661.

23. Nickels, McHugh, and McHugh, *Understanding Business*, p. 175.

are investing in a business that has a good chance of succeeding. Only a clear and logical business plan can persuade them.[24] Remember, business plans vary, but the following example includes most components, as in Figure 5–4. They are simply arranged differently. If you go to one of the many Internet search engines and type in "parts of a business plan," you will find all kinds of professional help for preparing a business plan—at a cost. However, your business plan will vary from the following example according to your specific needs or individual requirements of your lenders or investors.

Cover Sheet

The cover sheet acts as the title page of your business plan. It should include:

- name, address, and telephone number of the company
- names, titles, addresses, and telephone numbers of the owners and corporate officers
- month and year the plan was prepared
- name of the person who prepared the plan
- copy number of the plan[25]

Introduction

The introduction could also be referred to as the executive summary or statement of purpose. It sets out the business plan objectives. Use the keywords "who," "what," "where," "when," "why," "how," and "how much" to summarize:

- your company (who, what, where, when)
- who the management is and what each manager's strengths are
- what your objectives are and why you will be successful
- if you need financing, why you need it, how much you need, and how you intend to repay the loan or benefit investors

It is easier to write this section after you have completed your business plan, because it is a summary and reflects the contents of the finished plan.[26]

24. Betty J. Brown and John E. Clow, *Introduction to Business* (New York: Glencoe/McGraw-Hill, 2006), p. 75.
25. Linda Pinson, "Business Plan Outline," from *Anatomy of a Business Plan & Automate Your Business Plan* (2005), available at http://www.business-plan.com/outline.html.
26. Ibid.

Table of Contents

The contents page lists the major topics in your business plan. This makes it easier for lenders or investors to review your business plan quickly and find areas of particular interest.[27]

The Organizational Plan

This section includes two major areas: a summary description of your agribusiness and a more detailed description of the administrative parts of the company. This portion of the plan should include the following subsections.

Summary Description of the Business. In a couple of paragraphs, give a broad overview of the nature of your business, telling when and why the company was formed. Complete the summary by briefly addressing:

- mission—project short- and long-term goals
- business model—describe the model for your company and why it is unique to or appropriate for your industry
- strategy—give an overview of your business strategy, focusing on short- and long-term objectives
- strategic relationships—describe any existing strategic relationships
- SWOT analysis—identify and describe strengths, weaknesses, opportunities, and threats (both internal and external) that your company will face

Products or Services. If you are a manufacturer or wholesale distributor of a product, describe your products and briefly outline your manufacturing process. Include information on suppliers and availability of materials. If you are a retailer or an e-tailer, describe the products you sell. Include information about your sources of supply, handling of inventory, and order fulfillment. If you offer a service, describe your services and list any future products or services that you plan to provide.

Intellectual Property. Address issues concerning copyrights, trademarks, and patents. Back up your statements, assessments, and conclusions with supporting documents containing registrations, photographs, diagrams, and other pertinent materials.

27. Ibid.

Location. Describe your projected or current location, including advantages and disadvantages. Project costs associated with the location. In a section at the end of the plan, include copies of legal agreements, utilities forecasts, and other location-related supporting documents.

Legal Structure. Describe the legal structure of your company and why it is advantageous. List owners and corporate officers, describing their strengths and abilities. (Include a résumé for each person in the "Supporting Documents" section at the end of your plan.)

Management. List the people who are or will be running the business. Describe their responsibilities and abilities (include their résumés in the "Supporting Documents" section of your plan). Project their salaries.

Personnel. Detail how many employees you will have, and in what positions. What are the necessary qualifications for each position? How many hours will employees work, and at what wage? Project future needs for additional employees.

Accounting. What system will you set up for daily accounting? Who will you use as a tax accountant? Who will be responsible for periodic analysis of financial statements?

Legal. Who will you retain as an attorney?

Insurance. What kinds of insurance will you carry? (Consider property, liability, life, health, and disability, among other types.) What coverage amounts do you foresee needing? What will this coverage cost? Who will you use as a carrier (that is, who will issue the policies)? Will you use an insurance agent or broker?

Security. Address inventory control and theft of information, both online and offline, and project-related costs.[28]

The Marketing Plan

The marketing plan sets out all the elements and components of your marketing strategy. It should include the details of your market analysis, sales, advertising, and public relations campaigns. The plan should also combine traditional (offline) programs with new-media strategies, such as online marketing and affiliations, Web presence, and the like. The marketing plan should include the following areas.

Overview and Goals of Your Marketing Strategy. Show who your potential customers are and what kind of competition your business will face. Outline your marketing strategy and specify what makes the company unique. Do not underestimate the competition. Review industry size, trends, and the target market segment. Discuss the strengths and weaknesses of the product or service.

Market Analysis. Address the following areas:

- identify *target market* with demographics, psychographics, and niche market specifics
- describe major *competitors* and assess their strengths and weaknesses
- identify industry and customer *trends* in your target market
- describe methods of *market research* you have used and any database analysis you have undertaken; give a summary of the results

Marketing Strategy. Begin with a general description of budget percentage allocations (both online and offline), with expected returns on investment (ROIs).Then, include the following areas:

- method of sales and distribution—stores, offices, kiosks, catalogs, direct mail, Website
- packaging—quality considerations and type(s) of packaging
- pricing—price strategy and competitive position
- branding
- database marketing and customization
- sales strategies—direct sales, direct mail, e-mail, affiliate, reciprocal, and viral marketing
- sales incentives and promotions—samples, coupons, online promotions, add-ons, rebates, and so on
- advertising strategies—traditional, Web/new media, long-term sponsorships
- public relations—online presence, events, press releases, interviews
- networking—memberships and leadership positions

28. Ibid.

Implementation of Marketing Strategies. Detail what implementation responsibilities you will keep in-house, and what functions you will outsource. If you have contracts or relationships with advertising, public relations, marketing firms, ad networks, or other such entities, list them.

Customer Service. Describe your planned customer service activities and the expected outcomes.

Assessment of Marketing Effectiveness. This assessment is included in the business plan of an existing company after making periodic evaluations.[29]

Financial Documents

This section of the business plan is the quantitative interpretation of everything you set out in the organizational and marketing sections. Do not do this part of your plan until you have completed those two sections.

Financial documents are the evidence of past, current, and projected finances. The following are the major documents to include in your plan. Your work will be much easier if you do them in the order shown here because they build on each other, each using information from the ones previously developed.

Summary of Financial Needs. This outline briefly tells why you are applying for financing and how much capital you need.

Loan Fund Dispersal Statement. A dispersal statement describes how you intend to disperse (use, spend) the loan funds. Back up this statement with supporting data.

Pro Forma Cash Flow Statement (Budget). This document, which is also used for internal planning, shows cash inflow and outflow over a period of time; basically, what your business plan comes to in terms of dollars. This statement is one of the first things a lender will look at, because it shows how you intend to repay your loan. Cash flow statements show both how much and when cash must flow in and out of your business.

Three-Year Income Projection. A profit and loss (P&L or income) statement gives projections for the next three years of your company. Use the revenue and expense totals from the pro forma cash flow statement for the first-year figures and project for the next two years according to your forecasts and research on industry and economic trends.

Projected Balance Sheet. Your balance sheet shows a projection of assets, liabilities, and net worth of your company at the end of the next fiscal year.

Break-Even Analysis. The break-even point is where a company's expenses exactly match the sales or service volume. It can be shown in total dollars or revenue offset by total expenses, or by total units of production, the cost of which exactly equals the income derived from sales. This analysis can be done and presented either mathematically or graphically. Revenue and expense figures are taken from the three-year income projection.[30]

Appendix (Supporting Documents)

This section of your plan will contain copies of all the records that support, prove, or back up the statements you made in the three main parts of the business plan. The following are the most common supporting documents.

Personal Résumés. Include résumés for owners, managers, and other key persons. A résumé should be a single page that lists the person's work history, educational background, professional affiliations and honors, and his or her special skills relating to the position held in the company.

Owner's Financial Statements. Include a statement of personal assets and liabilities, unless you are a new business owner. In the latter case, this will appear in the financial section of your plan.

Credit Reports. Include credit reports, both business and personal, from suppliers, wholesalers, credit bureaus, and banks.

Contracts. Attach copies of all business contracts, both completed and currently in force. This includes copies of leases, mortgages, purchase agreements, and other such agreements between your

29. Ibid.

30. Ibid.

company and leasing agencies, mortgage companies, and other real estate or brokerage agencies, as well as any shipping or purchase contracts.

Letters of Reference. Letters of reference recommend you as being a reputable and reliable businessperson who should be considered a good risk. Include both business and personal references.

Other Legal Documents. The appendix is the place for copies of all legal papers pertaining to the legal structure of your company (articles of incorporation, limited partnership agreement, etc.), its proprietary rights (patents, copyrights, etc.), insurance, and so forth, if the document was not included in the "Contracts" category.

Miscellaneous Documents. This catch-all section is where you include all other documents that you referred to (but not set out) in the main body of the plan. (For example, include location plans, demographics, competition analyses, advertising rate sheets, cost analyses, and so on.)[31]

The Completed Plan

Your business plan should look professional, but the potential lender or investor needs to know that you yourself created it. These readers consider your business plan the best indicator by which to judge your potential for success.

Your plan should be no more than 30 to 40 pages long, excluding supporting documents. If you are seeking a lender or investor, include only the supporting documents that will be of immediate interest to the person examining your plan. Keep other documents with your own copy where you can retrieve them quickly.

Have a local print shop bind your plan neatly with a blue, black, or brown cover (available from stationery and office supply stores). Make a copy for each lender or investor you wish to approach. Do not give out too many copies at once, and keep track of each copy. If you are turned down for financing, be sure to retrieve your business plan from that investor or lender.

Your plan will be valuable only if you update it frequently to reflect what is actually happening within your business. Compare your projections with what really happens, and use the results to analyze the effectiveness of your operation. This will show you what changes to implement to gain a competitive edge and make your business more profitable.[32]

31. Ibid.

Major Areas of Agribusiness Management

Agribusiness management refers to the responsibility of a person to make decisions, organize resources to implement decisions, monitor the implementation of decisions, and evaluate the effects of decisions on the overall success of the operation. Agribusiness management has five major area of activity, as follows.

Planning

Planning means determining what is to be done and where, how, and when to do it. Planning is done on a day-to-day, year-to-year, and long-term basis. Planning involves the following functions:

- determining the present status of the business
- surveying the environment
- setting objectives
- forecasting future situations
- stating necessary actions and resources
- evaluating proposed actions
- revising plans in response to changing conditions
- communicating effectively[33]

A part of planning is goal setting. Refer to the section on goal setting later in this chapter for a further explanation of the importance of and procedure for planning.

Organizing

Organizing is the grouping together of activities, people, and other resources to implement a plan. Without proper organization, it is difficult to carry out or even follow a plan. Organizing involves coordinating activities and obtaining and coordinating resources so that the objectives of the business

32. Ibid.

33. *Agribusiness Management Lesson Plan Library,* Unit A, Problem Area 2, Lesson 1 (CAERT, Inc.), pp. 4–5.

can be met. To do this, the manager/leader breaks down the organizational plans into tasks to be performed. The manager/leader then decides what resources are needed to perform these tasks. More specifically, organizing involves the following responsibilities:

- identifying and defining required work
- breaking work into duties and tasks
- grouping duties by position
- defining the requirements of positions in an agribusiness
- grouping positions into manageable, effective units or departments
- assigning work to be performed, as well as accountability and authority for task performance
- revising and adjusting the organizational duties in response to changing conditions
- communicating duties, activities, and achievement expectations throughout the agribusiness[34]

If the needed resources are available within the agribusiness, the manager/leader must ensure that they are appropriately applied to the area and task when needed. If the resources are not available, the manager/leader finds out where they can be obtained.

Directing (Leading)

Directing or leading consists of providing instruction and guidance to employees. It is primarily concerned with the relationship between managers/leaders and employees. Directing/leading involves influencing, teaching, and guiding people to carry out their assigned tasks to meet the goals that have been set.

Good leaders understand and like people. Good leaders do not view workers just in terms of what they can produce or what problems they might cause. They recognize each person's potential within the organization and act accordingly. Directing includes the following responsibilities:

- communicating and explaining objectives to employees
- setting standards for performance
- providing motivation
- coaching employees to meet performance standards
- rewarding employees based on performance and achievement
- praising and criticizing constructively and fairly
- revising methods as necessary to respond to changes
- communicating throughout all processes[35]

Refer to Chapter 10 on human resource management for further discussion of leadership and personality types of employees.

Staffing

Staffing includes all activities involved in the recruitment, selection, training, and retention of personnel. According to Jack Frymier, former educator at the Ohio State University, hiring staff is the principal job of any leader. You cannot make silk out of a sow's ear. Hiring bad staff is analogous to having one rotten apple spoil the whole barrel. Good staff makes the workflow through the agribusiness go smoothly. Staffing involves the following responsibilities:

- determining human resource needs
- recruiting excellent employee candidates
- selecting excellent employees from those recruited
- training and developing employees
- revising the number of employees according to changing conditions of the agribusiness

Controlling

Controlling covers all the activities that are necessary to ensure that the policies of the agribusiness are being carried out. It focuses on whether the actions being taken are meeting expectations. More specifically, managers/leaders assign job responsibilities as part of their organizing function, based on their estimation of what they can accomplish with the resources available. Controlling involves setting standards for work (such as sales quotas, project goals, and/or quality requirements) and solving problems that prevent the completion of tasks according to these standards.

Controlling also involves evaluating or monitoring the activities employees undertake and the

34. Ibid.

35. Ibid.

procedures employees use in reaching these goals. Leaders are expected to review the way workers do their jobs. When they review employees, leaders may offer suggestions for improvement or may work with employees to set new goals. Controlling involves the following responsibilities:

- establishing standards (achievement, performance, quality, etc.)
- monitoring results and comparing them to standards
- correcting deviations from standards
- revising and adjusting methods in response to changes the agribusiness experiences, on an ongoing basis
- communicating necessary changes throughout the organization[36]

Goal Setting

As stated earlier in this chapter, people do not plan to fail—they simply fail to plan. Prior planning prevents poor performance. Setting goals is a necessity for your agribusiness. You must have a goal, because it is just as difficult to reach a destination you have not located as it is to come back from a place to which you have never been. You must have definite, precisely written, clearly set goals if you are going to realize the full potential of your agribusiness.

Setting Your Agribusiness Goals

There are some general rules that can help you set goals for yourself.

Write Down Your Goals. The best way to start thinking about the goals of your agribusiness is by writing them down. Initially, you need not worry about order or priority; just get your goals down on paper.

Organize Your Goals. You will have both specific goals that you want to reach in a few days, weeks, or months and goals toward which you will work for many years. In between, you will have goals that take a year or two to achieve. Arrange your goals according to these three groups: immediate, short term, and long term.

- *Immediate goals* are the goals that you would like to accomplish within a day, a week, or a month or two. As an entrepreneur, these immediate goals will probably include the first steps needed to get your business started.
- *Short-term goals* include the things you want to accomplish in a year or two. These goals often include the steps you need to build toward your long-term goals. For an entrepreneur, business expansion or marketing perfection would fit between immediate and long-term goals.
- *Long-term goals* are the ones toward which you intend to work for many years. They give you an idea of what you want to do with your business several years from now.

Reaching Your Goals

It is futile to set goals if they do not drive your actions. The following steps will help you start working to reach your goals.

Manage Your Time. Learning time management skills is a first step toward reaching your goals. This means understanding and maximizing the way you use the time you have. Everyone has the same number of hours in a day, a week, and a year; however, some people use these hours more wisely and more effectively than others. How do you spend the 24 hours that you have each day? Here are three techniques that will help you better use your time:

- *Avoid procrastinating.* A *procrastinator* puts off doing anything that can be done later. Good time managers, however, want to reach their goals quickly and use their available time to that end.
- *Judge your time.* Different tasks take different amounts of time to complete. Learning how to judge the time you need to accomplish a given task is a skill that you will develop as you gain experience.
- *Schedule your time.* Scheduling your time effectively is often a matter of balance. On the one hand, you want to allow enough time to complete a task well; on the other, you do not want to allow so much time that you finish early and waste time until your next appointment or project. Practice and experience will help you schedule effectively.

36. Ibid.

Immediate Goals

- Draw up a business plan to know exactly how much capital will be needed and how and when it will be repaid.
- Obtain capital and enlarge greenhouse, buy a refrigerating unit, and open a small shop.
- Produce enough income to ensure that the business will have an adequate and effective cash flow.
- Stay within a specified debt limit.
- Sell fresh-cut local floral arrangements and potted plants to people in the community.
- Locate suppliers for vases and other needed items.

Short-Term Goals

- Repay original loan.
- Capitalize expansion; perhaps include silk floral arrangements and garden plants.
- Provide flowers for weddings and add other floral-catering services.
- Have an income that provides a satisfactory profit over expenses.

Long-Term Goals

- Hire and train employees to expand the capacity of the company (and allow vacations).
- Broaden the market and price products reasonably to entice new clients from the larger town and surrounding area.
- Establish financial security.
- Sell the company and retire with a substantial profit.

Figure 5–5 If you do not know where you are going, you will not know when you get there. Nancy prepared these goals to measure the success of her floral shop.

Establish Priorities. Consider prioritizing your goals. You might start by listing all your goals on a piece of paper. Then number the most important goal "1," the second most important "2", and so on until all your goals are numbered in priority of importance. Then rewrite your list in order. Begin working toward the goals at the top of your list and continue, in order, until you have achieved them all. When you do this, you will find yourself working smarter, not harder.[37]

Breaking Goals into Manageable Units. Your best goals are often your biggest goals. They are probably also long-term goals—that is, they take the most time to accomplish. To avoid frustration over the time needed to achieve your long-term goals, it is wise to break them down into smaller, short-term goals. In this way, you experience the satisfaction of completing the short-term objectives while still progressing toward your long-term goal.[38]

Example of Business Goals

Nancy Baker is about to start a new business venture. The small town where she lives does not have a flower shop, and people in her community order flowers from a larger town 12 miles away. Nancy has always liked working with flowers, and her friends often compliment the floral arrangements in her home. For two years she has made a hobby of growing flowers in a small greenhouse that she built in her backyard. Recently, several friends have asked Nancy to make up floral arrangements for gifts, and she believes she can build a profitable business with her talent. Nancy prepared a list of goals for her venture. Refer to Figure 5–5 for a list of Nancy's goals.

37. Hamilton, Connelly, and Doster, *Agribusiness: An Entrepreneurial Approach.*

38. Ibid.

Problem Solving and Decisionmaking

When starting and organizing an agribusiness, the owner will have a number of decisions to make and problems to solve. How well the owner handles these issues will greatly affect the success of the business. Sometimes solutions require only minor changes in the business; at other times, a completely new or different course of action will be necessary. *Decisionmaking* is the term that describes the process used to choose the new and different action. Decisionmaking may involve setting new goals to solve a problem. In extreme cases, when a problem cannot be readily solved, the decision may actually be to do nothing at all.[39]

Skills Needed in Problem Solving and Decisionmaking

Lloyd J. Phipps suggests that individuals need to develop certain problem-solving and/or decision-making skills. Among these are the ability to:

- recognize problem situations
- clearly distinguish the problem from the problem situation
- clearly define goals and/or objectives
- develop creative, imaginative solutions to problems
- gather information related to possible solutions
- be open-minded about possible solutions offered by others
- carefully evaluate information before accepting or rejecting solutions
- work with others to solve problems
- avoid jumping to unwarranted conclusions (be flexible); accept the fact that you may make mistakes; and put aside opinions, feelings, emotions, and self-interests that may interfere with objective thinking
- understand different types of problems and techniques for solving them
- *understand,* and use, a systematic approach to problem solving and decisionmaking[40]

39. Robert N. Lussier, *Human Relations in Organizations: A Skill Building Approach* (Homewood, IL: Richard D. Irwin, 1990), p. 276.
40. Lloyd J. Phipps, *Handbook on Agricultural Education in Public Schools* (Danville, IL: Interstate Printers and Publishers, 1965), pp. 118–119.

Various Styles of Decisionmaking

Individuals make decisions in a variety of different ways. The criteria they use to make decisions may also differ from person to person and even time to time. *Criteria* are established standards or tests by which something is compared or judged. Notice the criteria used in these three decisionmaking styles.

Reflexive Decisionmaking. People with this style make decisions quickly. They typically do not spend a great deal of time considering all the different options and consequences of each action. They are interested in doing something immediately.

Reflective Decisionmaking. People with this style consider all the options and consequences before making a decision. They process as much information as possible to avoid poor decisions. They must be sure, however, that this careful thought process does not delay important decisions.

Consistent Decisionmaking. People with this style consider all the options and still make decisions in a timely manner. They combine the best characteristics of the reflexive and reflective decisionmaker to make the best decision possible.[41]

Steps in Problem Solving and Decisionmaking

A systematic approach to problem solving can be of great benefit when making important decisions. These steps are similar to the seven steps of the scientific method, which scientists use in their experiments:

- *Step 1, Recognize the Problem.* Remember, problems are inevitable and must be solved, not ignored, if you are going to be successful. Obviously, the first step toward solving a problem and making a decision is realizing that you have a problem that requires a decision.
- *Step 2, Determine Your Alternatives.* Once you have identified your problem, you need to determine your alternatives. Alternatives are the different courses of action you could take to solve your situation. There are likely to be many alternatives for solving any problem, and

41. Lussier, *Human Relations in Organizations,* p. 278.

you need to consider each one carefully before making a decision. It may be helpful to list your alternatives on paper to keep a clear focus.

- *Step 3, Gather Information.* Once you have listed your alternatives, you need to gather information relative to each one. Look at factual information. Relying only on opinions, emotions, and feelings may lead to hasty decisions that you will later regret. In gathering factual/objective information, ask yourself the following questions: What do I need to know about each alternative? What materials or information may be needed to implement each alternative? What costs will be involved? Is the alternative feasible in the first place?
- *Step 4, Evaluate the Alternatives.* Once you have gathered information for each of your alternatives, you will need to evaluate (and perhaps list) the advantages and disadvantages of each in relation to solving your problem and in relation to each other.
- *Step 5, Select a Workable Solution.* After you have evaluated each alternative and its possible results, you must choose the one that is the most practical, reasonable, and effective in solving your problem.
- *Step 6, Carry Out Your Solution.* Once you have determined your course of action, follow through. If you fail to implement your solution, you have wasted the time and effort spent to this point.
- *Step 7, Evaluate Your Results.* The problem-solving/decisionmaking process does not end when the proposed solution is carried out. It ends in the future, when you decide whether your problem has been solved, the same problem persists, or new problems have emerged. Evaluation may lead you to regard your solution as a good decision, make further adjustments to improve your solution, or discard your decision and start the process over again.[42]

Conclusion

Remember that you will encounter obstacles as you pursue your goals. Some of these obstacles can be foreseen. If you are prepared for the foreseen problems, you will have more time and energy to deal with the unexpected ones. If your business goals require a great deal of your time, you will want to consider how this will affect your other commitments, such as your family. Your goals may conflict with the goals of others. Communication is the key to human relations. Be sure that those who will be affected understand what your goals are.

Circumstances will change with your age, your health, and your family obligations, among other things. Be prepared to reevaluate your goals from time to time. You may find that you need to restructure them because your priorities or your values have changed. Do not be afraid to make decisions. Do not feel defeated if a decision turns out to be wrong—sometimes even the best decisions can have bad results. The most successful entrepreneurs are the ones who can profit from their mistakes and move ahead.

Develop a methodical approach to breaking your goals into manageable units and then pursuing those accomplishments one step at a time. It is up to you to make daily goals a habit as you pursue your dreams. You must think carefully about what kind of business you want. You are not likely to find everything you want in one business—easy entry, security, *and* reward. Choose those characteristics that matter the most to you; accept the absence of the others; plan, set your goals, and then go for it!

Summary

Even though most agribusiness students will never own their own business, it is important to understand the basic processes involved in planning and organizing a business. As we enter the world of agribusiness, we must decide whether to work for ourselves or work for someone else.

Small businesses, which include most agribusinesses, are not small in terms of the number of people they employ. Small businesses are really a big part of the U.S. economy. In fact, 90 percent of the nation's new jobs in the private sector are in small business.

Entrepreneurship is not for everyone. Entrepreneurs are people who have the initial vision, diligence, and persistence to follow through. An entrepreneur is a person who accepts all the risks pertaining to forming and operating a small business.

42. Cliff Ricketts, *Leadership, Personal Development, and Career Success* (Albany, NY: Delmar Publishers, 1997), pp. 313–316.

Entrepreneurs tend to be independent, self-confident, energetic, organized, persistent, optimistic, committed, self-nurturing, action-oriented, flexible, and emotionally stable; they are generally visionaries, problem solvers, and risk takers.

Being successful in a new agribusiness is a challenge. Three major challenges of entrepreneurship are total responsibility; long, irregular hours; and financial risks. The chances of a business reaching its fourth birthday are only about 50/50.

There are many potential reasons why agribusinesses fail, but most failures are attributable to bad management, improper use of labor, or misuse of financial resources. Many people go into business or expand without adequate planning, without analyzing the added cost of the added returns, and without analyzing potential risk. They also fail to realize that the strength of a business lies in its people. Hiring undependable and unqualified employees can quickly destroy your business.

Careful planning from the beginning is one way to ensure the success of your agribusiness. Every potential angle of your agribusiness venture has to be analyzed. Planning includes recognizing a need, considering startup factors, applying business fundamentals, and conducting a business survey.

A business plan is a written description of a new business venture that describes all aspects of the proposed business. It helps you focus on exactly what you want to do, how you will do it, and what you want to accomplish. The business plan is essential for the potential investors and financial institutions on whom you will depend for starting the business.

Agribusiness management refers to the responsibility of a person to make decisions, organize resources to implement decisions, monitor the implementation of decisions, and evaluate the effects of decisions on the overall success of the operation. Agribusiness management has five major areas: planning, organizing, directing, staffing, and controlling.

You must have definite, precise, clearly set, written goals if you are to realize the full potential of your agribusiness. You should write down your goals and organize them into immediate, short-term, and long-term goals. As you strive to achieve your goals, you must manage your time, establish priorities, and break your goals into manageable units. It is up to you to make daily goals a habit as you pursue your dreams.

The ability to solve problems and make decisions can mean the difference between excellence and mediocrity. Whether you are the chief operating officer of a large corporation or an individual agribusiness owner, your success may be determined by the ability to solve an individual problem or make an important decision. Those who succeed are most often those who plan, and those who plan are often accomplished goal setters, problem solvers, and decisionmakers.

End of Chapter Activities

REVIEW QUESTIONS

1. Define the Terms to Know.
2. What are nine reasons why people are willing to take the risks of starting a business?
3. What are 15 characteristics found most often in successful entrepreneurs?
4. What are three major challenges of entrepreneurship?
5. What are five responsibilities of the entrepreneur?
6. What are 13 management decisions that could cause your agribusiness to fail?
7. What are nine things to avoid as you manage your financial resources?
8. What are six questions to analyze when considering whether to start an agribusiness?
9. What are five advantages of buying an agribusiness rather than starting one?
10. What are eight disadvantages of buying an agribusiness rather than starting one?
11. List 14 fundamentals of business that entrepreneurs need to learn as they enter the agribusiness world as owners.

12. What are five basic questions that should be answered when doing a business survey?
13. What are the five major activity areas of agribusiness management?
14. What are three crucial steps to follow as you attempt to reach your goals?
15. List 11 skills needed in problem solving and decisionmaking.
16. Briefly explain the three distinct styles of decisionmaking.
17. Briefly explain the seven steps in the problem-solving and decisionmaking process.

FILL IN THE BLANK

1. When it comes to the number of jobs, small businesses are ___________ in the number of people they employ.
2. _________ percent of the nation's new jobs in the private sector are in small business.
3. There are about ________ full- and part-time home-based businesses in the United States.
4. Of all nonfarm businesses in the United States, almost ___________ percent are considered small by SBA standards.
5. Small businesses account for more than ___________ percent of gross domestic product.
6. It is not uncommon for an entrepreneur to work more than ___________ hours per week.
7. The chance of a business reaching its fourth birthday is ___________ percent.
8. Agribusiness failures usually fall into the category of________, ________, or _____.
9. The strength of a business is in its ___________.
10. A business succeeds only when it fills a/an ___________ need.
11. People do not plan to fail, they simply ___________ to ___________
12. Three techniques of successful time management are ___________, ___________, and ___________.

MATCHING

a. table of contents
b. marketing plan
c. appendix
d. introduction
e. organizational plan
f. financial documents
g. cover sheet

_____ 1. contains all the records that back up the statements and decisions in the business plan
_____ 2. includes a summary description of your agribusiness and the administrative areas of your agribusiness
_____ 3. is also referred to as the executive summary or statement of purpose
_____ 4. are records used to show past, current, and projected finances
_____ 5. serves as the title page of your business plan
_____ 6. includes the major topics in your business plan
_____ 7. includes analysis, sales, advertising, and public relations campaigns

Note: All the matching items relate to the parts of a business plan.

ACTIVITIES

1. Look through the Yellow Pages in your telephone book, or use any other source, and list all the agribusinesses in your area. Describe the product or service each sells. Identify whether the business is part of the agribusiness input sector or part of the agribusiness output sector.
2. The list on the following page shows categories of skills needed by most small business owners.[43] Read each one carefully and decide how adept you are at each one. Do this by placing a number from 1 to 5 in the space provided beside each item, with 1 meaning that you have no knowledge or expertise in the area. You will use these rankings in your personal inventory later. When you have completed each category, total the ratings within each section. Divide the total number by the number of items in each section. This will give the average rating in each section. Place the average score in the space provided for the category.

   When rating your personal skills, you can use any type of experience relating to the particular skill. For example, you might have organized a fundraiser for your club or organization. This would boost your ratings in the "Setting Up the Business" section. Do this with the other sections. Try to think of any personal experience that would relate to each item. If you do have such experience, then enter a number from 2 to 5 beside the item, depending on the amount. If not, place a 1 beside the item. Use your score as you consider starting an agribusiness. Compare your score with scores of your classmates.
3. Some entrepreneurs recognize a need in society and start a business to fill that need. Of the businesses you found in question 1, name three businesses that follow that pattern.
4. Some entrepreneurs create a need. They supply goods or services in the hope that the product is attractive enough that people will decide they "need" it. Of the businesses you found in question 1, list three companies that have offered new and unusual products.
5. Select an agribusiness that you believe would be successful. Answer the following questions.
   a. What product or service did you select?
   b. Is there a market for the product?
   c. What financial resources are needed to get started?
   d. What are the labor needs of the business?
   e. What skills or tasks would you have to learn how to do in order to succeed?
6. Complete a business plan for one of the agribusiness that you selected in question 5. Use the format in Figure 5-4, or follow the outline detailed in the chapter.
7. Using the same agribusiness as in questions 5 and 6, write three to five immediate goals, three to five short-term goals, and three to five long-term goals for that business. Refer to Figure 5-5 for an example.

43. *Entrepreneurship in Agriculture,* pp. 2–3.

A. Managing Money **Average Score___**
1. Borrowing Money
2. Keeping Business Records
3. Avoiding Losses
4. Handling Credit
5. Figuring Taxes

B. Managing People **Average Score___**
1. Hiring Employees
2. Supervising Employees
3. Educating Employees
4. Motivating Employees

C. Directing Business Operations **Average Score___**
1. Purchasing Supplies
2. Purchasing Equipment
3. Purchasing Merchandise
4. Managing Inventory

D. Directing Sales Operations **Average Score___**
1. Identifying Various Customer Needs
2. Developing Products for Different Customer Needs
3. Learning to Answer Customer Objections
4. Closing the Sale
5. Instructing Others in Selling Techniques

E. Marketing **Average Score___**
1. Developing New Ideas for Products or Services
2. Recognizing Community Needs
3. Recognizing Potential Customers
4. Creating Promotional Strategies
5. Designing Promotional Materials

F. Setting Up a Business **Average Score___**
1. Choosing a Location
2. Obtaining Licenses or Permits
3. Determining Initial Inventory
4. Obtaining Financing
5. Planning Long- and Short-Term Cash Flow
6. Choosing the Form of Ownership (e.g., Partnership)

Chapter 6

Types of Agribusiness

Objectives

After completing this chapter, the student should be able to:

- Compare proprietorships, partnerships, and corporations.
- Explain the characteristics of the single (sole) proprietorship.
- Explain the characteristics of partnerships.
- Discuss the different types of corporations.
- Explain the characteristics of limited liability companies.
- Explain the characteristics and value of cooperatives.
- Describe the characteristics of franchises.

Terms to Know

board of directors
common stock
cooperative
corporate charter
corporation
dividends
double taxation
franchise
franchisee
franchisor
general partnership
legal classification
legal entity
legal structure
limited liability
limited partner
limited partnership
marketing cooperatives
parent company
partnership
preferred stock
prepackages
single (sole) proprietorship
stock
stockholders
Subchapter C
Subchapter S
Subchapter T
unlimited liability

Introduction

As our society becomes increasingly complex, the process of selecting the most desirable type of agricultural business form, or **legal structure**, for an agribusiness becomes very important. The amount of regulation and restriction by government agencies, the tax structure, and other legal requirements are determined, to a large degree, by the **legal classification**, or type, of agribusiness. The type of business form or legal classification you choose may make the difference between success or failure. Although many specific legal structures are available, most legal specialists agree that the three major types of business organizations are: (1) the **single (sole) proprietorship**, (2) the **partnership**, and (3) the **corporation**. The three types of corporations are regular corporations (**Subchapter C**), family farms and small businesses (**Subchapter S**), and **cooperatives** (**Subchapter T**). Cooperative corporations are increasing in importance, and many authors list the cooperative as a fourth major type of business. Another corporation that is a special form of business organization is a **franchise**, which is the fifth type of agricultural business.

If you are going to operate a tractor repair shop, it is likely that you will be the sole owner. If you plan to have a manufacturing facility, you will probably have partners or form a corporation. If you want to own a fast-food chain, you will most likely have a franchise business. You need to assess all forms of legal classifications to decide which is best for your purposes. You can also change your initial decision about the type of business. As your agribusiness grows, changing conditions may require you to change your form of ownership.

Comparison of Proprietorships, Partnerships, and Corporations

Some agribusinesses are relatively easy to start, such as a lawn service or landscaping business. An organization that is owned, and usually managed, by one person is called a single (sole) proprietorship. It is the most common form of business ownership, with more than 19.7 million businesses in the United States using this structure.[1]

Many people lack the money, time, or desire to run a business on their own. They prefer to have another person or a group of people help them form an agribusiness. When two or more people make a legal agreement to become co-owners of a business, the organization is called a partnership.[2] In the United States, there are approximately 2.5 million partnership businesses.[3]

It is sometimes best to create a business that is separate and distinct from the owners. A corporation is a legal entity with authority to act that has liability separate from that of its owners.[4] Although there are only 2.8 million corporations in the United States, comprising only 17 percent of all businesses, they do 87 percent of the sales volume. Refer to Figure 6–1 for a comparison of proprietorships, partnerships, and corporations.

No single type of legal structure is best for all businesses. Therefore, when developing a new agribusiness, all structures should be considered relative to their advantages and disadvantages. Although single proprietorships are more numerous in all types of industries except manufacturing, they are especially dominant in production agriculture, including forestry and fisheries. Corporations are the least numerous for these groups, but they exceed partnerships in all other types of industries. We will now take a closer look at each of these legal classifications, or types, of agricultural businesses.

Single, or Sole, Proprietorship

The single proprietorship is the major type of legal structure in the agricultural industry, as well as in most other industries. It is the simplest type of business and is the easiest to organize. It is the type closest to the American dream, in which the

1. IRS Issues Fall 2006 Statistics of Income Bulletin, Dec. 13, 2006; available at http://www.irs.gov/newsroom/article/0,id=164959,00.html.
2. William G. Nickels, James M. McHugh, and Susan M. McHugh, *Understanding Business,* 5th ed.(Chicago: Irwin/McGraw-Hill, 1999), p. 134.
3. IRS Issues Fall 2006 Statistics of Income Bulletin.
4. Nickels, McHugh, and McHugh, *Understanding Business,* p. 134.

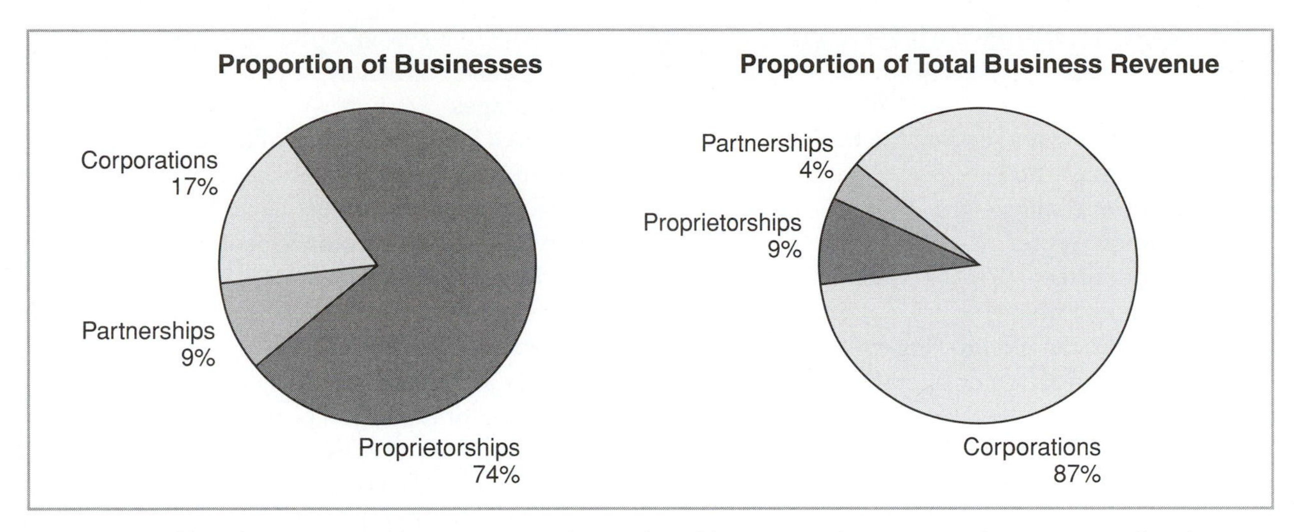

Figure 6–1 Although 74 percent of businesses are sole proprietorships, corporations account for 87 percent of total business revenue. Source: U.S. Internal Revenue Service, *Statistics of Income Bulletin*.

owner is in complete control. On average, sole proprietorships are the smallest businesses. As a result, sole proprietorships are especially suited to areas that require personalized service and are flexible enough to meet the specific needs of individual customers.

Advantages of Sole Proprietorship

Sole proprietorships have several advantages. As mentioned, they are simple to start. There are few government regulations or restrictions. Depending on local laws, the only requirement may be a *license,* which is a legal permit for doing business. Management and control are solely in the owner's hands. Therefore, no voting is necessary. This allows owners to make quick decisions, without waiting to talk with others or get their approval. Sole proprietors can choose their own products and set their own hours. Another advantage is that the owners receive all the profits from the business. Finally, sole proprietors pay taxes only once on the income from the company.[5] Refer to Figure 6–2 for a list of the advantages of a single or sole proprietorship.

Disadvantages of Sole Proprietorship

Even though we tend to stress the advantages of single proprietorships, they also have certain disadvantages. The major one is the claim that creditors can make on the owner's personal assets, as well as the business assets, for the payment of business debts. This relationship is known as **unlimited liability**. Another major disadvantage of one-owner firms is that it can be difficult to accumulate the large amounts of capital required to begin and operate many businesses today.

In some cases, owners may be expert in the "product" of the business (repairing tractors, for instance), but they may not have business skills in other areas, such as leadership, human and public relations, or tax preparation. Owners may have to seek outside help to keep the business going.[6] Another disadvantage of the sole proprietorship is its limited life. The business legally ends when the owner sells the business or dies. Someone may buy the business from the owner, but the success of the business may be based almost entirely on the former owner's special skills. Refer to Figure 6–2 for a list of the disadvantages of the sole proprietorship.

Partnerships

If starting a sole proprietorship sounds too risky for you, you may decide to share the responsibility and benefit from someone else's skills and find a business partner. A *partnership* is usually defined

5. Ibid., p. 135.

6. Ibid.

Single (Sole) Proprietorship	
Advantages	**Disadvantages**
■ Ease of starting and ending the business	■ Unlimited liability—the risk of losses
■ Being your own boss	■ Limited financial resources
■ Pride of ownership	■ Difficulty in management
■ Retention of profit	■ Overwhelming time commitment
■ No special taxes	■ Few fringe benefits
■ Simple and flexible	■ Limited growth
■ Independent decisionmaking	■ Limited life span
■ Change organization or operation quickly and easily	■ Business and owner are a single entity
■ Relatively low initial capital requirement	■ Restriction of expansion potential
■ Fewer government regulations	■ Owner assumes all risks
■ One owner in control	■ Owner is often tied to the business and may be unable to spend time away without closing it
■ Offers opportunity for personal advancement	

Figure 6–2 There are many advantages and disadvantages of a single (sole) proprietorship.

simply as a business association of two or more persons. Someone who wants to start a business but does not have the capital or management skills required usually seeks one or more individuals who meet these requirements. If these individuals can agree that a business is needed and a profitable arrangement can be made, the business is organized as a partnership. When formal agreements are made and recorded, it is easy to identify the partnership.

Business partnerships make it possible for persons to use their specialized skills to organize a company in which they can use their abilities fully. For example, three people might form a partnership to start an agricultural machinery dealership because one is a skilled mechanic, another is an excellent salesperson, and the third is an excellent manager.[7] Partnerships are found in many other areas of the agricultural industry as well. A few examples include agricultural supply stores, landscaping businesses, and hunting preserves. Many times the name of the business will give you a clue about how many partners are involved: a two-person partnership might be called "Walker and Jones Feed Store," and a three-person partnership might be named something like "Rodriguez, Rochelle and Jordan Meat Processors."

Types of Partnerships

General Partnership. **General partnerships** are formed when two or more people manage a business together as owners. General partnerships should be carefully structured using a formalized agreement. The agreement should specify all the terms for the partnership, including what resources each partner will contribute, who will have decisionmaking authority, and how profits will be divided. General partnerships are usually dissolved by mutual agreement, by bankruptcy, or by the death of a partner.

Voting and profit sharing in a general partnership are generally guided by the amount each member contributes to the partnership. If the partnership earns a profit, it is divided as specified in the formal partnership agreement. Legally speaking, each partner is completely liable for all activities of the partnership. The partnership itself does not pay taxes, though it must file a tax return. The partnership tax return is used by each of the partners in filing their individual returns.

Limited Partnership. There is a special type of partnership in which some partners are not completely liable for their partners' debts. This type is usually

7. Betty J. Brown and John E. Clow, *Introduction to Business* (New York: Glencoe/McGraw-Hill, 2006), pp. 87–88.

Partnerships	
Advantages	**Disadvantages**
■ More financial resources	■ Unlimited liability
■ Shared management and pooled knowledge	■ Division of profits
■ Longer survival	■ Disagreements among partners
■ Special skills of the partner	■ Difficult to terminate
■ Legal aspects of forming a partnership are relatively simple	■ Dissolved when a partner dies or leaves the partnership
■ Pay no taxes as a business	■ Size is limited by resources
■ May be ended anytime partners agree	■ Difficult to manage if there are too many partners
■ Limited government regulation	■ Lack of continuity
■ Greater management base than with one owner	■ Divided management authority
■ Partners, like sole proprietors, often feel pride in owning and operating their own company	

Figure 6–3 Business partnerships make it possible for persons to use their specialized skills to organize a company in which they can use their abilities to the fullest.

called a **limited partnership**, and most states recognize its legality. To be a limited partner, one must invest in the partnership, but may not participate in the day-to-day management of the business. The limited partner's name cannot even appear in the partnership name. Limited partnerships are for investment purposes only. "Limited partners are liable for partnership obligations only up to the amount of their investment in the partnership and may not be held personally liable."[8] This provides incentive for investing in a business without the fear of losing personal or other business assets. However, since passage of the Tax Reform Act of 1986, limited partners can no longer deduct partnership losses from other personal income. To be safe, limited partners should have legal documentation of their limited role in and relationship to the business, to protect them should creditors attempt to hold them responsible for business debts.

Limited Liability Partnership. A limited liability partnership gives the benefits of limited liability, in that with this business structure, partners can protect their existing personal assets. A **limited partner** risks whatever investment he or she makes in the firm, but has limited liability and cannot legally help manage the company. **Limited liability** means that limited partners are not responsible for the business's debts beyond the amount of their investment. Their liability (amounts that they may have to pay) is limited to the amount they invested in the company, and their personal assets are not at risk.[9]

Advantages of Partnerships

Partnerships offer an immediate advantage in obtaining capital, as two or more persons are the founders and therefore are sources of startup capital as well as operating funds. Decisionmaking is shared, as are all other aspects of the business. Startup costs are lower than for corporations. Like the income of the sole proprietorship, the income of a partnership is taxed only once.

Often it is much easier to own and manage a business with one or more partners. Your partner can cover for you when you are sick or go on vacation. Your partner may be skilled at inventory keeping and accounting, while you do the selling or servicing. A partner can also provide additional money, support, and expertise. Refer to Figure 6–3 for a list of the advantages of partnerships.

8. Neil E. Hard, *Farm Estate Business Planning*, quoted in N. Omri Rawlins, *Introduction to Agribusiness* (Englewood Cliffs, NJ: Prentice-Hall, 1980), p. 31.

9. Nickels, McHugh, and McHugh, *Understanding Business*, p. 136.

Disadvantages of Partnerships

The major disadvantage of a partnership—and its major distinguishing characteristic—is unlimited liability. This means that each partner is completely liable, legally, for all obligations of the partnership. The personal assets of either partner may be taken by a creditor in payment of a business debt; thus, a business partner should be selected very carefully. In fact, some businesspeople contend that this one disadvantage outweighs all the advantages of the partnership business form, and many refuse to take on the responsibility of this type of business structure.

Another major disadvantage of a partnership is its instability and the lack of continuity if one partner dies. Generally, if one partner dies, the remaining partners close the business and settle with the deceased partner's estate. However, the courts have recognized the rights of remaining partners to continue the business after one partner's death. In some cases, a new partnership is formed with the surviving family members and the business continues. "For federal tax purposes, a partnership does not terminate on death of a partner unless the partnership ceases to operate or there is a change of 50 percent or more in partnership capital and profit within a 12-month period."[10] Additional disadvantages are summarized in Figure 6–3.

Written Partnership Agreements

It is not hard to form a partnership, but it is wise for each partner to get the advice of a lawyer who is experienced with such agreements. Lawyers' services are usually expensive, so study and be prepared before calling a lawyer. To protect yourself, be sure that your partnership agreement is put in writing. The Model Business Corporation Act recommends that the following be included in a written partnership agreement:

- The name of the business. Many states require the partnership name to be registered with state and/or county officials if that name is different from any of the partners' names.
- Partners' names and addresses.
- The purpose and nature of the business, the location of the main offices, and any other locations where the partnership will conduct business.
- The date the partnership will start and the expected duration of the partnership. Will it exist for a specific period of time, or will it end with the death of a partner or when the partners agree to discontinue it?
- The contributions made (or to be made) by each partner. Will some partners contribute money and others provide expertise, labor, real estate, or personal property? When will the contributions be made?
- The responsibilities of management. Will all partners share equally in management, or will some be senior partners and others junior partners?
- The duties of each partner.
- The salaries and spending authority of each partner.
- Details on the sharing of profits or losses.
- Accounting procedures. Who will keep the accounts? What bookkeeping and accounting methods will be used? Where will the partnership books be located?
- Requirements for accepting new partners.
- Any special restrictions, rights, or duties of any partner.
- Provisions for a partner's retirement.
- Method for purchasing a deceased or retiring partner's share of the business.
- Grievance procedures.
- Details on dissolution of the partnership and distribution of the assets to partners.[11]

Corporations

Corporations are another way of doing business in the United States, especially where large investments of money are needed. Although the word *corporation* makes people think of big businesses, such as General Motors, Ford, IBM, FedEx, Exxon, and John Deere, a firm need not be big to incorporate. Obviously, many corporations are big.

10. Rawlins, *Introduction to Agribusiness*, p. 209.

11. Nickels, McHugh, and McHugh, *Understanding Business*, p. 139.

However, incorporating may be beneficial for small businesses, too.

Although agribusiness corporations are not as numerous as single proprietorships, they are increasing in number and importance. This increase is due primarily to the increasing requirements for large amounts of capital in modern agriculture and the limited liability characteristic of the corporate structure, which allows individuals to invest in a business without incurring personal liability beyond the amount of their investment.

Characteristics of Corporations

A corporation is an organization owned by many people but treated by law as though it were itself a person. A corporation is a **legal entity**, separate from the people who own it or work for it. It can own property, pay taxes, make contracts, sue and be sued in court, and do other things a "real" person can do. If you want to form a corporation, you issue **stock**, or shares in the ownership of your corporation. The new owners, called **stockholders**, pay a set price for their shares. For each share of stock purchased, the stockholder gets one vote in the major decisions made by the corporation. Some large corporations have more than a million stockholders.

Because a corporation often has numerous owners who are not involved in the management of the business, government has established strict regulations designed to protect these owners. Therefore, corporations must comply with more legal regulations than any other type of business. For example, each corporation must obtain a legal charter from a state government before it can operate; no other type of business is required to do this. Corporations can be chartered in any state. They must then elect a **board of directors**. Most state laws governing the formation of corporations are similar; the process generally begins with filing an application for articles of incorporation. If the articles conform to state law, the state will grant the business a **corporate charter**, which is a license to operate from that state.[12]

The stockholders of a corporation elect a board of directors, which is a group of individuals chosen to make the decisions for the company. They also appoint officers. The officers, such as the president, vice presidents, and treasurer, make most of the day-to-day decisions for the corporation.

Types of Corporations

There are three types of corporations. The Subchapter C is a regular corporation, which sells stock to investors. The Subchapter S is primarily for small businesses or a family, and Subchapter T is for cooperatives. Both Subchapter C and Subchapter S are profit-making corporations. Nonprofit corporations are formed under Subchapter T (for cooperatives).

Subchapter C (Regular) Corporation. The owners of a Subchapter C corporation sell stock to investors to raise capital for the business. When they buy stock, the investors each become owners of a portion of the company. They buy stock hoping that the business will do well and the price of the stock will increase. When a company makes a profit, its board of directors may choose to pay a certain amount of money to investors for each share of stock the investors own. This payment is called a **dividend**. If a company is consistently profitable, the price of its stock will rise because more investors will want to buy it. When stock prices increase, shareholders may keep their stock or sell it. When investors sell stock for more than the purchase price, they make a profit. Of course, the opposite is true also. Corporations that lose money do not issue dividends. They will experience a decrease in stock price, as fewer investors will want to purchase their stock. When investors sell stocks for less than the purchase price, they, too, lose money.

In a regular corporation, the board of directors and the shareholders make policy decisions. If a vote is required, each shareholder is entitled to one vote for each share of stock owned. The corporation is financially liable at all times.[13] Agribusiness corporations of this type include Ralston Purina, Case International, Ciba-Geigy, and many others. Refer to Figure 6–4 for a list of additional agribusiness corporations.

12. Brown and Clow, *Introduction to Business,* p. 89.

13. Robert J. Birkenholz and Ronald L. Plain, *Agricultural Management and Economics* (Columbia, MO: University of Missouri, Instructional Materials Laboratory, 1988).

Agribusiness Corporations		
Phillip Morris	Safeway	Monsanto
Procter & Gamble	RJR Nabisco	General Mills
Pepsico	Georgia-Pacific	H. J. Heinz
Conagra	Archer Daniels Midland	Eli Lilly
Kroger	IBP Packers	Campbell Soup
Dow Chemical	Winn-Dixie Stores	Farmland Industries
International Paper	Deere Ralston	Ralston Purina
Sara Lee	McDonald's	Kellogg
Quaker Oats	Tyson Foods	Hershey Food
Dole Food	Hormel Foods	Dean Foods

Figure 6–4 Agribusiness is big business. The agribusiness corporations listed here represent only a few examples of agribusinesses in the United States.

Subchapter S (Small Business or Family) Corporation. A Subchapter S corporation is a unique government creation that has the characteristics of a corporation but is taxed like a sole proprietorship or partnership. These corporations have the benefit of limited liability. The paperwork and details of Subchapter S corporations are similar to those of regular corporations. S corporations have shareholders, directors, and employees, but the profits are taxed as personal income for the shareholders, thus avoiding the double-taxation issue that regular corporations encounter.[14]

Regular corporations (Subchapter C) must first pay taxes on the profits of the firm, and then the income is taxed again as individual income when stockholders receive their dividends. This is commonly referred to as **double taxation**. In contrast, Subchapter S corporations are not taxed as regular corporations and thus do not pay the corporate tax. The opportunity to avoid the double corporate taxation was enough to persuade more than 1.6 million small U.S. businesses to operate as Subchapter S corporations.[15]

Not all businesses can become Subchapter S corporations. For a corporation to qualify for taxation under Subchapter S, it must meet the following requirements:

- The owners must be individuals or estates.
- All stockholders must be U.S. citizens or resident aliens.
- Only one type of stock is allowed.
- The owners must not be members of an affiliated group of corporations.
- The owners may not own 80 percent or more of the stock of another corporation.
- Complete agreement of all stockholders must be achieved to pass profits to the owners.
- No more than 25 percent of the gross income can come from rents, royalties, dividends, and interest.
- No more than 80 percent of gross income can come from outside the United States.

This type of corporation also allows small groups, such as production agriculturalists and small agribusinesses, to gain the advantages of incorporating without the disadvantage of double taxation. This type of corporation is very popular in the agriculture industry, and the number of Subchapter S corporations is increasing rapidly.

Subchapter T Corporations (Cooperatives). This is the third type of corporation. Because cooperatives are nonprofit and are very popular in the agricultural industry, they are discussed in detail later in this chapter.

14. Nickels, McHugh, and McHugh, *Understanding Business,* p. 143; Peter Weaver, "Heightened Scrutiny of S Corporations," *Nation's Business,* February 1995, p. 37.
15. Nickels, McHugh, and McHugh, *Understanding Business,* p. 143.

Corporations	
Advantages	**Disadvantages**
■ More money for investment	■ Initial cost
■ Limited liability—personal assets are protected	■ Paperwork
■ The right size to do needed things	■ Complicated to establish
■ Perpetual life	■ Complex organization
■ Ease of ownership change	■ Two tax returns
■ Ease of attracting talented employees	■ Double taxation
■ Separation of ownership from management	■ Size causes slow response time to market changes
■ Corporation is a legal entity	■ Difficulty of termination
■ Combined resources of shareholders	■ Possible conflict with the board of directors
■ Stable level of production	■ Must follow state laws
■ Owners (stockholders) do not have to devote time to the company to make money on their investment	■ Owners have limited control of business

Figure 6–5 An outstanding advantage of corporations is the ease of raising capital or money for expansion by issuing stock.

Advantages of Corporations

There are several advantages to the corporate form of business. Large corporations can issue more stock whenever they need to raise money to start or expand a business. Their shareholders assume only limited liability; that is, they cannot lose more money than they have invested. The corporation has the additional advantage of existing as a legal entity apart from its shareholders. This is important because the company does not dissolve if investors sell their shares of stock. They simply transfer ownership to another investor and business continues.[16] This ability to transfer stock is also advantageous in estate planning. Shareholders can give gifts of corporate stock to children (or others) before or at the time of death. In this way, large estate taxes may be avoided when the original owner dies.[17] See Figure 6–5 for a summary of the advantages of for-profit corporations.

Disadvantages of Corporations

Small agribusiness corporations may find it difficult to sell their stock. Such stocks usually represent a very small interest in a company that is not publicly traded. Also, in family-owned and small corporations, lenders may require personal guarantees from owners. This negates the limited financial liability afforded by the corporate structure. Forming a corporation is also a much costlier and more complicated process than forming most other business organizations. Filing fees, fees for drafting articles of incorporation, legal fees, and accounting fees are typical expenses for new corporations. In addition, corporations are burdened with complicated recordkeeping requirements and often with double taxation.

Ending a corporation may be complicated and expensive. For example, for tax purposes, land is valued at market value instead of its original value when a corporation is forced to dissolve. Because the corporation pays taxes on the difference between market value and original value, if the value of the land increases, a corporation will probably end up paying extra taxes. The other types of business structures do not face this problem. Refer to Figure 6–5 for other disadvantages of for-profit corporations.[18]

16. Brown and Clow, *Introduction to Business,* p. 89.
17. Birkenholz and Plain, *Agricultural Management and Economics,* p. II-5.
18. Ibid.

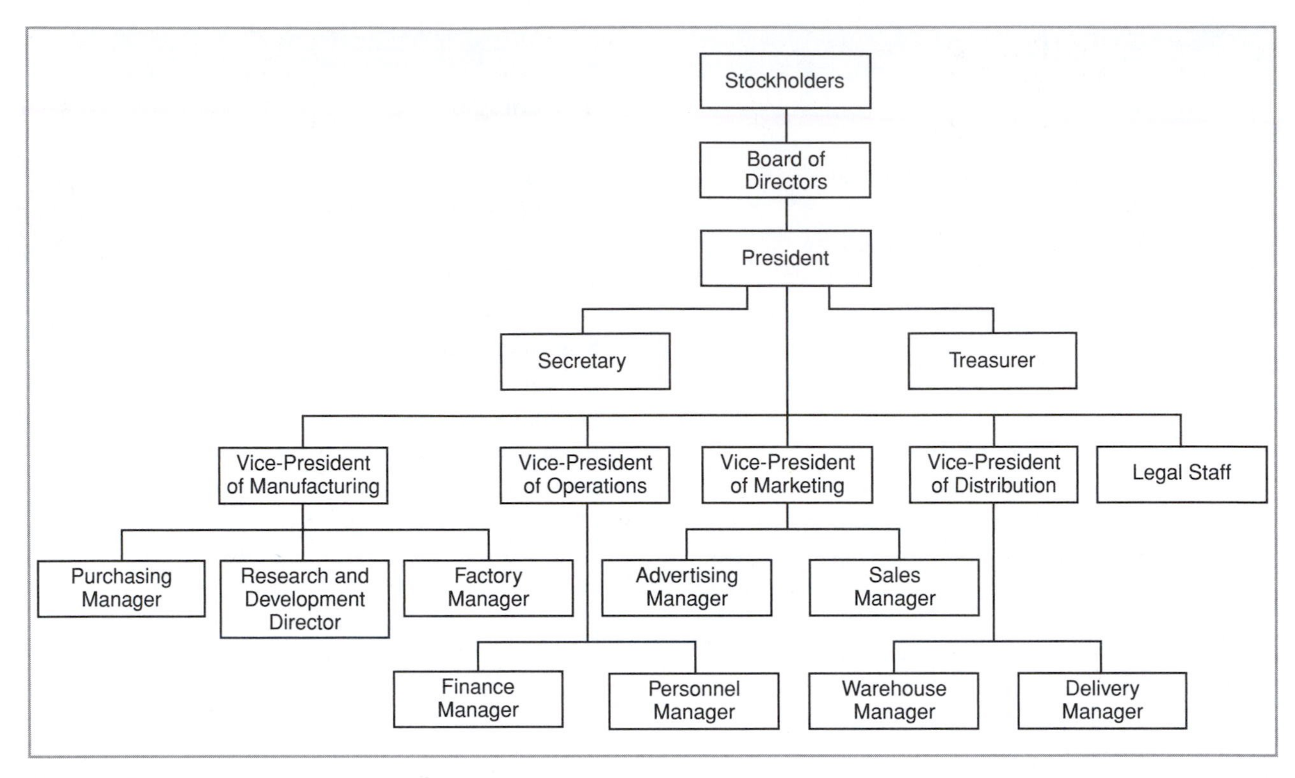

Figure 6–6 Many different structures may be used for a corporation. This schematic illustrates a typical corporate structure.

Establishing a Corporation

The process of forming a corporation varies somewhat from state to state. The articles of incorporation are usually filed with the secretary of state's office in the state in which the company incorporates. The articles contain:

- the name of the corporation
- the names of the people who incorporated it
- the purposes of the corporation
- the duration of the corporation (usually perpetual)
- the number of shares that can be issued, the voting rights attached to each type of stock, and any other rights of the shareholders
- the minimum capital of the corporation
- the address of the corporate office
- the name and address of the person responsible for legal service on the corporation
- the names and addresses of the first directors
- any other information the incorporators wish to make public[19]

Before a potential corporation can so much as open a bank account or hire employees, it needs a federal tax identification number. To apply for one, you must get an SS-4 form from the Internal Revenue Service (IRS).

In addition to its articles of incorporation, a corporation also has bylaws. These describe how the firm is to be operated, from both legal and managerial points of view. The bylaws include:

- how, when, and where shareholders' and directors' meetings are held, and how long directors are to serve
- specifics as to each director's authority
- the duties and responsibilities of officers and the length of their service
- how stock is issued
- other matters, including employment contracts[20]

A typical corporation will have a structure similar to that in Figure 6–6.

19. Stephen L. Nelson, "How Not to Incorporate," *Home Office Computing*, February 1994, pp. 28–30.

20. Ibid.

Limited Liability Companies

A limited liability company (LLC), like a corporation, is a legal entity that exists separate from its owners. Although a limited liability company is neither a partnership nor a corporation, it usually combines the corporate advantage of limited liability with the partnership advantage of pass-through (single) taxation.

A limited liability company is owned by its members. The members may directly manage the LLC, or they may appoint managers to do the direct, day-to-day management of the company for them.

The members may also assign duties amongst themselves as they see fit. Members may be appointed as company officers (such as president, vice president, or secretary) and take on the duties normally associated with those positions. In most states, the members can structure the company in any way they wish.[21]

Advantages of LLCs

The primary advantage of a limited liability company is that the liability of members is limited. Creditors of a sole proprietorship or partnership may seize the owners' or partners' personal assets to pay the business's debts. In contrast, members are not liable for the debts and obligations of the LLC unless they have given personal guarantees for those debts.

Additionally, a business formed as an LLC may avoid double taxation. The LLC's earnings may be treated like earnings from partnerships, sole proprietorships, and S corporations and "passed through" to members for treatment on their individual tax returns. In this way, the company income is not first subjected to a "company" tax. Also, the LLC structure is more flexible than the corporate form: for the most part, members can divide ownership and voting rights however they see fit and still have the LLC earnings qualify for pass-through taxation.[22]

Limited liability companies have all the flexibility of partnerships with the same kind of liability protection provided by corporations. It is relatively easy to switch from LLC status to C corporation status if you so desire (normally this is done so that money can be raised more easily).

Disadvantages of LLCs

The main disadvantage of the LLC business form is the work and expense involved in initial formation of the limited liability company. Most states also impose postformation recordkeeping requirements, which may be time-consuming and expensive to comply with.

Because limited liability companies have not existed in this country as long as corporations have, the laws regarding LLCs are still developing, particularly with regard to the rights of members. It is likely that laws resembling those regulating corporations will be developed in the future.[23]

Cooperatives

A cooperative is a corporation formed to provide goods and services to members either at cost or as near to cost as possible. Cooperatives (co-ops) are not formed to make profits, but to serve the people who own shares in the organization. Agribusiness cooperatives are very popular.

Kinds of Cooperatives

In the agricultural industry, there are three kinds of cooperatives: supply (purchasing) cooperatives, marketing cooperatives, and service cooperatives.

Supply (Purchasing) Cooperatives. **Supply cooperatives** or purchasing associations buy supplies, such as feed, seed, fertilizer, and fuel, in quantity for resale to their members (many of whom may be production agriculturalists). The big advantage is that by buying in large quantities, cooperative members are usually able to save money over what they would have paid individually.[24] In some cases, a supply cooperative manufactures its own supplies instead of buying from another company.

21. Milliken PLC, The LLC Link, "Limited Liability Company FAQ" (2006), http://www.4inc.com/llcfaq.htm.
22. Ibid.
23. Ibid.
24. Jasper S. Lee and Max L. Amberson, *Working in Agricultural Industry* (New York: Gregg Division/McGraw-Hill, 1979), pp. 95–97.

Marketing Cooperatives. For the most part, **marketing cooperatives** assist production agriculturalists in marketing their agricultural products by finding buyers who will pay the highest price. Some marketing cooperatives process agricultural products, such as milk and vegetables, and sell them directly to consumers and retailers. Examples of marketing cooperatives are fruit growers' cooperatives, such as Sunkist, and dairy marketing co-ops, such as Land O' Lakes and Mid-America Dairy.[25]

Service Cooperatives. Service cooperatives provide their members with a specific service, rather than a product, that members probably could not afford to obtain individually. Service cooperatives are not as numerous as marketing and supply cooperatives, but they provide valuable services to many farmers. Specific examples of service-type cooperatives include farm credit services, banks, rural credit unions, mutual irrigation cooperatives, dairy herd improvement associations, and artificial breeding cooperatives. Rural electric and telephone cooperatives once consisted primarily of farmers, but now they serve all rural residents, who are more nonfarmers than farmers in many areas. Therefore, these are now considered consumer cooperatives rather than agricultural cooperatives.

Statistics on Cooperatives

According to the National Cooperative Bank, there are more than 21,000 cooperatives nationwide, with both business and individual members.[26] A good example is the agricultural cooperative. Agricultural cooperatives have become a multibillion-dollar industry. For an idea of the size of the typical agricultural cooperative, consider Farmland Industries, Inc. In 1999, it was the country's largest agricultural cooperative, with annual revenues of $4.5 billion. Its 1,800 member cooperatives included 250,000 farmers in 19 states. Farmland Industries owns oil wells and refineries, fertilizer plants, feed mills, and manufacturing plants that produce everything from grease and paint to steel buildings. It also offers insurance, financial and technical services, and access to a network of warehouses.[27]

Cooperatives and Membership

Some co-ops serve the general public; others serve members only. Even with those that serve the general public, the major emphasis is still on the members. Voting stock and investment stock are separate in a cooperative, and only the former, also called **common stock**, gives a person the right to vote on business matters of the co-op. Investment stock, also called **preferred stock**, only gives a person the opportunity to invest in the business and, it is hoped, to receive a reasonable return on investment. In some cases, preferred stock is available only to holders of common stock, but some co-ops sell preferred stock to anyone who wishes to invest. In any event, members of the cooperative are encouraged to purchase preferred stock, whereas nonmembers are discouraged from doing so. The law currently limits the return on cooperative investment to an established percentage above the discount rate (the interest rate that the federal government charges for loans to member banks), regardless of how successful the business is in a given year.

Cooperatives and Control

Most cooperatives use a democratic system of control whereby each member has one vote only, regardless of how much business the member does with the co-op or how much investment stock he or she holds. Critics often contend that co-ops are run by a few "elite" members. In some cases this is true, but only because many members fail to exercise their right to vote during the annual stockholders' meetings.

Distinguishing Characteristics of Cooperatives

Three major distinguishing characteristics of cooperatives are:

- service at cost; any excess earnings are returned to patrons in the form of patronage dividends
- democratic control (one member, one vote)
- limited returns on investment

25. Jorean Huber, "All Together Now," *Entrepreneur,* April 1994, p. 14.
26. http://dev.ncb.coop//uploadedFiles/Co-op%20100(2).pdf (2006).
27. Nickels, McHugh, and McHugh, *Understanding Business,* p. 156.

Cooperatives	
Advantages	**Disadvantages**
■ Owners have limited liability	■ More legal formalities than sole proprietorship or partnership
■ Continues at death of shareholders	■ Members-shareholders have limited control over business
■ Benefits go to members	■ Expensive to form, maintain, and dissolve
■ Economic benefits for members	■ A few persons may gain excessive power
■ Members-shareholders share in direction of business	■ Restrictive charter requirements
■ Broad capital base	■ Lack of member understanding about cooperative (co-op) structure
■ Special tax advantages	■ Lack of member participation
■ Legal entity	■ Business community resentment against cooperatives
■ Special antitrust exemptions	■ Divided management authority

Figure 6–7 Agribusiness cooperatives are very popular. They are formed to provide goods and services to members at a reduced cost.

Cooperatives and Taxes

Many cooperatives are also exempt from paying a corporate tax, provided they meet minimum cooperative requirements. At least 50 percent of a co-op's business must be with members, and all profits must be returned to patrons in the form of patronage dividends. However, some cooperatives choose to pay the corporate tax and not be restricted by requirements imposed on tax-exempt organizations.

Advantages of Cooperatives

At their best, cooperatives provide distinct advantages to their members. Because the members own and control the cooperative, cooperatives tend to be member-friendly. For example, members' financial risk is limited only to the amount they have invested. Operating expenses of the cooperative are kept low to hold down member costs. Also, as previously mentioned, because a portion of the co-op's profits is returned to members (after certain requirements have been met), income taxes paid by cooperatives tend to be lower.[28] Refer to Figure 6–7 for other advantages of cooperatives.

Disadvantages of Cooperatives

There are some disadvantages for members of a cooperative. The cooperative withholds a portion of any patronage dividends distributed; therefore, members rarely receive the their full share of the profits. In addition, members are responsible for paying all of the taxes on the patronage dividends they receive. If the cooperative suffers business losses, either its customers pay higher prices for goods and services, or patronage dividends are withheld to cover the losses.[29] Refer to Figure 6–7 for other disadvantages of such cooperatives.

Franchises

Many of us eat often at fast-food restaurants. Therefore, we are already familiar with some of the most popular franchises, such as McDonald's, Wendy's, and Domino's. A *franchise* is a contract in which a **franchisor** sells to another business the right to use its name and sell its products. The **franchisee** (person purchasing the franchise) buys a system of operation that has proven successful.

28. Birkenholz and Plain, *Agricultural Management and Economics*, p. II-5.

29. Ibid.

Franchises	
Advantages	**Disadvantages**
■ Nationally recognized name and reputation	■ Large startup and franchise costs
■ Help with finding a good location	■ Additional cost may be charged for marketing
■ Management system with successful track record	■ A monthly percent of gross sales may go to the parent company
■ Successful methods for inventory and operations	■ Possible competition from other, nearby franchises
■ Financial advice and assistance	■ Little to no freedom to select decor or other design features
■ Training for owners and staff	■ Management regulations
■ National advertising and promotional assistance	■ Many rules and regulations to follow
■ Periodic management counseling	■ *Coattail* effect if other franchises fail nationwide
■ Proven record of success	■ Restrictions on selling
■ Personal ownership	

Figure 6–8 Franchising sales will soon account for nearly 50 percent of all retail sales in the United States.

There are about 600,000 franchised outlets in the United States.[30] McDonald's has more than 30,000 restaurants in more than 100 different countries; two-thirds of its new stores are outside the United States.[31] Government figures show that in 2004, franchising sales accounted for nearly 33 percent of all retail sales in the United States.[32]

Characteristics of Franchises

The **parent company** (franchisor) **prepackages** all the business planning. Prepackaging generally includes management training and assistance with advertising, selling, and day-to-day operations. In return, the franchisee agrees to run the business in a certain way.[33] Some people who want to own their own business are more comfortable entering into a franchise agreement and joining a company with a proven track record, rather than starting a completely new business. The business form of a franchise can be a sole proprietorship, partnership, or corporation.[34]

30. "How to See beyond the Burgers in Franchising," *The Globe and Mail*—Small Business, available at http://www.theglobeandmail.com/series/business/sb2/sbw/sbw_franchising.html.
31. McDonald's Corporation, *2006 Annual Report*; available at http://www.mcdonalds.com.
32. "How to See beyond the Burgers in Franchising."
33. Brown and Clow, *Introduction to Business,* p. 90.
34. Nickels, McHugh, and McHugh, *Understanding Business,* p. 149.

Advantages of Franchises

The most important advantage of a franchise is the support from the parent company, both during and after startup. This support comes in many forms. Training programs may be provided to teach the business and company operations. Help in choosing a location for the business and in obtaining financing may be available. Franchise owners benefit from the name recognition and advertising power of a nationally known franchise. Franchising also provides store owners with consistent, convenient, and cost-effective products and services. All of this support, experience, and technical expertise from the parent company increases the chance of success for the franchise owner. Refer to Figure 6–8 for other advantages of franchises.

Disadvantages of Franchises

If franchises are so great, why aren't all businesses franchises? Unfortunately, franchises do have drawbacks. First, it takes a large amount of money to purchase most franchises. Second, the owner must share sales or pay a predetermined yearly fee to the parent company. Also, the parent company places constraints on how the owner can manage the franchise. Refer to Figure 6–8 for other disadvantages of franchises. Refer to the Career Options section for a variety of careers in different types of agribusinesses.

Conclusion

There are several different types of legal structures for an agribusiness. You can start your own sole proprietorship, partnership, corporation, or cooperative, or you can buy into a franchise. Each business form has its own characteristics, advantages, and disadvantages. Before deciding on the legal structure that is appropriate for your business venture, carefully evaluate each alternative and its risks. Remember, everyone has to start somewhere. Businesses like J.C. Penney, John Deere, Levi Strauss, and (Henry) Ford are named after the real people who founded them. They started small, accumulated capital, grew, and became successful. Remember, you have to go out on a limb to get to the fruit.

Summary

The major types of agricultural businesses are single (sole) proprietorships, partnerships, and corporations. Although cooperatives are really a form of corporation, they are sometimes considered a fourth type of business. Another special form of business organization is the franchise, which is the fifth type of agricultural business.

No single type of legal structure is best for all agricultural businesses. Therefore, when developing a new agribusiness, consider the advantages and disadvantages of all structures. Although corporations constitute only 17 percent of U.S. business establishments, they do 87 percent of the sales volume.

The single proprietorship is the major type of legal structure in the agricultural industry, as well as in other major industries. It is the simplest type of business and the easiest to organize. The major disadvantage is that creditors can take the owner's personal assets as well as the business assets in payment of business debts.

Another type of agricultural business is a partnership, which is a business association of two or more persons. There are three types of partnerships: general, limited, and limited liability. General partners own and manage a business together, whereas a limited partner invests in the business

Career Options

Agribusiness Owner/Manager, Agribusiness Partnership, Agribusiness Corporation, Agribusiness Cooperative, Agribusiness Franchise

The manager of an agribusiness, whether it is a sole proprietorship, partnership, corporation, cooperative, or franchise, provides expert advice, supplies, or services to the agricultural industry. The agribusiness manager needs a good educational background in business and accounting procedures. A college degree in an agricultural discipline is often required. Education for agribusiness managers should begin at the high school level or earlier. The agribusiness is expected to have first-hand experience and detailed knowledge of procedures and techniques. Most agribusiness managers work their way up in the management system of the company or cooperative with which they are employed. As their management skills improve, they are given added responsibilities.

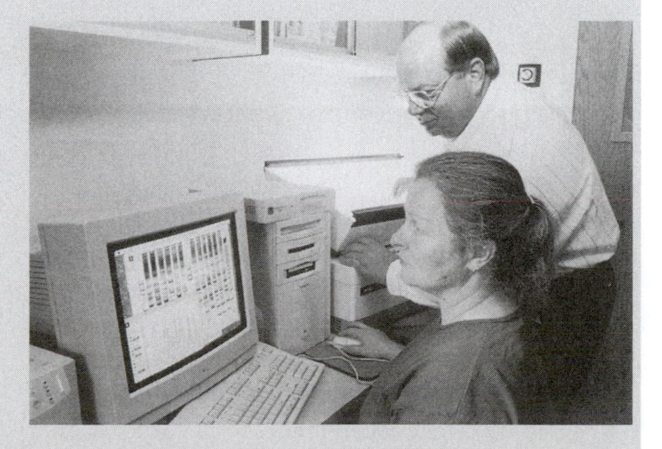

A Farm Credit officer should always thoroughly assess the potential borrower's financial position before making a loan. (Courtesy of USDA)

but may not participate in the management of the business. The major advantages of a partnership include greater resources and specialization of talent. The greatest disadvantage is its unlimited liability, though the limited liability partnership form helps partners avoid many of these problems.

Corporations are another way of doing business. There are three types of corporations: Subchapter C is a regular corporation, which sells stock to investors; Subchapter S is basically for small businesses or a family; and Subchapter T is for cooperatives. The major advantage of corporations is the ability to raise capital or money for expansion by issuing stock. The major disadvantage is that they are complicated and costly to organize. A limited liability company is not a partnership or a corporation, but it combines the corporate advantage of limited liability with the partnership advantage of pass-through (single) taxation.

Cooperatives are corporations formed to provide goods and services to members at cost or as near to cost as possible. There are three types of cooperatives in the agricultural industry: supply (purchasing) groups, which act as purchasing associations to buy supplies in bulk, at reduced rates; service cooperatives, which provide production agriculturalist members with a specific service rather than a product; and marketing cooperatives, which assist production agriculturalists in marketing and selling their agricultural products. The major advantage of cooperatives lies in being able to purchase goods and services at reduced cost. The major disadvantage is that members or shareholders will not receive the full amount of their shares of the profit (dividends), because a portion is withheld by the cooperative.

A franchise is a special business type in which a franchisor grants a franchisee the right to use its name and sell its products. The parent company prepackages all the business planning, management training, and assistance with advertising, selling, and day-to-day operations. The biggest advantage of a franchise is that the parent company helps new owners start their businesses. The biggest disadvantage, with most franchises, is that the initial cost to buy in is very high.

End of Chapter Activities

REVIEW QUESTIONS

1. Define the Terms to Know.
2. What are the three major types of agricultural business (business organizations)?
3. What are the three major types of corporations?
4. What are 10 advantages of single (sole) proprietorships?
5. What are 10 disadvantages of single (sole) proprietorships?
6. What are five recognizable characteristics of general partnerships?
7. What are four characteristics of limited partnerships?
8. What are nine advantages of partnerships?
9. What are nine disadvantages of partnerships?
10. Name five corporations.
11. Name five characteristics of corporations.
12. Name five agribusiness corporations.
13. What are eight requirements for a corporation to qualify for taxation under Subchapter S (small business or family corporations)?
14. What are 10 advantages of for-profit corporations?
15. What are 10 disadvantages of nonprofit corporations?
16. What are three advantages of forming a limited liability company?

17. What are two disadvantages of forming a limited liability company?
18. What are three major distinguishing characteristics of cooperatives?
19. What are nine advantages of cooperatives?
20. What are nine disadvantages of cooperatives?
21. What are five advantages of franchises?
22. What are five disadvantages of franchises?

FILL IN THE BLANK

1. Corporations constitute 17 percent of all business, but they do __________ percent of the sales volume.
2. A _________ is an organization owned by many people but treated by law as though it were one person.
3. There are ________ small businesses in the United States operating as Subchapter S corporations.
4. The articles of incorporation are usually filed with the ________ office in the state in which the company incorporates.
5. A _________ is a corporation formed to provide goods and services to members at cost or as near to cost as possible.
6. In the agricultural industry there are three kinds of cooperatives: _________ cooperatives, __________ cooperatives, and _________ cooperatives.
7. An example of a market cooperative for fruit growers is _________.
8. An example of a market cooperative for dairies is _________.
9. Nationwide, 2 million businesses belong to _________ cooperatives.
10. Farmland Industries has 1,800 member cooperatives, which include _________ farmers in 19 states.
11. Most cooperatives use a _________ system of control whereby each member has one vote only.
12. There are more than ________ franchised outlets in the United States.

MATCHING

a. corporation
b. Subchapter C
c. marketing cooperative
d. Farmland Industries, Inc.
e. single (sole) proprietorship
f. Subchapter T
g. agricultural cooperatives
h. franchise
i. Subchapter S
j. partnership

______ 1. 2.8 million of this type of business

______ 2. 1.4 million of this type of business

______ 3. 12 million of this type of business

______ 4. John Deere

______ 5. regular corporation that sells stock to investors

______ 6. small business or family corporations

______ 7. cooperatives

______ 8. act mostly as purchasing associations for such things as feed, seed, and fertilizer

______ 9. country's largest agricultural cooperative in 1999

______ 10. assist the production agriculturalist in selling agricultural products by finding buyers who are willing to pay the highest price

______ 11. buys the right from a parent company to use its name and sell its products

ACTIVITIES

1. Jake has his own business, and Maria is interested in becoming a part of it. They need your help in avoiding problems when forming their partnership. In the following table, list any problems they need to avoid as they form their partnership and list any steps you think they could take to avoid problems.

Problems to Avoid	Steps to Avoid Problems

2. Draw the following table on a separate sheet of paper and fill in the blanks to describe the characteristics of each type of business.

Type	Ownership	Startup Costs	Taxes
Single (sole) proprietorship			
Partnership			
Corporation			
Cooperative			
Franchise			

Type	Liability	Responsibility for Decisions	Major Advantages and Disadvantages
Single (sole) proprietorship			
Partnership			
Corporation			
Cooperative			
Franchise			

3. Answer each of the following:
   a. Have you ever considered starting your own business?
   b. What opportunities seem attractive?

   c. Write down the name of a friend (or friends) whom you might want as a partner.
   d. List all the financial resources and personal skills you need to start the business.
   e. Make a separate list of the capital and personal skills that you and your friend(s) might bring to your new venture.
   f. What capital and personal skills do you need but neither of you have?
4. Prepare a sample written partnership agreement. Refer to section in this chapter on partnerships, which lists everything you need to include.
5. Prepare sample articles of incorporation. Refer to the section in this chapter on corporations, which lists everything you need to include.
6. Prepare a sample set of bylaws. Refer to the section in this chapter on corporations, which lists everything you need to include.
7. Secure either a sample written partnership agreement, sample articles of incorporation, or sample set of bylaws; sources could include the library, a local business, or any other source. Share the document with the class.
8. Identify the franchises in your community. If the owner of one of the franchises is available, ask him or her to identify three advantages and three disadvantages of franchises. Share your results with the class.
9. Identify the agribusinesses in your community. For each, list whether it appears to be a single (sole) proprietorship, partnership, corporation, cooperative, or franchise. Also, identify whether it appears to use more than one legal structure.

Chapter 7

Financing the Agribusiness

Objectives

After completing this chapter, the student should be able to:

- Discuss the importance of farm credit.
- Explain three fundamentals of credit.
- List eight rational credit principles needed for effective decisionmaking.
- Describe three areas for which credit is needed.
- Differentiate the three lengths of financing terms.
- Discuss the three types of loans.
- Explain the components of a credit profile.
- Compute interest.
- List the agricultural credit sources for real estate and non–real estate loans.

Terms to Know

actuarial interest rate
add-on interest
amortize
annual percentage rate (APR)
appreciate
buyer's fever
collateral
contractual interest rate
debt-equity ratio
depreciable
discount (prime) rate
equilibrium price
equity capital
farm assets
financial assets
fixed expenses
foreclosure
interest
interest rates
lien
long-term credit
operating expenses
principal
secured
simple interest
speculative investments
startup expenses

Introduction

Decisions about credit often are the most important judgments that people in the agricultural industry must make. These decisions often determine whether individuals operating in the agribusiness input, production, or agribusiness output sectors will succeed in making a profit. Remember the old saying, "It takes money to make money." A business must have sufficient financial resources if it is to show a profit. Money is used in every area of the agricultural industry, including for land, buildings, equipment, livestock, crops, and operating expenses.

Questions about whether to use credit for specific purposes must be examined carefully. Refer to Figure 7-1. The ability to make sound business judgments comes from education, careful thought, and actual business experience.[1]

Figure 7-1 Credit decisions are some of the most important decisions a producer can make. Here a specialist (right) discusses effective farm management with a dairy cattle producer. (Courtesy of USDA, #NRCSCA 02053)

1. Jasper S. Lee and Max L. Amberson, *Working in Agricultural Industry* (New York: Gregg Division/McGraw-Hill, 1979), pp. 107–108.

Importance of Farm Credit

In the agricultural industry, credit is needed to overcome a shortage of **equity capital**. Restricted capital, changing interest rates, and lack of credit information are major problems facing many production agriculturalists. Credit allows production agriculturalists to:

- increase production
- improve the quality of what is produced
- revise operations to make them more profitable[2]

The money that is borrowed must, at the least, generate enough additional income to pay for the cost of the borrowed money (the **interest**) and to ensure that the **principal** is repaid according to the specified terms of the loan.[3]

Change in Credit Needs

One of the major changes in production agriculture during recent years has been the substitution of capital for labor. Around 1900, capital (including land) contributed about 25 percent to the farm production process and labor, 75 percent. Currently, however, about 90 percent of the production of food and fiber is attributable to capital, while labor accounts for only 10 percent. Although the rate of substitution has been reduced since 1990, the trend still continues. The total volume of capital needed has become so large that farm credit agencies now constitute a major supply of capital. As a result of increased borrowing and rising interest rates, interest payments have become one of the fastest growing costs of production agriculturalists during the past two decades.

Magnitude of Agricultural Credit

Agriculture and its products and services make up about one-fourth of the U.S. gross domestic product. The value of all **farm assets** has increased over time, from about $839 billion in 1990, to $957 billion in 1995, to more than $1.8 trillion in 2005. In 2005, real estate (land and buildings) accounted for $1.52 trillion; non-real estate assets (equipment, livestock, and crops) totaled $210

2. Randall D. Little, *Economics: Applications to Agriculture and Agribusiness*, 4th ed. (Danville, IL: Interstate Publishers, 1997), p. 131.
3. Ibid.

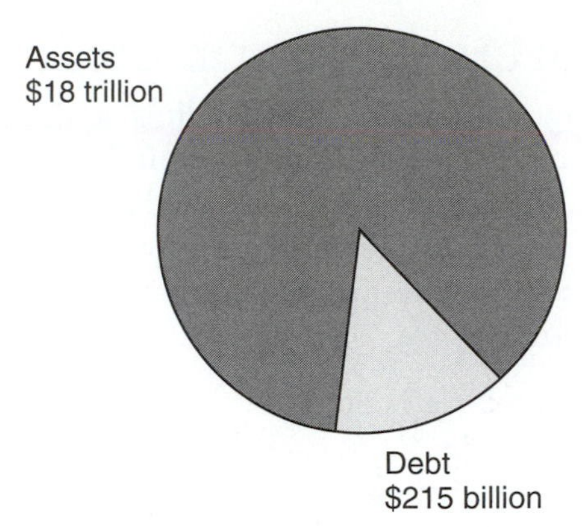

Figure 7–2 Total farm debt is only about 12 percent of total farm assets. (2005)

billion; and **financial assets** amounted to $67 billion. Total farm debt was about $215 billion, or 12 percent of assets.[4] Most of this debt is split between real estate and non-real estate loans. Refer to Figure 7–2.

Successful Farm Credit Objectives

To make appropriate financial decisions for an organization, one must recognize and understand factors such as capital, credit, and debt, and how these apply to the agribusiness. These aspects must be addressed when considering successful credit objectives. Specific objectives of successful farm financial management should include:

- seeing continuous growth in net worth (equity capital)
- earning a high return from farm investments
- being able to secure credit when needed
- reducing the interest cost of borrowed debt capital
- implementing risk management strategies to reduce potential financial losses
- maintaining both a strong liquidity and a solvency financial position for the farm business operation
- demonstrating a strong debt repayment capacity by the farming operation[5]

Achieving these objectives through effective financial management of an agribusiness is important for achieving the bottom line: maintaining and improving profits.

Three Fundamentals of Credit

Education has its traditional three "Rs." In credit, there are also three "Rs": returns, repayment, and risk.

Returns

The main reason for borrowing money is to increase net returns and make a profit. Will the net returns or profit be greater with the use of added assets (in the form of credit) than if credit were not used? A production agriculturalist must carefully choose among the best alternatives when considering credit. Money should not be borrowed based on quick, thoughtless decisions or **buyer's fever**. It is important to analyze each investment based on the ability to pay back the borrowed money.[6]

Repayment Ability

Agricultural lenders expect their money to be repaid in full plus interest. Production agriculturalists and others in the agricultural industry must determine their ability to repay loans, just as the lenders must. Because borrowers usually pledge their farms or agribusinesses as collateral for their loans, inability to repay may lead to **foreclosure**. Therefore, priority should always be given to loans that have earning capacity. For example, borrowing for dairy cows would take priority over borrowing for automatic feeders to replace hand feeding. The automatic feeder does not have any repayment ability, because it generates no direct income, unlike the sale of milk from the dairy cows. Repayment ability is also tied to sacrifices in standard of living in favor of additional investments. For example, it may be better to continue to drive a good used pickup and apply your funds to productive investments, such as the dairy cows, rather than to buy a new truck.[7]

Risk

Borrowers with strong assets can take on more risk than those with few assets. It is only reasonable that lenders tend to favor individuals in

4. United States Department of Agriculture Economic Research Service, (n.d.). Briefing Rooms: *Farm Income and Costs: Farm Sector Income Forecast*; available at http://www.ers.usda.gov/briefing/farmincome/data/Bs_t5.htm (accessed January 23, 2007).
5. Ronald Hanson, *Agricultural Economics 452 Lecture Study Notes* (Lincoln: University of Nebraska-Lincoln, 1999).
6. Little, *Economics,* p. 146.
7. Ibid., p. 146.

the agricultural industry who have enough stability to absorb a potential loss. Borrowers do not like **speculative investments**, because of their high risks. However, they do like to make loans to enterprises that can produce enough profit to ensure income stability or that **appreciate** in value.[8]

Rational Credit Principles

Agribusiness profitability is based on many different components, not the least of which is effective use of farm credit. The right combination of land, labor, equity, management, and credit is a valuable foundation for decisionmaking within the agribusiness. Applying the following principles wisely can lead to success and profitability for today's production agriculturalist:

- *Use credit for productive purposes*—purposes that increase income.
- *Limit borrowing on unfamiliar enterprises.* The ability to manage an enterprise should be tested before the enterprise is expanded through the use of borrowed funds, whenever possible.
- *Use credit where it will generate the highest return within reasonable risk limits.* Limited funds should be allocated first to the enterprise(s) expected to give the highest returns.
- *Know your own situation.* Keep records, evaluate performance, and regularly review resources.
- *Keep debt in line with income and repayment capacity.* Cash income must be adequate to meet operating expenses, debt retirement, and expansion; within family businesses, one must also meet living expenses, pay taxes, and save retirement funds.
- *Shop for loans and select a dependable lender.* Loan terms and conditions vary by lender. Do some research to find out what rates and terms are available.
- *Be businesslike, honest, and fair in credit dealings.* Be prepared to present your credit case.[9]

All these principles work together to provide a useful outline for production agriculturalists to use when making decisions regarding credit for their agribusinesses.

Three Areas of Credit Needs

In the agricultural industry, financing is needed in three areas: fixed expenses, operating expenses, and startup expenses.

Fixed Expenses

Fixed expenses are items that can be used over and over for a long period of time, incurring the same price (expense) each year. Examples of fixed expenses include land, buildings, machinery and equipment, tools, and fixtures. Usually, a large amount of money is borrowed for fixed expenses, but the repayment amount is fixed monthly or yearly for several years.[10] A brief discussion of the major fixed expenses follows.

Land. Land is a large, one-time expense item. Land is used not only by production agriculturalists, but also by those in the agribusiness input, output, and service sectors. Improvements to land can also be expensive. Examples of improvements include fencing, clearing land, leveling and ditching farmland, and preparing building sites.

Buildings. Buildings constitute a one-time, fixed expense that may be used for 25 or more years. Examples include dairy barns, hay sheds, grain elevators, livestock houses, and various other buildings for agribusiness.

Machinery and Equipment. Machinery and equipment usually have a shorter useful life than that of buildings. The useful life of a piece of machinery or equipment depends on how often it is used, care and maintenance, quality, and how complicated it is.

Fixtures and Tools. Cash registers, shelves, and refrigerators are some of the fixtures used as equipment in the agricultural industry. Tools include such things as wrenches, hammers, shovels, thermometers, and so on.

8. Ibid., p. 147.
9. *Agricultural Business Management*, (Stillwater, OK: Curriculam Instrutional Materials Center, Department of Carrer and Technology Education, 1998), Unit 8, pp.5-6.
10. Lee and Amberson, *Working in the Agricultural Industry*, pp. 108–109.

Figure 7–3 For production agriculturalists, operating expenses are usually figured on a yearly or seasonal basis. Operating expenses can fluctuate according to seasonal conditions. (Courtesy of USDA, #NRCSWY 02022)

Operating Expenses

The term **operating expenses** covers everything needed to run a farm, ranch, or agribusiness. The amount of money needed depends on the size of the operation. For production agriculturalists, operating expenses are usually figured on a yearly or seasonal basis. Refer to Figure 7–3. For example, a production agriculturalist who raises wheat would need money to purchase seed, fertilizer, and chemicals and to pay workers. In the agribusiness input, output, and service sectors, operating expenses would include communication, transportation, utilities and fuel, taxes, insurance, and advertising.

Startup Expenses

Startup expenses are payable before the business begins operation. Examples of startup expenses include attorneys' fees, incorporation expenses, and costs for the development of a site or product.[11] Startup expenses do not include the initial cost of land and other fixed items.

Length of Financing

Lending (financing, or credit) is important to any country's economic well-being. It stimulates economic activity by providing a means of purchasing that which would otherwise be impossible to obtain. In reality, a loan is a contract between the borrower and the lender. Loans usually fall into three borrowing time frames: short-term, intermediate, and long term. However, there are different types of loans within each category.

Short-Term Loans

Short-term loans are distinctive in that their terms are normally one year or less. The main use of the short-term loan is to finance operating inputs. Banks, individuals, merchants, and farm credit services are among the suppliers of short-term loans. Short-term operating loans are probably the most common in this category. They usually range anywhere from one month to one year in length. A typical operating loan is for six months, with a single payment retiring the loan amount at the end of the period.

In business applications, short-term or operating credit assists in purchasing items such as fuel, fertilizer, chemicals, and seed, and in meeting maintenance expenses. Short-term credit may be the most important type for the survival of an agribusiness, because it finances the everyday operations of the firm, which generate the cash flow in the business. To maintain good credit, these loans should be repaid upon receipt of money at harvest or auction time.[12]

Intermediate-Term Loans

Intermediate-term loans vary in length from 1 to 10 years. They finance assets that may be **depreciable** over their expected lives. Farm machinery and equipment, breeding livestock, irrigation systems, and any modernization of farm facilities are examples of these assets.

As technology increases in sophistication and complexity, the importance of intermediate-term

11. Ibid.

12. *Agribusiness Management and Marketing* (8720-B) (College Station, TX: Instructional Materials Service, 1988).

loans increases as well. Those in the agricultural industry must stay up with technology to remain competitive. Many of them have to borrow money to be able to purchase the newest technology. Because most of the technology takes the form of depreciable assets, intermediate-term loans assist in these purchases.

Commercial banks are responsible for most of the outstanding intermediate-term debt, followed by the Farm Credit Agency. Banks make both commercial and consumer loans on depreciable assets. They offer different types of loans according to the description of the asset, the borrower's needs, and loss security. They will want to see financial records and past credit history when considering a loan application.

Having good credit ratings and relationships with lending institutions will lower the cost of buying intermediate-term assets. A lender may require **collateral** when considering whether to grant a loan. A **lien** is the lender's right to take possession of the asset acting as collateral if the borrower fails to repay the loan.[13]

Long-Term Credit

Loans that extend over 10 years are possible with **long-term credit**. Purchases of land, buildings, and housing create the need for this type of credit. Typically, the credit instrument used in long-term financing is a mortgage. Businesses and individuals both use mortgages for long-term financing needs.

Long-term credit is especially important when starting an agribusiness. The manager will have to determine how much land and what types of buildings are suitable to the scale of operations. The startup of the agribusiness is, in effect, dependent on whether a lender is willing to extend this credit.

The growth of the agribusiness also depends on the ability to obtain long-term credit. As the agribusiness expands, it will need to purchase more land and storage space. Maintaining a good long-term credit source will lower the cost of borrowing in the expansion process. Repayment of loans of this type comes from funds left over after deducting all other expenses for the year.[14] Net income or retained earnings are examples of sources of repayment for long-term loans. Sources of loans for long-term credit include Farm Credit Services,

13. Ibid.
14. Ibid.

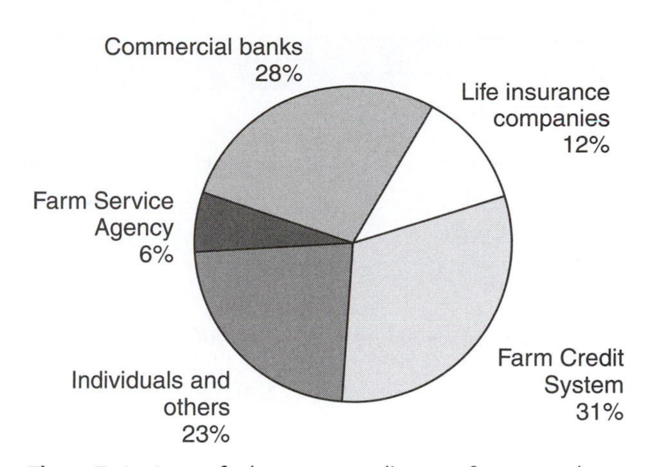

Figure 7–4 Loans for long-term credit come from several sources. Long-term credit is needed for land purchases, buildings, and housing.

Farm Service Agency (formerly FmHA), life insurance companies, commercial banks, and individuals. Refer to Figure 7–4.

Types of Loans

As mentioned in the previous section, loans generally have one of three time frames: short-term, intermediate, or long-term. However, different types of loans are found within each of these categories. The type of financing appropriate to each operation is determined by the needs of the agribusiness and what type of financing is desired.[15]

Line of Credit

A line of credit allows the borrower to acquire funding up to a maximum amount. This type of loan is generally used to buy production inputs such as fertilizer, feed, or feeder calves, and must be repaid completely within one year.

Revolving Line of Credit

A revolving line of credit again allows a producer to borrow up to a specified limit. The amount of a revolving line of credit loan fluctuates with seasonal credit requirements, and the loan need not be completely paid off as long as adequate collateral is available to secure the loan. These types of loans are generally more expensive than other types, but they are also more flexible. In addition, yearly renewals and increases are also available.

15. *Agricultural Business Management*, Unit 8, p. 7.

Debt-Equity Ratio

A. Asset Credit

Source of Credit	Asset Value (A)	Debt Outstanding (D)	Equity in Assets E=A-D	Max. D/E	Credit Capacity (E)(D/E)
Real Estate	$50,000	0	$50,000	2	$100,000
Non-Real Estate Assets	30,000	0	30,000	1	30,000
Total	$80,000	0	$80,000		$130,000

B. Income Credit

	Gross Value per Unit	Credit Rate	Credit per Unit	No. of Units	Credit Capacity
Crops	$200/acre	75%	$150	150	$22,500
Cattle	$250/head	85%	$213	300	$63,900
Total					$86,400

Figure 7–5 Examining debt-equity ratio is one way in which a lender may evaluate a borrower's credit capacity.

Term Loan

A *term loan* is a loan with a specified amount loaned for a specific amount of time. These loans are paid off with a single payment, or scheduled payments consisting of principal and interest. Detailed loan agreements generally accompany these types of loans. Term loans are often used for long-term growth, such as farmland improvements or new equipment, and are financed according to the expected rate at which the borrowed funds will generate cash.

Components of a Credit Profile

When reviewing a loan application, lenders will require a credit profile. Lenders want positive answers to the following questions—and more—before they will extend credit:

- **Personal characteristics.** What is the borrower's reputation for honesty, integrity, and judgment, as a person and as a manager?
- **Management ability.** Is the applicant a capable decisionmaker, with the appropriate experience, education, and background?
- **Financial position.** Does the applicant provide accurate and sufficient information with the financial statements? Do the firm's balance sheet, income statement, and cash-flow statement show loan repayment ability?
- **Loan purposes.** Why does the applicant want the loan? Does the length of the loan match the life of the asset?
- **Loan security.** Do *existing* assets amount to sufficient collateral?[16]

To illustrate one way a lender may evaluate a borrower's credit capacity, consider the following example. Assume that José's real estate lender will extend credit up to a **debt-equity ratio** of 2. In other words, José can have two times as much debt as equity in the business. The non-real estate lender allows a debt-equity (D/E) ratio of 1. José can obtain additional operating credit of up to 75 percent of expected gross crop production and 85 percent of expected gross cattle sales. Figure 7–5 shows that José's business has a total credit capacity of $216,400. Refer to Figure 7–6 for an example of a loan application.

16. *Agribusiness Management and Marketing.*

Form Approved - OMB No. 0560-0167

FSA 410-1
(03-31-97)

U.S. DEPARTMENT OF AGRICULTURE
Farm Service Agency

REQUEST FOR DIRECT LOAN ASSISTANCE

Federal Agencies may not conduct or sponsor, and a person is not required to respond to, a collection of information request unless it displays a currently valid OMB control number. Public reporting burden for this collection of information is estimated to average 60 minutes per response, including the time for reviewing instructions, searching existing data sources, gathering and maintaining the data needed, and completing and reviewing the collection of information. Send comments regarding this burden estimate or any other aspect of this collection of information, including suggestions for reducing this burden, to Department of Agriculture, Clearance Officer, OIRM (OMB No. 0560-0167), Stop 7630, Washington, D.C. 20250-7630. **RETURN THIS COMPLETED FORM TO YOUR FSA COUNTY OFFICE.**

INSTRUCTIONS TO APPLICANT: (For individuals, partnerships, or joint operations, show names, and trade names if any. Business entity applicants must provide additional information listed in Item 31. A husband and wife who want to apply for a loan together will be considered a joint operation. Either a husband or wife can apply as an individual.)

1. APPLICANT'S NAME
2. SPOUSE'S NAME
3. APPLICANT'S TELEPHONE NO.
4. APPLICANT'S ADDRESS
5. APPLICANT'S SOCIAL SECURITY NO. OR TAX IDENTIFICATION NO.
6. APPLICANT'S BIRTH DATE
7. SPOUSE'S BIRTH DATE
8. SPOUSE'S SOCIAL SECURITY NUMBER
9. TOTAL NUMBER OF HOUSEHOLD MEMBERS
10. TYPE OF OPERATION: ☐ INDIVIDUAL ☐ PARTNERSHIP ☐ JOINT OPERATION ☐ CORPORATION ☐ COOPERATIVE — ACRES OWNED ____ ACRES RENTED ____
11. MARITAL STATUS ☐ MARRIED ☐ SEPARATED ☐ UNMARRIED (INCLUDING SINGLE, DIVORCED, AND WIDOWED)

		YES	NO
12.	Have you or any member of your organization ever been in receivership, been discharged in bankruptcy, or filed a petition for reorganization in bankruptcy? If YES, please provide details in Item 31.		
13.	Are you, or any member of your organization, or the organization itself, involved in any pending litigation? If YES, provide details in Item 31.		
14.	Do you now, or have you ever, conducted business under any other name? If YES, give name in Item 31.		
15a.	Have you or any member of your organization ever obtained a direct or guaranteed farm loan from the Farm Service Agency (FSA) or Farmers Home Administration?		
15b.	If Item 15a is YES, did the government ever forgive any debt through a write-off, debt settlement, compromise, write-down, charge-off, adjustment, reduction, or bankruptcy? If bankruptcy, please provide details in Item 31. If Item 15a is NO, leave blank.		
16.	If you obtained a guaranteed loan, did the government pay the lender a loss claim? Leave blank if you did not obtain a guaranteed loan.		
17.	Are you or any member of your organization delinquent on any federal debt? If YES, provide details in Item 31.		
18.	Are you a citizen or permanent resident of the United States of America? If permanent resident, provide a copy of Form I-151 or I-551, "Alien Registration Receipt Card."		
19.	Are you a veteran? If YES, please indicate Branch and Dates of Service in Item 31.		
20.	Are you now, or have you ever farmed or ranched? If YES, provide the number of years and brief explanation in Item 31.		
21.	Are you an FSA employee or are you related to or closely associated with any FSA employee? If YES, please explain in Item 31.		

22. PURPOSE OF LOAN
23. APPROXIMATE AMOUNT OF LOAN NEEDED
24. NAME AND ADDRESS OF APPLICANT'S EMPLOYER
25. NAME AND ADDRESS OF SPOUSE'S EMPLOYER
26. APPLICANT'S APPROXIMATE ANNUAL INCOME
27. SPOUSE'S APPROXIMATE ANNUAL INCOME

28. FSA USE ONLY

A. DATE FORM FSA 410-1 RECEIVED
B. DATE APPLICATION COMPLETE
C. CREDIT REPORT FEE $
D. DATE RECEIVED
E. INITIALS
F. TYPE OF ASSISTANCE: FO ☐ OL ☐ EM ☐ SUBORDINATION ☐ OTHER (SPECIFY) ☐

Figure 7–6 A sample loan application (provided by the USDA Farm Service Agency). *(continued on the following page)*

Figure 7–6 *(continued)*

FSA-410-1 (03-31-97) Page 2

29. **FOR BUSINESS ENTITY APPLICANTS ONLY (COOPERATIVES, CORPORATIONS, PARTNERSHIPS, OR JOINT OPERATIONS)**

I. MEMBER INFORMATION - Business entity applicants must attach the following information regarding all members, stockholders, partners, and joint operators (If there are no individually owned assets, then husband and wife joint operations may submit one consolidated balance sheet):

A. Name, address, social security number, birth date, principal occupation, and a balance sheet not more than 90 days old.

B. The full name of each non-U.S. citizen.

C. Each members' ownership interest (expressed as a percent).

II. BUSINESS ENTITY INFORMATION - The business entity must provide:

A. Any Organizational and Operational Documents (e.g. Charter, Articles of Incorporation, Bylaws, Partnership or Joint Operation Agreements, etc).

B. Any evidence of its current registration with relevant state regulatory agencies (good standing).

C. A duly adopted resolution to apply for and obtain financing.

D. Tax identification number.

NOTE: Individual liability will be required regardless of the type of business organization.

30. **VOLUNTARY INFORMATION FOR MONITORING PURPOSES:**
The following information is requested by the Federal Government in order to monitor FSA's compliance with federal laws prohibiting discrimination against loan applicants on the basis of race, color, national origin, religion, sex, marital status, handicapped condition or age (provided that the applicant has the capacity to enter into a binding contract). You are not required to furnish this information, but are encouraged to do so. This information will not be used in evaluating your application or to discriminate against you in any way. However, if you do not furnish it, FSA is required to note your race/national origin and sex on the basis of visual observation or surname.

A. RACE/NATIONALITY: ☐ White ☐ Black ☐ Hispanic ☐ Asian or Pacific Islander ☐ American Indian or Alaskan Native

B. SEX: ☐ Male ☐ Female

31. ADDITIONAL SPACE FOR ANSWERS. Write the number to which each answer applies. If you need more space, use additional sheets of paper the same size as this page. On each sheet, write the applicant's name.

ITEM NUMBER	REMARKS

(continued on the following page)

Figure 7–6 *(continued)*

FSA-410-1 (03-31-97) Page 3

32. A signed and dated balance sheet not more than 90 days old is required. Business organizations must provide individual members' balance sheets. You may use this form or attach your own. If you have a balance sheet on file with FSA that is less than 90 days old, you need not complete this section at this time.

BALANCE SHEET **AS OF ____________**

CURRENT FARM ASSETS				$VALUE
Cash on hand	Checking	Savings		
$	$	$		
Other Investments:				
Time Certificates	Other			
$	$			
Accounts and Notes to be Received (Receivables)				
Crops and Feed	Units	Price Per Unit ($)		
Livestock to be Sold	No.	Unit Weight	Price Per Unit ($)	
Growing Crops	Acres	Cost/Acre ($)		
Supplies and Prepaid Expenses				
Leases				
Other				
			TOTAL CURRENT FARM ASSETS➤	

CURRENT FARM LIABILITIES				$AMOUNT
Farm Accounts and Notes Payable (Include Principal and Interest)				
Creditor	Payment Due Date	Interest Rate	Monthly or Annual Installment ($)	
CCC Loan:				
Type	Quantity	Due Date		
Current Portion of Principal Due on:				
Intermediate Liabilities				
Long Term Liabilities				
Accrued interest on:				
Intermediate Liabilities				
Long Term Liabilities				
Accrued Taxes on:				
Real Estate, Personal Property and Assessments				
Income Tax and Social Security				
Accrued Rent/Lease Payments				
Other (judgments, liens, etc.)				
			TOTAL CURRENT FARM LIABILITIES➤	

INTERMEDIATE FARM ASSETS			
Accounts and Notes to be Received beyond 12 months (Receivables)			
Breeding Livestock	No.	Price Per Unit ($)	
Machinery, Equipment and Vehicles			
Co-op Stock			
Cash Value, Life Insurance (Face Amount $)			
Farmer-Owned Reserve:			
Type	Quantity	Price/Unit ($)	
Other			
		TOTAL INTERMEDIATE FARM ASSETS➤	

INTERMEDIATE FARM LIABILITIES (portion due beyond 12 months)				
Creditor	Payment Due Date	Interest Rate	Amount Delinquent ($)	
Loans Secured by Life Insurance Policy(ies)				
Farmer-Owned Reserve				
Other				
			TOTAL INTERMEDIATE FARM LIABILITIES➤	

LONG TERM FARM ASSETS (Farm Real Estate)				$VALUE
Acres	Date Bought	Annual Tax	Cost	
			$	
			$	
Co-op Stock				
Equity in Partnerships/Corporations/Joint Operations/Cooperatives				
Other				
			TOTAL LONG TERM FARM ASSETS➤	
			TOTAL FARM ASSETS➤	

LONG TERM FARM LIABILITIES (portion due beyond 12 months)				$AMOUNT
Creditor	Payment Due Date	Interest Rate	Amount Delinquent ($)	
Other				
			TOTAL LONG TERM FARM LIABILITIES➤	
			TOTAL FARM LIABILITIES➤	

NONFARM ASSETS	
Household Goods	
Car, Recreational Vehicles, etc.	
Cash Value of Life Insurance	
Stocks, Bonds	
Nonfarm Business	
Other Nonfarm Assets	
Nonfarm Real Estate (Annual Tax $)	
TOTAL NONFARM ASSETS➤	
TOTAL FARM ASSETS➤	
TOTAL ASSETS➤	

NONFARM LIABILITIES				
Nonfarm Accounts and Notes Payable				
Creditor	Payment Due Date	Interest Rate	Monthly or Annual Installment ($)	
			TOTAL NONFARM LIABILITIES➤	
			TOTAL FARM LIABILITIES➤	
			TOTAL LIABILITIES➤	
			NET WORTH➤	
			TOTAL LIABILITIES AND NET WORTH➤	

(continued on the following page)

Figure 7–6 *(continued)*

FSA-410-1 (03-31-97) Page 4

33. SPECIAL PROGRAM INFORMATION

Certain FSA programs are, by law, designed to reach targeted applicants. If you are interested in any of the programs described below, or have questions about these programs and whether you may qualify for a specific program, the FSA office processing your application will help you.

A. **SOCIALLY DISADVANTAGED APPLICANTS:** A portion of FSA farm ownership and operating loan funds are, by law, targeted to applicants who have been subjected to racial, ethnic or gender prejudice because of their identity as a member of a group, without regard to individual qualities. Under the applicable law, groups meeting this condition are: Women, Blacks, American Indians, Alaskan Natives, Hispanics, Asians, and Pacific Islanders.

B. **BEGINNING FARMER ASSISTANCE:** FSA has the authority to assist beginning farmers and ranchers through the farm operating and ownership loan programs. A portion of FSA farm ownership and operating loan funds are, by law, targeted to beginning farmers and ranchers. In addition, FSA has a beginning farmer down payment program, which receives special funding. In some States, FSA has agreements with State beginning farmer programs to help meet the credit needs of beginning farmers and ranchers.

C. **LIMITED RESOURCE LOANS:** Limited resource farm ownership and operating loans are available to qualified FSA applicants. This program provides loans at reduced interest rates to low-income farmers and ranchers whose farm operations and resources are so limited that they cannot pay the regular rates for FSA loans. The program is also intended to provide beginning farmers with an opportunity to start a successful farming operation.

34. STATEMENT REQUIRED BY THE PRIVACY ACT

The following statements are made in accordance with the Privacy Act of 1974 (5 U.S.C. 552a): The Farm Service Agency (FSA) is authorized by the Consolidated Farm and Rural Development Act, as amended (7 U.S.C. 1921 et seq.), or other Acts, and the regulations promulgated thereunder, to solicit the information requested on its application forms. The information requested is necessary for FSA to determine eligibility for credit or other financial assistance, service your loan, and conduct statistical analyses. Supplied information may be furnished to other Department of Agriculture agencies, the Internal Revenue Service, the Department of Justice or other law enforcement agencies, the Department of Defense, the Department of Housing and Urban Development, the Department of Labor, the United States Postal Service, or other Federal, State, or local agencies as required or permitted by law. In addition, information may be referred to interested parties under the Freedom of Information Act (FOIA), to financial consultants, advisors, lending institutions, packagers, agents, and private or commercial credit sources, to collection or servicing contractors, to credit reporting agencies, to private attorneys under contract with FSA or the Department of Justice, to business firms in the trade area that buy chattel or crops or sell them for commission, to Members of Congress or Congressional staff members, or to courts or adjudicative bodies. Disclosure of the information requested is voluntary. However, failure to disclose certain items of information requested, including your Social Security Number or Federal Tax Identification Number, may result in a delay in the processing of an application or its rejection.

35. GENERAL INFORMATION

A. **RIGHT TO FINANCIAL PRIVACY ACT OF 1978 and TITLE XI, 1113(h) OF PUB. L. 95-630:** FSA has a right of access to financial records held by financial institutions in connection with providing assistance to you, as well as collecting on loans made to you or guaranteed by the government. Financial records involving your transaction will be available to FSA without further notice or authorization but will not be disclosed or released by this institution to another government Agency or Department without your consent except as required by law.

B. **THE FEDERAL EQUAL OPPORTUNITY ACT** prohibits creditors from discriminating against borrowers on the basis of race, color, religion, sex, handicap, familial status, national origin, marital status, age (provided the borrower has the capacity to enter into a binding contract), because all or a part of the borrower's income derives from any public assistance program, or because the borrower has in good faith exercised any right under the Consumer Credit Protection Act.

C. **FEDERAL COLLECTION POLICIES FOR CONSUMER DEBTS:** Delinquencies, defaults, foreclosures and abuses of mortgage loans involving programs of the Federal Government can be costly and detrimental to your credit, now and in the future. The mortgage lender in this transaction, its agents and assigns as well as the Federal Government, its agencies, agents and assigns, are authorized to take any and all of the following actions in the event loan payments become delinquent on the mortgaged loan described in the attached application: (1) Report your name and account information to a credit bureau; (2) Assess additional interest and penalty charges for the period of time that payment is not made; (3) Assess charges to cover additional administrative costs incurred by the Government to service your account; (4) Offset amounts owed to you under other Federal programs; (5) Refer your account to a private attorney, collection agency or mortgage servicing agency to collect the amount due, foreclose the mortgage, sell the property and seek judgement against you for any deficiency; (6) Refer your account to the Department of Justice for litigation; (7) If you are a current or retired Federal employee, take action to offset your salary, or civil service retirement benefits; (8) Refer your debt to the Internal Revenue Service for offset against any amount owed to you as an income tax refund; and (9) Report any resulting written-off debt to the Internal Revenue service as taxable income. All of these actions can and will be used to recover debts owed to the Federal Government, when in its best interests.

36. CERTIFICATIONS

A. RESTRICTIONS AND DISCLOSURE OF LOBBYING ACTIVITIES

1. **The loan applicant certifies that: if any funds, by or on behalf of the loan applicant, have been or will be paid to any person for influencing or attempting to influence an officer or employee of any agency, a Member, an officer or employee of Congress, or an employee of a Member of Congress in connection with the awarding of any Federal contract, the making of any Federal grant or Federal loan, and the extension, continuation, renewal, amendment, or modification of any Federal contract, grant, or loan, the loan applicant shall complete and submit Standard Form - LLL, "Disclosure of Lobbying Activities," in accordance with its instructions.**

2. The loan applicant shall require that the language of this certification be included in the award documents for all sub-awards at all tiers (including contracts, subcontracts, and subgrants, under grants and loans) and that all subrecipients shall certify and disclose accordingly.

3. This certification is a material representation of fact upon which reliance was placed when this transaction was made or entered into. Submission of this statement is a prerequisite for making or entering into this transaction imposed by 31 U.S.C. 1352. Any person who fails to file the required statement shall be subject to a civil penalty.

B. ABUSE OF CONTROLLED SUBSTANCES:

The loan applicant certifies that he/she as an individual, or any member, stockholder, partner or joint operator of an entity applicant, has not been convicted under Federal or State law of planting, cultivating, growing, producing, harvesting, or storing a controlled substance since December 23, 1985, in accordance with the Food Security Act of 1985 (Public Law 99-198).

C. TEST FOR CREDIT

The individual or authorized party certifies that the needed credit, with or without a loan guarantee, cannot be obtained by the individual applicant, or in the case of a business entity, the needed credit cannot be obtained considering all assets owned by the business entity and all of the individual members.

D. ACKNOWLEDGMENT

I, THE UNDERSIGNED LOAN APPLICANT, UPON SIGNING THIS LOAN APPLICATION, CERTIFY THAT I HAVE RECEIVED THE ABOVE NOTIFICATIONS AND ACCEPT AND COMPLY WITH THE CONDITIONS STATED THEREON. I CERTIFY THAT THE STATEMENTS MADE BY ME IN THIS APPLICATION ARE TRUE, COMPLETE, AND CORRECT TO THE BEST OF MY KNOWLEDGE AND BELIEF AND ARE MADE IN GOOD FAITH TO OBTAIN A LOAN. I UNDERSTAND THAT THE 60-DAY PROMPT APPROVAL PERIOD WILL NOT BEGIN UNTIL A COMPLETE APPLICATION HAS BEEN FILED. (WARNING: SECTION 1001 OF TITLE 18, UNITED STATES CODE PROVIDES FOR CRIMINAL PENALTIES TO THOSE WHO PROVIDE FALSE STATEMENTS ON LOAN APPLICATIONS. IF ANY INFORMATION ON THIS APPLICATION IS FOUND TO BE FALSE OR INCOMPLETE, SUCH FINDING MAY BE GROUNDS FOR DENIAL OF THE REQUESTED CREDIT.)

SIGNATURE OF LOAN APPLICANT OR AUTHORIZED REPRESENTATIVE	DATE

Computing Interest

Interest can be a major expense to agribusinesses that use debt to finance their operations. The **interest rates** that lending institutions charge actually represent the prices they charge for the use of their money.

Interest Rates

Normally, a lender will charge an interest rate tied to the current **discount (prime) rate** set by the Federal Reserve. Supply and demand factors for money influence the movement of these rates. Interest rates can, and often do, fluctuate over time.

It is often hard to think of money as a commodity that is exchanged for a certain **equilibrium price**. However, the interest rate charged on a loan is actually its selling price to the borrower. Figuring out the actual price, or interest rate, on loaned money is important in controlling the cost of capital to an agribusiness.[17]

Cost of capital is the difference between the actual amount paid for the loan and the actual amount received from it. For example, if a person decides to borrow $100 at 9 percent interest, the cost of capital would be $100 × 9% = $9. Thus, the person would receive $100 and pay back $109.

The **contractual interest rate** will not always be equal to the actual rate charged on a loan. This makes it important to know what a lender is charging for use of the money. Most borrowers try to choose a lender that offers the lowest **actuarial interest rate**.

The **annual percentage rate (APR)** is a common name for the actuarial interest rate. However, the lender with the lowest APR may not always suit your needs. You may need the lower monthly payments of a longer-term loan, which another lender may offer at higher rates.[18] The following are three ways to calculate the annual percentage rate.

Simple Interest. **Simple interest** applies to loans with a single payment. An example is a six-month, short-term operating note due in a single payment at the end of the period. The APR on the loan is the rate charged if no down payment or borrowing fees were required. In the example just presented, the APR is 9 percent.

Remaining Balance Method. This method applies to installment-type loans in which a series of installment payments go toward paying off the loan. Suppose that a person borrows $100 at 9 percent interest and will pay the loan off in four annual payments. The payment of principal will be $25 per year. The interest payments in each year are calculated as follows:

Year 1	$100 × .09 =	$9	+ $25 =	$34.00
Year 2	75 × .09 =	$6.84	+ $25 =	$31.84
Year 3	50 × .09 =	$4.50	+ $25 =	$29.50
Year 4	25 × .09 =	$2.25	+ $25 =	$27.25
Total		$22.59	+ 100 =	$122.59

In this example, the remaining balance is the basis on which the borrower pays interest. The contractual rate and the APR in this case would be the same.

Add-On Method. Often, lenders use **add-on interest** when making loans on automobiles, tractors, or other types of machinery. In computing the interest, you take the total interest paid on a loan and add it to the loan amount. After adding in the interest, you divide the total amount by the number of payment periods to get the installment payment amount.

The type of interest will be different from the contracted rate; in most cases, it will be higher.[19] The following example helps explain this point.

Suppose Mr. Shinn needs $10,000 to finance the purchase of a used truck. The bank offers to finance the truck for four years at 7 percent interest using the add-on method. Mr. Shinn's financial advisor tells him that the stated rate on the loan will not be his actual rate, or APR. She computes the APR as follows:

- Interest calculation: $10,000 × 7% × 4 years = $2,800. This is the finance charge.
- Then, she adds the interest to the loan amount: $10,000 + $2,800 = $12,800. This is the total amount to be repaid.
- Calculation of the monthly payment works out to: 4 years × 12 months = 48 monthly payments.
- Then, she divides $12,800/48 payments = $266.67 per month.

17. Ibid.
18. Ibid.
19. Ibid.

Use the following formula to calculate the APR for add-on interest.

$$R = \frac{2C}{L(P+A)} \times 100 \text{ where}$$

R = annual percentage rate
C = total interest cost
L = length of the loan in years
P = beginning principal of the borrowed amount
A = payment in each period

The payments do not have to be monthly in this equation. The variable *A* could apply to annual, semiannual, or quarterly payments as well. Substituting in the numbers from the example, the advisor computes the following APR.

$$\frac{2 \times \$2{,}800}{4(\$10{,}000 + \$266.67)} \times 100 = 13.64\%$$

Mr. Shinn is actually paying 13.64 percent interest, instead of 7 percent, as stated in the loan contract.

Amortization Tables

An *amortization table* is a chart used to calculate the constant payments needed to repay both the principal and interest on a sum of money. The term **amortize** refers to a loan that is set up with equal installment payments. These can be set up as annual, semiannual, quarterly, or monthly payments. The tables are set up most conveniently for yearly payments, but they can be used for other payment schedules as well. Many computer programs can provide the same information.

With equal payments, a larger amount is applied to interest cost and a smaller amount is applied to the payment of principal during the early stages of the loan. As the number of payments increases, the amount going to interest decreases and the amount going to principal increases. Interest is being paid only on the actual amount still owed. Because the total size of each payment is always the same, lenders and managers can more accurately calculate their cash flow.[20]

To use the amortization table in Figure 7-7, look at the far left column. It will be marked either in years or in time periods. Drop down to year five. Now cross over to 10 percent interest. The factor is .2637. So, if $10,000 is borrowed at 10 percent for five years, the annual payments would be $10,000 times .2637, or $2,637. The payment schedule is illustrated in Figure 7-8.

If semiannual payments are made, the same table can be used. If the loan is for five years with semiannual payments, the time period, or number of payment periods, is 10. The interest must be divided by two because half the interest twice a year is the same as all the interest once a year. Find the factor for 10 time periods (years) at 5 percent interest. That factor is .1295. The semiannual payment will be $1,295, or $1,000 times .1295. The total cost of the loan would be $12,950. It costs less than the other loan because the money is being paid back more frequently, which lowers the interest cost.

Feasibility

If an investment is feasible, it will generate enough after-tax income to pay for itself. The investment must have a positive cash flow during its economic life (the loan period). Many agricultural investments show a negative cash flow during the first year or two of activity. The difference must be made up from profits in other parts of the operation, savings, or short-term borrowing. Because such debt can be an excessive burden to an operation, it has become an important managerial concern.

Certain steps can be taken to help overcome this cash-flow problem. Prepare an estimated cash-flow statement for the investment item. Then, if the cash flow is negative in the first year or two, determine whether the operation can absorb any added costs of the new investment. Try to match the economic life, or length of loan, to the physical life, or how long the item will last. Also, the more money that can be invested at first, the easier it will be for the cash flow to become positive, because the payment will become smaller. A good example is a large machinery purchase.[21] Many times, the loan payment will initially be more than the money earned with the machine.

20. Veronica Feilner, *Agricultural Management and Economics* (Columbia, MS: Instructional Materials Laboratory, 1988), pp. I:32–I:33.

21. Ibid., pp. I:33–I:34.

Annual Payment per $1 of Loan at Given Interest Rates and Maturities (Amortization Table)

Year	4%	5%	6%	7%	8%	9%	10%	11%	12%	13%	14%	15%	16%	17%	18%	19%	20%
1	1.0400	1.0500	1.0600	1.0700	1.0800	1.0900	1.1000	1.1100	1.1200	1.1300	1.1400	1.1500	1.1600	1.1700	1.1800	1.1900	1.2000
2	.5301	.5378	.5454	.5530	.5607	.5684	.5761	.5839	.5916	.5994	.6072	.6151	.6229	.6308	.6387	.6466	.6545
3	.3603	.3672	.3741	.3810	.3880	.3950	.4021	.4092	.4163	.4235	.4307	.4379	.4452	.4525	.4599	.4673	.4747
4	.2754	.2820	.2885	.2952	.3019	.3086	.3154	.3223	.3292	.3361	.3432	.3502	.3573	.3645	.3717	.3789	.3862
5	.2246	.2309	.2373	.2438	.2504	.2570	.2637	.2705	.2774	.2843	.2912	.2983	.3054	.3125	.3197	.3270	.3343
6	.1907	.1970	.2033	.2097	.2163	.2229	.2296	.2363	.2432	.2501	.2571	.2642	.2713	.2786	.2859	.2932	.3007
7	.1666	.1728	.1791	.1855	.1920	.1986	.2054	.2122	.2191	.2261	.2331	.2403	.2476	.2549	.2623	.2698	.2774
8	.1485	.1547	.1610	.1674	.1740	.1806	.1874	.1943	.2013	.2083	.2155	.2228	.2302	.2376	.2452	.2528	.2606
9	.1344	.1406	.1470	.1534	.1600	.1667	.1736	.1806	.1876	.1948	.2021	.2095	.2170	.2246	.2323	.2401	.2480
10	.1232	.1295	.1358	.1423	.1490	.1558	.1627	.1698	.1769	.1842	.1917	.1992	.2069	.2146	.2225	.2304	.2385
11	.1141	.1203	.1267	.1333	.1400	.1469	.1539	.1611	.1684	.1758	.1833	.1910	.1988	.2067	.2147	.2228	.2311
12	.1065	.1128	.1192	.1259	.1326	.1396	.1467	.1540	.1614	.1689	.1766	.1844	.1924	.2004	.2086	.2168	.2252
13	.1001	.1064	.1129	.1196	.1265	.1335	.1407	.1481	.1556	.1633	.1711	.1791	.1871	.1953	.2036	.2121	.2206
14	.0946	.1010	.1075	.1143	.1212	.1284	.1357	.1432	.1508	.1586	.1666	.1746	.1828	.1912	.1996	.2082	.2168
15	.0899	.0963	.1029	.1097	.1168	.1240	.1314	.1390	.1468	.1547	.1628	.1710	.1793	.1878	.1964	.2050	.2138
16	.0858	.0922	.0989	.1058	.1129	.1203	.1278	.1355	.1433	.1514	.1596	.1679	.1764	.1850	.1937	.2025	.2114
17	.0821	.0886	.0954	.1024	.1096	.1170	.1246	.1324	.1404	.1486	.1569	.1653	.1739	.1826	.1914	.2004	.2094
18	.0789	.0855	.0923	.0994	.1067	.1142	.1219	.1298	.1379	.1462	.1546	.1631	.1718	.1807	.1896	.1986	.2078
19	.0761	.0827	.0896	.0967	.1041	.1117	.1195	.1275	.1357	.1441	.1526	.1613	.1701	.1790	.1881	.1972	.2064
20	.0735	.0802	.0871	.0943	.1018	.1095	.1174	.1255	.1338	.1423	.1509	.1597	.1686	.1776	.1862	.1960	.2053
25	.0640	.0709	.0782	.0858	.0936	.1018	.1101	.1187	.1275	.1364	.1454	.1546	.1640	.1734	.1829	.1924	.2021
30	.0578	.0650	.0726	.0805	.0888	.0973	.1060	.1150	.1241	.1334	.1428	.1523	.1618	.1715	.1812	.1910	.2008
35	.0535	.0610	.0689	.0772	.0858	.0946	.1036	.1129	.1223	.1318	.1414	.1511	.1608	.1707	.1805	.1904	.2003
40	.0505	.0505	.0664	.0750	.0838	.0929	.1022	.1117	.1213	.1309	.1407	.1505	.1604	.1703	.1802	.1901	.2001

Figure 7–7 An amortization table is used to calculate the constant payments needed to repay both the principal and interest on a given sum of money.

Payment Schedule

	Interest	Payment on Principal		Total Payment
Year 1	$1,000	$1,637		$2,637
2	836	1,801		2,637
3	656	1,981		2,637
4	458	2,179		2,637
5	218	2,419		2,637
	$3,168 +	$10,017*	=	$13,185

*Principal did not quite total $10,000 due to rounding.

Figure 7–8 Refer to Figure 7–7 and notice that a five-year loan at 10 percent interest has a factor of .2637. The payment schedule here shows that the yearly payments for a $10,000 loan are $10,000 × .2637 = $2,637.

Profitability Index

The most practical way to determine if something is profitable is to determine if the *benefits* (sales/income) outweigh the *costs* (inventory or investment). For a profitability index (PI), the benefits must be determined and divided by the costs. This determination is a useful management tool for making investment decisions.

The profitability index allows a comparison of investments with different rates of return and different terms. The most profitable investment can then be selected. The present value of each of the investments is used through the life of that investment. This puts the value of the investment in terms of today's dollar, which can easily be compared. If a person is currently earning $110 from a $100 investment, the profitability index would be figured as follows:

$$\frac{\$110 \text{ benefit}}{\$100 \text{ cost}} = 1.10 \text{ profitability index}$$

As long as the profitability index is greater than 1, the investment is profitable. When deciding between two investments, simply choose the one with the highest profitability index. Remember, money is a tool—nothing more and nothing less. If the tool is used correctly, it will benefit the operation. If the tool is used poorly, it might not be there to use tomorrow. Refer to Chapter 6 for further clarification on financial (profitability) ratios.

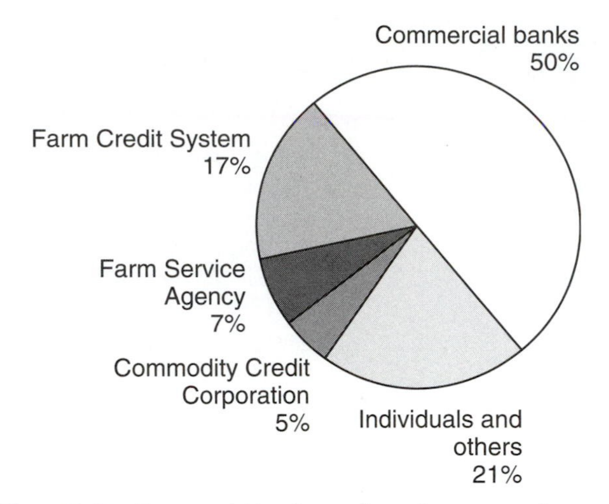

Figure 7–9 Commercial banks are the major source of non-real estate loans.

Agricultural Credit Sources for Real Estate and Non-Real Estate Loans

Production agriculturalists have access to all the credit sources that other businesses have, plus some special sources set up only for them. Agricultural lending is big business. Many lenders recognize this and compete for a share. The major sources of credit for those in the agricultural industry are the Farm Credit System, commercial banks, the Farm Service Agency (formerly the Farmers Home Administration or FmHA), the Commodity Credit Corporation, life insurance companies, and individuals.

Non-Real Estate Loans

Most of the money borrowed by production agriculturalists is used to finance farm expansion and higher-cost production items, such as farm machinery and motor vehicles. In 1995, money borrowed for non-real estate loans **secured** by farm assets totaled $72 billion. Commercial banks supplied 50 percent; the Farm Credit System, 17 percent; the Farm Service Agency, 7 percent; the Commodity Credit Corporation, 5 percent; and "individuals and others," 21 percent.[22] Refer to Figure 7–9.

22. Gail L. Cramer, Clarence W. Jensen, and Douglas D. Southgate, Jr., *Agricultural Economics and Agribusiness,* 8th ed. (New York: John Willey & Sons, 2001), p. 249.

Career Option

Farm Credit Officer

The Farm Credit officer fulfills many roles in helping to supply needed financial resources to farmers, their families, and their businesses, and to local communities that depend on healthy economic agricultural productivity. Often the Farm Credit officer acts as a salesperson, operating at the state and local levels to locate new prospects for loans and financial services. At other times the credit officer acts as a farm management advisor and resource person for the local producer.

Farm Credit is a cooperative with a national focus, but its districts are geographically divided so that loan officers can specialize in local business and farming trends. This local presence, coupled with a national interest and financial basis in agribusiness, allows the Farm Credit officer to provide broad-based yet personalized financial advice. Though some commercial banks offer loans to farm business enterprises, the benefit of Farm Credit agencies lies in their understanding of, and roots in, agriculture and their focus on the particular financial problems that farmers face. Farm Credit offers loans for homes in some nonagricultural areas, but the mission of the agency is to provide loans for agricultural development. This includes loans for farms, livestock, farm equipment, and farm operating expenditures. Repayment options are tailored to suit the specific agribusiness. Farm Credit officers often help small farms begin a new operation or rebuild or update operations, and they help part-time farmers become more efficient and productive by providing financial support and services. Often the Farm Credit officer can help steer family farm owners from a less productive enterprise to a more productive one, as the agency encourages new production ideas and supports local educational efforts. These efforts include training and workshops on topics specific to the agribusiness community.

A Farm Credit officer should always thoroughly assess the potential borrower's financial position before making a loan. (Courtesy of USDA)

Real Estate Loans

In 1998, the Farm Credit System supplied 32 percent of all outstanding real estate debt; life insurance companies supplied 11 percent; commercial banks, 31 percent; the Farm Service Agency, 5 percent; and individuals, 24 percent. Historically, individuals have been the major source of funds for land transfers. As can be seen, the Farm Credit System is the largest lender involved in the land mortgage field. The total real estate debt as of December 31, 1998, was $87.6 billion.[23]

Conclusion

Agricultural credit is big business. These financial services are a vital component of the agricultural

23. Ibid.

industry. Without credit, there would be a broken link in the chain of production agriculturalists' business. With it, they can acquire the credit needed to overcome the general shortage of equity capital; credit not only assists in paying business expenses, but also allows producers to finance and expand their agribusinesses. Major factors when considering a loan include the three "Rs" of credit: returns, repayment, and risk; what type of expenses a loan is needed for; the amount of the loan (including principal and interest) over its lifetime; the type of loan and time frame it can be financed over; the borrower's credit profile; and finally the source of the loan itself. Each of these components is an important piece of the overall financial puzzle. With careful planning, production agriculturalists can have a brighter and more successful future.

Summary

Decisions about credit often are the most important judgments that people in the agricultural industry must make. In the agricultural industry, credit is needed to overcome a general shortage of equity capital. Limited capital, fluctuating interest rates, and lack of credit information are major problems facing many production agriculturalists. To effectively address these issues, it is important for production agriculturalists to identify and utilize the specific objectives of successful farm financial management most appropriate to their situation.

The three "Rs" of credit are return, repayment, and risk. The main reasons for borrowing money are to increase net returns and make a profit. Agricultural lenders expect their money to be repaid in full plus interest. Borrowers with strong assets can assume more risk than those with few assets.

In the agricultural industry, financing is needed in three areas: fixed expenses, operating expenses, and startup expenses. Fixed expenses include items that have a fixed yearly price, such as land, buildings, machinery, and equipment. Operating expenses cover everything needed to run a farm, ranch, or agribusiness. These expenses depend on the size of the agribusiness, and may include seed, feed, fertilizer, and wages. Startup expenses, which are incurred before the business begins operation, include attorneys' fees, incorporation expenses, and costs for the development of a site or product.

Loans usually fall into one of three time frames: short-term, intermediate, or long-term. Short-term loans normally have a term of one year or less. Intermediate loans can range anywhere from 1 to 10 years and generally finance assets that are depreciable over their expected lives, such as farm livestock and equipment. Long-term loans extend over 10 or more years and are designed to finance acquisition or improvement of land, buildings, and housing. Within any of the three time frames, three different types of loans can be found: term loans, lined of credit, and revolving lines of credit.

When reviewing a loan application, lenders require a credit profile. Before extending credit, lenders want positive answers to questions about personal characteristics, management ability, financial position, loan purpose, and loan security. In addition to these questions, lenders also consider a borrower's debt-equity ratio. Depending on the type of lender, a ratio between 1 and 2 may be allowed.

Interest can be a major expense to agribusinesses that use debt to finance their operations. Normally, a lender charges an interest rate tied to the current discount or prime rate set by the Federal Reserve. The annual percentage rate is often used to determine what the lender is really charging for the use of its money. Three ways to calculate the APR are simple interest, the remaining-balance method, and the add-on method. An amortization table allows you to calculate constant payments needed to repay both the principal and the interest on a sum of money.

If an investment is feasible, it will generate enough after-tax income to pay for itself. The profitability index allows a comparison of the same investment with different rates of return and different lengths of time.

A wide variety of sources for both real estate and non–real estate loans are available to production agriculturalists. The major sources of credit for those in the agricultural industry are the Farm Credit System, commercial banks, the Farm Service Agency, the Commodity Credit Corporation, life insurance companies, and individuals.

End of Chapter Activities

REVIEW QUESTIONS

1. Define the Terms to Know.
2. Give three reasons why credit is important to production agriculturalists.
3. Name and briefly discuss the three fundamentals—"three Rs"—of credit.
4. Name the three areas of credit need.
5. What are four types of fixed expenses?
6. What items are typically purchased with short-term loans?
7. What is the key to maintaining good credit with short-term loans?
8. What items are typically purchased with (a) intermediate-term loans and (b) long-term loans?
9. What three types of financing can be found within any time frame?
10. Name and write down the questions asked by lenders when securing a credit profile on a borrower.
11. What would be the cost of capital for a $1,000 loan at 8 percent interest?
12. List and briefly describe the three ways to calculate annual percentage rate.
13. What are six sources of credit for production agriculturalists?

FILL IN THE BLANK

1. Agriculture and its products and services make up about ___________ of the U.S. gross domestic product.
2. ___________ loans are usually for one year or less.
3. When loans produce enough profit to ensure income stability, they ___________ in value.
4. The term of ___________ loans varies from 1 to 10 years.
5. A/an ___________ table allows calculation of payments needed to repay both the principal and interest on a sum of money.
6. If Tom has 1.5 times as much debt as equity in his business, his ___________ is 1.5.
7. If an investment is ___________, it will generate enough after-tax income to pay for itself.
8. ___________ may be the most important type of credit for survival of an agribusiness because it finances the everyday operations of the firm.
9. ___________ is determined by comparing the benefits with the costs.
10. The duration of ___________ loans is usually more than 10 years.
11. A lender may require ___________ when considering whether to grant a loan.
12. Current ___________ ___________ ___________ are set by the Federal Reserve.
13. As long as the profitability index is greater than ___________, the investment is profitable.
14. Often, lenders use ___________ ___________ when making loans on automobiles, tractors, or other types of machinery.

MATCHING

a. interest payments
b. "buyer's fever"
c. startup expenses
d. long-term
e. term loan
f. $1.8 trillion
g. amortize
h. operating expenses
i. Farm Credit System
j. foreclosure
k. cost of capital
l. loan
m. Farm Service Agency
n. profitability index
o. annual percentage rate (APR)
p. interest rates

______ 1. one of the fastest-growing costs of production agriculture in the past two decades
______ 2. value of all farm assets in 2005
______ 3. quick, thoughtless decisions
______ 4. seed, feed, fertilizer, chemicals, and wages
______ 5. expenses incurred before a business begins operation
______ 6. inability to repay loans may lead to this
______ 7. a contract between the borrower and the lender
______ 8. purchases of land, buildings, and housing are made possible through this type of credit
______ 9. the source providing 31 percent of the loans for long-term credit
______ 10. a loan with a specified amount loaned over a specific amount of time
______ 11. the prices lending institutions charge for use of their money
______ 12. the difference between the actual amount paid for the loan and the actual amount received from it
______ 13. actuarial interest rate
______ 14. a loan that is set up with equal installment payments
______ 15. allows a comparison of investments with different rates of return and different terms
______ 16. formerly the Farmers Home Administration

ACTIVITIES

1. Make a list of all the financial institutions in your community that make loans to those in the agricultural industry.
2. Suppose a lender requires a debt-equity ratio of 2. How much money could you borrow for agricultural purposes if you already had $150,000 equity in land?
3. Suppose a lender requires a debt-equity ratio of 1. How much money could you borrow for agricultural purposes if you had $24,000 in equity?

4. You want to buy a commercial lawn mower for your lawn service business. The mower cost $2,700, and the interest rate is 11 percent. How much simple interest will you have paid at the end of one year?
5. Suppose you are buying a used small tractor for your lawn service business. The used tractor cost $5,000, and you take out a loan for 8.5 percent interest to be paid off in five years. You set this up on the remaining-balance method of payment discussed in this chapter. Set up a payment plan showing the balance, principal, and interest paid yearly.
6. Calculate the price of the tractor described in Activity 5 by using the add-on method over five years at 8.5 percent interest. (a) What would be the total payment amount? (b) How much would the monthly payments be? (c) What would be the actual annual percentage rate (APR) of the loan?
7. Refer to the amortization table in Figure 7-7. You want to purchase a 100-acre farm for $3,000 per acre. The lender will charge a 9 percent interest rate on a loan to be paid off in 20 years. (a) How much will the payments be yearly? (b) How much will you pay for the farm over 20 years, including principal and interest?
8. What is the profitability index if a person earns $3,000 from a $2,500 investment?
9. Interview an agricultural loan officer in your community. List three things that the loan officer says will help you get a loan. List three things that will keep you from getting a loan.

Chapter 8

Personal Financial Management

Objectives

After completing this chapter, the student should be able to:

- Discuss earning money.
- Select a financial institution.
- Manage a checking account.
- Identify where your money goes.
- Plan and prepare a budget.
- Describe financial management and financial security tips and hints.
- Explain four ways to potentially retire as a millionaire.
- Explain key factors that make the millionaire plans work.
- Discuss how to achieve financial security.
- Understand the levels of financial management.

Terms to Know

APR
balancing a checkbook
brokerage
compatible
compound interest
enterprises
entrepreneurship
extrapolate
fidelity shares
financial security
fixed expenses
gross pay
individual retirement account (IRA)
investment portfolio
liquid income
locked into
mutual funds
net pay
part-time (avocational) enterprises
perseverance
plastic prosperity disease
prospectus
reconcile
risk management
tax-deferred
tax-deferred savings
term life insurance
therapeutic activity
tax shelter account (TSA)
universal life insurance
utilization of resources
variable expenses

Introduction

As the opening ceremony of a Future Farmers of America meeting notes, "George Washington was better able to serve his country because he was financially independent." The first step toward managing the finances of a business in the agricultural industry is becoming better able to manage your personal finances. People can lead better and receive greater respect when they have control of their money. Washington could better lead the country because he had no monetary problems. He could not be influenced financially by outside forces and would not be distracted from his presidential duties by money problems. Washington used the same sound management techniques that made him successful in his personal life to further the status and well-being of his country.

You have many decisions to make in life. The decisions you make about how to handle money, however, will have a profound impact on your life. This chapter shows you how to manage your money in the most effective and responsible way. You should seek the help and advice of a trustworthy financial advisor as the need arises. Also, the College for Financial Planning publishes a "High School Financial Planning" program. Some banks also have high school financial planning curriculums. In the meantime, this chapter can open your eyes to financial opportunities that you might not have been aware of previously.

Everyone needs to be a student of money management. J. Paul Getty, the billionaire oil magnate, once said, "No man's opinions are better than his information." We hope that the information in this chapter will help you form accurate, well-founded opinions about money management and wealth accumulation. It includes discussion of how to budget, tips on money management, how to potentially retire as a millionaire, and achieving financial security through agricultural **entrepreneurship**.

Earning Money

The career or type of work you choose, and how you choose to work once you get your job, can be paramount to your financial well-being. You must plan your work and then work your plan. Select a job for which you have a natural aptitude and then do that job extremely well. The ideal job is one in which you have a vacation as a vocation.

Working hard at something you are good at will, sooner or later, put you where you want to be. In most careers, you can simply outwork 80 percent of your peers and outsmart (with superior knowledge) 15 percent of the rest; that alone puts you in the top 5 percent, which always pays very well.[1] If you are good at something, it normally makes you more intense.

Gross versus Net Pay

This leads us to the point of receiving the paycheck. There are hidden costs to earning money. These include taxes and optional employee benefits, which are deducted from each individual's paycheck. The total earned before deductions is called **gross pay**. The amount of money left after deductions is called **net pay**, or "take-home" pay. Figure 8–1 illustrates a typical pay statement showing gross versus take-home pay.

Mandatory Deductions

Money is withheld from an employee's pay to cover the cost of government-mandated taxes and programs. These vary from state to state, and include:[2]

- **Income taxes (federal, state, and local).** These taxes are based on the amount of money earned on a job or through investment income. Income taxes are used by federal, state, and local governments to provide public goods and services for the benefit of the community as a whole. This includes funding for education, military, roads, the legal system, and police and fire departments. State and city taxes vary, based on where you live and work. For example, seven states impose no income tax.
- **Social Security tax.** Taxes required by the Federal Insurance Compensation Act (FICA) are used to provide retirement or disability income for individuals or their survivors who have contributed to the fund.

1. Dave Ramsey, *Financial Peace* (Nashville, TN: Lampo Press, 1992), p. 43.
2. *Money Skills: A Resource Book for Teaching Personal Economics* (Atlanta, GA: NationsBank Corporation, 1994), pp. 31–32. (*Money Skills* was produced for NationsBank by Martin Kidd Associates, Nashville, Tennessee.)

PAY STATEMENT						
			AGRIBUSINESS SUPPLY Main Street and First Avenue Our Town, USA 54321			
Employee: ARTIE F. DAVIS						Pay period: 9/22 to 9/28
HOURS		EARNINGS				
Regular	Overtime	Regular	Overtime	Bonus	Other	Gross Pay
20	-	100.00	-	-	-	100.00
			DEDUCTIONS			
FICA	Federal With. Tax	State With.Tax	Health Ins.	Retirement	Other	Net Pay
7.20	10.80	3.60	-	-	-	$78.40
Year to Date	Regular	Overtime	Gross Pay	FICA	Federal With. Tax	State With. Tax
1988	1,920.00	-	1,920.00	144.00	216.00	72.00

Figure 8–1 Pay statement illustrating gross versus take-home pay.

- **Unemployment insurance.** This insurance provides income for qualified individuals whose employment has been terminated.
- **Workers' compensation insurance.** This insurance protects individuals in the event of injury or illness that happens on, or because of, a job.

Employee Benefits

Most employers offer some benefits, which help them attract and keep good employees. These benefits may be paid for by the employer, by the employee, or by a combination of the two. Sometimes the benefits are optional. Benefits may include:[3]

- *Life insurance:* pays a beneficiary (designated by the employee; usually a family member) in the event of an employee's death.
- *Long-term disability insurance:* provides income for employees who become disabled and are unable to work.
- *Medical insurance:* covers some portion of employee and/or family medical and dental expenses, often including preventive, diagnostic, basic, and major medical and orthodontic services.
- *Retirement savings plan:* provides employees with an opportunity to save money on a **tax-deferred** basis. This is discussed further later in this chapter.
- *Profit sharing:* allows employees to share a portion of the profits the company makes.
- *Other:* some employee benefits may not require pay deductions. These include paid holidays, vacation time, sick pay, performance bonuses, and use of a company vehicle.

Taxes and Benefits Summary

Federal, state, and local taxes are deducted by withholding. This is simply a means of spreading taxes out over the year, rather than waiting until the April 15 filing deadline to pay all the money owed. Benefits cost an employer between 25 and 50 percent of an employee's salary. Although the employee cannot spend them, benefits are worth money because they reduce out-of-pocket expenses for the employee. Refer to Figure 8–2 for a summary of taxes and benefits.

3. Ibid., pp. 3.2–3.3.

Taxes and Benefits Summary		
Mandatory taxes and benefits		
Deduction	***What is it?***	***Who pays?***
Federal income tax	Funds services provided by the federal government, such as defense, human services, and the monitoring and regulation of trade.	Employee
State income tax	Funds services provided by state government, such as roads, safety, and health. (Not all states levyan income tax.)	Employee
Local income tax	Funds services provided by the city or other local government, such as schools, police, and fireprotection. (Not all areas levy an income tax.)	Employee
Social Security tax (FICA)	Provides income for retired or disabled employees or their survivors.	Employee and employer
State unemployment insurance	Provides income and other benefits to qualified people who have lost their jobs.	Employer
Worker's compensation insurance	Provides protection in the event that an employee sustains an injury or illness in the course of a job, or because of it.	Employer
Employee benefits*		
Benefit	***What is it?***	***Who pays?***
Life insurance	Pays a beneficiary in the event that an employee dies.	Employer or employee, or shared
Long-term disability insurance	Provides benefits in the event that an employee is completely disabled.	Employer or employee, or shared
Medical insurance	Employee and family insurance coverage for medical care expenses, including hospitalization, physician service, surgery, and major medical expenses.	Employer or employee, or shared
Dental insurance	Employee and family insurance coverage for dental care expenses, including preventative, diagnostic, basic, major, and orthodontic services.	Employer or employee, or shared
Retirement savings plan	A tax-deferred savings plan for retirement.	Employee (employer may contribute)
Profit sharing	A distribution of company profits to employees.	Employer
*Note: Whether these benefits are offered, and who will fund them, varies by company.		

Figure 8–2 Taxes and benefits summary. (Courtesy of NationsBank)

Selecting a Financial Institution

Financial institutions help consumers manage, protect, and increase their money. People have a variety of financial needs, which vary at different stages of life.

Types of Institutions

There are several different types of institutions:

- banks
- credit unions
- savings & loans (S&Ls)
- brokerage firms

In the past, each type of institution offered a specific, limited range of services. Banks took deposits in the form of checking accounts, savings accounts, and certificates of deposit (CDs), and they granted credit to qualified individuals. S&Ls offered savings accounts and home mortgages. Credit unions made low-interest loans available to their members. **Brokerage** firms bought and sold stocks and bonds on behalf of their customers. Because of deregulation of the financial services industry, though, the delineations among the different types of financial institutions have begun to blur. For instance, many banks now sell stock and bond mutual funds, while many credit unions, brokerages, and S&Ls now offer accounts that are very similar to checking accounts.[4]

Factors to Consider in Choosing a Financial Institution

Consider the following when choosing a financial institution:

- *Types of service:* Do they offer the financial services I need?
- *Convenience:* What are their hours? Is there a location convenient to me? How many locations do they have? Can I do business with them from a remote location (by mail, over the phone, or by computer)?
- *Costs:* How do the costs compare with those of other institutions? What are the charges for the product or service I need? Examples include monthly service charges, annual fees, transaction charges, and interest rates.
- *Relationships:* Do I already have a relationship with this institution? Do I like the people? Do they know my financial history? Will it be to my advantage to do more business with them, since we already have a relationship?
- *Security:* Is my money insured? Will I get my money back if something happens to the institution?
- *Customer qualifications:* Am I qualified to do business with this institution? Does it require membership or have other criteria to do business?[5] For example, to buy a mutual fund, some companies require a $2,000 minimum investment to open the account.

Figure 8-3 offers an overview of financial services. Study this chart as you select the financial institution that is appropriate for you.

Checking Accounts

Use Caution

Keep your checkbook properly updated and balanced, with all checks, deposits, and other charges recorded. This is simple, but many people fail to do it. Bankers tell horror stories of people who bring in checking accounts so far out of balance that the only thing that can be done is close them out and start over with a new account. The only math involved in keeping and balancing a checkbook is basic addition and subtraction.[6] Some students may wonder why checking account information is included here; after all, you may have had a checking account for several years. The authors' experience, however, is that many high school and college students do better in algebra and trigonometry than they do in balancing a checkbook.

The biggest caution about keeping a checkbook concerns being rushed. People become hurried and forget to record the proper amount, especially when they are in a grocery store or some other busy checkout line (Figure 8-4). When they get home, they cannot remember the right

4. Appreciation is extended to Dr. Joyce Harrison, Certified Financial Planner and faculty member of the Department of Human Sciences at Middle Tennessee State University, for input on the first portion of this chapter.

5. *Money Skills*, p. 5.2.

6. Ramsey, *Financial Peace*, p. 163.

Overview of Financial Services	
DEPOSIT SERVICES	
Checking accounts	The convenience and safety of paying by check instead of cash
Savings accounts and certificates of deposit (CDs)	Safe places to let your money grow
Automated teller machines (ATMs)	Easy access to your money from multiple locations, 24 hours a day
Direct deposits and automatic withdrawals	The ability to deposit money or pay bills automatically
Deposit insurance (such as FDIC)	The knowledge that your deposits are insured by the federal government for up to $100,000 per depositor. Agencies that provide this insurance are FDIC (banks) SAIF (savings & loans), or NCUA (credit unions)
CREDIT SERVICES	
Credit cards	The ability to access credit conveniently up to the amount of your approved credit limit
Installment loans and credit lines	The opportunity to borrow for major items such as a new or used vehicle, education, home improvement, and other personal or household items
Mortgages	The opportunity to borrow for a home purchase
Home equity loans	The ability to borrow against the equity in your home
Student loans	The ability to borrow at below-market rates to pay for a college education
Small business loans	The ability to borrow for financing the growth of a small business
INVESTMENT SERVICES	
Retirement accounts (IRAs, SEPs, KEOGHs)	The ability to save money on a tax-deferred basis toward retirement
Stocks, bonds, and mutual funds	The ability to invest in corporations and governments to meet your financial needs for the future
TRUST SERVICES	
Living trusts, testamentary trusts, estate planning and administration	The ability to ensure that your property will pass smoothly to your heirs with a minimum amount of tax
OTHER MONEY MANAGEMENT SERVICES	
Safe deposit boxes	Safekeeping for important documents and valuables
Wire service	The ability to transfer money quickly and safely over long distances
Traveler's and cashier's checks	The safety of checks and the convenience of their accepted by parties that usually require cash
Remote banking	The ability to conduct banking transactions by mail, over the phone, or by Internet hookup

Figure 8–3 This overview of financial services will help you select the appropriate financial institution. (Courtesy of NationsBank)

Figure 8–4 Even though you feel rushed in a checkout line, be patient and take the time needed to record your purchase amount in your checkbook register.

amount. Sometimes the check is recorded, but the balance is never brought forward.

If you have trouble recording your checks, try using duplicate checks. Most banks sell an NCR-paper or carbon check, which automatically records what you write as you fill out a check. However, you still have to carry your balance forward and **reconcile** your checkbook to the statement each month.[7]

Reasons for Writing Checks

Writing a check is perhaps the simplest way of withdrawing money from your account—maybe even too simple. We tend to spend more money when we write checks than when we use cash. However, checks are needed for many purchases. They are a convenient way to pay bills and protect money from loss or theft, and they are usually accepted as legal proof of payment. Many people use checks as their accounting procedure to determine expenditures, especially in production agriculture. A properly written check will have the type of expense written on the check. As expenses are being calculated for income tax preparation, the checks are used as official records for the year, along with the cash receipts.

7. Ibid., p. 164.

Points to Remember When Writing Checks

Checks must be written accurately because the bank needs to understand exactly how much you want to give to whom. Also, a poorly written check can be tampered with. Some points to remember when writing checks include:

- Write in ink.
- Write clearly.
- Fill in all blanks.
- Use the correct date.
- Place the decimal correctly.
- Write the amount close to the dollar sign so that it cannot be changed.
- Write the amount in words as far to the left as possible on the space given; fill in the unused portion with a wavy line.
- Do not cross out the name or erase anything on the check. If you make a mistake, tear up the check and write another.
- Never sign a blank check.
- Keep your checkbook in a safe, protected place.[8]

Steps in Writing a Check

- Write the correct date.
- Write the name of the person or company you intend to pay.
- Enter the amount of the check in numbers, including a decimal point to show cents. Start the numbers as close to the dollar sign as possible.
- Enter the amount of the check in words. Start writing from the far left side of the line. Follow the dollar amount by the word "and"; then write the cents amount as a fraction, over 100. (If there are no cents, use 00/100 or the word "exactly.") Draw a wavy line from the end of your writing to the end of the line, so there is no room to insert additional words or numbers.
- Sign your check the same way you signed the signature card when you opened your account.

8. *Development of Financial Skills* (Successful Living Skills Series) (Stillwater, OK: Oklahoma Department of Vocational and Technical Education, Curriculum and Instructional Materials Center, 1989), p. 238.

Completed Personal Check

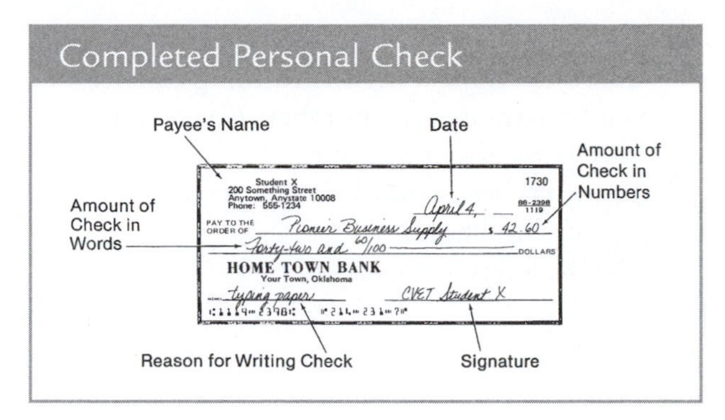

Figure 8–5 Completed personal check. (Courtesy of CIMC, Oklahoma Department of Vocational and Technical Education)

Checkbook Register

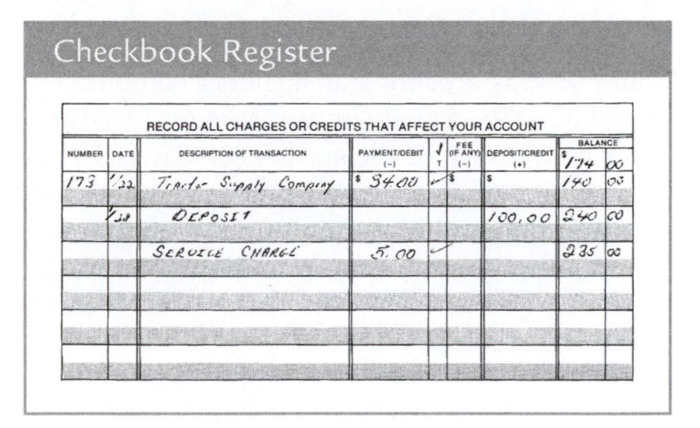

RECORD ALL CHARGES OR CREDITS THAT AFFECT YOUR ACCOUNT

NUMBER	DATE	DESCRIPTION OF TRANSACTION	PAYMENT/DEBIT (−)	✓ T	FEE (IF ANY) (−)	DEPOSIT/CREDIT (+)	BALANCE
							$ 174 00
173	1/22	Tractor Supply Company	$ 34 00	✓	$	$	140 00
	1/28	DEPOSIT				100.00	240 00
		SERVICE CHARGE	5.00	✓			235 00

Figure 8–6 Checkbook register with an example of an entered check, deposit, and service charge.

- Write down the purpose of the check. You may also use this space to write the account or invoice number, if you are paying a bill.

Figure 8-5 shows a check completed using the preceding steps.

Keeping a Checkbook Register

A *checkbook register* is the small booklet that you keep with your checkbook for recording deposits made and checks written. It allows you to tell at a glance the amount of money you have in your account. There is nothing difficult about keeping a checkbook register. You simply must take the time to do it. Remember being rushed in the checkout lane? Reasons for keeping a check register properly are that it:

- shows your balance at all times
- helps avoid overdrafts
- keeps a record of checks written
- keeps a record of deposits made
- shows dates of checks and deposits
- provides a record to compare with bank statements[9]

Figure 8-6 shows a checkbook register with an example of an entered check, deposit, and service charge.

Entering Checks Written. By referring to Figure 8-6, you can see the following steps:

- Write the check number (173).
- Write the date (1/22).
- Write the name of the person or business to whom the check will be written (Tractor Supply Company).
- Write the amount of the check in the correct column ($34.00).
- Subtract the check from the balance, using scratch paper or a calculator.
- Write in the balance ($140.00).
- Repeat each of these steps each time you write a check.

Entering Deposits. By referring to Figure 8-6, you can see the following steps:

- Write the date (1/28).
- Write the word *deposit* on the line.
- Write the amount of the deposit in the proper column ($100.00).
- Add the balance, using scratch paper or a calculator.
- Write in the balance ($240.00).
- Repeat each of these steps each time you make a deposit.

Entering Service Charges. Service charges can be found on the monthly bank statement. These may be monthly fees for bank services or (we hope not!) overdraft charges. By referring to Figure 8-6, you can see the following steps:

- After locating any service charges on the bank statement, write the service charge on a line in the check register.
- Write the amount of the service charge in the check column ($5.00).
- On scratch paper or with a calculator, subtract the service charge from the balance.
- Write the new balance ($235.00).[10]

9. Ibid., p. 246.

10. Ibid., pp. 247–251.

Balancing (Reconciling) Your Checkbook

Your checking account statement is a complete record of all the activity in your account (deposits and withdrawals) for the previous 30-day cycle, or period. You will want to take time each month to compare, or reconcile, this information with your checkbook register, to make sure your records agree with those of the bank. This process is called **balancing a checkbook**.

Importance of Balancing Your Checking Account Monthly. Although banks do make mistakes, the monthly account statement is an accurate picture of how much money is really in your account and available to be spent. If your checkbook is not balanced, you are in danger of spending more money than you have available. Other dangers include:

- There may be a steep financial penalty for "overdrawing" your account.
- If you overdraw your account often, it could affect your ability to get credit; the bank may even ask you to close your account.
- Unauthorized use of your automated teller machine (ATM) card will go unnoticed and unreported.
- Bank errors can go unnoticed. It is your responsibility to notify the bank of these errors in a timely manner.
- You may actually have money available of which you were not aware.[11]

Common Reasons Why Checkbook Registers and Account Statements Differ.

- Checks you have written have not yet been cashed.
- Checks you have deposited have not yet cleared and been added to your balance.
- Interest has been automatically deposited to your account (on interest-bearing accounts).
- Fees have been charged to your account, such as monthly service fees and fees for check printing.
- You have forgotten to record checks or deposits in your checkbook register (this is especially likely to happen with ATM transactions, direct deposits, and automatic withdrawals).
- You have made an error in addition or subtraction when calculating the running balance, or you have transposed numbers (for example, $89.30 instead of $98.30).
- There has been unauthorized use of your ATM card or checkbook.
- The bank has made an error.[12]

Steps in Balancing (Reconciling) a Bank Statement. Compare the bank statement and your checkbook register. When your checkbook total and the bank statement total agree (after subtractions and additions have been made), you have reconciled your checkbook with your bank statement. Here are the steps to balancing your checking account:

- Subtract from the checkbook register any service charges that are shown on the bank statement. Refer to Figure 8–7.
- Arrange canceled checks numerically (170, 171, 172, etc.).
- Compare the amounts on canceled checks and deposits in the bank statement with the amounts written in the checkbook register. (Figure 8–8 shows a simplified bank statement.)
- Check off all canceled checks and deposits in the checkbook register. (Refer to Figure 8–7.)
- List and add total outstanding checks (those not yet returned to the bank or checked in your register) on the back of your bank statement (Figure 8–9). Be sure to list the check number and the amount of the check.

Checkbook Register

RECORD ALL CHARGES OR CREDITS THAT AFFECT YOUR ACCOUNT

NUMBER	DATE	DESCRIPTION OF TRANSACTION	PAYMENT/DEBIT (−)	√ T	FEE (IF ANY) (−)	DEPOSIT/CREDIT (+)	BALANCE $206.00
170	1/16	MT. JULIET FFA ALUMNI	$10.00	√			196.00
171	1/18	SAM'S WAREHOUSE	10.00	√			186.00
172	1/20	MOTO PHOTO	12.00				174.00
173	1/22	TRACTOR SUPPLY SERVICE	34.00	√			140.00
	1/28	DEPOSIT				100.00	240.00
		SERVICE CHARGE	5.00				235.00

Figure 8–7 Checkbook register showing subtracted service charges.

---

11. *Money Skills*, p. 92.

12. Ibid.

Checkbook Register

Date of Statement: January 31, ____				
DATE	#	CHECK	DEPOSITS	BALANCE
1-17	170	$10.00		$196.00
1-19	171	$10.00		$186.00
1-23	173	$34.00		$152.00
1-29		$5.00 SC		$147.00
				LAST AMOUNT IN THIS COLUMN IS YOUR BALANCE

Figure 8–8 Bank statement showing checks, check numbers, and date on which each check cleared the bank. This bank statement also shows a $5.00 service charge.

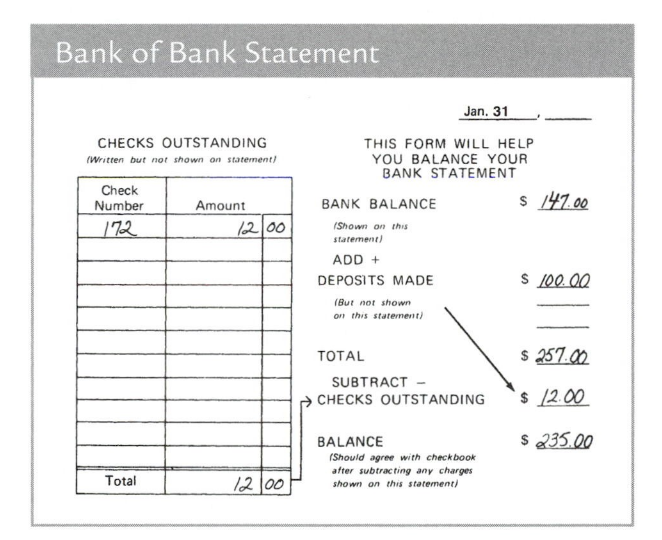
Bank of Bank Statement

Jan. 31, ____

CHECKS OUTSTANDING
(Written but not shown on statement)

Check Number	Amount
172	12 00
Total	12 00

THIS FORM WILL HELP YOU BALANCE YOUR BANK STATEMENT

BANK BALANCE (Shown on this statement) $ 147.00
ADD + DEPOSITS MADE (But not shown on this statement) $ 100.00
TOTAL $ 257.00
SUBTRACT – CHECKS OUTSTANDING $ 12.00
BALANCE (Should agree with checkbook after subtracting any charges shown on this statement) $ 235.00

Figure 8–9 Back of bank statement used in balancing (reconciling) your checking account.

- List and add total deposits.
- Add total outstanding deposits to your balance ($100.00).
- Subtract total outstanding checks from the total of the bank balance ($147.00) and outstanding deposits ($100.00); in our example, $257.00 – $12.00 = $235.00.
- Make sure that the adjusted bank balance and adjusted checkbook balance are equal.
- Circle the checkbook total to show that you agree with the bank.[13] If you forget how to reconcile your statement, check the back of the statement for instructions, or ask your bank for help.

Where Your Money Goes

You must gain control over your money, or your money will gain control over you. Money is active. Time, interest rates, amounts, cash flows, inflation, and risk all intermingle to create a current that is ever-flowing. Dave Ramsey, in his book titled *Financial Peace*, said that money is similar to a beautiful horse. It is very powerful and always moving and in action, but if not trained when young, it will become an out-of-control, dangerous animal when it grows to maturity.[14]

Handling Money Takes Discipline

Money and finances are not static. They must be managed, and you must do the managing. If you do not, money and finances will always manage you. Handling money takes discipline; however, you will achieve peace of mind and avoid financial stress in your life if you are willing to make the effort to become a disciplined money manager.[15] You discipline your spending with both fixed and variable expenses.

Fixed Expenses

A **fixed expense** is a set amount of money due on a set date. As a rule, fixed expenses must be paid when due. Although we do not have control over the types of fixed expenses, we do have some control over their amounts. Several examples of fixed expenses follow. Because you are still a student, many of the examples will not apply to you; however, they will certainly apply in the next one to five years.

Mortgage or Rent Payments. Obviously, this is an unavoidable expense once you begin living on your own. Housing expense could also include property tax and insurance if these are a part of your house payment. Your rent or house payments should not exceed 35 percent of your income, at most.

13. *Development of Financial Skills*, pp. 251–255.

14. Ramsey, *Financial Peace.*

15. Ramsey, *Financial Peace*, pp. 16–17.

Other Real Estate Payments. These could include a second mortgage, home improvement loan, storage rental, pasture for horses, or some other item.

Income and Property Taxes. This refers to taxes not taken directly out of your paycheck. Examples are federal and state taxes, past taxes, and property taxes not included in your house payments.

Installment Contract Payments. Many financial counselors advise against incurring debt. (This is discussed fully later in this chapter.) Examples include payments on contracts for purchase of cars, furniture, appliances, and many other items.

Insurance. Some insurance is not taken directly out of the paycheck and not included in the house payment. These payments can be real "budget busters," as they are often due every three months. An envelope technique (discussed later in this chapter) can help manage these budget busters. Examples include real property (fire, theft, liability), personal property (homeowner's, renter's, auto), life, and health insurance.

Regular Contributions. Examples are contributions to the United Way or other charities. Many people believe that making charitable donations is the humanitarian thing to do. Steel magnate Andrew Carnegie spent the first half of his life attaining wealth and the second half giving it away. St. Ambrose said, "Just as riches are an impediment to virtue in the wicked, so in the good they are an aid of virtue."

Dues. The type of job you have determines whether dues will be a fixed expense. Teachers, doctors, lawyers, and real estate agents pay dues to professional organizations. Many hourly workers pay dues to labor unions.

Savings. More is discussed about savings later in this chapter. You must develop the conviction that savings constitute a fixed expense, even though you could argue that they do not. If you do not build an emergency fund, you are destined for pain, embarrassment, and potential disaster.

Variable Expenses

Variable expenses vary in amount and frequency and offer the best opportunities for adjustments in a money plan. Be conservative with your variable expenses, or they will quickly get out of hand. Carefully consider your motives for buying. Why do you want or think you need this item? Could you live without it? Do you want it for selfish reasons, like showing up the neighbors, or would the item be truly useful to you? You must develop power over your purchases rather than letting your purchases gain power over you.

Utilities. Among utility charges, you probably have the most control over gas and electricity bills. Moderating thermostat settings, conserving electricity, and making maximum use of insulation can make a big difference in monthly expenses. Telephone and water bills can also be monitored. Garbage, sewage, and cable TV are other possible utility charges.

Charge Accounts. Credit cards are everywhere today. You can obtain credit cards for department stores and gas stations. You can easily qualify for cards from major credit companies that will allow you to purchase just about any product or service imaginable. However, many financial experts warn that credit cards are an unwise and dangerous financial tool. They are so convenient to use that you can easily be tempted to purchase items that you would never consider buying otherwise. The convenience is increased by the fact that no actual cash passes from your hand to the vendor, so you never have the emotional realization that you have spent money. The emotional detachment experienced when you use a credit card is referred to as "**plastic prosperity disease.**"[16]

Medical/Dental Bills Not Covered by Insurance. Even if you have insurance, most plans only cover 80 to 90 percent of medical bills. My son recently had surgery on his elbows. He checked into the hospital at 6:00 a.m. and got out the same day at 3:00 p.m. The total bill for the hospital, doctor, and related services was $14,200 for his nine-hour stay. Even with insurance, unless you have an emergency fund set aside, medical and dental bills can cause financial hardship. Prescriptions, office calls, and eyeglasses are also included in this category.

---

16. Ibid., p. 102.

Transportation. Car payments are a fixed expense; however, gas, oil, repairs, license plates and registration, your driver's license, and servicing must also be considered when you are drawing up a budget.

Household Maintenance and Repair. Some expenses in this area include gardening, cleaning supplies, appliance and TV repairs, electrical repairs, and the dreaded unexpected expenses, such as a new roof or water heater.

Child Care. Besides food and clothing, there will be some babysitting expenses. Depending on your lifestyle, there could also be day care expenses. Make sure that a second income is not an economic myth. What may appear on the surface to be an economically advantageous thing to do may prove, with some serious budgeting, to be infeasible or not worthwhile (Figure 8–10).

Food. This category also includes all nonfood items that are part of your supermarket bill. Eating out may be included in this category, or you may put it in the recreation/entertainment category. Regardless of where you place groceries and eating out in your budget, consider funding them with actual cash. When you write a check or use a credit card, you register very little emotion; however, you certainly register an emotion when you spend cash. The more of it you spend, the more emotion you experience. As a result, you tend to spend less when you use cash.

Second Income ($18,000 Annual Salary)		
Monthly	**Yearly**	
$1,500	$18,000	Income
		Minus:
-550	-6,600	Taxes and payroll deduction
=$950	**=$11,400**	**Take-home pay**
		Minus:
-100	-1,200	More extensive wardrobe
-45	-540	Extra dry cleaning
-50	-600	Extra mileage depreciates car
-50	-600	Extra maintenance on car
-560	-6,720	Day care, two kids, $65 each per week
-50	-600	Extra meals out due to fatigue
=$95	**$1,140**	**Real Net Income**

Figure 8–10 In some cases, earning a second income may not be the economically correct thing to do, as shown by this real-life example. (Courtesy of Dave Ramsey, *Financial Peace*, Lampo Group, Nashville, TN: 1993)

Funding the food and eating-out categories with cash has been found to reduce spending by more than $100 per month.[17]

Personal Maintenance. A part of self-concept includes making ourselves look and feel good, and helping to project a professional image. Examples of such expenses are clothing purchases, laundry, barber and beauty salon visits, and grooming and health products.

Self-Improvement/Education. We all want to stay up-to-date and know what is happening. Leaders continually try to improve themselves and learn. Examples of educational expenses are books, magazines, newspapers, seminars, music lessons, tuition, and travel and conference expenses.

Recreation/Entertainment. Recreation and entertainment are very important and should be included in a budget. Examples of expenses in this area are movies, sports, restaurant visits, vacations, parties, CDs, and photographs.

Other Expenses. There are always unexpected expenses, so budget for them. Expect the unexpected. It can come in the form of your car breaking down, children, injury, loss of work, and so on. There are gifts (graduation, birthday, wedding), traffic tickets, and holiday decorations and presents, among many other expense items.

Figure 8–11 shows a pie chart giving the portion of your income for fixed and variable expenses. It is a compilation drawn from several sources. These are only recommended percentages, and they will change dramatically accordingly to your income and lifestyle.

I hope this section on fixed and variable expenses did not depress you, but rather motivated you to give special attention to the next section, which covers budgets and budgeting.

Planning and Preparing a Budget

A *budget* is a plan for spending and saving money. A good budget is an essential tool in every household. Most of us know this, yet 90 to 95 percent of American households operate without a detailed,

17. Ibid., pp. 168–169.

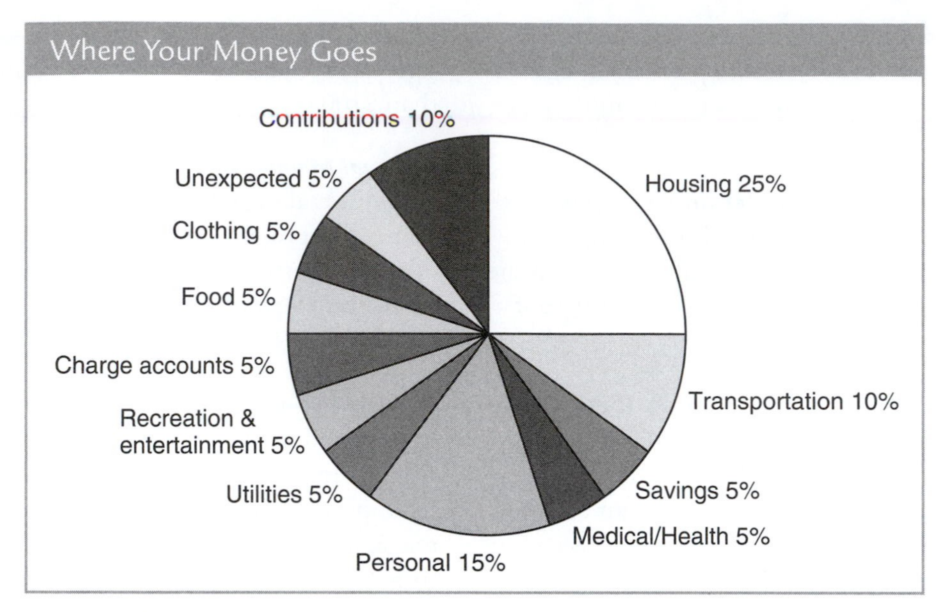

Figure 8–11 A compilation from several sources showing recommended percentages of fixed and variable expenses. Obviously, these percentages will change somewhat according to your income.

accurate written outline of income and expenses (a budget). By failing to plan, we are planning to fail, and there will always be too much *month* left at the end of the *money*.[18]

Benefits of a Budget

Simply put, everyone needs a written budget. A written budget helps you remember to pay all your bills and protects you from unexpected expenses. At times, your budget may even tell you that you cannot afford to eat out or that your vacation will have to wait.

Budgets may appear to restrict you, and in the beginning may seem like torture to prepare. With time, however, the process gets easier, and you will find that your money goes farther. You may also find that you gain confidence in your ability to take control of your finances. Devoting the time to write a simple budget will actually give you more free time and more money to do the things you enjoy.

Here are some other benefits of a budget:

- A budget puts you in control of your financial future.
- A budget ensures that you do not spend more than you earn.
- A budget helps you prepare for major periodic expenses, such as car insurance, medical insurance, and vacations.
- A budget helps you save money to prepare for unpredictable expenses, such as medical bills or major car repairs, as well as for special opportunities, such as a sale on needed furniture or a special family event in another city.[19]

Budget while Still Living at Home

While you are still living at home, you obviously do not have as many expenses. However, you should start preparing a personal budget immediately. Figures 8–12 and 8–13 show two simple examples. Figure 8–12 shows a weekly budget, and Figure 8–13 shows a budget that is more appropriate as a monthly or even yearly budget.[20]

Living on Your Own Budget

Once you move out of your family's home to attend college, or simply start living on your own, your household and monthly expenses ratchet up another notch. Many things that you formerly took for granted now are major expense items. Examples are rent, electricity, telephone, laundry and dry cleaning, groceries, and even medicine-cabinet supplies. If you have a roommate, some of

18. Ibid., pp. 157–158.

19. *Money Skills*, p. 141.

20. *High School Financial Planning Program* (Denver, CO: College for Financial Planning, 1988), pp. 54–57.

Personal Budget
One Week

Income (after taxes)	
Part-time job	$53.00
Allowance	12.00
Total	$65.00

Savings & Spending	
Savings	$15.00
Food	8.00
Clothing	13.00
Entertainment	10.00
Gifts & contributions	6.50
Transportation (bus)	7.50
Miscellaneous	5.00
Total	$65.00

Figure 8–12 Example of a personal weekly budget while still living at home.

Personal Budget

NAME: ______________________

TIME PERIOD: ______________________

INCOME (after taxes)	
1	$ ________
2	________
Allowance	________
Other ____________	________
TOTAL	$ ________

EXPENDITURES	
Savings (emergency/opportunity)	$ ________
Food	________
Clothing/shoes	________
Entertainment	________
School supplies	________
Gifts and contributions	________
Transportation	________
Insurance	________
Loans/credit payments	________
Other ____________	________
Miscellaneous	________
TOTAL	$ ________

Figure 8–13 Example of personal monthly or yearly budget while still living at home.

the expenses can be shared. Figure 8–14 gives an example of items to consider in a budget for people living on their own. It also includes columns in case you have any roommates.

Family Budgets

For those who choose to start a family, the budget is somewhat different because of added responsibility. A good budget changes as your life changes. You will need to make periodic adjustments. You may have budgeted too little for some areas and be strained; you will need to adjust. In other areas you will have budgeted too much. It will take you a few minor adjustments to bring the plan to a realistic level.

A budget is not intended to complicate your life; on the contrary, when you begin to know where your cash is going, it will make life easier. You cannot possibly realize that you are spending too much on a category if you do not track your spending.

Figure 8–15 offers an example of a family budget. A few items require some explanation. Charitable gifts are technically optional, but they may not be an option to many people psychologically. Many people value helping humankind in some fashion. Emergency and retirement funds are discussed later in this chapter.

What is the discretionary income item? Plan to blow, waste, or not account for some portion of your money. Even if you do not plan for this, you will do it anyway.[21] Most people desire a discretionary income category in their budget. The problem is that for some people, their entire budget is discretionary. However, a budget that is too regimented is unrealistic. Allowing some discretionary income for flexibility or as a cushion in your budget is wise and will ensure that the budget process works for you.[22]

Financial Management and Security Tips

Here are 16 tips to help you better manage your money. These "sweet 16" money tips can change your financial life if you adhere to them.

Limit Debt

Do not believe that debt is a way of life. Except for a home mortgage, there are ways to stay out of debt if you develop the proper mindset. Unfortunately, we want our dessert first. We borrow as if we had forgotten that we have to pay it back.

21. Ramsey, *Financial Peace*, p. 166.
22. Ibid.

Living on Your Own Household Expenses								
STARTUP EXPENSES								
Expense	**Cost**				**Cost per roommate**			
					#1	**#2**	**#3**	**#4**
First/last month's rent								
Security deposit								
Telephone deposit								
Moving expenses								
Refrigerator, if needed								
Small appliances, if needed								
Total								
MONTHLY EXPENSES								
Expense	**Monthly cost**	**Annual cost**	**Fixed, flexible, discretionary?**	**Shared? (yes/no)**	**Cost per roommate**			
					#1	**#2**	**#3**	**#4**
Housing								
Rent								
Gas, oil, and water								
Telephone								
Renter's insurance								
Food								
Groceries								
Dining out								
Transportation								
Car payments								
Gasoline, oil								
Car insurance								
Car maintenance and repair								
Public transportation								
Personal								
Clothes								
Dry cleaning, laundry								
Education (tuition and books)								
Personal allowances								
Health								
Doctors and dentists								
Prescriptions								
Health insurance								
Savings and investment								
Savings								
Recreation								
Vacation								
Movies, theater, cable TV								
Subscriptions								
Other								
Gifts and contributions								
Credit card payments								
Bank charges								
Total expenses								

Figure 8–14 Examples of a budget of household expenses for those living on their own or with roommates. (Courtesy of NationsBank)

Family Budget

	Amount Paid Monthly	Total Annual Amount
Charitable Gifts		
SAVINGS		
Emergency fund		
Retirement fund		
College fund		
HOUSING		
Mortgage/Rent		
Furnishings		
Property taxes		
Electricity		
Heating		
Water		
Garbage collection		
Telephone		
Yard work		
Repairs/Maintenance		
Remodeling		
Cable		
TOTAL		
FOOD (Toiletries, Cosmetics, etc.)		
CLOTHING		
TRANSPORTATION		
Car payments		
License plates		
Gas and oil		
Maintenance/Repairs		
Parking		
New tires		
TOTAL		
ENTERTAINMENT		
RECREATION		
Eating out		
Babysitters		
Periodical subscriptions		
Vacations/Holidays		
Clubs		
Hobbies		
TOTAL		

	Amount Paid Monthly	Total Annual Amount
MEDICAL EXPENSES		
Hospitalization Ins.		
Physicians		
Dentists		
Optometrists		
Pharmacists		
TOTAL		
INSURANCE		
Home		
Life		
Disability		
Automobile		
TOTAL		
CHILDREN		
School lunches		
Allowances		
Tuition		
Lessons		
Club dues		
Child care		
TOTAL		
GIFTS		
Graduations		
Holidays		
Birthdays		
Anniversaries		
Weddings		
TOTAL		
MISCELLANEOUS		
Organizational dues		
Cleaning/Laundry		
Pet care		
Beauty/Barber		
Credit Card 1 (Hopefully 0)		
Credit Card 2		
Credit Card 3		
Discretionary $$		
TOTAL		
TOTAL LIVING EXPENSES		

Figure 8–15 Example of a budget for those who choose to start a family. Special features of this budget include savings for an emergency fund, retirement fund, and college fund. There is also a discretionary item.

One of the biggest challenges in managing money is impulse spending. When we see something, we want it and we buy it. The next time you're tempted to make an impulse purchase, take the time to think it through overnight. Talk it over with a family member. Calculate the true cost, including the interest you will have to pay if you use a credit card for the purchase. You may be shocked at how much the item really costs when you factor in interest.

Once you have determined what you want to buy, do some comparison shopping. *Consumer Reports* regularly rates items based on quality, price, durability, and so on. Additionally, many retailers now have Websites that allow you to compare prices and even purchase items online. Allow for shipping costs (and sales tax where applicable) when you compare online prices and prices at your local store. Some time spent on research and comparison shopping can save you hundreds of dollars.[23]

If you had no debt, how much money could you save every month? We will learn later in this chapter about the magic of compound interest. With a totally freed-up budget, you could be wealthy—wildly wealthy—within just a few years. However, we are strapped with debt. An old Middle Eastern proverb says, "The rich rule over the poor and the borrower is servant to the lender."

Do Not Use Credit Cards

The typical consumer owns from five to seven credit cards. As previously mentioned, many financial advisors warn that credits cards are unwise and dangerous financial tools. Their convenience makes you buy "stuff" you cannot afford and gives you a false sense of prosperity. This "plastic prosperity disease" worsens because you do not experience the emotional reality that comes from spending actual cash.

People generally contend that they can control their credit card use. Addiction therapists who counsel people with spending compulsions point out that the first level of treatment involves dealing with the addict's denial. That is, the addicts do not believe they have a problem.

Consider the following:

- About 20 percent of Americans have charged the maximum on their credit cards.
- About 25 percent of adults in the United States have had credit problems.
- Average credit card debt in America is $8,400 per household.
- In the United States, about 24 percent of personal expenditures are made with bank credit cards, retail cards, and debit cards.
- In the first quarter of 2002, total credit debt was $660 billion; of that, total credit card debt was about $60 billion.
- Approximately 185 million Americans have at least one credit card.
- 1.3 million credit card holders declared bankruptcy in 2002.
- The average interest rate on credit card balances is 18.9 percent.
- The average household pays $83.33 in monthly credit card interest.
- On average, the typical credit card purchase is 112 percent higher than if cash were used.
- More than 40 percent of American families spend more than they earn.
- As of 1995, 92 percent of American family disposable income is spent on paying debts.
- An $8,000 debt, at an 18 percent interest rate, will take more than 25 years to pay off and cost more than $24,000.[24]

The credit card marketing companies want you to just sign and spend; they will not help educate you on the responsible use of a credit card. The problem is that most students have not been properly educated as to the harsh realities of credit—and there are many consequences to bad credit. Consider the following:

- You cannot open a checking account.
- You cannot get another credit card.
- You cannot move into an apartment.
- You cannot buy furniture, computers, or anything on credit.
- You cannot get a mortgage.
- You will pay double the interest on car loans, and excess dealer fees.

23. http://creditcounselingbiz.com/education.htm.

24. http://creditcounselingbiz.com/credit_counseling_statistics.htm.

- In some cases, you cannot get a job, if employers run credit checks.
- You may not be able to rent a car.

Credit Card Alternative. A great credit card alternative is called a *check card.* A check card (also called *debit card*) is a plastic card that looks and acts like a credit card, but the transaction amount is deducted from your checking account, just as though you were using your ATM card in the grocery store checkout. When you make a purchase at a restaurant or store with a check/debit card, the merchant treats it as though it were a credit card; however, when the transaction is completed, the money is taken directly out of your checking account. With a check card, you incur no interest because the money comes right out of your bank account. This can save you thousands of dollars annually because it forces you to pay for purchases on the spot, and you can never spend more than you have in your account. You are prevented from running up a high credit card debt. This is a great alternative to secured credit cards. Just make sure you save all your receipts, and don't forget to enter them in your checkbook register. Also, be careful not to spend too much money with a check card. It's not like running up a balance with a credit card. When you use a check card, you are actually running down your checking account, so be careful not to overdraw.[25]

Credit Scores. Credit scores usually range from 300 to 900—the higher the better. Credit scores are created by credit bureaus when they put together your credit report. They do it using software from the Fair Isaac Company; hence the name *FICO score*. Calculation of a credit score is very complicated, as it is based on more than 100 parameters in your credit file, including length of credit history, number of open accounts, loans, mortgages, public records, and other items.

Other companies may also generate credit scores, but most lenders use the Beacon FICO score. For lenders, your credit score is a predictor of your ability to pay. The higher your score, the more likely it is that you will pay back your creditors on time. According to some analysts, 51 percent of all people with a credit score from 550 to 599 will default on their debts.[26]

How does a low credit score affect you? Your credit score is the single most important factor determining whether you will get approved for a home or car loan, a refinance loan, or a credit card, and what your annual percentage rate (**APR**) will be. If your credit score is low, you will be offered only very high interest rates, up to 23 percent. Most people do not realize that your credit score also affects how much you will pay for car insurance, too. Many insurance companies run a credit check before selling insurance to a new client.

Maintaining a high credit score should be an ongoing process, not a task you hurriedly begin when you need to apply for new credit, a home loan, or a car loan. That is the worst time: Banks will reject your loan application if you have a current dispute as to any items on your credit report. It can take 60 days to clean up your credit score, so do not apply for any new or more credit until all disputes are resolved.[27]

What should your credit score be for credit and loan approval? If your credit score is above 680, you are considered a "prime borrower" and will be eligible for a good APR on your loan or credit card. If your credit score is below 680, you are "sub prime," and will pay a much higher APR on any loans. If it is below 550, you most likely will not be approved for a home loan, car loan, or credit card, regardless of the interest rate.

You can find your credit score online at Equifax, Consumerinfo, or TrueCredit. However, there will probably be a charge to do so.[28]

Protect Your Credit Rating. Although you have been warned not to use credit cards, if you do, never, ever have more than one. Even then, set a low limit, such as $500, in case of an emergency. If you still insist on keeping your card, here are some tips that will help keep you on track in protecting your credit rating:

- Start an accurate budget spreadsheet and always stay within your budget.
- If you max out your card, do not start using another credit card; pay off your current one.

---

25. http://www.debtwizards.com/creditcards.html.

26. http://www.debtwizards.com/creditreports.html.

27. Ibid.

28. Ibid.

- Pay your balance in full when you get your statement, to avoid paying interest.
- Never pay just the minimum payment; always send more, or it will take years to pay off.
- Never use your credit card for cash advances. You will incur transaction fees and pay 19.3 percent interest (or more).
- NEVER, EVER take out a cash advance to pay another credit card bill.
- Avoid frequent trips to an ATM if it is not affiliated with your bank. Many charge fees of more than $1.50 just to use them. Avoid high-fee ATMs at highway rest stops, night clubs, airports, gas stations, and so on.
- When dining with friends, do not collect cash from them and pay the bill with your credit card. That subjects the whole amount to an 18 percent interest rate.
- When buying clothes and CDs or dining out, use cash or a debit card instead of a credit card. Do not use your credit card to buy groceries.
- If you cannot afford to make your purchase with cash, you should not be buying it on a credit card.
- If the credit card company sends you checks to "pay your income tax" with, do not use them. They are treated as cash advances.
- Always read the fine print on balance transfer checks. Be sure they are not treated as cash advances.
- Leave plenty of room on your card for emergencies.
- NEVER, EVER go to a "credit doctor." Most are scams and should be avoided.
- Do not file bankruptcy; it stays on your record for 10 years. Talk to a nonprofit credit counseling service first.[29]

Save Big Money on Home Mortgages

If you must borrow, borrow on short terms and only on items that go up in value. That means you never borrow on any consumer items except homes, land, or livestock. The terms are very important. If you are able, buy less so you can pay the loan off faster. Make sure you get the lowest possible interest rate. You may pay more than three times the amount borrowed on a 30-year mortgage if you pay normal payments for the full term. Less than 10 percent of your payments are applied to the principal for the first 15 years at an interest rate of approximately 10 percent. Furthermore, it takes approximately 24 years to reduce your mortgage to one-half the amount financed. Consider the following examples:[30]

15-Year Rather than 30-Year Loans. If you were to finance $80,000 on a home at 10 percent, here are two ways to pay:

30 years: 360 payments at $702 per month = $252,720 total
15 years: 180 payments at $860 per month = $154,800 total

The total difference is $158 more in payments monthly, but savings of $97,920.

Pay an Additional Payment per Year. The effect of paying one additional payment per year on your home mortgage substantially reduces the principal.

30 years: $75,000 at 10.75 percent; monthly payment = $700.11
(360 payments = $252,740 principal and interest)

Assume that from each monthly payment of $700.11, $534.00 is interest, and $157.11 is paid on the principal. By paying one additional payment per year of $700.11, you will pay four and one-half months ($700.11 divided by $157.11) of principal, thus reducing your mortgage payoff in direct proportion. In other words, you can pay off your loan in approximately one-third less time by making one extra payment per year. You save approximately $84,000.

Divide the Total into Two Monthly Payments. By dividing the monthly amount into two monthly payments rather than one lump sum, you will have savings similar to the additional payment per year just discussed. However, all loans are not set up this way. If your principal can be reduced by extra payments, then it can be reduced quicker by the divided payments.

29. Ibid.

30. Cliff Ricketts, Science IA (Agriscience) (Nashville, TN: Tennessee Department of Education, 1990), pp. 8–19.

Interest Rates	Total Cash
At 0%	$ 81,213
At 5.5%	$106,880
At 8%	$121,969
At 10%	$135,988
At 12%	$152,036

Figure 8–16 By paying your home off early and then continuing to put the same payment into some kind of savings account, you can accumulate a substantial amount of money.

Increasing Retirement Benefits. To retire with comfort, use one of the three techniques just discussed. This will lead to a 30-year mortgage being paid in full 9 years and 8 months early. Because your budget is adjusted to this payment, using self-discipline, continue until the remainder of the proposed mortgage length is paid off, but place this monthly payment in an investment. Depending on the rate of interest earned at the end of 30 years, not only will your home be paid for, but more than $100,000 will now be available for your retirement. If your original payments were $700.11 per month and you put that amount in investments for the remainder of the original mortgage term, you will see what happens by examining Figure 8–16.

Reduce Debt by Using the Debt Snowball Technique

Rationale. If you accumulate debt, what is the best way to get out of it? The best way is the "debt snowball" technique. The first step in this technique is to put your debts in ascending order, with the smallest remaining balance first and the largest last. Do this regardless of the interest rate or payment. These will be paid from top to bottom. This works because you get to see some success quickly and are not trying to pay off the largest balance just because it has a high rate of interest. Consider the example in Figure 8–17.

Procedure. Follow this strategy as you refer to Figure 8–17. Suppose you get an extra job or have a garage sale and can pay off the Sears card in the first month. Next, do not spend the $60 per month you used to spend on the Sears card; instead, add $60 to the next payment on the list. You are then paying Castner Knott $130 per month until paid. Castner Knott will be paid off in the seventh month. Next, add $130 to the $200 Den Charge payment to make your payments $330. Because you have already been paying on the Den Charge for seven months, it will be paid off in the eighth month (the next month).

Continue the "snowball" by adding $330 to your Medicount payment of $250, making your Medicount payment $580. This will cause Medicount to be paid off in 17 more months, only 25 months into the program. Continue this process for the Credit Union to pay it off in four additional months.

Debt Snowball Payoff Technique

ITEM	BALANCE	PAYMENT	RATE
Sears	400	60	18%
Castner Knott	700	70	18%
Den Charge	1200	200	18%
Medicount	6500	250	12%
Credit Union	7000	123	9%
House	60,000	540	9%

Figure 8–17 The debt snowball payoff technique works by arranging debts in ascending order, with the smallest remaining balance first and the largest last, regardless of interest rate. Once the first bill is paid off, apply that amount of money to the next bill. Continue this procedure until all bills are paid, except, perhaps, your mortgage.

What about the House? Since you are having so much fun, you may be wondering how to pay off the house. Now you have $1,243 per month with which to pay on the house, and it will be paid for in five more years.

I Need a Car. As an alternative, you could save $275 per month of the $1,243 to purchase a car with cash. When you pay the remaining $968, the house will be paid off in just seven years, while the $275 per month saved will grow to $11,490 in just three years (at 10 percent) for your cash car purchase.[31]

31. Ramsey, *Financial Peace*, p. 118.

Attitude toward Debt. As these bills are paid off, you will change your attitude about financial matters. You will start using credit very cautiously and incurring debt very sparingly. Depression and misery do not have to control you. Do not get into a debt mess again. If you do, remember you will instantly become the servant of the lender.

Develop Power over Purchases

Learn to manage your money, or it will manage you. Realize that money is amoral—it is neither good nor bad. Remember that the popular expression says that the *love* of money, not the money itself, is the root of all evil. Your abundance of money or your lack of money will reveal a great deal about your character.[32]

Learn the concept of power over purchase. Do not get a case of "buyer's fever," where you want something so badly that you rationalize that you need it or you deserve it. When you get this fever, you lose patience, perspective, and bargaining power. If you find yourself in this situation, determine that you will sleep on it; maybe even wait 48 hours before deciding to purchase an item. If you do not make the purchase, chances are that there will be another "one-time" opportunity waiting for you. Good spending habits develop when you have power over your purchasing.

Live within and below Your Income

Avoid the lifestyles of the rich when you are not rich. Henry David Thoreau said, "Almost any man knows how to earn money, but not one in a million knows how to spend it."[33] You must limit your style of living, figure out what your income is, and live far below that mark.

Paying Cash for a Car versus Going into Debt

In his book *Financial Peace,* Dave Ramsey gives the following example.

Suppose you were to purchase a new car for $16,000 by going into debt to be paid off at $300 per month. What if, instead, you bought a preowned car for $5,400 at $100 per month and saved the other $200 per month at 10 percent for seven years? You would have $24,190 at the end of seven years. By then, either car will be nearly worn out, but let's say you saved that $24,190 and went out and bought a $16,000 car for cash from that savings, leaving $8,190.26.

You now have a new car with no car payments and $8,190 in the bank—but let's think ahead for yet another seven years. You have no car payment, so instead of the $100 payment and the $200 savings you were putting aside every month, now you decide just to save $100 per month, freeing up $200 per month. At the end of seven more years, that leftover $8,190, plus $100 per month and the 10 percent interest, will grow to $28,539. Again, you have a seven-year-old car that is worthless, so you buy another $16,000 car for cash, leaving $12,539 in savings. For the next seven years you will have no car payment, and though you contribute no more to savings, the $12,539 at 10 percent will grow to $24,436.[34]

Negotiate and Buy Only Big Bargains

Another useful skill to develop is comparison shopping. Some people spend more time deciding which brand of laundry detergent to buy than they do in selecting which major household appliance to buy. When you comparison-shop, especially for large purchases, you may be able to purchase an item with more desirable features and still save hundreds of dollars. When buying, learn three essential things: how to negotiate, how to find bargains, and how to have patience.

Impulse is the buyer's enemy. Take your time and look for a bargain. You will find that most often you can negotiate on price. Most people try to avoid confrontation, so you may gain an advantage in negotiations if you are willing to just ask for a better deal. Strive to create a "win-win" situation in the negotiation. You win by getting a better price or better terms, and the seller wins by getting a necessary. Here are seven basic principles of negotiating:

1. *Always tell the truth.* There is never enough money to be made, or lost, on a deal to warrant a lack of integrity.
2. *Use the power of cash.* People get silly when they see cash. There is something highly emotional and powerful about the effect of flashing cash when making a purchase.

32. Ibid., p. 20.
33. Ibid., p. 70.
34. Ibid., p. 71.

3. *Understand and use "walk-away" power.* You must be prepared to walk away and not make the purchase. If the seller senses (and he or she usually can) that you are married to that purchase, you will receive no discounts. Sometimes this is not a bluff, which means you must simply walk away to buy another day.
4. *Talk sparingly.* When negotiating a purchase, we may sense the inherent confrontation, get nervous, and talk too much. Just make small comments and let the opposing negotiator rattle. For example, say, "But it seems your price may be too high." Then be quiet and see just how far the seller will come down.
5. *"That's not good enough."* Everyone has some flexibility to expand or reduce anything, and the same principle holds true in negotiating. When the price is given, reply with, "That's not good enough—what can you really do?" You will see the price drop, sometimes lower than you would have offered. Instead of giving a lower offer, try using that statement.
6. *Good guy–bad guy.* A parent or another person who is not with you should always seem mean to the other side. "My mom would kill me if I took that price," or "You know, my dad wouldn't like it if I came home with that if I didn't get a good buy." Note: When you use this technique, you must be telling the truth.
7. *If I give, I also take.* When you reach a point where you must give up something (money, for example), be sure you take something at the same time. You should say, for example, "If I give you $2,500 for that entertainment center, then you have to throw in the cabinet [or something else] and the sales tax at no extra charge."[35]

Another skill to master in getting great bargains is to look for hidden treasures. Rather than shopping only at retail stores, consider buying from individuals, garage sales, public auctions, and flea markets. Use coupons and classified ads to your advantage. If you are looking for real estate, read the foreclosure notices in your local newspaper. Bargains abound if you just look, and if you accept the premise that even good things can be bought at a bargain.

Even with knowledge of where to look for bargains and how to negotiate price, you must still learn to have patience. Without the discipline of patience, you can miss a true bargain. The old adage says that "patience is a virtue." There must be something to that, as the saying has stood the test of time. We must begin to think like vultures. Vultures are not very pretty, but the patience they display in the wild is something we should emulate.

Build an Emergency Fund

A good financial planner will tell you that from the start, you should have three to six months of income in savings that are available (liquid) for emergencies. An example of **liquid income** is a simple bank savings account or a money market fund on which you can write checks. If you make $24,000 annually, you should have $8,000 to $12,000 located where you can easily get to it before you do any other investing. As mentioned previously, you have to expect the unexpected.

Manage Your Budget Busters

Payments that are made on a nonmonthly basis can destroy your budget unless they are planned for. If you convert them to a monthly basis and use the "envelope system" (discussed next) to set money aside each month, you can avoid strain or borrowing when these payments come due. Refer to Figure 8–18 for examples of nonmonthly payments. When you prepare your own sheet of nonmonthly payments, adjust it to your budget. If you make a payment quarterly, then adjust the payments monthly for this sheet.

Envelope System

The time-honored envelope system of cash management is a budgeting system recommended by most good financial counselors. Items for which you use a credit card or a check, for the supposed convenience, will likely destroy your budget if you continue the practice.

If you get paid once a month, put cash in an envelope for each budgeted category. If your grocery and food budget is $300 monthly, put $300 cash in an envelope for food each payday. As you buy food, take the money from that envelope, and from nowhere else. This provides you with instantaneous cash management, in that you will almost

35. Ibid., pp. 81–86.

Nonmonthly Payments for Which to Plan

ITEM NEEDED	ANNUAL AMOUNT		MONTHLY AMOUNT
Real estate taxes	______	/12 =	______
Howeowner's insurance	______	/12 =	______
Home repairs	______	/12 =	______
Replace furniture	______	/12 =	______
Medical bills	______	/12 =	______
Health insurance	______	/12 =	______
Life insurance	______	/12 =	______
Disability insurance	______	/12 =	______
Car insurance	______	/12 =	______
Car repair/tags	______	/12 =	______
Replace car	______	/12 =	______
Clothing	______	/12 =	______
Tuition	______	/12 =	______
Bank note	______	/12 =	______
IRS (Self-employment)	______	/12 =	______
Vacation	______	/12 =	______
Gifts	______	/12 =	______
Other	______	/12 =	______
Other	______	/12 =	______

Figure 8–18 This example will, we hope, help you manage your budget. Without proper planning, bills that are not on a monthly cycle can destroy your budget.

never spend more than you allotted. If you have had a hard week and you "deserve" to go out to eat on Friday night, simply pull out the "Food" envelope and look into it to see if you can afford to do what you want to do.

Again, refer to Figure 8–18 for the nonmonthly budget busters. Prepare an envelope for each of these items. You can put cash in the envelope, or, for security, you can put play money in the envelope and put the real money in a savings account. However, make sure there is enough real money in your savings account to cover all the play money in the envelopes. Then, as the payments become due, the money will be in place and you will be able to manage the payments.

Have Enough Insurance

Every person should buy the insurance needed to financially secure the well-being of themselves and their families. You may think that you cannot afford it, but you cannot afford to be without it. Just do it, even if paying the premiums makes the budget tight. Make sure you have life insurance, health insurance, disability insurance, auto insurance, and homeowner's insurance. Without these, you have great potential for true financial disaster, including bankruptcy. By having enough insurance, you will have greater control over your destiny and show others that you are competent and capable.

Master the Magic of Compound Interest

You do not have to have a fortune to invest to enjoy the magic of **compound interest**. You can accumulate a substantial amount of money by "investing" in a small way.

If you invest the money used to purchase your afternoon soft drink, see Figure 8–19. If you quit smoking cigarettes, see Figure 8–20. If you save $2.40 a day, see Figure 8–21. You see, it does not take a lot of money to begin enjoying the magic of compound interest. It truly is amazing.

Use an Automatic Savings Plan

If you have the opportunity, use a payroll deduction savings plan. If you do not receive the money, you will hardly miss it. The accumulation in your savings will motivate you to do even more. If you get a raise, put a percentage of your raise in savings. This is an excellent way to build your emergency fund.

Take Advantage of Matching Stock Purchases

Some larger corporations offer their employees an opportunity to buy stock in their companies. Some companies match employee contributions dollar for dollar; some match 50 cents per dollar, and there are many other options. In reality, you will get 100 percent return on your investment if every dollar is matched dollar for dollar. Do not miss this opportunity if it is provided to you.

Use Your Hobby to Make Money

You can have a hobby that costs money, or you can have a hobby that makes money and have just as much fun. We discuss this further in "The $2,000

$.50/day @ 5 days/week = $10.00 month							
	10 yrs.	20 yrs.	30 yrs.	40 yrs.	50 yrs.	60 yrs.	70 yrs.
5%	$1,559	$4,127	$8,357	$15,323	$26,797	$45,695	$76,820
10%	$2,065	$7,656	$22,793	$63,767	$174,687	$474,952	$1,287,781

Figure 8–19 Example showing the magic of compound interest through saving 50 cents per day that you would spend on a soft drink.

$2.40/day for 30 days = $72.00 month		
Years	**5%**	**10%**
10	$11,178.92	$14,745.79
20	$29,593.08	$54,664.39
30	$59,922.59	$162,721.41
40	$109,877.56	$455,224.63
50	$192,157.16	$1,247,011.58
60	$327,677.84	$3,390,326.69
70	$550,890.58	$9,192,139.44

Figure 8–20 Example of the magic of compound interest for a person who quits smoking.

$1/day @ 5 days/week = $20.00 month		
Years	**5%**	**10%**
10	$3,118	$4,131
20	$8,254	$15,313
30	$16,714	$45,586
40	$30,647	$127,535
50	$53,595	$349,375
60	$91,390	$949,904
70	$153,640	$2,575,561

Figure 8–21 Example of the magic of compound interest for a person who saves $1.00 a day, five days a week, or who saves $20.00 per month.

Work-Unit Approach" later in this chapter. Examples are coins, antiques, books, stamps, phonograph records, autographs, baseball cards, old car restoration, undeveloped plots of land, metals, bonds, art, stock, and rental property.

The first step is to become knowledgeable about the type of investments in which you want to put your money. The next step is to commit a certain amount of your time and energy (perspiration equity) toward making your investments prosper. The third step is to get started. Explore a variety of investments, find mentors to guide you, constantly upgrade your investments—and then enjoy them. You can use the profits from your hobby money to accumulate great wealth, as discussed in the next section.

Retiring as a Millionaire

Is it possible for an agribusiness manager or worker to retire as a millionaire? It is very possible. Although nothing in life is guaranteed, things certainly will not happen unless you make them happen. If you think you can, you can. If you think you cannot, you cannot. Either way, you are correct. Your attitude determines your altitude. Being a millionaire may not be a goal of yours. However, I believe most people would like to know how it could be done.

The four following plans assume that you can earn a 12 percent interest rate. This will be discussed later.

Plan One: $2,000 a Year for Six Years

Starting at age 22, put $2,000 a year into a tax-free retirement account, **individual retirement account (IRA)**, or **tax shelter account (TSA)** under 403(b), for six years, and then stop. At age 65, $12,000 will compound to $1,348,000 if a 12 percent interest rate is attained. Figure 8–22 illustrates this. Notice the importance of time and consistency. If you wait until age 28 to start, you will have to put in $2,000 a year until age 65 ($76,000) to get the same amount of money.

Plan One: $2,000 a Year for 6 Years (12 Percent Interest Rate)

	EXAMPLE A		EXAMPLE B	
Age	Payment	Accumulation End of Year	Payment	Accumulation End of Year
22	$2,000	$2,240	0	0
23	2,000	4,749	0	0
24	2,000	7,559	0	0
25	2,000	10,706	0	0
26	2,000	14,230	0	0
27	2,000	18,178	0	0
28	0	20,359	$2,000	$2,240
29	0	22,803	2,000	4,749
30	0	25,539	2,000	7,559
31	0	28,603	2,000	10,706
32	0	32,036	2,000	14,230
33	0	35,880	2,000	18,178
34	0	40,186	2,000	22,599
35	0	45,008	2,000	27,551
36	0	50,409	2,000	33,097
37	0	56,458	2,000	39,309
38	0	63,233	2,000	46,266
39	0	70,821	2,000	54,058
40	0	79,320	2,000	62,785
41	0	88,838	2,000	72,559
42	0	99,499	2,000	83,507
43	0	111,438	2,000	95,767
44	0	124,811	2,000	109,499
45	0	139,788	2,000	124,879
46	0	156,563	2,000	142,105
47	0	175,351	2,000	161,397
48	0	196,393	2,000	183,005
49	0	219,960	2,000	207,206
50	0	246,355	2,000	234,310
51	0	275,917	2,000	264,668
52	0	309,028	2,000	298,668
53	0	346,111	2,000	336,748
54	0	387,644	2,000	379,398
55	0	434,161	2,000	427,166
56	0	486,261	2,000	480,665
57	0	544,612	2,000	540,585
58	0	609,966	2,000	607,695
59	0	683,162	2,000	682,859
60	0	765,141	2,000	767,042
61	0	856,958	2,000	861,327
62	0	959,793	2,000	966,926
63	0	1,074,968	2,000	1,085,197
64	0	1,203,964	2,000	1,217,661
65	0	1,348,440	2,000	1,363,780

Figure 8-22 "Plan One" for potentially retiring as a millionaire involves saving $2,000 per year for 6 years, starting at age 22, and getting a 12 percent interest rate on your investments. Example B shows the importance of time.

You can see the importance of starting early: it's $12,000 if you start at age 22, but $76,000 if you start at age 28, to have $1 million by age 65.

Because of the impact of compound interest over a long period of time, it pays big dividends to start saving early in life. The following example again proves the point:

> Rose scrapes and saves every penny she can, and at age 19, she's able to stash away $2,000 in a long-term investment that earns 10 percent compounded annually. She does that for eight years. At age 27, Rose simply cannot afford to save any more because of family responsibilities and other expenses. But the $16,000 investment made early in life enables her to retire as a millionaire. Refer to Figure 8-23.

> Nancy has a good time while she's young and spends her money on a fancy car. At age 27, she figures it's time to settle down, so she starts saving $2,000 per year until she is 65. It, too, is invested at 10 percent compounded annually. Because Nancy got a later start, she retires on only $805,185. As you can see from Figure 8-23, that is $213,963 less than Rose's retirement. Nancy had to work a lot harder at it—saving $2,000 per year for 39 years for a total of $78,000.

Obviously, these are hypothetical examples. There are not many Roses who will save $2,000 per year. The principle of compounding interest remains the same, regardless of the amount of money involved. There is nothing wrong with enjoying life by spending some of your hard-earned money while you are young, but you need to weigh those expenditures against their impact on your future. In the long run, Nancy's new car cost her $213,963!

You can give your children financial security. Give your child $2,000 when he or she is born, with the understanding that it remains untouched until age 65. Invested at a rate of 10 percent compounded annually, that $2,000 birthday gift will be worth $980,740 at retirement!

Plan Two: Buy a Used Car Instead of a New One and Invest the Difference

Give up on owning a new car before age 23 and buy a used car instead. Invest the difference between a new car and a used car in an IRA or TSA. Consider the following illustration:

> Jeff, a 17-year-old senior in high school, bought a nice pickup with payments of $250 per month for five years. If $166.66 monthly were put into an IRA or TSA, $83.34 would still be left to pay monthly for a used vehicle. By analyzing Figure 8-22, we can **extrapolate** and see that the new truck cost Jeff $2.6 million. The Rule of 72 states that by dividing the interest rate into 72 (72 ÷ 12 = 6), one can determine how long it will take for one's money to double. The additional $1,300,000 in six more years would double to $2.6 million. The Rule of 72 is discussed later in this chapter.

Do not let the allure of a new car cause you to make a serious financial mistake. Despite what advertisers lead you to believe, driving a new car will not make you feel better, look better, or be more successful. Success is not defined by the kind of car you drive. In fact, financing a new car may have the opposite effect on you and your family: it may actually make you a slave to a piece of metal! Not only do you have to make a monthly payment on the vehicle, but you still have to fuel it, clean it, and service it. The average car payment today is somewhere between $350 and $450 per month. This kind of debt can be a crushing weight on the American family. It threatens to destroy the people it supposedly helps. What most people fail to realize is that, depending on the make and model, a new car depreciates (goes down) in value by approximately 25 percent as soon as you drive it off the lot.[36]

What about the Repairs? You should always have a trustworthy and qualified mechanic check out a used vehicle before you buy it. This may prevent you from making a bad purchase. To further protect yourself from costly repairs, you might want to purchase a used car maintenance warranty. Not only might it save you money, it might actually be a positive selling point if you trade the car before the warranty expires. Even if you do have to make some repairs to your used car, you will most likely find them cheaper than repairs on a

36. Art L. Williams, *Common Sense* (Atlanta, GA: Parklake Publishers, 1986).

What You Don't Know Can Cost You

	ROSE		NANCY	
AGE	Investment	Total Value	Investment	Total Value
19	$2,000	$2,200	0	0
20	2,000	4,620	0	0
21	2,000	7,282	0	0
22	2,000	10,210	0	0
23	2,000	13,431	0	0
24	2,000	16,974	0	0
25	2,000	20,871	0	0
26	2,000	25,158	0	0
27	0	27,574	0	0
28	0	30,442	2,000	$2,200
29	0	33,486	2,000	4,620
30	0	36,834	2,000	7,210
31	0	40,518	2,000	13,431
32	0	44,570	2,000	16,974
33	0	49,027	2,000	20,971
34	0	53,929	2,000	25,158
35	0	59,322	2,000	29,874
36	0	65,255	2,000	35,062
37	0	71,780	2,000	40,768
38	0	78,958	2,000	47,045
39	0	86,854	2,000	53,949
40	0	95,540	2,000	61,544
41	0	105,094	2,000	69,899
42	0	115,603	2,000	79,089
43	0	127,163	2,000	89,198
44	0	139,880	2,000	100,318
45	0	153,868	2,000	112,550
46	0	169,255	2,000	125,005
47	0	186,180	2,000	140,805
48	0	204,798	2,000	157,086
49	0	225,278	2,000	174,994
50	0	247,806	2,000	194,694
51	0	272,586	2,000	216,363
52	0	299,845	2,000	240,199
53	0	329,830	2,000	266,419
54	0	362,813	2,000	295,261
55	0	399,094	2,000	326,288
56	0	439,003	2,000	351,886
57	0	482,904	2,000	400,275
58	0	531,194	2,000	442,503
59	0	584,314	2,000	488,953

(continued on the following page)

Figure 8–23 Another example of the importance of getting started early and being consistent with your savings. This example is based on a 10 percent interest rate on your savings.

What You Don't Know Can Cost You *(concluded)*				
60	0	642,745	2,000	540,048
61	0	707,020	2,000	596,253
62	0	777,722	2,000	658,078
63	0	855,494	2,000	726,086
64	0	941,043	2,000	800,995
65	0	1,035,148	2,000	883,185
TOTAL EARNINGS		**$1,019,148**		**$805,185**

new car. Always keep in mind that the purpose of an automobile is to get you from one place to another. The style and luxury you enjoy driving down the road depends on your commitment to stay out of the debt trap!

Plan Three: The 25-Plus-10 Plan

For the first job you take at the age of 22, put $25 per month into an IRA or TSA during the first year. Each year thereafter, increase the monthly contributions by $10. In other words, the second year on the job, you would be putting $35 monthly into your IRA or TSA. Assuming a return of 12 percent, this will amount to a total of $1,830,000 by age 65. Most years, because of inflation, your pay raises will stay ahead of your initial $120 investment, even though you will be investing much money later in life. Thus, the percentage of your salary you are investing will hardly ever be more than when you first started.

Plan Four: The 80-10-10 Plan

Some people's beliefs and values cause them to allot the first 10 percent of a paycheck to charitable contributions. Another suggested strategy is an 80-20 or 90-10 plan. By following this plan, 10 percent goes to charity or as a gift, 10 percent is invested in an IRA or TSA, and 80 percent is used for living expenses. If you do not allocate 10 percent to charity or as a gift, then you could follow a 90-10 plan. Using this plan, by following a budget and buying accordingly, one can retire as a multimillionaire. This can be achieved with little disruption of your lifestyle.

You're probably thinking: "I can't even live off 100 percent of my income. How can I live off 80 percent?" By using the money management tips discussed earlier in this chapter, you can. Plan your work (budget), and work your plan. It is my belief that Americans are somehow programmed to spend 3 to 10 percent more than they make, no matter what their income level. Simply budget as discussed earlier, and you can do it. Buy the lower-priced automobile, get a house with a one-car garage rather than a two-car garage, or do whatever else it takes to live within or below your income level.

Key Factors in Making the Millionaire Plans Work

A farmer does not simply plant a seed and let it grow, hoping for an abundant harvest. He or she has to work on it. Depending on the cropping method and the region of the country, the crop has to be cultivated, fertilized, sprayed for pesticides, and watered, either naturally or by irrigation. The same is true for dairy and livestock animals. Monthly, weekly—and for dairy, daily—care must be given. The same is true with the millionaire plans. They have to be worked at. You do not simply start something and let it grow without nurturing it. Knowledge of the following factors will help you reach the millionaire goal.

Manage Investments to Achieve a 12 Percent Interest Rate

Every plan is simply a mathematical process based on a 12 percent return on your investments. It is an achievable production goal. Ten pigs per litter, 100 bushels of corn per acre, and 18,000 pounds of milk per cow are viable production goals. They are challenges, but they can be achieved, just as the 12 percent interest can be achieved. It may not be achieved every year, but including very good and very bad years, this is our goal.

For a person who is able to save $2,000 per year for 43 years (ages 22 to 65), a 1 percent increase in the interest rate would amount to $316,946. Look at the following figures:

8 Percent Interest

- Ages 22 to 65
- $2,000 per year for 43 years at 0.08 interest compounded monthly
- $839,054

9 Percent Interest

- Ages 22 to 65
- $2,000 per year for 43 years at 0.09 interest compounded monthly
- $1,156,000

Difference in 1 Percent (0.01) Interest Rate. At $2,000 per year for 43 years, a 0.01 (1 percent) interest rate difference equals $317,946. Also, refer to Figure 8–21 for a comparison of returns at 5 percent and 10 percent interest.

The Rule of 72

The "Rule of 72" refers to the process of dividing an interest rate into 72 ($72 \div 12 = 6$) to determine how long it will take for one's money to double. Money that yields 12 percent interest will double in six years. For example, $1,000 yielding 18 percent interest would double in four years. If 36 percent were the interest rate, $1,000 would double to $2,000 in just two years. Figure 8–24 further illustrates this.

Invest in Mutual Funds to Achieve a 12 Percent Interest Rate

There are many types of investment strategies. Take courses, read, and seek advice from financial counselors. Do not simply take our word. However, we believe the novice (inexperienced) investor would do well to select **mutual funds** among many other investment possibilities.

Mutual funds were invented to make investment expertise available to ordinary people. A mutual fund company makes investments solely on behalf of others. All the profits (or losses) go to the fund's shareholders. In existence in the United States since 1924, there are now more than 3,400 different mutual funds holding more than $2 trillion for about 70 million shareholders. This amount rises yearly.

What you get from a mutual fund, at relatively low cost, is professional management of your money by people who devote their full time and attention to investment decisions. Even professionals sometimes get poor results, but the average fund does better than you probably can do for yourself, according to studies by Marshall Blume, chairman of the finance department at the University of Pennsylvania's Wharton School. With a mutual fund, you get a portfolio that usually consists of 50 or more stocks or other securities.

Select and monitor your mutual funds. There are many publications that give information on mutual funds: *Money* magazine publishes an annual yearbook that gives mutual fund data; the

ORIGINAL INVESTMENT	72	% INTEREST	YEARS TO DOUBLE	ENDING AMOUNT
$1,000	72	2	36	$2,000
$1,000	72	4	18	$2,000
$1,000	72	6	12	$2,000
$1,000	72	8	9	$2,000
$1,000	72	10	7.2	$2,000
$1,000	72	12	6	$2,000
$1,000	72	18	4	$2,000
$1,000	72	36	2	$2,000

Figure 8–24 The "Rule of 72" refers to the process of dividing the interest rate into 72 (72 divided by 12 equals 6) and determining how long it takes for one's money to double.

Wall Street Journal and many local newspapers from larger cities list performances of mutual funds each day. Most of the funds have toll-free numbers to call for information. By law, you have to be given a **prospectus** before you can invest with any mutual fund. Consider the following as you select and monitor your funds:

- *Select a mutual fund on the basis of whether the fund has a sales charge* (no-load fund). Many companies charge 4 percent or more just to invest. There are enough good mutual fund companies that do not charge anything.
- *Select a mutual fund based on performance. Money* magazine, among a number of other magazines and newspapers, publishes mutual fund rankings each month, which show the funds that have gained the most during the past month, the past year, and the past five years. Do not select a mutual fund unless it has a lifetime track record of more than 12 percent.
- *Monitor your mutual fund closely.* Remember the earlier analogy of planting seeds? When you purchase a mutual fund, you are only planting the seed; then you have to nurture it.
- *Diversify your mutual funds.* Do not put all your eggs into one basket. You may want to have three to five different mutual funds with different investment objectives. The minimum investment for many mutual funds for IRAs is only $500. If you have $2,000, for example, deposit $500 in each of four different mutual funds.[37]

Types of Mutual Funds. As you select and monitor investments, be aware of the types of mutual funds. Most funds carry a risk rating of high, medium, or low. Furthermore, mutual funds carry this risk to some extent within each of the following investment strategy categories:

- *Aggressive growth:* Strive for maximum capital gains, buying the **fidelity shares** of fast-growing companies.
- *Long-term growth:* Own the least erratic issues of better-known and still rapidly growing firms.
- *Growth and income:* To achieve a growth-and-income combination, funds buy bonds and stocks of large companies that are big dividend payers and also offer some prospect for capital gains.
- *Income:* Funds with this objective attract investors seeking high yields.
- *Sector:* A sector fund specializes in stocks of one industrial sector or several related industries, such as health care, high technology, or energy.[38]

Are Mutual Funds Risky? The term **risk management** refers to the use of various methods to deal with potential personal or financial loss. Practicing risk management means that an individual acknowledges the existence of potential losses, knows their possible magnitude, and uses various methods to control them.[39] Even though mutual funds are riskier than some investments, they are less risky than others. In reality, use of mutual funds is a type of risk management, in which many different stocks and securities are managed by a professional financial team. Even then, do not forget the old saying: "Nothing in life is guaranteed except death and taxes."

Are Mutuals the Only Investment? There are several types of investment, from low-risk bank savings (CDs) to high-risk commodities. Of course, stocks are also risky. Mark Twain summed it up well when he said, "October." This is one of the peculiarly dangerous months in which to speculate in stocks—the others are July, January, September, April, November, May, March, June, December, August, and February. Space in this chapter does not permit an explanation of each investment type and the associated risk. However, Figure 8-25 shows various types of investment alternatives and their risk and earnings rankings. Mutuals are in-between in terms of risk.

More Information. For a directory of more than 3,400 funds, their addresses, telephone numbers, investment objectives, fees charged, and other information, contact the Investment Company

37. Ricketts, *Science IA (Agriscience)*, pp. 8-16.

38. Ibid.

39. *High School Financial Planning Program*, pp. 67-80.

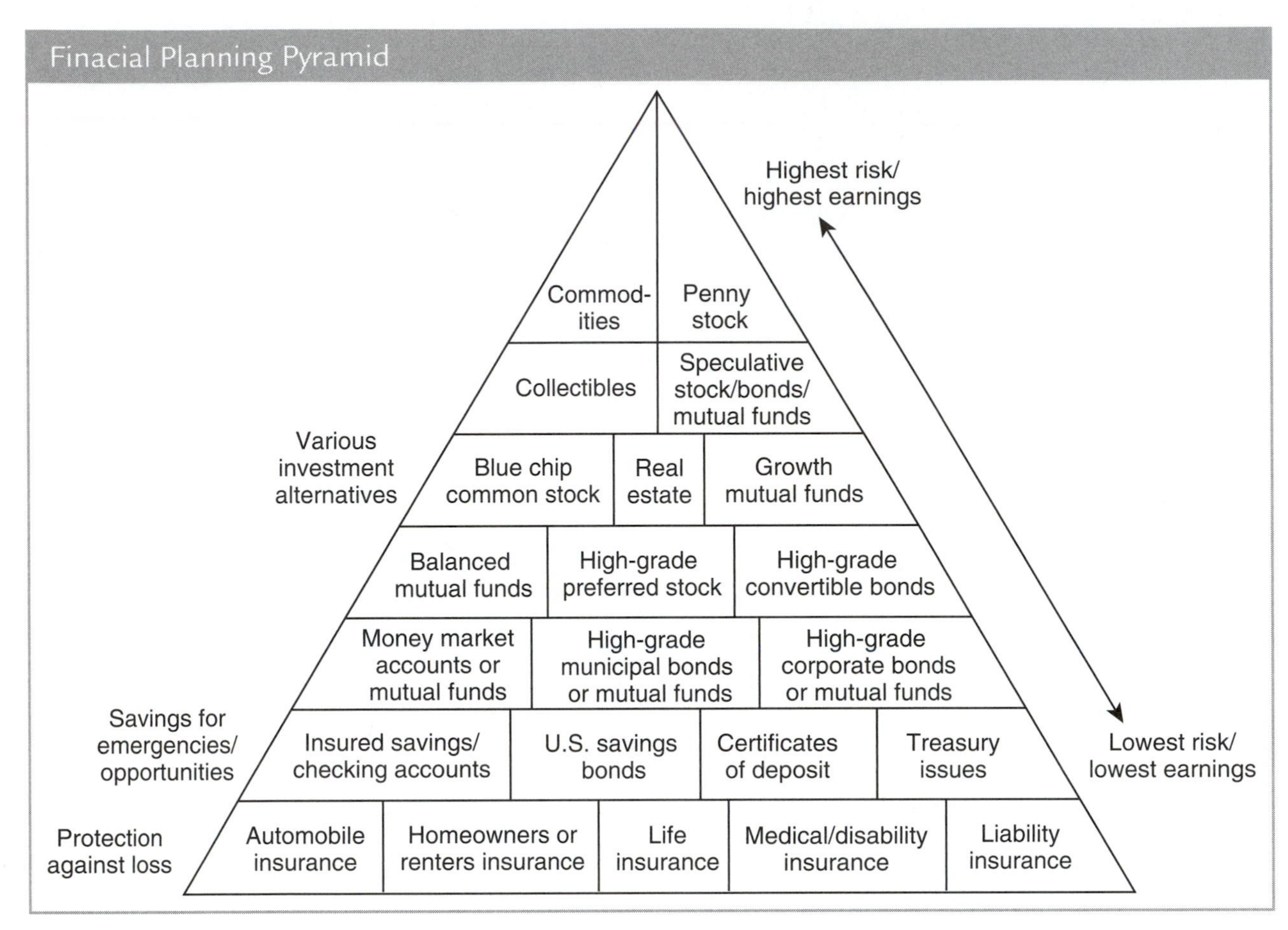

Figure 8–25 Financial planning pyramid showing the risk on various investment alternatives. (Courtesy of National Endowment for Financial Education)

Institute, P.O. Box 66140, Washington, DC, 20035-6140. The institute is the national association of the investment company industry. Its mutual fund members represent approximately 95 percent of the industry's assets and have about 36 million shareholders.

Tax-Deferred Savings

The four millionaire plans are **tax-deferred savings** plans. Tax-deferred savings allow an individual to postpone taxes while building an investment. Examples of tax-deferred savings include:

- Individual retirement accounts (IRAs), available to everyone who has earned income, with contributions limited to $2,000 per year
- 401(k) tax-sheltered plans, which are available to government employees and the self-employed
- 403(b) tax-sheltered plans, which are available to teachers and hospital employees

Contributions are allowed for a percentage of your income. In some situations, you can put up to 20 percent of your income into a 403(b) plan.

Tax-Free Earnings

The latest type of IRA is the Roth IRA. With the Roth IRA, tax is paid on the front end, not when you withdraw it. You can put $3,000 yearly into a Roth IRA. Consider the following advantages of a Roth IRA:

Amount Invested. In a regular IRA you receive only part of the benefit of your investment earnings, because part will end up going to the government in the form of taxes. With a Roth IRA, every dollar is working for you. This is the primary reason some financial analysts have been able to project significantly more wealth accumulation for people who can maximize their contributions to a Roth IRA. If you do not contribute the maximum amount to your IRA, this factor is less important.

Required Distributions. The rules for regular IRAs require you to start taking distributions when you reach age 70½. If you do not need the money at that time, this rule reduces your ability to continue investing in a tax-free vehicle. The required distribution rules do not apply to the Roth IRA. As a result, the Roth IRA may enhance your postretirement investment earnings.

Estate Tax Reduction. Suppose you have a Roth IRA with a value of $1 million when you die. Your heirs will not pay income tax on their benefits from the Roth IRA, so the benefit to your heirs is the same as the amount included in your estate: $1 million. Now suppose you have a regular IRA instead. The balance is $1 million, but your heirs will pay $250,000 income tax when they receive the benefits.

Early Distributions. No matter how well you plan, you may need to withdraw money from your IRA before retirement. The Roth IRA makes this easier because of a rule that says the first dollars out are considered a return of your contributions. You do not pay tax (or penalty) on the amounts you withdraw until you have taken out all your contributions and start to withdraw the earnings. This is not true of a regular IRA. If your regular IRA includes any earnings, your distributions will be partly taxable (and possibly subject to penalty) even if all your contributions were nondeductible.

When Do I Get My Money? Withdrawals from such plans are designed to occur at retirement. As you withdraw your money, you pay taxes on your withdrawals according to your tax bracket. Actually, you can take your money out at any time, but you will pay taxes and heavy penalties on money withdrawn from these plans before you reach age 59½.

Financial Security

You are probably saying, "I don't care about retirement, I want to know how to get rich quick." But we cannot tell you how to do that by ethical, legal, or moral means. The best way to get rich quick is to get rich slowly. We have heard it said that the safest way to double your money quickly is to fold it over once and put it back into your pocket. However, you can reach financial security by the young adult years through knowledge gained in your agricultural education program and your supervised agricultural experience program (SAEP). Nevertheless, you have to work and use "sweat equity."

Financial Security Defined

Financial security involves being able to meet monthly expenses (house payments, car payments, utilities, groceries, medical, clothing, etc.) with pay from your regular full-time job, plus accumulating (saving) approximately $100,000 in tax-free municipal or government bonds (with a return of 8 percent, which would be $8,000 tax-free earnings yearly). The interest on the savings is to be used only for travel, entertainment, and other nonbudgeted items. This money is not to be used or confused with money from your full-time regular job intended for living expenses, house payments, and so on. We realize that you may have a different definition of financial security, but this is the foundation for the discussions in this book.

Assumptions for Reaching Financial Security

It is possible for an agribusiness manager or worker to become financially secure by age 30 to 35. The age depends on whether a student continues his or her education after high school before pursuing other plans. For this to happen, there are some beliefs, assumptions, or criteria that must be taken into consideration.

1. *Think part-time (avocational).* The key is **part-time (avocational) enterprises**.
2. *Emphasize marketing.* The key to a profit in avocational **enterprises** is marketing. As much emphasis must be placed on marketing as on production.
3. *Develop new ways of thinking.* Traditional thinking and values may have to be discarded, or at least carefully examined. For example, bigger may not always be better. Enormous debt, big machinery, and large amounts of land are not necessary to make profits in agribusiness, production agriculture, and avocational enterprises.
4. *Manage investments.* As much thought and planning will have to be devoted to an **investment portfolio** as to an avocational enterprise.

Career Options

Avocational (Part-Time) Production Agriculture, Agribusiness Recreation, Agricultural Mechanic, Nursery Production

Because of emerging technologies, many agriculture enterprises can now be done as an avocation or part-time endeavor. For example, 60 years ago, hay was harvested loose and put in hay lofts by a rope, spear, and pulley; 40 years ago, almost every hay producer harvested with a square baler; today, most hay producers use a round baler to harvest their hay. The invention of the round hay baler permits many cow-calf producers to manage a 60- to 70-cow herd on an avocational basis; 60 years ago, this would have been a full-time job. Refer to Figure 8-26 for several other avocational enterprises.

Another popular avocational enterprise near large metropolitan areas is strawberry production. With new strawberry varieties, using plastic and irrigation, some strawberry producers are able to gross $10,000 per acre. Besides using workers to pick the strawberries, some strawberry producers have pick-your-own operations. The photo shows strawberry growing using special varieties, plastic, and drip irrigation.

Some strawberry producers are able to gross $10,000 per acre using new varieties, plastic, and drip irrigation. Raising strawberries as an avocational enterprise can be very advantageous if the market is available and the necessary skills can be mastered. (Courtesy of Cliff Ricketts)

5. *Get tough and stay focused.* Self-discipline, hard work, and **perseverance** are prerequisites for anyone to achieve challenging goals. This is especially true for the goal to achieve financial security through avocational enterprises. Refer to the Career Options section for further explanation and examples of avocational enterprises.

Selecting Part-Time Agricultural Enterprises

There is a story about the farmer who sold his farm and went in search of diamonds. Many agriculture students go searching for diamonds and gold, and they do not realize that they are already sitting on gold mines. Many agriculture students have resources available that most nonagriculture students do not have.

The $2,000 Work-Unit Approach. The "$2,000 work-unit approach" simply involves selecting avocational enterprises that will net you $2,000 on an acre or less. No more than five acres of land, including home and buildings, will be needed. In many cases, such as real estate investing, no land, equipment, or supplies are needed. By selecting just four different avocational enterprises that will do this, you can achieve financial security if you are self-disciplined.

Figure 8-26 lists more than 50 examples. Remember, these are not automatic. However, if you study them, use all the approved practices, visit people (working models) who have succeeded at that particular avocational enterprise, and become a "student of that enterprise," these 50 enterprises will meet the "$2,000 work-unit" criteria in most cases for most years. Remember, these can also be agricultural mechanical enterprises. In reality, they do not have to be agricultural all the time. Hobbies such as coin collecting, stamps, crafts, and so on can be used; as long as the $2,000 per year is net profit, it can be considered a $2,000 work-unit.

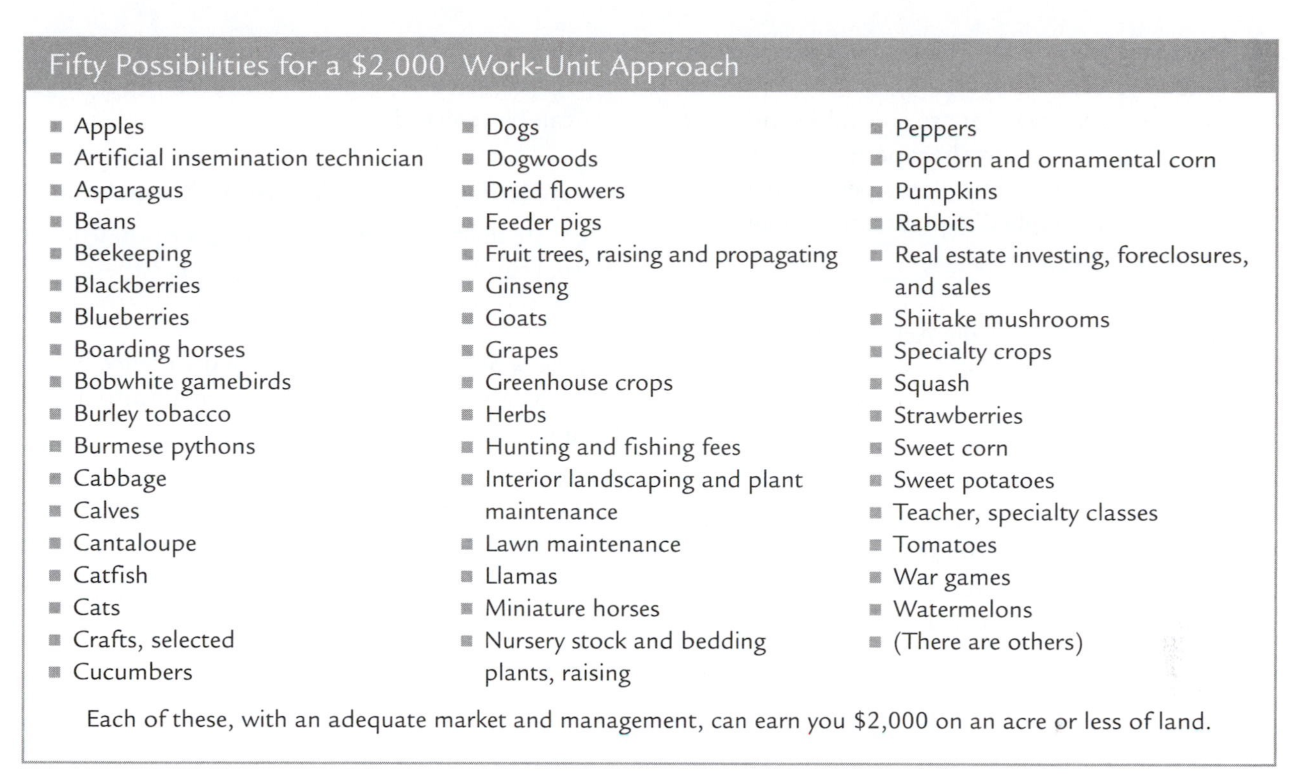

Fifty Possibilities for a $2,000 Work-Unit Approach

- Apples
- Artificial insemination technician
- Asparagus
- Beans
- Beekeeping
- Blackberries
- Blueberries
- Boarding horses
- Bobwhite gamebirds
- Burley tobacco
- Burmese pythons
- Cabbage
- Calves
- Cantaloupe
- Catfish
- Cats
- Crafts, selected
- Cucumbers
- Dogs
- Dogwoods
- Dried flowers
- Feeder pigs
- Fruit trees, raising and propagating
- Ginseng
- Goats
- Grapes
- Greenhouse crops
- Herbs
- Hunting and fishing fees
- Interior landscaping and plant maintenance
- Lawn maintenance
- Llamas
- Miniature horses
- Nursery stock and bedding plants, raising
- Peppers
- Popcorn and ornamental corn
- Pumpkins
- Rabbits
- Real estate investing, foreclosures, and sales
- Shiitake mushrooms
- Specialty crops
- Squash
- Strawberries
- Sweet corn
- Sweet potatoes
- Teacher, specialty classes
- Tomatoes
- War games
- Watermelons
- (There are others)

Each of these, with an adequate market and management, can earn you $2,000 on an acre or less of land.

Figure 8–26 Fifty enterprises that potentially can net $2,000 per acre per year with an adequate market and management. However, market conditions can change rapidly.

This approach is not needed if your family is established in agriculture production and you already have land, facilities, and equipment. Therefore, dairy, crop, and livestock producers (except in specialty situations) are excluded from this approach. However, students in many agricultural education programs in the United States do not come from farms, and if they do, their parents are generally not full-time production agriculturalists, or there is not enough economic opportunity for all their brothers and sisters to join the farming operation after graduation. Therefore, the $2,000 work-unit approach allows them to get into or stay with agribusiness and agriscience production. Also, by needing only five acres, few facilities, and little equipment, this approach addresses and avoids the factors that have caused farms to periodically endure hard times:

- high interest payments
- high machinery costs
- depreciated land value
- inadequate market

Considerations for selecting part-time avocational enterprises that meet the criteria of the $2,000 work-unit approach include the following:

- Select enterprises that do not interfere with your full-time job. Remember, this is only your avocation. Your full-time job has priority.
- Select an enterprise that you will not get **locked into**. This would mean you could not get out of what you were doing because your debts exceeded your assets.
- Start small and capitalize on the **utilization of resources**.
- Make sure that there is a market for the enterprise you select. All else is in vain without a market. If there is no market, abandon your plan.
- Select enterprises that are **compatible** with each other in relation to land, labor, and capital. For example, strawberry growing is intense during one particular month, so select an enterprise that is intense in another month, like blackberry growing.
- Select avocational enterprises that will require one acre or less per enterprise.

- Select enterprises that make a net profit of $2,000 yearly.
- Select enterprises that are enjoyable and serve both as a leisure and **therapeutic activity**. Make this your hobby. Some people like animals; some people like plants; some like mechanical things.

Procedures for Potentially Achieving Financial Security by Age 30 to 34

- After graduation from high school or college, select one avocational enterprise that is compatible with your full-time job. Remember, marketing is the key! The Business Research Center (or Small Business Development Center) at many universities can provide assistance in determining if there is a market for many avocational enterprises.
- Select an enterprise that will make a net profit of $2,000. Let us use strawberries as an example: one acre of properly managed strawberries will net $2,000.
- The second year after graduation, another enterprise can be added to net another $2,000. Add an enterprise that will distribute your workload throughout the year. Since strawberries naturally produce in one month, an acre of blackberries can be added, since they naturally produce in another month.
- The third year after graduation, add another enterprise. The possibilities are unlimited. Some include blueberries, catfish, tomatoes, squash, nursery stock, bedding plants, floriculture crops, and agricultural mechanics projects. Refer to Figure 8–26 for further selections.

There are a thousand ways in which something can fail to work, but only one way it can. Select enterprises that will work for you. Part-time jobs and working overtime are other $2,000 work-unit alternatives.

- The fourth year, add another $2,000 enterprise. (Another approach is to expand one of the first three enterprises. It may be best to add another acre of strawberries or blackberries.) Refer to Figure 8–27 for a summary of the generated income. Although the goal was $100,000 by age 30 to 34, Figure 8–27 shows that you can accumulate more than $120,000. As you can see, even with bad years, $100,000 can be achieved.

Invest Earnings in Tax-Free Bond Mutual Funds

There are many tax-free bonds. Many of these are mutual fund tax-free bonds. The goal of these funds is to achieve 8 percent interest.

Change in Strategy. Why is there a difference in strategy in the goal of 12 percent on a mutual fund investment for the millionaire plan compared to the goal of an 8 percent tax-free bond for the financial security plan? There are three reasons:

1. Only $2,250 can be put into an IRA ($2,000 if your spouse also has an IRA).
2. There is less paperwork and recordkeeping, because interest income from tax-free bonds is, obviously, not taxable.
3. There is less risk in tax-free bonds, especially utility bonds.

Limitations on Potentially Achieving Financial Security through Avocational Enterprises

Is this plan a sure thing? No, there are limitations that should be taken into consideration. Remember, if this were easy, everyone would do it. The limitations are as follows:

- *This is only a plan.* Few things in life are absolute. However, if you do not try you surely will not succeed.
- *Market conditions change yearly.* You will need to constantly seek and evaluate your markets.
- *Hard work and perseverance are taken for granted.* Many people do not have the maturity or the ability to work as hard as this plan will require.
- *Life requires tradeoffs.* You cannot hunt, fish, play ball, attend ball games every weekend, and still maintain four acres of avocational enterprises. However, there will still be plenty of time for recreation.
- *Self-discipline* is needed to keep from withdrawing any of the principal or interest before the goal of $100,000 is achieved.
- *The stock market can crash,* as it did in 1929, 1987, and 1998. This can drastically affect the value of one's investments.

Four Enterprises at 8 Percent Interest
To Age 30

First Year Out of High School	Age 19	Strawberries	2,000.00
Second Year Out of High School	Age 20	Blackberries	2,000.00
		Strawberries	2,000.00
		Interest at 8%	320.00
			$4,320.00
Third Year Out of High School	Age 21	Crafts	2,000.00
		Blackberries	2,000.00
		Strawberries	2,000.00
		Previous Savings and Interest	4,320.00
			$10,320.00
Fourth Year Out of High School	Age 22	Catfish	2,000.00
		Crafts	2,000.00
		Blackberries	2,000.00
		Strawberries	2,000.00
		Previous Savings and Interest	11,134.80
			$19,134.80
	Age 23	Four Enterprises	8,000.00
		Previous Savings and Interest	20,642 58
			$28,642.58
	Age 24	Four Enterprises	8,000.00
		Previous Savings and Interest	30,933.99
			$38,933.99
	Age 25	Four Enterprises	8,000.00
		Previous Savings and Interest	42,048.71
			$50,048.71
	Age 26	Four Enterprises	8,000.00
		Previous Savings and Interest	54,052.61
			$62,052.61
	Age 27	Four Enterprises	8,000.00
		Previous Savings and Interest	67,016.82
			$75,016.82
	Age 28	Four Enterprises	8,000.00
		Previous Savings and Interest	81,018.17
			$89,018.17
	Age 29	Four Enterprises	8,000.00
		Previous Savings and Interest	96,139.62
			$104,139.62
	Age 30	Four Enterprises	8,000.00
		Previous Savings and Interest	112,470.78
			$120,470.78

Figure 8–27 Example of how selecting four enterprises that net $2,000 per year per acre each can potentially accumulate enough money to make you financially secure.

- *Environmental conditions can change.* Environmental conditions for agricultural enterprises, especially crop-related enterprises, are often beyond one's control; a crop can fail due to drought, freeze, flood, and pests.

Levels of Financial Management

The more levels of financial security you can attain, the better off you will be at retirement. Consider the following levels of financial management:

1. *Live within a budget.* Follow good money management practices to limit debt and live within your means.
2. *Social Security.* This is required by law. Let us hope that the system is still in existence when you retire.
3. *Retirement benefits through your job.* This is generally required by most places of employment.
4. *Life insurance.* **Universal life insurance** can also become a savings plan. If you do not reach retirement, your family will be protected financially if insured adequately. Some financial advisors advise using **term life insurance**. With term insurance, you pay a yearly premium, but the premium never accumulates toward a cash value. Because term insurance is much cheaper than universal or whole-life, many advise you to buy term and invest the difference in mutual funds.
5. *One of four plans to retire as a millionaire.* Select the plan that best fits your needs, as discussed earlier in this chapter.
6. *Financial security by age 30 to 34, through part-time avocational enterprises.* This is the $2,000 work-unit approach.
7. *Stock market or futures (commodity) trading from interest only on principal from the previous level.* These were not discussed due to space, and they have a high risk. Refer to Chapter 18 for a discussion of commodity (futures) marketing.

Conclusion

Few things in life are guaranteed. There are few absolutes. Thus, there is no promise that these financial ideas will work for you. However, we believe that we can say with 100 percent accuracy that if you do not try them or some other financial plan, you will not reach your financial goals. Many people have accumulated financial wealth, and they may not have been smarter than any of the rest of us. However, they have been educated to know how money works, to exercise restraint and self-control, to follow many of the suggestions made in this chapter, and to understand the power of compound interest. A part of being successful is learning to have power over our money rather than letting our money have power over us.

Summary

Decisions about how to handle money will have a profound impact on your life. You must manage your money in an effective and responsible way. You must plan your work and then work your plan.

When we earn money, we need to understand how to interpret our paychecks. There is a difference between gross pay and net pay. Deductions include income taxes (federal, state, and local), Social Security tax (FICA), unemployment insurance, and worker's compensation. Employee benefits include life insurance, long-term disability, medical insurance, retirement savings plan, and profit sharing, as well as several benefits that do not require deductions, such as paid holidays and vacations.

You must select a financial institution to help manage, protect, and increase your money. There are several types of institutions, including banks, savings and loans, credit unions, and brokerage firms. Some factors to consider in choosing a financial institution include types of service, convenience, cost, relationships, security, and customer qualifications.

Use caution with your checking account. Keep checks properly recorded, and your checkbook updated and balanced. Checks must be written accurately because banks need to understand exactly how much you want to give to whom. Remember to record your deposits. As you balance your checkbook, compare the bank statement with your checkbook register.

You must gain control over your money, or your money will gain control over you. Money is active. If you do not take action continually with your money, or lack of it, it will take action with you. You will have conflict, worry, shortages, and a general lack of fun and funds until you achieve some discipline in handling your money. You can discipline your spending with both fixed and variable expenses.

You need to prepare a budget, which is a plan for spending and saving money. A good budget is an essential tool in every household. Budgets help us avoid unexpected expenses and provide greater freedom. A budget is not meant to complicate your life; on the contrary, when you begin to know where your cash is flowing, it will make life easier.

Some financial management and security tips include the following: limit debt, do not use credit cards, save on home mortgage by varying the payments, reduce debt by using the debt snowball technique, develop power over purchases, live within and below your income, pay cash for a car rather than going into debt, negotiate and buy only big bargains, build an emergency fund, manage budget busters, use the envelope system, have enough insurance, master the use of compound interest, use automatic saving plans, take advantage of matching stock purchases, and use your hobby to make money.

Is it possible for an agribusiness manager or worker to retire as a millionaire? Although nothing in life is guaranteed, things certainly will not happen unless you make them happen. If you think you can, you can. If you think you cannot, you cannot. Either way, you are correct. Your attitude determines your altitude. The four plans that hinge on a 12 percent interest rate are: Plan One, $2,000 a year for six years; Plan Two, buy a used car instead of a new car and invest the difference; Plan Three, the 25-plus-10 plan; and Plan Four, the 80-10-10 plan.

There are key factors to making the millionaire plan work. It has to be worked at. You must manage investments to achieve a 12 percent interest rate; be aware of the Rule of 72; study, select, invest, and monitor the best mutual funds; and be aware of tax-deferred savings.

Financial security can be attained through knowledge gained in agricultural education and your supervised agricultural experience program. However, you have to work and use "perspiration (sweat) equity." To do this, you need to select four avocational enterprises that will net $2,000 dollars each on an acre of land or less. This is called the "$2,000 work-unit approach." Invest the earnings in tax-free bond mutual funds. Of course, the plan has limitations due to changing markets, environmental conditions, and sudden downturns in the stock market. However, you can either make excuses or you can work hard, be persistent, stay focused, get tough, and just do it. Remember, if it were easy, everyone would do it.

End of Chapter Activities

REVIEW QUESTIONS

1. Define the Terms to Know that relate to personal financial management.
2. What mandatory deductions are taken from your paycheck?
3. List 10 employee benefits that may be paid by the employer (some are optional).
4. What are six factors to consider in choosing a financial institution?
5. Name four reasons for writing checks.
6. List 10 points to remember when writing checks.
7. What are six reasons for keeping a checkbook register?
8. What are five reasons for balancing your checking account monthly?
9. List eight common reasons why checkbook registers and account statements differ.
10. Give eight examples of fixed expenses.
11. Give 10 examples of variable expenses.
12. What are six benefits of a budget?
13. List 16 financial management and security tips.
14. List 10 potential dangers of credit cards.
15. What are eight consequences of bad credit due to credit card misuse?

16. Explain the use of a credit card alternative.
17. What does a low credit score mean to you?
18. Although you have been warned not to use credit cards, if you do, what are 18 tips that will help keep you on track in protecting your credit rating?
19. What are three ways to save big money on home mortgages?
20. What are the seven basic principles of negotiating?
21. What are five types of insurance that everyone should have?
22. Briefly discuss the four ways to potentially retire as a millionaire.
23. Why were mutual funds selected as the investment strategy most likely to get a 12 percent interest rate?
24. What are the five types of mutual funds?
25. Explain the $2,000 work-unit approach.
26. List eight considerations for selecting a part-time avocational enterprise.

FILL IN THE BLANK

1. The math involved in keeping and balancing a checkbook is basic __________ and __________.
2. You must gain control over your money, or your money will gain __________ over you.
3. By failing to plan, we are planning to fail, and there is always too much __________ left at the end of the money.
4. Creating a __________ guards you from being blindsided by an unexpected expense or a forgotten bill.
5. Almost any person knows how to earn money, but not one in a million knows how to __________ it.
6. The __________ of money is the root of all evil.
7. One of the biggest challenges in managing money is __________ __________.
8. If your credit score is above __________, you are considered a "prime borrower."
9. Payments you make on a non-__________ basis can be budget busters if they are not planned.
10. You can get __________ percent interest on your money if your employer matches investments dollar for dollar.
11. If you do not see the money, you will hardly miss it—this is the benefit of a __________ savings plan.
12. The amount of interest that must be achieved to make the millionaire plans work is __________ percent.
13. Besides the interest rate, "Plan One," illustrated in Figure 8-22, shows the importance of __________ and __________.
14. If a person were to invest $2,000 per year for 43 years, the difference in total yield between 8 percent and 9 percent would be __________.

MATCHING

a. savings and loan
b. credit cards
c. emergency fund
d. brokerage firms
e. envelope system

f. credit unions
g. banks
h. debt snowball
i. variable expense
j. fixed expenses
k. 8 percent
l. 12 percent

_____ 1. primary purpose is to buy and sell stocks and bonds on behalf of the customer
_____ 2. primary purpose is to offer checking accounts, savings accounts, and certificates of deposit (CDs)
_____ 3. primary purpose is to offer savings accounts and home mortgages
_____ 4. primary purpose is to make low-interest loans available to members, who usually have an affiliation with the company
_____ 5. home mortgage
_____ 6. household maintenance and repair
_____ 7. can be very addictive
_____ 8. technique to reduce bills quickly
_____ 9. three to six months of income in savings
_____ 10. budgeting technique by using cash
_____ 11. interest goal of millionaire plan
_____ 12. interest goal of financial security plan

ACTIVITIES

1. You have just accepted a position that pays an annual salary of $20,000. You will be paid in equal installments on the 15th and the last day of each month. Here is a summary of the deductions on your paycheck:
   - federal income tax–10.60%
   - state income tax–3.00%
   - Social Security tax (FICA)–7.65% (employee portion; employer matches as required by law)
   - medical insurance, family coverage (shared costs: employee pays $30 per pay period)
   - dental insurance, family coverage (shared costs: employee pays $3.50 per pay period)

   What will be the amount of your check per pay period?
2. Contact a person at a financial institution that handles your money matters or a financial institution that you are considering and determine how many of the 18 financial services listed in Figure 8-3 that institution offers.
3. Write out the following check on a copy provided by the teacher. Use the following information.

   Date: Today's date
   Payee: Red Stafford's Feed Supply
   Amount: $132.47
   Amount written: One hundred thirty-two and 47/100 [dollars]
   Memo: Horse feed
   Signature: Your name

4. Fill out a checkbook register provided by the teacher.

1/7	Carter Hardware	$50.00	#1731
1/10	Farmer's Co-op Supply	$23.52	#1732
1/12	Wal-Mart	$8.48	#1733
1/14	Mires Rental	$35.00	#1734
1/18	Kroger Grocery	$18.50	#1735
1/21	Jones Supply	$12.36	#1736
1/6	Deposit	$69.35	
1/8	Deposit	$161.40	
1/15	Deposit	$337.80	
1/22	Deposit	$126.00	
Service charge		$4.82	

5. At the end of the month, you will receive a statement from the bank. It will tell you how much money you have in your checking account. You must reconcile (balance) your bank statement every month. To complete this involvement activity, you will need to use the checkbook register you completed in activity 4. Reconcile your checkbook register with the bank statement illustrated here. Complete the reverse side of the bank statement.
6. Plan a budget for your present living and financial situation: living at home, living on your own (with roommates), or family budget. You may write it out as in Figure 8-12, or your teacher will provide you with a blank sheet of either Figures 8-13, 8-14, or 8-15.
7. Select one of the "sweet 16" financial management and security tips. Write a brief report on it and present it to the class. Use real-life examples or any other technique to enhance your presentation. Use today's prices and interest rate if different from those in the text.
8. Select a hobby that you feel would make you money. Research it and write a report, and present the report to the other class members.
9. Select the millionaire plan that you believe would work best for you. Write a short essay defending your selection and present the report to the other class members.
10. A person offers you a job in July. The conditions are that you work for a penny the first day, and your salary will be doubled each day for the next 31 days. If you take the job, how much money will you make in 31 days (considering you work 31 days without taking off for the weekends or Independence Day)?
11. Look in the advertisements of a financial magazine or newspaper. (Some cable TV stations advertise mutual funds.) Select three mutual funds. Call the toll-free numbers and have them send you a prospectus of the mutual fund that you have selected. Once you get the prospectus, determine the fund's past performance and any other pertinent information, such as type and diversity of investments. Make a short presentation to the class of your findings. Note: The purpose of this exercise is to find mutual funds that have a track record of 12 percent.
12. Using the Rule of 72, determine the cash accumulated to age 61, assuming a one-time investment of $1,000, beginning at age 25.

    a. Age 25: $1,000, interest 4%
    b. Age 25: $1,000, interest 6%
    c. Age 25: $1,000, interest 12%

13. From the list in Figure 8-26, select 10 "$2,000 work-units" that you believe would meet the criteria for selecting avocational enterprises.
14. Choose one of the 10 "$2,000 work-units" that you selected in activity 13, or come up with one that is not on the list in Figure 8-26. Write a paper justifying your selection. Remember, it has to meet the criteria for part-time enterprises (as stated for the $2,000 work-unit approach). Convince the class why you believe your selection will work. If you find while writing that you do not believe it will work, make another selection. Budgets will enhance your defense of your selection.

Chapter 9

Agribusiness Recordkeeping and Accounting

Objectives

After completing this chapter, the student should be able to:

- Differentiate bookkeeping from accounting.
- Complete budgets.
- Describe the single-entry bookkeeping system.
- Describe the double-entry bookkeeping system.
- Complete journals and ledgers.
- Complete a trial balance.
- Explain basic accounting considerations.
- Prepare an income statement, balance sheet, and statement of cash flows.
- Prepare a statement of owner equity.
- Analyze a financial statement.
- Describe the impact of computer technology on agribusiness recordkeeping.
- Explain the benefits of using computer software to complete tax returns.

Terms to Know

accounting cycle
accounts payable
accounts receivable
accrued
assets
balance sheet
budget
budgeting
capital
capital expenditures budget
cash-flow budget
chronological
cost of sales
credit
creditor
debit
deficits
depreciation
double-entry bookkeeping system
equities
erroneous
incidental expenses
income statement
inventory
journals
ledger

(Continued)

liabilities
net profit
net worth
note payable
operating budget
operating expenses
owner equity
posting
receivables
revenue
salvage value
single-entry bookkeeping system
solvency
statement of cash flows
taxable income
trial balance
write-off

Introduction

Planning is a big part of success. Part of planning in an agribusiness includes budgeting, financial planning, and keeping good records. Keeping records takes time and effort, but it is essential. We should work as hard at our recordkeeping as at any other phase of the business.

Your attitude determines your success. Your agribusiness can achieve new heights if you develop a positive attitude toward recordkeeping and what it will do for your business. In reality, good records save time and money. Records must be kept accurately and in an orderly fashion. We need records to:

- know our cash balance
- know who is in debt to our business
- keep track of our business debts
- file correct and timely tax reports
- know our financial position
- make business decisions[1]

Role of the Financial Manager

Whether an agribusiness is large or small, one person is usually responsible for the bookkeeping, accounting, and financial management of the business. In a small agribusiness, the owner may be the person responsible for the finances. Therefore, the owner has to make sure the agribusiness can meet its financial obligations. To do this, the financial manager determines how money comes into the business and how the agribusiness uses that money. Refer to Figure 9-1.

Sources of Income. Sources of income include revenues from the sale of products and services, profits reinvested in the business, loans and credit received from outside the business, and the **owner equity**, which is the money the owner invests in the business. The owner or financial manager must decide the best way to use these sources of income.

Expenses. Operating expenses include wages and salaries, cost of maintaining the machinery and equipment needed to produce products and services, advertising, rent, electricity, and insurance. To meet unexpected expenses, the owner or financial manager needs to have a reserve or emergency fund of cash that can be made available quickly.[2]

Accounting versus Bookkeeping

The distinction between accounting and bookkeeping is often confusing. *Bookkeeping* refers to the actual recording of business transactions. This clerical recordkeeping function may be done manually or by computer. *Accounting* goes far beyond bookkeeping. The purpose of an accounting system for a business is to allow preparation, review, and understanding of reports.[3]

1. William H. Hamilton, Donald F. Connelly, and D. Howard Doster, *Agribusiness: An Entrepreneurial Approach* (Albany, NY: Delmar Publishers, 1992), p. 104.
2. Betty J. Brown and John E. Clow, *Introduction to Business* (New York: Glencoe/McGraw-Hill, 2006), pp. 273–275.
3. *Advanced Agribusiness Management and Marketing: Accounting* (8709-B) (College Station, TX: Instructional Materials Service, 1990).

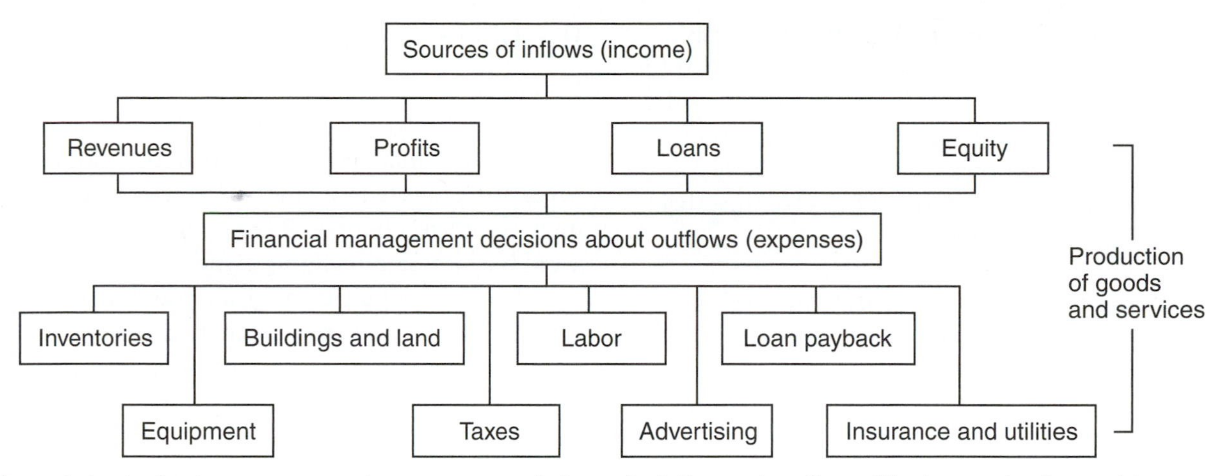

Figure 9–1 Agribusiness owners and managers try to balance the inflow and outflow of funds to maintain a stable business.

Bookkeeping

If you were a bookkeeper, you would divide all the business's paperwork into meaningful categories, such as sales documents, purchasing receipts, and shipping documents. This information must then be recorded into record books called **journals**. Journals are the books where accounting data are first entered.

Accounting

Accountants categorize and summarize the data provided by bookkeepers. They interpret the data and report the results to the business owners or management. Accountants also suggest strategies for enhancing the agribusiness's financial condition and progress.[4] Accounting is the basis for the financial statements and income tax preparation that must be done for any business. As you can see, an understanding of basic accounting is critical to the successful control of every agribusiness.

The Six-Step Accounting Cycle

The sequence of accounting procedures used to record, classify, and summarize accounting information is often termed the **accounting cycle**. The accounting cycle generally requires the work of both the bookkeeper and the accountant. It begins with the preparation of formal financial statements summarizing the effects of transactions on the business's **assets, liabilities**, and owner equity. The objective of this chapter is to teach you each of the following six steps of the accounting cycle so that you can perform them properly when the need arises.

Analyze and Categorize Documents. You will need a system to get your financial data and records organized and keep them organized. This system may be as simple as putting incoming bills or documents in a shoebox or as complex as a very structured filing system or computer database system.

Record the Information into Journals. As each business transaction occurs, it is entered into a journal, thereby creating a **chronological** record of events.

Post the Preceding Information into Ledgers. The **debit** and **credit** changes in account balances are posted from the journal to the **ledger** accounts. This procedure classifies the effects of the business transactions in terms of specific assets, liability, and owner equity accounts.

Prepare a Trial Balance. A **trial balance** is a summary of all the data in the ledger accounts, which shows whether the figures are correct and the books balance. If some data are not correct, they must be corrected before the income statement and balance sheet are prepared.

Prepare an Income Statement, Balance Sheet, and Statement of Cash Flows. The **balance sheet** reports the financial condition of the business on

4. William G. Nickels, James M. McHugh, and Susan M. McHugh, *Understanding Business,* 5th ed. (Chicago: Irwin/McGraw-Hill, 1999), p. 534.

a specific date. The **income statement** records revenues, costs, expenses, and profits (or losses) for a specific period of time; thus, it shows the results of business operations during that time period. It is helpful to make notes disclosing any facts necessary for proper interpretation of the balance sheet and income statement.[5]

Analyze the Financial Statements. The sixth step is to analyze the financial statements and evaluate the financial condition of the firm. Bookkeeping and accounting are not done just as a formality or tradition. Besides being needed for income tax purposes, these financial statements can help you evaluate the success of your business so that you can make any necessary adjustments in a timely and informed manner. These statements will also let you know whether you can expand or allocate profits to other investments.

A set of financial statements consists of four related accounting reports that summarize in a few pages the financial resources, obligations, profitability, and cash transactions of a business. A complete set of financial statements includes:

- A balance sheet, giving the financial position of the company at a specific date by showing the resources it owns, the debts it owes, and the amount of owner equity (the owner's investment) in the business. Refer to Figure 9-16.
- An income statement, which shows the profitability of the business over the preceding year (or other time period). Refer to Figure 9-15.
- A statement of owner equity, which explains certain changes in the amount of the owner's investment (equity) in the business. (In businesses that are organized as corporations, the statement of owner equity is replaced by a statement of retained earnings.) Refer to Figure 9-19.
- A statement of cash flows, summarizing the cash receipts and cash payments of the business over the same time period covered by the income statement. Refer to Figures 9-3, 9-17, and 9-18.

5. Ibid., pp. 535-536.

Budgets

If something does not work on paper, it will not work once the agribusiness opens. After an agribusiness owner or manager has planned for customers' needs and projected income and expenses, he or she must complete a budget. Proper planning can help avoid business failure. Some owners of an agribusiness do only mental planning, and can give only an approximation of where they are financially. Others put their financial plans on paper in an orderly manner. This technique is called **budgeting**, and the plan is called a **budget**. Three types of budgets must be prepared: an **operating budget**, a **cash-flow budget**, and a **capital expenditures budget**.[6] A discussion of each follows.

The Operating Budget

The operating budget summarizes the anticipated sales, production activities, and related costs for the year (or other relevant period). It is an estimate of sales and income plus the expected expenses, both fixed and variable, for the year. The operating budget is a positive statement of what management expects to do during the coming year to maximize profits.[7] Refer to Figure 9-2 for an example of an operating budget.

The Cash-Flow Budget

The cash-flow budget shows, in summary form, the amount of income that will flow into and out of the business during the year, and when it will do so. Income normally comes from sales and services, from borrowing, from the sale of capital items, and from accounts receivable payments collected. Expenses can include payments for goods and services purchased from others, principal and interest payments on debt, taxes, salaries, and payments on capital purchases, such as a down payment on a truck. The cash-flow budget shows:

- when cash receipts should be available to the agribusiness during the year
- when the agribusiness will need to make cash payments

6. James G. Beierlein, Kenneth C. Schneeberger, and Donald D. Osburn, *Principles of Agribusiness Management,* 2d ed. (Prospect Heights, IL: Waveland Press, 1995), p. 78.
7. Ibid., p. 80.

Operating Budget Green's Nursery & Landscaping Co. for the Year	
Income:	
Plants	50,000
Soil, fertilizer, etc	10,000
Accessories	6,000
Gross Sales Income	66,000
Cost of Goods	33,000
Gross profit	33,000
Landscaping fee income	20,000
Lawn care fee income	7,000
Total Income	60,000
Variable Expenses:	
Labor	30,000
Advertising	2,000
Office expense	1,000
Total	33,000
Fixed Expenses:	
Equipment depreciation and lease	5,000
Rent	4,000
Insurance	3,000
Total	12,000
Total Expenses	45,000
Net income before income tax	15,000
Income tax expense	5,000
Net Income	$10,000

Figure 9–2 The operating budget summarizes the expected sales or production activities and related costs for the year.

The cash-flow budget differs from the operating budget in that the cash-flow budget leaves out such items as:

- credit sales that do not generate a cash flow during the year
- depreciation and other noncash expenses
- noncash portions of any capital purchases or sales

The cash-flow budget may tell the owner little about the profitability of the business. Its purpose is to help the owner determine whether the funds on hand will be adequate to meet obligations. If not, money may have to be borrowed. If there is excess cash during certain periods, it can be invested.[8] Refer to Figure 9–3 for an example of a cash-flow budget. Notice how the cash inflows and outflows are divided by quarter.

The Capital Expenditures Budget

Each agribusiness must make regular capital investments in equipment, because wear and tear will eventually take their toll and contribute to depreciation of the item. The capital expenditures budget shows how money to be spent on capital expenditures is allocated among the various divisions or activities of the agribusiness. The capital expenditures budget is a list of projects (equipment purchases, etc.) that management believes will be needed, along with the anticipated cost of each.[9] Ranking the items or projects helps ensure that the money spent on capital expenditures will be allocated efficiently. Refer to Figure 9–4 for an example of a capital expenditures budget.

Relationships between the Operating, Capital Expenditures, and Cash-Flow Budgets

The sales part of the operating budget supplies the basic information for preparation of the cash-flow budget. Only the sales that are actually expected to generate cash during the budget year appear in the cash-flow budget. The capital expenditures budget provides the information needed regarding large cash outflows.[10]

Single-Entry Bookkeeping System

The simplest way to keep one's books is with the **single-entry bookkeeping system**. This is because the entry is made only in the checkbook or the checkbook and the journal. Therefore, one of the first activities when starting your agribusiness is to establish a checking or other bank account for your business.

Separate Business Accounts from Personal Accounts

Business funds and accounts should always be kept separate from all personal funds and accounts. This is best for tax records. Mixing

8. Ibid., pp. 80–81.
9. Ibid., p. 83.
10. Ibid., p. 82.

Cash Flow Budget
Green's Nursery & Landscaping Co.
for the Fiscal Year Ending December 31

	Quarter			
	1	2	3	4
Cash Balance, beginning of first quarter	10,000	10,075	27,250	16,725
Income:				
Plants	5,000	20,000	15,000	10,000
Soil, Fertilizer, etc.	1,000	4,000	3,000	2,000
Accessories	600	2,400	1,800	1,200
Gross Sales Income	6,600	26,400	19,800	13,200
Cost of Goods	3,300	13,200	9,900	6,600
Gross Profit	3,300	13,200	9,900	6,600
Landscaping Fee Income	2,000	8,000	6,000	4,000
Lawn Care Fee Income	700	2,800	2,100	1,400
Total Income	6,000	24,000	18,000	12,000
Variable Expenses:				
Labor	3,000	12,000	9,000	6,000
Advertising	200	800	600	400
Office Expense	100	400	300	200
Total Variable Expenses	3,300	13,200	9,900	6,600
Fixed Expenses:				
Lease Payments	625	625	625	625
Rent	1,000	1,000	1,000	1,000
Insurance	750	750	750	750
Total Fixed Expenses	2,375	2,375	2,375	2,375
Total Expenses	5,675	15,575	12,275	8,985
Net Income before Income Tax	325	8,425	5,725	3,025
Income Tax Expense	1,250	1,250	1,250	1,250
Net Inflow	(925)	7,175	4,475	1,775
Capital Expenditures			(15,000)	
Loan Repayments				(2,000)
Additional Borrowings		10,000		
Additional Investment	1,000			
	$10,075	$27,250	$16,725	$16,500

Figure 9–3 The cash-flow budget summarizes the amount and timing of income that will flow in and out of the business during the year.

personal accounts with business accounts will lead to problems. A sense of job security can develop, leading to unnecessary withdrawals for personal reasons from the business account. If the owner does draw money from the business account for personal expenses, these funds should be paid back as soon as possible. Also, whenever funds are withdrawn from a business account, use a check to provide the appropriate documentation.

Sales Slips Plus Checking Account

Besides the checking account, a small agribusiness should use sales slips. The checking account and use of sales slips will provide the information

Capital Expenditures Budget Green's Nursery & Landscaping Co. for the Year Ending December 31			
Capital Item	**Priority**	**Estimated Cost**	**Total Cash Outlay**
Computer Software and Printer	1	$3,000	$3,000
Trucks and Trailers	1	$15,000	$3,000
Bobcat Loader	2	$8,000	$2,000
John Deere Mower	1	$5,000	$5,000
(3) Small Trimmer Mowers	2	$1,000	$1,000
(4) Weedeater Trimmers	3	$1,000	$1,000

Figure 9–4 The capital expenditures budget is a list of projects and/or equipment needed for the agribusiness.

needed to file income taxes. Use a sales slip for each and every sale, and process all money through the checking account. Refer to Figures 9–5 and 9–6. A checkbook and a sales slip book can document revenues and expenditures. The single-entry system is basically a cash method of accounting, as the payments for goods and services are recorded when they are made or received. The single-entry system must allow the owner of the agribusiness to:

- ensure that the business will pay all the taxes it owes
- prevent errors in **accounts receivable**
- prevent erroneous payments for goods receivable
- have an up-to-date cash position
- make projections for the future[11]

Green's Nursery & Landscaping Co.
Rose Garden Drive
Murfreesboro, TN, 37000

Customer:
Name
Address
Phone #

Quantity	**Description**	**Unit Price**	**Extended Amount**

For Office Use:
Date Completed: _____ Completed by: _______
Time Required: ___ Workers Used: __________

Figure 9–5 Use a sales slip for every sale to provide information needed to file income taxes.

Double-Entry Bookkeeping System

It is possible to make a mistake when recording a financial transaction. For example, you could easily write $12.97 as $12.79. Because of this, a bookkeeper records all transactions in two places. The bookkeeper can check one list against another list to make sure all the transactions add up to the same amount in both places. If they do not, the bookkeeper knows that a mistake was made.

Recording the Transaction in Two Places

The practice of writing every transaction in two places is called **double-entry bookkeeping**. This process involves journals and ledgers. For example,

11. Hamilton, Connelly, and Doster, *Agribusiness: An Entrepreneurial Approach,* p. 118.

Checking Account Register

Check Number	Date	Description of Transactions	G/L Acct #	Payment Amount	Deposit Amount	Balance $850.000
1001	6/15/xx	McMinnville Farm	501	500.00		350.00
	6/15/xx	Daily Deposit	401		1,200.00	1,550.00
1002	6/16/xx	Carl's Truck Repair Center	551	100.00		1,450.00
1003	6/16/xx	First Tennessee Bank—Payroll A/C	510	500.00		1,950.00
	6/16/xx	Daily Deposit	401		1,500.00	$2,450.00

Figure 9–6 By running all money through the checking account, information is recorded that will later be needed to file income taxes.

Simplified Example of Double Entry Bookkeeping System

Assets		=	Liabilities	+	Owner's Equity
Cash in Bank	25,000		20,000		80,000
Accounts Receivable	5,000				
Inventory	20,000				
Equipment	50,000				
	$100,000	=	$20,000	+	$80,000

Figure 9–7 The practice of writing every transaction in two places is called double-entry bookkeeping. Debits and credits are always kept in balance.

if the owner of an agribusiness puts $10,000 of his or her own money into the business, the bookkeeper (who may be the owner) would debit (make an entry into) the cash account. Refer to Figure 9–7. To keep the transaction in balance, a corresponding credit entry is recorded in the capital (owner equity) account. Therefore, debits and credits are always in balance. A *debit* refers to a bookkeeping entry on the left side of the account. A *credit* means a bookkeeping entry that is recorded on the right side of the account. This is illustrated by the accounting equation:

Assets = Liabilities + Owner's equity
or
(Resources = Claims)

Keeping the Books Balanced

Accountants insist on maintaining the balance of resources (assets) and claims or liabilities. Accountants want to identify and track each transaction and entry to ensure this. The balance sheet in Figure 9–8 is another example of double-entry bookkeeping. Notice that the assets equal the liabilities, and the books "balance."

Journals and Ledgers

Journals

The single-entry and double-entry systems of bookkeeping could be considered journals, but not in the world of accountancy. In an actual accounting system, information is initially recorded in an account record called the *journal,* as shown in Figure 9–9. After a transaction has been recorded in the journal, the debit and credit changes in the individual accounts are entered in the ledger. Because the journal is the accounting

Green's Nursery & Landscaping Co. Balance Sheet as of December 31		
	Assets	**Liabilities**
Cash in Bank	25,000	
Accounts Receivable	5,000	
Inventory	20,000	
Equipment	50,000	
Total Assets	$100,000	
Accounts Payable		10,000
Bank Note		10,000
Surplus		80,000
Total Liabilities		$100,000

Figure 9–8 The balance sheet is another example of double-entry bookkeeping. The assets equal the liabilities.

record in which transactions are recorded initially, it is sometimes called the *book of original entry.*[12]

Use of the Journal. The journal is a chronological (time-ordered) record of business transactions. The information noted about each transaction includes the date of the transaction, the debit and credit changes in the appropriate ledger accounts, and a brief explanation. Again, refer to Figure 9–9.

Why Use a Journal? It is technically possible to record transactions directly in the ledger, so why is the journal necessary? The unit of organization for the ledger is the transaction, but the unit of organization for the ledger is the account. Having both a journal and a ledger gives three advantages over using a ledger alone:

1. The journal shows all the information about a transaction in one place and also provides an explanation of the transaction.
2. The journal provides a chronological record of all the events in the life of a business.
3. The use of a journal helps to prevent errors.[13]

The Ledger

At convenient intervals, the debit and credit amounts recorded in the journal are transferred (*posted*) to the accounts in the ledger. The updated ledger accounts can then be used to prepare the balance sheet and other financial statements. The ledger is an accounting system that has a separate record for each item. For example, a separate record is kept for asset cash, showing all the increases and decreases in cash that result from the many transactions in which cash is received or paid. A similar record is kept for every other asset, for every liability (such as telephone, fuel, office supplies, etc.), and for the owner equity. Therefore, the form of record used to show increases and decreases is a single balance-sheet item called an *account* or a *ledger account.* All the separate accounts grouped together are called a *ledger.*[14]

Parts of the Ledger. Very simply, a ledger account page has only three parts:

1. A title, consisting of the name of the particular asset, liability, or owner equity
2. A left side, which is called the debit side
3. A right side, which is called the credit side

Students and people unfamiliar with accounting often have mistaken ideas about the meaning of the terms *debit* and *credit.* For example, many people think that a credit is somehow more favorable or desirable than a debit. This is not correct. The terms *debit* and *credit* simply mean *left* and *right,* and do not have any hidden meaning or value connotations.[15]

Posting. **Posting** is the process whereby debits and credits are transferred from the general journal into the ledger account. The debits from the journal are recorded on the left side (debit side) of a ledger account. Likewise, the credits

12. Robert F. Meigs, Mary A. Meigs, Mark Bettner, and Ray Whittington, *Accounting: The Basis for Business Decisions*, 10th ed. (New York: McGraw-Hill, 1996), p. 60.

13. Ibid., pp. 60–61.
14. Ibid., pp. 51–52.
15. Ibid., p. 52.

General Journal

Date		Account Titles and Explanation	LP	Debit	Credit
Jan	2	Cash		70,000.00	
		Owner's Equity			70,000.00
		Original Investment			
Jan	20	Cash		20,000.00	
		Bank Note			20,000.00
		Loan from bank			
Jan	31	Inventory		10,000.00	
		Cash			10,000.00
		Purchase of inventory			
Feb	10	Equipment		25,000.00	
		Cash			25,000.00
		Purchase of equipment			
Feb	15	Accounts Receivable		2,500.00	
		Sales Income			2,500.00
		Income from landscaping for Ideal Realty			
Feb	20	Cash		1,500.00	
		Accounts Receivable			1,500.00
		Partial payment, Ideal Realty			
Feb	25	Cash		2,750.00	
		Sales Income			2,750.00
		Cash sales of plants			

Figure 9–9 This is an example of a general journal showing several entries. Each of these entries will later be recorded in a ledger account.

from the journal are recorded on the right side (credit side) of a ledger account. Posting methods may vary with the preference of the individual, but the sequence shown in Figure 9-10 is commonly used:

1. In the ledger, locate the first account named in the journal entry.
2. In the debit column of the ledger account, enter the amount of the debit as shown in the journal.

Posting from General Journal to Ledger Account

General Journal

Date		Account Titles and Explanation	LP	Debit	Credit
Jan	2	Cash	1	70,000.00	
		Owner's Equity	2		70,000.00
		Original Investment			

Ledger

Cash						Debit
Date		Explanation	LP	Debit	Credit	Balance
Jan	2	Original Investment	1	70,000.00		70,000.00

Owner's Equity						Credit
Date		Explanation	LP	Debit	Credit	Balance
Jan	2	Original Investment	2		70,000.00	70,000.00

Figure 9–10 An example of posting (transferring) the debits and credits from the general journal to the ledger account.

3. Enter the date of the transaction in the ledger account.
4. In the reference column of the ledger account, enter the number of the journal page from which the entry is being posted.
5. The recording of the debit in the ledger account is now complete; as evidence of this fact, return to the journal and enter in the ledger page (LP) column the number of the ledger account and page to which the debit was posted.
6. Repeat the posting process described in the preceding five steps for the credit side of the journal entry.[16]

16. Ibid., p. 63.

Trial Balance

The proof of the equality of debit and credit balances is called a *trial balance*. Because equal dollar amounts of debits and credits are entered in the accounts for every transaction recorded, the sum of all the debits in the ledger must be equal to the sum of all the credits. Refer to Figure 9–11.

Before the account balances are used to prepare a balance sheet, the bookkeeper should check the figures to ensure that the debits equal the credits. This is a trial balance. The usual form for a trial balance is a two-column schedule listing the names and balances of all the accounts in the order in which they appear in the ledger; the debit balances appear in the left-hand column and the credit balances appear in the right-hand column.[17] The totals of the two columns should agree, as shown in Figure 9–11. Notice the correlation between the trial balance in Figure 9–11, the entries in the ledger accounts in Figure 9–12, and the balance sheet in Figure 9–8.

Limitations of the Trial Balance

A trial balance does not prove that transactions have been recorded in the proper accounts, or that they were recorded in the correct account; it also cannot indicate that something was completely omitted from the ledger. In reality, the trial balance proves only that debits and credits are equal. Nevertheless, the trial balance is very helpful. It not only shows that the ledger is in balance, it also serves as a convenient source during later preparation of the required financial statements.[18]

Green's Nursery & Landscaping Co. Trial Balance as of February 28		
Cash	59,250	
Accounts Receivable	1,000	
Inventory	10,000	
Equipment	25,000	
Prepaid Expenses	4,750	
Accounts Payable		10,000
Bank Note		10,000
Surplus		80,000
Total Liabilities	$100,000	$100,000

Figure 9–11 The sum of all the debits in the ledger must be equal to the sum of all the credits. These sums will then be used to prepare a trial balance.

17. Ibid., p. 66.

Basic Accounting Considerations

Before we proceed to the next step (balance sheet, income statement, and statement of cash flows), let us discuss a few basic accounting considerations: assets and liabilities; capital and owner equity; sales, cost of sales, and net profit; operating and incidental expenses; depreciation; principal and interest payments; inventory; and profit and loss.

Assets and Liabilities

An *asset* is something of value that is owned. Assets are economic resources that are owned by a business and are expected to benefit future operations. Some examples of assets are cash, **receivables**, inventory, investments, equipment, buildings, and prepaid accounts.

Liabilities are amounts that are owed, or debt. Examples of liabilities include accounts payable, notes payable, and mortgages. Even the larger agribusinesses find it convenient to purchase merchandise and supplies on credit rather than to pay cash at the time of each purchase. Liability arising from the purchase of goods and services on credit are called **accounts payable**. The person or company to whom the account payable is owed is called a **creditor**.

When an agribusiness borrows money for any reason, a liability is incurred and the lender becomes a creditor of the business. The form of liability when money is borrowed is usually a **note payable**, which is a formal written promise to pay a certain amount of money, plus interest, at a definite future time.

Capital and Owner Equity

Capital is the investment that the owner has made or put into the business. For example, money used to start the agribusiness is capital. Capital also includes machines, tools, and buildings. *Owner equity*

18. Ibid., p. 67.

Ledger

Ledger

Cash						*Debit*
Date		**Explanation**	**LP**	**Debit**	**Credit**	**Balance**
Jan	2	Original Investment		70,000.00		70,000.00
Jan	20	Bank note		20,000.00		90,000.00
Jan	31	Purchase inventory			10,000.00	80,000.00
Feb	10	Purchase equipment			25,000.00	55,000.00
Feb	20	Payment, Ideal Realty		1,500.00		56,500.00
Feb	25	Cash sales		2,750.00		59,250.00

Accounts Receivable						*Debit*
Date		**Explanation**	**LP**	**Debit**	**Credit**	**Balance**
Jan	2	Sales to Ideal Realty		2,500.00		2,500.00
Jan	20	Payment, Ideal Realty			1,500.00	1,000.00

Inventory						*Debit*
Date		**Explanation**	**LP**	**Debit**	**Credit**	**Balance**
Jan	31	Purchase inventory		10,000.00		10,000.00

Equipment						*Debit*
Date		**Explanation**	**LP**	**Debit**	**Credit**	**Balance**
Feb	10	Purchase equipment		25,000.00		25,000.00

Figure 9–12 A trial balance is a two-column schedule listing the names and balances of all the accounts in the order in which they appear in the ledger. The totals of the two columns should agree.

is the difference between total assets and total liabilities. Legally, the claims of creditors come first. If you are the owner of an agribusiness, you are entitled to whatever remains after the claims of the creditors have been fully satisfied (that is, after they have been paid what they are owed).

Revenue, Cost of Sales, and Net Profit

Revenue is the value of what is received from goods sold and services rendered, plus what comes in from other financial sources. Be careful not to confuse the terms *revenue, sales,* and *income.* Most revenue (money coming into the agribusiness) comes from sales. However, there could be other sources of revenue, such as rents received, interest earned, or income from other sources.

Cost of sales (also known as *cost of goods sold*) is the expense and labor that are directly associated with the production or acquisition of products held for sale. This includes the purchase price plus any packaging or freight charges paid to bring in the goods, plus the cost associated with storing the goods, less the cost of any item returned.[19]

Net profit is revenue minus the cost of sales. Net profit is also called the *profit margin.* In an agribusiness service, there may be no cost of goods sold; therefore, net sales could equal gross margin.

Operating and Incidental Expenses

Operating expenses are any regular expenses associated with operation of the business. Obvious ones include rent, salaries, supplies, utilities, insurance, and depreciation of equipment. **Incidental expenses** are any other expenses, including ones that seldom occur but are business related. Examples are unexpected expenses incurred because of emergencies, accidents, or weather damage.

Inventory

An important aspect of financial accounting is the development of inventory for the agribusiness. An **inventory** is a physical count of all the assets of a business along with their estimated worth.

When to Inventory. Although inventories can be done at any time, January 1 is a good date because it corresponds with the tax year of many businesses. The inventory date must be the same each year, to allow accurate comparisons of financial performance. The ending inventory for the year is always the beginning inventory for the next year.

Determining Inventory Value. There are two methods of putting a dollar value on agribusiness inventories: book value and market value. *Book value* is the original cost of an item minus depreciation and any other adjustments. Typically, book value is used for income tax purposes and for estate planning. *Market value* is book value minus any variable cost (for example, for such things as transportation or sales commissions). This method is typically used for establishing credit and determining insurance needs.

There are many variations of these two methods. No matter which method a manager selects to value inventory, it is important to use the same one consistently.

Depreciation

Depreciation is the systematic **write-off** of the value of a tangible asset over the estimated useful life of that asset. Agribusinesses are permitted to treat depreciation as an expense of business operations. This loss in value occurs because of normal use, weathering, and/or changes in technology. Depreciation of machinery, buildings, equipment, and storage structures constitutes a major agribusiness expense.

Real Estate Depreciation. With regard to real estate, *depreciation* is the loss in value of an asset (building) over time due to wear and tear, physical deterioration, and age. The cost of reproducing an income property can be recovered over the useful life of the asset (the useful life is determined by law). Depreciation is treated as an expense and is a line item on an income statement. Depreciation can be applied only to the building, not to the land, because land does not wear out over time. Residential income property must be depreciated over a 27.5-year period using straight-line depreciation. Commercial income property must be depreciated over 39 years using straight-line depreciation. The straight-line depreciation method stipulates that an asset must be depreciated by an equal amount in each year of its useful life. The depreciation deductions that you write off in any

19. Nickels, McHugh, and McHugh, *Understanding Business,* pp. 541–542.

year reduce your taxable income, thus increasing your profit for that year.

Capital improvements are subject to the same depreciation laws. Capital improvements include such things as a new roof, a new furnace, an addition to a building, siding, and the like. All depreciation amounts that you write off in each year for the building and any capital improvements reduce your adjusted basis for the property, thus increasing the taxable profit you must declare when you sell that asset.

Calculating Depreciation. There are various methods of calculating depreciation. Straight-line depreciation is the simplest and easiest to calculate. To calculate straight-line depreciation, compute the annual depreciation by dividing the original cost of the asset, less any **salvage value**, by the expected useful life of the asset. For example, a $25,000 asset with a $1,000 salvage value is determined to have an expected life of six years. The depreciable value would thus be $24,000 and the depreciation amount is $4,000 per year. The depreciation amount remains the same over the next five years. Refer to Figure 9–13.

Many depreciation programs are available on the Internet. Just go to your favorite search engine and type in "calculate depreciation." Remember that some materials have a charge associated with them, whether for purchase of the program or online use of the depreciation calculator.

Principal and Interest Payments. Many people are surprised to learn that the amounts one pays toward interest and principal vary dramatically over time. This is because mortgage loans work so that the early payments go primarily for interest; later payments go primarily toward paying the principal. Although your monthly payment remains the same throughout the life of the loan (if it has a fixed interest rate), in the beginning most of your payment only pays interest. Lenders long ago developed what are known as amortization tables. These tables make it fairly easy to calculate how much of each payment is for interest and how much pays down the principal balance.

For example, let's calculate the principal and interest for the very first monthly payment on a 30-year, $100,000 mortgage loan at 7.5 percent interest. According to the amortization tables, the monthly payment on this loan will always be $699.21. The first step is to calculate the annual interest by multiplying $100,000 × .075 (7.5%). This equals $7,500, which we then divide by 12 (for the number of months in a year), which equals $625. If you subtract $625 from the monthly payment of $699.21, you can see that $625 of the first payment is interest, while $74.21 of the first payment goes toward paying off the principal. Next, if we subtract $74.21 (the first principal payment) from the $100,000 of the loan, we come up with a new unpaid principal balance of $99,925.79. To determine the next month's principal and interest amounts, we just repeat the steps already described. Thus, we now multiply the new principal balance ($99,925.79) by the interest rate (7.5%) to get an annual interest payment of $7,494.43. Divided by 12, this equals $624.54. So, of the second month's payment, $624.54 is interest and $74.67 goes toward the principal. Refer to Chapter 8 for ways to reduce this debt more quickly and save thousands of dollars.

Straight-Line Depreciation Example

	Annual Depreciation	Accumulated Depreciation	Book Value
Year 1	$4,000	$4,000	$21,000
Year 2	4,000	8,000	17,000
Year 3	4,000	12,000	13,000
Year 4	4,000	16,000	9,000
Year 5	4,000	20,000	5,000
Year 6	$4,000	$24,000	$1,000

Figure 9–13 An asset costing $25,000 with a salvage value of $1,000 and a useful life of six years will depreciate by $4,000 yearly.

Profit and Loss

Profit and loss are the difference between sales minus all expenses. In an income statement (discussed later in this chapter), profits can be noted as three different amounts: gross profit, operating income, and net income.

Gross profit is the term used when the cost of goods sold is deducted from sales revenue. The gross profit is a subtotal. *Operating income* is the result when operating expenses are deducted from gross profit. This is also a subtotal. Some call this *income from operations*. *Net income* is determined by subtracting nonoperating costs from operating income. Many call this the "bottom line."

Preparing an Income Statement, Balance Sheet, and Statement of Cash Flows

Agribusiness owners and managers need up-to-date financial information to make decisions about future operations. As discussed earlier, the financial information for an agribusiness is recorded in ledger accounts. Transactions occurring during a fiscal period change the balance of ledger accounts. These changes should be reported periodically in the form of financial statements.

Financial statements summarize the financial information contained in ledger accounts. These financial statements provide adequate disclosure of the condition of the agribusiness. A financial statement showing the revenue, expenses, and net income or net loss of a business is called an *income statement.* A financial statement showing all assets and **equities** of a business on a specific date is known as a *balance sheet.*

A balance sheet and an income statement can be prepared whenever the financial information is needed. However, agribusinesses should always prepare these two statements at the end of a fiscal period. An income statement shows financial information over a specific period of time. A balance sheet shows financial information on a specific date. The financial progress of an agribusiness in earning a net income (or incurring a net loss) is shown by the income statement. The financial condition of an agribusiness on a specific date is shown by the balance sheet. Refer to Figure 9-14.

Income Statement

An income statement measures the profit gained by an agribusiness over a given time period. This time period is usually one year; however, monthly or quarterly reports may be desired. Other names for the income statement are *operating statement* or *profit and loss statement.* The four major components of the income statement are revenue, expenses, taxes, and "other." Refer to Figure 9-15.

Revenue. Many agribusinesses earn revenue from a variety of sources. Revenue comes in the form of cash receipts from product sales, services, and

Comparison of Income Statement and Balance Sheet

The income statement reports financial progress from the beginning of the month until the end; from January 1st to January 31st.

For Month of January

S	M	T	W	T	F	S
	1	2	3	4	5	6
7	8	9	10	11	12	13
14	15	16	17	18	19	20
21	22	23	24	25	26	27
28	29	30	31			

The balance sheet reports the balances carried as of the end of a day; for example, January 31st.

As of January 31

S	M	T	W	T	F	S
	1	2	3	4	5	6
7	8	9	10	11	12	13
14	15	16	17	18	19	20
21	22	23	24	25	26	27
28	29	30	31			

Figure 9-14 Agribusiness owners should prepare these two statements at the end of every fiscal period.

Green's Nursery & Landscaping Co. Income Statement for the Year			
Revenue			
Plants	200,000		
Soil, Fertilizer, etc.	50,000		
Accessories	30,000		
Gross Sales Income		280,000	
Cost of Goods		140,000	
Gross Profit			140,000
Landscaping Fee Income			20,000
Lawn Care Fee Income			7,000
Total Income			167,000
Variable Expenses:			
Labor	30,000		
Advertising	4,000		
Office Expense	1,000		
Total Variable Expenses		35,000	
Fixed Expenses:			
Equipment Depreciation and Lease	5,000		
Rent	4,000		
Insurance	3,000		
Total Fixed Expenses		12,000	
Total Expenses			47,000
Net Income before Income Tax			120,000
Income Tax Expense			36,000
Net Income			

Figure 9–15 An income statement measures the profit for an agribusiness over a given time period.

interest and dividend income. The gain or loss on the sale of intermediate- and long-term assets also counts as revenue. Wages earned from a job unrelated to the business, noncash adjustments in the value of inventories, and **accrued** revenue due from others for products already sold are counted as revenue, too.[20]

Expenses. Total agribusiness expenses include cash operating expenses, interest expenses, and depreciation and other noncash adjustments. Depreciation of property and equipment is a business expense even though no actual cash payment is made at the time. Accounts payable accrued expenses, cash investments in the business, and prepayments during the year may also require expense adjustment.[21]

Taxes. The difference between revenue and expenses represents the **taxable income**, or loss, of an agribusiness. The income statement typically uses a separate category to account for payment of income and self-employment taxes. This category does not include property taxes or employee taxes—they count as business expenses.[22]

20. *Advanced Agribusiness Management and Marketing* (8709-D) (College Station, TX: Instructional Materials Service, 1998), p. 2.

21. Ibid.

22. Ibid.

Other. The final component of an income statement captures the gain or loss from unusual events. For example, a freeze that destroys an insured crop results in an insurance payment that would go into the "other" category.

Cash or Accrual? All small businesses need to choose one of two methods of accounting: the cash method or the accrual method (sometimes called *cash basis* and *accrual basis*). It is important to understand the basics of these two principal methods of keeping track of the income and expenses of a business. In a nutshell, these methods differ only in the timing of when transactions, including sales and purchases, are credited or debited to the accounts. The accrual method is more commonly used.

With the accrual method, transactions are counted when the order is made, the item is delivered, or the services are rendered, regardless of when the money for them (receivables) is actually paid or received. In other words, income is counted when you make the sale, and expenses are counted when you receive the goods or services. You need not wait until you see the money, or until you actually pay money out of your checking account, to record a transaction.

With the cash method, income is not counted until cash (or a check) is actually received, and expenses are not counted until they are actually paid. For example, your computer installation business finishes a job on November 30, 2007, but you do not get paid until January 10, 2008. Using the cash method, you would record the payment in January 2008. Using the accrual method, you would record the income in your books in November 2007. Consider this example. You purchase a new laser printer on credit in May and pay $1,000 for it in July, two months later. Using cash method of accounting, you would record a $1,000 payment for the month of July, the month when you actually disburse the money. Using the accrual method, you would record the $1,000 payment in May, when you take the laser printer and become obligated to pay for it.

Determining the Transaction Date. Sometimes it is not completely clear when a sale or purchase has occurred. The key date with the accrual method is the job completion date. You do not put the income down in your books until you finish a service, or deliver all the goods a contract calls for. Likewise, you do not record an item as an expense until the service is completed or all goods have been received and installed, if necessary. (If a job is mostly completed but it will take another 30 days to add the finishing touches, technically the income and expenses related to that job do not go on your books until the 30 days have passed.)

Balance Sheet

The three major components of a balance sheet are assets, liabilities, and **net worth**. The name *balance sheet* references the fundamental accounting operational axiom that "assets equal liabilities plus net worth." This equation always holds true. Any transaction changing one side of this equation results in an equal change to the other.

The balance sheet normally has two columns. The left side lists the assets; liabilities and net worth appear on the right side. Assets and liabilities have direct values, whereas net worth has an indirect value because it is calculated from total assets minus total liabilities. The three classifications of both assets and liabilities are current, intermediate, and long-term. Refer to Figure 9–16.

Current Assets and Liabilities. *Current assets* are those assets that an agribusiness manager could convert to cash within a 12-month period without disrupting business operations. Current assets include such items as cash, accounts receivable, inventories, and prepaid expenses. Current liabilities are debt payments due within 12 months of the balance-sheet date. Those liabilities include such items as accounts payable, accrued expenses, and any payments due on loans. Any tax due (contingent) on the sale of current assets is also a current liability.

Intermediate Assets and Liabilities. *Intermediate assets* are those that are normally in service for more than 1 year but less than 10. During this time frame, management either depreciates, liquidates, or replaces the asset. This group includes machinery, equipment, vehicles, and business furnishings.

Intermediate liabilities also fit within a time frame from 1 to 10 years. They primarily involve notes that mature within a 10-year

Green's Nursery & Landscaping Co.
Balance Sheet
December 31

Assets			
Current Assets:		**Current Liabilities:**	
Cash	69,000	Accounts Payable	10,000
Short-Term Investments	20,000	Short-Term Notes Payable	2,000
Accounts Receivable	20,000	Current Portion of Long-Term Debt	2,000
Inventory	34,000	Accrued Rent	500
Supplies	5,000	Accrued Payroll	1,500
Prepaid Expenses	2,000	Accrued Taxes	1,000
Total Current Assets	150,000	Total Current Liabilities	17,000
Property and Equipment:		**Long-Term Liabilities:**	
Land	10,000	Mortgages net of current portion	20,000
Buildings	25,000	Notes Payable net of current portion	15,000
Machinery and Equipment	20,000	Total Long-Term Liabilities	35,000
Total Original Cost	55,000		
Accumulated Depreciation	5,000	**Total Liabilities**	52,000
Total Property and Equipment	50,000		
Long-Term Assets:		**Owner's Equity:**	
Long-Term Investments	15,000	Original Investment	70,000
Notes Receivable	4,000	Retained Earnings—Jan. 1st	14,000
Deposits	1,000	Current Year Net Income/Loss	84,000
Total Long-Term Assets	20,000	**Total Owner's Equity (net worth)**	168,000
Total Assets	**$220,000**	**Total Liabilities & Owner's Equity**	**$220,000**

Figure 9–16 The three components of a balance sheet are assets, liabilities, and net worth. The balance sheet has two columns. The left side lists the assets and liabilities; net worth appears on the right side.

period and taxes contingent on the sale of intermediate assets.

Long-Term Assets and Liabilities. *Long-term* or *fixed assets* have an expected life or maturity exceeding 10 years. These assets include land, buildings, and other permanent improvements.

Long-term or fixed liabilities have maturity periods of more than 10 years. They include mortgages on land and buildings and other long-term obligations.

Net Worth. *Net worth* is the difference between total assets and total liabilities. If this value is positive, the business has **solvency** (is solvent). This means that if management converted all the assets to cash, there would be more than enough to cover all liabilities.

Statement of Cash Flows

In 1988, the Financial Accounting Standards Board required that businesses replace the statement of changes in financial position with the statement of cash flows.[23] The **statement of cash flows** reports cash receipts and disbursements related to the major activities of the agribusiness: operations, investments, and financing. Refer to Figure 9–17. Generally Accepted Accounting

23. Nickels, McHugh, and McHugh, *Understanding Business,* p. 544.

Green's Nursery & Landscaping Co. Statement of Cash Flows for the Year Ended December 31		
Cash flows from operations:		
Sales	307,000	
Cash payments from Accounts Receivable	15,000	
Inventory	(140,000)	
Cash expenses for operations	(42,000)	
Income taxes paid	(36,000)	
Net cash from operations		104,000
Cash flows from activities other than operations:		
Investments	(24,000)	
Notes receivable	2,000	
Mortgage payments	(40,000)	
Proceeds from notes payable	15,000	
Net cash from nonoperations activities		(47,000)
Net change in cash		57,000
Cash at beginning of year		12,000
Cash at end of year		$69,000

Figure 9–17 The statement of cash flow reports cash receipts and disbursements related to operations, investments, and financing.

Principles (GAAP) mandate inclusion of the statement of cash flows as one of the four basic financial statements. The other three required financial statements are the income statement, balance sheet, and statement of owner equity.

Comparison of Cash-Flow Statement and Statement of Cash Flows. The statement of cash flows is an analysis and planning tool. The cash-flow statement is a projection tool.

A *cash-flow statement* highlights the financial arrangements necessary to cover cash requirements of the agribusiness. The previous year's cash-flow statement shows when cash surpluses and **deficits** occurred. The surpluses and deficits largely determine the credit needed and payback schedules. For example, a loan may be due in March, but your best month of sales is April. In this situation, you could go bankrupt, because improper planning has not arranged to cover the fluctuation in cash flow, even though money would be available within the fiscal year. Refer to Figure 9–18.

A *statement of cash flows* relates to evaluation and planning. Questions that are answered by the statement of cash flows include:

- Where did the cash flows come from?
- How were dollars spent?
- Is the change in cash flow consistent with the balance sheet's cash position?
- Was cash used to buy stocks, bonds, and other investments?
- Were some investments sold that brought in cash?

Sources of Cash Flows. Cash and cash equivalents are the only types of entries allowed on the statement of cash flows. Permitted items include checking account balances, actual cash on hand, savings accounts, time certificates, government payment certificates, and so on. As stated earlier, entries to the statement of cash flows fall into three categories: operating, investing, and financing activities.[24]

24. National Council for Agricultural Education, *Decisions and Dollars* (Alexandria, VA: Author, 1995).

Green's Nursery & Landscaping Co. Quarterly Cash Flow Statement for the Year Ended December 31					
Categories	**1st Qtr.**	**2nd Qtr.**	**3rd Qtr.**	**4th Qtr.**	**Total**
Cash inflows from operations:					
1. Beginning Cash	12,000	28,500	16,500	26,000	12,000
2. Plants Receipts	30,000	70,000	70,000	30,000	200,000
3. Soil and Fertilizer Receipts	10,000	25,000	30,000	15,000	80,000
4. Landscaping Fees	2,000	6,000	8,000	4,000	20,000
5. Lawn Care Fees	500	2,000	3,500	1,000	7,000
	54,500	131,500	128,000	76,000	319,000
Cash outflows from operations:					
6. Inventory	5,000	90,000	40,000	5,000	140,000
7. Operating Expense	8,000	13,000	13,000	8,000	42,000
8. Income Tax Payments	9,000	9,000	9,000	9,000	36,000
	22,000	112,000	62,000	22,000	218,000
Cash Flows from Operations	32,500	19,500	66,000	54,000	101,000
Cash flows from other activities:					
9. Accounts Receivable	10,000	10,000	(30,000)	25,000	15,000
10. Investments	(4,000)	(20,000)			(24,000)
11. Notes Receivable		2,000			2,000
12. Mortgage Payments	(10,000)	(10,000)	(10,000)	(10,000)	(40,000)
13. Proceeds from Notes Payable		15,000			15,000
	(4,000)	(3,000)	(40,000)	15,000	(32,000)
Ending Cash Balances	$28,500	$16,500	$26,000	$69,000	$69,000

Figure 9–18 A quarterly cash-flow statement highlights the financing arrangements necessary to cover the cash requirements of the agribusiness.

The values within the "Operating Income and Expenses" section should correspond to the accrual income statement. An example is entries related to the production of goods and services that include consumable supplies. Another example is withdrawals for family/personal living and investments. This entry is made if the agribusiness does not pay a wage or salary as a business expense for labor and management contributions of family members.[25]

Information as to why purchases and sales of capital assets occurred is provided in the "Investments" section. It is important to know if

25. Ibid.

Green's Nursery & Landscaping Co. Statement of Owner's Equity for the Year Ended December 31		
1. Beginning owner's equity		84,000
2. Net income	84,000	
3. Gifts and inheritances	5,000	
4. Additions to paid-in capital, including investments of personal assets into the business	10,000	
5. Distributions of dividends, capital, or gifts made (cash or property)	(5,000)	
6. Withdrawals for family living, gifts made, and investments into personal assets	(10,000)	
7. Total change in contributed capital and retained earnings (Line 2 + Line 3 + Line 4 - Line 5 - Line 6 = Line 7)		84,000
8. Change in valuation of equity		0
9. Ending owner's equity		$168,000

Figure 9–19 The primary role of the statement of owner equity is to conduct a final check on the accuracy of the change in owner equity. The statement of owner equity utilizes information from the other three basic financial statements: balance sheets, income statement, and statement of cash flows.

assets are being sold to cover operating cash deficiencies. Nonbusiness investments are also included within this category. A negative cash flow requires either the operating or financing area to cover it, whereas a positive cash flow requires managers to scrutinize the records to determine if cash is being used in the best interest of the business. Examples of investments include land, buildings, machinery/equipment, personal loans to others, retirement accounts, and so on.[26]

Borrowing funds or other injections of equity constitute the primary cash inflows. Loan repayments or capital lease repayments are the major cash outflow entries. The "Financial Activities" section provides the business operator with a summary of debt financing. Both operating and term debt are accounted for. This information is an invaluable aid when making decisions about acquiring and repaying loans.[27]

Statement of Owner Equity

The primary role of the *statement of owner equity* is to allow a final check on the accuracy of the change in owner equity between the beginning and ending balance sheets with net income and other possible changes in owner equity. Owner equity is the most fundamental financial measure of a firm's financial position and performance. Verifying that the base financial statements, balance sheet, income statement, and statement of cash flows are in agreement is the goal of the statement of owner equity. It also makes it possible to establish and understand what occurred financially within the business during the financial year. This statement is a powerful addition to the basic financial statements, and it offers exceptional insight into the financial status of a business.

As previously mentioned, the statement of owner equity utilizes information from the other three, basic financial statements. Beginning and ending owner equity amounts come from the balance sheets. Net income is calculated in the income statement, and other values are derived from the statement of cash flows.

Explanation of the Statement of Owner Equity

Figure 9–19 shows nine entries within the statement of owner equity financial statement. An explanation follows for each of the nine entries.

26. Ibid.
27. Ibid.

1. Beginning Owner Equity. This is the amount of equity at the beginning of the year. It is the same amount shown on the balance sheet ("Total Owner Equity") for the previous year on December 31.

2. Net Income (Accrual). This amount is gross revenue minus expenses. It is the same amount as the "Net Business Income" total taken from the income statement.

3. Gifts and Inheritances. This includes any amount given to you outside the revenue-producing ability of the agribusiness. An example would be a $5,000 inheritance that you invested in the agribusiness.

4. Additions to Paid-in Capital, Including Investments of Personal Assets into the Business. If money is taken from investments in personal accounts, stocks, or any other personal assets and put into the agribusiness, this is where it is recorded. These contributions to the agribusiness are helpful, but their origin must be accurately noted to more accurately reflect the stability of the business.

5. Distributions of Dividends, Capital, or Gifts (Cash or Property). These values are usually found directly on corporation or partnership financial distribution forms. These values are recorded in the statement of cash flows. However, few beginning agribusiness sole (single) proprietorships would need to use this type of entry.

6. Withdrawals for Living Family, Gifts Made, and Investments into Personal Assets. As indicated, this entry represents money taken from the agribusiness for family living, gifts, and investments into personal assets. The amount is taken from the statement of cash flows.

7. Total Change in Contributed Capital and Retained Earnings. Retained earnings is the amount of money put back into the business. This practice of reinvesting income back in the business is common in most firms where growth of the business is a goal. Also, in Figure 9-19 notice that Line 2 + Line 3 + Line 4 − Line 5 − Line 6 = Line 7.

8. Change in Valuation Equity. If the beginning balance sheet and ending balance sheet are done correctly, this amount is found by subtracting the amount in "Owner equity: retained earnings, contributed capital, personal net worth and valuation equity" in the beginning balance sheet from the amount in that same category in the ending balance sheet.

9. Ending Owner Equity. This amount is calculated as follows: Line 1 +/− Line 7 +/− Line 8 = Line 9 ("Ending owner equity").
Line 9 and owner equity on the ending balance sheet should be the same.[28]

Analyzing Financial Statements

The last step in the accounting cycle is to analyze the financial statements. Balance sheets, income statements, and cash-flow statements provide information to agribusiness owners and managers on how well the business is doing. An agribusiness manager can obtain a clear picture of the financial position and performance of the business only by analyzing the components of all three financial statements together.

Compare All Financial Statements with Each Other

Agribusiness managers can expect problems in evaluating financial position and performance when only one or two of the statements are available. For example, a solid net income in one year does not necessarily mean that the business is doing well, as profits might be the result of the sale of intermediate- or long-term productive assets. An examination of the balance sheet would reveal if total assets were increasing or decreasing and what was happening to net worth. Also, a strong cash position on the balance sheet does not necessarily mean that the business is in good shape.[29]

Furthermore, neither the balance sheet nor the income statement show critical financial periods during the year. Only the cash-flow statements show when financial stress occurs during the year and the steps that were taken to improve the situation.

28. Ibid.
29. Ibid.

Objective of Financial Analysis

The objective of financial analysis is to estimate financial position and monitor financial performance. A financial analysis concentrates on the solvency, liquidity, and changes in the net worth of the agribusiness.

Methods of Financial Analysis

Several methods are available for analyzing financial statements: historical analysis, comparative analysis, and ratio analysis. *Historical analysis* makes comparisons over time. *Comparative analysis* compares events within the year with what happened in other, similar firms. *Ratio analysis* expresses key financial relationships in percentage terms. This allows management to assess liquidity ratios, leverage (debt) ratios, profitability (performance) ratios, and activity ratios as compared to previous years or similar operations. These ratios provide valuable information to agribusiness owners, managers, and lenders. A brief description of each follows.

Liquidity Ratio. Creditors are interested in a company's ability to repay its short-term debts. This ability is known as the firm's *liquidity ratio*. Liquidity may be measured by determining the company's current ratio and its acid-test ratio.[30] The current ratio is calculated by dividing the firm's current liabilities by its current assets. This ratio will appear on the company's balance sheet.

For example, suppose that Hall's Auto Repair has $25,000 of current assets and $10,000 of current liabilities. Its current ratio is thus 2.5. This means the company has $2.50 of current assets for every $1 of current liabilities.

$$\textit{Current ratio} = \frac{\textit{Current liabilities}}{\textit{Current assets}} = \frac{\$25{,}000}{\$10{,}000} = 2.5$$

Leverage (Debt) Ratios. Lenders and stockholders are interested in the amount of borrowed money that a company needs to operate. This is the company's leverage (debt) ratio. The amount of debt a company has may influence whether loans are repaid or dividends are issued. A company's debt ratio is calculated by dividing its total liabilities by the owner equity.

$$\textit{Debt of owner equity ratio} = \frac{\textit{Total liabilities}}{\textit{Owner equity}} = \frac{\$150{,}000}{\$275{,}000} = 54.5\%$$

Ratios above 1 (100 percent) alarm lenders and stockholders because it means that a company actually has more debt than equity. Comparing a company's past and current debt ratios can be useful in identifying trends within the firm. It may also be useful to compare debt ratios of similar companies, as acceptable debt amounts vary by industry.[31]

Profitability (Performance) Ratios. In business, management is often evaluated by how effectively it uses available resources to make a profit. This is called the profitability (performance) ratio.[32] Three important profitability ratios are earnings per share, return on sales, and return on equity. An example of each follows.

$$\textit{Earnings per share} = \frac{\textit{Net income}}{\textit{Number of common shares outstanding}} = \frac{\$60{,}000}{\$150{,}000} = \$0.12 \textit{ earnings per share}$$

$$\textit{Return on sales} = \frac{\textit{Net income}}{\textit{Net sales}} = \frac{\$25{,}000}{\$250{,}000} = 10\% \textit{ return on sales}$$

$$\textit{Return on equity} = \frac{\textit{Net income}}{\textit{Total owner equity}} = \frac{\$25{,}000}{\$75{,}000} = 33\% \textit{ return on equity}$$

30. Nickels, McHugh, and McHugh, *Understanding Business,* p. 548.
31. Ibid., p. 549.
32. Ibid., p. 549.

Activity Ratios. Management's effectiveness may also be measured by how well it uses the firm's various assets to generate profit. This is called the company's *activity ratio*.

Inventory Turnover Ratio. The *inventory turnover ratio* is a measure of how quickly a firm moves inventory through its systems and converts inventory into sales. Better inventory control will yield greater profit. The inventory turnover ratio is calculated by dividing the cost of goods sold by the average inventory.

$$\textit{Inventory turnover ratio} = \frac{\textit{Cost of goods sold}}{\textit{Average inventory}} = \frac{\$320{,}000}{\$116{,}000} = 2.75\ \textit{turns}$$

If a company purchases poorly or stocks unwanted inventory, it will have an inventory turnover ratio that is lower than industry standards. If the company has an inadequate amount of inventory and loses sales, it will have an inventory turnover rate that is higher than industry standards.[33]

Computer Technology and Agribusiness Records

Many industries, such as tractor companies, supply computer software programs for their dealers to use. Certified representatives can train the workers at the dealership to execute most business transactions. Many things can be done with these computer programs, including:

- sales and billing
- customer accounts
- weekly payroll
- receipt of products into inventory
- monthly statement printing
- financial records and reports
- quarterly tax calculations
- placement of monthly bills in correct accounts
- journal balancing

Besides helping keep financial records, these same computer software programs can assist with ordering parts, locating part numbers, inventory, and various other functions the agribusiness needs. We hope that the information in this chapter will give you a better understanding of the information that such computer programs can provide.

Filing Taxes with Computer Software

There are numerous computer software programs with which to complete and file tax returns. If you have a personal computer, any of the leading software packages will be extremely helpful in calculating and filing your taxes. These tax programs have reached a high state of proficiency. They assist in the calculation of sums, percentages, and tax liabilities. However, it is best not to rely completely on the computer program alone. Mistakes made during the input phase can cause the output to be incorrect. Always double-check your work!

The tax software programs are designed to be easy to use, even for a beginner. These programs will lead you through all the steps necessary to prepare and file the return. The programs are accurate and fast, and all the leading programs provide technical support in addition to the computer program prompts and menus. However, the support people are not trained in accounting or tax law and will refer the taxpayer to the Internal Revenue Service for questions concerning taxes. The leading software programs do checks after the return is complete to verify the accuracy and consistency of the return. You can access an audit report that shows any possible errors in the return. If you have an Internet connection, you can transmit your tax return electronically, rather than printing it out and mailing it (there may be an additional charge for this). There are benefits to filing electronically, including acknowledgment that the tax return was received; some programs even allow you to specify that the refund should be deposited directly into a checking account.[34]

Many careers are available in agribusiness accounting and recordkeeping. Refer to the Career Options section for further discussion of careers in this area.

33. Ibid., p. 551.

34. Margaret Wagner, *Personal Tax Edge* (Hiawatha, IA: Parsons Technology, 1997), pp. 19–22.

Career Options

Accountant, Bookkeeper, Financial Manager, Management Analyst

An important aspect of managing a sound business is the ability to plan ahead financially. Production agriculturalists today must spend their time managing crops and livestock, so the task of preparing financial paperwork must fall to someone other than the farm operator. A bookkeeper can handle most of the simple, day-to-day recording of business transactions. A bookkeeping position requires good math skills along with accounting classes at the high school level, and a postsecondary education at a community college often enhances employment opportunities.

Accountants perform a broad range of accounting, auditing, tax preparation, and consulting activities for their clients. Accountants have gained professional recognition through a certification/licensure test (upon passing this test, one is designated a Certified Public Accountant, or CPA). This is a rigorous four-part, two-day test, and only about one-fourth of all test takers pass the first time. Each of the four sections of the test can be taken separately and passed within a specified time period. The minimum requirement for licensure is a bachelor's degree in accounting or a related field. A master's degree, in addition to familiarity with computer software for accounting and an area of accounting expertise, are an advantage in today's job market. Financial management is an area of specialization within the accounting field.

A management analyst interprets financial information so that a business owner can make sound financial decisions. A management analyst prepares stockholder reports, financial statements for banks and other creditors, and government paperwork and tax forms. He or she can help you to plan for future growth by creating a budget for your business. The prerequisites for this occupation are similar to those for the accountant. A bachelor's degree in accounting or business management is preferred, although expertise in both areas is a plus. Management analysts are listed on the Department of Labor, Bureau of Statistics, 1994–2005 National Industry Occupation Employment Matrix as among the top 25 fastest growing occupations requiring a bachelor's degree or higher for the period 1994 through 2005. Cashier, billing clerk, and accounting clerk are entry-level positions, which can be attained with good grades in math and a high school diploma.

Computer programs for daily, weekly, and monthly recordkeeping for accountants, bookkeepers, financial managers, and management analysts are abundant. Most businesses also file their taxes with the aid of computer programs. (Courtesy of Cliff Ricketts)

Conclusion

Planning is a big part of success. Proper budgets, recordkeeping, reports, and financial analysis are a part of sound business conduct and operations. Proper and accurate recordkeeping lets you know the financial status of your business. Financial reports show you whether you can reinvest in the business or make other investments, and help you time your loan payments to match your cash flow. Almost all recordkeeping can be done by computer, which makes the process much easier.

Summary

An important, yet often overlooked, part of planning in an agribusiness involves budgeting, planning, and keeping good records. We need to work as hard at recordkeeping as at any other phase of the business.

Often, the distinction between accounting and bookkeeping is misunderstood. Bookkeeping refers to the actual recording of business transactions. An accounting system includes the preparation, review, and comprehension of reports. There are six steps in the accounting cycle: analyze and categorize documents; record the information in journals; post the information in ledgers; prepare a trial balance; prepare an income statement, balance sheet, and statement of cash flows; and analyze the financial statements.

If a business will not succeed on paper, it will not succeed once it is open. After an agribusiness owner or manager has planned for customers' needs and projected income and expenses, he or she must complete a budget. Three types of budgets must be prepared: an operating budget, a cash-flow budget, and a capital expenditures budget.

The simplest bookkeeping system is the single-entry bookkeeping system, in which the entry is simply made in the checkbook or in the checkbook and the journal. Besides a checking account, a sales slip should be used in this bookkeeping system. These two things will provide the information needed to file income taxes.

In double-entry bookkeeping, every transaction is recorded in two places. The two places (columns) must be the same; that is, in balance. This process involves journals and ledgers. To keep the transactions in balance, debits and credits must always equal each other. A debit is a bookkeeping entry on the left side of the account, and a credit is a bookkeeping entry on the right side of the account.

Information from sales or purchases is recorded in an account record called a journal. After the transaction has been recorded in the journal, the debit and credit changes in the individual accounts are posted in the ledger. The proof of the equality of debit and credit balances is called a trial balance. A trial balance is a two-column schedule listing the names and balances of all the accounts in the order in which they appear in the ledger.

Every agribusiness owner should be familiar with basic accounting terminology. Among other things, the owner or manager should understand assets and liabilities; capital and owner equity; sales, cost of sales, and net profit; operating and incidental expenses; depreciation; inventory; and profit and loss.

Agribusiness owners and managers need up-to-date financial information to make decisions about future operations. Financial statements summarize the financial information contained in ledger accounts. A financial statement showing the revenue, expenses, and net income or net loss of a business is called an income statement. A financial statement showing the asset and equities of a business on a specific date is known as a balance sheet. The statement of cash flows reports cash receipts and disbursements related to the agribusiness's major activities: operations, investment, and financing.

The primary role of the statement of owner equity is to allow a final check on the accuracy of the change in owner equity between the beginning and ending balance sheets with net income and other possible changes in the owner equity. Owner equity is the most fundamental financial measure of the firm's financial performance.

The last step in the accounting cycle is to analyze the financial statements. Balance sheets, income statements, and cash-flow statements provide information to agribusiness owners and managers on how the business is doing. However, all three financial statements must be analyzed together to obtain a clear picture of the financial position of the agribusiness.

After studying the necessary reports and creating the good financial records needed to run an agribusiness, one can appreciate the value of computers in the accounting process. Computers can record and analyze data and print out financial reports, providing continuous financial information for the agribusiness.

End of Chapter Activities

REVIEW QUESTIONS

1. Define the Terms to Know.
2. What are six reasons for keeping records?
3. What is the role of a financial manager within an agribusiness?
4. What are four potential sources of income for an agribusiness?
5. What are six potential expenses of an agribusiness?
6. Explain the differences between bookkeeping and accounting.
7. What are four reasons why accountants or accounting are vital to an agribusiness?
8. What are the six steps in the accounting cycle?
9. A set of financial statements consists of what four account reports?
10. Explain the relationship between the operating, capital expenditures, and cash-flow budgets.
11. Why should business accounts and personal accounts be kept separate?
12. What are five things the single-entry system must provide the agribusiness owner?
13. What information is recorded in a journal?
14. What are three advantages of using both a journal and ledger rather than the ledger alone?
15. What are three parts of a ledger account page?
16. What are the six steps of posting in a ledger?
17. What are three limitations of trial balances?
18. Do *revenue, sales,* and *income* describe the same thing? Explain.
19. What are six examples of operating expenses?
20. Explain the two methods of determining inventory values.
21. Explain and give an example of how to calculate straight-line depreciation.
22. Name and briefly discuss the four parts of an income statement.
23. Discuss the difference between the cash and accrual methods of accounting.
24. What are the four basic financial statements required by the Generally Accepted Accounting Principles (GAAP)?
25. List three reasons why a cash-flow statement is important.
26. What five questions are answered by a statement of cash flows?
27. Name five sources of cash flow shown on a statement of cash flows.
28. Name and briefly discuss the three areas used to categorize entries on the statement of cash flows.
29. What is the primary role of the statement of owner equity?
30. List four contributions of the statement of owner equity.
31. Name and briefly explain the nine entries on a statement of owner equity.
32. Explain why all the financial statements should be analyzed together.
33. What is the objective of financial analysis?
34. Name and briefly explain the three methods of financial analysis.
35. Why should an agribusiness owner consider using computer technology for the agribusiness records?

36. List six reasons for filing taxes with computer software.
37. Name and briefly explain the four types of ratio (financial) analysis.

FILL IN THE BLANK

1. The purpose of the cash-flow budget is to help the owner determine whether there will be adequate _________ to meet obligations.
2. In a single-entry bookkeeping system, a checking account and _______________ would provide the information needed to file income taxes.
3. ________ means that a bookkeeping entry was made on the left side of the account.
4. _________ means that a bookkeeping entry was recorded on the right side of the account.
5. When the assets equal the liabilities, the books are said to _________.
6. The unit of organization for the ledger is the ________, but the unit of organization for the journal is the _________.
7. The _________ is an accounting system that includes a separate record for each item.
8. The sum of all the debits in the ledger must be equal to the sum of all the _________.
9. Examples of incidental expenses would be unexpected emergencies due to _________ or _________ damage.
10. The ending inventory for the year is always the beginning _________ for the next year.
11. An _______________ shows financial information over a specific period of time.
12. A _______________ shows financial information on a specific date.
13. The three major components of a balance sheet are _________, _________, and _________.
14. ________________ is the most fundamental financial measure of a firm's financial position and performance.

MATCHING

a. capital expenditures budget
b. double-entry bookkeeping
c. gross profits
d. operating budget
e. single-entry bookkeeping
f. net income
g. exceeding 10 years
h. 1 to 10 years
i. operating income
j. cash-flow budget
k. 1 year or less

______ 1. summarizes the expected sales or production activities and related costs

______ 2. summarizes the amount and timing of income that will flow into and out of the business during the year

______ 3. list of projects (equipment, etc.) that management believes are worthwhile, with the estimated cost of each

______ 4. this bookkeeping system may only involve a checkbook and a journal

______ 5. practice of writing every transaction in two places

______ 6. cost of goods sold is deducted from sales revenues

______ 7. results when operating expenses are deducted from gross profits

______ 8. many call this the "bottom line"

______ 9. current assets and liabilities

______ 10. intermediate assets and liabilities

______ 11. long-term assets and liabilities

ACTIVITIES

1. In one paragraph, discuss why keeping good financial records is essential to the success of an agribusiness.
2. Name and discuss the three types of budgets used in planning for a business.
3. Illustrate the fundamental accounting equation.
4. Calculate the yearly principal and interest payments for a business that you bought for a cost of $275,000 at an 8.5% interest rate with a 30-year loan.
5. Using the same facts as in activity 4, calculate the payments on a 15-year loan. How much money was saved?
6. Describe journals and ledgers and explain the process of posting from journals to ledgers.
7. Assets and liabilities are grouped by length of time to maturity. Name the groups and discuss the time factors of each.
8. Using the following account balances, prepare a balance sheet for Sumner Nursery, December 31.

Current Assets:		**Current Liabilities:**	
Cash	$35,000	Accounts payable	$5,000
Short-term investments	$10,000	Short-term notes payable	$1,000
Accounts receivable	$10,000	Current portion of long-term debt	$1,000
Inventory	$17,000	Accrued rent	$500
Supplies	$2,500	Accrued payroll	$750
Prepaid expenses	$1,000	Accrued taxes	$500
Property and Equipment:		**Long-Term Liabilities:**	
Land	$5,000	Mortgages net of current portion	$10,000
Buildings	$12,500	Notes payable net of current portion	$7,500
Machinery and equipment	$10,000	**Owner Equity:**	
Accumulated depreciation	$2,500	Original investment	$35,250
Long-Term Assets:		Retained earnings—Jan. 1st	$7,000
Long-term investments	$7,500	Current year net income/loss	$42,000
Notes receivable	$2,000		
Deposits	$500		

9. Using the following account balances, prepare an income statement.

Income:		**Fixed Expenses:**	
Plants	$100,000	Equipment depreciation and lease	$2,500
Soil, fertilizer, etc.	$25,000	Rent	$2,000
Accessories	$15,000	Insurance	$1,500
Cost of goods sold	**$10,000**	**Income Tax Expense**	**$18,000**
Landscaping fee income	**$10,000**		
Lawn care fee income	**$7,000**		
Variable Expenses:			
Labor	$15,000		
Advertising	$2,000		
Office expense	$500		

Chapter 10

Managing Human Resources

Objectives

After completing this chapter, the student should be able to:

- Discuss six considerations when employees are being hired for an agribusiness.
- Discuss five characteristics an agribusiness manager should possess to work successfully with employees.
- Explain six theories that agribusiness managers may use in motivating employees.
- Discuss six things to consider when evaluating employees.
- Describe laws affecting human resources management.

Terms to Know

alien
goal-setting theory
Hawthorne Effect
human resource management
hygiene factors
impairments
inherent
job description
job enlargement
job rotation
Medicare
motivators
nonexempt employees
performance appraisals
résumé
self-actualization
self-esteem
social maturity
statutes

Introduction

Agribusinesses need to be aware of how important it is to manage human resources; that is, the employees who work for the business. **Human resource management** includes finding the right people, motivating them by providing incentives in an ideal environment, and evaluating them. Agribusinesses know that the best way to improve is to hire the right people and train them appropriately. Employees who are well trained, motivated, and satisfied help their employer businesses be successful.

Considerations as Employees Are Hired for Agribusiness

The owners or managers of an agribusiness should consider several things as they seek people to work for them. A brief discussion of each follows.

Determine Job Requirements

Agribusiness managers need to determine what jobs have to be filled in the company so they can choose and hire the right people. After the various job requirements are assessed, a **job description** for each position must be written. Job descriptions list the duties and responsibilities of each job as well as the educational level, skills, and special characteristics required for that post. Refer to Figure 10–1 for an example of a job description.

Determine Cost of Labor

Agribusiness managers need to calculate the labor cost of their employees. Besides the wage that the employee makes, the manager has to consider benefits (such as health insurance and life insurance) and the payroll taxes to be withheld for each employee. The size of the agribusiness will influence what benefits it can offer.

Recruitment

Agribusiness managers can take advantage of several sources of prospective employees. From that pool of available candidates, the manager must then recruit the best person to fill the job. Consider the following sources:

- local or drop-in applicants
- former employees or part-time employees
- family and friends of existing employees
- vocational-technical schools, colleges, or universities
- newspaper advertisements
- private or state employment agencies

Selection

When selecting employees, the agribusiness manager must know and follow federal and state regulations. Equal opportunity laws prohibit discrimination against potential employees on the basis of sex, age, race, or ethnic background. Selection involves hiring the employee with the best potential who appears to be the most likely to succeed. One concern that is often overlooked is hiring someone who is overqualified. This could be as much of a problem as hiring someone who is underqualified. The selection process usually involves an application, a résumé, and an interview.

Application. An application helps the agribusiness manager compare job applicants with one another. Applications also help the manager see the pride and care that each candidate takes in preparing

Agribusiness Computer Technician

- Tests, adjusts, and repairs office equipment.
- Performs periodic preventive maintenance on computer systems, using test equipment and protocols.
- Maintains computer systems, including software backups.
- Assists operators when technical problems occur.
- Checks for common causes of trouble, such as loose connections or obviously defective components.
- Reviews manufacturers' specifications that show connections and provide instruction on how to locate problems.
- Installs passwords and security features and keeps them updated.

Figure 10–1 Well-written job descriptions are needed so that agribusiness managers can hire the right people for specific jobs.

APPLICATION FOR EMPLOYMENT
(PRE-EMPLOYMENT QUESTIONNAIRE) (AN EQUAL OPPORTUNITY EMPLOYER)

PERSONAL INFORMATION

DATE April 15, 2000

NAME Fisher Ronald R. SOCIAL SECURITY NUMBER 351-44-5751

PRESENT ADDRESS 6428 Valley Rd. (STREET) Cambden (CITY) Ohio (STATE) 67423 (ZIP)

PERMANENT ADDRESS (STREET) Same (CITY) (STATE) (ZIP)

PHONE NO. 627-353-2761 ARE YOU 18 YEARS OR OLDER? Yes X No ☐

ARE YOU EITHER A U.S. CITIZEN OR AN ALIEN AUTHORIZED TO WORK IN THE UNITED STATES? Yes X No ☐

EMPLOYMENT DESIRED

POSITION engine + powertrain mechanic DATE YOU CAN START May 1 SALARY DESIRED Open

ARE YOU EMPLOYED NOW? Yes IF SO MAY WE INQUIRE OF YOUR PRESENT EMPLOYER? Yes

EVER APPLIED TO THIS COMPANY BEFORE? No WHERE? WHEN?

REFERRED BY Ken Jenkins

EDUCATION	NAME AND LOCATION OF SCHOOL	NO. OF YEARS ATTENDED	DID YOU GRADUATE?	SUBJECTS STUDIED
GRAMMAR SCHOOL	Cambden Elementary School District #95	8	yes	general curriculum
HIGH SCHOOL	Cambden High School	4	yes	vocational curriculum automotive
COLLEGE	Hillside Community College	2	yes	automotive technology
TRADE, BUSINESS OR CORRESPONDENCE SCHOOL	—			

FORMER EMPLOYERS (LIST BELOW LAST THREE EMPLOYERS, STARTING WITH LAST ONE FIRST)

DATE MONTH AND YEAR	NAME AND ADDRESS OF EMPLOYER	SALARY	POSITION	REASON FOR LEAVING
FROM June 19– TO present	Goodman's Tire + Auto Center – Cambden, OH	6.55 hr.	service technician	currently employed
FROM June 19– TO May 19–	Hunter's Auto Repair – Eastbrook, OH	4.30 hr.	general auto repair	part-time only
FROM Aug. 19– TO May 19–	Texacon Service Station – Cambden, OH	3.75 hr.	auto maintenance	co-op student learner
FROM TO				

WHICH OF THESE JOBS DID YOU LIKE BEST? Hunter's Auto Repair

WHAT DID YOU LIKE MOST ABOUT THIS JOB? engine diagnosis

REFERENCES: GIVE THE NAMES OF THREE PERSONS NOT RELATED TO YOU, WHOM YOU HAVE KNOWN AT LEAST ONE YEAR.

NAME	ADDRESS	BUSINESS	YEARS ACQUAINTED
1 Earl Thompson	Goodman's Tire + Auto 219 E. Sycamore Cambden, OH	service manager	1.5
2 Archie Hunter	Hunter's Auto Repair 2025 N. Walnut Eastbrook, OH	owner/manager	2.0
3 Frank Hopkins	Hillside Community College – Eastbrook, OH	auto instructor	2.0

Figure 10–2 Sample of a correctly completed job application form.

the application. Most job applications ask for the following information:

- the applicant's name, address, telephone number, and e-mail address.
- the applicant's education, training, and work history.
- references giving recommendations and providing information about the applicant's qualifications and skills.

Refer to Figure 10–2 for an example of a job application.

Résumé. Besides what is included in an application, the **résumé** provides detailed information about work history, personal traits, and skills. The main purpose of a résumé is to highlight a person's qualifications and abilities. As with job applications, the way a résumé is completed can tell the agribusiness manager a great deal about to the pride and attention to detail of the potential employee.[1] Refer to Figure 10–3 for a sample résumé.

1. Betty J. Brown and John E. Clow, *Introduction to Business* (New York: Glencoe/McGraw-Hill, 2006), p. 337.

Interview. No one should be hired without a personal interview. During an interview, the agribusiness manager can learn about the potential employee's attitude, experience, interests, and personal motivation. Even if the candidate does well in the initial interview, the manager should follow up with the person's references before making a final decision.

Orientation and Training

After new employees are selected, they need to know several things about the agribusiness company for which they will be working. They need to know, specifically, what they will be doing; how they will be paid and how deductions will be made; what benefits they are eligible for and what paperwork they must do regarding benefits; and what their work, meal, and break schedules are. They should also be introduced to the other employees.

Most agribusiness managers use on-the-job training. For each new employee, the manager will explain and demonstrate the job. After new employees have begun to do their jobs, the manager will give feedback on their performance and make any necessary adjustments. Other employees can also help train new hires.

Another training method that agribusiness managers can use is **job rotation**, a system in which new employees move from job to job until they learn the various tasks. Each rotation could last an hour, days, or weeks. Because it provides variety, job rotation helps prevent boredom and boosts morale. The exposure to other jobs and cross-training also makes the new employees more valuable, as they can fill in when needed for people who are out sick or on vacation.

Working with Employees

To work well with employees, agribusiness managers need special characteristics. Consider each of the following:

- *Adaptability.* An agribusiness manager has to work with many different employees in a variety of situations. The manager has to adapt his or her approach and actions depending on the situation.

SAMPLE RESUME
FOR A RECENT HIGH SCHOOL GRADUATE

JOAN SMITH
123 Some Street, City, State, Zip
Home (123) 980-2576, Work (123) 985-8973

OBJECTIVE: Seeking full-time office position involving data entry and/or computer operations.

EDUCATION: Big City High School, Big City, MO
Certificate: Computer/data entry (June, 1992)

Related course work:
-Introduction to computer literature
-Business accounting
-Computer operations, mainframe
-Records management
-Computer operations, micro
-Office machines

CAPABILITIES:
-Accurate and efficient worker.
-Work well without supervision.
-Pay attention to details.
-Computer experience: Wordperfect, Lotus 1-2-3, Q&A
-Typing 65 WPM, alpha and numerical filing.
-10 key calculator operation.

EXPERIENCE: Office worker, Ace Apple Barn, Small Town, RI (1986-1988)
Worked in records department.
-Processed work orders.
-Checked inventory.

Switchboard Operator, Jane's Telephone Exchange, Summersville, CA (1988).
-Operator for message center.
-Answered phone system and recorded messages.

(continued on the following page)

Figure 10–3 (A) Sample résumé of a recent high school graduate, including objective, education, capabilities, and experience. (B) Sample résumé of a person with considerable experience. (Source: R. A. Wolf and T. A. Wolf, *Job Hunter's Secret Handbook for Success* (St. Joseph, MI: WINSS, 1993). Used with permission.)

Figure 10–3 *(concluded)*

SAMPLE RESUME
CONSIDERABLE EXPERIENCE

BILL ROBERTS
507 Happy Valley St., Some City, IL 44548
Home (456) 453-6432, Car (456) 412-2222

OBJECTIVE: Seeking full time position as Building Supervisor and/ or Foreman

EDUCATION: University of Michigan, Bachelor of Science, Drafting Major.
Seminars Completed
"Managing Workers Today"
"Insight Seminars/Training for Construction"
"Advanced Constructors Framing"

EXPERIENCE: General Builders, Detroit, MI.
-Building duties: framing, roofing, trimming, overseeing subcontractors, plumbers, electricians, masons, painters, overseeing up to 30 laborers.
-Arranged for purchase of all supplies and equipment.
-Conducted training of many apprentice carpenters.
-Residential and commercial experience.

CAPABILITIES: -Attention to details.
-Excellent problem-solving ability.
-Mechanically inclined.

MILITARY: U.S. Army
Second Lieutenant.
Honorable Discharge.

REFERENCES: Available on request.

- *Human relationships.* Managers should be sensitive to the needs and feelings of their workers and respond to them with respect.
- *Emotional and* **social maturity**. Agribusiness managers must accept their own feelings and control their own emotions. They must work with many types of people even if they do not like a particular person.
- *Insight.* Good agribusiness managers have to have—or learn to develop—insight, a sort of sixth sense regarding other people. They must be able to analyze a complicated situation and see the relationships, causes, and effects so that they can reach appropriate solutions.
- *Self-motivation.* "If it's going to be, it's up to me." Good agribusiness managers motivate themselves. They have a great determination to reach their goals and objectives.[2]

2. William G. Ouchi, *Theory Z: How American Business Can Meet the Japanese Challenge* (Menlo Park, CA: Addison-Wesley, 1981).

Motivating Employees

One of the keys to success as an agribusiness manager is whether you can motivate your employees to do their best. People like to know and be told that they are appreciated and that their work is appreciated. Some say that people cannot be motivated unless they want to be; it is said that you can "lead a horse to water but you cannot make him drink." This author does not agree with this statement. If you give horses a heavy dose of salt in their feed, they will seek water and drink. Several theories of "salting" employees, to ensure their motivation, are discussed here.

Hawthorne Effect

The **Hawthorne effect** refers to people's tendency to behave differently when they know they are being studied. This recognition stemmed from research done in the 1920s. Employees in large production plants were placed in two different groups to see if better workplace lighting for one group would result in greater productivity. In fact, both groups increased their productivity by more than 50 percent. Several other research studies produced similar results. The researchers finally determined that the increased productivity was the result of human relation factors, rather than the physical factors being researched.

The reason for increased productivity was that the employees felt needed and appreciated. Besides the attention of the researchers, the employees liked the informal atmosphere, the ability to talk freely, and the opportunity to interact with their employers. This caused them to feel special, so they worked harder. The employees' ideas were respected, and they were involved in decisionmaking for the company. This increased job satisfaction led to greater productivity.

Maslow's Hierarchy of Needs

More than a century ago, Abraham Maslow determined that humans have both physiological and psychological needs. Physiological needs include air, water, food, and shelter. Once these basic physiological needs are met, we begin to pay attention to psychological needs. Psychological needs include safety, love and acceptance, recognition, and **self-actualization**, which is a person's need to accomplish personal goals and develop to his or her full potential.

To relate Maslow's "hierarchy of needs" to the agribusiness workplace, we use the "dangling the carrot" theory. If you give a carrot to a rabbit, he immediately eats it and thus meets his need for food. If the carrot is obviously out of reach, the rabbit gives up without trying. However, if the carrot is dangled just close enough so it can be reached with a fair amount of extra effort, the rabbit is motivated to keep trying until he gets it.

This, is essence, is the theory of motivation in the workplace. Don't make the goals too easy, don't make them too hard. Make the goals challenging enough so that people must stretch to achieve them, but make them realistic enough that they can be reached with effort and ingenuity. You can probably relate this to some of your teachers: a teacher whose class is too easy is not appreciated; nor is a teacher who sets impossibly difficult tasks. Most students tend to appreciate a teacher who encourages them, using stimulating, challenging tasks, to reach their full potential. This is the way agribusiness managers should treat their employees in the workplace. Refer to Figure 10-4 and Figure 10-5 for Maslow's hierarchies of physiological and psychological needs.

McGregor's Theory X and Theory Y

Agribusiness managers try to motivate employees according to their own attitude. In 1960, the late Douglas McGregor, while a professor at MIT, published a book called *The Human Side of Enterprise*. In this book, McGregor set forth what were to become his famous Theory X and Theory Y. Although some people have generalized and applied these ideas

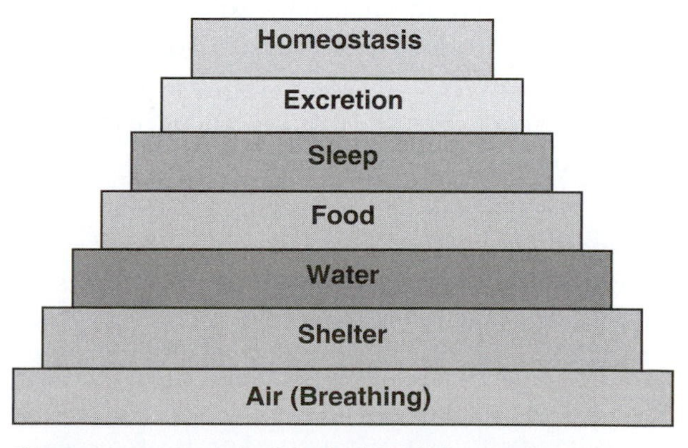

Figure 10-4 Our physiological needs have to be met before we even become concerned about our psychological needs.

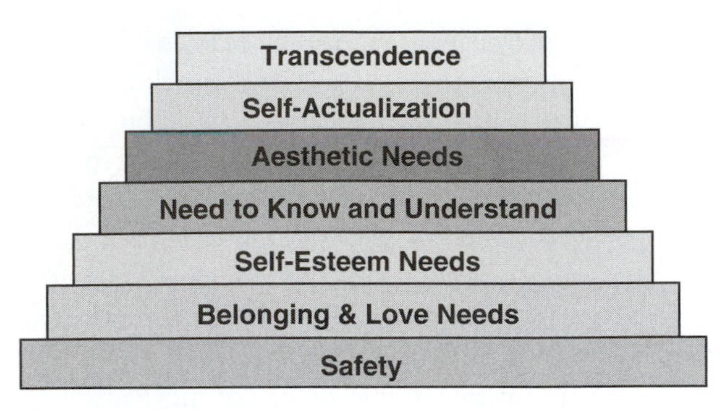

Figure 10–5 Once psychological needs on one level are met, most people feel the urge to meet the next need level.

inappropriately, a brief review at this point will be helpful in achieving better understanding of how agribusiness managers go about motivating their workers.

Theory X. Each theory, as McGregor defined it, is based on a set of assumptions made about people in general—a way some of us look at humanity. Theory X is traditional and quite familiar. Here are the main characteristics of a Theory X person:

- The average human being has an **inherent** dislike of work and avoids it if possible.
- Because of this human characteristic of dislike for work, most people must be coerced, controlled, directed, or threatened with punishment to get them to work toward the achievement of organizational objectives.
- The average human being prefers to be directed, wishes to avoid responsibility, has relatively little ambition, and wants security above all.

Those who hold such views expect their employees or co-workers to be lazy, to "goof off" at every opportunity, and thus to require close control. The man in Figure 10–6 appears by his facial expression to be a Theory X individual.

Theory Y. Theory Y reflects a totally different set of values and expectations of people. Here is McGregor's Theory Y description:

- The expenditure of physical and mental effort in work is as natural as in play or rest.
- External control and threat of punishment are not the only means to stimulate effort toward organizational goals.

Figure 10–6 What characteristics of this person would lead one to believe that he has a Theory X philosophy? (Courtesy of USDA #k4830-2)

- The average human being learns, under proper conditions, not merely to accept but to actively seek responsibility.
- Under the conditions of modern industrial life, the intellectual potential of the average human being is only partially used.

Theory Y states that people work because it is natural for them to do so, and, under the proper conditions, they will want to achieve the goals of the group of which they are a part. Generally, the Theory Y approach offers a leader more options in working with people. By their nonverbal expression, the two workers in Figure 10–7 appear to be Theory Y individuals.

Theory X and Theory Y may seem somewhat simplistic. However, they are important because they dramatize and summarize how we may feel about employees and co-workers. If you view your co-workers as lazy people who must be coerced and controlled so that you can perform your leadership functions properly, this attitude will affect your behavior toward them.

You may unconsciously be causing mistrust, suspicion, rebellion, and many other forms of nonproductive behavior. In contrast, demonstrating acceptance of and confidence and trust in co-workers shows that you expect them to want to do a good job. It adds to their **self-esteem**, self-respect, and motivation to do a good job,

Figure 10–7 What characteristics of these two people would lead one to believe that they have a Theory Y philosophy? (Courtesy of USDS #k-4563-14)

thus contributing to group effectiveness. Refer to Figure 10–8 for other assumptions about Theory X and Theory Y leadership.

In reality, the way an agribusiness manager treats workers is also based upon his or her own personality. Therefore, it is wise for managers to learn to be flexible in acting according to either Theory X or Theory Y. Employees are also different from each other, and need to be treated differently. Some employees do better with direction; some employees do better with more freedom. Agribusiness managers should use common sense and act appropriately for each situation.

Theory Z. A somewhat different approach for agribusiness managers to use is Theory Z. The key to Theory Z is teamwork. Agribusiness managers can rely on the efforts of teams to enhance motivation and productivity. Although recognition of employees is based on their individual contributions, the manager emphasizes working together and sharing networks. Other characteristics and foundations of Theory Z include:

- Employee involvement is the key to increased productivity.
- Employee control is implied and informal.
- Employees prefer to share responsibility and decisionmaking.
- Employees perform better in environments that foster trust and cooperation.
- Employees need guaranteed employment and will accept slow evaluations and promotions.[3]

Herzberg's Motivating Factors

Besides the styles of management just discussed, psychologist Frederick Herzberg studied how the content of a job could increase motivation. In other words, what can agribusiness managers do with employees' job to motivate employees? When employees were asked to rank various job-related factors as to which best motivated them, created

3. Ibid.

Motivating Subordinates (Two Sets of Assumptions about People)	
Traditional X (Authoritarian)	**Progressive Y (Democratic)**
People are naturally lazy; they prefer to do nothing.	People are naturally active; they set goals and enjoy striving.
People work mostly for money and status rewards.	People seek many satisfactions in work: pride in achievement, enjoyment of process, sense of contribution, pleasure in association, and the stimulation of new challenges.
The main force keeping people productive in their work is fear of being demoted or fired.	The main force keeping people productive is the desire to achieve their personal and social goals.
People remain children; they are naturally dependent on leaders.	People normally mature beyond childhood; they aspire to independence, self-fulfillment, and responsibility.
People expect and depend on direction from above; they do not want to think for themselves.	People close to the situation see and feel what is needed and are capable of self-direction.
People need to be told, shown, and trained in proper methods of work.	People who understand and care about what they are doing can devise and improve their own methods of doing work.
People need supervisors who will watch them closely enough to be able to praise good work and reprimand errors.	People need a sense that they are respected and are mature enough to do good work without constant supervision.
People have little concern beyond their immediate, material interests.	People seek to give meaning to their lives by identifying with nations, communities, churches, unions, companies, or causes.
People need specific instruction on what to do and how to do it; larger policy issues are not any of their business.	People need ever-increasing understanding; they need to grasp the meaning of the activities in which they are engaged; they have cognitive hunger as extensive as the universe.
People appreciate being treated with courtesy.	People crave genuine respect from their peers.
People are naturally compartmentalized; work demands are entirely different from leisure activities.	People are naturally integrated; when work and play are too sharply separated, both deteriorate.
People naturally resist change; they prefer to stay in the old ruts.	People naturally tire of monotonous routine and enjoy new experiences; everyone is creative to some degree.
Jobs are primary and must be done; people are selected, trained, and fitted to predefined jobs.	People are primary and seek self-realization; jobs must be designed, modified, and fitted to people.
People are formed by heredity, childhood, and youth; as adults, they remain static; old dogs don't learn new tricks.	People constantly grow; it is never too late to learn; they enjoy learning and increasing their understanding and capability.
People need to be "inspired" (pep talk), pushed, or driven.	People need to be encouraged and assisted.

Figure 10-8 "Traditional X" represents the old beliefs about viewing and leading others. It is the authoritarian way. "Progressive Y" represents a newer way of leading, the democratic way. Some situations, however, may require the traditional X perspective. (Adapted from *Horace Small Manufacturing Training Manual*, Nashville, TN: Horace Small Mfg. Co., 1992)

enthusiasm, and made them work to their fullest potential, the following factors were mentioned, in order of importance:

1. sense of achievement
2. recognition earned
3. interest in the work itself
4. opportunity for growth
5. opportunity for advancement
6. ability to demonstrate responsibility
7. peer and group relationships
8. pay
9. supervisor's fairness
10. company policies and rules
11. status
12. job security
13. supervisor's friendliness
14. working conditions[4]

Motivators. The preceding list could be divided into two categories: motivators and hygiene factors. The first seven in the list are **motivators**, factors that provide satisfaction and stimulate people to work. The best way to motivate workers is to make the job interesting, help employees achieve their objectives, and recognize achievement through advancement and/or added responsibility.

Hygiene Factors. The **hygiene factors** are the last seven items in the preceding list. These have to do mostly with the job environment, and may cause dissatisfaction if they are missing, but hygiene factors would not necessarily increase motivation if they were improved or increased.

Money Isn't Everything. It is interesting to find that one of the hygiene factors is pay. Maybe pay is taken for granted, and it certainly is a factor when selecting a job. Nevertheless, research has shown that once a person is on the job, pay is not a top motivator. One study found that 80 percent of employees, when asked to pick the most important factor in the workplace, chose recognition of good work.[5]

Another study, by Jim Barlow, reported similar results. When employees were asked what makes them content with their current jobs, more then half listed open communication with their supervisors, the nature of the work, the quality of management, their own control over work content, the opportunity to gain new skills, the quality of co-workers, and the opportunity to do intellectually stimulating work. Only a third mentioned salary as important.[6]

Job Motivation

Job motivation strategies emphasize making jobs more interesting, challenging, and rewarding. Work is assigned to employees so that they have the opportunity to complete a task from beginning to end; that is, the worker is responsible for the successful completion of a whole task. The following characteristics of work are believed to be important in affecting individual motivation and performance:

- *Skill variety*—the extent to which a job demands that the worker use different skills.
- *Task identity*—the degree to which the job has an identifiable, visible outcome with a beginning and an end.
- *Task significance*—the impact the job has on the lives or work of others in the company, and/or its importance to the company mission, goals, and objectives.
- *Autonomy*—the degree of freedom, independence, and discretion the employee has to schedule work and determine procedures.
- *Feedback*—the amount of direct, specific information received about job performance.

The first three—skill variety, task identity, and task significance—contribute to the meaningfulness of a job. The last two, autonomy and feedback, give employees a feeling of responsibility and contribute to a feeling of achievement and recognition.[7]

Two other things that contribute to employees' motivation as they do their work are **job enlargement** and job rotation. With the job enlargement strategy, several tasks are combined into one job to

4. Frederick Herzberg, *Work and the Nature of Man* (Cleveland, OH: World Publishers, 1966).
5. "Loyalty Surprise," *Boardroom Reports*, March 15, 1994, p. 10; "Brainstorming," *Boardroom Reports*, April 15, 1994, p. 15.
6. Jim Barlow, "Company Loyalty: The Feeling Is Mutual," *Washington Times*, April 18, 1994, p. E13.
7. William G. Nickels, James M. McHugh, and Susan M. McHugh, *Understanding Business*, 5th ed. (Chicago: Irwin/McGraw-Hill, 1999), pp. 299–300.

Figure 10–9 When employees are properly motivated and challenged, it makes their job more interesting and pleasurable, leading to greater productivity.

make that job more challenging, interesting, and motivating. As previously mentioned, job rotation is an enrichment strategy that involves moving employees from one job to another, so that no employee becomes bored with or "stale" in a position with which he or she is already intimately familiar. The results are that tasks are more interesting, employees learn new things and develop new perspectives, jobs are more challenging and thus more motivating, and morale is increased.[8] Refer to Figure 10–9.

Setting Goals and Objectives

Many people start out the day, especially on weekends, with a "Things to Do" list. Every time you mark something off the list, it makes you feel good, motivates you, and gives you a sense of accomplishment. In reality, a to-do list is a list of goals and objectives.

8. Ibid., p. 300.

This simple mechanism can be taken to a higher lever in the workplace. The **goal-setting theory** is based on the fact that the setting of specific, attainable goals creates high levels of motivation and performance if the goals are accepted by managers and accompanied by feedback. It makes sense if both the agribusiness manager and the workers have some agreement about the overall goals and objectives of the agribusiness.

Agribusiness managers should formulate goals and objectives in cooperation with their employees, commit to those goals and objectives, support employees' performance of tasks to reach the goals, monitor work results, and reward accomplishment. Doing so generates feelings of involvement, motivation, and responsibility; improves collaboration; and reinforces loyalty to the team and company.

Listening to Employees

Listening to the people who have "boots on the ground" is a simple but too-often-overlooked concept. Agribusiness managers must listen to their employees, and each manager should create an environment in which employees feel free to make suggestions. Because the workers are the ones doing the jobs and routine tasks, they may discover things or have ideas that would improve the effectiveness and efficiency of the company. When approached with suggestions for improvement, the agribusiness manager should listen; employees see things from a different perspective, and often come up with creative solutions to problems that management did not even realize were problems.

Agribusiness managers must create an environment that honors and rewards employee suggestions. Employees must feel free to say anything to their managers, even if it is merely a complaint. The agribusiness manager should encourage dialogue and constantly show workers that their ideas and opinions count. The agribusiness management should follow up by providing feedback, adopting employee suggestions whenever possible, and rewarding employees even if their suggestions are negative or cannot be implemented for some reason.

Evaluating Employees

Like any company, agribusinesses must evaluate their employees. Employees like to know how they are doing—or at least how their supervisors perceive they are doing. Evaluations, also called **performance appraisals**, should be done routinely for every job and employee. Agribusiness managers must be able to determine whether their employees are doing an effective job with a minimum of errors and disruptions. When appraising or evaluating an employee's performance, consider following these six steps:

1. *Establish performance standards.* Standards must be understandable, measurable, and reasonable. The best standards are achievement oriented, so that employees are evaluated on their achievement rather than mere performance of an activity.
2. *Communicate the standards and expectations.* Employees must be told clearly and precisely what the standards and expectations are and how they are to be met. Managers cannot assume that the workers know.
3. *Evaluate performance and achievement.* Over a specified time period, evaluate the employee's behavior to see if his or her activities and performance have met the standards and achieved the expectations.
4. *Discuss the results of the evaluation with employees.* Discuss each employee's successes and areas in which improvement is needed. This meeting is an opportunity to be understanding and helpful and to guide the employee to better performance. This is also a good time to get employee suggestions on how a particular task could be better performed.
5. *Take any necessary corrective action.* The primary purpose of conducting a performance appraisal is to make suggestions to increase employee performance—even if the employee is already meeting or exceeding the standards.
6. *Use the evaluation results to make personnel decisions.* Decisions about promotions, salary increases, additional training, and retention are all based on performance evaluations.[9]

The whole purpose of evaluation is improvement. The agribusiness manager needs excellent human relations skills when talking with employees about performance. In many instances, it is not what you say but how you say it. Consider the following when evaluating an employee's performance:

- *Never* attack an employee personally. Critically but objectively evaluate the employee's work. Fairness and honesty are crucial.
- Allot sufficient time, without distractions, for each appraisal. Take the phone off the hook and close the office door.
- Do not make employees feel uncomfortable or uneasy. If you have properly communicated the performance standards, employees should know what to expect during a review. *Never* conduct an appraisal where other employees are present.
- Include employees in the evaluation process as much as possible. For example, you might ask the employee prepare a self-improvement program.
- Do not wait until the appraisal to deal with problems in the employee's work that have been developing or apparent for some time. Employees who are having trouble meeting expectations need immediate feedback, support, coaching, or training.
- Always end the appraisal/evaluation meeting with positive suggestions for employee improvement.[10]

Laws Affecting the Management of Human Resources

Legislation has changed and expanded the role and the challenge of managing human resources. It has made hiring, promoting, firing, and managing workers very complex, as there are many restrictions on and legal implications of most personnel actions.

Since the 1930s, legislation and legal decisions have greatly affected all areas of human resource management, including hiring, firing, and working conditions. The following is a brief overview of the most important federal **statutes** governing employees and employment that may apply to your agribusiness. The number of people your

9. Ibid., pp. 329–330.

10. Ibid., p. 330.

Laws Affecting Human Resource Management

- Age Discrimination in Employment Act (ADEA)
- Americans with Disabilities Act
- Consolidated Omnibus Budget Reconciliation Act (COBRA)
- Employee Retirement Income Security Act (ERISA)
- Equal Pay Act
- Fair Labor Standards Act (FLSA)
- Family and Medical Leave Act (FMLA)
- Federal Insurance Contributions Act (FICA)
- Federal Unemployment Tax Act (FUTA)
- Immigration Reform and Control Act (IRCA)
- Medicare
- Occupational Safety and Health Administration Act (OSHA)
- Older Worker Benefit Protection Act (OWBPA)
- Pregnancy Discrimination Act
- Social Security Act
- Title VII of the Civil Rights Act
- Worker Adjustment and Retraining Notification Act (WARN)

Figure 10–10 Laws have changed and expanded the role and the challenge of managing human resources.

agribusiness employs makes a difference in what laws pertain to your business. You may wish to consult with an attorney for more detail about the laws that affect your specific organization. Refer to Figure 10–10 for a list of these laws.

Laws That Apply to Agribusinesses of Any Size

Whether your agribusiness has 1 or 100 employees, the following laws apply.[11] Some of them are so basic that you probably will forget they are even legal requirements. You can find more information about any of these statutes by typing the name of the law into an Internet search engine. Government agency Websites are also good places to start a search for authoritative, current information.

Fair Labor Standards Act. The Fair Labor Standards Act (FLSA) addresses minimum wage and overtime. It establishes a standard 40-hour work week. This means that you must pay time and a half to **nonexempt employees** who work more than 40 hours per week. However, the FLSA exempts certain classifications of employees from the overtime pay requirements. This act also sets the minimum age for employment and the minimum hourly wage rate and minimum wage for most workers.

Social Security. The Social Security law established several social programs that meet the material needs of individuals and families. The programs include retirement insurance, survivor's insurance, disability insurance, hospital and medical insurance for the aged, disability income, and supplemental security income, among others. A report of earnings must be filed annually by every employer that is required to withhold income tax from wages and/or who is liable for FICA taxes (also called Social Security and Medicare taxes).

Each employer must keep a record of the name and Social Security number of each employee as these items are shown on the employee's Social Security card. This information is needed for the earnings report. Also, each employer must get an employer identification number (EIN) from the Internal Revenue Service (IRS). This number must be used on the employer's tax returns and earnings report. You can get an EIN by filing an application on Form SS-4, which may be obtained from any IRS or Social Security office. The *Social Security Handbook* is available online, at http://www.ssa.gov/OP_Home/handbook/ssa-hbk.html, if you need more information.

11. http://www.smallbusinessnotes.com/operating/legal/laborlaws.html.

Federal Insurance Contributions Act. The Federal Insurance Contributions Act (FICA) actually imposes two separate taxes: one is the Social Security tax and the other is the Medicare tax. These taxes are paid jointly, by both employer and employee, and are withheld from wages. To learn more about paying these taxes or doing withholding, visit "*employment taxes*" at the Internal Revenue Service or search the FICA Website.

Medicare. **Medicare** is the U.S. national health insurance program that was established under the Federal Insurance Contributions Act. Taxes are paid by both employer and employee; the employee's portion is withheld from wages. For more information, visit "*employment taxes*" at the Internal Revenue Service Website or go to http://www.medicare.gov.

Equal Pay Act. The Equal Pay Act prohibits discrimination in pay on the basis of sex when jobs are performed under similar conditions and require equal skill, effort, and responsibility. However, it permits pay differentials between men and women where such differences are based on or created by seniority systems, merit systems, wage incentive plans, or other factors other than sex.

Immigration Reform and Control Act. The Immigration Reform and Control Act (IRCA) requires employers to verify that applicants for employment are authorized to work in the United States. It also sets civil and criminal penalties for those who knowingly employ unauthorized **aliens**. This act also prohibits discrimination based on national origin or citizenship if the alien is authorized to live and work in this country. This act is enforced by the Department of Justice and the Immigration and Naturalization Service. For more information about the IRCA, go to http://www.dol.gov/esa/regs/compliance/ofccp/ca_irca.html.

Federal Unemployment Tax Act. Under the Federal Unemployment Tax Act (FUTA), workers can receive payments for a given period of time or until they find new jobs, if they lost their jobs or their jobs were terminated through no fault of their own. Unemployment insurance is administered according to federal and state systems, with employees receiving a credit against federal tax for state taxes paid. A mix of federal and state laws determine which employees are eligible for compensation, the amount they receive, and how long benefits are paid.

Laws That Apply to Agribusinesses with More than 10 Employees

Occupational Safety and Health Administration Act. The Occupational Safety and Health Administration Act (OSHA) requires all employers to provide a workplace that is free from recognized hazards that cause, or are likely to cause, death or serious physical harm to employees. This law established the Occupational Safety and Health Administration, which is responsible for setting workplace safety standards, issuing regulations for various industries, and enforcing employer compliance.

Laws That Apply to Agribusinesses with More than 14 Employees

Title VII of the Civil Rights Act. One of the most important pieces of legislation ever passed by Congress was the Civil Rights Act of 1964. With much debate and contention, this act was amended 97 times before it finally passed. Title VII of the Civil Rights Act of 1964 injects government directly into the management of human resources.[12]

Title VII prohibits discrimination in hiring, firing, promotion, compensation, or terms, conditions, and privileges of employment on the basis of race, color, sex, religion, or national origin. It does permit employment discrimination based on race, religion, sex, or national origin when any of these factors is a bona fide occupational qualification necessary to the operation of the enterprise. This act also permits bona fide seniority, merit, or incentive systems that have the effect of discriminating, provided that such systems are not the result of an intention to discriminate. This act is enforced by the Equal Employment Opportunity Commission.[13]

Americans with Disabilities Act. The Americans with Disabilities Act (ADA) prohibits employment discrimination on the basis of physical or mental **impairments** that limit one or more major life activities. These impairments may be

12. Nickels, McHugh, and McHugh, *Understanding Business,* p. 337.
13. http://www.smallbusinessnotes.com/operating/legal/laborlaws.html.

real or perceived. The ADA also requires that reasonable accommodation be made for individuals with handicaps.

Pregnancy Discrimination Act. The Pregnancy Discrimination Act prohibits employment discrimination based on pregnancy, childbirth, or related medical conditions. It applies to employers with 15 or more employees, unions with 25 or more members, employment agencies, and federal, state, and local governments.

Laws That Apply to Agribusinesses with More than 19 Employees

Age Discrimination in Employment Act. The Age Discrimination in Employment Act (ADEA) prohibits discrimination against individuals over the age of 40 with respect to hiring, compensation, and terms, conditions, and privileges of employment, if such discrimination is based solely or primarily on age. Furthermore, the ADEA eliminates mandatory retirement age for most employees (with some exceptions for jobs in which age is considered a true job qualification). This act applies to employers with 20 employees, unions with 25 or more members, employment agencies, and federal, state, and local governments. The Equal Employment Opportunity Commission also enforces this law.

Older Worker Benefit Protection Act. The Older Worker Benefit Protection Act (OWBPA) is an amendment to the Age Discrimination in Employment Act. It prohibits discrimination with respect to employee benefits on the basis of age. This act regulates early retirement incentive programs.

Consolidated Omnibus Budget Reconciliation Act. The Consolidated Omnibus Budget Reconciliation Act (COBRA) included provisions pertaining to continuation of group health insurance when employment is terminated. It requires the employer to continue to include ex-employees in the employer's group health plan for a certain amount of time. However, the full cost of the health insurance is paid by the former employee. The act requires the employer to provide information about COBRA rights to employees on the date they are terminated. COBRA is administered by the U.S. Department of Labor's Pension and Welfare Benefits Administration.

Laws That Apply to Larger Agribusinesses

Family and Medical Leave Act. The Family and Medical Leave Act (FMLA) requires that employers provide up to 12 weeks of unpaid leave, within any 12-month period, to employees for the care of a newborn or adopted child, for the care of a seriously ill family member, or for treatment and care of the employee's own serious medical condition. It applies to employers with 50 or more employees and is enforced by the Wage and Hour Division of the U.S. Department of Labor.

Worker Adjustment and Retraining Notification Act. The Worker Adjustment and Retraining Notification Act (WARN) requires that employers give 60 days' advance notice to employees of impending plant closure or layoffs of 50 or more employees. This act applies to employers with 100 or more employees.

Employee Retirement Income Security Act. The Employee Retirement Income Security Act (ERISA) governs the operation of pensions and retirement benefits provided by employers to their employees. It does not require that employers provide such benefits, but regulates the conduct of employers that do provide such plans. This act is enforced by the Pension and Welfare Benefits Administration of the U.S. Department of Labor. This act also applies to employers with 100 or more employees.[14]

Conclusion

Managing human resources involves evaluating human resource needs, finding people to fill those needs, and getting the optimum performance from each employee by providing the right incentives and job environment. All of this is done by the agribusiness manager with the goal of meeting the objectives of the business. Our time in this life is relatively short. The majority of each person's hours, when they are awake, are spent at work. Therefore, the agribusiness manager should strive to treat employees in a way that makes their work lives as pleasant and happy as possible.

14. Ibid.

Summary

The owner or manager of an agribusiness must do several things when they seek people to work for the business. These include determining job requirements and determining the cost of labor and recruitment. When selecting employees, the agribusiness manager must follow federal and state regulations in interviewing and hiring. The selection process usually involves submission of an application and a résumé, and a personal interview. After new employees are selected, they should receive orientation and training.

To work successfully with employees, agribusiness managers need special characteristics, including adaptability, ability to form relationships, emotional and social maturity, insight, and self-motivation.

One of the keys to success for an agribusiness manager is motivating employees to do their best. Some of the motivation theories underlying business management are the Hawthorne effect, Maslow's hierarchy of needs, McGregor's Theory X and Theory Y, Theory Z, and Herzberg's motivating factors. Management techniques based on these theories include job motivation, setting goals and objectives, and listening to employees.

Agribusinesses need to evaluate their employees. Employees like to know how they are doing. When appraising or evaluating an employee's performance, consider following these steps: establish performance standards, communicate the performance standards and expectations, evaluate performance, discuss evaluation results with employees, take corrective action, and use the evaluation results to make decisions. The purpose of evaluation is improvement.

Legislation has changed and expanded both the role and the challenge of managing human resources. It has made hiring, promoting, firing, and managing workers very complex and imposes many legal requirements. Some of the laws that affect agribusinesses include the Fair Labor Standards Act, Equal Pay Act, Immigration Reform and Control Act, Federal Unemployment Tax Act, Occupational Safety and Health Administration Act, Title VII of the Civil Rights Act, Americans with Disabilities Act, Pregnancy Discrimination Act, Age Discrimination in Employment Act, Older Worker Benefit Protection Act, Consolidated Omnibus Budget Reconciliation Act, Family and Medical Leave Act, Worker Adjustment and Retraining Notification Act, and Employee Retirement Income Security Act.

End of Chapter Activities

REVIEW QUESTIONS

1. Define the Terms to Know.
2. Briefly discuss five things the owner or manager of an agribusiness should consider when finding and hiring people to work for the business.
3. Besides labor cost, what are three other costs the employer incurs in relation to employees?
4. Name six sources through which agribusiness managers could recruit prospective employees.
5. What three things does the selection process involve?
6. After employees are selected and hired, what are seven things they should know about the agribusiness?
7. List five special characteristics that agribusiness managers need to work successfully with employees.
8. According to the Hawthorne effect, what are four reasons for increased productivity?
9. List four psychological needs described by Abraham Maslow.

10. List four physiological needs described by Abraham Maslow.
11. List three characteristics or beliefs of a Theory X leader.
12. List four values and expectations of Theory Y leaders.
13. What are five characteristics of Theory Z leaders?
14. According to Frederick Herzberg, what are 14 job-related factors, in order of importance, that best motivated employees, created enthusiasm, and stimulated employees to work to their fullest potential?
15. Explain motivators according to Herzberg.
16. Explain hygiene factors according to Herzberg.
17. List five characteristics of work believed to be important in affecting individual motivation and performance.
18. Why is it important for agribusiness managers to listen to their employees?
19. What are six steps to follow when appraising or evaluating employee performance?
20. List three "do's" and three "don'ts" when evaluating an employee's performance.

FILL IN THE BLANK

1. The main purpose of a résumé is to highlight a person's ____________.
2. One of the keys to success for an agribusiness manager is ____________ employees to do their best.
3. The ____________ refers to people's tendency to behave differently when they know they are being studied.
4. Abraham Maslow determined that humans have both ____________ and ____________ needs.
5. Theory ____________ leaders expect their co-workers to be lazy, to require close control, and to "goof off" at every opportunity.
6. Theory ____________ leaders believe that people work because it is natural for them, and that, under the proper conditions; people want to achieve the goals of the group.
7. The key to Theory Z is ____________.
8. ____________ are factors that provide satisfaction and motivate people to work.
9. ____________ factors deal mostly with the job environment and could cause dissatisfaction if they were missing.
10. The ____________ theory is based on the fact that the setting of specific, attainable goals is related to high levels of motivation and performance.

MATCHING

a. Federal Insurance Contributions Act (FICA)
b. Immigration Reform and Control Act (IRCA)
c. Occupational Safety and Health Administration Act (OSHA)
d. Title VII of the Civil Rights Act
e. Fair Labor Standards Act (FLSA)
f. Older Worker Benefit Protection Act (OWBPA)
g. Medicare
h. Age Discrimination in Employment Act (ADEA)

i. Consolidated Omnibus Budget Reconciliation Act (COBRA)
j. Social Security
k. Family and Medical Leave Act (FMLA)
l. Worker Adjustment and Retraining Notification Act (WARN)
m. Equal Pay Act
n. Pregnancy Discrimination Act
o. Americans with Disabilities Act (ADA)
p. Employee Retirement Income Security Act (ERISA)

_____ 1. U.S. national health insurance program

_____ 2. Provides for payments for a given period of time or until a former employee finds a new job

_____ 3. Prohibits discrimination in employment on the basis of physical or mental impairments

_____ 4. Includes the Medicare tax

_____ 5. Prohibits discrimination in employment based on pregnancy or childbirth

_____ 6. Pertains to continuation of group health insurance when employment is terminated

_____ 7. Prohibits discrimination against individuals over the age of 40 with respect to hiring, compensation, and privileges of employment

_____ 8. Requires employers to verify that applicants for employment are authorized to work in the United States

_____ 9. Prohibits discrimination with respect to employee benefits on the basis of age

_____ 10. Prohibits discrimination in hiring, firing, and promotion

_____ 11. Requires all employers to provide a safe workplace

_____ 12. Addresses minimum wage and overtime compensation, and discrimination on the basis of race, color, sex, religion, or national origin

_____ 13. Prohibits discrimination in pay on the basis of sex

_____ 14. Establishes a number of social programs that meet the material needs of individuals and families

_____ 15. Requires that employers provide up to 12 weeks of unpaid leave under certain circumstances

_____ 16. Requires employers to give 60 days' advance notice to employees of impending plant closings or layoffs

_____ 17. Governs the operation of pensions and retirement benefits provided by employers to their employees

ACTIVITIES

1. Prepare a résumé. Refer to the example in this chapter or go to your favorite search engine for examples.
2. Explain the "dangling the carrot" theory.
3. Which do you believe are the most important, physiological needs or psychological needs? Defend your answer and share it with the class.
4. Which do you believe would most contribute to depression, not meeting your physiological needs or not meeting your psychological needs?

5. Examine yourself and determine whether you are a Theory X or Theory Y person. Share your answer with the class.
6. According to research on worker satisfaction, only one-third of the workers surveyed said that salary was the most important. Discuss this with your classmates and share your ideas on why salary is not ranked highest.
7. Make out a "Things to Do" list for the rest of the day. How does this list compare with goals and objectives?
8. Pretend that you are an agribusiness manager and you have to evaluate one of your employees. However, you are the employee. How would you evaluate yourself?

Chapter 11

Farm Management

Objectives

After completing this chapter, the student should be able to:

- Define management and list important management skills.
- Discuss the importance and challenges of farm management.
- Choose appropriate farming enterprises.
- Accurately calculate depreciation.
- Perform accurate machinery cost calculations.
- Explain the different types of budgeting used for farm operations, and develop working examples of each.
- Develop an effective farm recordkeeping system and discuss how and why important financial statements are employed.
- Discuss nonfinancial farm management issues.
- Describe whole-farm planning.

Terms to Know

accelerated cost recovery system (ACRS)
accrual adjustments
accumulated depreciation
break-even point
cash-flow budget
cash-flow statement
complete budget
current ratio
debt-to-asset ratio
depreciable asset
depreciation
economic depreciation
enterprise
enterprise budget
financial efficiency ratios
fixed costs
income statement
Industrial Revolution
liquidity
management
managers
modified accelerated cost recovery system (MACRS)
net farm income
net farm income from operations
net worth statement
nondepreciable asset
operating costs
ownership costs
partial budget
profitability
repayment capacity
salvage value
solvency
whole-farm planning
working capital

Introduction

Although farms are only one aspect of today's agribusiness, these centers of production continue to be one of the most important. Today's farms not only provide a glimpse into the most traditional agriculture landscape, but also serve as the point of origin for a majority of agricultural production.

Importance of Farm Management

Effective farm management is important because it provides structure and guidance for successful farms. With the constantly advancing developments in society and technology, effective farm management is becoming ever more important, for the following reasons:

- Farm operation size has increased.
- Farming operations are more specialized.
- Higher levels of technology are being used.
- Capital is more and more frequently being substituted for labor.
- Larger capital investments are required.
- More farmers are borrowing more (incurring larger amounts of debt).
- Inflation keeps pushing up the costs of production.
- Profit margins are narrower.[1]

Each of these factors clearly illustrates the need for effective farm management. Refer to Figure 11-1. This chapter takes you through the basics of farm management, and provides the most basic tools you will need to manage your farm enterprise effectively.

Figure 11-1 Alternating strips of alfalfa with corn on the contour protects this crop field in northeast Iowa from soil erosion. (Photo by Tim McCabe, courtesy of USDA)

Basics of Farm Management

What Is a Manager?

Within any type of organization, a variety of individuals assist in getting things done: servers, customer representatives, managers, executives, and so on. All these people work together to achieve the goals and improve the bottom line of the organization. Particularly important to this process are **managers**—individuals who make sure that things get done.[2] They can be found in any level of an organization and may have a variety of titles, including director, foreman, chief, or principal. Whatever their titles, managers generally do not perform the work personally, though they are responsible for its timely and correct completion.

Before the late 1800s, there was little need for managers, as a majority of businesses were operated by their owners and had only a few employees. With the advent of the **Industrial Revolution**, advances

1. Ronald Hanson, *Agricultural Economics 201 Lecture Study Notes* (Lincoln: University of Nebraska-Lincoln, 1997).

2. James G. Beierlein, Kenneth C. Schneeberger, and Donald D. Osburn, *Principles of Agribusiness Management,* 2d ed. (Prospect Heights, IL: Waveland Press, 1995), p. 22.

- Ability to make timely and judicious decisions
- Knowledge and use of effective communication strategies
- Thorough contextual knowledge, according to farming operation
- Proficiency in technical management skills and areas:
  - Marketing
  - Projecting and forecasting
  - Budgeting
  - Accounting
  - Finance
  - Personnal (handling employees and human resources matters)
- Basic knowledge of leadership skills, such as delegation and motivation
- Willingness to take appropriate risks
- Ability to align and guide all business elements for business profitability

Figure 11–2 Effective management skills needed by farm managers.

in technology allowed many firms to become more efficient and productive. Enhanced communications and transportation allowed businesses to expand sales into larger areas, thereby increasing sales and the need for employees. With this expansion, it became necessary to separate the ownership and management functions, leading to the need for appropriate, competent managers.

Management itself is a relatively new discipline. Introduced into college curriculums in the 1930s, **management** is most simply defined as accomplishing tasks through people.[3] It utilizes a variety of skills and characteristics from other disciplines, as illustrated in Figure 11–2. Managers perform a variety of tasks, often unique to the context in which they are managing, and use a variety of skills. Farm managers may handle everything from budgeting farm fuel for the entire year to hiring seasonal workers to help with harvest.

The ultimate task of all managers is to effectively combine the components of a business (human, financial, and physical) to maximize profitability. To accomplish this, managers must not only possess the skills already mentioned (Figure 11–2), but also have to apply just the right combination of technical expertise, logic, and judgment needed in each situation.

Three Basic Functions of Management

Specifically, there are three basic functions of management:

1. *Planning*: formulating a course of action to take or procedure to use. Determining an appropriate course of action is an important first step.
2. *Implementing*: taking steps to put the selected plan of action into operation. It is necessary to acquire and organize the essential resources to effectively carry out any plan of action.
3. *Analyzing*: observing the results to determine whether specific goals are being achieved. This includes keeping accurate records so that results can be measured.[4]

The best managers are not merely technicians; they are organizers and facilitators who enable a business to be more than the sum of its parts.

The Challenge of Farm Management

As mentioned earlier, today's ever-changing environment creates unique problems for farm managers. Today's farm managers have to deal with all the business issues that other firms and industries must face; in addition, they must also deal with matters such as weather uncertainties, crop and animal diseases, perishability of goods, and changes in government farm and trade policies. All of these aspects of business create a challenging environment, to say the least. For managers to effectively address each of these issues, they need the management skills listed in Figure 11–2, as well as a strong knowledge of the production of food and fiber and the ability to quickly adapt and make important decisions in response to changing market conditions. Successful farm managers effectively combine the technical skills of management with the practical knowledge of their particular enterprise.

3. Ibid., p. 23.

4. Hanson, *Agricultural Economics 201,* p. 2C.

Enterprise Selection

Now that we have discussed management, the skills needed by managers, and the current farm environment, it is time to address enterprise selection.[5] On an average farm within the United States, numerous production activities often occur simultaneously. A good example of this is the farmer who produces corn, soybeans, hay, and cattle all at the same time. Each of these production activities is considered an **enterprise**.[6] An important initial consideration for any farm manager is selection of the enterprise most appropriate for the agribusiness.

Although enterprise selection is often considered a relatively basic decision, following several specific steps can make the decisionmaking process easier and more effective.

Determine Your Goals

The first step in enterprise selection is setting goals specific to your operation. Almost any agribusiness operation has the broad, overall objective of achieving financial success and security; however, goals must be more specific. Goals should involve a measurable action and be set within a definite time frame. Identifying precise goals not only assists in measuring the success of your business, but also helps in developing an effective implementation plan (refer to Figure 11-3). The following important questions should be answered when selecting an enterprise:

- What other activities are you involved in, and what are the priorities of the other activities relative to the farm business?
- Do you want to devote full-time effort to the farm, or would you prefer to do farming as a part-time activity or an avocation?
- What is the desired period between initial investment and cash returns?

5. Karen Klonsky and Patricia Allen, *Family Farm Series Publications: Farm Management* (Davis: University of California-Davis, Small Farm Center, 2005).
6. H. Evan Drummond and John W. Goodwin, *Agricultural Economics* (Upper Saddle River, NJ: Prentice-Hall, 2001), p. 199.

Figure 11–3 These two agricultural productionists are discussing multiple range conservation practices, such as watering developments, pipelines, and fences. (Photo by Bob Nichols, courtesy of USDA)

Inventory Your Resources

Ultimately, the availability of resources will guide and limit your decision as to which enterprises are most profitable and appropriate. Resource requirements vary according to the enterprise chosen. Almost all enterprises require land, labor, and capital; however, there are other important resource factors that are often overlooked, such as access to markets and market potential. Access to markets could possibly be the most important resource available to your business—just because you can grow something does not mean you can sell it, and simply because you can sell something does not mean that your operation will be profitable. Consider the following questions:

- What is the market potential for your business?
- How long will it take to firmly establish your business?
- What is your cash-flow situation?

List the Physical, Financial, and Management Factors Involved

Physical Factors. Physical factors are the basis for your farming operation. Significant physical factors include land, climate, irrigation water, farm structures, and machinery and equipment. Each of these factors is important when selecting an enterprise for your business. Consider the following:

- How much land do you have available?
- What are the soil texture, drainage capability, and nutrient levels?
- What other crops have been grown on the land?
- What is the average rainfall in your area, and when are the rainy periods?
- When are the first and last frost dates?
- Where does your water come from, and how much does it cost?
- What is your water quality?
- Do you have water rights? Are you within an irrigation district?
- What type(s) of buildings exist on the property?
- Are any fences structurally sound?
- What farm implements are available?
- Would it be possible and cost-effective to lease or rent some equipment?
- What are the possibilities of contracting for custom operations in your area?

Financial Factors. Financial factors include those resources you are willing to invest in your operation. The type and amount of financial resources you are willing and able to invest in your operation provide a loose outline for the type of enterprise(s) you can undertake as the basis for your agribusiness. Important questions to consider as you determine the financial factors are:

- How much capital are you willing and able to invest?
- Are you able or willing to borrow capital?
- What is your cash-flow situation?
- Are you willing to consider risky enterprises?

Management Factors. Management factors are the final component in outlining the factors for a successful agribusiness. These factors describe important skills related to less tangible aspects of the business, such as personnel and information-gathering capacity. Specific items in this area include personal skills, information access, and labor and marketing factors. Consider these important questions:

- What are your mechanical skills?
- Do you consider your management skills (e.g., recordkeeping, personnel management, budgeting, etc.) to be adequate?
- Are you willing to learn new skills if they are required?
- Are you familiar with agricultural information delivery systems?
- What are your labor needs on a monthly basis?
- Are you planning to use family or mostly hired labor?
- Have you considered the opportunity cost of using your own labor?
- What is your proximity to various potential markets?
- How much time you are willing to spend marketing your products?
- Are you familiar with marketing regulations for the enterprises you are considering?

Develop a List of Potential Enterprises

Once you have identified your goals and resources, the next step is to develop a list of potential enterprises that might fit your needs, goals, and resources. These products or production activities could be something as simple as raising cattle and corn, like many of the other production agriculturists in your area, to something as innovative as producing organic squash and beans for a local restaurant. Refer to Figure 11-4. In many parts of the country, agricultural tourism is a viable option; examples in this area include corn mazes and pick-your-own pumpkin fields. (Refer to Chapter 8 for other examples.) Brainstorming what enterprises interest you and asking a few key questions will help you develop a good starting list.

- Which enterprises are predominant in your area?
- For the enterprises that interest you, have other agribusinesses with similar soil and climate conditions been successful with such enterprises?
- What crops or livestock have been raised on your land in the past?

Figure 11–4 Canola has emerged as a viable alternative or specialty oil crop, not only for its products, but also for its potential to diversify cropping systems. (Photo by Bob Nichols, courtesy of USDA)

Compatibility with Resources

Once you have brainstormed a list of possible enterprises for your business, the next step is to determine which of the enterprises are compatible with the resources you have available. This can be done by systematically comparing the individual resources needed for each enterprise to the resources available. Identifying resource requirements for each enterprise may require a little bit of homework, but a good place to begin is by talking with other growers or farmers in your area who already work with the enterprise of interest. The Internet, your Cooperative Extension Service office, or your local library are other good resources, too. Important questions to consider include:

- What types of resources are needed for the enterprise of interest?
- If the resources are not available, are they obtainable? At what cost?
- Are the resources available more effectively used for several enterprises, or would it be more cost-efficient to stick with a single enterprise?

Growing Considerations

Once you have narrowed down your enterprise options, it is important to take growing requirements into consideration (particularly if you have chosen crops, vegetables, fruit, trees, etc., as your enterprise). Important considerations include the general crop situation, climate, rotation, equipment requirements, irrigation, pollination, pest management, labor, operation and investment capital, harvesting, marketing, and profitability. This is a lot to think about and weigh, but the choices can be made more straightforward by answering a few important questions.

- What is your personal experience with the crop?
- What is the research basis for the crop under consideration?
- What are the effects of spacing on yield and quality?
- Is the crop adaptable to the climate during the growth period in your location?
- What is the crop's tolerance for rainfall, frost, high/low temperatures?
- How does the crop fit into rotation with other crops planned?
- How much time elapses from planting through total harvest period?
- Will you need special materials or equipment (e.g., harvesting equipment, storage facilities, etc.)?
- How often will you need to irrigate?
- Does the quality of water affect production?
- Will there be pollination requirements (for fruiting crops)?
- Will this require you to purchase or rent beehives? What will the cost be?
- What are the important pest problems with this crop?
- Are crop varieties available that are resistant to major diseases? If so, do they have a high yield with high-quality characteristics?
- How many acres of the crop can you handle with the available labor resources?
- Is seasonal labor available?
- How many harvests will be needed to obtain an economic yield?

- What is the harvest interval for the crops being considered?
- How long will harvest take with the available labor?
- What are the total production costs?
- What kinds of yields do you expect?
- What are the expected gross and net incomes?
- How much money will you need to invest in growing the crop?
- Where will you get the capital?
- Will you need to take out any loans? If so, what amounts and terms are available?
- Are you familiar with the market history and market trends of the crop?
- Have you explored different types of market outlets for your product?
- How does this crop compare to other crops? How does this crop compare to livestock returns?

Compatibility among Enterprises

The final step in enterprise selection is considering the relationship among enterprises. When selecting more than one enterprise, or looking to diversify your current agribusiness, it is important to consider how the enterprises of interest will fit together (or work against each other). For example, you may currently have plenty of labor resources to efficiently produce one enterprise—but you may not if you begin another labor-intensive enterprise.

Still, there are several advantages to diversifying your agribusiness. First and foremost, by adding another enterprise you are reducing risk. Total production failure, price crashes, and pest annihilation are less likely when you have several commodities. In addition, your cash flow and profitability may be more stable from year to year when you are relying on more than one enterprise.

Furthermore, diversification can allow you to spread fixed costs over several commodities, as well as using resources more evenly throughout the year. This may allow you to offer seasonal help longer work periods and enable you to better regulate cash flow. Another advantage to diversifying your business is that you may have greater access to available markets when you produce a range of products. Many buyers are interested in producers who can offer a variety of commodities, rather than having to deal with several different producers.

A final advantage of adding enterprises to your business involves pest control and soil quality. When done correctly, crop rotation and crop mix have been shown to effectively control pests as well as increase soil fertility (an example is alternating corn and beans every other year). This allows the use of fewer pesticides and reduces the need for extra nitrogen and fertilizer (depending on the crops rotated), all of which helps to decrease the inputs and costs of production.

Though all these advantages are important factors to consider when determining the appropriateness of adding enterprises to your agribusiness, it is still more important to consider the interaction between enterprises as part of your final decision. Refer to Figure 11-5. Introducing a new enterprise is sensible only when it is not going to hinder the production and profitability of your current enterprise.

After following this step-by-step process, go back and compare your initial goals to your enterprise selection(s). This enterprise selection process may seem pretty intensive, but it is important to take all the steps listed. This is a major decision, one that helps to form the foundation for your agribusiness. Without effective enterprise selection, you may find yourself experimenting in an area for which you are not adequately prepared.

Depreciation

Another important factor in business decisions is depreciation. **Depreciation** is an estimate of the annual dollar loss in value associated with the ownership of a depreciable asset, and is done to recover the costs of "using up" farm assets.[7] This process is based on value and is calculated over the useful life of the asset. The total amount of depreciation claimed (to date) during the ownership period of an asset is termed **accumulated depreciation**.[8]

What Qualifies as a Depreciable Asset?

Determining what qualifies as a **depreciable asset**—an asset that is important to the farm and loses

7. Hanson, *Agricultural Economics 201,* p. 171.
8. Ibid.

Figure 11–5 These landowners decided to diversify with the production of 100 acres of kiwi, peaches, and walnuts. (Photo by Bob Nichols, courtesy of USDA)

value over time—is key to calculating depreciation. To be depreciated, an asset must not retain its value (as land does) or already be accounted for within another class of farming expenses (such as Christmas trees, breeding stock, and raised dairy animals). Types of depreciable assets include purchased livestock, purchased machinery, and real estate improvements (e.g., building a barn, buying a tractor, improving fences, etc.). Land, breeding livestock, and anything raised on the farm are considered **nondepreciable assets**; these are things that do not depreciate (lose value) or are not kept for more than one year.

Reasons for Using Depreciation

There are three main reasons for using depreciation within an agribusiness operation: (1) it is needed for income tax purposes, (2) it assists in calculating asset value, and (3) it helps to distribute fixed costs.[9] To calculate depreciation, the following information must be available for each asset:

1. Original cost
2. Depreciation method
3. **Salvage value**, or the value of the asset at the end of its useful life[10]

The depreciation method is chosen according to the purpose behind the depreciation charged. The salvage value may or may not be used, depending on the depreciation method employed.

9. Patricia A. Duffy, Ronald D. Kay, and William Edwards, *Farm Management,* 5th ed. (Europe: McGraw-Hill Education, 2003).
10. Jack Elliot, *Agribusiness Decisions & Dollars* (Albany, NY: Delmar Publishers, 1998).

Methods of Depreciation

There are three methods of depreciation: straight-line depreciation, the accelerated cost recovery system, and the modified accelerated cost recovery system. A brief discussion of each follows.

Straight-Line Depreciation. Often considered the most common type of depreciation, straight-line depreciation makes planning easier because the asset is depreciated equally over every year of its useful life. The equation for this is

$$\frac{\text{Original cost of the asset} - \text{Salvage value}}{\text{Useful life of the asset}}$$

This method of depreciation is most often used to depreciate property equally over several years; it is also commonly used for intangible property.

Accelerated Cost Recovery System. With the **accelerated cost recovery system (ACRS)** depreciation method, the asset is depreciated more in the early years of its useful life. The depreciation rules for ACRS were set by a 1981 tax act, and may be applied to property placed into service after 1980 and before 1987. In ACRS, salvage value is always zero.

Modified Accelerated Cost Recovery System. As with the ACRS, assets depreciated under the **modified accelerated cost recovery system (MACRS)** are depreciated more at the beginning of their lives. However, with MACRS, the years of an asset's useful life are mandated by the 1986 Tax Reform Act, and therefore apply to any property placed into service after 1987. Under this depreciation method, the cost of specific assets can be recovered over a set number of years (3, 5, 7, 10, 15, or 20) depending on the type of asset involved (refer to Figure 11-6). Under MACRS, salvage value is also always zero.

MACRS Depreciation Property Classificationss

Years Depreciated	Property
3 years	Breeding hogs
5 years	Dairy, breeding cattle, breeding sheep, farm trucks, computers
7 years	Machinery, horses
10 years	Single-purpose buildings
15 years	Tile, LD improvements
20 years	General-purpose buildings

Figure 11-6 As a general rule, property not found in this MACRS depreciation list is considered to fall into the seven-year classification.

Machinery Cost Calculations

More likely than not, if you are managing a farming operation, you will employ some type of machinery to assist in planting, harvesting, feeding, or carrying out a variety of other activities within the operation. Furthermore, although a majority of the financial investment is made up front with the purchase cost of the machinery, there are many other costs associated with equipment ownership. To make effective management decisions regarding the purchase, trade, or lease of farm machinery, you must be able to accurately determine the costs of owning and operating each specific piece of equipment.

The costs of owning and operating machinery are generally divided into two categories: fixed and variable costs. **Fixed** (or **ownership**) **costs** are those that occur regardless of the use of the machinery. **Operating** (or **variable**) **costs** are those that depend on how much the machinery is used.[11]

Ownership Costs

Ownership costs include depreciation, interest, taxes, and outlays for housing/storage and maintenance facilities. Depreciation, in this context, applies to the decline in the value of a machine due to age or wear. This **economic depreciation** differs from the depreciation described earlier (which is used primarily for tax purposes). To calculate economic depreciation, one follows a declining-balance schedule for the remaining value of a machine or piece of equipment where the amount of depreciation remains constant for the life of the machine. Another ownership cost is interest cost, or the annual average interest rate over the useful

11. "How to Estimate Your Machinery Costs," *Doane's Agricultural Report*, vol. 58, no. 25-6 (1995).

life of a machine. Finally, you have the ownership costs of taxes, insurance, and housing. These costs will vary slightly according to your operation and state of residence, but you can figure an average annual cost of approximately 3 percent of the remaining value (depreciated value) for taxes, insurance, and housing.

Operating Costs

Operating costs include expenses such as repair and maintenance, fuel and lubrication, labor costs, and any other costs associated with the operation of machinery on your farm. Repair and maintenance costs can be estimated, but will depend largely on the type of machinery, as well as usage, soil type, terrain, climate, and equipment care. As with any piece of equipment, a farm manager who uses good maintenance and operating techniques should have lower costs.

One way to estimate fuel and lubrication costs is with the information found in Figure 11-7. Lubrication costs generally run 10–15 percent of fuel costs. A final necessary operating cost is the labor cost. This is the cost of your time and effort in operating the machine itself. As a general rule of thumb, as the capacity of a machine increases, the labor cost per acre decreases.

It is important for any farm manager to consider both the ownership and the operating costs of machinery. Not only does this information assist in planning activities such as cash-flow budgeting, but it also helps in making major capital-investment decisions regarding purchases or leases.

Agribusiness Budgeting

Budgeting is a large and important part of a farm manager's responsibilities. It assists in making the farming operation more efficient, as well as providing a foundation on which to base future decisions. However, not all budgeting is the same. Complete budgeting is necessary for the overall picture of the business, but it is not necessary to revise or redo a complete budget for every decision to be made. Partial, enterprise, and cash-flow budgeting each have their own special purpose within the planning system.

Complete Budgeting

A **complete budget**, sometimes called *whole-business budgeting* or an *operating budget*, is the most comprehensive of all budgets. Very simply, a complete budget is a financial plan for all segments of the business—literally, the "complete" business.

Importance. Budgeting allows farm managers to be more precise, reduces wasteful activity, and helps to decrease overlapping and duplicative efforts in the farming operation. To make purchases, a farm manager must have some idea of future profits; to borrow money, there must be some indication of the business's profitability. Each of these indications, as well as many other financial projections, is made possible by developing a budget for your operation.

Purpose. The complete budget summarizes planned sales, production activities, and other

Fuel Consumption Center for Tractor Engines

Engine	Average Fuel Consumption (gallons per hour per max. PTO hp)
Gasoline	0.068
Diesel	0.044
LP-gas	0.080

To calculate fuel consumption, use the formula:
hp × average fuel consumption × hours used × fuel price
Example: You have a 150-horsepower tractor, diesel fuel costs $2.30 per gallon, and you use the tractor 500 hours annually. Your fuel cost would be $7,590 per year:
150 hp × 0.044 × 500 hrs. × $2.30 = $7,590

Figure 11-7 The average fuel consumption factor, with the formula, is one way to estimate machinery operating cost.

related costs for a specific period. This type of budget is unique because it provides the most complete overall picture of the agribusiness, rather than highlighting just one specific section or aspect. The first step in developing a complete budget is to determine the expected sales over the projected period. This will require information from past sales as well as forecasts of future (potential) sales. The manager should also factor in expected changes in competition or the economy.[12]

It is important to determine the costs or expenses expected over the budgetary period. This requires looking at both fixed and variable expenses, again both historically (past cost data) and forthcoming (current price information). Complete budgets are not merely a projection: they are a statement of what the farm manager expects to accomplish in the upcoming period. This budget is a significant planning document, because it acts as a reference tool to assist the farm manager in making effective business decisions and therefore maximizing farm profits.[13]

Partial Budgeting

Changes within a farming operation often affect only a specific area or enterprise within the operation. Although complete budgeting is invaluable for managing a successful farming operation, it would be time-consuming and unnecessary to alter this detailed budget every time you consider a change within your operation. Thus, many important farming decisions can be made more promptly and easily by using a **partial budget**.

A partial budget, by definition, measures the change in net farm income that can be anticipated from a projected change in a farming operation. Examples of this could be a change in enterprise production, farm expansion, or an increase or decrease in livestock investment. Specific advantages of partial budgeting include the ability to analyze and decide on proposed changes that affect only a specific portion of or enterprise within the farming operation; the capability to measure every return and cost (increase or decrease) in the farm business that is affected by the projected change; and the capacity to illustrate what effect various prices, costs, and yields will have on the change in net farm income if the projected investment is made.[14]

Calculating a Partial Budget. A partial budget will show that a proposed change can have one of two effects on a farm operation: either an income-increasing effect or an income-decreasing effect. Income-increasing effects occur as a result of additional returns and reduced costs. Income-decreasing effects occur as a result of reduced returns and additional costs. Changes as part of a partial budget can be calculated relatively simply, through plugging pertinent information into a form such as the one illustrated in Figure 11-8.

Enterprise Budgeting

As defined earlier, an *enterprise* is part of the farm production process: a specific crop or livestock commodity that produces a marketable product. Because what your farming operation produces directly affects its profitability, enterprises can be considered the basic building blocks of an agribusiness. By looking at and evaluating the revenues and expenses associated with various enterprises on your farm, you can determine which enterprises should be expanded and which should be cut back and possibly eliminated. This process can also be used to compare the profitability of one production technique to another. It should be kept in mind, however, that any enterprises maintained as part of the farming operation should meet the overall vision and goals of the operation.

Explanation. An **enterprise budget** is an inventory of all the estimated income and expenses associated with a specific enterprise, which in turn help the manager estimate the profitability of that enterprise.[15] This type of budget differs from a partial budget in that it always involves only a single enterprise; a partial budget could focus on a specific aspect of the farming operation (such as custom

---

12. Beierlein, Schneeberger, and Osburn, *Principles of Agribusiness Management,* p. 80.
13. Hanson, *Agricultural Economics 201,* p. 59A.
14. Colorado State University Cooperative Extension, *Partial Budgeting;* available at http://www.coopext.colostate.edu/ABM/abmprt.htm (accessed August 27, 2007).
15. Colorado State University Cooperative Extension, *Enterprise Budgeting;* available at http://www.coopext.colostate.edu/ABM/abmenterbudget.htm (accessed August 27, 2007).

Partial Budget Form	
Additional Costs ($)	Additional Returns ($)
Reduced Returns ($)	Reduced Costs ($)
A. Total annual additional costs and reduced returns (A) ______	B. Total annual additional returns and reduced costs (B) ______ (A) ______
Net change in income (B minus A) ______	

Figure 11–8 Many important farm management decisions can be made more quickly and easily by using a partial budget.

harvesting or machinery investment) that affects many enterprises. Given this definition, an enterprise budget can be considered a type of partial budgeting.

Purpose. An enterprise budget may be developed for a variety of reasons. It could be created for each current or potential enterprise associated with the farming operation. Several budgets could be created and merged into a single overall budget to illustrate different combinations of inputs and outputs. Whatever the purpose, each budget should be developed with a small, common unit of measurement (e.g., one acre of corn, one acre of soybeans, one acre of wheat, etc.), so that there is a basis for future comparisons.

Developing an Enterprise Budget. An enterprise budget generally has four components: (1) income/receipts, (2) variable or operating costs, (3) fixed costs, and (4) net receipts. For the first step of determining income and receipts, a manager should estimate the total production (or output) as well as the output price expected from the individual enterprise. When making these estimates, assume normal conditions, and be as realistic as possible.

Building upon that work, the second step requires the manager to estimate variable costs.

Sample Enterprise Budget for Corn Production (1 Acre)

Item		Value per Acre
Income:		
175 bushels @ $2.50 per bushel		$525.00
Variable costs:		
Seed	$40.00	
Fertilizer and lime	50.00	
Chemicals	22.45	
Machinery fuel and expense	17.30	
Labor @ $10 per hour	50.00	
Miscellaneous	8.00	
Interest on variable costs (10% for 6 months)	12.75	
Total variable costs		$200.50
Income above variable costs		$200.50
Fixed costs:		
Machinery depreciation/interest, taxes, and insurance	$42.36	
Total fixed costs		$42.36
Total costs		$242.86
Estimated profit		$282.14

Figure 11–9 The enterprise budget can help the farm manager determine which enterprises should be expanded, cut back, or possibly eliminated.

Variable costs, as noted, are the costs that change according to the amount produced, and are always incurred if the enterprise is undertaken. Variable costs include (but are not limited to) repairs, feed, fuel, and seed.

Once variable costs are estimated and recorded, the manager must consider the fixed costs associated with the enterprise. Fixed costs differ from variable costs in that they will occur (and stay relatively constant) regardless of whether the enterprise is undertaken. A few examples of fixed costs are taxes, insurance, and depreciation.

The final step involves calculating net receipts, which constitute the income left for living expenses. An example of an enterprise budget is shown in Figure 11–9.

Break-Even Analysis. A break-even analysis is a helpful tool when undertaking enterprise analysis. The **break-even point** is when total receipts equal total costs (or when your expenditures equal your income). This figure helps the farm manager to determine what the break-even price is for a specific yield and vice versa. By studying break-even points, a farm manager can make important decisions regarding the production changes needed to obtain a price/yield combination that will cover total costs. Specific break-even formulas are:[16]

$$\text{Break-even yield} = \frac{\text{Total costs}}{\text{Total production}}$$

$$\text{Break-even sale price} = \frac{\text{Total costs}}{\text{Sale price}}$$

Cash-Flow Budget

Yet another type of budget is the **cash-flow budget.**[17] Used to assist in planning the use of

16. William Edwards, *Twelve Steps to Cash Flow Budgeting* (ISU Extension Publication FM 1792) (Ames, IA: Iowa State University Extension, 2000).
17. Ibid.

capital within a farm business, a cash-flow budget can be an invaluable tool to the farm manager who needs to plan and track cash flow within the business. Specifically, a cash-flow budget is an estimate of all cash receipts and expenditures expected to occur during a particular period (a month, a quarter, a year).

Importance. Cash-flow budgets concern only cash movement within the business, and do not show solvency or profitability. This type of budgeting is particularly useful because it not only requires farm managers to think through farming plans for the entire year, but also provides a guide against which to compare the actual cash flows (refer to the discussion of the statement of cash flows mentioned later in this chapter). Finally, and maybe most importantly, it assists in determining the operating line of credit the farm business will need for the next year.

Developing a Cash-Flow Budget. Though the process is relatively straightforward, developing a cash-flow budget can be time-consuming (especially the first time). The process requires attention to many details, so it is probably easiest to prepare a cash-flow budget for your operation using a step-by-step approach:

1. Outline your tentative plans for livestock and crop production.
2. Take an inventory of livestock on hand and crops in storage.
3. Estimate feed requirements for the proposed livestock program.
4. Estimate feed available.

Now begin drawing up the actual cash-flow budget:

5. First, estimate livestock sales (based on production and marketing plans).
6. Plan sales of nonfeed crops and excess feed.
7. Estimate income from other sources, such as income from off-farm work, interest, and/or government payments.
8. Project crop expenses and other farm expenditures.
9. Estimate operating surplus (use projected cash inflows/outflows).
10. Consider capital purchases, such as machinery, equipment, or land.
11. Summarize debt repayment.
12. Calculate the cash-flow surplus (or deficit).[18]

Refer to Figure 11–10 for a sample cash-flow budget.[19]

Analyzing Your Budget. Once you have completed the preceding 12 steps, you must analyze your budget. If you find that the projected net cash flow is negative (at a deficit), you will need to make some adjustments to the budget and your plans. The manager usually has options such as selling more current assets, financing capital expenditures with credit (taking out a loan), or refinancing short-term debt as intermediate- or long-term debt. One note of caution, however: Although undertaking any of these actions will assist in reducing the cash-flow squeeze for this year, make sure the decisions being made are not going to cause more severe problems in the future. You may need to do projections to help you pick different, more profitable enterprises.

Farm Recordkeeping

Working as a farm manager requires you to use a few primary tools so that you can make informed decisions. One of the most important tools for effective decisionmaking is accurate farm records. Farm records are generally kept for two different reasons: income tax reporting and business management. Our interest here is management of the farm business.

Farm managers face many questions in today's environment. Which livestock enterprises should be expanded? Which should be reduced or possibly eliminated? How can you use farm credit more profitably? By themselves, records will not answer these questions or solve these problems. However, with effective recordkeeping, a farm manager can make more informed decisions that *will* help to solve these problems.

Selecting an Appropriate Farm Record System

Keeping accurate records begins with the selection of an appropriate farm record system. A useful system incorporates the statements and records that

18. John R. Schlender, *Farm Record Keeping* (Manhattan, KS: Kansas State University, 1991).
19. Hanson, *Agricultural Economics 201*, p. 105.

Cash-Flow Planning Form

Date Completed ______, 20____
For period from ______ **to** ______

Name of Operation: Bob & Cindy Baker
Address: Farmerville, OK

	TOTAL	Jan.–March	April–June	July–Sept.	Oct.–Dec.
1. Beginning cash balance	$ 1,200	$ 1,200	$ 1,700	$ 1,900	$ 1,485
Operating Sales 2. Crop and hay sales	80,000	4,000	6,000	20,000	50,000
3. Market livestock sales	138,000	17,000	19,000	44,000	58,000
4. Livestock product sales					
5. Custom work	9,000			6,000	3,000
6. Other (ag. program payments, etc.)	17,000	9,000		3,000	5,000
7. Capital sales					
8. Breeding livestock	12,000		12,000		
9. Machinery & equipment	16,000			16,000	
10.					
11.					
12. Personal income	20,000	5,000	5,000	5,000	5,000
13. Total cash available (add lines 1–12)	293,200	36,200	43,700	95,900	122,485
Operating Expenses 14. Car & truck expenses	400	100	100	100	100
15. Chemicals	4,000		4,000		
16. Conservation expenses					
17. Custom hire					
18. Feed purchased	14,000	4,000	4,000	3,000	3,000
19. Fertilizers & lime	12,000		10,000		2,000
20. Freight & trucking	1,200			1,200	
21. Gasoline, fuel, & oil	3,000	700	1,200	600	500
22. Insurance	500	500			
23. Labor hired, benefits, etc.	2,000	400	1,000		600
24. Rents & leases					
25. Repairs & maintenance	2,800		300	2,200	300
26. Seeds & plants	4,200		4,200		
27. Storage & warehousing					
28. Supplies	400	200	100	100	
29. Taxes (real estate & personal property)	4,000	2,000		2,000	
30. Utilities	500	100	100	140	160
31. Veterinary, breeding fees, & medicines	5,400	1,400	1,300	1,300	1,400
32. Feeder livestock (purchased for resale)	116,000	56,000			60,000
33.					
34.					

(continued on the following page)

Figure 11–10 A cash-flow budget is a tremendously useful tool for planning and tracking cash flow within the farming operation, and for estimating the operating line of credit needed for the next year.

Cash-Flow Planning Form *(concluded)*

	TOTAL	Jan.–March	April–June	July–Sept	Oct.–Dec
Capital Purchases 35. Breeding livestock	34,000		34,000		
36. Machinery & equipment	15,700	6,700		9,000	
37.					
38. Family living withdrawals	10,000	2,500	2,500	2,500	2,500
39. Income tax & social security (self-employment)	3,200	3,200			
Term loan payments 40. *Principal*	28,000	15,000	10,000	3,000	
41. *Interest*	38,000	11,700	14,000	12,300	
42. Total cash required (add lines 14–41)	299,300	104,500	86,800	37,440	70,560
43. Net cash available (lines 13–42)	-6,100	-68,300	-43,100	58,460	51,925
44. Operating loan borrowings (if line 43 is negative)	115,000	70,000	45,000		
Operating loan payments					
45. *Principal*	98,000			53,000	45,000
46. *Interest*	9,050			3,975	5,075
47. Ending cash balance (43 + 44 - 45 - 46)	1,850	1,700	1,900	1,485	1,875
48. Ending operating loan balance					
Principal	17,000	70,000	115,000	62,000	17,000
Accrued interest	—0—	1,750	4,625	3,525	

are directly involved in your farming operation. For example, it is not necessary to maintain a landlord-tenant agreement or partnership settlement if you are not renting any property or are operating a sole proprietorship.

Farm Record Uses. A good set of farm records can be used in three ways:

1. As a *service tool* for:
   - income and Social Security tax returns
   - employee wage and Social Security reports
   - balance sheet, income statement, and cash-flow statement for improved credit access and relationships with creditors
   - landlord-tenant relationships and partnerships (records of past agreements and settlements)
   - estate planning
   - appraisal of land values
2. As an indication of business profitability and growth, through the:
   - income statement—present, past, and change over time
   - balance sheet and net worth—present, past, and change over time
3. As a farm business analysis tool, to indicate:
   - sources of farm profits and losses
   - weak and strong parts of the business
   - facts for important decisions
   - bases for future/forward planning[20]

This chapter focuses on several of the more useful (and sometimes more complex) forms: net worth statement (balance sheet), income

20. William Edwards, *Your Net Worth Statement* (ISU Extension Publication FM 1791) (Ames, IA: Iowa State University Extension, 2000).

statement, cash-flow statement, and statement of owner equity. Refer to Chapter 9 for information on similar forms used in running an agribusiness.

Net Worth Statement (Balance Sheet)

The net worth statement is one of the most basic components in any farm record system.[21] The **net worth statement** (sometimes called a *balance sheet*) is a listing of property owned (assets) and debts owed (liabilities). The difference between assets and liabilities is considered the net worth of the business. The majority of farm operations today comprise a combination of land, livestock, crops, and machinery acquired by debt or contributed by the operator. The net worth statement is a snapshot of assets and liabilities at a given time. By comparing end-of-the-year net worth statements over several years, you can get an accurate measurement of the progress of your business. In addition, you can use this statement to determine your ability to pay off current debts and take on additional ones. Refer to Figure 11–11 for a sample net worth statement form.

Valuing the Assets. The first step in developing a net worth statement is valuing the assets of your farming operation. Assets are generally listed on the left-hand side of the statement, and are divided into current and fixed categories. *Current assets* include cash, crops, livestock, and supplies that will normally be used by the end of the year. When listing current assets, it is important to accurately estimate the number, as well as the value, of items on hand (e.g., accurate estimates of bushels, tons, bales, etc. on hand). It is important to differentiate between cost value (actual cost) and market value (includes inflation). Fixed assets are also used in farm production; however, these assets are not converted directly into marketable products. Fixed assets include breeding and dairy livestock, machinery, equipment, and vehicles.

Operational Liabilities. After valuing your assets, you must assess the liabilities of your operation. These are generally listed on the right-hand side of the statement (according to the length of time before they are due), and are also broken down into current and fixed categories. Current liabilities are those debts due within the next 12 months. Such liabilities include accounts payable (or unpaid open accounts), contractual obligations, and debts such as operating or feeder livestock notes. *Fixed liabilities* include debts that are payable more than one year into the future. Items falling into this category include loans for breeding stock, land, or machinery and mortgages or contracts on real estate. For example, a $60,000 loan for machinery is payable in four annual installments of $15,000 each, plus interest at 10 percent on the unpaid balance. You would show the $15,000 due this year as a current liability and the remaining $45,000 as a fixed liability.

Farmer's Equity. The difference between total farm assets and total farm liabilities is the net worth of the business. This is also considered the farmer's equity, or current value of your own investment within the operation. Adding the net worth to total liabilities should give you an amount equal to total assets; this is a good check on your calculations.

Usefulness. A great deal of useful information can be garnered from an examination of your operation's net worth statement. As an initial consideration, look at and compare each major liability listed to see if there is a corresponding asset. The corresponding item will generally be in the same category (current or fixed). If this is not the case, one of two things has happened: you might have forgotten to list something, or an asset purchased with borrowed money might already have been sold or used up. This is a caution signal to you as a manager. It means that you must find funds from somewhere else in the business to pay off this debt.

Other useful aspects of the net worth statement include financial ratios and year-to-year comparisons.

Debt-to-asset ratio. The **debt-to-asset ratio** is equal to total liabilities divided by total assets. This illustrates the amount of total capital in your business supplied by creditors. A low debt-to-asset ratio is desirable, because it usually indicates less year-to-year variability in net farm income.[22]

21. Ibid.

22. William Edwards, *Your Farm Income Statement* (ISU Extension Publication FM 1816) (Ames, IA: Iowa State University Extension, 2000).

Net Worth Statement

Name____ Date______________

Farm Assets	Cost Value	Market Value	Farm Liabilities	Market Value
Checking and savings accounts			Accounts payable (Sched. N)	
			Farm taxes due (Sched. O)	
Crops held for sale or feed (Sched. A)			Current notes and credit lines (Sched. P)	
Investment in growing crops (Sched. B)			Accrued interest - short (Sched. P)	
Commercial feed on hand (Sched. C)			- fixed (Sched. Q)	
Prepaid expenses (Sched. D)				
Market livestock (Sched. E)			Due in 12 months - fixed (Sched. Q)	
Supplies on hand (Sched. F)				
Accounts receivable (Sched. G)			Other current liabilities	
Other current assets				
Total Current Assets			***Total Current Liabilities***	
Unpaid coop. distributions (Sched. H)			Notes and contracts, remainder (Sched. Q)	
Breeding livestock (Sched. I)			Machinery	
Machinery and equipment (Sched. J)			Land	
Buildings/improvements (Sched. K)				
Farmland (Sched. L)				
Farm securities, certificates (Sched. M)				
Other fixed assets			Other fixed liabilities	
Total Fixed Assets			***Total Fixed Liabilities***	
a. Total Farm Assets			***b. Total Farm Liabilities***	
c. Farm Net Worth (a – b)			$\frac{\text{Current Assets (market)}}{\text{Current Liabilities}}$ = ______ Current ratio	
d. Farm Net Worth Last Year			$\frac{\text{Total Liabilities}}{\text{Total Assets (market)}}$ = ______ Debt-to-asset ratio	
e. Change in Farm Net Worth (c – d)				

Personal Assets		Personal Liabilities	
Bank accounts, stocks, bonds		Credit cards, charge accounts, other loans	
Automobiles, boats, etc.		Automobile loans	
Household goods, clothing		Other loans, taxes due	
Real estate		Real estate, other long-term loans	
f. Total Personal Assets		***g. Total Personal Liabilities***	
h. Total Personal Net Worth (f – g)		$\frac{\text{Total Personal Liabilities}}{\text{Total Personal Assets}}$ = ______ Debt-to-asset ratio	
i. Total Net Worth, Market Value (c + h)			

Figure 11–11 The net worth statement is a snapshot of property owned (assets) and debts incurred (liabilities) at a specific point in time.

Current ratio. By dividing total current assets by total current liabilities, you get the **current ratio** of the business. This measures **liquidity**, or the ability to pay debts as they come due. As a general rule, a current ratio of close to 1.0 or lower indicates potential cash-flow problems; a ratio of 2.0 or greater is considered good.

Working capital. A final useful figure, called **working capital**, represents the difference between current assets and current liabilities. This figure indicates the amount of cash available for meeting daily operating costs.

Year-to-year comparisons. Comparing one year to another is a good way to determine the financial progress of your farm business. Maintaining such records allows a farm manager to compare current net worth statements with those from previous periods. Two figures are important in this comparison: the change in cost-value net worth and the change in market-value net worth. The former shows the growth or loss due to net income earned from the farm business and consumption from one year to the next. In contrast, the change in market-value net worth shows the change in market value of your equity share in the farming operation. Both help to reveal the financial progress of your farming operation, as well as providing a sound base or foundation from which to make financial decisions.

Farm Income Statement

The farm income statement could be the easiest way to answer the question "How much did your farm earn last year?" A farm **income statement** is an overall summary of the income and expenses incurred by your farm over a certain period, usually the calendar year. Operationally speaking, it is the value of what your farm produced over the year, and how much it cost to produce it.[23]

A farm income statement is normally divided into two areas: income and expenses. Both of these areas are then further divided into a section for cash entries and a section for accrual (noncash) adjustments. Much of the information needed to prepare your income statement can be found in other records, such as beginning and ending net worth statements and various tax forms. Refer to Figure 11–12. If you use a computer for recordkeeping and/or tax preparation, much of this information will be available immediately from your saved records.

Cash Income. The first section of an income statement concerns cash income. This is where you list your sources of income, such as crop or livestock sales, government payments, and tax credits and refunds. Do not include proceeds from outstanding loans; sale of land, machinery, or other depreciable assets; or income from nonfarm sources.

Accrual Adjustments. Following the cash-income section are the **accrual adjustments**, or adjustments to income. It is not possible to account for all farm income through cash sales. Changes such as fluctuations in inventory value can increase or decrease your farm's income for the year. In addition, changes in the value of inventories of market or breeding livestock, feed, and grain could result from increases or decreases in the quantity of items on hand or changes in unit values. For all these reasons, it is necessary to make adjustments for accurate recordkeeping. Finally, changes in the market value of land, buildings, machinery, and equipment should not be included (unless such items are sold); however, accounts receivable are included because they are considered income earned but not yet received.

Cash Expenses. You also have cash expenses, which are the expenses incurred for operation of the farm business during the year. Cash expenses include almost anything purchased, insurance, interest paid on loans (but not the principal), rent or lease payments, and a number of other expenses. Not included are death losses of livestock, the purchase of capital assets (such as machinery or equipment), or land.

Adjustments to Expenses. Adjustments to expenses are used in situations in which cash expenses are paid in one year for items used in the following year. Such situations may occur regarding feed inventories, prepaid expenses, or growing crop investments. To adjust for these, subtract the ending value from the beginning value for the net adjustment. Other expense adjustments include accounts payable (e.g., farm taxes due or accrued interest) or services that have been rendered but

23. Ibid.

SCHEDULE F (Form 1040)

Department of the Treasury
Internal Revenue Service

Profit or Loss From Farming

▶ Attach to Form 1040, Form 1040NR, Form 1041, Form 1065, or Form 1065-B.

▶ See Instructions for Schedule F (Form 1040).

OMB No. 1545-0074

2007

Attachment Sequence No. **14**

Name of proprietor | **Social security number (SSN)**

A Principal product. Describe in one or two words your principal crop or activity for the current tax year. | **B Enter code from Part IV** ▶

C Accounting method: **(1)** ☐ Cash **(2)** ☐ Accrual | **D Employer ID number (EIN), if any**

E Did you "materially participate" in the operation of this business during 2007? If "No," see page F-2 for limit on passive losses. ☐ Yes ☐ No

Part I **Farm Income—Cash Method.** Complete Parts I and II (Accrual method. Complete Parts II and III, and Part I, line 11.) Do not include sales of livestock held for draft, breeding, sport, or dairy purposes. Report these sales on Form 4797.

1	Sales of livestock and other items you bought for resale	1	
2	Cost or other basis of livestock and other items reported on line 1	2	
3	Subtract line 2 from line 1		3
4	Sales of livestock, produce, grains, and other products you raised		4
5a	Cooperative distributions (Form(s) 1099-PATR)	5a	**5b** Taxable amount 5b
6a	Agricultural program payments (see page F-3)	6a	**6b** Taxable amount 6b
7	Commodity Credit Corporation (CCC) loans (see page F-3):		
a	CCC loans reported under election		7a
b	CCC loans forfeited	7b	**7c** Taxable amount 7c
8	Crop insurance proceeds and federal crop disaster payments (see page F-3):		
a	Amount received in 2007	8a	**8b** Taxable amount 8b
c	If election to defer to 2008 is attached, check here ▶ ☐		**8d** Amount deferred from 2006 8d
9	Custom hire (machine work) income		9
10	Other income, including federal and state gasoline or fuel tax credit or refund (see page F-3)		10
11	**Gross income.** Add amounts in the right column for lines 3 through 10. If you use the accrual method, enter the amount from Part III, line 51 ▶		11

Part II **Farm Expenses—Cash and Accrual Method.** Do not include personal or living expenses such as taxes, insurance, or repairs on your home.

Line	Description		Line	Description	
12	Car and truck expenses (see page F-4). Also attach **Form 4562**	12	25	Pension and profit-sharing plans	25
13	Chemicals	13	26	Rent or lease (see page F-6):	
14	Conservation expenses (see page F-4)	14	a	Vehicles, machinery, and equipment	26a
15	Custom hire (machine work)	15	b	Other (land, animals, etc.)	26b
16	Depreciation and section 179 expense deduction not claimed elsewhere (see page F-5)	16	27	Repairs and maintenance	27
			28	Seeds and plants	28
			29	Storage and warehousing	29
17	Employee benefit programs other than on line 25	17	30	Supplies	30
			31	Taxes	31
18	Feed	18	32	Utilities	32
19	Fertilizers and lime	19	33	Veterinary, breeding, and medicine	33
20	Freight and trucking	20	34	Other expenses (specify):	
21	Gasoline, fuel, and oil	21	a		34a
22	Insurance (other than health)	22	b		34b
23	Interest:		c		34c
a	Mortgage (paid to banks, etc.)	23a	d		34d
b	Other	23b	e		34e
24	Labor hired (less employment credits)	24	f		34f

35 **Total expenses.** Add lines 12 through 34f. If line 34f is negative, see instructions ▶ 35

36 **Net farm profit or (loss).** Subtract line 35 from line 11.
- If a profit, enter the profit on **Form 1040, line 18,** and **also** on **Schedule SE, line 1.** If you file Form 1040NR, enter the profit on **Form 1040NR, line 19.**
- If a loss, you **must** go to line 37. Estates, trusts, and partnerships, see page F-6.

36

37 If you have a loss, you **must** check the box that describes your investment in this activity (see page F-7).
- If you checked 37a, enter the loss on **Form 1040, line 18,** and **also** on **Schedule SE, line 1.** If you file Form 1040NR, enter the loss on **Form 1040NR, line 19.**
- If you checked 37b, you **must** attach **Form 6198.** Your loss may be limited.

37a ☐ All investment is at risk.
37b ☐ Some investment is not at risk.

For Paperwork Reduction Act Notice, see page F-7 of the instructions. Cat. No. 11346H **Schedule F (Form 1040) 2007**

(continued on the following page)

Figure 11–12 Schedule F of IRS Form 1040 is the tax form farm managers use when paying their taxes each year; it shows the profit or loss from farming.

Schedule F (Form 1040) 2007 Page 2

Part III **Farm Income—Accrual Method** (see page F-7).
Do not include sales of livestock held for draft, breeding, sport, or dairy purposes. Report these sales on Form 4797 and do not include this livestock on line 46 below.

38	Sales of livestock, produce, grains, and other products		38	
39a	Cooperative distributions (Form(s) 1099-PATR)	39a	39b Taxable amount	39b
40a	Agricultural program payments	40a	40b Taxable amount	40b
41	Commodity Credit Corporation (CCC) loans:			
a	CCC loans reported under election			41a
b	CCC loans forfeited	41b	41c Taxable amount	41c
42	Crop insurance proceeds			42
43	Custom hire (machine work) income			43
44	Other income, including federal and state gasoline or fuel tax credit or refund			44
45	Add amounts in the right column for lines 38 through 44			45
46	Inventory of livestock, produce, grains, and other products at beginning of the year	46		
47	Cost of livestock, produce, grains, and other products purchased during the year	47		
48	Add lines 46 and 47	48		
49	Inventory of livestock, produce, grains, and other products at end of year	49		
50	Cost of livestock, produce, grains, and other products sold. Subtract line 49 from line 48*			50
51	**Gross income.** Subtract line 50 from line 45. Enter the result here and on Part I, line 11 ▶			51

*If you use the unit-livestock-price method or the farm-price method of valuing inventory and the amount on line 49 is larger than the amount on line 48, subtract line 48 from line 49. Enter the result on line 50. Add lines 45 and 50. Enter the total on line 51 and on Part I, line 11.

Part IV **Principal Agricultural Activity Codes**

CAUTION *File Schedule C (Form 1040) or Schedule C-EZ (Form 1040) instead of Schedule F if **(a)** your principal source of income is from providing agricultural services such as soil preparation, veterinary, farm labor, horticultural, or management for a fee or on a contract basis, or **(b)** you are engaged in the business of breeding, raising, and caring for dogs, cats, or other pet animals.*

These codes for the Principal Agricultural Activity classify farms by their primary activity to facilitate the administration of the Internal Revenue Code. These six-digit codes are based on the North American Industry Classification System (NAICS).

Select the code that best identifies your primary farming activity and enter the six digit number on page 1, line B.

Crop Production

111100 Oilseed and grain farming
111210 Vegetable and melon farming
111300 Fruit and tree nut farming
111400 Greenhouse, nursery, and floriculture production
111900 Other crop farming

Animal Production

112111 Beef cattle ranching and farming
112112 Cattle feedlots
112120 Dairy cattle and milk production
112210 Hog and pig farming
112300 Poultry and egg production
112400 Sheep and goat farming
112510 Aquaculture
112900 Other animal production

Forestry and Logging

113000 Forestry and logging (including forest nurseries and timber tracts)

Schedule F (Form 1040) 2007

have yet to be paid for. A final adjustment consideration is depreciation, the amount by which the value of machinery, buildings, and other capital assets has declined over the year.

Summarizing. Once all of the income and expenses have been accounted for, it is time to summarize. By subtracting gross farm expenses from gross farm revenue, you get the net income generated from the production and marketing activities of the operation. Another name for this figure is **net farm income from operations**. After figuring capital gains or losses (from the sale of capital assets such as land, equipment, or machinery), add or subtract this figure from the net farm income from operations to determine the **net farm income**. The net farm income represents the overall income earned from the farm manager's capital and labor. Refer to Figure 11-13 for a farm income statement form.

Cash-Flow Statement

The third financial statement of interest within this chapter is a **cash-flow statement**. Though not as commonly used as the net worth and farm income statements, this financial statement still provides useful financial information. A statement of cash flows summarizes all of the cash receipts and cash expenditures throughout the year.[24] Though it does not illustrate an operation's profitability like the net income statement, it does show the sources and uses of cash within the farm. A statement of cash flows has five sections:

- cash income and cash expenses
- purchases and sales of capital assets
- new loans received and principal repaid
- nonfarm income and expenses
- beginning and ending cash on hand[25]

As mentioned previously in this chapter, a cash-flow budget is the precursor to the more formal statement of cash flows. Both are useful in effective planning and budgeting. Refer to Figure 11-14 for an example of a cash-flow and owner equity statement.

24. Ibid.
25. Ibid.

Statement of Owner Equity

The final financial record of interest is the statement of owner equity. This statement connects net farm income with the change in net worth. The net worth of an operation can increase during the year, depending on three factors: (1) net farm income, (2) net nonfarm withdrawals (nonfarm income minus nonfarm expenditures), and (3) adjustments to the market value of capital assets. If these items are recorded correctly and added to the beginning net worth of the operation, they should equal the ending net worth.[26] Refer to Figure 11-14 for a combined cash-flow and owner equity statement.

Employing Farm Records for Analysis

All of the financial statements detailed within this section provide valuable information regarding the financial condition of the operation. This information is critical for farm managers, because it provides a foundation from which to make effective decisions. Several measures of financial performance can be described through farm recordkeeping. These measures not only help the farm manager to assess profitability, debt capacity, and the financial risk currently taken on by the business, but also to describe the financial health of the farming organization.

Financial Performance Measures. Five different types of financial measures are detailed through good recordkeeping.[27]

Liquidity describes the degree to which debt obligations due within the next year can be paid with cash, or with assets that will be turned into cash. This is determined through your current ratio and the amount of working capital, which can be found on the farm's statement of net worth.

Solvency refers to the mix of equity and debt capital used by the farm, or the degree to which all debts are secured. Financial ratios based on an operation's assets, liabilities, and net worth, and their respective relationships, are often used to measure solvency. One of the most common ratios used is the total debt-to-asset ratio, which can also be found on the statement of net worth.

26. William Edwards, *Financial Performance Measures for Iowa Farms* (ISU Extension Publication FM 1845) (Ames, IA: Iowa State University Extension, 2000).
27. Ibid.

Net Farm Income Statement

Name____ Year__________

Income

Cash Income		Income Adjustments	Ending	Beginning
Sale of livestock, other bought for resale		Crops held for sale or feed (Sched. A)		
Sales of market livestock, grain, produce		Market livestock (Sched. E)		
Cooperative distributions paid		Accounts receivable (Sched. G)		
Agricultural program payments		Unpaid coop. distributions (Sched. H)		
Crop insurance proceeds		Breeding livestock (Sched. I)		
Custom hire income				
Other cash income		**Subtotal of Adjustments**	b.	c.
Sales of breeding livestock		***d. Value of Home Used Production***		
a. Total Cash Income		***e. Gross Farm Revenue (a + b – c + d)***		

Expenses

Cash Expenses		Expense Adjustments	Beginning	Ending
Car and truck expenses		Investment in growing crops (Sched. B)		
Chemicals		Commercial feed on hand (Sched. C)		
Conservation expenses		Prepaid expenses (Sched. D)		
Custom hire		Supplies on hand (Sched. F)		
Employee benefits			**Ending**	**Beginning**
Feed purchased		Accounts payable (Sched. N)		
Fertilizer and lime		Farm taxes due (Sched. O)		
Freight, trucking		Accrued interest (Sched. P, Q)		
Gasoline, fuel, oil		**Subtotal of Adjustments**	g.	h.
Insurance				
Interest paid		***i. Depreciation (Sched. J, K)***		
Labor hired		***j. Gross Farm Expenses (f + g – h + i)***		
Pension and profit-share plans				
Rent or lease payments				
Repairs, maintenance		***k. Net Farm Income from Operations (e–j)***		
Seeds, plants				
Storage, warehousing				
Supplies purchased		***l. Sales of Farm Capital Assets***		
Taxes (farm)		***m. Cost Value of Items Sold (Sched. J, K, L)***		
Utilities		***n. Capital Gains or Losses (l – m)***		
Veterinary fees, medicine, breeding				
Other cash expenses				
Livestock purchased				
f. Total Cash Expenses		***o. Net Farm Income (k + n)***		

Figure 11–13 By subtracting gross farm expenses from gross farm revenue, you get net income generated by the production and marketing activities of your farming operation.

Statement of Cash Flows

Name________________________________ Year__________

Cash Farm Income and Expenses (Operating)	Cash Inflows	Cash Outflows
Total Cash Income (Line a, Net Farm Income Statement)		
Total Cash Expenses (Line f, Net Farm Income Statement)		
Capital Assets (Investing)		
Sales of Capital Assets (Line I, Net Farm Income Statement)		
Cost of Purchases and Trades (Sched. J, K, L)		
Loans (Financing)		
New Loans Received		
Principal Paid		
Nonfarm		
Nonfarm Income (wages, rents, interest, etc.)		
Nonfarm Expenditures (family living, income tax, etc.)		
Cash on Hand (farm and nonfarm cash, checking, savings)		
Beginning of Year (Net Worth Statement)		
End of Year (Net Worth Statement)		
Total		

If all cash transactions are included correctly, the totals for the two columns will be approximately equal.

Statement of Owner Equity

Name________________________________ Year__________

	Cost Value	Market Value
a. Farm Net Worth, Last Year		
(Line d, Net Worth Statement)		
b. Change in Market Value of Capital Assets (net of depreciation)		
(Line e, Net Worth Statement, market value ________ minus cost value ________)		
c. Net Farm Income		
(Line o, Net Farm Income Statement)	***same for cost and market***	
d. Net Nonfarm Withdrawals		
(Statement of Cash Flows, nonfarm income ________ minus nonfarm expenditures ________ , minus value of home used production ________ , line d, Net Farm Income Statement)	***same for cost and market***	
e. Farm Net Worth, This Year		
(Line c, Net Worth Statement)		

Line e should approximately equal the sum of lines a, b, c, and d.

Figure 11-14 A statement of cash flows summarizes all of the cash receipts and cash expenditures during the year; the statement of owner equity connects net farm income with the change in net worth.

Profitability is a third financial performance measure. *Profitability* is the difference between an operation's income and expenses. One of the most important measures of this is net farm income. Other important ratios, as well as net farm income, can be found within the statement of net worth.

Financial efficiency ratios are financial figures that illustrate the percent of gross farm revenue that went to pay interest, operating expenses, and depreciation, and how much was left for net farm income. An important measure of financial efficiency is the asset turnover ratio, which measures the amount of gross income that was generated for each dollar invested in livestock, equipment, and other various assets.

Repayment capacity is the final type of financial performance measure; it illustrates the adequacy of cash generated by the farm to pay the principal and interest on loans as they come due. Both financial efficiency ratios and repayment capacity figures can be determined from information found within your farming operation's statement of net worth, as well as the farm income statement.

Other Farm Management Issues

As a farm manager, you will face a wide variety of issues not directly related to financial matters. These issues are often just as important as financial issues, and in the end may exert their own influence on financial decisions. Thus, it is important for today's farm manager to pay attention to these operational issues as they pertain to the farming operation.

Managing Risk and Uncertainty

Uncertainty and risk are both pervasive features of the farm environment. All businesses have to deal with these to a degree, but they uniquely affect the farming community. *Uncertainty* is defined as an imperfect knowledge of the future, which of course is highly relevant to managerial decisionmaking done in the course of planning and running a farm operation.[28]

28. Food and Agriculture Organization of the United Nations (FAO), "Planning and Managing Farm Systems under Uncertainty," in *Farm Management for Asia: A Systems Approach* (FAO Farm Systems Management Series 13, 1997); available at http://www.fao.org/docrep/w7365e/w7365e0e.htm.

Uncertainty. Farm managers must deal daily with uncertainties such as weather, environment, and supply and demand; in addition, they undertake risks with regard to capital investments, loans, and other production decisions. For these reasons, it is important not only to recognize the presence of these factors within the farm environment, but also to make a conscious effort to incorporate their possibility into farm management decisions.

Although they are considered here within the same section, risk and uncertainty are slightly different. Uncertainty, in and of itself, is always present.

Risk. Risk is present only when the uncertain outcomes of a decision are regarded as significant or worth worrying about. Therefore, if similar decisions are made frequently or have historically been successful, risk may be considered small or negligible.

Risk experienced within a farming operation generally comes from one of two sources: (1) the external environment, as it affects the operation, and (2) the internal operating environment of the farm.[29]

External sources of risk include (but are not limited to):

- *Natural environment*: things such as weather, natural disasters, market supply (indirectly), prices
- *Economic environment*: matters such as uncertainty about supply and demand, inflation, interest, productivity through new technology
- *Social environment*: such things as lifestyle and consumer habit changes, farm labor availability, other societal issues
- *Governmental (policy) environment*: issues of policy relating to commodity prices and marketing, credit availability, water rights, environmental standards, labor laws, and the like
- *Political environment*: any marked changes in political ideology[30]

These types of risk exist for any farming operations, because they are an inherent part of the external environment.

29. Ibid.
30. Ibid.

Internal sources of risk, by contrast, affect only the individual farming operation. Major internal sources of risk include:

- the health of the farm family
- interpersonal relations between farm household members
- the farm manager's decisionmaking approach toward farm resources, farm credit, and intergenerational transfer of the farm[31]

It is important to note that although there are a wide variety of approaches to handling and managing farm risk, the utility of these approaches varies depending on the situation. They can only provide guidance; no system can give absolute answers or direction. Still, as long as a farm manager recognizes the existence of uncertainties and risks, includes them in planning considerations, and uses a little intuition and good judgment, the decisions made should prove effective.

Income Tax Management

Income tax management is a potentially complex aspect of business operation; nevertheless, farm managers should be aware of tax issues and know the basics of tax management. When they know these basics, farm managers can benefit substantially because they will be able to recognize a tax problem or opportunity. Most major business decisions have income tax implications, including business expansion, capital expenditures, retirement planning, estate planning, and many others. Effective tax management can allow farm managers to save in taxes and apply these savings elsewhere in the business.

Although saving on taxes is important, the overall objective of effective tax management is to maximize income after taxes over time. Keeping this objective in mind is essential when making tax decisions, because it keeps you focused on the impact of your decisions upon the entire operation. Bad decisions may result if you make decisions based only upon tax savings, and do not consider the future and goals of the farm operation. A good example of this is postponing the sale of a commodity because of declining prices. This may help the farm in the short term—at that given point in time—but the shift in revenue could badly hurt the operation in the future.

31. Ibid.

The basics of good income tax management include selecting a good accounting system and the appropriate accounting period (monthly, quarterly, annually). In addition, an informed farm manager should be aware of the general tax law provisions, be able to recognize benefits gained from effective tax decisionmaking, seek competent professional help as necessary, and know when to employ basic tax management strategies.[32]

Whole-Farm Planning

Through looking at the wide variety of areas and factors involved in farm management, we have gotten a pretty good picture of the environment for today's farm manager. However, it is necessary and important to really study the overall picture, which can be done through **whole-farm planning**.

Family Planning

Farm managers work in a distinctive environment (one much different from those of most other managers), mainly because their job includes their families, livelihood, and other unusual considerations. It is important to take each of these considerations into account when doing long-term planning. *Whole-farm planning,* also known as comprehensive farm planning, is a process used by the farm family to balance important life aspects, such as their quality of life, with important business aspects, such as the farm's resources, the need for production and profitability, and long-term stewardship.[33] Through this process, farm managers have been able to more effectively manage their farms, while at the same time providing more financial stability, improved stewardship, and a higher quality of life for their families.[34]

32. Gayle S. Willett, Larry K. Bond, and Norman Dalsted, *Federal Income Tax Management for Farmers and Ranchers* (WREP 0148) (Western Farm Management Extension Committee, Washington State University, 1995); available at http://www.extension.usu.edu/files/agpubs/taxpub.pdf.
33. Whole Farm Planning Working Group, *Whole Farm Planning: What It Takes* (St. Paul: Minnesota Department of Agriculture, 1997).
34. David Mulla, Les Everett, and Gigi DiGicomo, *Whole Farm Planning: Combining Family, Profit, and Environment* (BU 06985) (Minnesota Institute for Sustainable Agriculture, University of Minnesota Extension Service, 1998); available at http://www.extension.umn.edu/distribution/businessmanagement/components/6985a1.html.

Long-Term Vision

Whole-farm planning differs from other farm planning because it focuses on the "big picture." It includes all of the planning done on a daily basis and allows the farm manager to base that planning on the long-term goals of the operation. Specifically, whole-farm planning can allow a farm manager to create a long-term vision for the farm, extending well into the future; increase the profitability and efficiency of farm operations; increase community respect for work done to protect the environment; and make the farm a better place for all family members to live.

There are four main steps in whole-farm planning: (1) setting goals, (2) making an inventory and assessment of farm resources, (3) developing and implementing an action plan, and (4) monitoring on-farm progress toward goals.[35]

Setting Goals. As with many other planning approaches, the first step in whole-farm planning is to create goals and a long-term vision for your farming operation. The vision is an overall view of where you want your farm to be in the future; your short- and long-term goals will help you to work toward that vision. Both the vision and goals should be based on three considerations: (1) the quality of life you want, (2) your desires for the future of your farm, and (3) how the farming operation will provide the income and living environment you need. For each consideration, there are also resource areas that should be addressed: human and social resources, environmental and natural resources, and economic and financial resources. The manager and his or her family should discuss both short- and long-term goals, addressing each of these considerations and resource areas.[36]

Making an Inventory and Assessment of Farm Resources. Once you have developed a vision and short- and long-term goals for your farming operation, it is appropriate to inventory and assess your resources. Resources of interest include natural resources, human resources, financial and capital assets, and crops and livestock systems. To effectively assess these resources, you will need information such as farm financial data, animal management histories, and soil test results, to name just a few. Going through this process may help to identify problems with asset allocation or management, highlight strengths or weaknesses in your financial or capital status, or reveal that your labor or time is being used inefficiently.[37]

Developing and Implementing an Action Plan. Using the inventory and assessment results, you can now consider, evaluate, and choose management alternatives, and develop them into an action plan. The number and kind of management alternatives you select depend on the issues facing your particular farming operation. It is important to note, however, that a broader range of alternatives creates a greater likelihood that you will find options that address specific goals and further your overall vision.

Management alternatives can come from many different places. Other farm managers, family, extension publications, and even farm journals are just a few of the sources that can provide viable alternatives. For example, you may want to consider an enterprise change from a continuous cropping system to a combination livestock-cropping system. Initially, you need to look at how the change would affect your income, quality of life, and natural resources, and assess whether it would help you to achieve your goals more effectively. You may want to study specific changes, such as the effect of manure (as opposed to chemical fertilizer) on the remaining crops, or the effect of building a feedlot on the quality of your family's drinking water or your neighbor's quality of life. As part of the action plan, you could even evaluate changes from a family member going back to school, or a child graduating and going away to college.

Once you have carefully considered the available management alternatives, choose the most effective to incorporate into your action plan. As you make your decisions, though, it is important to continuously compare alternatives to the goals and vision you developed in step 1. If they are not properly aligned with the farming organization's goals and vision, certain management alternatives that may seem appropriate on paper may prove ineffective upon implementation.[38]

35. Ibid., http://www.extension.umn.edu/distribution/businessmanagement/components/6985b.html.
36. Ibid.
37. Ibid.
38. Ibid.

Monitoring Farm Progress toward Goals. Once you have a functional action plan for your farming operation, it is important that you monitor progress toward these goals. During implementation of the whole-farm plan, make sure to keep good records and use them to check the progress of your farm toward the goals that you set. If you find that the plan is not progressing as it should, or that something is not working out as planned, you may have to revise the overall plan. You may even find that you need to revise those goals or your overall vision as your farming operation and your family changes over time.[39]

Conclusion

Farm managers of today have more complex and more diverse responsibilities than managers of the past. Because of factors such as technology, changes in society, and environmental transformations, today's farm managers must be able to balance the decisions made with the impacts of those decisions. The bottom line for any manager is to maintain and increase business profitability; however, farm managers must balance moneymaking with other direct effects of the business, both on their families and beyond.

As the manager of a modern farming operation, it is important for you to consider all of the aspects of farm management brought up in this chapter, as well as other issues in modern agriculture. Agriculture within the United States has gone through many changes within the past few decades. During the "golden age" of agriculture (the 1950s and 1960s), agriculture was just coming into its own—plentiful water supplies, cheap energy, and extensive pesticide and fertilizer use allowed large productivity increases while decreasing labor needs and land use. More recently, many of these inputs have become more expensive; this, coupled with the potential adverse effects of pesticides and fertilizers, an ever-increasing population (and therefore, food demand), and limited land availability, create a much different environment for today's farm manager. Having more highly educated and knowledgeable consumers further complicates the situation. Other modern issues include (but are certainly not limited to):

- land exhaustion
- nitrate runoff
- soil erosion
- agricultural fuel
- animal slurry
- habitat destruction
- water quality
- environmental contamination

These issues, and many others, create an environment for farm managers much more complex than ever before. To be an effective farm manager, it is imperative to focus on producing high yields, while at the same time protecting soil productivity and maintaining environmental quality.[40] To do this, it is important to remain current on the issues affecting modern agriculture and how those issues affect the decisionmaking process, particularly within your farming operation. Today's farmland is an exhaustible resource—you must find the delicate balance between being an effective manager and protecting this resource for the future.

Summary

To be an effective manager, you must understand the role of a manager within the farming operation. As farms increased in size and complexity, it became necessary to hire extra help and diversify the services and products offered. To effectively control growth and diversification, and increase profitability, farmers needed to become farm managers. Important skills for farm managers include timely decisionmaking, appropriate risk taking, contextual knowledge, effective communication, and the ability to combine and oversee all the various parts of the farm operation. The challenge for today's farm managers is to maintain the profitability of their organizations while dealing with the uncertainties that will always affect their businesses.

One of the most basic responsibilities of a farm manager is enterprise selection. Determining the most appropriate (and profitable) products and

39. Ibid.

40. Nancy M. Trautmann, Keith S. Porter, and Robert J. Wagenet, *Modern Agriculture: Its Effects on the Environment*; available at http://pmep.cce.cornell.edu/facts-slides-self/facts/mod-ag-grw85.html.

services your farming operation can produce is an important foundation on which to build a business. Seven steps should be considered when selecting an enterprise: (1) determine your goals, (2) inventory your resources, (3) list the physical, financial, and management factors involved, (4) develop a list of potential enterprises, (5) investigate compatibility with resources, (6) look at growing considerations, and (7) determine compatibility among enterprises. Although the work needed for each of these steps may seem intensive, this basic but important decision could make or break your business.

Two calculations important to farm budgeting and accounting are depreciation and machinery cost calculations. Both involve calculations that directly affect the decisions made within a farming operation. Depreciation allows one to determine the value of an asset over its useful life. Depreciation can be determined by three methods: straight-line depreciation, ACRS, or MACRS. Machinery cost calculations involve a number of different costs associated with owning and operating machinery (including depreciation). Ownership costs include depreciation, interest, taxes, and maintenance facilities; operating costs include anything associated with the use of machinery, such as repair and maintenance, fuel and lubrication, and labor costs. Both of these calculations are important in an effective farm manager's planning process.

Another important planning tool is budgeting. Budgeting allows farm managers to be more precise, reduces wasteful activity, and helps to decrease overlapping or duplicative efforts in the farming operation. A complete budget is a financial plan for all segments of the business. Other useful budgets include the partial budget (which outlines a specific area of the operation), the enterprise budget (which relates to a specific enterprise), and the cash-flow budget (which tracks the projected cash flow within an operation). Each of these budgets provides useful information within different contexts of the farming operation.

Farm recordkeeping is another fundamental aspect of the farm manager's job. Recordkeeping itself involves several different business forms, calculations, and formulas. The first step in effective recordkeeping is selecting the system most appropriate to your farming operation. Whether general or detailed, computer-based or written, good farm records allow a farm manager to determine anything from profitability to solvency, from liquidity to financial efficiency. Important records include the net worth statement (balance sheet), farm income statement, statement of cash flows, and statement of owner equity. From each of these statements, the manager can draw five key financial measures that assist in illustrating the financial health of the business: liquidity, solvency, profitability, financial efficiency ratios, and repayment capacity. By keeping accurate records, a farm manager can plan better and make more effective decisions.

There are other management issues that are not related to budgeting, recordkeeping, or business calculations. Both income tax management and management of risk and uncertainty fall into this category. The need to deal with risk and uncertainty is not unusual; when working as the manager of a farming operation, these issues are a way of life. Weather, fluctuations in supply and demand, and ever-changing farm policies are just a few of the issues that continue to play a major role in farming operations. Risk can come from one of two sources: external (natural, economic, social, governmental, and political) or internal (family health, interpersonal relations, and decision-making approach). Farm managers must also deal with income taxes. Though tax issues often are confusing or difficult to understand, effective tax management can help farm managers save on taxes and apply these savings elsewhere in the business.

Whole-farm planning, in addition to keeping current with farm issues, should both be included in a farm manager's decisionmaking process. Considering the "big picture" is important because this overall planning takes all aspects of the farming operation into account. This allows development of a long-term vision and goals that move the operation toward that vision. The four main steps in whole-farm planning are: (1) setting goals, (2) inventorying and assessing farm resources, (3) developing and implementing an action plan, and (4) monitoring progress toward goals. Each of these steps assists in bringing the farming operation closer to achievement of the overall vision.

A final consideration, which is an aspect of whole-farm planning and other planning within the farming operation, is maintaining relevance within the environment of modern farming. It is difficult to maintain and improve profitability, competitiveness, and productivity in the best of situations; to be an effective farm manager, you must work toward these goals while also taking into account issues such as land exhaustion, soil erosion, and water quality. Effective farm management can be a tough yet rewarding job. By considering all of the issues discussed in this chapter, farm managers can ensure that they are on the road to success.

End of Chapter Activities

REVIEW QUESTIONS

1. Define the Terms to Know.
2. What role do managers play in business operations?
3. Why is management important in today's farming operations?
4. Name and define the three basic functions of management.
5. List the steps involved in enterprise selection.
6. What management factors are involved in enterprise selection? Name at least three and list pertinent questions regarding each.
7. What is the difference between depreciation and economic depreciation?
8. What are the three main reasons for using depreciation within an agribusiness operation?
9. List four fixed costs associated with owning machinery.
10. What are specific advantages to a partial budget?
11. Which of the budgets outlined within this chapter is the best for making overall management decisions? Why?
12. What is the difference between partial and enterprise budgeting?
13. What is a cash-flow budget, and how is it related to the statement of cash flows?
14. What are three major uses for a good set of farm records?
15. Which of the four financial statements mentioned in this chapter is best for determining the progress of your farming operation? Why?
16. Define the difference between current and fixed liabilities (in the balance sheet) and give three examples of each.
17. List and define the four major sections of a farm income statement.
18. Define the five financial performance measures and include a specific measurement ratio or gauge for each.
19. What are the major external sources of risk for today's farm manager? Include an example of each.
20. What are the basics of good tax management?
21. When setting goals in whole-farm planning, the goals and vision for your operation should be based on what three considerations?
22. Besides those mentioned within the text, what are some other issues affecting today's farm manager?

FILL IN THE BLANK

1. Financial, ____________, and ____________ considerations are the basic components of Step 3 of enterprise selection.
2. A good set of farm records can be used as a service tool, an indication of ____________________ and ____________, and a ______________________________ tool.

3. __________ costs generally run 10–15% of __________ costs.
4. A ____________________ is a financial plan for all segments of the business.
5. Income-increasing effects occur as a result of __________ returns and __________ costs.
6. A(n) __________ budget could be considered a type of __________ budget.
7. Three components of an enterprise budget are __________, ____________________, and ____________________.
8. Farm records are generally kept for two different reasons: __________ and __________.
9. A good set of farm records can be used as an indication of ____________________ and ________.
10. The ______________________________ best answers the question "How much did your farm earn last year?"
11. ____________________ is the part of a farm income statement where you list your sources of income.
12. The statement of owner equity connects ______________________________ with ______________________________.
13. The __________ of the farm family is considered a(n) __________ source of risk.
14. Effective ____________________ can allow farm managers to save in taxes.
15. ______________________________ focuses on the "big picture."
16. Plentiful __________, __________ energy, and extensive __________ and __________ use were all characteristics of the "golden age" of agriculture.

MATCHING

a. economic depreciation
b. current assets
c. liabilities
d. Industrial Revolution
e. vision
f. enterprise
g. net worth statement
h. goals
i. current ratio
j. balance sheet
k. current liabilities
l. adjustment to expenses
m. debt-to-asset ratio
n. profitability
o. cash-flow budget
p. uncertainty
q. action plan
r. solvency
s. whole-farm planning
t. "golden age" of agriculture

______ 1. type of depreciation used in machinery calculations
______ 2. one of the catalysts of major change in agriculture
______ 3. production activities
______ 4. these should be measurable over a specific period of time
______ 5. a statement of where you "see" your agribusiness in the future
______ 6. operational expenses
______ 7. one of the best financial statements for measuring profitability
______ 8. total current assets/total current liabilities
______ 9. balance sheet
______ 10. expenses due in the next 12 months
______ 11. cash, crops, livestock, and supplies used by the end of the year

______ 12. total liabilities/total assets

______ 13. a section in the farm income statement for cash expenses that were paid in one year for items used the next year

______ 14. precursor to the statement of cash flows

______ 15. difference between an operation's income and its expenses

______ 16. degree to which all debts are secured

______ 17. imperfect knowledge of the future

______ 18. a process used to balance life aspects with business aspects

______ 19. a plan developed to move toward the farming operation's vision and goals

______ 20. 1950s to 1960s

ACTIVITIES

1. Complete the following activity using the computer program Quicken. Develop pages within the program duplicating the pages you find within any generalized record book, or develop your own. Input the information provided below and calculate totals for the year.

    Last year, you filled out your financial records and showed a pretty good profit from your productive enterprises. They were dairy cows, swine breeding, and strawberries.

    This year, when you close out your record book on December 31, you find out that you have done even better than last year. Your program is very innovative and you did not have any help from anyone. Last year, you thought it was best to rent rather than buy everything, so you found a 25-acre farm with a dairy barn on it already equipped with Grade A milking facilities. When you finish the year, you have 20 cows valued at $1,000 each, 7 brood sows and 1 boar valued at $200 each, and one acre of strawberries that you feel is worth approximately $400. However, at the beginning of the year, January 1, you valued your 20 dairy cows at $900 per head, your sows and boar at $175 each, and your strawberries at only $300, because the plants had not had time to establish themselves very well. You pay $200 per month for the farm and you considered $150 to go for the dairy cow enterprise, $30 for the swine breeding enterprise, and $20 for the strawberry enterprise.

    Your expenses were astronomical because you bought all of your concentrates and roughages. On November 2, you bought a trailer-load of alfalfa hay that weighed 15 tons at $95 per ton. On December 2, you bought another 15 tons, and this cost $98 per ton. On January 2, you bought another load weighing 15 tons and this load cost $100 per ton. You bought three more loads of 15 tons each for $105 per ton on February 2. You bought Purina Cow Ration from a local dealer monthly for $9.85 per hundredweight. You bought 90 hundred-pound bags every month for 12 months. You bought barn supplies from Haley and Rigsby's Milk Supply Company twice. The first time was on January 25 and you bought $157 worth of supplies. On June 16, you bought more supplies valued at $143.72. Your vet bill for the year totaled $231.45. The vet bill was for the dairy cows only.

    Your 20 cows averaged 45 pounds of milk per day for 12 months, averaging Class I and II milk and totaling out for haul bill, insurance, etc. You got $12.50 per hundredweight for your milk. You also sold 12 heifer calves for $100 per head on September 23, and 8 bull calves on October 5 for $50 per head. Your cows were artificially inseminated; therefore, it cost $12 each to service your cows.

    Labor was the only expense for your strawberries.

On March 21, Artie Davis cleaned off your patch in 21 hours at the cost of $2.50 per hour. On April 15, Jeff Crafton hoed the berries in 18 hours for $2.75 an hour. Joey Willis pulled weeds from the berries on May 10, for $3.00 per hour. You had people pick the berries on halves; thus, you had no labor cost for picking them, yet you still sold 175 crates of strawberries at $16.00 per crate.

You sold the pigs from your sows as feeder pigs. On April 19, you sold 60 pigs that brought $1.20 per pound. The pigs averaged 60 pounds per head. On December 9, you sold feeder pigs again. This time you sold 52 pigs that averaged 45 pounds per head for $1.35 per pound.

Your hog feed was shorts, corn, and hog supplement. You bought 15 bags of corn each month at $4.50 per hundredweight, 4 bags of shorts each month at $8.50 per hundredweight, and 4 bags of hog supplement each month at $12.00 per hundredweight. All bags were hundred-pound bags.

You spent numerous hours on the enterprises. For the whole year, you spent 1,885 hours on your dairy operation, 365 hours on your swine enterprise, and 75 hours on your strawberries (this is self labor). Your hourly rate is $3.00 per hour.

You inherited $30,000 from a family member a couple of years ago. After you bought your dairy cows and swine, you bought a used Ford diesel tractor for $2,000, a grader blade for $500, a manure spreader for $1,000, a farm pickup for $3,000, a goose-neck trailer for $2,000, a front-end loader for $750, and a bushhog for $1,100. You set these up on a straight-line 10-year depreciation (10 percent for the next 10 years).

Fill out this information on the record pages (within Quicken) and find out what your net farm profit and your investment in farming are. Print the pages out to illustrate the recordkeeping entries, as well as the final calculations.

2. Describe how the agricultural environment has changed since the 1950s.
3. List five technical management skills important for a farm manager and provide an example of each.
4. Develop and design your own enterprise selection worksheet. After reviewing the enterprise selection section in this chapter, outline the major steps involved and the questions you feel are most important to the selection process. Take this information and organize it into a reader-friendly formatted worksheet.
5. Describe the difference between partial budgeting and enterprise budgeting and develop an example of each.
6. You have a 150-horsepower tractor, diesel costs $2.50 a gallon, and you use the tractor 700 hours annually. What is your estimated annual fuel cost?
7. Define the difference between depreciable and nondepreciable assets and make side-by-side columns exhaustively comparing examples of each. (Exhaustive comparison means listing examples of each until you cannot think of any more.)
8. Research the 1986 Tax Reform Act, and develop a detailed explanation of the connection between this act and MACRS.
9. Using poster board or large sheets of paper, draw a flowchart illustrating the step-by-step development of a cash-flow budget. Once complete, make a presentation to the class outlining the steps, using the flowchart as an illustration.
10. Choose one of the financial statements discussed in this chapter to take home and calculate using a farming operation familiar or available to you. Once you have done this, discuss your results with the farm manager of the operation.

11. Carry out the whole-farm planning process step-by-step using a farm familiar to you (or make one up). Be sure to include the development of short- and long-term goals, a vision, and specific action steps.
12. Using the library and/or Internet, research one of the farming issues affecting today's farm manager. Write a two-page discussion on the chosen topic, concluding with viable options for addressing the issue.

Chapter 12

Production Economics

Objectives

After completing this chapter, the student should be able to:

- Define economics and the role it plays in the U.S. economy.
- Explain how economics is characterized within the United States (including the food and fiber system).
- Illustrate the interaction of the product and resource market sectors within the economy.
- List the six basic concepts of economics and explain their significance.
- Explain the production function relationship and give a working example of a production function.
- Accurately calculate TP, AP, and MP.
- Identify and characterize the three stages of production.
- Use the decision rule for the principle of diminishing returns–input basis to determine the most efficient amount of inputs into a production process.
- List and accurately calculate short-run costs (TC, ATC, MC).
- Compare the concepts of returns to scale and economies of scale within an agricultural setting.
- Effectively illustrate resource substitution and describe the various types of substitutions.
- Correctly determine the least-cost combination of inputs.
- Accurately calculate profit maximization using both the total revenue–total cost method and the marginal revenue–marginal cost method.

Terms to Know

cost function
demand
demand curve
demand schedule
economics
economies of scale
efficiency
elasticity
equilibrium price
externalities
isocost line
isoquant curve

(Continued)

least-cost combination of inputs
margin
marginal product (MP)
market
opportunity cost
perfect complements
perfect substitutes
point of equilibrium
principle of diminishing returns–input basis
principle of resource substitution
production function
profit maximization
quantity supplied
resources
scarcity
supply
total cost (TC)
total fixed costs (TFC)
total product (TP)
total variable costs (TVC)

Equations to Know

average fixed costs (AFC)
average product (AP)
average total costs (ATC)
average variable costs (AVC)
marginal costs (MC)
marginal factor cost (MFC)
marginal rate of substitution (MRS)
marginal revenue–marginal cost method
marginal value product (MVP)
total revenue–total cost method

Introduction

The field of economics is an important one that in some way affects all members of society, whether they are producers, consumers, or just part of the general population (refer to Figure 12-1). **Economics** itself can be generally defined as "a social science that studies how consumers, producers, and societies choose among the alternative uses of scarce resources in the process of producing, exchanging, and consuming goods and services."[1] Production economics is particularly concerned with the aspects of economics that directly affect cost and production within the production process.

Economics in Society

Within the United States, economics is expressed through our capitalistic economy. More specifically, the United States uses the marketplace to determine prices for both the resources used in production and the good/services sold or traded. Thus, the study of economics gives us important insight into how the U.S. economy and other capitalistic societies operate. Much of economics involves the study of:

- how prices are actually determined in the market for goods/services
- how price signals are communicated between buyers and sellers, or producers and consumers
- how institutions should be organized to facilitate efficient operation and function of the markets for goods/services
- the appropriate role of government, in both the private and public sectors of the economy[2]

This chapter more thoroughly explains several of these economic fundamentals, as well as outlining essential aspects of basic production economic theory.

The U.S. Food and Fiber System. The phrase *food and fiber system* is used to represent all business units (including services) that are involved, either directly or indirectly, in the production, processing, and distribution of food and fiber products to U.S. consumers. Sometimes characterized as the movement of a product from the "farm gate to the store shelf," this system comprises four major sectors:

1. the farm input supply sector
2. the farm business sector
3. the processing and manufacturing sector
4. the wholesale and retail sector

The food and fiber system is particularly important in the U.S. economy, as it employs more than one-fifth of the nation's workforce overall.[3]

1. Ronald Hanson, *Agricultural Economics 141: Lecture Study Notes* (Lincoln: University of Nebraska-Lincoln, 1996), p. 1-4.

2. Ibid., p. 1-7.
3. Ibid., p. 2-1.

Figure 12–1 One illustration of how the economy affects producers is the cattle market, which has fluctuated considerably over the past few years because of various environmental considerations. (Courtesy of USDA, #NRCSA 02031)

Comprehensive View of the Economy. Two primary entities interact with each other to shape the overall workings of the economy: households and business firms. *Households* are what individuals and families call home; they are the consumers of all goods and services and own the economic resources of the country. On the other side, *business firms* are the companies that produce all goods and services; they purchase or rent the economic resources available within the economy, and are generally organized as sole proprietorships, partnerships, corporations, or cooperatives. Both households and business firms play a role in the two sectors that constitute a comprehensive view of the economy: the product and resource market sectors.[4] Refer to Figure 12–2.

Product market sector. This sector represents the market for finished goods, and illustrates the flow of finished goods from producer (business) to consumer (households). Consumers pay producers for specific goods and services. The product market establishes the prices that regulate the quantity and quality of goods produced and consumed within the economy.

Resource market sector. This sector represents the movement of economic resources owned by consumers to business firms. These resources are land, labor, capital, and management. Within this sector, the producer (business) pays the consumer (household) for the desired resources. The resource market determines the prices (cost paid for resources used) that regulate the flow of resources from consumer to producer.

In general, households constitute the selling side of the resource market, with businesses being

4. Ibid.

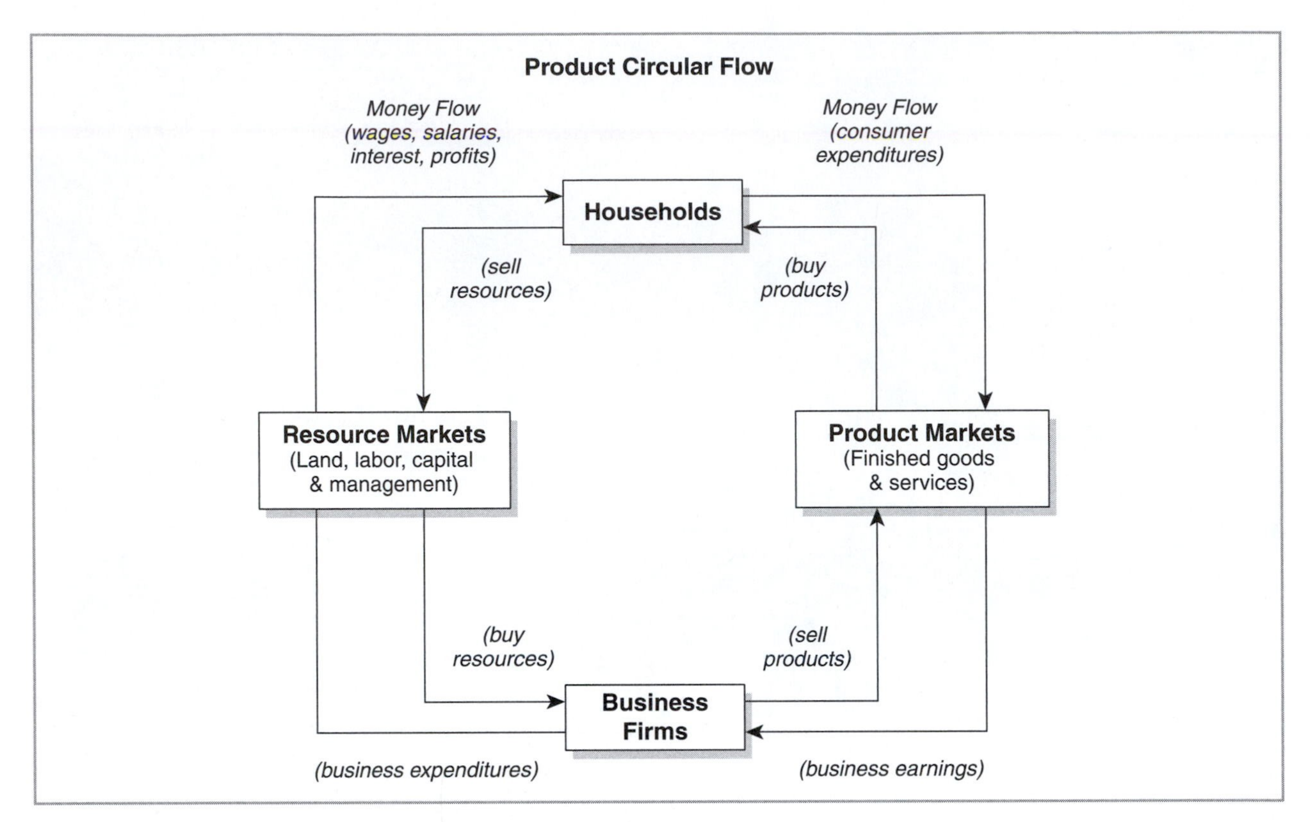

Figure 12–2 Households constitute the selling side of the resource market, with businesses being the buying side; in the product market, households are the buying side and businesses are the selling side.

the buying side. Conversely, households are the buying side of the product market, with businesses being the selling side. The overall market (or economy) is in equilibrium when the flow of funds through both market sectors is exactly the same. Disequilibrium results when the flows of funds through these sectors differ.

Economics Basics

Economics is an interesting and multifaceted field. Fundamentally, it is based on making the best possible decisions on how to spend our income and how best to use the limited resources we have. Because we—individually or collectively (as a society)—cannot have everything we desire, we must make appropriate choices and decisions. For example, agricultural producers must decide how to allocate their time between work and home. Therefore, economics as a science evolved from the collective problem of **scarcity**: the relationship between the amount of a good or service that is available and the amount of that item that is truly desired. If the desire or need for an item exceeds the availability of that item, the item is considered scarce.[5]

Scarcity of Resources

For the purposes of this text, **resources** are defined as inputs provided by nature that are transformed by human labor and technology to produce goods and services. The resources needed for production are called *factors of production*. Because resources themselves are scarce, many goods and services are also scarce. Therefore, choices must be made among alternative uses of those scarce resources:

- *Alternative uses in production*—example: using land for soybean rather than oat production
- *Alternative uses in consumption*—example: using native grassland for conservation purposes rather than as pastureland

5. Ibid.

- *Alternative uses regarding time*—example: working more overtime rather than having additional leisure time[6]

These alternatives allow managers to make the decisions and choices most appropriate to their operations, with profitability and efficiency generally being the primary goals.

Basic Economic Concepts

As previously mentioned, the field of economics is based on making the best possible decisions in various conditions and situations. This effort includes the evaluation of alternative choices and decisions regarding the same situation. Six basic concepts underlie nearly all economic decisions, and form the basis from which economists approach decisionmaking:

1. supply and demand
2. opportunity cost
3. diminishing returns
4. marginality
5. costs and returns
6. externalities[7]

Following is a more detailed discussion of each of these concepts.

Supply and Demand. Supply and demand is both one of the best-known and one of the most misunderstood economic concepts. Together, supply and demand work to determine price within economic markets. Individually, **demand** can be defined as the quantity of a good or service that buyers are willing to purchase at different price points; **supply** is the quantity of a good or service that sellers are willing to offer, also at various price points. Both occur within a given market, over a specific period of time. A **market** can be defined as "the interaction between potential buyers and potential sellers of a good or service."[8]

Demand. The most important aspect of demand is that it is a relationship between quantities and prices: specific *quantities* will be *demanded* at different *prices.* Therefore, demand can be viewed as a

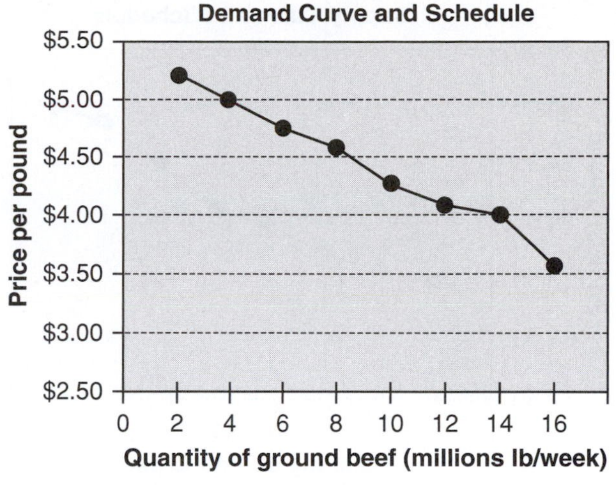

Figure 12–3 A demand curve (a) is a graphical illustration of the points with a demand relationship. A demand schedule (b) is a table outlining the various price and quantity combinations within a specific demand relationship.

two-dimensional concept, with the relationship between the two dimensions of quantity and price being demand. *Demand* itself refers to buyers' specific desires or intentions (not necessarily the purchases they actually make). A **demand schedule** is a table outlining the various price and quantity combinations that exist within a specific demand relationship. Finally, a **demand curve** is a graphical illustration of the points within a demand relationship.[9] Both of these aspects are clearly illustrated in Figure 12–3.

Supply. Like demand, *supply* also describes a relationship between quantity and price. In this case, however, *quantity* refers to the amount sellers are willing to offer, at various prices. Again, supply describes a relationship, not a specific amount or quantity. The **quantity supplied** is the amount of a good or service a supplier is willing to provide at a particular price. Supply itself is the relationship between that quantity and the correlated price. Supply can also be detailed through supply schedules and supply curves, both of which illustrate individual correlating points of quantity and price within the supply relationship.[10] Refer to Figure 12–4.

Price determination. The question remains: "How are quantity and price determined through supply and demand?" This is where the interaction

6. Ibid.
7. H. Evan Drummond and John W. Goodwin, *Agricultural Economics* (Upper Saddle River, NJ: Prentice-Hall, 2001), pp. ix–xii.
8. Ibid., pp. 26, 27–28.
9. Ibid., pp. 27–29.
10. Ibid., p. 29.

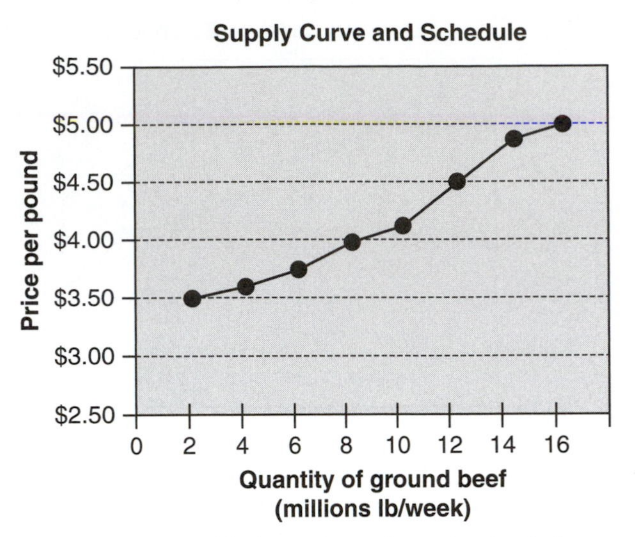

Figure 12–4 As with demand, supply can be detailed through a supply schedule and supply curve. Both illustrate individual correlating points of quantity and price in the supply relationship.

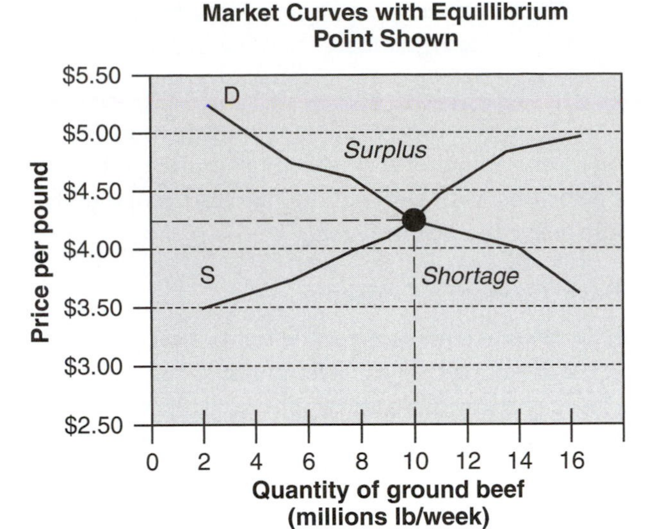

Figure 12–5 If you overlay the demand curve with the supply curve for the same product, you can find the point of equilibrium at the spot where the curves intersect.

between supply and demand becomes fundamental. As long as a perfectly competitive market is allowed to operate under no controls or regulation, the price will adjust to the point where quantity demanded equals quantity supplied—the **equilibrium price**. If you superimpose the demand curve on the supply curve for the same product, the point where they intersect is considered the **point of equilibrium**. At this price and quantity, the amount supplied equals the amount demanded. Anything below this point (on the graph) is considered a shortage of the good, whereas anything above the point of equilibrium is regarded as a surplus.[11] Refer to Figure 12-5 for further illustration of these points.

Elasticity. A final and essential consideration in the study of supply and demand is the concept of **elasticity**, which is the price responsiveness of an item. When the supply of or demand for a good is very price responsive, it is considered to be *elastic*. More specifically, with regard to a good for which the supply-demand relationship is very elastic, a small change in price will bring about a reasonably large response in the quantity either supplied or demanded. Conversely, with regard to a good for which the relationship is considered to be *inelastic*, a large change in price will bring about a relatively small response in the quantity supplied or demanded.[12]

Opportunity Cost. A second fundamental concept of economics is that of opportunity cost. In any situation, choices must be made regarding the use of scarce resources. When a resource is used for one activity, it cannot be used for another. The value of what is given up (or decided against) is considered the **opportunity cost** of the chosen activity. Technically, *opportunity cost* can be defined as a measure of how much of an earning opportunity is relinquished by using a resource in its current employment.[13]

This can be effectively illustrated through an example. Agricultural producers often have to make choices regarding allotment of time, which is a valuable, limited resource. Does a producer choose to spend extra time working to make the agribusiness more efficient, or does he or she spend it with her family? The opportunity cost is what is lost by the decision that is made—either production time or family time. Often opportunity cost cannot be translated into a specific dollar

11. Ibid., pp. 31–33.

12. Ibid., pp. 36–37.
13. Ibid., p. x.

amount, but it may nevertheless have a significant impact on the situation and the choices made.

Diminishing Returns. A third fundamental concept within the field of economics is that of diminishing returns. As you add more of something while holding everything else constant, the added benefit received from each additional unit input will eventually begin to decline. When the benefits begin to decrease, you have reached the point of *diminishing returns.* Eventually, you could reach a point where you get no added benefit from adding additional units. A good example of this is eating a pizza. The first piece is mouth-watering, and the second is also good, but as you begin to fill up you tend to get less satisfaction out of each subsequent piece consumed. Sooner or later you become full, and cannot consume any more pizza. This is the point at which another piece has no added benefit for you. This concept applies in a variety of settings: eating, sleeping, working out, and even fertilizing crops all are subject to the rule of diminishing returns.[14]

Marginality. It has been said that "in economics, all the action is on the margin."[15] In economics, the **margin** is an additional or incremental unit of something. It is the marginality of something that often influences decisionmaking. Economic theory relating to the agribusiness is based on marginal revenues and marginal costs. If you look at economic theory from the viewpoint of the consumer, it is based on the marginal utility (use) and marginal cost of a good. Many investment decisions are based on marginal returns versus marginal costs. In each of these cases, marginality provides a measurement by which to guide effective decisionmaking. This is what makes marginality one of the fundamental concepts of economics.[16]

Costs and Returns. As previously introduced in relation to both marginality and opportunity cost, costs and returns are another cornerstone of economic theory. Much of economics deals with the tradeoff between costs and returns. Taken separately, determining costs in a given situation is relatively straightforward (particularly when the concept of opportunity cost is understood). In contrast, determining the returns in the same situation can be a bit more complex. Depending on the situation, you can have a variety of terms—returns, benefits, utility, etc.—and accurate measurement of these terms requires making some assumptions. Nonetheless, finding the bottom line of any economic situation requires looking at the costs and returns of alternative decisions and determining which decision will most effectively reach the desired goal.[17]

Externalities. Finally, externalities are an economic concept that must be considered, but generally cannot be captured or realized. Often, these have to do with the context of (the environment surrounding) the economic situation. Much of the study in economics focuses on transactions; in any transactions, the parties do not usually bear all of the costs or receive all of the returns associated with that transaction. These forgone costs and returns are considered **externalities** to the situation.[18]

An excellent example of this is taxes. In certain instances, agribusiness operators can purchase specific farm items tax-free. Thus, the government absorbs some of the costs of such a transaction for the operator. However, when the same operator markets and sells his goods within the economy, the taxes added on to the price of the good go directly back to the government, imparting some of the returns of that sale transaction to the government.

Given these basics, you can see that it is important to consider all aspects of any specific economic situation.

The Production Function

As one of the most basic concepts of economic theory, a **production function** can be defined simply as a relationship summarizing the process of the conversion of inputs into a particular output. This relationship is determined by the specific

14. Ibid.
15. Drummond and Goodwin, *Agricultural Economics,* p. xi.
16. Ibid., pp. x–xi.
17. Ibid., p. xi.
18. Ibid.

technology used in the production process. One of the most straightforward production functions is called the "factor-product" model because one variable factor works to produce one product. Within this model, the production function describes a single variable input and a single product working with any combination of other inputs that are fixed or preset. For example, let's say that fertilizer is the variable input, land is the fixed input, and soybeans are the product. The production function describes the relationship between the amount of fertilizer (variable input) used per acre and the quantity of soybeans produced (product or output).

Factors of Production

Specifically, four types of factors are involved in the production process:

1. *Land*—the physical earth (or soils) and its mineral deposits
2. *Labor*—the services or physical energy provided by workers
3. *Capital*—the machinery, equipment, buildings, inventories, and financial assets invested in production
4. *Management*—the entrepreneurial and decision-making skills necessary for effective production[19]

Each of these factors plays an important role in the production process of any agribusiness.

Inputs

As noted, inputs are important in the production function. An *input* is a material or good used in creating a final product. Inputs help to determine the fixed and variable costs associated with the production of goods and services. As with other agribusiness concepts, there are both fixed and variable inputs. *Fixed inputs* exist regardless of production. These are available to a farm manager in a set amount for the production period. Agricultural fixed inputs may include things such as buildings, land, machinery, taxes, and even insurance. *Variable inputs* are inputs over which the farm operator has control; they change according to the amount of production. By adjusting the amount of variable inputs used, the operator can work on maximizing profits. Examples of variable agricultural inputs include feed, fertilizer, seed, chemicals, fuel, and repairs. A cost comparison is used to determine the most efficient inputs to use in a given production situation.[20]

Outputs

The other component in the production relationship is the *output,* or the good or service produced. The amount of output produced is the result of varying one (or sometimes more) of the input amounts used. Extending our example from the preceding section, the amount of soybeans produced would be the product (output) that results from varying the amount of fertilizer used. To determine the cost of production, use the following generalized equation:

Fixed Costs + Variable Costs = Total Output Costs

Production Equations

Several key components work together to mathematically describe production relationships. The following are some of the fundamental relationships used in production equations.

Total Product. The direct relationship between inputs and outputs can be effectively described using common sense. If you begin with a small amount of fertilizer per acre, that acre can be expected to produce a small amount of soybeans per acre. As you increase the amount of fertilizer used per acre, the amount of soybeans produced per acre will also increase. This relationship between fertilizer/acre and soybeans/acre will continue to increase until the point at which an additional unit of fertilizer may burn or harm the soybean plants; that is, when an additional unit of fertilizer will actually assist in producing fewer soybeans per acre. At this point, the total product curve levels off and begins to decline. Specifically, **total product (TP)** is the level of output (yield) produced within the production process corresponding to the amount of the variable input added. Overall, TP illustrates the relationship between a variable input and output. This relationship is illustrated in Figure 12-6.

19. Hanson, *Agricultural Economics 141*, p. 1-2.

20. Deere and Company, *Farm and Ranch Business Management* (Moline, IL: Author, 1992).

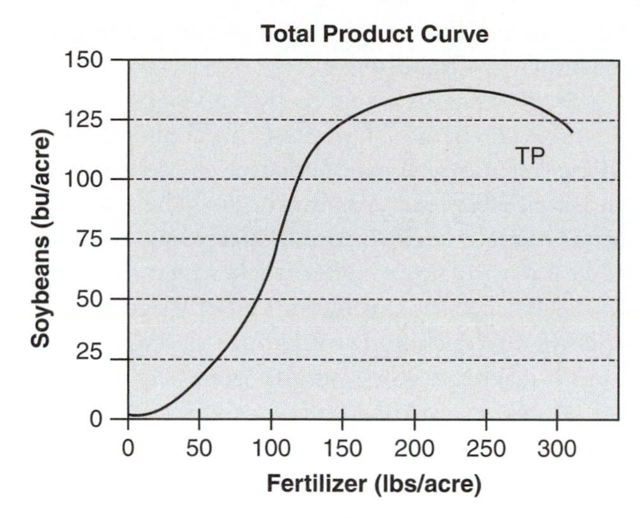

Figure 12-6 The relationship between fertilizer/acre and soybeans/acre will continue to increase until the point where an additional unit of fertilizer may burn or harm the soybean plants.

As you can see from this curve, as fertilizer/acre increases, soybeans/acre initially increases at an increasing rate. After a specific point (where the marginal product peaks—to be explained shortly), soybeans/acre increases at a decreasing rate, and finally the function reaches a maximum point, beyond which it decreases. This is typical of a normal production function.

Average Product. Another component of the production process, **average product (AP)**, describes the amount of output produced by each individual unit of the variable input in the production process. This amount can be found by dividing the TP by the number of input units corresponding to the level of production.

$$AP = \frac{TP}{\#\text{ input units}}$$

Marginal Product. The preceding discussion not only illustrates the important concept of total product, but also introduces us to **marginal product (MP)**. As defined earlier, *marginal*, as used in the field of economics, refers to an additional unit. When discussing marginal product, we are referring to the added production (i.e., increases in yield) associated with each unit increase in variable input. This helps the producer to answer the question, "If I added one more unit of input, how much additional output would I receive?" For our preceding example, the question would become, "If I added one more unit of fertilizer per acre, how much would my soybean production per acre increase?"[21]

Other Production Equations. Two other important considerations for the production process are marginal factor cost and marginal value product. **Marginal factor cost (MFC)** is the additional cost of adding one more unit of a variable input to the production process. To calculate the MFC, divide the change in total cost by the change in number of input units employed. Conversely, the **marginal value product (MVP)** demonstrates the change in total returns received through adding one more unit of input. Explicitly, MVP is the additional value received from the production process through adding one more variable input unit to the production process.

$$MFC = \frac{\Delta\text{ TC}}{\Delta\,\#\text{ input units}} \qquad MVP = \frac{\Delta\text{ total revenue}}{\Delta\,\#\text{ input units}}$$

These equations work together to assist the producer in making decisions regarding the production process. To determine the most appropriate amount of input in relation to the individual production situation, the principle of diminishing returns-input basis is utilized.

Three Stages of Production

Before we address the principle of diminishing returns-input basis, though, it is important to understand how all the components work together within the production process. To do this, we will further examine the relationships between the TP, AP, and MP curves, within the various stages of production. There are three specific stages of the production function:

Stage I = Increasing marginal returns
Stage II = Decreasing marginal returns
Stage III = Negative marginal returns

These can be identified according to the relationship between the TP curve and the MP curve. Refer to Figure 12-7 for an illustration.

Stage I. The process begins with Stage I, in which there are relatively low levels of input use but the

21. Drummond and Goodwin, *Agricultural Economics*, p. 202.

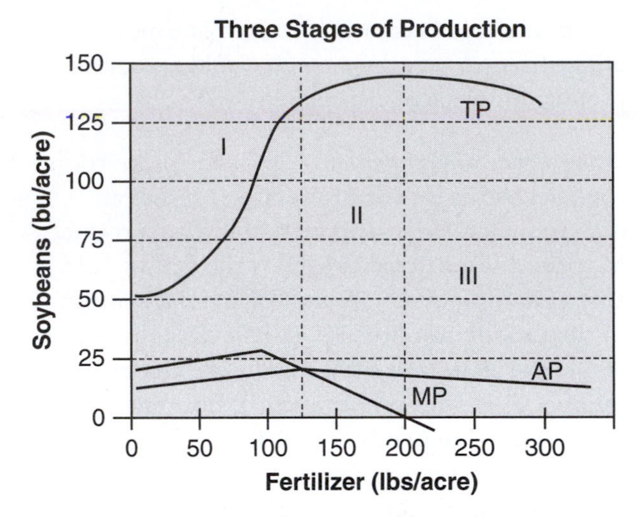

Figure 12-7 This diagram illustrates the three stages of production: Stage I, increasing marginal returns; Stage II, decreasing marginal returns; and Stage III, negative marginal returns.

output rises at an increasing rate. This stage is considered an irrational stage of production, as using more of the variable input results in more output being produced. The AP continues to increase as more variable input units are added. Production does not take place during this stage, as the level of maximum input efficiency (peak point of the AP curve) has not yet been reached—production (consideration) always begins at the point of maximum input efficiency in the production process.[22]

Stage I begins where zero units of the variable input are added and extends to the point where the AP curve reaches its maximum. It is at this point (the peak of the AP curve) that the marginal product curve intersects the average production curve. Individually, the TP curve first increases at an increasing rate; it then increases at a decreasing rate. The AP curve increases throughout this stage until it reaches its maximum point; the marginal product curve first increases to its peak, then begins to decrease. Increasing marginal returns are realized throughout this stage.

Stage II. The second stage of production begins once the maximum point of the AP curve is reached. This is considered the rational stage of production, the stage where production always occurs. During this stage, the producer decides how much of the variable input to add in order to maximize profitability.

Stage II begins where the MP and AP curves intersect (at the AP curve peak), and culminates where the marginal product curve equals zero and the TP curve is at its maximum. The total product curve itself continues increasing, though at a decreasing rate, until it reaches its maximum level. AP decreases throughout all of Stage II, and MP also declines throughout this stage, until it reaches zero. As an important side note, MP equals zero at the same point where TP is at its highest point (maximum) within the production process. Decreasing marginal returns are realized within this stage.

Stage III. Within this final stage, even though more units of input may be added, the output (TP) has reached its maximum level and begins to decrease. Therefore, there is no benefit to adding additional input units. This is also considered an irrational stage of production, as TP is decreasing and MP is actually negative.

This stage begins where the MP curve is zero and becomes negative, and the TP curve peaks and begins decreasing—something known as *negative marginal returns*. AP also continues to decline. Net returns within the production process can never be maximized during this stage, even if the input is completely free.

Principle of Diminishing Returns–Input Basis

The **principle of diminishing returns–input basis** takes into consideration the relationship among product curves, and assists producers in determining the most effective production level for their operations. By definition, this principle is concerned with varying the amount of one input while keeping all other inputs constant within the production process. A good example of this was introduced in the preceding section: Different amounts of fertilizer can be applied in soybean production while keeping the amounts of seed, herbicide, water, labor, and machinery constant. With the continued addition of a variable input, eventually a decline in output will be experienced. As illustrated through the different stages of production, as inputs continue to be added, TP will first increase at an increasing rate; then it will increase at a decreasing rate; and, finally, total

22. Hanson, *Agricultural Economics 141*, p. 4-2.

product will actually decrease (i.e., total yield per acre will decrease) if further inputs are added to the production process.[23]

Purpose of Diminishing Returns. The purpose of exploring the different stages of production as well as the principle of diminishing returns-input basis is relatively simple: to maximize net returns above the cost of the input (that is, to maximize profitability). By focusing on one variable, and holding all others constant, the producer can determine the amount of input to add to maximize the net returns realized.

A decision rule for this economic principle involves comparing marginal value product to marginal factor cost. Net returns above the cost of the input will be maximized when the added return from using one more additional unit of input (MVP) is equal to or just greater than the added cost of adding one more unit of input.[24]

Decision Rule: MVP $\geq$ MFC

Additional input units will continue to be added to the production process as long as this rule holds true, because the added return from using an additional unit of input is greater than the added cost of using that same input. If this equation become untrue, net returns would begin decreasing. As a result, production would have to be reduced (and less variable inputs added) until MVP and MFC were once again equal.

Returns to Scale

Returns to scale describes production efficiency in terms of an agribusiness's input and output. As addressed later in the chapter (in the section on the cost function), in the long run, all factors (and costs) of production are variable. Therefore, taking returns to scale into consideration is an important step for the producer who is choosing the optimal level of inputs into the production process.

1. *Increasing returns to scale:* As inputs are increased by a specific proportion, production (output) increases by a greater proportion.
2. *Decreasing returns to scale:* As inputs are increased by a specific proportion, output increases by a lesser proportion. At the point where the production process begins showing decreasing returns to scale, the business becomes less efficient.
3. *Constant returns to scale:* As inputs are increased by a given proportion, production increases by the same proportion.[25]

Each of these concepts and definitions is used to describe how efficiently the production process is performing within the agribusiness. Producers can use returns-to-scale figures to adjust the amounts of inputs they are employing to make production more efficient—and result in a higher profit.

Production Efficiency. When a production process can be made to yield increasing economy of scale, you would expect larger agribusinesses to drive smaller ones out of business, as they can produce more output per unit of input. **Efficiency** is defined as this ratio of output per unit of input. Looking at the larger picture, industries (collections of agribusinesses/firms) with increasing returns to scale would most likely be composed of a few larger, more efficient firms, rather than several small, less efficient ones. With constant returns to scale, both large and small agribusinesses are equally efficient and can peacefully coexist. Studies have revealed that U.S. agriculture, with the exception of extremely small farms, is characterized by constant returns to scale.[26]

The Cost Function

Like the production function, the **cost function** plays an important role in production economics. Specifically, the cost function provides valuable information that an agribusiness manager needs to determine the level of output for profit maximization. To accurately illustrate this function, we will break costs down into short-run and long-run costs.

Short-Run Costs

Short-run costs describes a function that illustrates the minimum possible costs of producing at each level of output, when variable factors are used

23. Ibid., p. 4–1.
24. Ibid.
25. Drummond and Goodwin, *Agricultural Economics*, p. 200.
26. Ibid.

in a cost-minimizing manner. This time frame is often used by producers for planning purposes, to characterize some inputs as fixed inputs. Short-run costs include total costs, average total costs, and marginal costs.

Total Costs. Whether they are fixed or variable, inputs have costs. The **total cost (TC)** of production in the short run includes both fixed and variable costs. Like the variable and fixed inputs described earlier, **total fixed costs (TFC)** are those that do not change according to output—they remain the same regardless of the level of production. Total fixed costs include machinery costs, many utility costs, and depreciation. **Total variable costs (TVC)** are costs that change according to the level of production, such as those inputs that vary with the level of output (e.g., cost of seed, fertilizer, etc.). Q is the notation for the number of units of output.

$$\text{ATC} = \text{AFC} + \text{AVC} \quad \textbf{OR} \quad \text{ATC} = \frac{\text{TC (Q)}}{\text{Q}}$$

Average Total Costs. Average total costs (ATC) include two different factors: average fixed costs (AFC) and average variable costs (AVC).

- *Average Fixed Costs:* TFC divided by the number of units of output.

$$\text{AFC} = \frac{\text{TFC}}{\text{Q}}$$

- *Average Variable Costs:* **TVC divided by the number of units of output.**

$$\text{AVC} = \frac{\text{TVC}}{\text{Q}}$$

Marginal Costs. **Marginal costs (MC)** are considered to be the costs of producing each additional unit of output.

$$\text{MC} = \frac{\Delta\text{VC}}{\Delta\text{Q}}$$

Relationships between Costs. To accurately illustrate the association between costs, relationships must be found among all of the aforementioned costs. Given that fixed costs (TFC, AFC) do not vary according to the output, as the amount produced increases, fixed costs decline. When output increases, both AVC and MC initially decline, then reach a minimum and begin to increase. Finally, the MC curve intersects the ATC and AVC curves at their minimum points. Refer to Figure 12–8.

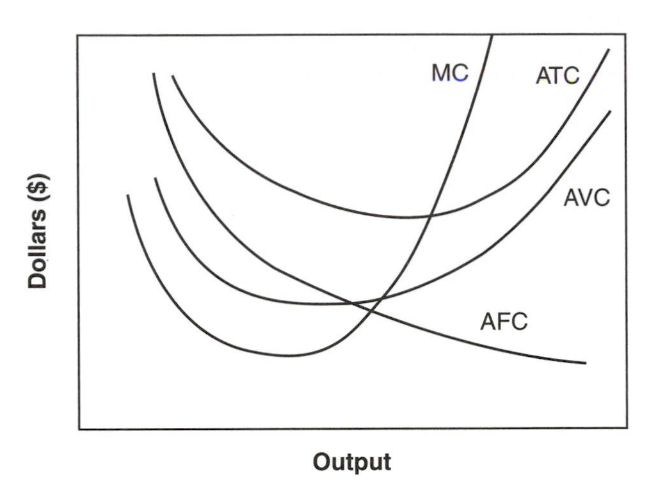

Figure 12–8 This diagram illustrates the relationship between marginal cost (MC), average total cost (ATC), average variable cost (AVC), and average fixed cost (AFC).

Long-Run Average Costs

In contrast to short-run costs (where certain inputs can be considered fixed), in the long run all costs are considered variable, as the producer can adjust the level of inputs. Thus, long-run average costs is another function that assists the producer in making production decisions. Specifically, a graph of these costs is a curve that defines the minimum average cost of producing at different levels of output. The long-run average cost curve allows the producer to choose the optimal level of fixed and variable factors of production.[27]

Economies of Scale. Directly related to the concept of returns to scale (discussed earlier), the concept of **economies of scale** also addresses efficiency, but from a slightly different angle. Whereas returns to scale look to adjust efficiency according to the amount of inputs used in the production process, economies of scale look at production efficiency from the cost perspective. The link between the two concepts lies in the obvious fact that most inputs have a cost, which inevitably affects the overall profitability of the business. The long-run average

27. "The Cost Function"; available at http://www.people.memphis.edu/-vfarber/manecon9.htm (accessed February 21, 2006; no longer available at this URL).

cost curve brings three important aspects into consideration:

1. *Economies of scale:* Initially, an expansion allows the agribusiness to produce at a lower long-run average cost. Increasing the size of the business makes it continuously more efficient, thus decreasing the minimum average cost.
2. *Diseconomies of scale:* At a certain point, additional increases to output lead to an increase in average cost. At this point, the agribusiness starts to become less efficient.
3. *Constant returns to scale:* This occurs when long-run average costs remain constant, even as output is increased.

Factors of economies of scale. Two of the most important factors when working to achieve economies of scale are improving the division of labor and specialization within a business. Other factors that can be considered are lower input costs (buying in bulk), more efficient use of costly or specialized inputs, and more effective use of organizational and learning inputs. As tasks are performed better and more efficiently, time and money are saved while production increases. Therefore, as an agribusiness grows and production units increase, the business may have a greater opportunity to decrease its costs. Often when economies of scale are realized within a business, economic growth is also achieved.[28]

Economies of scale in agriculture. As agriculture has progressed through the decades, the size of the American farm has become larger and larger. When more units of a good can be produced on a larger scale—with less input cost (and thus greater overall efficiency)—economies of scale are said to have been attained. However, diseconomies of scale also exist. Diseconomies may arise when the inputs are out of proportion to a business's production. This may be caused by things such as overhiring or deteriorating transportation networks, for example. Usually this is an indication of inefficiencies within the business, or rising average costs of production.

To apply this theory to practice, we will use an example of swine manure utilization. Refer to Figure 12–9. Within most of today's modern swine finishing facilities, producers look for ways to maximize profits by achieving economies of scale, and thus minimize the costs of production. With larger operations, producers can spread fixed costs over a greater number of animals and take advantage of technologies that increase production efficiency. More advanced systems (generally used only by larger producers) help in minimizing the cost of manure storage and handling. Another factor in successful economies of scale in this example is concentrating the swine finishing facilities at one site; this helps to reduce feed storage, manure storage, and infrastructure costs. With these techniques, overall production costs can be reduced, improved economies of scale can be achieved, and, ultimately, the producer can reap greater economic profit.[29]

Resource Substitution

The preceding sections have focused on the variability of only one input, while keeping the other inputs in the production process fixed. This was done primarily to simplify complex production equations and to help you understand how producers use them to make informed decisions. However, while appropriate for study and useful in theory, real life offers few opportunities to vary only a single input. In reality, all inputs are variable, so agribusinesses can substitute between inputs. The **principle of resource substitution** addresses two or more variable inputs within the production equation. Specifically, the purpose of this principle is to determine the combination of two or more variable inputs (that can be substituted for each other in varying amounts) that will produce a specific amount of a given product with the least cost of production.[30]

If we have two variable inputs (X_1 and X_2), there are four different combinations that can be substituted for each other:

- Use all of X_1
- Use all of X_2
- Use more of X_1 and less of X_2
- Use more of X_2 and less of X_1

28. Reem Heakal, "What are Economies of Scale?" (January 27, 2003); available at http://www.investopedia.com/articles/03/012703.asp (accessed August 27, 2007).

29. Raymond E. Massey, John A. Lory, John Hoehne, and Charles Fulhage, "Economies of Scale in Swine Manure Utilization"; available at http://www.p2pays.org/ref/21/20986.htm (accessed August 27, 2007).

30. Hanson, *Agricultural Economics 141*, p. 5–1.

Figure 12–9 This state-of-the-art swine facility is completely automated and temperature controlled. (Courtesy of USDA, #NRCSGA 02039)

Using the principle of resource substitution, the producer can answer the question, "What combination of inputs would be most useful within the production process?" In the agricultural context, this can be applied in numerous situations. A good example is using alfalfa and corn to feed feeder calves. What combination of these two inputs would be the least expensive (least cost) feed ration for feeder calves to produce 150 pounds of weight gain?

Marginal Rate of Substitution

An equation to assist in making decisions regarding resource substitution is the **marginal rate of substitution (MRS)**. By definition, MRS is the amount of one input that is replaced by an additional unit of another input; this can be illustrated through the following equation:

$$\text{MRS} = \frac{\Delta X_2}{\Delta X_1} = \frac{\Delta \text{ Input Replaced}}{\Delta \text{ Input Added}}$$

Isoquant Curve

The **isoquant curve** is a graphical illustration of the principle of resource substitution. It illustrates the set of all pairs of inputs (X_1 and X_2) that can be used to produce a specific output (Y). Additionally, it indicates the amount of one input that can be replaced by another input, while sustaining the same level of output. In general, the further an isoquant curve is from the origin, the higher the level of production. Isoquant curves do not cross, and they always slope downward.

Types of Substitutes. The shape of the isoquant curve depends on the substitutability of the inputs. There are three types of substitutes that can effectively illustrate resource substitution: perfect substitutes, perfect complements, and imperfect substitutes. If inputs are **perfect substitutes**, isoquants are straight lines; if they are **perfect complements**, the "curve" is L-shaped.

When one has a *perfect substitute,* one input (X_1) always replaces a consistent amount of another input (X_2). For example, two pounds of alfalfa always replace one pound of corn. In this situation, the MRS between these two inputs is always constant (which is why this isoquant curve is really a straight line). Refer to Figure 12–10 for an illustration.

With a *perfect complement,* adding more of one input (X_1) will not change or replace the amount of the other input (X_2) used. The opposite of perfect substitutes, perfect complements exist when inputs cannot be substituted for each other. As the L-shaped graph shows, it takes a certain amount of each input to produce the required output. Thus, MRS between these inputs is always zero. Refer to Figure 12–11 for an illustration.

Imperfect substitutes, illustrated by the final isoquant curve in Figure 12–12, exist when additional units of one input replace less and less of the other input (resulting in a negatively sloping graph). For example, as more corn is added to the overall ration, it replaces less alfalfa. The MRS within this situation is always decreasing.

Isocost Line

Another aspect to address when considering the principle of resource substitution is the **isocost line**. This line takes into consideration the price or cost of the inputs added and replaced. Specifically, it illustrates the different combinations of two inputs that can be purchased with a specific amount of money.

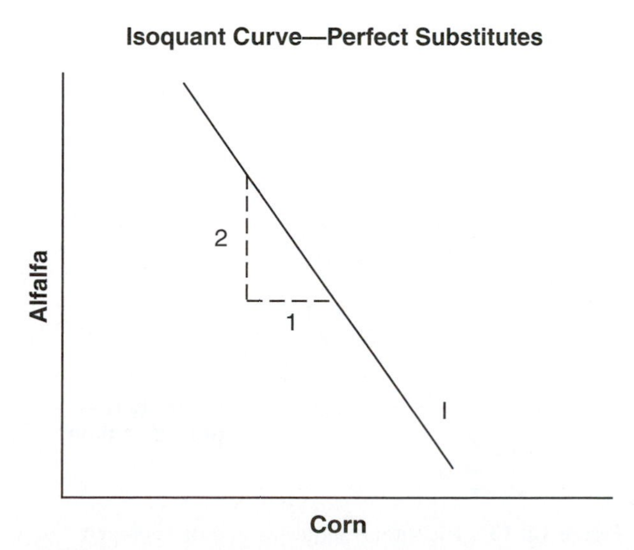

Figure 12–10 With a perfect substitute, one input (here X_1, two pounds of alfalfa) always replaces a consistent amount of another input (here X_2, one pound of corn).

The isocost line also shows the amount of one input that would cost the same if another unit of the other input was purchased. The slope of this line can be illustrated through the following price ratio:

$$\text{Price Ratio} = \frac{\text{Price Input Added}}{\text{Price Input Replaced}} = \frac{P_1}{P_2}$$

Extending our previous example, let's assume that the price of alfalfa (P_2) is \$75 per ton, and that the price of corn (P_1) is \$60 per ton.

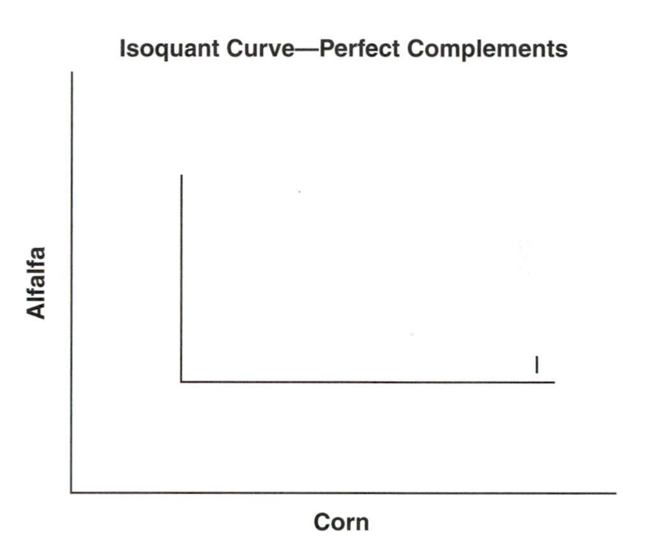

Figure 12–11 Perfect complements are the direct opposite of perfect substitutes. These inputs cannot be substituted for each other. Thus, as the L-shaped graph illustrates, it takes a certain amount of each input to produce the required output.

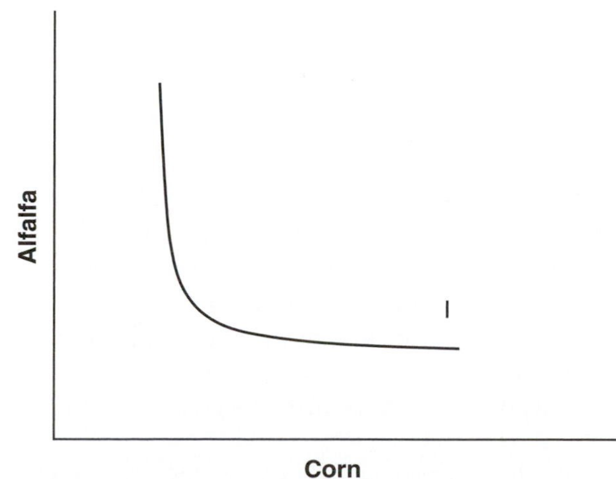

Figure 12–12 With an imperfect substitute, additional units of one input (here, corn) replace less and less of the other input (here, alfalfa), resulting in a negatively sloping graph.

Alfalfa = \$75/ton (3.7 cents/lb)
Corn = \$60/ton (3 cents/lb)

One pound of corn costs the same as 0.81 pound of alfalfa.

One pound of alfalfa costs the same as 1.23 pounds of corn.

$$\text{Price Ratio} = \frac{P_1}{P_2} = \frac{.03}{.037} = 0.81$$

Therefore, if a cattle feeder had \$50 to spend on ration to produce 100 pounds of grain for feeder calves, the feeder could feed 200 pounds of alfalfa at the same cost of feeding 246 pounds of corn.

Least-Cost Combination of Inputs

So, how do we decide what combination of inputs is most efficient in a particular production process? The **least-cost combination of inputs provides the decision rule for producers to work with:**

Decision Rule: MRS ≥ Price Ratio

$$\frac{\Delta X_2}{\Delta X_1} \geq \frac{\Delta P_1}{\Delta P_2}$$

Cost of Input Replaced ≥ Cost of Input Added

According to this rule, the producer should continue to add more of one input (X_1) to replace another input (X_2) as long as the cost of the input being replaced (cost saved by replacing X_2) is greater than or equal to the cost of the input being added (cost added by adding X_1).

Profit Maximization

Regardless of all the production considerations, variables, and equations discussed in this chapter, the bottom line remains how best to answer one question: "How do I maximize my profits as an agribusiness operator?" **Profit maximization** is one of the single most important goals of any operation. By definition, profit maximization is the process by which a firm determines the price and output level that returns the greatest profit.[31]

31. Wikipedia, "Profit Maximization"; available at http://en.wikipedia.org/wiki/Profit_maximization (accessed August 27, 2007).

Though there are several approaches to this issue, two of the most useful are the **total revenue–total cost** method and the **marginal revenue–marginal cost** method.

The Totals Approach

The total revenue–total cost approach is based on the fact that total revenue minus total cost equals profit: *TR − TC = Profit*. By plotting data, we can illustrate the relationships between revenue, cost, and profit more effectively. Refer to Figure 12–13 for a visual representation of these relationships.

To find the profit-maximizing output, initially we look for the amount of output (quantity) where the profit reaches its maximum. At this point, the output is represented by Q_1 on the graph; the peak point (or maximum) of the profit curve is represented by A.

Graphically, Q_1 is determined to be the optimal point of production in two different ways. As mentioned previously, we see that the profit curve is at its maximum (A). In addition, at point B, the tangent of the TC curve is parallel to the TR curve, showing us that the surplus of revenue net of costs (the line between B and C) is the greatest. Because profit equals total revenue minus total costs, the line from B to C should equal the line between A and Q_1.

Profit Maximization—The Totals Approach

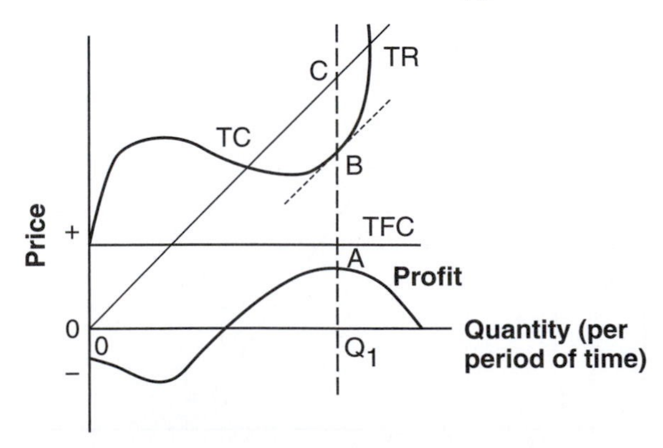

Figure 12–13 The total revenue–total cost approach is defined by the fact that total revenue minus total cost equals profit: TR − TC = profit. Profit maximization is represented by profit curve A.

The Marginal Approach

In some situations, total revenue and cost figures are difficult to obtain. In these situations, it is often easier to use the marginal cost-marginal revenue method. This method is based on the premise that for each unit sold, marginal profit equals marginal revenue minus marginal cost: *MR − MC = Marginal Profit*. This equation shows that if marginal revenue is greater than marginal cost, (marginal) profit is negative. When MR and MC are equal, marginal profit is zero. Finally, if marginal revenue is greater than marginal cost, marginal profit is positive. This is best explained and illustrated graphically; refer to Figure 12-14.

It is important to locate the intersection of the marginal revenue (MR) line and the marginal cost (MC) curve on this graph. This point, denoted by A, is the point at which MR is equal to MC. This is critical, because when marginal revenue equals marginal cost, marginal profit is zero. Because total profit increases when marginal profit is positive and decreases when marginal profit is negative, total profit must reach a maximum where marginal profit is zero—that is, where MR and MC intersect (point A). Total economic profits are represented by the area outlined by points A, B, C, and D. The optimum quantity is the same as in Figure 12-13 and is again designated Q_1.

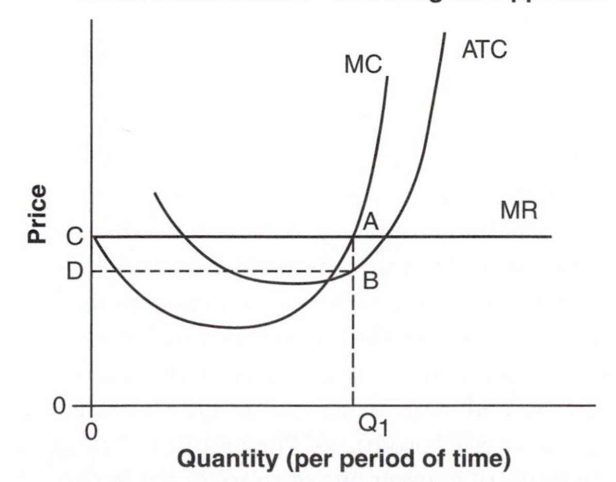

Figure 12-14 Because the total profit increases when marginal profit is positive and decreases when marginal profit is negative, total profit must reach a maximum where marginal profit is zero; that is, where MR and MC intersect (point A).

Conclusion

Production economics plays an important role within society, and is particularly important to today's farm managers and producers. Effective farm managers must take a wide variety of factors into account when planning for the success of their agribusinesses. Production economics provides the tools to achieve this success. Production economic basics, knowledge of the different market sectors of the economy, and an understanding of how the production and cost functions operate are all important to the manager and producer. The principle of resource substitution provides a framework within which producers can consider alternative inputs to their production processes. Finally, they can use the profit maximization framework to determine the profitability of an agribusiness—the bottom line for any manager.

Summary

Production economics plays an important role in society, and is particularly relevant in the agricultural domain. Economics can be defined generally as how consumers, producers, and societies choose to use scarce resources in production, exchange, and consumer situations—each very important to the process of production agriculture. More specifically, the study of economics includes examining how prices are determined for goods and services, how price signals are communicated between buyers and sellers, how institutions can efficiently facilitate the operation of markets, and the appropriate role of government, among many other considerations.

The U.S. economy and society are capitalistic. The United States uses the marketplace to determine prices for production resources as well as for the goods to be sold or traded. The U.S. food and fiber system is comprised of all business units that are involved in the production, processing, and distribution of food and fiber within the country. Sometimes characterized as the movement from the "farm gate to the store shelf," this system is comprised of four major sectors: (1) the farm input supply sector, (2) the farm business sector, (3) the processing and manufacturing sector, and (4) the wholesale and retail sector.

The U.S. economy is basically constituted of two entities: households and business firms.

Households represent the consumers within our society; business firms, which may be organized as proprietorships, partnerships, corporations, or cooperatives, are the producers. Both groups play an important role in the product and resource market sectors. The product market is the market for finished goods, in which finished goods flow from producer to consumer. The resource market represents the movement of economic resources from consumers to producers. In general, the market is in equilibrium when the flow of funds through both markets is exactly the same.

Several economic basics help us to define the field of economics and apply its theories. Because the resources needed for production are scarce, many goods and services are also scarce. Therefore, agriculturalists need to decide between alternative uses regarding production, consumption, and time. These alternatives allow managers to make decisions most appropriate to their situation. Other considerations (and basic economic concepts) include supply and demand, opportunity cost, diminishing returns, marginality, costs and returns, and externalities. Each of these aspects also plays an important role within economics.

Supply and demand—Supply and demand determine price within economic markets. Demand is the quantity of a good or service that buyers are willing to purchase; supply is the quantity of a good or service that sellers are willing to offer. Within a perfectly competitive market, the price will adjust to the point where quantity demanded equals quantity supplied, also called the equilibrium price. By overlaying the supply curve with the demand curve, you discover the point of equilibrium (the point where the curves intersect). The concept of supply and demand also involves elasticity, which is the price responsiveness of an item. A good that has high price responsiveness is considered very elastic; with low price responsiveness, the good is called inelastic. For a very elastic item, a small change in price will bring about a large change in the quantity supplied or demanded. For an inelastic item, the opposite is true.

Opportunity cost—Opportunity cost refers to the value of what is given up when a resource is employed for one activity and not another.

Diminishing returns—As you add more and more of something while holding everything else constant, the added benefit in returns received from each additional unit will eventually begin to decline. The point at which the benefits begin to decrease is defined as the point of diminishing returns.

Marginality—In economics, the margin is an additional or incremental unit of something. A large part of economic theory is based on marginal revenues and marginal costs. Thus, marginality provides a measurement to aid in effective decisionmaking.

Costs and returns—Costs and returns help to show the bottom line for the producer. Usually, the bottom line of any economic situation requires investigation of the costs and returns of alternative possible decisions and determining which action will more effectively reach the desired goal.

Externalities—In most economic situations, the parties generally do not bear all of the costs or receive all of the returns involved in a transaction. These factors, which affect gain and loss, are called externalities, and must be considered when making economic decisions.

A foundational component of production economic theory is the production function, which is a relationship summarizing the process of the conversion of inputs into a specific output. One of the most basic production functions is the factor-product model, which describes how one variable factor works to produce one product.

Four factors are involved in the production process: land, labor, capital, and management. Any of these factors may be an input or an output of the production process. Inputs are any material or good used in creating the final product. Fixed inputs remain the same regardless of whether production takes place; these include things such as buildings, land, and machinery. Variable inputs change according to the amount of production that takes place; these include feed, fertilizer, seed, chemicals, and many other things. Outputs are the goods or services produced. The amount of output is the result of varying one or more of the input amounts used. To determine the costs of production, the fixed and variable costs are added together; the sum equals total output costs.

To mathematically describe production relationships, several production equations are used:

Total Product (TP)—the level of output produced within the production process corresponding to the amount of variable input added.
Average Product (AP)—the amount of output produced by each individual unit of a variable input in the production process.
Marginal Product (MP)—the added production associated with each unit increase in variable input.
Marginal Factor Cost (MFC)—the additional cost of adding one more unit of a variable input to the production process.
Marginal Value Product (MVP)—demonstrates the change in total returns received by adding one more unit of input.

All of these components work together during the three stages of production. These stages are identified according to the relationships between the AP, TP, and MP curves. Stage I begins with zero units of variable input, and extends to where the AP curve reaches its maximum point. Increasing marginal returns are realized throughout this stage. Stage II begins where the MP and AP curves intersect, and ends where the TP curve is at its maximum. Decreasing marginal returns occur within this stage. During Stage III, TP has reached its maximum level and thereafter begins to decrease. At this point, there is no benefit to adding more input units, and negative marginal returns are realized. Thus, Stage II is the only rational stage of production.

The principle of diminishing returns-input basis takes the stages of production into account, and provides a standard for farm managers and producers to work from. At some point, as inputs are continually added, eventually adding one more input unit will not be useful, and will actually be detrimental to the production process. This is the definition underlying the principle of diminishing returns-input basis, and helps farm producers to maximize profitability. The decision rule for this principle is: $MVP \geq MFC$.

The concept of returns to scale also describes production efficiency in terms of input and output. Agribusinesses may experience increasing returns to scale, decreasing returns to scale, or constant returns to scale. This principle is used to describe the efficiency of the production process, and assists producers in adjusting the amount of inputs they are employing to make production more efficient.

The cost function is as important as the production function. The cost function itself provides valuable information in determining the level of output for profit maximization. This function can be broken down by time frame into short-run costs and long-run costs. Short-run costs include total costs, average total costs, and marginal costs, as follows:

Total Costs (TC)—the total costs of production in the short run; includes both fixed and variable costs.
Total Fixed Costs (TFC)—costs that do not change according to output.
Total Variable Costs (TVC)—costs that change according to the level of production.
Average Total Costs (ATC)—include both fixed and variable costs.
Average Fixed Costs (AFC)—TFC divided by the number of units of output.
Average Variable Costs (AVC)—TVC divided by the number of units of output.
Marginal Costs (MC)—the costs of producing each additional unit of output.

All of these costs are related. As production increases, fixed costs (TFC, AFC) decline. Furthermore, as output increases, both AVC and MC decline initially, reach a minimum, and then increase. Finally, the MC curve intersects the ATC and AVC curves at their minimum points.

Long-run costs are determined differently. In the long run, all costs are considered variable. The long-run average cost curve allows the producer to choose the optimal level of both fixed and variable factors of production. A concept describing long-run costs is economies of scale. With a constant return to scale, a business has long-run average costs that remain constant as output is increased. To achieve this, a producer needs to focus on improving the division of labor and specialization within the business. Other considerations include lower input costs, costly or specialized inputs, and the use of organizational or learning inputs.

The preceding concepts and decision rules are based on the factor-product model, which assumes a single input and a single output. The principle of resource substitution addresses two or more variable inputs within a production process. Some inputs can be substituted for each other in varying amounts, which can be determined by the least-cost method of production. Four different input combinations are available for substitutions: use all of X_1, use all of X_2, use more of X_1 and less of X_2, or use more of X_2 and less of X_1.

Several other concepts assist in the resource-substitution decisionmaking process. The marginal rate of substitution (MRS) is the amount of one input that is replaced by an additional unit of another input. This provides an equation for producer use. The isoquant curve is a graphical illustration of the principle of resource substitution; the curve can illustrate perfect substitutes, perfect complements, or imperfect substitutes. An isocost line shows the price or cost of the inputs to be added and replaced. For the producer, this illustrates the varying combinations of two inputs that can be purchased with a specific amount of money. Finally, the least-cost combination of inputs provides the decision rule that ties all these components together: MRS $\geq$ Price Ratio. All of these aid the producer or manager in answering the question, "What combination of inputs would be most profitable in my production process?"

A final consideration of production economics is profit maximization, the process by which a firm determines the price and output levels that return the greatest profit. The two approaches addressed in this text are the total revenue-total cost method and the marginal revenue-marginal cost method. The total revenue-total cost approach is defined by the equation TR − TC = Profit. If total revenue and cost figures are difficult to obtain, the marginal approach is used. The marginal revenue-marginal cost method is defined by the equation MR − MC = Marginal Profit. Both of these approaches allow the agribusiness producer to determine how profitable a business is currently, and how profitable it could be in the future.

End of Chapter Activities

REVIEW QUESTIONS

1. Define the Terms to Know.
2. List each equation in the Equations to Know section.
3. What are important economic considerations within a capitalistic society?
4. What four sectors constitute the U.S. food and fiber system?
5. How do the product and resource market sectors interact to create a comprehensive view of society?
6. How does scarcity affect economics?
7. What role does the market play in supply and demand?
8. How does the principle of diminishing returns work?
9. List the four main factors in a production process.
10. What is the difference between fixed and variable inputs? Provide two examples of each.
11. Outline total product (TP) through each of the three production stages.
12. Outline marginal product (MP) through each of the three production stages.
13. Outline average product (AP) through each of the three production stages.
14. How is the principle of diminishing returns-input basis applied within the production process?
15. What is the basic principle behind returns to scale as that concept is applied in an agricultural setting?
16. How is production efficiency determined?

17. Outline the various costs defined as short-run costs.
18. What is the relationship between AVC, ATC, and MC?
19. Outline the major components of long-run costs.
20. What are the major factors of economies of scale?
21. Using the two variable inputs X_1 and X_2, list the four different resource substitution combinations.
22. What role do isoquant curves play within the principle of resource substitution?
23. What is the relationship between isoquant curves and isocost lines?
24. Outline the principle of the least-cost combination of inputs, and describe how it assists in a producer's decisionmaking process.
25. Compare and contrast the total revenue–total cost method and the marginal revenue-marginal cost methods of profit maximization.
26. Besides those mentioned within the text, what are some other economic issues affecting today's producer?

FILL IN THE BLANK

1. One aspect of economics involves the study of how ________ are determined in the ____________ for goods or services.
2. The ____ is sometimes characterized as the movement of a product from the "farm gate to the store shelf."
3. The ______________ establishes the prices that regulate the quantity and quality of goods produced and consumed within an economy.
4. The overall market is in __________ when the flow of funds through both market sectors is the __________.
5. When allowing for the scarcity of resources, alternative uses in production, ____________, and ______ ________________ should be considered in relation to the operation.
6. To determine the most efficient inputs to use within a production situation, a __________________ is used.
7. Fixed costs + __________________ = Total __________________.
8. Whenever the ___________ curve is increasing at a decreasing rate, the ___________ curve must be decreasing.
9. The boundary point between Stages II and III occurs when the ___________ curve is equal to zero, and the ___________ curve is at its maximum point.
10. The decision rule for applying diminishing returns is ________ ≥ ________.
11. Production will never take place in Stage III, because _________ is decreasing as additional input units are added.
12. Production always occurs at the point of ______________________________.
13. Whenever the _______ curve is decreasing, the _______ curve will be negative.
14. Production never occurs in Stage I, because the ___________ curve continues to increase throughout this entire stage.
15. The decision rule for the principle of diminishing returns is: ___________ ≥ MFC.
16. In the ______________________, all costs are variable.

17. Lower input costs, more efficient use of costly or specialized units, and more effective use of __________ or ____________________ are important considerations when working to achieve economies of scale.
18. TFC, TVC, and MC are all considered ____________ costs.
19. Resource substitution is concerned with the variability of ___________.
20. The three types of substitutes possible in resource substitution are: ____________________, _________ complements, and imperfect ___________.

MATCHING

a. irrational
b. MFC
c. marginal value product (MVP)
d. AP
e. scarce
f. U.S. food and fiber system
g. households
h. market
i. agribusiness firms
j. resource market
k. margin
l. variable inputs
m. efficiency
n. returns
o. factor-product model
p. returns to scale
q. TP
r. perfect substitutes
s. marginal
t. isocost line
u. constant returns to scale
v. isoquant curve

_____ 1. the additional return from adding one more unit of input
_____ 2. average product
_____ 3. the cost of adding one more unit of input to a production process
_____ 4. Stages I and III are considered in this stage of production
_____ 5. employs more than one-fifth of the nation's workforce
_____ 6. producer
_____ 7. consumer
_____ 8. sector that represents the movement of resources from consumers to agribusiness firms
_____ 9. term for an item when the need for the item exceeds its availability
_____ 10. the interaction between potential buyers and potential sellers of a good or service
_____ 11. an additional or incremental unit of something
_____ 12. synonym of benefits, utility, revenue, etc.
_____ 13. one of the most straightforward production functions
_____ 14. feed, fertilizer, seed, chemicals, fuel, and repairs
_____ 15. illustrates the relationship between variable input and output
_____ 16. describes production efficiency in terms of input and output
_____ 17. ratio of input per unit of output
_____ 18. when long-run average costs remain constant and output is increased
_____ 19. graphical illustration of resource substitution

_____ 20. illustrated when the isoquant curve is a straight line

_____ 21. shows the amount of one input that would cost the same if another unit of the other input were purchased

_____ 22. profit maximization approach used when total revenue and cost figures are difficult to obtain

ACTIVITIES

1. Describe the U.S. economic system and how it operates. Include specific examples.
2. Dryland corn yields with respect to changes in the amount of fertilizer (variable input) applied are presented in the following table:

Units of Fertilizer (F_x)	Yield or TP	Adjusted TP	AP	MP	MFC	MVP
0	60	0	—	—	—	—
1	68	8		8	$7.00	$
2	80		10		$7.00	$
3	87		9		$7.00	$
4	92	32		5	$7.00	$
5	95		7		$7.00	$
6	97	37			$7.00	$
7	98			1	$7.00	$
8	95		4.38		$7.00	$

a. Assuming that corn sells for $2.50 per bushel and fertilizer costs $7 per unit, fill in the blanks in the table.

b. To reach the point of maximum input efficiency, __________ units of fertilizer should be applied per acre.

c. If corn sold for $2.50 a bushel and fertilizer cost $7.00 a unit, the producer should apply __________ unit(s) of fertilizer per acre to maximize net returns at $__________ per acre above the input cost.

d. If corn now sells for $2.00 per bushel and fertilizer costs $12.00 per unit, __________ units of fertilizer should be applied to maximize net returns at $__________ per acre above the cost of fertilizer.

e. Regardless of the market price for corn or the cost per unit of fertilizer, a producer would never apply fewer than __________ units or more than __________ units of fertilizer per acre.

f. To justify producing a yield of 97 bushels, assuming that fertilizer costs $6 per unit, what would the minimum price of corn have to be per bushel to justify applying 6 units of fertilizer? $__________

3. Define and explain the six basic economic concepts introduced at the beginning of the chapter. Tie them all together by providing an overall agricultural example applying each concept.

UNIT 3

The Agribusiness Input (Supply) Sector

Chapter 13
Supplies, Machinery, and Equipment

Chapter 14
Economic Activity and Analysis

Chapter 15
Agricultural Policy and Governmental Agribusiness Services

Chapter 16
Private Agribusiness Services

Chapter 13

Supplies, Machinery, and Equipment

Objectives

After completing this chapter, the student should be able to:

- Describe the changes in the agribusiness input sector.
- Discuss the size of the agribusiness input sector.
- Name the major types of agribusiness inputs.
- Discuss the history, market structure, and careers in the feed industry.
- Explain the various types and consumption of feed.
- Discuss the use of fertilizer and the history of and career opportunities in the fertilizer industry.
- Discuss the history, types, benefits, and drawbacks of pesticides.
- Describe the role and types of farm supply stores.
- Explain the market structure of the farm machinery and equipment industry.
- Analyze the diversification of the farm machinery and equipment industry.
- Evaluate the impact of foreign trade on the farm machinery and equipment industry.
- Describe trends in the farm machinery and equipment industry.
- List various farm machinery and equipment products.
- Discuss possible careers in the farm machinery and equipment industry.
- Describe the role of farm machinery and equipment wholesalers.
- Discuss farm machinery and equipment dealership.
- Explain the types of jobs in farm custom and contracting services.

Terms to Know

base mixes
biopesticides
broaches
bulk feed
castings
centralization
complete feed
complete fertilizer
concentrates
consolidation
cost-effective
custom grinding
defoliants
desiccants
economy of scale

Continued

farm contracting
forge shop
formulation
foundry
full-line companies
fumigants
fungicide
hatcheries
herbicide
incomplete fertilizer
insecticide
lathes
long-line companies
machine shop
milling machines
nonruminant
pesticide
premix
roughages
ruminant
short-line companies
straight fertilizer
supplements
synthetic chemicals
target customers
vertically integrated
wholegoods

Introduction

Until fairly recently, one of the most neglected areas of the agricultural industry, in terms of recognition and understanding, was the agribusiness input (supply) sector. Agricultural economists erroneously assumed that this sector was adequately covered in general business courses and programs. Part of this neglect is due to various pieces of national legislation. Agricultural education programs were originally funded under the Smith-Hughes Act of 1917. However, it was not until the Vocational Act of 1963 was passed that agribusiness input (and output) became part of the instructional content of federally funded agricultural education programs. More specifically, the 1963 act allowed students to take vocational agriculture courses (now called agricultural education) if those courses offered Supervised Agricultural Experience Programs (SAEPs) in areas other than production agriculture. Refer to Figure 13-1.

Figure 13-1 Students working in garden centers or nursery supply businesses were not considered to be participating in legitimate Supervised Agricultural Experience Programs (SAEP) until the Vocational Act of 1963. (Courtesy of USDA)

Changes in the Agribusiness Input Sector

Historically, the pattern of agribusiness input has changed. Many years ago farmers produced agricultural products for their own use. They produced their own energy, used family labor, raised their own food, and produced their own clothing and shelter. A small percentage of agricultural products produced on the farm was sold to finance the purchase of supplies.

Industrialization in the United States led farmers to become more specialized, both in how they worked and in what they produced. Farmers also became more able to afford production inputs from other sources. Currently, most input products are purchased rather than produced on farms.[1]

To further illustrate the changes in the agribusiness input sector, consider the following: In 1910, labor accounted for 75 percent of the total production expenses. Today, however, labor accounts for less than 10 percent of production expenses. However, capital items for the production of agricultural products (which includes buildings, livestock, machinery, equipment, operating inputs, etc.) accounted for only 17 percent of input in 1910. Today, more than 80 percent of production costs are for capital items.[2]

1. Randall D. Little, *Economics: Applications to Agriculture and Agribusiness*, 4th ed. (Danville, IL: Interstate Publishers, 1997), pp. 9–20, 397; Ewell P. Roy, *Exploring Agribusiness* (Danville, IL: Interstate Printers & Publishers, 1980), pp. 21–23.
2. Little, *Economics,* p. 286.

Size of the Agribusiness Input Sector

Presently, about $200 billion is spent on agribusiness input by American production agriculturalists, and in 2005 more than 6 million people had jobs filling the demand in the agribusiness input sector.[3]

To be more specific, farm equipment (including tractors, other motor vehicles, and machinery) cost approximately $10 billion and required 150,000 employees to produce.[4] Fuel, lubricants, and maintenance costs for machinery and motor vehicles were $11 billion. Other expenditures for agricultural inputs include $7 billion for seed, $30 billion for feed, $12 billion for livestock, $12.5 billion for fertilizer and lime, $17 billion for hired labor, and $21 billion for depreciation and other consumption of farm capital.[5] (For the most current statistics, use your favorite Internet search engine to find the latest USDA agricultural statistics and other relevant information.) Many other examples could be added, but it should already be clear that the agribusiness input sector is big business. In fact, the agribusiness input sector handles a wide range of products manufactured or processed by hundreds of firms.

Types of Agribusiness Inputs

Major Agribusiness Inputs

The agribusiness input sector supplies many and varied products to production agriculturalists. The four biggest inputs are feed, fertilizers, agricultural chemicals, and farm machinery and equipment.

Minor Agribusiness Inputs

Many other agribusiness inputs are required by production agriculturalists. Petroleum and petroleum products, such as diesel fuel, gasoline, motor oil, and transmission and hydraulic oil, are needed to keep tractors, trucks, and self-propelled machinery going. Seed and lime are necessary for crop producers. Veterinary supplies are needed to maintain the health of farm animals. Containers, bags, sacks, cartons, and crates are needed for shipping and transporting agribusiness supplies to production agriculturalists. Lumber and building materials are necessary to produce shelters for humans, animals, and plants in the production sector.

Agribusiness Inputs That Are Often Overlooked

Many agribusiness input suppliers are taken for granted or not even considered significant, yet production agriculturalists could not function without them. Hardware, iron, steel, and related products are needed for the construction and daily maintenance of agricultural buildings. Seeds, plants, and trees are needed for crop and forestry production. Utilities, including electricity, water, gas, and telephone services, are integral parts of the production of agricultural items. Credit, insurance, and other private and government services (which are discussed in Chapters 7, 15, and 16) are also crucial input services in the production sector.[6]

Feed Industry

Historical Development

The feed industry is relatively new, dating back less than 100 years. The **formulation** of animal feeds began shortly before the turn of the 20th century, and scientific animal agriculture had its real beginning in the first two decades of the 20th century. Until that time, the by-products of corn and other grain refineries were considered industrial wastes and were dumped in the rivers. The soybean, which today has become the most important source of feed protein, was a botanical curiosity grown only in greenhouses. Today, among the numerous expenditures of production agriculturalists, the purchase of livestock feed is the largest. It represents about 14 percent of total farm expenses.

Much of the manufactured or processed feed previously produced at feed mills is now being produced on the farm. Based on cost, about 60 percent of the feed used comes from the manufactured feed industry. The other 40 percent is stored, ground, mixed, and enhanced with supplements onsite at the particular livestock operation. This

3. United States Department of Agriculture, *Agricultural Statistics 2005* (Washington, DC: U.S. Government Printing Office, 2005).
4. Ibid.
5. Ibid.

6. Roy, *Exploring Agribusiness*, pp. 35–36.

can be done because livestock operations have become larger, making it **cost-effective** to invest in feed-processing equipment.[7] A prime example is the poultry industry, which is **vertically integrated** through ownership or a contract with a parent company. The integrated firms, such as Tyson Foods, mix their own feed.[8]

Feed Production and Consumption

Production agriculturalists spent $30 billion on feed in 2005. Manufacturers in the United States produced more than 130 million tons of primary feed in 2005 and 3.3 million tons of secondary feed, which represents a total of more than 133 million tons of commercial feed. (Refer to the latest USDA agricultural statistics on the Internet for information on the amount of feed currently produced.)

Today, approximately 8,000 animal-feed processing plants are in existence, making this one of the largest manufacturing industries in the United States. The feed industry employs more than 100,000 workers. This figure does not include local feed mill workers, custom grinders and mixers, at least 2,000 wholesale feed dealers, more than 20,000 retailers, and several thousand **hatcheries** that sell feed.[9]

Types of Feed

Feed is not produced for livestock in general, but for a particular type of livestock. Four major types of feed are used in the poultry industry: chick starter, broiler grower, turkey grower, and laying feed. There are also different types of feed for swine, beef, and dairy. **Ruminant** and **nonruminant** animals, in particular, consume different types of feed. In reality, there are some 1,500 different combinations of ingredients. Refer to Figure 13-2 for the amounts consumed by each type of animal. (Conduct an Internet search to compare the 1997 data in Figure 13-2 with the latest USDA agricultural statistics.)

7. Kevin Kimbe and Marvin Hayenga, *Agricultural Input and Processing Industries* (Report RD-05) (Ames, IA: Iowa State University, Department of Economics, 1992).
8. Little, *Economics,* pp. 288-289.
9. Marcella Smith, Jean M. Underwood, and Mark Bultmann, *Careers in Agribusiness and Industry*, 4th ed. (Danville, IL: Interstate Publishers, 1991), p. 119.

Species	Percentage of Total
Broilers	30.9
Beef/sheep	15.7
Dairy	13.4
Hogs	12.5
Starter/grower/layer/breeder	12.7
Turkey	7.8
Other	6.9
Total	100.0

Figure 13-2 It may surprise many to know that broilers (poultry) consume more purchased feed than any other animal. Broilers are followed by beef/sheep, dairy, hogs, laying hens, and turkeys. (Source: *Feedstuffs*, 1988 Reference Issue, vol. 60, no. 31, July 1988)

Commercial feeds are of three general types: complete feeds, supplements, and premixes. **Complete feeds** require no additional treatment and can be fed directly to animals. About 80 percent of feed produced is of this type. Another 18 percent of the feed produced comes in the form of **supplements**. These are formula feeds that add protein, vitamins, and minerals to regular feed. A specified amount of grain must be added to a supplement before feeding it to stock. The final 2 percent of the feed produced is in the form of **premixes**. Premixes contain only vitamins and minerals; they are used at a ratio of less than 100 pounds of premix per ton of feed.[10]

Market Structure

The production and market structures of the feed industry are complex and interrelated. The general production and marketing channel for commercial feed is shown in Figure 13-3. The process begins with the ingredient and premix suppliers, who sell the feed inputs to formula feed manufacturers. The manufacturers produce complete feeds, premixes, **base mixes**, and/or supplements. As previously mentioned, about 80 percent of formula feed is manufactured as complete feed, with the remainder consisting of supplements, concentrates, base mixes, and premixes.

The manufacturers ship feed to warehouses for storage or send it to commercial feedlots or

10. Little, *Economics,* pp. 289-290.

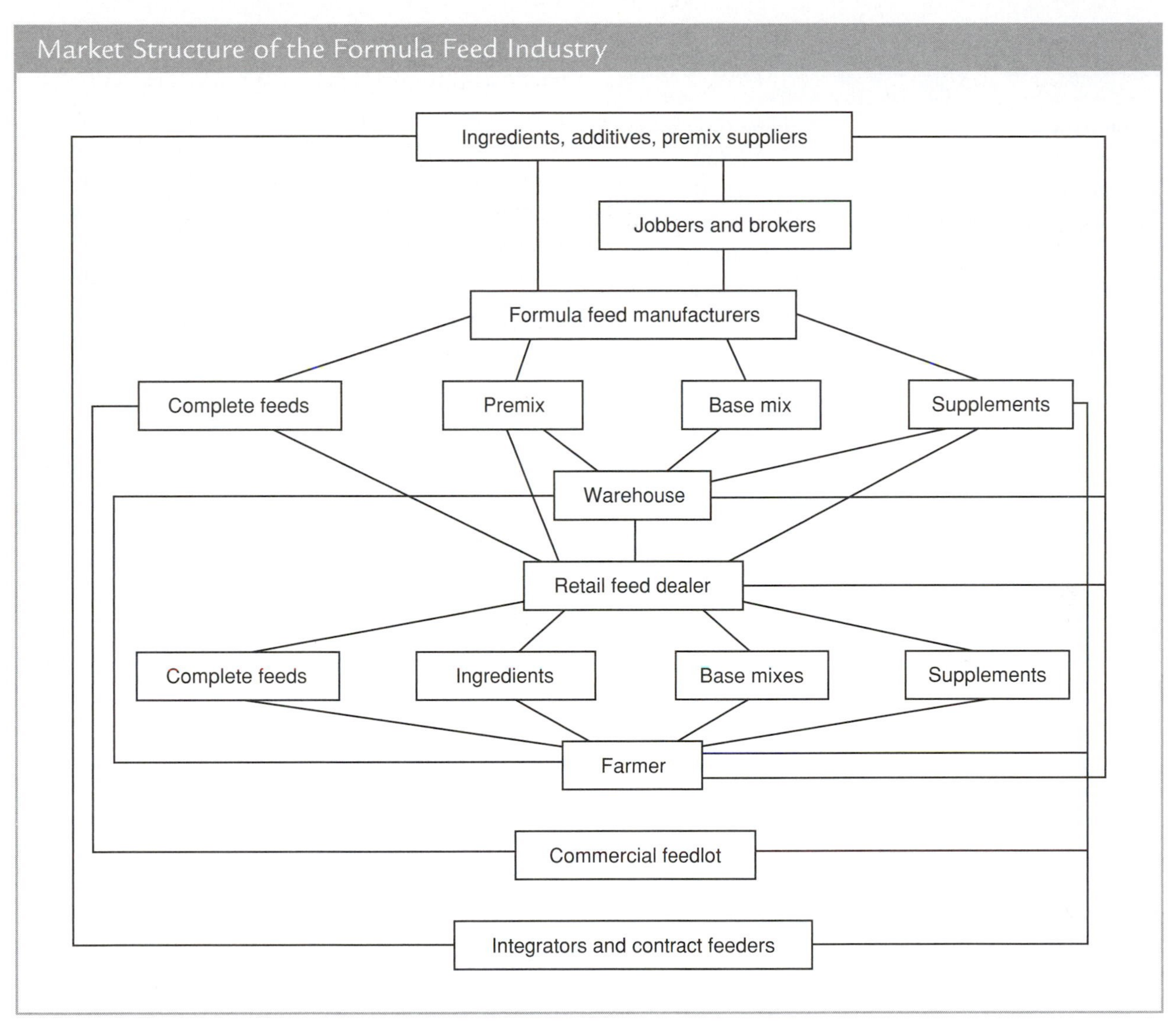

Figure 13–3 The flow of feed and market channel of the formula feed industry is complex and interrelated, as shown by this diagram. (Source: *Feedstuffs*)

feed retail stores. According to Rawlins's book, *Introduction to Agribusiness,* about 20,000 feed dealers account for 60 percent of retail feed sales. The remaining 40 percent is handled by direct-selling feed firms or sold directly to large feeders. The volume handled by retail feed stores varies considerably. About 25 percent of the stores handle 75 percent of the volume. Until recently, the major volume of feed was sold in bags. However, the trend has shifted to favor **bulk feed**, and now more than 60 percent of formula feed is sold in bulk.

The feed industry is dominated primarily by the 10,000 manufacturing firms, although retail firms outnumber them two to one. This is due primarily to the large volume of feed produced by these manufacturers. The top four manufacturers are international in scope, whereas the remaining top twenty firms sell feed only on a national and regional basis. A listing of the top feed producers in the United States and the volume produced, as of 1994, is shown in Figure 13–4. Use the Internet to find the latest USDA agricultural statistics and compare feed manufacturing firms and their current annual production.

An analysis of the top 20 feed manufacturing companies indicates that most of them are for-profit corporations, although 4 of the top 20 firms are cooperative corporations. Most of the firms also make products other than feed, and the products of the top four firms are highly diversified. Summary descriptions of the top four firms are given in Figure 13–5.

Estimated Annual Production of Top Feed Manufacturing Firms in the United States

Company	Volume (000 tons)
Purina Mills	7500
Cargill, Inc.	7000
PM Agricultural Products	3500
Consolidated Nutrition LC	3000
Land O'Lakes, Inc.	2500
Kent Foods, Inc.	2000
Farmland Industries	1820
Continental Grain Co. (Allied Mills)	1500
Southern States Co-Op	1290
Cactus Feeders, Inc.	1200
Agway, Inc.	850
Hubbard Milling Co.	800
MFA, Inc.	796
Mark II Plan	770
Friona Industries, L.P.	680
Morenman Mfg. Co.	600
Quality Liquid Feeds, Inc.	600
SF Services Inc.	600
Star Milling	513
Pennfield Corp.	500

Figure 13–4 Feed manufacturing is a big part of agribusiness. By multiplying an average of $160 per ton by the number of tons produced, one can get an even more vivid view of the impact of the feed industry. (Source: N. Omri Rawlins, *Introduction to Agribusiness,* Murfreesboro, TN: Middle Tennessee State University, 1999)

Careers in the Feed Industry

There are many opportunities for employment in the feed industry. Graduates with a strong background in agribusiness and the animal sciences and a desire to work closely with production agriculturalists will find employment opportunities in the feed industry. The following are some specific job types: buyer for feed manufacturer; plant manager, assistant, or plant operator for a feed manufacturing plant; sales representative; feed wholesaler; researcher; manager or employee of an agricultural cooperative; manager of a poultry hatchery; feed specialist with a chemical or pharmaceutical company; manager or employee of a retail feed and supply store; and similar positions in many related industries.[11] Refer to Figure 13-6 for the home office addresses of some of the major food companies.

11. Roy, *Exploring Agribusiness,* pp. 264–267.

Fertilizers

The first mixed fertilizers sold in the United States for commercial purposes were manufactured in Maryland in 1849. Factories were soon built in other areas and production expanded rapidly, primarily in the eastern states. Some of the major events and dates in the fertilizer industry in the United States are summarized in the following list:

1851–first record of fertilizer analysis by state officials
1856–first state fertilizer control law (Massachusetts)
1865–potash first imported into the United States
1928–first direct use of anhydrous ammonia in mixed fertilizers
1929–first shipment of synthetic nitrate of soda, from Hopewell, Virginia
1931–first commercial shipment of potash, from Carlsbad, New Mexico
1939–first experiments with the use of anhydrous ammonia on nonirrigated land (direct)
1942–first commercial use of anhydrous ammonia directly in soil
1950–centennial of the first manufacture of mixed fertilizers in the United States[12]

Fertilizer Analysis

Commercial fertilizers are made by combining selected plant food materials to obtain specific ratios and quantities of plant nutrients. The three major plant food nutrients found in fertilizers are nitrogen, phosphorus, and potassium. Because these elements are rarely found in pure, elemental form, they must be combined with other substances to be used as fertilizers. The mining, manufacture, and distribution of these materials constitute the basis of the fertilizer industry.

A **complete fertilizer** is one that contains nitrogen (N), phosphoric acid (P_2O_5), and potash (K_2O). Fertilizers containing only two of the three materials are known as **incomplete fertilizers**. A fertilizer containing only one of the primary materials is sometimes called a **straight fertilizer**.

Fertilizer distribution in the past was done primarily in 100-pound bags, so the analyses were

12. Malcolm McVickar, *Using Commercial Fertilizer* (Danville, IL: Interstate Printers and Publishers, 1952), pp. 17–20.

A Summary of Selected Characteristics of the Four Major Feed Manufacturers						
Company	Central Office	Scope	Major Feed Brand	Type of Feed	Other Products	Other Information
Ralston Purina	St. Louis, MO	International profit corporation	Checkerboard Chow	Beef, swine, dog, cat, dairy, poultry, horse, other pets, catfish, shrimp	Poultry products Cereals and snacks Sea foods (Chicken of the Sea) Chemical specialties Animal health products Protein products Candy and beverages Institutional foods Mushrooms, shrimp	Major feed producers in United States Owns and operates fast-food restaurant chain—Jack in the Box Owns and operates specialty dinner houses—Hungry Hunter, Boar's Head, Tortilla Flats Owns and operates Keystone Ski Resorts Owns and operates shrimp and mushroom farms
Allied Mills	Chicago, IL	International profit corporation	Wayne Feeds	Beef, swine, poultry, dog, horse, other pets, lab animal, zoo animal	Poultry products Pork products Soybean and oil meal Dehydrated alfalfa	Subsidiary of Continental Grain Company Second largest feed producer Sixth largest poultry processor Owns Baronet Corp., a major producer of leather goods
Central Soya	Fort Wayne, IN	International profit corporation	Master Mix and Provima	Beef, swine, dairy, poultry	Animal health products Poultry products Consumer foods: Mrs. Filbert's margarine Super beef mix Grain marketing	Third largest feed producer Fourth largest broiler processor Expanding rapidly into major food lines to promote "food-power"
Cargill	Minneapolis, MIN	International profit corporation	Nutrina	Beef, swine, dairy, other animals	Poultry products Food and feed grains Food and oil seeds Chemical products Farm seeds P-A-G brand Fertilizers, metals, plastics, and resins	Fourth largest feed producer Second largest grain exporter Major transportation and warehousing system More than 250 plants and offices in North America and 60 in 36 other countries

Figure 13–5 The four major feed companies are involved in more than just feed production, as you can see.

Feed Companies

A. E. Staley Manufacturing Co.
2200 East Eldorado Street
P.O. Box 151
Decatur, IL 62525
Tel.: 217-423-4411/Fax: 217-421-2936

Agway Grain Marketing
P.O. Box 4933
Syracuse, NY 13221-4933
Tel.: 315-449-6272/Fax: 315-449-5499

Archer Daniels Midland Co.
P.O. Box 1470, 4666 Faries Parkway
Decatur, IL 62425-1820
Tel.: 217-424-5450/Fax: 217-424-5447

Bluebonnet Milling Co.
100 South Mill, P.O. Box 2006
Ardmore, OK 73402
Tel.: 405-223-3010/Fax: 405-223-3546

Cargill
9300-MS48 Minneapolis, MN 55440
Tel.: 612-742-6212

Central Soya Co., Inc.
P.O. Box 1400
Fort Wayne, IN 46801
Tel.: 219-425-5100, 1-800-422-7692
Fax: 219-425-5254

Conagra Feed Co., Inc.
980 Molly Pond Road
Augusta, GA 30901
Tel.: 706-722-6681/Fax: 706-722-4561

Dekalb Feeds, Inc.
105 Dixon Avenue, P.O. Box 111
Rock Falls, IL 61071
Tel.: 815-625-4546/Fax: 815-625-4896

Dickey-John Corp.
P.O. Box 10
Auburn, IL 62615
Tel.: 217-438-6012, 1-800-637-2964
Fax: 217-438-6012

Doane Products Company
West 20th Street & State Line Road,
P.O. Box 879 Joplin, MO 64802
Tel.: 417-624-6166

Farmland Industries
P.O. Box 7305
Kansas City, MO 64116-0005
Tel.: 816-459-6000/Fax: 816-459-3980

General Mills, Inc.
54 S. Michigan Avenue
Buffalo, NY 14203-3086
Tel.: 716-856-6100/Fax: 716-857-3689

Hubbard Milling Company
P.O. Box 338
Ft. Pierre, SD 57532
Tel.: 605-223-2557

Kent Feeds Inc.
1600 Oregon Street
Muscatine, IA 52761
Tel.: 319-264-4211

Land O'Lakes Inc.
P.O. Box 64089
St. Paul, IN 55164
Tel.: 612-451-5412/Fax: 612-451-5407

Monsanto (Agricultural Group)
800 North Lindbergh Boulevard
St. Louis, MO 63167
Tel.: 314-694-1000/Fax: 314-694-3900

Moorman Manufacturing Co.
1000 N. 30th Street
Quincy, IL 62305-3115
Tel.: 217-222-7100/Fax: 217-222-4069

Murphy Farms Inc.
P.O. Box 759, U.S. Highway 117 S.
Rose Hill, NC 28458-0759
Tel.: 919-289-2111/Fax: 919-289-6400

Nutrena Feed-Nutrena Feed Division
15407 McGinty Road, P.O. Box 5614
Minneapolis, MN 55440-5614
Tel.: 612-475-7495/Fax: 612-475-6391

Purina Mills, Incorporated
1401 S. Hanley Road
St. Louis, MO 63144
Tel.: 314-768-4100/Fax: 314-768-4188

Walnut Grove
201 Linn Street
Atlantic, IA 50022
Tel.: 712-243-1030/Fax: 712-243-6360

Wayne Feed
10 S. Riverside Plaza
Chicago, IL 60606
Tel.: 312-930-1050

Gold Kist, Inc.
244 Perimeter Center Parkway, N.E.,
P.O. Box 2210 Atlanta, GA 30301
Tel.: 404-393-5000/Fax: 404-393-5347

Figure 13-6 Feed companies are a source of potential employment. The home office addresses of the companies are listed here.

always given in the same order and the numbers represented the weight percentage of nitrogen, phosphoric acid, and potash. For example, a 100-pound bag of fertilizer labeled 4-12-12 contained 4 pounds of nitrogen, 12 pounds of phosphoric acid, and 12 pounds of potash, a total of 28 pounds of plant food. (The remaining 72 pounds in the bag was "filler" or "conditioner.") Federal law requires that each bag of fertilizer have a label or tag stating the minimum amount of N, P_2O_5, and K_2O contained in the bag. However, most fertilizer today is sold in bulk by the manufacturers and mixed at the retail level through a process known as *bulk blending.*

Nitrogen. Nitrogen fertilizer manufacturing uses a process that combines nitrogen from the atmosphere with hydrogen from some natural source to produce synthetic ammonia. Natural gas is the primary source of hydrogen, although fuel oil, coke, naphtha, refinery gas, and other sources are also used. Nitrogen may be applied as a liquid, gas, or solid, although solids and liquids are most commonly used. The major gas source of nitrogen is anhydrous ammonia, whereas aqua-ammonia is the major liquid source. The three major sources of nitrogen available in solid form are ammonium nitrate, ammonium sulfate, and urea. The amount of nitrogen currently produced for fertilizer is about 10 million tons per year.

Phosphate. Phosphate fertilizers are made by treating rock deposits, certain iron ores, or bones with sulfuric or phosphoric acid. The phosphate content of fertilizer is expressed as phosphoric acid (P_2O_5); pure, elemental phosphorus cannot be used as a plant nutrient. The major sources of phosphates include normal superphosphate, concentrated superphosphate, ammonium phosphate, and phosphoric acid (H_3PO_4). Currently, more than 32 million tons are produced each year. The major phosphate deposits in the United States are found in Florida, North Carolina, Tennessee, Idaho, Montana, Utah, and Wyoming.

Potassium. Potassium in its pure state is a light-gray element that bursts into flame when exposed to air. Thus, it must be combined with other elements before it can be used as fertilizer. Potassium in fertilizer is commonly called *potash* (K_2O). The major sources of potash in commercial fertilizers are muriate of potash (KCl), sulfate of potash (K_2SO_4), and nitrate of potash (KNO_3). The major potash deposits in the United States are found in New Mexico, Utah, and California, which currently produce more than 2 million tons of potash each year. In addition, about 3.5 million tons are imported, primarily from Canada.

Fertilizer Use

Production of basic fertilizer products in the United States is carried out by some 40 to 50 manufacturers. The fertilizer is further processed at more than 10,000 local marketing centers, which supply granular mixed fertilizer and liquid mixed fertilizer bulk blends.[13] The average-size fertilizer plant has a storage capacity of 6,041 tons and an annual distribution volume of 15,400 tons. Almost half the fertilizer plants are single (sole) proprietorships; approximately 42 percent are organized as cooperatives, and about 9 percent are corporations.[14]

Production agriculturalists in the United States spend approximately $11 billion yearly on fertilizers. Fertilizer usage increased substantially during the 1960s and 1970s, and peaked in 1981. Fertilizer use has been relatively stable since the mid-1980s, ranging from 20.1 to 23.8 million tons per year. The amount of fertilizer used is affected by the number of acres in production, rates of application (often reduced to cut costs), and exports, which are approximately $3 billion annually.[15]

Environmental Policy. Environmental policy will likely affect the fertilizer industry in the future. Environmental regulations on certain aspects of fertilizer production and use will become increasingly strict, and compliance will be closely monitored. Local, state, and federal regulations concerning groundwater contamination by nitrogen fertilizers will certainly affect the industry. If contamination of underground water supplies becomes worse, regulation will almost certainly increase.

Interrelationships among Fertilizers, Pesticides, and Other Chemicals

The market structures of pesticide (discussed in the next section) and fertilizer producers and distributors are very closely related. Most of the major pesticide producers also produce some type of fertilizer, and vice versa. The chemical companies are also very closely associated with the petroleum industry, as farm chemicals are usually derivatives of oil products. The chemical industry is highly concentrated, with the processors or chemical manufacturers in control. The market is also highly integrated, with the processors normally owning the mines, processing

13. N. Omri Rawlins, *Introduction to Agribusiness,* 3rd ed. (Murfreesboro, TN: Middle Tennessee State University, 1999), p. 93.

14. Little, *Economics,* p. 292.

15. United States Department of Agriculture, *Agricultural Statistics 1998* (Washington, DC: U.S. Government Printing Office, 1998).

plants, and wholesale operations. The retailers are primarily cooperative or independent farm supply stores.

The chemical industry is also highly diversified. Chemical processors handle a wide variety of products, including plastics, metals, and pharmaceuticals. Many produce numerous brands of food products. In the farm pesticide market, Novartis ranks first in chemical sales, and Monsanto and Zeneca rank second and third, respectively. Some of the major fertilizer producers include DuPont, Agrevo, Bayer, Rhone-Poulenc, Dow Agrosciences, Cyanamid, and BASF. A list of the top 10 chemical companies is shown in Figure 13-7. You can use the Internet to identify the top chemical companies today and compare their sales volumes.

Careers in the Fertilizer Industry

All parts of the agricultural supply sector use conventional business services, such as accounting and auditing, control of production cost, distribution cost, selling costs, data processing and management, office management, sales and sales promotion, credit and collections, communications, publicity, advertising, and customer relations. Manufacturer representatives work with chemists, botanists, and other agricultural scientists. Manufacturing plants hire many workers, as do wholesale distribution operations. Field service contracting includes applying fertilizer and spreading lime for farm customers. Refer to Figure 13-8 for the names of a few of the nation's leading fertilizer and pesticide companies.

Rank	Company	Sales (Million $)
1	Novartis	4,199
2	Monsanto	3,126
3	Zeneca	2,674
4	DuPont	2,518
5	Agrevo	2,352
6	Bayer	2,254
7	Rhone-Poulenc	2,202
8	Dow Agrosciences	2,200
9	Cyanamid	2,119
10	BASF	1,855

Figure 13-7 The top 10 agricultural chemical companies based on sales, 1997. (Courtesy of AEROW, *World Crop Protection News*, 1998)

Pesticides

Since the beginning of time, humankind has competed with insects and other animals for the fruits of the plant world. There has also been a constant battle between desirable and undesirable plants. Until the 20th century, we used natural means and simple machines to control weeds, insects, and diseases. However, because of successful efforts in chemical and biological research, weeds, insects, and diseases are now controlled primarily with **synthetic chemicals**. It is estimated that there are 10,000 species of harmful insects, more than 1,500 diseases caused by fungi, 1,800 different weeds, and approximately 1,500 different types of nematodes that cause damage to crops and livestock.

Types of Pesticides

Pesticide is a broad term that refers to all chemicals used to control weeds, insects, and diseases that affect crops, livestock, or people. The three major types of pesticides are herbicides, insecticides, and fungicides. Other types of pesticides include **fumigants, defoliants, desiccants**, and growth regulators. An **herbicide** (80 percent of pesticides sold) is one or more chemicals used to control undesirable plants, whereas an **insecticide** (15 percent of sales) is used to control various insects. A **fungicide** (5 percent of sales) is one or more chemicals used to control the fungi that attack plants and animals.

Historical Background

The foundation for modern pesticides can be traced to the mid-19th century. In 1859, Julius Sacks, a German botanist, wrote articles on the translocation of growth-regulating substances in plants.

Around 1900, three scientists–Bonnet of France, Schultz of Germany, and Bolley of the United States–found that copper salt solutions would kill broad-leaved plants. These scientists were all working independently at about the same time. In 1907, a German scientist, Hans Fitting, found that the swelling of the ovary and the fading

Fertilizer and Pesticide Companies

Abbott Laboratories, Chemical & Agricultural Products Division
1400 Sheridan Road
North Chicago, IL 60064
Tel.: 708-937-6737/Fax: 708-937-3697

Allied Products Corporation
10 South Riverside Plaza
Chicago, IL 60606
Tel.: 312-454-1020/Fax: 312-454-9608

American Cyanamid Co.
1 Cyanamid Plaza
Wayne, NJ 07470
Tel.: 201-831-2000/Fax: 201-831-4470

BASF Corporation
P.O. Box 13528
Research Triangle Park,
NC 27709-3528
Tel.: 919-361-5300/Fax: 919-361-5722

Cargill Fertilizer Division
P.O. Box 9300
Minneapolis, MN 55440
Tel.: 612-475-7167/Fax: 612-475-7313

Ciba-Geigy Corporation
P.O. Box 18300
Greensboro, NC 27419-8300
Tel.: 910-632-6000

Conoco/Du Pont
P.O. Box 4784-TA2136
Houston, TX 77210-4784
Tel.: 713-293-3987

Dow Elanco
U.S. Crop Protection
9330 Zionsville Road
Indianapolis, IN 4668
Tel.: 317-337-3000

E. I. du Pont de Nemours & Co.
Barley Mill Plaza, EA/WM5-167
Wilmington, DE 19880
Tel.: 302-992-3022

Farmland Industries
P.O. Box 7305
Kansas City, MO 64116-0005
Tel.: 816-459-6000/Fax: 816-459-3980

Fermenta Animal Health Company
10150 North Executive Hills Boulevard
Kansas City, MO 64153-2314
Tel.: 816-891-5500/Fax: 816-891-0663

Gold Kist Inc.
244 Perimeter Center Parkway, N.E.,
P.O. Box 2210
Atlanta, GA 30301
Tel.: 404-393-5000/Fax: 404-393-5347

Mobil Oil Canada
300-5th Avenue S.W.
Calgary, Alberta Canada T2P 2J7
Tel.: 403-260-7910/Fax: 403-260-4277

Monsanto (Agricultural Group)
800 North Lindbergh Boulevard
St. Louis, MO 63167
Tel.: 314-694-1000/Fax: 314-694-3900

Occidental Chemical Corporation, Ag. Products—Feed Products Division
5005 LBJ Freeway
Dallas, TX 75244
Tel.: 214-404-3800/Fax: 214-404-3200

Potash & Phosphate Institute
655 Engineering Drive, Suite 110
Norcross, GA 30092
Tel.: 404-447-0335

The Upjohn Company
700 Portage Road
Kalamazoo, MI 49001
Tel.: 616-323-4000/Fax: 616-323-4000

The Vigoro Corporation
225 N. Michigan Avenue
Chicago, IL 60601
Tel.: 312-819-2020/Fax: 312-819-2027

Figure 13–8 Fertilizer and pesticide companies are a source of potential employment. The home office addresses of the major companies are listed here.

of the flower after pollination were caused by a substance that developed during pollination.
In 1908, Bolley of North Dakota reported effective weed control in wheat using sodium chloride, iron sulfate, copper sulfate, and sodium arsenite. Bolley predicted that chemical weed control would be one of the most significant developments in agricultural research, and current trends and usage indicate that he was correct.

Discovery of Herbicides. In 1941, Pokorny dis covered a compound called 2,4-D. Other researchers later tried it as a fungicide and insecticide and found it ineffective. However, a year later Zimmerman and Hitchcock of the United States discovered that 2,4-D was a plant growth substance. In 1944, Garth and Mitchell established the selectivity of 2,4-D for broad-leaved plants and Hamner and Tukey first used it successfully in field weed control. In 1945, Templeman of England established the preemergence principle of soil treatment for selective weed control. Since that time, the development of chemical herbicides has expanded rapidly.

Discovery of Fungicides and Insecticides. The earliest fungicides and insecticides were inorganic compounds developed from copper, mercury, lead, and arsenic, many of which are still being used. The natural organic compounds of nicotine, pyrethrum, and rotenone were also used in limited amounts during the 19th century. The modern pesticide industry is based primarily on synthetic organic compounds, which made their first major impact during the 1940s. The compounds that had the greatest impact were DDT, an insecticide, and the herbicide 2,4-D.

DDT. The value of DDT as an insecticide was discovered in 1939 by Dr. Paul Muller, for which he received the 1948 Nobel Prize in physiology and medicine. DDT was the major compound used to control malaria-carrying mosquitoes during World War II, and the effectiveness of this compound encouraged researchers to develop other, similar compounds, such as chlordane, aldrin, dieldrin, and other valuable insecticides. Ironically, many of the most successful pesticides, which showed so much promise in earlier years, have come to be viewed as enemies of the environment. Currently, DDT, aldrin, and dieldrin are illegal, and numerous others are being evaluated.

Benefits of Pesticides

Pesticides represent more than 4 percent of expenditures of production agriculturalists, who spend about $10 billion yearly on herbicides, insecticides, and fungicides to assure that crop production is not lost to pests.[16] Nevertheless, each year losses of about $3 billion are caused by hordes of insect pests that chew, suck, bite, and bore away at food, food crops, and livestock. Bacteria, viruses, fungi, nematodes, and weeds destroy another $7 billion, making for a total production loss of about $10 billion.[17]

Today, pesticides do a great deal to reduce crop losses. In addition to controlling insects and killing weeds, pesticides are used to fight plant diseases, treat seed, clear brush, sterilize and fumigate soil, and regulate harvest times. Production agriculturalists now have literally thousands of government-approved pesticides at their disposal.[18]

Concerns about Pesticides

Although pesticides are very beneficial, they do cause some environmental damage. Research is being conducted to find effective pesticides that will not be harmful to water, air, soil, wildlife, humans, animals, or the food we consume. One example is the development of **biopesticides**, which will be safer to use.

Because of potential harm to the total environment, pesticide production and use are closely regulated by the Office of Pesticide Programs of the U.S. Environmental Protection Agency. The first major effort to control pesticides was the Federal Insecticide, Fungicide, and Rodenticide Act (FIFRA) of 1947. Until 1970, the USDA was responsible for administering FIFRA; thereafter, that authority was transferred to the Environmental Protection Agency.

Government restrictions make it extremely difficult to develop new and more effective pesticides. The registration of a new pesticide takes an average of 5 to 10 years and costs anywhere from $5 million to $10 million. Future projections indicate that newly developed pesticides will take even longer to receive approval and be much more expensive. However, scholars contend that if the world's population is to be fed, more and better pesticides must be found.

Careers

Just as in the fertilizer industry, the pesticide industry needs many conventional business services. Jobs are available in manufacturing, with wholesale distributors, with researchers, and in sales. Many new jobs are opening in the area of environmental policy, regulation, and control. Many colleges are now offering college degrees in environmental science and technology or have programs with similar names. Refer to Figure 13–8 for a list of some of the major pesticide manufacturers (listed along with the top fertilizer companies). Figure 13–9 lists animal health companies, which also offer many career opportunities. Refer to the Career Options section for a further explanation of possible careers in the agribusiness supply industry.

Farm Supply Stores

Most agribusiness input supplies get to the production agriculturalist by way of farm supply stores. There are approximately 36,000 farm

---

16. Ibid.
17. Ibid., p. 220.
18. Smith, Underwood, and Bultmann, *Careers in Agribusiness and Industry*, pp. 218–219.

Animal Health Companies		
A. L. Laboratories Inc. 1 Executive Drive Fort Lee, NJ 07024 Tel.: 201-947-7774	Eli Lilly and Company Lilly Corporate Center Indianapolis, IN 46285 Tel.: 317-276-2000/Fax: 317-276-2000	Pfizer, Incorporated 235 E. 42nd Street New York, NY 10017 Tel.: 212-573-2323/Fax: 212-573-7851
American Cyanamid Co. 1 Cyanamid Plaza Wayne, NJ 07470 Tel.: 201-831-2000/Fax: 201-831-4470	Fermenta Animal Health Company 10150 North Executive Hills Boulevard Kansas City, MO 64153-2314 Tel.: 816-891-5500/Fax: 816-891-0663	SmithKline Beecham Animal Health Whiteland Business Park 812 Springdale Drive Exton, PA 19341 Tel.: 215-363-3100/Fax: 215-363-3285
Animal Health Institute P.O. Box 1417-D50 Alexandria, VA 22313-1480 Tel.: 703-684-0011	Merck Ag. Vet, Division of Merck & Co., Inc. P.O. Box 2000 Rahway, NJ 07065-0912 Tel.: 908-855-6709/Fax: 908-855-9366	The Upjohn Company 7000 Portage Road Kalamazoo, MI 49001 Tel.: 616-323-4000/Fax: 616-323-5251
Elanco Animal Health Lilly Corporate Center Indianapolis, IN 46285 Tel.: 317-276-3000/Fax: 317-276-9430		

Figure 13–9 Animal health companies, which offer many career opportunities, also sell products beyond the agricultural industry. The home office addresses of several of these companies are listed here.

supply stores throughout the United States.[19] They sell seed, feed, fertilizer, pesticides, animal health products, farm machinery and equipment, petroleum products, and hardware. In the spring, many flower and garden supplies are also offered. The majority of small stores are organized either as sole (single) proprietorships or as partnerships. The larger, chain-type farm supply stores are either corporations or cooperatives.

The recent trend with many farm supply stores is toward **consolidation** and **centralization**, which means a complete line of farm production supplies and services is available. Some of the supplies and services offered include bagged and bulk fertilizers, bulk blending, delivery and spreading of fertilizers, free or reduced-cost use of fertilizer spreaders when the company's fertilizer is used, use of a feed mill to grind and mix **roughages** and/or **concentrates**, **custom grinding** and custom mixing, and soil testing. Many of these businesses also offer technical information and furnish credit. Products for tractors, trucks, and automobiles are available, such as diesel, gasoline, propane, oil, tires, batteries, and accessories. Some stores even perform truck and car service and repair.

Types of Farm Supply Stores

The types of farm supply stores can be divided into three major categories: farmer cooperatives, chain-store corporations, and small sole-proprietorship feed and supply stores. The legal makeup of each of these businesses forms is discussed in Chapter 6. The following is a brief description with examples of each.

Farmer Cooperatives. Farmer cooperatives sell nearly 28 percent of all supplies purchased by production agriculturalists. More specifically, farmer cooperatives supply about 45 percent of fertilizer and lime, 23 percent of feed, 13 percent of seed, and 32 percent of farm chemicals used by production agriculturalists today.[20]

Farmer Supply Chain Stores. Another major source of supplies for production agriculturalists is farmer supply chain stores. These are corporations that are organized and operate like other corporations, such

19. Ibid., p. 222.

20. Little, *Economics,* p. 296.

Career Options

Animal Nutritionist, Feed Formulator, Pesticide Applicator, Pesticide Specialist, Seed Analyst, Fertilizer Chemist

There are many jobs in the agribusiness input (supply) sector. Careers in animal nutrition are varied and interesting. One may work essentially as an organic chemist or technician in a laboratory, where complex equipment is used for research on and analysis of animal needs and the nutritional values of feedstuffs. One may also work selling feeds or at an agribusiness that produces feedstuffs. Careers in nutrition may be in the basic or applied sciences. They may focus on fish, small animals, pets, equines, poultry, livestock, dairy, or wild animals.

Tens of thousands of chemical formulations have been developed to control insects, diseases, weeds, nematodes, rodents, and other pests. We must use chemicals to help control pests. However, if not properly used, these materials are likely to be hazardous to the applicator, the farm operator, other people, plants, animals, or the environment.

Pesticide applicators are employed by production agriculturalists, lawn service companies, farm and garden supply firms, termite control companies, highway departments, and railroads, and often are self-employed persons. Special training and licensing are required for handling most pesticides.

Most of the major agricultural food and fiber crops are grown from seeds. Because large amounts of quality seed are needed each growing season, a giant seed industry has developed. Many seed analysts work for large seed companies. They test the seed the company plans to sell to determine how well the seed can be expected to perform in the field and to ensure high quality. Some seed analysts work with government agencies that are responsible for making sure that commercial seed meets the quality standards set by federal laws. Seed analysts require various degrees of technical or college-level training prior to employment. They must be patient individuals who are able to perform many repetitions of the same task.

There are many jobs with fertilizer companies. Almost all production agriculturalists use some type of fertilizer. All inorganic fertilizers are manufactured from some raw material. A fertilizer chemist then tests, analyzes, and makes recommendations for the appropriate mixes. This position requires various degrees of technical or college-level training prior to employment.

At the Yakima Agricultural Research Laboratory in Wapato, Washington, technician Tom Treat applies a test pesticide to a rapeseed variety being grown for canola oil production. (Courtesy of USDA)

as Wal-Mart or K Mart, except that their **target customers** are full-time and part-time production agriculturalists. Many other consumers also shop at these chain stores because of the hardware, lawn and garden, and work clothing available there.

There are several farm supply chain stores, each of which is popular in a different region of the country. They sell small tractors, fertilizer, feed, seed, tractor supplies, auto and truck supplies, lawn and garden equipment and supplies, hardware, shop supplies, clothing, and various other products. The largest farm supply chain store is Tractor Supply Company (TSC), with national headquarters in Nashville, Tennessee. TSC has more than 300 stores throughout the country, and also makes catalog sales. Each TSC store employs approximately 10 to 20 full-time and part-time workers, depending on the store size and hours of operation. The home office of TSC also employs approximately 350 people; the total employment by TSC is more than 9,100 people.

Single (Sole) Proprietorship Feed and Supply Stores. There are thousands of smaller, sole-proprietorship operations throughout the United States. They tend to market a name-brand feed, such as Purina, Master Mix, Nutrena, Moorman, or Wayne, along with other, locally grown feeds. They also carry seed, fertilizer, and miscellaneous farm supplies. The smaller stores are usually family managed, though they may hire some extra help during rush seasons.

Farm Machinery and Equipment

Market Structure

Farm machinery companies may be classified into three types based on the kind and variety of equipment sold: full line, long line, or short line. **Full-line companies** produce and sell tractors as well as a wide variety of other equipment. **Long-line companies** produce and sell a wide variety of general farm equipment, including self-propelled combines, but no tractors. **Short-line companies** produce highly specialized equipment, such as planters and cultivators, forage equipment, and milking machines.

Full-Line Companies

The farm machinery industry is similar to the automobile industry in terms of market concentration. According to Paine Webber's "Agricultural Dealer Surveys," four full-line companies account for more than 90 percent of total farm machinery sales in the United States. Deere, Inc., which is the largest producer, makes more than one-third of total sales. New Holland is second, followed by Case and AGCO. Full-line companies constitute the major basis of the farm equipment industry. They are the smallest in number, but they have the largest sales volume and are international in scope, with manufacturing plants and distribution centers throughout the world.

Figure 13–10 Vermeer is an example of a long-line company. Vermeer produced the first round hay baler. (Courtesy of Vermeer Manufacturing Company)

Long-Line Companies

The second major group is the long-line companies. They are more numerous, but they have less dollar volume than full-line companies. Most long-line companies operate in a national market. The major long-line companies include Gehl and Vermeer. Several firms that previously were long-line companies have expanded or merged into full-line companies. Refer to Figure 13–10.

Short-Line Companies

Short-line companies have the least amount of impact on the industry, because of their lower volumes of sales. Most short-line companies are regional in scope. They typically supply items for full-line companies under some type of production agreement. Full-line companies may buy parts and fully assembled machines from short-line companies, foreign-based subsidiaries, and/or foreign markets.[21] Refer to Figure 13–11.

21. Ibid., p. 293.

Figure 13–11 Bush Hog is an example of a short-line company. (Courtesy of Bush Hog Corporation)

Size of the Farm Machinery and Equipment Market

Approximately 1,500 firms are involved in producing farm machinery in the United States.[22] These firms franchise nearly 7,000 retail outlets to distribute their products. In 1993, the average expenditure for machinery was slightly more than $7,000 per farm; currently, annual total sales for farm machinery and equipment exceed $14.6 billion. As you might expect, about 30 percent of the total amount is spent on tractors and self-propelled vehicles.[23]

In many cases, a retail dealer holds a franchise for more than one type of tractor or other types of equipment. This trend emerged as a result of the consolidation the farm equipment industry experienced during the 1980s. By having two to three franchises, a dealer can generate sufficient sales volume to maintain a full service and parts operation and to provide adequate income for the dealership. An average full-line dealer is required to have a minimum net worth of $250,000 to $500,000 to be considered for award of a franchise. Approximately two-thirds of a dealership's business volume comes from the sale of new and used equipment, with the remainder coming from sale of parts, service, and small farm supply items.

22. Rawlins, *Introduction to Agribusiness*, p. 75.
23. Little, *Economics*, p. 292.

Diversification of Farm Machinery and Equipment Companies

The major farm equipment companies are highly diversified and do not limit their products to farm equipment. A major portion of recent expansion has been made in nonfarm product lines. For example, J.I. Case is a major producer of backhoes and other industrial equipment; Deutz-Allis, purchased by AGCO, was a major producer of electric generators; and White Motor Company, also purchased by AGCO, was a major producer of industrial trucks. In their early days, both Case and International Harvester made automobiles, but they discontinued production soon after they began. Ford is still a major producer of farm pickup trucks and cars. The Ford tractor was bought by the Fiat Company in Italy and is now sold under the New Holland name. Refer to Figure 13–12.

Most of the full-line companies have expanded their consumer product lines to attract urban homeowners. The most attractive consumer line consists of riding lawn mowers and garden equipment, but many other consumer products are offered, from chainsaws and snowmobiles to bicycles. This area of the industry is expected to expand even more as the urban population increases and the number of farmers decreases. Refer to Figure 13–13.

Figure 13–12 Original farm machinery and equipment companies have recently expanded into nonfarm product lines such as industrial equipment. (Courtesy of Cliff Ricketts)

Figure 13–13 Full-line companies have expanded to attract urban homeowners, so as to maintain a strong market for their products. (Courtesy of John Deere)

Foreign Trade

In the 1980s, the farm machinery and equipment industry in the United States recognized a growing need in the international market. Prior to that time, the United States and Canada were the primary markets, and buyers there usually needed larger types of machinery and equipment. Foreign countries, in contrast, were more interested in smaller items, especially smaller tractors. As the need for these smaller tractors grew in Western Europe, and then eventually in the United States, major producers of farm machinery responded. As a result, domestic producers contracted with foreign firms to manufacture smaller tractors under the domestic producers' names and brands.[24]

Exports

Because most tractor manufacturers now have an international scope, foreign trade plays a major role in the economic success of the farm machinery and equipment industry. From 1972 to 1981, the value of U.S. farm equipment exports increased at an annual rate of 24 percent. However, the export market peaked in 1981 and has declined by about 8 percent per year since then.[25] Most of the decrease was due to declining markets for high-priced machinery such as large tractors and self-propelled combines.

Imports

Because of inflationary pressures in the farm machinery and equipment industry during the 1970s, American farmers became increasingly interested in the tractors made in and imported from other countries. Imported tractors tend to be 10 to 20 percent cheaper, depending on their size. The major foreign tractors available to American farmers included the former Deutz-Allis (Germany), Long (Romania), Satra Belarus (Russia), Same (Italy), and numerous small to medium-size tractors from Japan, such as the Kubota and Satoh.

Although exports have declined since 1981, imports have continued to increase. The sharp increase in the demand for small tractors for nonagricultural uses has increased the market for small tractor imports. Low-power (under 40 horsepower) tractors account for about 30 percent of all farm machinery imports. The value of all tractors and parts imported amounts to more than $1 billion. The United States imports about 56,000 low-power tractors, and about 90 percent of these come from Japan.[26]

The major purchasers of U.S.-made farm machinery are Canada, Western Europe, and Central America. The major sellers of farm machinery to the United States are Canada, Japan, the United Kingdom, and West Germany. Japan and Germany account for almost 80 percent of all foreign tractors sold in the United States.[27]

Trends in the Farm Machinery and Equipment Industry

The farm equipment industry was the first resource group to experience the effects of the farm financial crisis of the 1980s. As incomes declined, farmers responded by repairing older farm equipment instead of buying new, thus reducing domestic demand. In addition, the increasing value of the dollar relative to other currencies reduced foreign demand. Increasing imports reduced demand even more for farm machinery made in the United States. As a result, farm expenditures for tractors and farm equipment peaked at $11.7 billion in

24. Ibid., p. 293.
25. Rawlins, *Introduction to Agribusiness,* p. 77.
26. Ibid.
27. Ibid.

1970 and declined throughout the 1980s. In 1987, the industry experienced a slow recovery and sales volume increased to a peak in 1990 before experiencing another slight reduction. An "Agricultural Dealer Survey" conducted by Paine Webber in 1997 showed that since 1992 tractor sales have slowly increased.

Mergers

The large reduction in demand has resulted in several mergers of farm equipment manufacturers, a reduction in the number of dealerships, and a change in the types of tractors produced. Two major mergers of full-line companies occurred, and several long-line and short-line companies were purchased by full-line companies. In 1984, J.I. Case purchased the tractor division of International Harvester, creating Case-International, and in 1986 Steiger Equipment Company was purchased. Deutz-Allis was created in 1985 when Klockner-Humboldt-Deutz of West Germany purchased Allis-Chalmers. In 1980, a U.S. management team bought Deutz-Allis and formed AGCO. In 1991, AGCO acquired White Tractor Division, which earlier had acquired Oliver. AGCO also purchased Massey-Ferguson. Ford Tractor Company purchased New Holland Equipment Company from Sperry Rand Corporation in 1986 and was called Ford-New Holland for a while. Presently, Ford tractors are sold under the New Holland name. Ford also purchased Versatile in 1987 to expand its large-tractor line. Allied Products, a newcomer to the farm equipment industry, purchased White Equipment Company and several short-line companies, including Bush Hog, Kewanee, and New Idea. Refer to Figure 13–14 for an illustration of other mergers.

Products of the Farm Machinery and Equipment Manufacturers

Although we have concentrated on tractors and related tractor equipment up to this point, many other items are produced by the farm machinery and equipment industry. The agricultural industry includes the agribusiness input sector and agribusiness output (marketing) sector, as well as the agricultural production sector, so many different types of farm machinery and equipment are needed. Besides tractors, field tillage and planting equipment, and harvesting equipment, the following products are sold:

- farm structures and equipment (including prefabricated structures)
- greenhouses for ornamental horticulture
- forestry and logging equipment
- lawn, garden, and landscaping equipment
- light industrial power machines
- grain handling and storage equipment
- livestock feeding and handling equipment
- dairy feeding and milking equipment
- cotton ginning and compressing equipment
- fertilizer and pesticide equipment
- irrigation equipment

As you can see, manufacturers produce a tremendous variety of equipment for production agriculturalists, contractors, gardeners, and landscape specialists. Billions of dollars are spent on farm machinery and equipment. Production agriculturalists now own more than $2.5 billion worth of machinery and equipment.[28] Each year more nonfarm users purchase products from the farm machinery and equipment industry, which today also sell to landscaping firms, construction contractors, highway departments, parks, golf courses, rural and suburban homeowners, and home gardeners. Refer to Figure 13–15 for a list of major farm machinery and equipment manufacturers.

Careers in Farm Machinery and Equipment Manufacturing

Careers are available in farm machinery and equipment manufacturing businesses for both high school and college graduates.

Factory Jobs

Factory jobs in farm machinery and equipment manufacturing businesses include: the **foundry**, where molten metal is poured into molds to make

28. United States Department of Agriculture, *Census of Agriculture* (Washington, DC: U.S. Government Printing Office, 1997).

castings; the **forge shop**, where heavy hammers forge the red-hot steel billets into shape; the **machine shop**, with its **lathes**, **milling machines**, cutters, planes, **broaches**, and automatically controlled (and in many cases computer-controlled) machine tools; the paint shop, the assembly line; and the shipping department.[29]

Office and Professional Jobs

Several large divisions and departments of the farm machinery and equipment industry provide specialized, essential business services. Many of these jobs go to college graduates with a major in agribusiness or a related major. Refer to Figure 13-16. Although the names and job titles vary from one manufacturing business to another, the following departments or divisions are potential sources of employment:

- purchasing department
- sales and marketing division
- finance department
- engineering department
- consumer relations division
- human resource department
- parts department
- distribution department
- product planning division
- export division
- administrative (or executive) department[30]

Farm Machinery and Equipment Wholesalers

Once the farm machinery and equipment leaves the manufacturer, it goes to the wholesaler, which links the manufacturers to the dealerships. Because the manufacturers are often located far from the dealerships, the dealers cannot get prompt delivery or quick service. Dealerships want to be close to a main source of supply to be assured of readily available **wholegoods**, repair parts, and prompt service. The following are some important wholesale services:

- assembling a large inventory of equipment, machinery, and parts at a central point near dealerships in a region
- representing manufacturers and establishing good working relationships with retail outlets
- securing new retail dealers as necessary and contracting with them to sell the manufacturer's equipment and parts
- fulfilling the manufacturer's service and warranty obligations
- assisting retailers with merchandising methods, displays, and showroom arrangements
- helping retailers do sales training, sales promotions, advertising, and demonstrations
- assisting dealers in arranging service shops and parts departments, training mechanics, and inventorying parts
- providing management counseling to retail dealers
- assisting dealers with financing plans and programs[31]

Careers in Wholesaling

Many of the same jobs are available in wholesaling as in the manufacturing office and professional areas. Postsecondary and college training will be an asset in getting many of these jobs. Regardless of your formal education, you should gain knowledge and experience in crop production methods, the principal farming enterprises, harvesting and storage of crops, animal production, and farm power and equipment. Your knowledge and skills should also include marketing, economics, sales and sales promotion, accounting, communication, and public speaking. Especially important is a knowledge of the language of the production agriculturalist. Specifically, the following areas offer potential employment opportunities in wholesaling:

- purchasing agent
- publications preparation
- territory or zone manager
- traffic and transportation manager
- liaison officer
- marketing surveying
- business services
- salesperson

29. Smith, Underwood, and Bultmann, *Careers in Agribusiness and Industry*, p. 189.

30. Ibid., pp. 190-192.

31. Ibid., pp. 193-194.

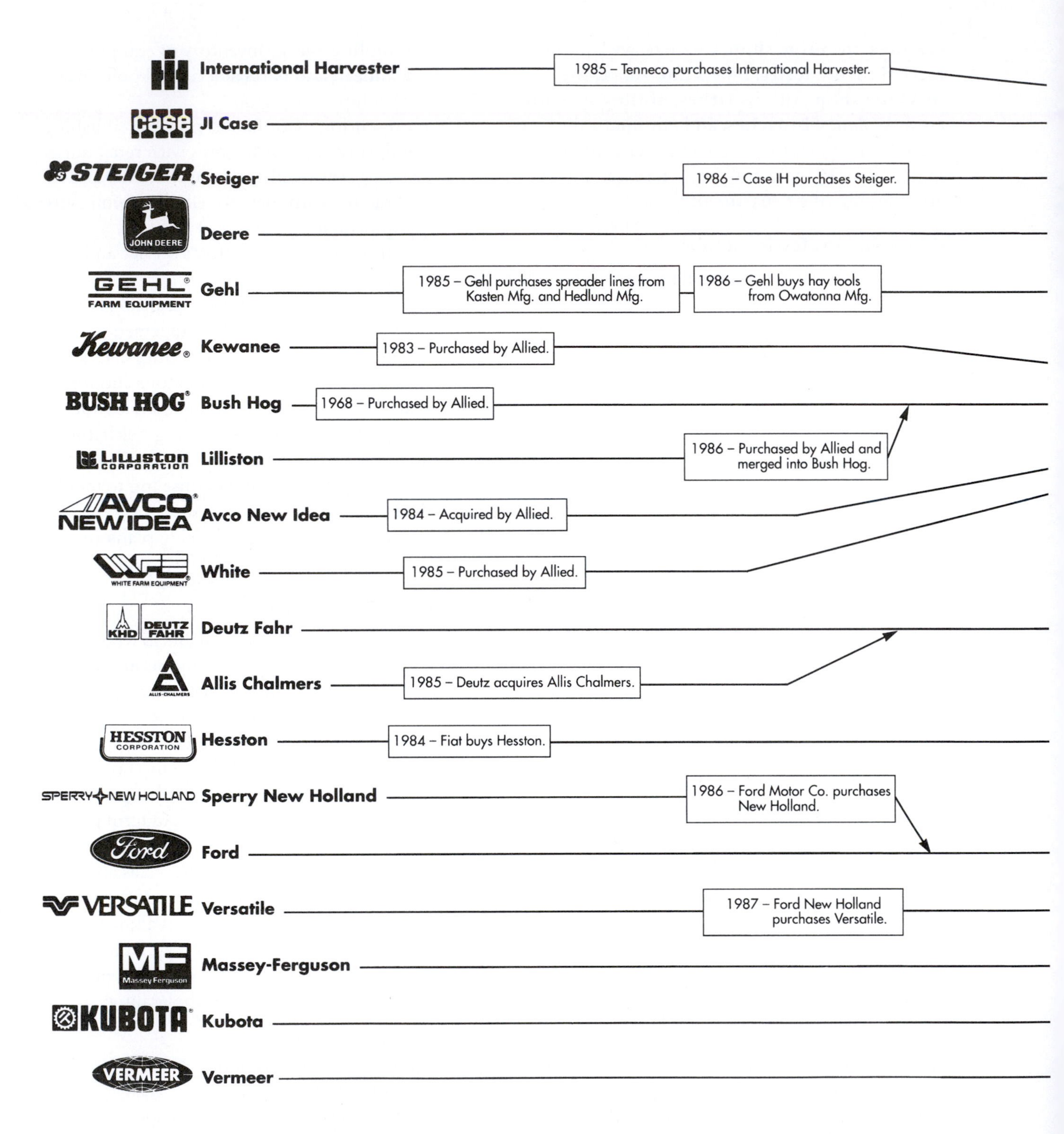

Figure 13–14 Many mergers within the farm machinery and equipment industry have occurred in the past 30 years. (*Courtesy of Agri-Marketing Magazine*, October, 1991)

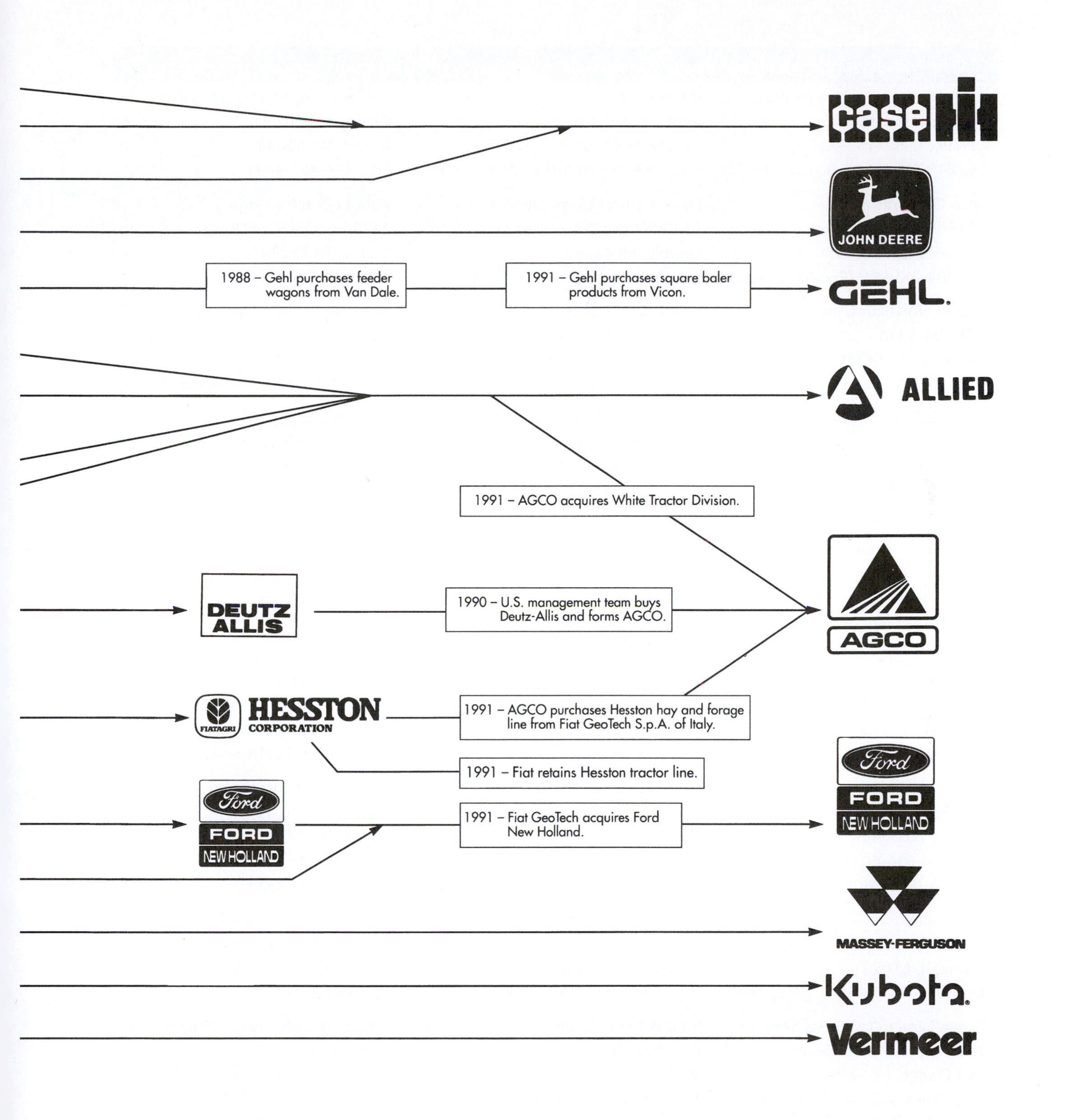

case IH
JOHN DEERE
1988 – Gehl purchases feeder wagons from Van Dale.
1991 – Gehl purchases square baler products from Vicon.
GEHL.
ALLIED
1991 – AGCO acquires White Tractor Division.
DEUTZ ALLIS
1990 – U.S. management team buys Deutz-Allis and forms AGCO.
AGCO
FIATAGRI
HESSTON CORPORATION
1991 – AGCO purchases Hesston hay and forage line from Fiat GeoTech S.p.A. of Italy.
1991 – Fiat retains Hesston tractor line.
Ford
FORD
NEW HOLLAND
1991 – Fiat GeoTech acquires Ford New Holland.
Ford
FORD
NEW HOLLAND
MASSEY-FERGUSON
Kubota.
Vermeer

Farm Machinery and Equipment Manufacturers

A. O. Smith Harvestore Products Inc.
345 Harvestore Drive
DeKalb, IL 60115
Tel.: 815-756-1551/Fax: 815-756-7821

Acco Corporation
5295 Triangle Parkway
Norcross, GA 30092
Tel.: 404-447-5546/Fax: 404-246-6118

Badger Northland, Inc.
P.O. Box 1215
Kukauna, WI 54130
Tel.: 414-766-4603

Briggs & Stratton Corporation
P.O. Box 702
Milwaukee, WI 53201
Tel.: 414-259-5333
Fax: 414-259-5338

Bush Hog Corporation
P.O. Box 1039
Selma, AL 36702
Tel.: 205-872-6261/Fax: 205-872-0168

Butler Manufacturing Co.
7400 East 13th Street
Kansas City, MO 64126
Tel.: 816-968-6141

Case J.I.
700 State Street
Racine, WI 53404
Tel.: 414-636-6011, 636-7394
Fax: 414-636-7809

Caterpillar, Inc.
100 Northeast Adams Street
Peoria, IL 61629
Tel.: 309-675-1000/Fax: 309-675-6155

Cenex/Land O'Lakes Ag. Services
2827 Eighth Avenue South
Fort Dodge, IA 50501
Tel.: 515-576-7311
Fax: 515-576-2685

Cummins Engine Company, Inc.
500 Jackson Street, P.O. Box 3005
Columbus, IN 47201
Tel.: 812-377-3695

Deere & Co.
John Deere Road
Moline, IL 61265
Tel.: 309-765-8000/Fax: 309-765-5772

Detroit Diesel Corporation
13400 Outer Drive, West
Detroit, MI 48239
Tel.: 313-592-7642

Deutz Corporation
5295 Triangle Parkway
Norcross, GA 30092
Tel.: 404-564-7100

Ford New Holland Americas, Inc.
500 Diller Avenue
New Holland, PA 17557
Tel.: 717-355-1371
Fax: 717-355-3192

Vermer Manufacturing Company
One Mile E. New Sharon Road
P.O. Box 200
Pella, IA 50219
Tel.: 515-628-3141/Fax: 515-628-7734

Gehl Co.
143 Water Street
West Bend, WI 50395
Tel.: 414-334-9461/Fax: 414-334-1565

Homelit Div. of Textron, Inc.
14401 Carowinds Boulevard
Charlotte, NC 28273
Tel.: 704-588-3200
Fax: 704-588-0926

J. I. Case-A Teneco Company
700 State Street
Racine, WI 53404
Tel.: 414-636-7843

John Blue Company
P.O. Box 1607
Huntsville, AL 35807
Tel.: 205-721-9090

Keystone Steel & Wire Co.
7000 S.W. Adams Street
Peoria, IL 61641
Tel.: 309-697-7020/Fax: 309-697-7422

Kohler Company/Power Systems
Group 444 Highland Drive
Kohler, WI 53044
Tel.: 414-457-4441

Kubota Tractor Corp.
13780 Benchmark Drive
Dallas, TX 75234
Tel.: 214-241-5900

Massey-Ferguson Operations
5295 Triangle Parkway
Norcross, GA 30092
Tel.: 404-447-5546/Fax: 404-409-6040

McCullough & Co.
4717 E. 119th Street
P.O. Box 9824
Kansas City, MO 64134
Tel.: 816-761-2857/Fax: 816-761-2845

Morton Buildings, Inc.
P.O. Box 399, 252 W. Adams Street
Morton, IL 61550
Tel.: 309-263-7474

New Holland, Incorporated
500 Diller Avenue, P.O. Box 1895
New Holland, PA 17557
Tel.: 717-355-1121/Fax: 717-355-1826

Patz Sales, Incorporated
Highway 14 S., P.O. Box 7
Pound, WI 54161
Tel.: 414-895-2251
Fax: 414-897-4312

Robison & Grisson/MT
Route 2, Box 36
Guntown, MS 38849
Tel.: 601-869-1028

White-New Idea Farm Equipment Company
123 W. Sycamore Street
Coldwater, OH 45828
Tel.: 419-678-5311/Fax: 419-678-5496

Figure 13–15 There are many full-line, long-line, and short-line farm machinery and equipment manufacturers, which have retail dealerships throughout the country. The names and home office addresses of several of these companies are listed here.

Figure 13–16 Many careers are available in farm machinery and equipment manufacturing. Specialized training is needed in many areas, and an essential skill for almost everyone is the ability to use computer technology. (Courtesy of USDA)

Dealerships

A major consolidation of farm machinery dealerships has occurred in the past 30 or so years. The number of dealerships fell from 16,700 in 1967 to 9,500 in 1972, then to 7,900 in 1989. The number dropped even further over the following decade. However, average annual equipment sales per dealer increased from about $2 million in 1979 to more than $3 million in 1988. USDA reports show that farm equipment companies sold $4.6 billion worth of farm machinery during 1995.[32] Search the Internet and compare the most recent sales figures with those of 20 years ago. Many of the companies have added or switched their emphasis to lawn and garden equipment, including items for use on golf courses and for other nonfarm or industrial use.

To survive, dealers have cut costs in every way possible, increasing in size to gain **economy of scale**, expanding the variety of farm equipment they carry, and in some cases expanding into nonfarm equipment lines. Some analysts predict that future farm equipment dealerships will resemble supermarkets, offering a wide variety of tractors and equipment.

A Complex Business

Dealerships are a very complex type of business because of the many different pieces of equipment handled, the parts inventory maintained, the servicing done, and the equipment repairs offered. In addition to the capital outlay required to start a business, the fact that many sales are made on credit means that availability of funds is often delayed. Manufacturers of farm equipment also provide consumers with financing, which is processed by or through the dealership. Challenges that add to the complexity of dealerships include:

- sales of used machinery that reduce sales of new machinery
- increasing demands for better service and quicker repairs
- increased credit needs of farmers
- increased popularity among farmers of leasing farm equipment
- need for more automated livestock production equipment
- increased specialization in the handling and sale of equipment
- safety and environmental issues, such as oil disposal[33]

Dealership Careers

Although dealerships are complex, they offer a good place to start a career in the farm machinery and equipment industry. A wealth of experience can be gained from such an experience. By being on the ground floor, where you can see the whole business, you can learn a great deal about farm markets, management, customer relations, finance, product promotion, and sales. Company training is also available. For example, Deere has an excellent program, which includes study of marketing, parts, credit, services, and management. Specific jobs available at a dealership include:

- retail operations
- purchasing
- facility management
- sales
- parts department
- service department
- advertising
- finance, credit, collections
- accounting
- management[34]

32. United States Department of Agriculture, *Agricultural Statistics 1997* (Washington, DC: U.S. Government Printing Office, 1997).

33. Roy, *Exploring Agribusiness*, p. 33.
34. Smith, Underwood, and Bultmann, *Careers in Agribusiness and Industry*, pp. 197–198.

Refer to the Career Options section for the skills needed to be an agricultural equipment dealer.

Farm Custom and Contracting Services

Custom and contracting services are another segment of the farm machinery and equipment industry. They are vital services that benefit both the operator and the customer.

Custom Farming

Production agriculturalists benefit from the work of custom farmers. Tilling soil, planting and cultivating crops, applying pesticides, harvesting crops, grinding and mixing feed, baling hay and straw, picking cotton, transporting milk, and picking and packing fruits and vegetables are all types of jobs performed by custom farmers.[35] Refer to Figure 13-17.

One major reason that production agriculturalists use custom farmers is that it is economically wise. For example, a new combine may cost more than $100,000, but it may be used for only two months per year. Rather than invest such a large sum of money in equipment, the production agriculturalist may pay a custom farmer to harvest a crop. The cost is lower than purchasing new equipment and may result in a faster harvest if the custom farmer owns several combines. A faster harvest prevents weather damage and many other types of loss.[36]

Custom farmers charge for their services based on the job. For instance, those who harvest grain or hay may charge by the bushel or bale; tillage may be priced by the acre; and so on. As long as the costs are reasonable, both the custom farmer and the production agriculturalist benefit: the former gets extended use from the equipment (so that the purchase price is amortized more quickly), and the latter saves money on equipment and gets the job completed more quickly.[37]

35. Jasper S. Lee, James G. Leising, and David E. Lawver, *Agrimarketing Technology: Selling and Distribution in the Agricultural Industry* (Danville, IL: Interstate Publishers, 1994), pp. 162–163.
36. Ibid., pp. 162–163.
37. Ibid., p. 164.

Figure 13–17 Certain farmers perform a variety of jobs that are necessary for production agriculturalists, offering several economic advantages. (Courtesy of Holland North America, Inc.)

Farm Contracting Services

Another career area in the farm machinery and equipment industry is **farm contracting** services. This field includes special jobs done by individuals with specialized equipment. Specific skills are needed, such as mechanical operation, maintenance and repair; effective cost control through estimating; bidding; and business management. Some of the jobs that can be very profitable are:

- clearing land
- leveling and reforming land surfaces
- constructing irrigation ditches
- installing drainage systems
- terracing and constructing water courses
- developing watersheds
- doing aerial spraying[38]

Conclusion

The agricultural input (supply) industries have developed mostly since the end of World War II. The main reason the agribusiness input sector has flourished is because the supplies it sells are cheaper and more effective than what production agriculturalists could provide for themselves. The development of the agribusiness supply sector has permitted production agriculturalists to concentrate more on production, and in the process has made the supply firms a major force—and major

38. Smith, Underwood, and Bultmann, *Careers in Agribusiness and Industry,* pp. 202–205.

Career Option

Agricultural Equipment Dealer

A career as an agricultural equipment dealer involves daily interactions with production agriculturists, mechanics, technicians, salespeople, customers, office staff, and the entire farm community. During years when production agriculture is profitable, the business is likely to do well, but during low-profit years for farms, an agricultural equipment dealership often takes financial losses. The success of the business is tied directly to the economic conditions of those in production agriculture.

Many agricultural equipment dealers first entered the business as salespeople or mechanics and only later were able to generate enough credit to buy into the ownership of the agribusiness. Large amounts of capital are needed to maintain a dealership's inventories of agricultural equipment and parts. An agricultural equipment dealer typically has a huge investment in the business. In addition to the new and used farm equipment kept in inventory, replacement parts must be purchased and stocked for resale; service and repair departments must be staffed and operated properly as well.

Success in an agricultural equipment dealership requires a good understanding of the machines that are used in producing crops. An agricultural background is useful in understanding the kinds of conditions in which agricultural machines are operated. Formal education and a strong background in business practices are also important in managing agricultural equipment dealerships. Good public relations skills and a real interest in providing services to people are important for success in this agricultural business.

An agricultural equipment dealer must follow up and assist the producer to make sure new equipment is functioning properly. (Courtesy of USDA)

employers—in agribusiness today. The agribusiness input sector provides approximately 75 percent of the total inputs used by production agriculturalists. The remaining 25 percent includes items, such as hay, livestock, and grains, which are produced on the farm for use as an input to other production of agricultural products.[39]

39. James G. Beierlein and Michael W. Woolverton, *Agribusiness Marketing: The Management Perspective* (Englewood Cliffs, NJ: Prentice-Hall, 1991), pp. 105–106.

Summary

Historically, the pattern of agribusiness input has changed. As America gradually industrialized, the production of agricultural products became more specialized, and it became more economical for farmers to purchase production inputs from others. The agribusiness input sector is big business. It handles many products, which are manufactured or processed by hundreds of firms. About $200 billion is spent on agribusiness inputs by American production agriculturalists, and more than 6 million people have jobs filling the demand in the agribusiness input sector.

The four biggest products of the agribusiness input sector are feed, fertilizer, agricultural chemicals, and farm machinery and equipment. Many other inputs are also required for production agriculturalists, such as seed, lime, diesel fuel, gasoline, motor oil, transmission oil, hydraulic oil, veterinary supplies, and many other products.

Among the numerous agribusiness supply expenditures of production agriculturalists, feed for livestock is the largest, accounting for about 14 percent of total farm expenses. Production agriculturalists spend more than $30 billion per year on purchased feed. There are over 1,500 different combinations of feed ingredients for poultry, swine, beef, and dairy animals. The production and market structures of the feed industry are complex and interrelated: feed starts with premix suppliers, goes through several processes, and ends up at the retail store for consumer purchase. There are many opportunities for employment in the feed industry.

Fertilizer was first manufactured for sale in Maryland in 1849. The three major plant food nutrients found in fertilizers are nitrogen, phosphorus, and potassium. Production of basic fertilizer products in the United States is carried out by about 50 manufacturers and fertilizer is further processed in more than 10,000 local marketing centers. Production agriculturalists in the United States spend approximately $11 billion yearly on fertilizers. Environmental policy will likely affect the fertilizer industry in the future, adding to the already abundant career opportunities.

Due to successful efforts in chemical and biological research, weeds, insects, and diseases are now fairly well controlled in the United States, primarily by use of synthetic chemicals. It is estimated that there are 10,000 species of harmful insects, more than 1,500 diseases caused by fungi, 1,800 different weeds, and approximately 1,500 different types of nematodes that cause damage to crops and livestock. Production agriculturalists spend about $10 billion yearly on pesticides. Although pesticides are very beneficial, they do cause some environmental damage. Research is being conducted to find effective pesticides that will not be harmful to water, air, soil, wildlife, humans, or the food we consume.

Almost all agribusiness input supplies get to the production agriculturalist by way of farm supply stores. There are approximately 36,000 farm supply stores, which sell the inputs and items discussed in this chapter. The three major categories of farm supply stores are: farmer cooperatives; chain-store corporations; and small, single (sole) proprietorship feed and supply stores. Many job opportunities are available with farm supply stores.

Farm machinery companies are usually classified as full-line, long-line, or short-line. The full-line companies dominate the industry and are highly concentrated. The top 2 companies account for 57 percent of all sales; the top 6 companies make 93 percent of all sales. Most of the full-line companies are international in scope, with manufacturing plants and distribution centers throughout the world. Most long-line companies are national in scope, whereas short-line companies generally serve regional and local markets. About 1,500 firms are involved in producing farm machinery in the United States. These firms sell their products through approximately 7,000 retail outlets, which are franchised by the major manufacturers.

The major farm equipment companies are highly diversified and do not limit their products to farm equipment. Many have expanded heavily into nonfarm product lines. Changes made in the farm machinery and equipment industry in the 1980s also resulted in the industry shifting from being primarily domestic to being competitive internationally. Although exports have declined since 1981, imports have continued to increase.

The trend since the 1980s has been for farm equipment manufacturers to merge, the number of dealerships to decrease, and the types of tractors produced to change to smaller models. Each year, more nonfarm users purchase products from the farm machinery and equipment industry, which now sells to landscaping firms, construction contractors, highway departments, parks, golf courses, rural and suburban home owners, and home gardeners. As a result, many careers are available in the farm machinery and equipment manufacturing industry.

Once a piece of farm machinery or equipment leaves the manufacturer, it goes to a wholesale company. These companies link the manufacturers to the dealerships. Dealerships conduct a very complex type of business because they must handle many different pieces of equipment, parts inventory, and the servicing and repair of equipment. Many dealerships have been consolidated, resulting in fewer dealerships, but the annual equipment sales and dollar volume per dealer have increased substantially.

Farm custom and contracting services form another segment of the farm machinery and equipment industry. These are vital services of benefit to both the operator and the customer. The farm machinery and equipment industry has raised the productivity and profit of the American production agriculturalist. Furthermore, purchasing patterns established in the farm machinery and equipment industry spread to other agricultural inputs and American business in general.

End of Chapter Activities

REVIEW QUESTIONS

1. Define the Terms to Know.
2. What changes have occurred in the agribusiness input sector since the beginning of the 20th century?
3. Name the four major agribusiness input products.
4. Name 10 minor agribusiness inputs.
5. Name seven agribusiness inputs that are often overlooked.
6. What is unique about the feed industry among the industries that provide supplies to production agriculturalists?
7. Name and explain the three major categories of commercial feeds.
8. Name 10 potential careers available within the feed industry.
9. What are the three major plant food nutrients found in fertilizers?
10. What is contained in a 100-pound bag of fertilizer labeled 4-12-12?
11. How is nitrogen fertilizer manufactured?
12. What are the three major sources of nitrogen in solid form?
13. How are phosphates made?
14. What are the four major sources of phosphates?
15. Where are the major sources of phosphates found in the United States?
16. What are the major sources of potassium in commercial fertilizer?
17. Where are the major potassium deposits in North America?
18. What three things affect the amount of fertilizer used?
19. Describe the interrelationships among fertilizer, pesticides, and other chemicals.
20. Briefly discuss the discovery of herbicides.
21. Briefly discuss the discovery of fungicides and insecticides.
22. What are seven benefits of pesticides?
23. What are 15 types of supplies and services available at large consolidated and centralized farm supply stores?
24. Name 10 items sold by farm supply chain stores.
25. What are the four major full-line tractor companies?

26. Compare the trend in tractor size manufacturing in both the United States and foreign countries.
27. What has been the trend for export of tractors?
28. What has been the trend for import of tractors?
29. Describe four mergers of major tractor companies that have occurred in the past 30 years.
30. List 16 products produced by the farm machinery and equipment industry.
31. Name eight nonfarm user groups of products produced by the farm machinery and equipment industries.
32. What are eight factory-type jobs that could lead to a career in the farm machinery and equipment industry?
33. What are 11 office and professional jobs that could lead to a career in the farm machinery and equipment industry?
34. List nine important services of wholesalers.
35. What are 12 areas of knowledge, experience, and skills that should be attained by people preparing for careers in farm machinery and equipment wholesaling?
36. List eight areas that offer potential employment opportunities in farm machinery and equipment wholesaling.
37. Name five factors that make dealerships complex businesses.
38. What are seven challenges that add to the complexity of dealerships?
39. List 10 specific jobs available at dealerships.
40. What are the advantages to a production agriculturalist of using a custom farmer?
41. List seven types of jobs that custom farmers perform.
42. List seven types of jobs of those who do farm contracting services.

FILL IN THE BLANK

1. In 1910, labor accounted for ___________ percent of the agriculturalist's total production expenses.
2. Today, capital items such as buildings, livestock, machinery, equipment, and operating inputs account for ___________ percent of production costs.
3. In 2005, more than ___________ billion dollars was spent on agribusiness inputs by American production agriculturalists.
4. Production agriculturalists' largest expenditure on inputs is for livestock feed, representing about ___________ percent of farm expenses.
5. About ___________ percent of the feed used comes from the manufactured feed industry.
6. The United States produces about ___________ percent and consumes about ___________ percent of the world's feed grains.
7. About 25 percent of feed retail stores sell ___________ percent of the feed.
8. More than ___________ percent of formula feed is sold in bulk.
9. The first mixed fertilizers sold in the United States for commercial purposes were manufactured in Maryland in the year ___________.
10. ___________ ___________ will likely affect the fertilizer industry in the future.

11. One of the first and most popular insecticides, which is now illegal, is ___________.
12. There are approximately ___________ farm supply stores.
13. Approximately ___________ percent of the supplies purchased by production agriculturalists come from farm cooperatives.
14. Production agriculturalists purchase about ___________ percent of their fertilizer and lime, ___________ percent of their feed, ___________ percent of their seed, and ___________ percent of their agricultural chemicals from farmer cooperatives.
15. Four full-line tractor companies account for ___________ percent of all farm machinery sold in the United States.
16. There are approximately ___________ firms producing farm machinery through approximately ________ retail outlets.
17. Japan and Germany account for almost ___________ percent of all foreign tractors sold in the United States.

MATCHING

a. 8,000
b. 32 million
c. 1,500
d. 5.5 million
e. 11 billion
f. 6 million
g. 100,000–150,000
h. 10 million
i. 10,000
j. 30 billion
k. 1,800
l. 10 billion
m. International Harvester
n. White Motor Company
o. J.I. Case

_____ 1. number of jobs in the agribusiness input sector
_____ 2. amount production agriculturalists spend yearly on purchased feed
_____ 3. number of animal-feed processing plants
_____ 4. number of people the feed industry employs
_____ 5. number of different combinations of feed ingredients
_____ 6. number of tons of nitrogen produced in the United States yearly
_____ 7. number of tons of phosphate produced in the United States yearly
_____ 8. number of tons of potassium produced yearly in North America
_____ 9. amount spent yearly by production agriculturalists on fertilizer
_____ 10. estimated number of harmful insects causing damage to crops and livestock
_____ 11. number of different weeds causing damage to crops and livestock
_____ 12. amount spent yearly on pesticides
_____ 13. major producer of backhoes and other industrial equipment
_____ 14. major producer of industrial trucks
_____ 15. once made automobiles

ACTIVITIES

1. Suppose that someone denigrates or makes light of the agricultural industry. Given some facts and figures from this chapter, what can you tell them about the significance of agribusiness in the U.S. economy?
2. Visit the largest farm feed and supply/support store in your community. List as many products as you can that it sells. Your teacher may wish to set a limit on the number or assign product categories to various students (example: animal health, pesticides, etc.).
3. Visit your local feed store. Find out its parent company (Purina, Cargill, etc.) and locate the home office address in Figure 13–6. Write this down and keep it for future reference if you decide to seek employment with the company.
4. Follow the same process as in Activity 3 for fertilizers, pesticides, and animal health products. Refer to Figures 13–8 and 13–9.
5. Suppose you are working the sales desk of a local farm supply store and a customer orders a ton of 15–15–15 fertilizer. She wants to know how many actual pounds of nitrogen she is buying. What is your answer?
6. The agricultural industry is constantly monitoring the use of fertilizers, pesticides, and other agricultural chemicals. Write a paper or locate an article in a newspaper or magazine about a particular environmental concern. Present this to the class.
7. The agricultural industry is often criticized unjustly regarding matters of health or the environment. Write a report or locate an article citing an example. Present this to the class.
8. Several agribusiness input supply companies were listed in this chapter. Select five companies that interest you as potential future employers. Give reasons for your selection of each and share these with your class.
9. Refer to Figure 13–15 for a list of the various farm machinery and equipment manufacturers. Identify each as a full-line, long-line, or short-line company.
10. Identify three different tractors owned by you, a neighbor, or someone in your community. Match each one with one of the four major tractor companies that exist today. Refer to Figure 13–14 for help.
11. List the farm machinery and equipment businesses in your community (if any) that are either manufacturers, wholesalers, or dealerships. Using this chapter as a guide, list the potential jobs available at those businesses.
12. State whether you think it is better to buy equipment for all farm jobs or better to hire custom farmers for certain jobs. Defend your answer. (You might want to debate this in front of the class.)
13. Identify five farm contracting services in your community. Select one of these and give a report as to its profit potential.

Economic Activity and Analysis

Objectives

After completing this chapter, the student should be able to:

- Discuss economic activity in the United States.
- Measure economic activity through economic indicators.
- Explain the function of money and banks.
- Explain the role of the federal government in promoting economic stability.
- Explain the basis of economic policy.
- Discuss the nation's economic goals.
- Describe various types of economic analysis.
- Explain ways to attain economic efficiency.

Terms to Know

artificial insemination
astute
bank money
barter
base year
business cycle
cloning
commodity money
consumer price index (CPI)
contraction
deflation
depression
discouraged workers
economic fluctuations
economic indicators
economic policy
economic recovery
equal product curve
Federal Reserve System
fiat money
fiscal policy
gross domestic product (GDP)
hypothesis
implicit GDP price deflator
inflation
law of increasing cost
marginal analysis
marginal benefit
marginal cost
market basket
monetary policy
opportunity cost
peak
producer price index
production possibilities
prosperity
recession
scenario
standard of living
stock prices
trough
value judgments
variables

Introduction

It is sometimes hard to separate the area of agricultural economics from the fields of study of other social sciences, such as sociology, anthropology, political science, and psychology. Both agricultural economics and these other social sciences investigate individual and social behavior. Agricultural economics focuses on behaviors that affect the way individuals and societies produce and consume goods and services.[1] A part of understanding agricultural economics is understanding economic activity and its relationship to federal, state, and local governmental actions. For example, it is difficult to understand what federal, state, and local governments do if we do not comprehend the *economic* circumstances underlying their actions. In reality, government budgets are economic documents. Taxation and government spending are economic tools that a political system uses to achieve economic, as well as political and social, objectives. To understand issues such as the national debt, budget deficits, and the welfare system, we must also understand economics.[2]

The role of the family in society and how the family behaves as an economic unit are also an integral part of agricultural economics. To some extent, even when and whom we marry, the number of children we have, and interpersonal relationships within the family are determined by economics. Therefore, an understanding of economic activity, economic goals of the nation, and methods of analyzing these goals and activities are crucial to understanding the U.S. economic system.

Economic Activity in the United States

Many factors are involved in maintaining a strong economy. These factors include efficient production of agricultural products, which frees many workers for industrial and nonagricultural work; efficient production of goods and services; and good wages so consumers can purchase goods, services, and entertainment. The cycle is continuous: the more money spent, the stronger the economy.

Purpose of the Economist

Obviously, the job of economists is to alert us when the factors making up our economy are beginning to get unbalanced. An unbalanced economy can cause a snowball effect, leading to unemployment, which in turn leads to people buying fewer goods and services. If fewer goods and services are being purchased, the industrial segment of the economy produces less, causing higher unemployment and resulting in even less money available to purchase goods and services. This snowball cycle is also continuous.

Every day, economists and business leaders provide forecasts, reports, and predictions about the state of the economy. These reports appear as often as reports about the weather.

The Great Depression

Why does the state of the economy require daily attention? What is so important about economic information? The answer is fear of another Great Depression. History has taught the world a great deal about the effects of changes in the state of the economy. The Great Depression, which occurred in 1929 and lasted for years thereafter, painfully illustrated the results of a severe economic decline.

In the summer of 1929, there was a peak of economic prosperity. A few months later, on a day that has become known as Black Thursday (October 24, 1929), **stock prices** collapsed. From Black Thursday in 1929 to about 1939, the United States was in a **depression**. Investors lost millions of dollars and the downward economic snowball cycle began. Great declines in production and employment occurred. Factories shut down. Businesses and banks failed. As businesses failed, more and more people lost their jobs. By 1932, more than 12 million people were unemployed, or 25 to 30 percent of the workforce.[3]

Business Cycles and Patterns

If you study the economic changes that have taken place in the past, you will notice an irregular cycle of

1. Fred M. Gottheil, *Principles of Economics* (Cincinnati, OH: South-Western College Publishing, 1996), pp. 6–7.
2. Ibid.
3. Betty J. Brown and John E. Clow, *Introduction to Business* (New York: Glencoe/McGraw-Hill, 2006), p. 43.

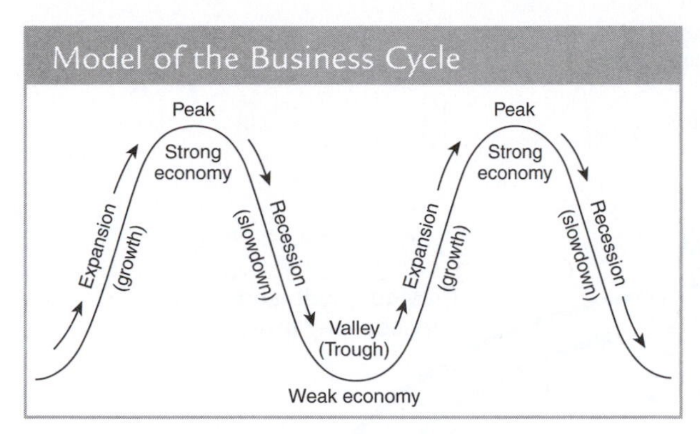

Figure 14–1 A business cycle is the repeated rise and fall of economic activity over time. A business cycle has four phases: expansion, peak, recession, and valley.

ups and downs. The ups and downs in economic activity are called **economic fluctuations**. The repeated rise and fall of economic activity over time is called a **business cycle**.[4]

Economic Recovery. Figure 14–1 illustrates a business cycle. A rise in business activity is **economic recovery**. The rise indicates an expansion of the economy (economic growth). A *recovery* is a period of economic growth or expansion following a **recession** or depression.

Peak. When economic activity reaches its highest level during a business cycle, it has hit a **peak**. During a peak, **prosperity** is evident: the economy expands rapidly, new businesses are started, unemployment rates are low, and goods and services are produced at high rates. Normally, inflation is moderate and numerous jobs are available.[5]

Contraction. When economic activity slows down during a business cycle, the resulting drop in the growth of the economy is referred to as a **contraction**. Contraction may be a sign that the unemployment rate is increasing or that prices of goods and services are stable or beginning to fall.

Trough. When economic activity reaches its lowest level in a business cycle, it is in a valley or **trough**. When the economy declines over a period of time to the point that spending dramatically decreases, there is less demand for goods and services, and unemployment rates rise sharply, the economy is said to have entered a *recession*. A recession that persists for several years is referred to as a *depression*.[6]

Circular Flow of the Economy

Figure 14–2 shows the circular flow of economic activity. This diagram illustrates how the economy's resources, money, goods, and services flow between households and firms through the resource and product markets.

In Figure 14–2, people appear as both consumers and producers. They are consumers in the sense that they purchase and use goods and services bought on the product market. At the same time, they are producers in that they provide their resources (their labor, property, money, and management) to firms to use in creating more goods and services for the product market. Entrepreneurs, hoping to make a profit, influence and provide direction for a firm's production of goods and services.[7]

Specific Explanation of Circular Flow. Figure 14–2 illustrates the relationship of the different parts of our economic system. Individuals sell the factors of production (land, labor, capital, and management) to businesses and, in return, gain income such as wages, rents, dividends, and profits. Businesses purchase the factors of production in order to produce and sell goods and services. Using some of the money that businesses generate, they pay taxes to the government. The government, in turn, provides public services and other payments to individuals. Likewise, individuals use their wages to purchase good and services, and they pay income taxes on their earnings. They also provide labor and land to the government. Thus, the cycle is completed.[8]

Your Effect on the Circular Flow. Suppose you get a job selling flowers and shrubs at a lawn and garden center. You are a factor of production to the garden center because you sell your labor to it. You receive a paycheck from the garden center and the money flows back to you. Some of your paycheck flows to the government, in the form of taxes, to provide roads and other services. With some of your weekly

4. Ibid., p. 40.
5. Ibid., p. 41.
6. Ibid., pp. 41–43.
7. Gottheil, *Principles of Economics,* p. 9.
8. Brown and Clow, *Introduction to Business,* p. 40.

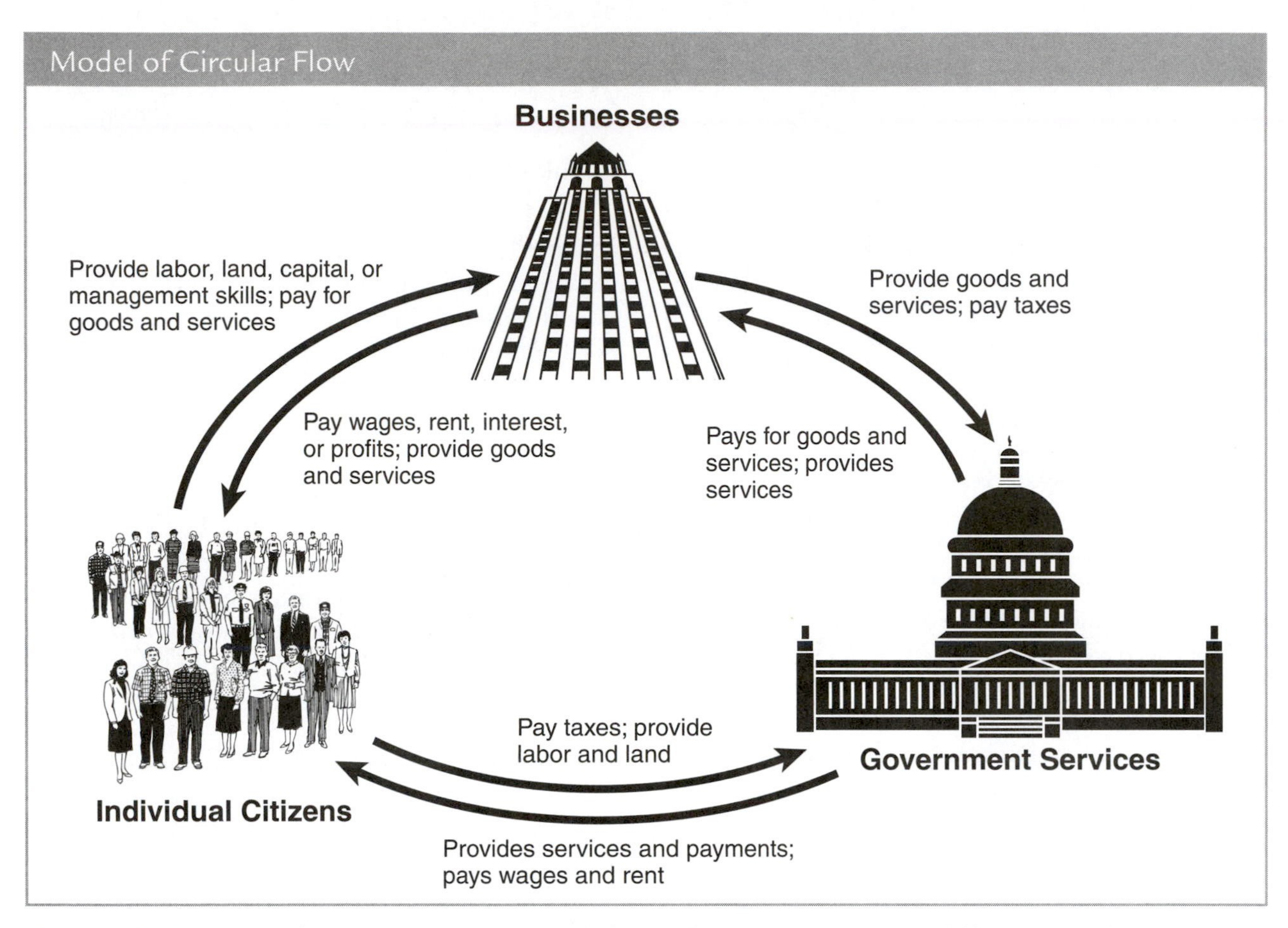

Figure 14–2 The circular flow model illustrates the high degree of dependence among the different parts of our economic system.

paychecks, you buy a lawnmower for your own lawn care business. Money flows from you to the agribusiness that sells you the mower. Taxes on the sale of the lawnmower and the business's profits flow to the government, completing the circle.

Interdependence. High degrees of dependence among the different parts of an economic system are illustrated by the circular flow. Consider the following:

> If people become concerned about the stability of the economy, they may reduce their spending. The reduction in spending leads to a reduction both in the amount of goods and services needed and in the number of resources needed to produce them. The final result is a reduction in the amount of wages earned and issued. In contrast, the sequence can be totally reversed if people have a positive view of the economy. They spend more money; more goods, services, and resources are needed; and more income is earned and distributed.

Measuring Economic Activity via Economic Indicators

The illustration of the circular flow of the economy showed us how economists study and define economic activity. However, it does not reveal the future state of the economy. This is done by **economic indicators**. Economic indicators measure business cycles and economic activity through a variety of data and statistics. These indicators normally change before the rest of the economy does. There are many economic indicators, but three major economic indicators are discussed here: gross domestic product, inflation, and unemployment.

Gross Domestic Product

One indicator of how well our economy is performing is how many goods and services it produces during a certain period of time, usually a year. **Gross domestic product (GDP)** is the total value of goods and services produced in a country in a given year. To calculate the gross domestic product, economists add to find a total of goods and services. There are four main types of good and services:

- consumer goods and services
- business goods and services
- government goods and services
- goods and services sold to other countries[9]

Determining Economic Growth. A dollar value is assigned to the goods and services to find the sum. The gross domestic product in 1970 was more than $1 trillion; in 1980 it was just over 2.7 trillion. In 1997, the gross domestic product was approximately $7 trillion. Evaluating the change in gross domestic product from one year to the next is one way to determine whether there has been economic growth.[10]

Economists track variations in the growth of the GDP on an annual basis to determine the health of the economy. If the GDP grows steadily or rapidly, economic policies are generally left as they are. If the GDP fails to grow at a significant pace, the economy may be experiencing problems that require governmental intervention to correct.[11]

Standard of Living. The health of a country's gross domestic product may also measure the **standard of living** of its people—that is, the amount of goods and services available to and utilized by individuals.[12] A nation will have a high standard of living when its economy can meet its citizens' basic needs with affordable goods and services.

Inflation

Rising or falling prices affect the dollar value of the gross domestic product. A protracted rise in the general price level of goods and services is called **inflation**. Business and government leaders consider the inflation rate to be an important general economic indicator. For example, last year a gallon of milk may have cost $2.00. This year, it may cost $2.25. The physical output (the gallon of milk) has not changed; only its money value has changed.

Inflation and the Purchasing Power of Money. Inflation reduces the "purchasing power" of money. Simply put, your money buys less. Sometimes people describe inflation as a time when "a dollar is not worth a dollar anymore." The purchasing power of a dollar is measured by the amount of real goods and services that it can buy. For example, if your income remains the same, but the price of goods increases, you cannot buy as much, and the purchasing power of your dollar has decreased. Conversely, if the price of goods declines over a marked period of time, through a process called **deflation**, the purchasing power of your dollar increases. However, deflation rarely occurs.[13]

Losing from Inflation. Perhaps more than any other single group, people on fixed incomes, such as retirees, have reason to worry about inflation. Retired people cannot count on increases in income as prices rise. Because of declining purchasing power, a person on a fixed income cannot keep up with rising prices. Many retired people have fewer goods and services to enjoy because of the rising prices that signal inflation. Refer to Figure 14-3 for a **scenario** on the plight of retirees during times of inflation.

Fixed Income and Inflation

Perhaps more than any other single group, people living on fixed incomes, such as retirees, have to worry about inflation. Back in 1958, Maria and her husband bought a deferred annuity that cost them $100 monthly. In 1998, it started to pay them $700 a month in retirement benefits. They were excited about the prospect of living comfortably in retirement on the savings they had accumulated. After all, back in 1958, when they put their retirement plans together, their rent was $135 per month; a new automobile cost $1,500; milk was 25 cents a quart; and a first-run movie ticket cost 35 cents. What they failed to factor into their plans, however, was inflation.

When Maria and her husband retired in 1998, they began to receive their $700 each month, yet their retirement dreams were shattered. The $700 does not come close to covering their apartment rent, and a new car—now at $17,000—is simply out of the question. They, along with millions of other people who live on fixed incomes, are big losers under inflation.

Figure 14-3 Inflation is especially difficult for retirees living on a fixed income.

9. Ibid., p. 37.
10. Ibid.
11. Sanford D. Gordon and Alan D. Stafford, *Applying Economic Principles* (New York: Glencoe/McGraw-Hill, 1994), p. 220.
12. Brown and Clow, *Introduction to Business,* p. 37.
13. Roger LeRoy Miller, *Economics Today and Tomorrow* (New York: Glencoe/McGraw-Hill, 1995), pp. 338-339.

Landlords worry about inflation when their rental income is determined by long-term leases. Workers who accept union-negotiated long-term fixed wages can lose due to inflation. For example, a contract negotiated during the 1980s, when the inflation rate was 4.6 percent, allowed you to buy perhaps $100 worth of groceries. By 1990, with that same wage you could buy only $63.70 worth of groceries (assuming that the price of groceries went up at the same rate as the price level).[14] Savers can lose if the inflation rate exceeds the interest rate they are getting. However, if compound interest is taken into account over several years (as discussed in Chapter 8), savers probably will never lose.

Gaining from Inflation. Not everyone loses from inflation. Homeowners, landowners, and investors gain from inflation. In the early 1960s, you could buy a brick homes with three bedrooms, a kitchen, den, and one bathroom for about $15,000. The same house today is worth approximately $75,000, or more depending on location. Farms purchased in certain regions of the Southeast for $200 per acre in the early 1960s are now selling for $3,000 or more per acre. Typically, a rise in inflation causes investments to increase in value. Obviously, not every investment is a winner, but an **astute** businessperson will constantly monitor investments, and buy and sell, to get the most profit possible.

Measuring Inflation. Prices in different sectors of the economy rise at different rates. Therefore, the government uses several measures of inflation. The three most commonly used are the **consumer price index (CPI)**, the **producer price index**, and the **implicit GDP price deflator**. Only the consumer price index is discussed here.

The consumer price index is based on a survey of several hundred goods sold in approximately 21,000 outlets.[15] These items, collectively called the **market basket**, include about 400 goods and services, such as food, housing, transportation, clothing, entertainment, medical care, and personal care.[16] Refer to Figure 14-4 for the consumer price index for several years.

Consumer Price Index	
1985	107.6
1986	109.6
1987	113.6
1988	118.3
1989	124.0
1990	130.7
1991	136.2
1992	140.3
1993	144.4
1994	148.4
1995	152.4
1996	156.9
1997	160.5
1998	161.6
1999	164.3
2000	163.8
2001	175.1
2002	177.1
2003	181.7
2004	185.2
2005	190.7
2006	198.7

Figure 14-4 The consumer price index is a major indicator of inflation. (Source: Bureau of Labor Statistics)

The Federal Bureau of Labor Statistics compiles the consumer price index. The prices for the goods and services are set in relation to a **base year**. The average prices for 1982, 1983, and 1984 are the base-year prices used by many economists; they are given a value of 100. Base years are chosen because they are considered typical of most years. Change in price levels cause the consumer price index to either rise or fall.[17]

Calculating Inflation from the Consumer Price Index. Let us consider two examples. The consumer price index for later years indicates the percentage that the market basket price has risen since the base year (100).

Example 1: If the index for the current year is 140, it means that prices have gone up 40 percent

14. Gottheil, *Principles of Economics,* p. 584.
15. Gordon and Stafford, *Applying Economic Principles,* p. 272.
16. Miller, *Economics,* p. 339.
17. Ibid., pp. 339–340.

since the base year (140 − 100 for the base year = 40 percent increase).

Example 2: The CPI also can be used to calculate inflation from year to year. Refer to Figure 14-4. At the end of 1993, the CPI is 144.4. At the end of 1992, it was 140.3, which is a difference of 4.1 (144.4 − 140.3 = 4.1). If we use 1992 as the base year, we can find out by what percentage average consumer prices rose from 1992 to 1993. We do this by dividing 4.1 by 140.3, which gives us 0.0292 (4.1/ 140.3 = .0292 × 100 = 2.92). The .0292 is multiplied by 100 to give the result as a percentage (2.92%).[18]

Unemployment

The rate of unemployment is another significant economic indicator. Each month the U.S. Bureau of Labor and Statistics collects data to measure the employment/unemployment rate of the economy. Roughly 50,000 households are questioned each month.[19] These households are drawn from all geographic, ethnic, social, and occupational groups so that they will represent a cross-section of the American population. The unemployment rate in February 2006 was 4.8 percent. In 1982 and 1983, the United States had an unemployment rate near 10 percent, which is considered high.

Calculating the Unemployment Rate. The unemployment rate measures the percentage of the total civilian labor force that is currently unemployed. The formula is as follows:

$$\text{Unemployment rate} = \frac{\text{Number of people unemployed}}{\text{Number of people in the civilian labor force}}$$

Types of Unemployment. Some types of unemployment are a greater problem than others. Four specific, different types have been identified by economists: structural, frictional, seasonal, and cyclical unemployment.

Structural unemployment—the most serious—refers to workers whose skills are no longer in demand because of technological advances or discoveries of natural resources. *Frictional unemployment* refers to workers who are between jobs, and who will quickly find new ones. This kind of unemployment is due to firings, layoffs, voluntary searches for new jobs, or training. *Cyclical unemployment* refers to workers who have lost their jobs because of a recession or a downturn in the economy. *Seasonal unemployment* occurs when demand for labor varies over the year, as with the harvesting of crops.

Discouraged workers are those who are not actively searching for work. They are not considered a part of the civilian labor force and therefore are not counted among the unemployed. Essentially, these are workers who have given up looking for a job.

Why Unemployment Matters as an Economic Indicator. High rates of unemployment usually signal an unhealthy economy. When unemployment rates are high, human resources are wasted, standards of living are reduced, families can be devastated, and individuals suffer from low self-esteem. Given the seriousness of these consequences, keeping a low unemployment rate is a major goal in maintaining a healthy economy.[20]

Value of Economic Indicators

Although economic indicators are not always exact, they are precise enough to show what is happening generally in a segment of the economy. Various groups and agencies use and study the economic indicators. The government uses the indicators to control the rate of inflation. Businesspeople check the economic indicators and adjust their production of goods and services accordingly. Investors watch the economic indicators to determine when to buy or sell stock. If inflation is moderate and the unemployment rate is low, investors might buy stock in hopes of making a profit during future economic prosperity.

18. Ibid., p. 340.
19. Gordon and Stafford, *Applying Economic Principles,* p. 268.
20. Miller, *Economics,* p. 438.

Money and Banks

To an economist, *money* is an item used to obtain something that has value. The actual item used as money normally has little value of its own. However, people in a society agree to recognize it as having value.[21]

Different Types of Money

Economists differentiate three different types of money: fiat money, commodity money, and bank money. A brief discussion of each follows.

Fiat Money. When a government issues an order that something has value and can be used to satisfy debt, it establishes **fiat money**. The dollar bill is an example of fiat money. The paper has little value in itself, but because the U.S. government has recognized it as "legal tender"; it has value to pay debts and must be accepted (under most circumstances).[22]

Commodity Money. **Commodity money** is a *good* that is used as money. Items such as gold, gems, and cattle are all examples of things that may act as commodity money. Not only can they be used as a medium of exchange (money), they also have intrinsic value as goods. For example, gemstones have been used in some countries like money, to purchase things. In our country, however, they have value only as a good—that is, jewelry.[23]

Bank Money. **Bank money** consists of the bank credit that banks extend to their depositors. Transactions made using checks drawn on deposits held at banks involve the use of bank money. Fiat money or currency makes up only 20 to 25 percent of the total money supply in the United States; bank money (checkbook dollars) constitutes the remaining 75 to 80 percent of the money supply.

Functions of Money

Money is often defined in terms of the three functions or services that it provides. Money serves as a medium of exchange, as storage value, and as a standard of value.

Medium of Exchange. The most important function of money is as a medium of exchange to facilitate transactions. Buyers and sellers use money as a means of payment for goods and services. Money is the "go-between" that facilitates each exchange. Without money, all transactions would have to be conducted by **barter**, which involves the exchange of one good or service for another. Barter is a clumsy and time-consuming method of exchange, and is not always achievable between the parties to a transaction.

Storage Value. Money has the ability to hold its value over time—it does not spoil, rot, or disintegrate.[24] This is called *storage value*. Money can be accumulated and saved, one hopes without losing value. If there were no inflation, the value of a dollar today would be the same as its value was three years ago.

Standard of Value. Money also functions as a *unit of account*, providing a common measure of the value of goods and services being exchanged. It helps us compare the value of different goods and services. In this country, we measure economic value in terms of our basic unit of money, the dollar.[25] In Japan, it is the yen; in much of Europe, it is the euro.

How Banks Create Money

Consider what happens when a bank receives a $100,000 deposit from one of its depositors. The bank is required to set aside 10 percent of this deposit (in this case $10,000) as reserves. Refer to Figure 14-5. It then lends out the excess; in this case, the remaining $90,000 of the initial deposit.

For the sake of better understanding, suppose that all borrowers redeposit their loan funds in the same bank. The bank thus receives another $90,000 in new deposits, of which the bank sets $9,000 aside as required reserves and lends out all its excess funds. Let us further pretend that all borrowers again redeposit their loan funds in the same bank. The bank again sets aside 10 percent of these deposits, and again lends out the remainder, which is redeposited in the bank to continue the cycle.

If you were to follow this multiple-deposit expansion to the end, the result would be that the

21. Gordon and Stafford, *Applying Economic Principles,* p. 292.
22. Ibid.
23. Miller, *Economics,* pp. 365–366.
24. Brow and Clow, *Introduction to Business,* p. 313.
25. Ibid.

Multiple Expansion of Deposits

Round	New Deposits	New Reserves	New Loans
1	$100,000	$10,000	$90,000
2	90,000	9,000	81,000
3	81,000	8,100	72,900
4	72,900	7,290	65,610
5	65,610	6,561	59,049
6	59,049	5,904	53,145
7	53,145	5,314	47,831
8	47,831	4,783	43,048
9	43,048	4,304	38,744
10	38,744	3,874	34,870
11	34,870	3,487	31,383
12	31,383	3,138	28,245
13	28,245	2,824	25,421
14	25,421	2,542	23,879
15	23,879	2,387	21,492
16	21,492	2,149	19,343
17	19,343	1,934	17,409
18	17,409	1,740	15,669
19	15,669	1,566	13,103
20	13,103	1,310	11,793
21	11,793	...	...

Figure 14–5 With an original deposit of $100,000, a bank could create $1 million if borrowers redeposited all of their borrowed money there.

bank's deposits would increase by $1 million, its loans would increase by $900,000, and its reserves would increase by $100,000, all because of the initial deposit of $100,000.[26] Again, refer to Figure 14–5.

In reality, loan recipients do not deposit all of their loan funds into a bank; they hold a portion of their loan funds as currency. When some loan funds are held as currency, money is said to "leak" out of the bank system, and the 10 percent of new deposits from the loans would not occur.[27] However, remember what we learned earlier: 75 to 80 percent of the money flow in the United States is deposited in banks and checking accounts.

26. John Duffy, *Economics* (Lincoln, NE: Cliffs Notes, 1993), pp. 67–68.

27. Ibid., p. 69.

The Federal Government and Economic Stability

Economic stability is one of our economic goals. However, some changes in our business and economic world are expected. Earlier we discussed the Great Depression. The threat of another Great Depression keeps the federal government involved in managing the economy. When the economy is in a recession, the government tries to prevent a depression and bring about a recovery. When economic times are good, the government attempts to maintain the status quo. The government also tries to control the rate of inflation.

Government economic policies designed to influence economic activity are **fiscal policy** and **monetary policy**. Fiscal policy involves using government spending and taxation to influence the economy. Monetary policy involves controlling the supply of money and credit to influence the economy. The goals of both the fiscal and monetary policies are the same: to promote price-level stability, full employment, and the achievement of a strong gross domestic product.

Fiscal Policy

Two branches of government have the authority to carry out fiscal policy in this country: the legislative branch and the executive branch. Fiscal policy is implemented mainly through government expenditures and taxation.[28]

Government Expenditures. At times it is beneficial for the government to slow down the economy. When inflation reaches high rates, the government may decide to decrease its own spending. As a major consumer of goods and services in the country, the government can naturally slow the economy by reducing the amount it spends.

Taxes. Another way the federal government tries to either stimulate or dampen the economy is through taxation. Cutting taxes indirectly increases consumer spending. When the government cuts taxes, workers have more take-home pay. By increasing their spending on goods and services, consumers help spend the economy out of a recession.

28. Ibid., p. 73.

Monetary Policy

Monetary policy concerns the management of the money supply and interest rates. Monetary policy is conducted by a nation's central bank. In the United States, monetary policy is implemented by the **Federal Reserve System**.

Slowing Down the Economy. The Federal Reserve is a major source of money. It has the power to increase or decrease the amount of money in the economy as needed. One way the Federal Reserve can subtract money is by increasing the interest rate on the money it lends to banks. Banks pass such increases on to businesses, causing the economy to slow as businesses borrow less.[29]

Speeding up the Economy. The Federal Reserve can also "speed up" the economy by lowering the interest rates it charges banks. This can be especially helpful when unemployment rates are high. The Federal Reserve, in essence, is able to boost the economy because the lower interest rates stimulate spending and business growth. As businesses grow, they hire more people. As more people are put to work, the unemployment rate decreases.[30]

Independence of the Federal Reserve System. The Federal Reserve System operates independently of the president and Congress. It has the goal of keeping the economy growing without causing inflation. As discussed earlier, it does this by managing the money supply and interest rates.

Basis of Economic Policy

An **economic policy** is a course of action that is intended to influence or control the behavior of the economy. Economic policies are typically implemented and administered by the government. Examples of economic policies include decisions made about government spending and taxation, about the redistribution of income from rich to poor, about the supply of money, and about what to include in the farm bill. The effectiveness of economic policies can be accessed in one of two ways: positive economics (truth) and normative economics (**value judgments**).

Positive Economics

Positive economics examines how consumers and producers "are" behaving and how they "might behave." No value judgments are made in terms of policy issues or questions. Only explanations and predictions are offered. For example, the **hypothesis** that "an increase in the supply of money leads to an increase in prices" belongs to the realm of positive economics because it can be tested by examining the data on the supply of money and the level of prices.[31] Economists call these things *positive statements* because they can be tested and proven or disproven. (Refer to Chapter 2 for more discussion of positive economics.)

Normative Economics

Normative economics suggests and studies how things "ought to be" or how things "should be." Value judgments are made as to the effectiveness of the economy and economic policy. Hypotheses and theories cannot be confirmed by normative economics. For example, when experts state that the "inflation rate is too high," they are working with normative economics; such statements are based on a value judgment that cannot be tested or proven.[32] (Refer to Chapter 2 for more discussion of normative economics.)

The Nation's Economic Goals

Nations have values, and they set goals based on these values. These goals are evident in government policies and in the actions of people. The values and goals of a nation determine the kind of economic system it chooses (if there is a choice). We learned earlier that the United States tends toward a market or capitalistic economic system.

Before we take a trip, we usually decide where we want to go and plan the best route to get there. After all, how can we know when we have arrived if

29. William G. Nickels, James M. McHugh, and Susan M. McHugh, *Understanding Business,* 5th ed. (Chicago: Irwin/McGraw-Hill, 1999), p. 62.
30. Ibid.
31. John B. Penson, Rulon D. Pope, and Michael L. Cook, *Introduction to Agricultural Economics* (Englewood Cliffs, NJ: Prentice-Hall, 1986), p. 8.
32. Duffy, *Economics,* pp. 2–3.

we do not know where we are going? As a nation, we also need to know where we are going economically, so that we can plan the best route. We do this by setting some specific goals and periodically evaluating our progress.

Over the years, several general economic goals have been established for the nation as a whole. Some of the major ones include economic growth, economic stability, economic justice, economic security, and economic freedom. A brief discussion of each follows.

Economic Growth

Economic growth means an expansion of the economy to produce more goods, services, jobs, and wealth. Economic growth represents an effort to raise our standard of living. It also involves levels of employment or unemployment and numerous other economic concepts that affect society directly. High rates of growth may cause some problems, such as environmental pollution, but low growth tends to cause more problems.

Before the 1930s, we assumed that the private sector of the economy would accomplish this goal without the intervention of government. Today, we hold the federal government responsible for setting and implementing appropriate monetary and fiscal policies to attain an acceptable level of economic growth.

Economic Stability

The goal of economic stability is to reduce extreme ups (inflation) and extreme downs (deflation) in the economy. The most desirable goal is a low rate of inflation; usually 2 to 4 percent is considered acceptable. Some inflation is desirable because it tends to stimulate economic growth. However, when it exceeds about 10 percent, economists and others become concerned.

Production Agriculturalists and Economic Stability. Economic instability is a major problem for production agriculturalists, much more so than for those in other sectors of the economy. Because of the large number of U.S. producers and the impact of world production on American agriculture markets, agricultural product prices are constantly changing. For most nonfarm business, when prices change, they usually go up. For production agriculturalists, prices go down about as often as they go up. This results in "boom and bust" production agriculture. It is very difficult to plan future production when you have little to no control over the price you will receive for your products. During the first half of the 1980s, most production agriculturalists received no returns for their labor and got very little return on their investments. To reduce the impact of market price variation and help achieve economic stability, the government has provided minimum price guarantees for a wide variety of agricultural products.

Economic Justice

One of the most difficult, yet most important, economic goals to accomplish is to divide the economic pie so that everyone gets a fair share. When a society consists of millions of people and each has a different perspective as to what is fair and just, it is almost impossible to achieve the goal of economic justice. However, in all societies, the members of that society must feel that the system is at least acceptable, or a new system will be developed. Many revolutions have been fought for just this reason.

Ways to Attain Economic Justice. In an effort to provide equal opportunity for everyone, the following programs have been developed: a free public school system, special programs for the handicapped and disabled, minimum wage laws, price support programs, and laws prohibiting discrimination against individuals based on sex, race, religion, and age.

Progressive Income Tax. In addition, the United States has a progressive income tax system, under which those who have more income are required to pay a larger share for public services. We have an inheritance tax so that increasing amounts of wealth will not accumulate over time in the hands of a few families. As our concept of what is fair changes, new programs are developed to accomplish the continuous goal of economic justice. However, we are not likely to find a system that will provide complete economic justice for everyone.

Economic Security

Security means protecting people against poverty and supplying them with the means to survive emergencies, whether created by medical crises or natural disasters. The United States attempts to provide security through an increasing number of government

Figure 14–6 Even though production agriculturalists have economic freedom, there are many government restrictions on who can produce tobacco and how much they can produce. (Courtesy of Jamie Mundy)

social programs directed, among others, to the elderly. One of the best examples of government efforts to provide economic security for citizens is the social security program. This program is designed primarily to provide financial assistance to citizens over the age of 65 (or 67, depending on the year of birth), and to ensure financial support for those who cannot work because of some medical problem.

Economic Freedom

The goal of economic freedom allows each member of society to enjoy the freedoms of entrepreneurship, choice, and private property. Economic freedom also allows individuals to make their own decisions in the marketplace.

Economic freedom in America means that individuals can select the career of their choice. They can establish a business of their own or work for someone else. However, all economic activity has restrictions. There are zoning rules that restrict where certain types of businesses can be located. There are health restrictions to protect workers and consumers. In fact, the government establishes thousands of economic rules that specify what individuals can and cannot do.

Production Agriculturalists and Economic Freedom. Even production agriculturalists, who work in a sector that is usually associated with a high degree of economic freedom, do not have complete freedom to do as they please. For example, they are often told what they can produce and how much they can produce. Only producers who have grown tobacco in the past can produce and sell tobacco today, unless they buy an allotment from someone else who formerly raised tobacco. Today those who produce tobacco are also constrained by government programs regulating how much they can sell. Those who are not restricted by government are restricted by the marketplace. Refer to Figure 14–6. Any production agriculturalist who expects to make a profit must produce the type and amount of products that buyers are willing to purchase. However, even with all these restrictions, Americans still have more economic freedom than any society in the world.

Types of Economic Analysis

There are many ways to analyze economic situations. Whole textbooks have been written on economic analysis. However, our purpose here is to briefly introduce six economic analysis concepts that have been developed to help us understand more clearly what economics is and how it is applied.

The six economic concepts treated here are: marginal analysis, *ceteris paribus* (all else held constant), production possibilities, opportunity cost, law of increasing cost, and equal product curve.

Marginal Analysis

In **marginal analysis**, one examines the consequences of adding to or subtracting from the current state of affairs. For example, suppose you are a small agribusiness owner who operates a lawn and garden center. You are considering hiring a new worker. You determine the marginal benefit of hiring the additional worker as well as the marginal cost.

Marginal Benefit. The **marginal benefit** of hiring a new worker is the value of the additional goods or services that could be produced by hiring, for example, a new lawn and garden center employee. In other words, can more product (such as supplies, shrubs, and flowers) be produced by hiring the new employee?

Marginal Cost. The **marginal cost** is the additional wages the employer will have to pay the new lawn and garden center employee. (Some businesses might include the additional paperwork necessary for another employee and any supervisory or training time as part of marginal cost.)

Economic Decisions. An economic analysis of the decision to hire a lawn and garden center employee involves weighing the marginal benefits against the marginal cost. If the marginal benefits are greater than the marginal costs, it makes sense for the owner of the lawn and garden center to hire a new employee. If not, then the new worker should not be hired. In other words, if the cost of the new worker were $8,000 per year, net profits from the sale of trees, flowers, shrubs, and so on by the lawn and garden center would have to exceed $8,000 per year.

Ceteris Paribus ("All Else Held Constant")

In performing economic analysis, it is sometimes difficult to separate out the effects of different factors, decisions, or outcomes. For example, let us say sales of the lawn and garden center doubled after the new worker was hired. However, the same week the new worker was hired, television advertising began, an ad was placed in the newspaper, and a new sign was erected in front of the lawn and garden center.

Determining the Cause of an Increase in Sales. Four things (**variables**) could have caused the sales to double: new worker, television advertising, newspaper advertising, or new sign. If we really wanted to determine what increased the sales, we could hold constant three of the variables. Suppose that we stop the television and newspaper advertising and remove the sign. If sales remain doubled after these three variables were held constant, our economic analysis will have identified ("separated out") the new worker as the reason for the doubled sales. The variable that was allowed to vary was the new worker. Of course, we could have held constant any of the four variables, but in the final analysis, we would have found that the new worker is what made the difference in increased sales.

Production Possibilities

Production possibilities is another type of economic analysis. The total amount of products produced by a society, or any business that produces products, is limited by the type and amount of resources available, in combination with a specific level of technology. To simplify this concept, let us assume that the resources available to the lawn and garden center owner allow her to produce only two products, poinsettias and chrysanthemums. The owner can produce all poinsettias, all mums, or several combinations of the two. A hypothetical example of several production possibilities is shown in Figure 14–7.

Figure 14–7 is used to make a specific economic point. In modern terms, it shows that you can't have your cake and eat it too. Specifically, in order to produce poinsettias, the lawn and garden center owner must give up some mums. To gain the first 25 six-inch pots of poinsettias, 23 six-inch pots of mums must be sacrificed. This relationship

Selected Production Options for Poinsettias and Mums

Production Possibilities	Poinsettias (6″ pots)	Mums (6″ pots)
A	0	270
B	25	247
C	50	215
D	75	170
E	100	105
F	125	0

Figure 14–7 Only a certain quantity can be produced by any one individual or firm. To produce more of one item, you must produce less of one or more other items.

Production Possibilities Curve for Lawn and Garden Center Owner

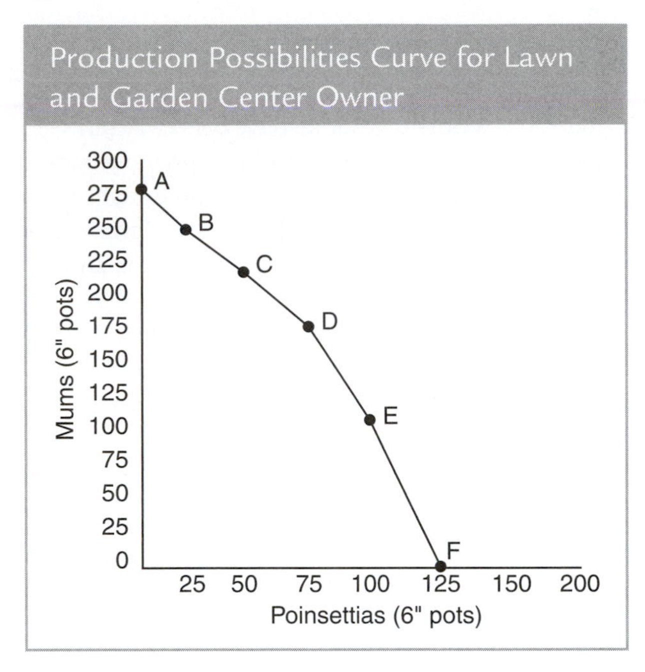

Figure 14–8 This curve shows how much of one product can be produced relative to production of another product.

holds throughout the example; if the owner wants to produce 125 six-inch pots of poinsettias, she cannot produce any mums. Conversely, if the owner wants to produce more mums, she will have to give up some poinsettias. This concept applies not only to the lawn and garden center owner, but also to society as a whole. All societies are limited by what they can produce. Therefore, to produce more of one item we must give up the production of one or more other items.

This concept is often illustrated with a graph as well as a table. The graphic illustration is called a *production possibilities curve*. This curve shows all production possibilities of two products. This type curve is sometimes called a *production possibilities frontier* or a *transformation curve*. Refer to Figure 14–8.

Opportunity Cost

One of the most important economic analysis concepts is that of opportunity cost. Economists define **opportunity cost** as the value of a service or product that must be given up in order to obtain another good or service. To illustrate this concept, refer again to Figure 14–8, the production possibilities curve. The opportunity cost of producing poinsettias is the value of the mums the owner has to give up. Specifically, the opportunity cost of producing the first 25 six-inch pots of poinsettias is the value of the 23 six-inch pots of mums the owner cannot produce and has to give up. The opportunity cost of producing 125 six-inch pots of poinsettias is the value of 270 six-inch pots of mums.

The concept of opportunity cost applies to all aspects of our lives. There is an opportunity cost to every decision we make. For example, the total cost of a college education is not just the direct cost of tuition, books, and room and board. It also includes an *opportunity cost*–the amount you could have made if you had worked full-time instead of going to college. From an economic analysis point of view, the best decision is one in which the alternative you select (college) provides a greater return than your opportunity cost (money loss in wages while obtaining college).

Law of Increasing Cost

We have established the economic relationship between the production of two products using limited resources: as you produce more of one, you have to produce less of the other. However, a closer look at the illustration shows an even more significant relationship; namely, an unequal rate of substitution between the two products. The **law of increasing cost** states that to produce equal extra amounts of one product or service, an increasing amount of another product or service must be given up. Figure 14–9 shows that the first 25 six-inch pots of poinsettias require a sacrifice of only 23 six-inch pots of mums. However, the second 25 six-inch pots of poinsettias require a sacrifice of 32 six-inch pots

Law of Increasing Cost

Production Possibilities	Poinsettias (6″ pots)	Mums (6″ pots)	Mums Given up for Poinsettias
A	0	270	–
B	25	247	23
C	50	215	32
D	75	170	45
E	100	105	65
F	125	0	105

Figure 14–9 The law of increasing cost states that to produce an extra amount of one product (poinsettias), an increasing amount of another product (mums) must be given up.

of mums. To add 25 more six-inch pots of poinsettias, the owner of the lawn and garden center would have to give up 105 six-inch pots of mums.

You may be wondering, at this point, why the sacrifice in mum production increases rather than remaining constant or decreasing. The answer is because there is not perfect flexibility or interchange ability among the resources used to produce both crops. For example, poinsettias can be inserted into a rotation, making use of surplus labor, spreading out the growing seasons, and using greenhouse space that is more efficient for poinsettia than mums. However, these advantages decline progressively as more poinsettias are produced. This also explains why a production possibilities curve is bowed rather than a straight line. Again, refer to Figure 14-8.

Equal Product Curve

Figure 14-8 illustrated a production possibilities curve showing that one product can be substituted for another, though not at a constant rate. However, different *resources* can also be substituted for one another in the production of a given product. There are numerous combinations of land, labor, and capital that will produce the same amount of product. This concept is often expressed by an **equal product curve**. Refer to Figure 14-10. An equal product curve shows the combinations of two sets of resources that will produce the same amount of total output. It should be noted that the curve is bowed in rather than bowed out. This is because we have resources instead of products on the axes, and because there is not perfect substitutability between resources.

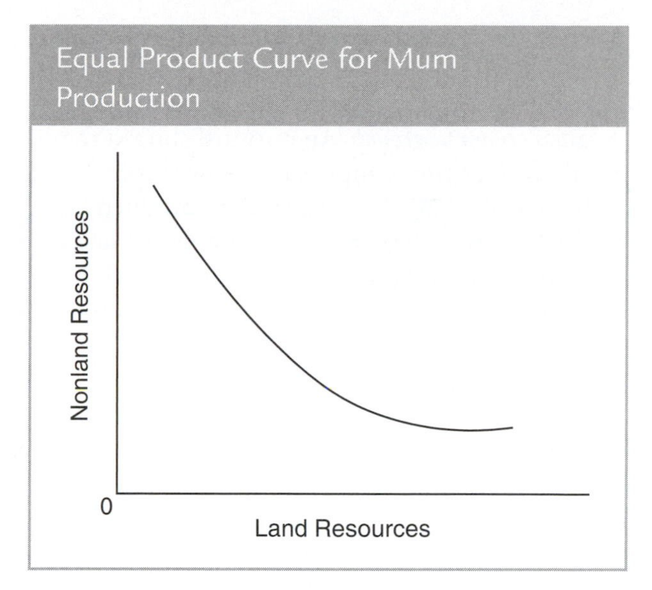

Figure 14–10 An equal product curve shows the combinations of two sets of resources that will produce the same amount of total output.

Productive Power of Advanced Technology and Innovations

Ideas, more than any factor of production, are the most revolutionizing force in shifting the production possibilities of an economy. Ideas can shift the curve out almost beyond imagination.[33] Who would have thought, just a century ago, that we could breed thousands of cows from one bull (by **artificial insemination**), or that we could produce thousands of new rose plants from one rose (by **cloning**)? Economists have a term for ideas that eventually take the form of new applied technology: they are called *innovations*.[34]

Earlier we emphasized the fact that a production possibilities curve is fixed, because of limited resources; in general, a producer or society cannot produce beyond their production possibilities curves. However, this limitation is not absolute, because of new technologies and innovations. We do have some flexibility. We can substitute one

33. Gottheil, *Principles of Economics*, p. 36.
34. Ibid.

product for another. Best of all, we can even shift the production possibilities curve outward. An outward shift of the production possibilities curve is called *economic growth*. An outward shift in the equal product curve implies the same thing. How is this possible? We can improve the productivity of our resources through new technologies and by changing the combination of resources used.

For example, through the use of hybrid seed corn, we have increased the average yield of corn from 74 bushels per acre in 1965 to more than 120 bushels per acre today. We have seen similar results through cross-breeding, artificial insemination, and embryo transfer in cattle. The use of computers has improved the productivity of resources in all phases of our economy. The substitution of capital for labor also shifts the production possibilities curve outward.

Economic levels of countries that apply a large amount of capital relative to labor tend to be much higher. We also classify countries as developed or underdeveloped relative to their level of economic growth. Figure 14–11 shows a production possibilities curve using food and nonfood products for a developed country (B) versus an underdeveloped country (A). Note that B can produce more of both food and nonfood items.

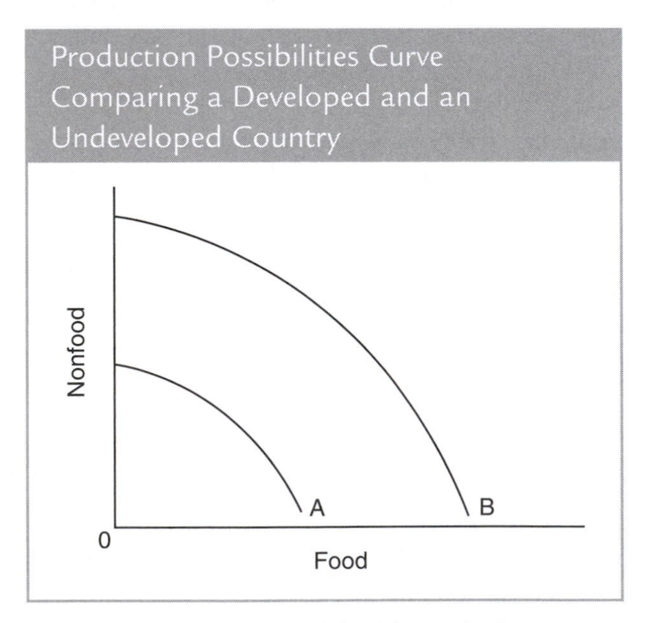

Figure 14–11 An outward shift of the production possibilities curve is called economic growth. With new technologies and greater amounts of capital relative to labor, developed countries have a greater outward shift of the production possibilities curve.

Common Pitfalls in Economic Analysis

Two pitfalls should be avoided when conducting economic analysis: the fallacy of composition and the false-cause fallacy.

Fallacy of Composition. *Fallacy of composition* is the belief that if one individual or firm benefits from some action, all individuals or all firms will benefit from the same action. Although this may in fact be the case, it is not necessarily so. For example, suppose that a purebred Angus breeder decided to start embryo transfer as a management technique to increase profits. It worked, and the news spread throughout the country. However, when other cattle breeders implemented the new technology, they did not experience the same profits, because the market for their calves produced from embryo transfer was not as good as that for the calves of the purebred Angus breeder. The others did not have as good a quality of cattle from the start. Hence, they did not realize increased profits from use of the new technology.

False-Cause Fallacy. A similar pitfall in economic analysis is called the *false-cause fallacy*. The false-cause fallacy often arises in economic analysis when two actions or events appear to be correlated. One must not automatically conclude that because two actions or events are correlated (they occur together), one must have caused the other. To avoid the false-cause fallacy, realize that *correlation does not imply causation*.

For instance, imagine a scenario in which the price of new tractors has increased over a period of time. During the same period of time, the sales of new tractors also increased. It may be tempting to conclude that the increase in price in new tractors caused an increase in sales of new tractors. However, that conclusion is an example of the false-cause fallacy. Though both the rise in price and the rise in sales occurred together, this correlation does not imply that one caused the other. Other factors, such as inflation, increasing production costs, and increasing consumer incomes, must be examined to explain why both events are taking place at the same time.[35]

35. Duffy, *Economics*, p. 9.

Career Options

Ag Banker, Financial Manager, Agriscience Loan Officer, Financial Analyst, Marketing Specialist

The management of finances is the most important single function of any career associated with agribusiness. Possessing a great deal of technical knowledge and know-how is very important. However, this know-how will not make you a success in your career field unless you are also a competent money manager.

As a financial manager, you might work for a bank or a credit agency as an agriscience loan officer, or for a company that sells supplies to producers and growers. As credit manager, loan officer, financial analyst, or marketing specialist, you could utilize your knowledge of business and finance as well as your knowledge of and experience with production, processing, or distribution. Careful planning of finances is probably more important than planning any other aspect of an agribusiness. Some very good agribusinesses have failed because the manager was not a good money manager.

Many jobs are available in banks in the position of agricultural loan officer. An overall knowledge of business and finance is necessary, as well as knowledge about the agribusiness input, production, and output sectors. (Courtesy of Cliff Ricketts)

Attaining Economic Efficiency

Economic growth is the result of improving output relative to the limited resources provided by nature. How can we change an input-output ratio to improve economic efficiency? Here we discuss five specific steps that will improve the input-output ratio at any point in time. Let us assume that it takes 500 units of resources to produce 100 units of outputs (for example, gallons of milk). How can we improve this input-output ratio of five to one? There are at least five ways:

Strategy 1: increase output with the same input
Result: more milk from the same number of cows

Strategy 2: same output with less input
Result: same amount of milk with fewer cows

Strategy 3: increase output with less input
Result: more milk with fewer cows

Strategy 4: increase inputs and increase output but increase output more than input
Result: 5 percent more cows results in 10 percent more milk

Strategy 5: decrease inputs and decrease output but decrease inputs more than outputs
Result: using 10 percent less feed results in 5 percent less milk

All of these possibilities are difficult, some more than others, but they are all realistic alternatives. As one or more of these alternatives are accomplished by agribusinesses or society in general, our economic standard of living is improved. Some societies have been very successful with this process; others cannot seem to make any progress with it. Likewise, some industries have been more successful than others. For example, output per hour of labor for farm workers has increased approximately 30 percent since 1979, whereas output per hour of labor for nonfarm workers increased only 10 percent during the same period.

Several careers are related to economic activity and analysis. The Career Options section explains some of these in the area of agribusiness.

Conclusion

Economics is a complex process involving land, labor, capital, and management. Much of the research, and many of the policies and laws that are in place today, are the result of the Great Depression. The Great Depression was so devastating to the American people, both economically and psychologically, that efforts are continuously being made to avoid another such catastrophe. In addition to the memory and lessons of the Great Depression, the desire of the American people to have a comfortable standard of living is another reason we study economic activity and do economic analyses.

Summary

A part of understanding agricultural economics is understanding economic activity and its relationship to federal, state, and local governmental action. Also, the role of the family in society and how the family behaves as an economic unit are an integral part of agricultural economics. Therefore, an understanding of economic activity, economic goals of a nation, and methods of analyzing these goals and activities are crucial to understanding the economic system of the United States. Economics is a complex process involving land, labor, capital, and management. The reasons we study economic activity and conduct economic analysis include the impact of the Great Depression and the desire of the American people to have a comfortable standard of living.

Many factors are involved in maintaining a strong economy. These factors include efficient production of agricultural products, efficient production of goods and services, and good wages. The job of the economist is to alert us when the factors in the economy are beginning to get unbalanced. Two things that economists study are business cycles and patterns, and the circular flow of the economy.

Economic indicators are data or statistics that measure economic activity and business cycles. These indicators normally change before the rest of the economy does. There are many economic indicators, but the three major ones are gross domestic product, inflation, and unemployment.

Economists view money as a claim on something that has value. Money usually has little value of its own, but it represents value. Economists differentiate three different types of money: fiat money, commodity money, and bank money. Banks create money by making loans, keeping some in reserves, and reloaning money from deposits of the loaned money.

Economic stability is one of our national economic goals. When the economy is in a recession, the government tries to prevent a depression and bring about a recovery. When economic times are good, the government attempts to maintain it. Government economic policies designed and used to influence economic activity are fiscal policy and monetary policy. Fiscal policy involves using government spending and taxation to influence the economy. Monetary policy involves controlling the supply of money and credit to influence the economy.

An economic policy is a course of action that is intended to influence or control the behavior of the economy. The effectiveness of economic policies can be assessed in one of two ways: positive economics (truth) and normative economics (value judgments). Positive economics focuses on testable, provable, "what is" and "what would happen if" questions and policy issues. Normative economics uses value judgments to assess the performance of the economy and economic policies.

Nations have values and set goals based on these values. These goals are evident in government policies and in the actions of people. Several general economic goals have been established for the United States as a whole. Some of the major ones include economic growth, economic stability, economic justice, economic security, and economic freedom.

There are many ways to analyze economic situations. Six economic analysis concepts in particular help us understand more clearly what economics is and how it is applied. These six economic analysis concepts are marginal analysis, *ceteris paribus* (all else held constant), production possibilities, opportunity cost, law of increasing cost, and equal product curve. However, there are two pitfalls to be avoided when conducting economic analysis: the fallacy of composition and the false-cause fallacy.

Economic efficiency consists of getting the most profit possible from a product or economy by having the most optimum input-output ratio. Economic growth describes the results of improving output relative to the limited resources provided by nature.

End of Chapter Activities

REVIEW QUESTIONS

1. Define the Terms to Know.
2. List four family matters that are affected by economics.
3. List three things that are crucial to understanding the economic system of the United States.
4. What are three factors involved in maintaining a strong economy?
5. What is the purpose of economists?
6. Why does the state of the economy require daily attention?
7. Name the four parts of a business cycle and give two characteristics of each.
8. Differentiate between a recession and a depression.
9. After examining Figure 14-2, explain the circular flow of the economy.
10. What are three economic indicators?
11. In calculating the gross domestic product, economists add the sum of what four main areas of goods and services?
12. What four groups can lose from inflation?
13. What are three groups that gain from inflation?
14. What are three ways to measure inflation?
15. Name and define the four types of unemployment.
16. Why does unemployment matter as an economic indicator?
17. What are three values of economic indicators?
18. Name and describe the three functions or services of money.
19. Explain how banks create money.
20. What are two economic policies that the government uses to stabilize the economy?
21. Name and explain two fiscal policies the federal government can establish to influence the economy.
22. Name and explain two ways in which the Federal Reserve System can influence the economy through monetary policies.
23. State four examples of economic policies.
24. Name and briefly describe five of the nation's economic goals.
25. List six economic analysis concepts.
26. Briefly explain the economic analysis concept of marginal analysis.
27. Briefly explain the economic analysis concept of *ceteris paribus*.
28. Briefly explain the economic analysis concept of production possibilities.
29. Briefly explain the economic analysis concept of opportunity cost.
30. Briefly explain the economic analysis concept of law of increasing cost.
31. Briefly explain the economic analysis concept of equal product curve.
32. Explain how technology and innovations can shift the production possibilities curve outward.

FILL IN THE BLANK

1. In reality, government budgets are _________ documents.
2. During the Great Depression, in 1932, the unemployment rate reached _________ to _________ percent of the workforce.
3. The gross domestic product in 1970 was more than $1 trillion, in 1980 it was over $2.7 trillion, in 1997 it was $_________ trillion.
4. By measuring the changes in the growth of the gross domestic product annually, economists are able to track the health of the _________.
5. Inflation reduces the _________ of money.
6. The _________ _________ _________ _________ _________ compiles the consumer price index.
7. Workers who have given up looking for a job are referred to as _________ _________ and are not included in the unemployment rate.
8. _________ usually has little value of its own, but it represents value.
9. Bank money or checkbook dollars make up _________ to _________ percent of the money supply.
10. In the United States, monetary policy is carried out by the _________ _________ _________.
11. An outward shift of the production possibilities curve is called _________ _________.
12. More so than any other factor of production, _________ _________ _________ are able to shift the production possibilities of any economy.

MATCHING

a. Federal Reserve System
b. dollar bills
c. normative economics
d. checks
e. *ceteris paribus*
f. legal tender
g. positive economics
h. fallacy of composition
i. monetary policy
j. fiscal policy
k. false-cause fallacy
l. gold coins

_____ 1. must be accepted to satisfy a debt
_____ 2. example of fiat money
_____ 3. example of commodity money
_____ 4. example of bank money
_____ 5. involves using government spending and taxation to influence the economy
_____ 6. involves controlling the supply of money and credit to influence the economy
_____ 7. independent of the president and Congress
_____ 8. economic truths
_____ 9. economic value judgments
_____ 10. all else held constant
_____ 11. correlation implies causation
_____ 12. belief that if one individual or firm benefits from some action, all will benefit from the same action

ACTIVITIES

1. Prepare a 100-word essay outlining a set of events that could cause another Great Depression. (Extra reading and library research may be needed but are not necessary.)
2. Write a paragraph examining your effect on the circular flow of the economy.
3. Refer to Figure 14–4. Calculate the rate of inflation from 1985 to 1990.
4. Refer to formula for calculating unemployment in this chapter. Calculate the unemployment rate if there are 100 million people in the workforce and 5 million are unemployed.
5. Suppose you make a $10,000 deposit in the bank. Using the example in Figure 14–5, show graphically how much money the bank can create from your loan.
6. Opportunity costs face us daily. Give an example of an opportunity cost that you have to decide upon. Remember, it need not involve money. Share this with the class in a large-group discussion.
7. Suppose you went to the grocery store to buy milk, blank CDs, and paper towels. You also consider buying a two-liter soda, but you notice that the price is 30 cents higher than last week. The prices of the other items seem to be the same as always. Is the soda price an example of inflation? Why or why not?
8. Work in a group of three to five people to find graphs, tables, and charts showing the gross domestic product and other economic indicators over the past few years. Use the library and any computer resources you can. Create a poster of your findings. What phase of the economic cycle are we in now? What groups of people need to know such information? Share the findings of your group with the class.
9. Consider the two statements: "Fifteen percent of our people live below the poverty line," and "Too many people live below the poverty line." Distinguish the positive economic statement from the normative economic statement.
10. Read an editorial from your local newspaper. Cite at least one positive economic statement from the editorial and at least one normative economic statement. Explain how you know which is positive and which is normative.
11. What does *ceteris paribus* mean? Give an example showing why the concept is useful to economists.
12. Identify one of the most widely debated economic policies or issues and discuss the various viewpoints with your classmates and teacher. Choose a position and be ready to defend it by researching facts and figures to support it.

Chapter 15

Agricultural Policy and Governmental Agribusiness Services

Objectives

After completing this chapter, the student should be able to:

- Discuss the historical perspective of public agribusiness services.
- Explain why agricultural policies are needed.
- Discuss the forces that cause policy change.
- Explain the conditions leading to and reasons for government involvement in agriculture.
- Discuss the legislative process of agricultural policy development.
- Describe the influence of groups and farm organizations on agricultural policy development.
- Discuss the history of farm legislation.
- Explain the future of agricultural policy.
- Explain why governmental services are provided.
- Describe the organizational structure of the USDA.
- Describe the following divisions of the USDA and each of the major agencies:
    - food, nutrition, and consumer services
    - marketing and regulatory programs
    - farm and foreign agricultural services
    - rural development
    - natural resources and environment
    - research, education, and economics
    - food safety
- Discuss the role of the state departments of agriculture.
- Explain the land grant system.
- Explain non-land grant agricultural programs.
- Describe the sea grant program.
- Discuss agricultural education programs.

Terms to Know

agricultural policy
agricultural statistics
attachés
consortium
distance learning
economic indicators
economic policy
eradicate
flex time
germ plasm
grants
income stabilization
inflation
initiatives
interdisciplinary
liaison
policy
policy analysis
price support
production controls
public policy
teleconferencing

Introduction

American agriculture has not become the world's greatest food producer by accident, but by planning and hard work on the part of agriculturalists and nonagriculturalists within a conducive social, economic, and political environment. American leaders have long recognized the importance of strong agriculture, possibly because many of our political leaders had their roots in farming. According to Vivian Whitehead, 23 presidents have had close farm ties.[1] Included in this list are George Washington, Thomas Jefferson, Andrew Jackson, Abraham Lincoln, Theodore Roosevelt, Harry S. Truman, Lyndon Johnson, and Jimmy Carter. In addition, numerous members of Congress were part-time farmers before and during their terms of service.

Government plays its major role in agriculture at the federal level. However, state and local governments are also heavily involved in agricultural development. Agriculturalists obtain information from teachers of agricultural education, their county agents, and their state colleges of agriculture. American governments have influenced agriculture directly through the following mechanisms: research, education, and extension programs; health regulation programs and regulations; **price support** laws and **production controls**; and the collection and distribution of **agricultural statistics**. In addition to the general policies of the nation as a whole, agriculture is specifically affected by **inflation**, unemployment, and foreign policy. Agriculturalists do very little today that is not affected, either directly or indirectly, by government action.

The government's role in agriculture has always been controversial, some aspects more than others. For example, price supports and production controls have been extremely controversial. Many production agriculturalists who are supposed to benefit from price supports and production controls dislike governmental interference in the free market system; other production agriculturalists and farm organizations demand higher price supports and either more or fewer production controls. The major controversy surrounding research, education, and regulation is how much should be provided. Agriculturalists want more research and educational services and less regulation, whereas consumer groups want more regulation and less funding for research and education.

Historical Perspective

When the term *government* is used without qualification, we usually assume that it refers to the federal government, and most involvement of government in agriculture has indeed occurred at the federal level. However, state and local governments also play a major role in agricultural development. The first federal aid to agriculture in the United States was granted in 1839, when Congress appropriated $1,000 to collect farm statistics and to collect and distribute seeds of various types. An agricultural

1. Vivian Whitehead, "White House Farmers," in USDA, ed., *The American Farmer* (Washington, DC: USDA, Economic Research Services, 1976).

division was established in the Patent Office to carry out these objectives. Ten years later, the Patent Office was shifted from the Department of State to the newly created Department of Interior, and the agriculture division went with it. On May 15, 1862, President Lincoln signed the act that created the U.S. Department of Agriculture (USDA). The purpose of the new department was "to acquire and diffuse among people of the United States useful information on subjects connected with agriculture."[2] The administrators of the newly created Department of Agriculture were called commissioners until 1889, when Congress raised the head of the department to the rank of Cabinet secretary. Isaac Newton was the first of seven commissioners, and Norman Colman was the first Secretary of Agriculture.

Why Agricultural Policy?

The specific rationales that will govern action, and will guide present and future decisionmaking in achieving agreed-upon goals and objectives in a given environment, are called **policy**. One focus of government is to conduct the public business. This public business consists of numerous issues and activities, such as security, defense, economic stability, civil rights, and the welfare of citizens. All types of people and groups, including farm organizations, provide input as to what should be done to get desired results in each of these areas. All of the input is considered as government officials formulate various policies.

Public policy can be defined as the steps that the government will follow to achieve specific objectives or to solve pressing problems that affect the general public. To implement public policy, laws, rules, and regulations are established.[3]

Economic policy dictates the way the government will manage the national economy. For example, if an administration has an economic policy that favors free trade, it will oppose reductions on imports and exports, and it will try to eliminate barriers to trade with other countries.[4]

Agricultural policy outlines the steps that will be taken to reach certain goals in the food and fiber economy. Typically such policies affect the resources, production, and markets related to agricultural products and services. They often are concerned with the safety, consumption, and nutritional value of food. Living conditions in rural areas may also be addressed. Agricultural policy, like most national-level policies, is influenced by economic, foreign, and environmental policies and considerations.[5]

Goals of Present and Future Agricultural Policy

Specifically, goals for U.S. agricultural policy include:

- Maintaining a profitable, viable, efficient, and environmentally safe agricultural production sector capable of meeting demands for food and fiber while providing satisfactory incomes to production agriculturalists for use of their land, labor, capital, and management
- Providing for an efficient, profitable, and dynamic agribusiness sector, including input suppliers and the agricultural output subsector
- Providing consumers with an abundant, varied, and safe supply of food and fiber at the lowest possible cost consistent with other goals
- Operating a food and fiber economy within the framework of a democratic society, relying on the free market system as much as possible, consistent with other goals
- Maintaining and enhancing the competitiveness of U.S. agricultural products in the global market[6]

Forces That Cause Policy Change

U.S. agricultural policy is constantly changing. Major revisions occur every few years, but smaller adjustments are made much more frequently. The

2. USDA, *A Guide to Understanding the United States Department of Agriculture* (Washington, DC: USDA, Office of Personnel, 1970).
3. Randall D. Little, *Economics: Applications to Agriculture and Agribusiness*, 4th ed. (Danville, IL: Interstate Publishers, 1997), pp. 383–384.
4. Ronald D. Knutson, J. B. Penn, Barry L. Flinchbaugh, and Joe L. Outlaw, *Agricultural and Food Policy*, 6th ed. (Upper Saddle River, NJ: Prentice-Hall, 2006), p. 1.
5. Ibid.
6. Little, *Economics*, p. 403.

following forces may cause changes in policy: instability, globalization, technology, food safety, environment, industrialization, politics, and unforeseen events.[7] Each of these is briefly addressed here.

Instability

One major problem faced by production agriculturalists is instability of prices and incomes. Changes in supply and demand can and do affect farm prices. This is common when government chooses to intervene only a little, or infrequently, in the agricultural sector.

Globalization

The cultures, politics, and economies of countries around the world have become more interdependent. This process is referred to as *globalization*. Increased globalization requires that all governments consider the effects their actions and policies will have on other countries.

Technology

Advances in technology have forced changes in agricultural policy, because of increased efficiency in the production of agricultural products and services and increased quality of those products and services. Through various research projects, land grant and agricultural colleges and universities have played (and continue to play) a significant role in advancing technology. Some examples of major technological advances include mechanization and hybridization, the creation of new and better pesticides, and improved commercial fertilizers.[8]

Food Safety

Since the early 1900s, the safety of the nation's food supply has been of major concern. This concern intensified during the 1980s when *E. coli* bacteria, salmonella, and listeria were found in food products, and has remained high. More recently, bovine spongiform encephalopathy (BSE or mad cow disease) caused alarm not only in Europe and Canada, but also in the United States. Agricultural policy has consistently had to be modified to ensure a safe food supply.

Environment

Agricultural policy is often changed to respond to issues related to the environment. In the 1930s, the main concern was soil conservation. Today, there is continuing concern about water quality in the United States. In other countries, the use of pesticides and genetically altered organisms sparks controversy. Some believe that future concerns may center on how agriculture affects air quality.

Industrialization of Agriculture

The multitude of changing conditions in agriculture and food systems may be referred to as *agricultural industrialization*. Technological advances, the growth of large agribusinesses, the need for supply-chain management, and the effects of corporate ownership, acquisitions, and contraction of agricultural functions all contribute to the industrialization of the agricultural sector. Policy focus in this area is on how the traditional family farm is affected by these sweeping changes.

Politics

Politics plays a significant role in determining all policies, including agricultural policy. As one might expect, politics also affects which programs are chosen to implement agricultural policy. A common misconception is that farmers have little political influence today. Although it is true that commercial farms are decreasing in number (and increasing in size), it is also true that farmers, as a minority group, continue to successfully lobby for legislation to benefit them and the agricultural sector in general.

Unforeseen Events

Significant unforeseen events can force changes in policy in nearly every sector of society. One need only consider how individual lives, the economy, the government, and even USDA programs were affected by the terrorist attacks of September 11, 2001. Other notable examples that have had similar effects include the potato famine in Europe at the beginning of the 20th century, Hitler's regime in Germany, the attack on Pearl Harbor, the collapse of the Soviet Union, the fall of the Berlin

7. Knutson, Penn, Flinchbaugh, and Outlaw, *Agricultural and Food Policy*, p. 2.
8. Ibid., p. 3.

Wall, the 2004 tsunami in the Indian Ocean, and the damage and flooding caused by Hurricane Katrina in 2005.[9]

Conditions Leading to and Reasons for Governmental Involvement

It was once thought that governmental involvement in agriculture was rooted in the beliefs that agriculture is the most basic of occupations, that numerous small family farms and farmers are essential to a democratic system, and that rural life is simply superior to urban life. Although these sentiments are still well received in parts of rural America (especially in election years), there are more fundamentally important reasons underlying the government's role and involvement in agriculture. Some of these have a solid basis in the economics of agriculture. A brief discussion of six of these reasons follows.

Price and Income Stability

In the past, farm incomes lagged far behind nonfarm incomes. The government intervened to help reduce or eliminate the disparity. Today, farm incomes are significantly higher than incomes in the nonfarm sector. Calculations based on USDA statistics indicate that since 1999, farm families have averaged earnings in excess of $60,000 annually.[10] For commercial farms, this figure is at least double. In comparison, the average nonfarm household income has been roughly $44,000 annually.[11]

Importance of the Food Supply

Food is a necessity for the survival of people in any society. Throughout history, one can find numerous examples of governments that eventually failed because they could not feed their people. A recent example was the government of the former Soviet Union.

Importance of Food Safety

Each day, Americans trust the government's regulatory bodies to ensure that our food is safe to eat. This confidence has developed over many years. In the early 1800s, the public became very concerned about the safety of meat and milk. This led to the formation of public policies designed to monitor the quality of foods. As a result, regulatory offices were created; they are now known as the Food and Drug Administration (FDA) and the Food Safety and Inspection Service (FSIS). Other state and local agencies with similar missions have also been created.

Poverty (Economic Justice)

A basic tenet of American democracy is that people should be treated fairly. In addition, there is a strong belief that there should be economic justice; that is, that citizens should be able to live above the poverty level. Data reveals that some 12 percent of American households do not have an adequate amount of food. These households represent more than 38 million people, including almost 14 million children.[12] USDA programs that tackle the problems of hunger and malnutrition are immensely popular with the public and are typically well respected as governmental programs.

Environmental Externalities

Environmental externalities are created when the market supply and demand do not include all the production costs of an item. For example, a farmer who has a cash-flow problem may decide to accept a high level of soil erosion even though that erosion is detrimental to the environment. The resulting runoff may pollute streams. This pollution is considered an *externality* to fishermen, to outdoorsmen who enjoy streams, and to cities that use surface water. Externalities may thus create markets that fail to allocate resources in the best interests of society as a whole.[13]

Other Public Concerns and Expectations

The general public today does not have to be concerned about growing food or raising stock

9. Ibid., pp. 4–7.
10. http://www.ers.usda.gov/Briefing/FarmIncome (accessed August 27, 2007).
11. http://www.whitehouse.gov/fsbr/income.html (accessed August 27, 2007).
12. http://www.centeronhunger.org/facts.html (accessed August 27, 2007).
13. Knutson, Penn, Flinchbaugh, and Outlaw, *Agricultural and Food Policy*, p. 17.

to put food on the table. Thus, people examine agricultural issues with a more critical eye. There are concerns over humane slaughter practices, animal welfare, genetically modified products, and wildlife and habitat preservation. We can expect these and other related concerns to increase in the future.

Development of Agricultural Policies

When policies are developed, a need is defined and transformed into a goal by an individual, a legislator, or an interest group. The next step is making the public aware of the issue and educating the public about the merits and benefits of the proposed goal. After public awareness is raised, the next step is public acceptance and general agreement that there is a need for action.

The next step is to determine the importance of reaching the goal; this determination is based on values. Then, costs and benefits of alternative plans or potential policies for addressing the need are analyzed in depth by any number of various individuals and groups. Following analysis, a proposed policy can be drafted as a bill, introduced into the legislature, and enacted.

Once enacted, a policy must be implemented or carried out. During this process, many different levels of government may become involved. A number of agencies and interest groups must agree that the policy meets their basic requirements before the policy can be fully implemented. Once implemented, the effects of the policy must be evaluated. Evaluation is not difficult when the objectives and the expected results of the policy are clearly specified.

Even when a policy has been drafted, enacted, implemented, and evaluated, the public may still have a variety of reactions to it. Some may want it repealed because it is burdensome; others may be dissatisfied with the policy and call for revisions.[14] Refer to Figure 15–1 for the steps in the legislative process by which a policy becomes law.

14. Little, *Economics,* p. 385.

Influence of Groups and Farm Organizations on Agricultural Policy Development

- There are numerous agricultural *interest groups*, which include general farm organizations, commodity groups, and agribusiness groups. They have special interests or agendas that lead them to suggest new policies or advocate changes to existing ones. Leaders within these groups often are able to field lobbying efforts to get policies enacted by Congress.
- *General farm organizations* represent nearly all sectors of agriculture and the food industry. They take broad policy positions on farm bills, taxes, and environmental policy. There are two major general farm organizations today. The largest, and more conservative, is the American Farm Bureau Federation (AFBF). The more liberal and populist group is the National Farmers Union (NFU).
- *Commodity organizations* exist for most agricultural commodities and/or groups of commodities. These organizations, such as the National Association of Wheat Growers, National Corn Growers Association, American Soybean Association, and National Cattlemen's Beef Association (NCBA), focus on narrowly drawn issues relating to a specific commodity or activity. Some organizations are made up of members of more than one agricultural sector. For instance, the NCBA represents both beef cattle ranchers and feeders. The National Cotton Council, known as one of the most cohesive lobbying groups, represents the entire cotton production and marketing chain, including farmers, millers, vendors, and exporters.
- *Agribusiness organizations* represent the agricultural sector from the farm to the consumer; members include input suppliers, commodity and food processors, marketers, and exporters. For example, the Fertilizer Institute is conservation oriented and is concerned with modern commercial farming systems. The North American Export Grain Association focuses on sale of grain abroad. The agriculture bankers division of the American Bankers Association represents lenders that specialize in farm loans. The International Dairy Foods Association (IDFA) represents milk processor interests

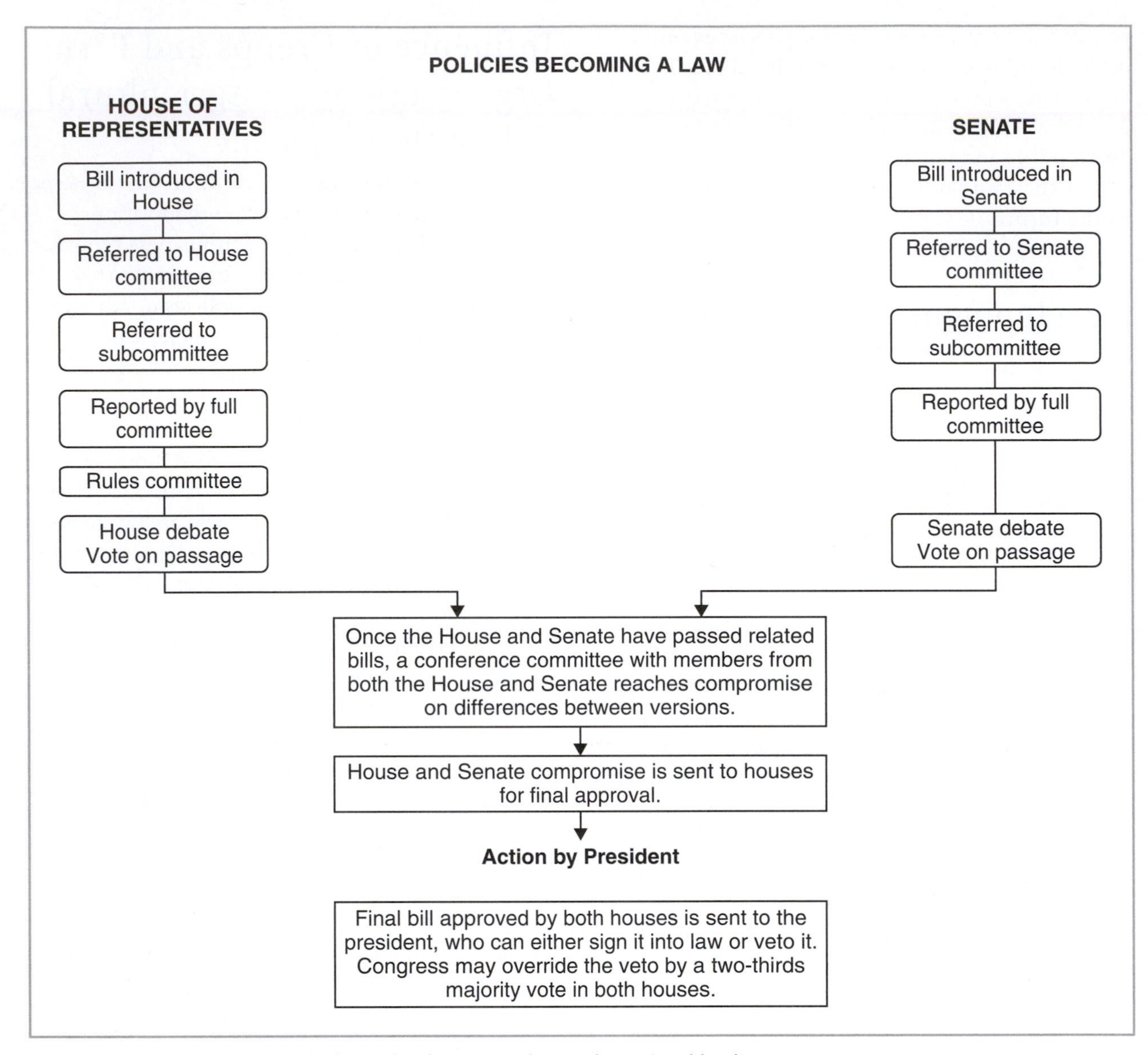

Figure 15–1 The process by which a policy becomes a law at the national level.

regarding pricing, marketing regulations, product quality, labeling regulations, and advertising programs. The IDFA seeks to offset the dairy farmer group, the National Milk Producers Federation (NMPF), in its attempts to raise milk prices. The agriculture and food sector also recognizes the Grocery Manufacturers of America, the Food Marketing Institute, and the Food Products Association. These groups are primarily concerned with food safety, labeling, and nutrition.[15]

15. Knutson, Penn, Flinchbaugh, and Outlaw, *Agricultural and Food Policy*, pp. 30–31.

Any of these groups has the opportunity to influence agricultural committees, in both houses of Congress, to support or oppose passage of laws that affect farm income, trade relations, conservation and environmental programs, technology, food safety, and nutrition. Typically, these areas are revisited in a comprehensive farm bill that is introduced every four to seven years.

Agricultural policy receives attention from the executive branch of government as well. The USDA is the lead agency working with the White House to ensure that agricultural policies under development coincide with the president's overall agenda. This might involve coordinating efforts with congressional committees and special interest groups to draft acceptable policies.[16]

Refer to Figure 15-2 for a list of many commodity/farm/agribusiness special interest organizations. For information on each of these organizations, use your favorite search engine or go to http://usextension.org/directory/farmorgs.htm and click on the desired site.

History of Farm Legislation

Agricultural policy tends to evolve over time and in response to changes in the economy, changes in the perceived roles of government, and changes in beliefs, values, and objectives. As we have seen, agricultural policy must also evolve as a result of outside forces such as globalization, technology, and politics. All the while, agricultural policy makers and those they represent must continually review the effectiveness of their work in solving current issues.

Settlement Period (1776–1929)

In the settlement period, U.S. farm policy focused on developing a capacity for research, education, and dissemination of information. The Homestead Act granted parcels of land to farmers who were willing to develop and cultivate that land. The USDA, warmly referred to as the "people's department," was created to advance scientific study and share information related to agriculture. In every state, land grant universities, extension services, and experiment stations furthered the knowledge and education of agriculturalists while also improving their lifestyles and productivity.[17]

During this same period, the U.S. government was heavily involved in agricultural policy. This was because during the early 18th and 19th centuries, society was basically agricultural in nature. Agriculture provided food and jobs; it influenced trade policy, land distribution, and even banking regulations. Policy makers tended to be those interested in these issues, although they typically were not agricultural specialists. Over time, the need for specialized agricultural policies and representatives became apparent, particularly when agriculture was no longer the dominant occupation in the country. Although some new policies were proposed and even passed by Congress during the 1920s, it was not until the 1930s that most agricultural policies were actually enacted and implemented.[18]

Official Farm Legislation Passed by Congress

Today, agriculture is influenced not just by its own policy **initiatives**, but also by general economic policy and policies regarding other economic sectors. Nevertheless, legislation related to agriculture is far-reaching, and encompasses many areas other than just production agriculture, such as food consumption, environmental regulation, trade relations, and subsidies. The area of subsidies alone involves credit programs, crop insurance, irrigation initiatives, grazing of livestock on public lands, marketing cartels, disaster relief, and price and income programs.[19] Figure 15-3 briefly outlines important U.S. farm legislation.

Future of Agricultural Policy

Most likely, total federal budget expenditures for agriculture will not increase or decrease significantly in the near future. Funding changes that do occur will probably shift monies to address various issues and concerns of the moment. Given these uncertainties, the following are educated predictions as to how agricultural policy will change during the next few years:

- Globalization will continue to influence the call for freer markets and fewer international trade restrictions. Import and export policies will be at the forefront. Modifications to the U.S. agricultural commodity mix will be needed.
- Economic and political policies will focus on protecting against terrorism.
- Agricultural policy will be directly influenced by the General Agreement on Tariffs and Trade (GATT), the Uruguay Round Agreements Act and negotiations, and activities of the World Trade Organization.

16. Ibid., p. 31.
17. Ibid., p. 87.
18. Daniel A. Sumner, in *Historical Statistics of the United States,* millenial ed., ed. Susan Carter et al. (Cambridge, UK: Cambridge University Press, 2006).
19. Ibid.

Commodity, Farm, and Agribusiness Organizations

- ABS Global (ABS)
- Accelerated Genetics
- AgriTech Analytics
- AgSource Cooperative Services
- Alta Genetics
- American Angus Association (AAA)
- American Association of Bovine Practitioners (AABP)
- American Association of Equine Practitioners (AAEP)
- American Association of Small Ruminant Practitioners (AASRP)
- American Association of Swine Veterinarians (AASV)
- American Belgian Blue Breeders (ABBB)
- American Brahman Breeders Association (ABBA)
- American Dairy Goat Association (ADGA)
- American Egg Board (AEB)
- American Farm Bureau Federation (AFBF)
- American Feed Industry Association (AFIA)
- American Gelbvieh Association (AGA)
- American Guernsey Association (AGA)
- American Hereford Association (AHA)
- American Honey Producers Association (AHPA)
- American Horse Council (AHC)
- American Human Association (AHA)
- American-International Charolais Association (AICA)
- American Jersey Cattle Association (AJCA)
- American Maine Anjou Association (AMAA)
- American Mean Institute (AMI)
- American Milking Shorthorn Society (AMSS)
- American Salers Association (ASA)
- American Sheep Industry Association (ASI)
- American Shorthorn Association (ASA)
- American Simmental Association (ASA)
- Animal Agriculture Alliance (AAA)
- Associated Milk Producers Inc. (AMPI)
- Association of Avian Veterinarians (AAV)
- Ayrshire Breeders Association (ABA)
- Beefmaster Breeders United (BBU)
- Brahman Breeders Association—Area 13
- Brown Swiss Association (BSA)
- California Dairy Herd Improvement Association (CDHIA)
- California Milk Advisory Board (CMAB)
- Cooperative Resources International (CRI)
- Dairy Farmers of America (DFA)
- Dairy Lab Services, Inc. (DLS)
- Dairy Management, Inc. (DMI)
- Dairy Records Management Systems (DRMS)
- DHI-Provo
- Equity Cooperative Livestock Sales Association
- Genex Cooperative
- Holstein Association USA
- National Animal Interest Alliance (NAIA)
- National Association of Animal Breeders (NAAB)
- National Cattlemen's Beef Association (NCBA)

Figure 15-2 Commodity, farm, and agribusiness organizations.

(continued on the following page)

Commodity, Farm, and Agribusiness Organizations *(concluded)*

- National Dairy Council (NDC)
- National Dairy Herd Improvement Association (NDHIA)
  - Quality Certification Services, Inc. (QCS)
- National Farmers Organization (NFO)
- National FFA Organization (FFA)
- National Forage Testing Association (NFTA)
- National Grange
- National Institute for Animal Agriculture (NIAA)
- National Lamb Feeders Association (NLFA)
- National Livestock Producers Association (NLPA)
  - National Producers Service Company (NPSC)
- National Mastitis Council (NMC)
- National Meat Association (NMA)
- National Milk Producers Federation (NMPF)
- National Pork Board (NPB)
- National Pork Producers Council (NPPC)
- National Sheep Industry Improvement Center (NSIIC)
- National Swine Improvement Federation (NSIF)
- National Turkey Federation (NTF)
- North American Lab Managers Association (NALMA)
- North American Limousin Foundation (NALF)
- NorthStar Cooperative
- Pennsylvania DHIA
- Professional Dairy Heifers Growers Association (PDHGA)
- Red & White Dairy Cattle Association
- Red Angus Association of America (RAAA)
- Select Sires
- Simmental USA
- Taurus-Service
- Texas Longhorn Breeders Association of America (TLBAA)
- United Braford Breeders (UBB)
- United Egg
- United Soybean Board (USB)
- U.S. Poultry and Egg Association (USPEA)
- Wisconsin Milk Marketing Board, Inc. (WMMB)
- World Dairy Expo
- World Wide Sires

- Limited water supplies and competition for farmland will be major challenges. Agriculture's potential to provide sources of energy will influence research.
- Governmental regulation will increase because of concerns over the environment and food safety; litigation and heightened regulation will particularly affect the livestock and poultry sectors.
- Farming will come to be viewed as a public service; thus, private property rights and interests in land may be restricted.
- As the "baby boomers" age, and the proportion of older people in U.S. society increases, adjustments will have to be made to policies, programs, and eligibility requirements to ensure that older citizens are adequately fed.[20]

20. Knutson, Penn, Flinchbaugh, and Outlaw, *Agricultural and Food Policy*, pp. 176–177.

U.S. Farm Legislation

Agricultural Adjustment Act of 1933

- Established first major price support and acreage reduction program
- Set parity as goal for farm prices
- Achieved acreage reduction through voluntary agreements with producers
- Regulated markets by voluntary agreements with processors and others
- Offset program cost through processing taxes (Agricultural Adjustment Act Amendments of 1935)
- Allowed president to impose quotas if imports interfered with adjustment program

Soil Conservation and Domestic Allotment Act of 1936

- Provided soil conservation and soil building payments

Agricultural Adjustment Act of 1938

- Reenacted modified Soil Conservation and Domestic Allotment Act
- Provided acreage allotments, payment limits, protection for tenants
- Introduced comprehensive price support legislation with nonrecourse loans
- Established marketing quotas for several crops

Stegall Amendment of 1941

- Required support for many nonbasic commodities at 85 percent of parity or higher (increased to 90 percent in 1942)

Agricultural Act of 1948

- Shifted price supports from fixed to flexible
- Modernized parity formula

Agricultural Act of 1949

- Completed fundamental federal farm program structure started in 1938 act; legislation had no expiration date
- Superseded 1948 act and postponed flexible price supports
- Cushioned impact of new parity formula

Agricultural Act of 1954

- Established flexible price supports, beginning in 1955
- Authorized a Commodity Credit Corporation reserve for foreign and domestic relief

Agricultural Trade Development and Assistance Act of 1954 (Food for Peace)

- Established the primary U.S. overseas food assistance program
- Made U.S. agricultural commodities available through long-term credit at low interest rates and provided food donations

Agricultural Act of 1956

- Began Soil Bank program for long- and short-term removal of land from production

Emergency Feed Grain Program of 1961

- Launched voluntary acreage reduction program with payment-in-kind (PIK) program

Food and Agricultural Act of 1962

- Gave president power to impose mandatory production controls

Figure 15–3 Major U.S. farm legislation.

(continued on the following page)

U.S. Farm Legislation *(continued)*
Agricultural Act of 1964 ■ Created two-year voluntary marketing certificate program for wheat ■ Established PIK for cotton
Food Stamp Act of 1964 ■ Established the basis for the Food Stamp Program ■ Supplemented the buying power of eligible low-income households
Food and Agricultural Act of 1965 ■ First multiyear farm bill ■ Extended voluntary acreage controls to wheat and cotton ■ Authorized Class I milk base plan for 75 federal milk marketing orders
Agricultural Act of 1970 ■ Introduced cropland set-aside program ■ Amended and extended authority of Class I milk base plan
Agriculture and Consumer Protection Act of 1973 ■ Established target prices, deficiency payments to replace price support payments ■ Lowered payment limits to $20,000 for all program crops ■ Provided for disaster payments, disaster reserve inventories ■ Emphasized expanded production to meet world demand
Trade Act of 1974 ■ Provided the president with tariff and nontariff trade-barrier negotiating authority for the Tokyo Round of multilateral trade negotiations
Food and Agriculture Act of 1977 ■ Raised price and income supports ■ Created farmer-owned reserve for grains ■ Set up two-tiered peanut program
Federal Crop Insurance Act of 1980 ■ Expanded national crop insurance program to cover majority of crops
Agriculture and Food Act of 1981 ■ Continued programs in effect since the 1930s ■ Set target prices for four-year length of bill ■ Lowered dairy supports, eliminated rice allotments and marketing quotas
Agricultural Programs Adjustment Act of 1984 ■ Froze target price increases ■ Authorized paid land diversions for feed grains, upland cotton, and rice ■ Provided a wheat payment-in-kind program for 1984
Food Security Act of 1985 ■ Lowered price and income supports ■ Established dairy herd buyout program ■ Created Conservation Reserve Program

(continued on the following page)

U.S. Farm Legislation *(continued)*

Omnibus Budget Reconciliation Act of 1986

- Required advance deficiency payments to be made to producers of 1987 wheat, feed grains, upland cotton, and rice crops at a minimum of 40 percent for wheat and feed grains and 30 percent for rice and upland cotton

Farm Disaster Assistance Act of 1987

- Provided assistance to producers who experienced crop losses from natural disasters in 1986

Agricultural Credit Act of 1987

- Provided credit assistance to farmers
- Strengthened Farm Credit System

Omnibus Budget Reconciliation Act of 1990

- Required USDA to calculate deficiency payments for 1994 and 1995 wheat, feed grain, and rice crops using as 12-month average market price instead of the 5-month average required under previous law

Food, Agriculture, Conservation and Trade Act of 1990

- Moved agriculture further in market-oriented direction
- Froze target prices
- Allowed increased planting flexibility
- Established Rural Development Administration within USDA
- Extended and improved food stamp and other domestic nutrition programs

Food, Agriculture, Conservation and Trade Act Amendments of 1991

- Allowed Farm Credit Bank for Cooperatives to make loans for agricultural experiments
- Established new handling requirements for eggs to prevent food-borne illness

Agricultural Credit Improvement Act of 1992

- Created new Farmers Home Administration loan programs for beginning farmers

North American Free Trade Agreement Implementation Act of 1993

- Eliminated all nontariff trade barriers to agricultural trade
- Maintained provisions of U.S.-Canada Free Trade Agreement

Federal Crop Insurance Reform and Department of Agriculture Reorganization Act of 1994

- Supplements the federal crop insurance program with a new catastrophic coverage level (CAT) available to farmers; producers must purchase crop insurance coverage at the CAT level or above to participate in commodity support programs

Uruguay Round Agreements Act of 1994

- Approved, implemented trade agreements
- Permitted reduction of tariffs and government subsidies on agricultural products

Federal Agriculture Improvement and Reform Act of 1996

- Eliminated price-sensitive deficiency payments and most acreage-use restrictions.
- Provided seven years of predetermined direct payments to farmers
- Suspended farmer-owned reserve program
- Reduced spending on commercial agricultural export programs
- Extended conservation and wetland reserve programs
- Authorized new Fund for Rural America
- Modified farm credit and agricultural commodity promotion programs

(continued on the following page)

U.S. Farm Legislation *(concluded)*

The Farm Security and Rural Investment Act of 2002 (Farm Bill)

- Replace federal legislation passed in 1996
- Set national farm legislation for the next six years
- Included provisions on commodities, conservation, trade, nutrition, credit, rural development, research, forestry, and energy
- Included higher loan rates for most crops
- Provided direct payments for wheat, feed grains, cotton and rice
- Expanded eligibility of direct payments to oilseed producers and counter-cyclical payments
- Available loan deficiency payments when market prices are lower than commodity loan rates
- Updated crop based acres using the past four years of planting history to calculate direct and counter-cyclical payments
- Provided opportunity for livestock producers to obtain cost share dollars to meet current and future environmental regulations
- Expanded the Conservation Reservation Program from 36.4 million acres to 39.2 million acres and allowed for limited haying and grazing
- Continued Wetlands Reserve Program along with the Environmental Quality Incentive Program which increased funding from $200 million to $400 million
- Included a Grassland Reserve Program for 1 million acres and a Conservation Security Program to help producers adopt practices on private agricultural land and in incidental forests
- Established a national dairy market loss payments (DMLP) whenever the national Class I base price falls below $13.69 per hundredweight
- Note: Tobacco was not included in the Farm Bill

Reasons for Governmental Services

Production

Various governmental agencies and services are made available to help production agriculturalists solve some of their farm-related problems. Though many believe that governmental assistance to production agriculturalists is a recent development, various services and forms of assistance have existed for more than 150 years. One of the important early agriculture-related laws passed by the federal government was the Homestead Acts of Congress, which aided farmers in settling and farming land to help feed America. In 1862, President Abraham Lincoln signed an act of Congress that established land grant colleges. This enabled better education on farming technology and was intended to increase production. In 1890, an act of Congress provided support for land grant colleges for blacks.[21]

Education

The land grant colleges, as well as other colleges offering courses on agriculture subjects (non-land grant colleges), educate students about agribusiness inputs, agribusiness services, production agriculture, agriscience, and the agribusiness output sector. *Extension specialists* are graduates of these universities who make agricultural information available to production agriculturalists, processors, handlers, farm families, communities, and consumers. The Extension Service of the USDA also administers the 4-H program. The government grants funds to state and local districts to help finance instruction in agricultural education; these funds are administered through the U.S. Department of Education.[22] The agricultural education teachers provide instruction for production agriculturalists of all ages, as well as for school-age students. Agricultural education teachers also act as advisors to the Future Farmers of America (FFA).

21. Alfred H. Krebs and Michael E. Newman, *Agriscience in Our Lives*, 6th ed. (Danville, IL: Interstate Publishers, 1994), p. 57.

22. Ibid., p. 60.

Services

The government offers many services that we take for granted. It is ironic that often those who criticize government spending on agriculture also complain about some lack of service, product, or consumer production. Most of the following governmental services are administered through divisions and programs of the U.S. Department of Agriculture (USDA), as discussed later in this chapter. Through the USDA, the government:

- manages the national forest system
- makes loans to farmers, cooperatives, and rural electric systems
- improves the marketability of crops
- encourages fair trade practices
- maintains fair and open competition in marketing, livestock, poultry, and meat
- prevents discrimination against producers
- requires truthful labeling
- extends patent-type protection to developers of plants
- helps prevent loss of stored products
- inspects commodity exports and imports
- expands foreign markets for agricultural products
- provides food assistance for those in need
- inspects food to assure safety and quality
- assists states in controlling insects, weeds, and plant diseases that are a serious threat
- keeps foreign diseases from entering the country
- **eradicates** diseases that do get into the country
- fights major domestic animal disease outbreaks
- mandates humane treatment of animals
- regulates all futures trading
- protects natural resources
- provides comprehensive information services for food and agricultural sciences
- conducts research in the areas of livestock, plants, soil, water, air quality, energy, lead safety and quality, food processing and marketing, nonfood agricultural products, and international development
- sets aside land not suitable for farming to be used for forests and reservations[23]

To carry out and accomplish all these services, thousands of workers are needed. In the sections that follow, we discuss the departments within the USDA that perform most of these services.

Organizational Structure of the USDA

Because the USDA plays such an important role in the agricultural sector, a student of agribusiness should have a basic understanding of its organizational structure and component agencies. The USDA is not one agency, but a rather complex organization of individual agricultural agencies, in addition to several administrative ones. The specific number changes frequently as agencies are reorganized and new ones added; changes are especially likely with installation of each new administration. However, the USDA normally consists of 15 to 20 agencies. Many divisions and agencies within the USDA changed names and functions after passage of the Department of Agriculture Reorganization Act of 1994.

The USDA is administered by a secretary, deputy secretary, and several assistant secretaries and undersecretaries. Each assistant secretary or undersecretary is responsible for overseeing and administering several agencies that relate to specific areas of agriculture. Currently, the major divisions include Food, Nutrition and Consumer Services; Marketing and Regulatory Services; Farm and Foreign Agricultural Services; Rural Development; Natural Resources and Environment; Research, Education, and Economics; and Food Safety. In addition, there are numerous offices that provide administrative services. According to the USDA's *Agriculture Fact Book, 2002,* the USDA currently operates with a budget of more than $65 billion and employs more than 88,000 people.[24] Figure 15-4 shows the organization of the USDA. The following sections discuss the major divisions and most of the agencies that provide a specific service for the clients of the USDA.

23. Ibid., pp. 58-60.

24. United States Department of Agriculture, *Agriculture Fact Book, 2001–2002;* available at http://www.usa.gov/factbook/2002factbook.pdf.

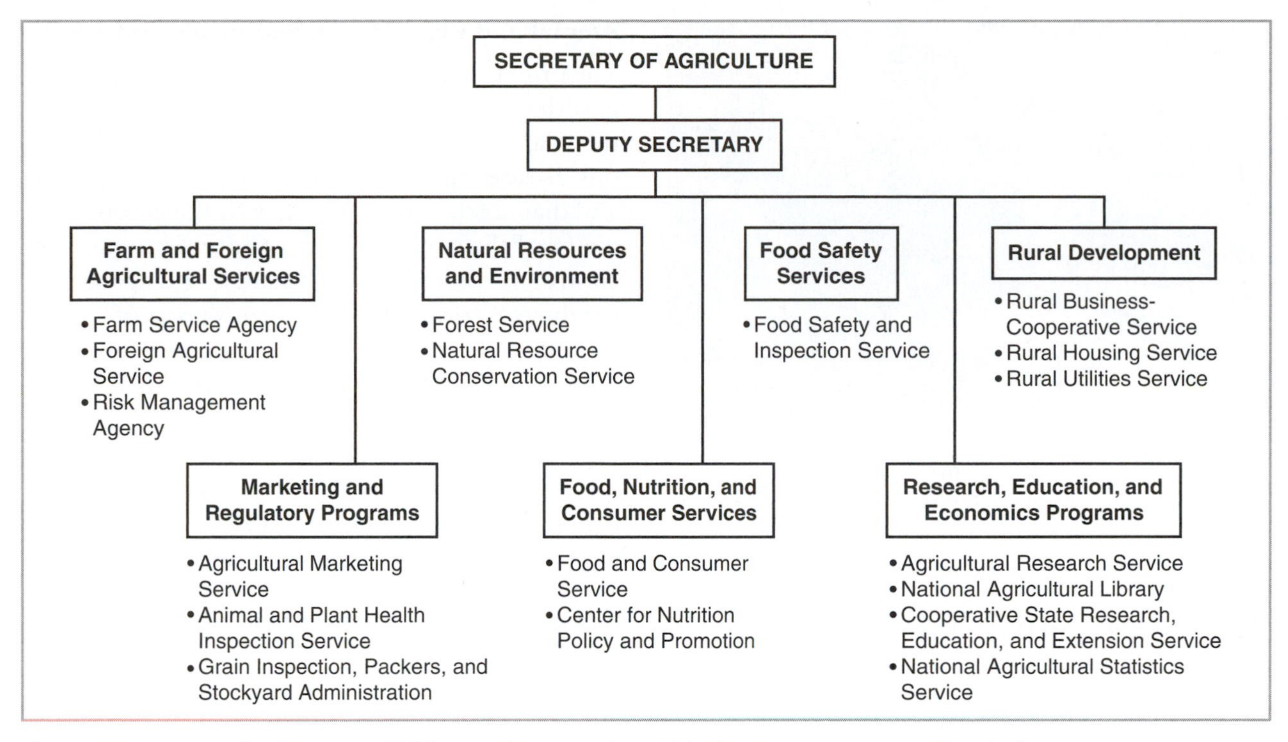

Figure 15–4 Presently, there are 7 divisions and 19 agencies within the U.S. Department of Agriculture.

Food, Nutrition, and Consumer Services

Until 1977, food and consumer services functions were carried out through Agricultural Marketing Services, but they are now in a separate division of their own. The Food, Nutrition, and Consumer Services division now coordinates the largest budget of any division in the USDA. The FNCS administers food assistance and nutrition programs that account for 40 to 70 percent of the USDA budget, depending on the size of farm program expenditures.[25] The major responsibilities of this division date back to the 1960s, when the USDA had tons of surplus food with no markets. Under Public Law No. 480, programs were developed to distribute this food to needy individuals as a part of the War on Poverty. Currently, emphasis is also being directed toward consumer education on all aspects of nutrition. The agencies currently assigned to this division are the Food and Nutrition Service and the Center for Nutrition Policy and Promotion.

25. Ibid.

Food and Nutrition Service

The Food and Nutrition Service administers a number of programs that provide food assistance to individuals through state and local agencies. These programs are designed to meet the food needs of families, individuals with special nutritional requirements, and persons in certain institutions. One of the largest of these programs, in terms of people served, is the school lunch program, which provides more than 27 million students in the nation's schools with subsidized lunches. Other initiatives administered by the Food and Nutrition Service include the Food Stamp Program, the most expensive of the nutrition programs, which serves about 19 million people; nutrition assistance programs; the Child and Adult Care Food Program; the Special Milk Program; food distribution programs; the Special Supplemental Nutritional Program for Women, Infants, and Children; the Commodity Supplemental Food Program; and the Emergency Food Assistance Program. Although the Food Stamp Program is the largest program in terms of the amount of funds disbursed, more than 5 million people are reached by the other programs as well.[26] Refer to Figure 15–5.

26. Ibid.

Figure 15–5 The USDA employees shown here are reviewing the National Nutrient Database for the Child Nutrition Program. (Courtesy of USDA)

Marketing and Regulatory Programs

The Marketing and Regulatory Programs division administers one of the largest number of agencies and has the overall responsibility of coordinating the activities of firms involved in moving food and fibrous products from the farm to the final consumer. These agencies provide protection for farmers, on the one hand, and consumers, on the other. The specific agencies in this division include the Agricultural Marketing Service, the Animal and Plant Health Inspection Service, and the Grain Inspection, Packers, and Stockyards Administration.

The Agricultural Marketing Service

The Agricultural Marketing Service promotes the orderly and efficient marketing and distribution of agricultural commodities. The AMS's Federal-State Marketing Improvement Program studies problems at the local and state levels. It is involved in improving marketability of crops, finding new markets, improving market efficiency, and improving market information.[27] Specific activities administered by the Agricultural Marketing Service include a market news service, egg products inspection, development of quality standards for commodity grading, and marketing orders and agreements for certain commodities. Participation in most of these programs is initiated by the producer and/or marketing firm, which may pay a fee to the AMS for services rendered.

27. Ibid.

Animal and Plant Health and Inspection Service

The primary purpose of the Animal and Plant Health and Inspection Service (APHIS) is to protect plants and animals from economically dangerous diseases and pests. The agency initiates survey and diagnostic activities and, with the governments of the affected states, conducts programs to control or eradicate these hazards. This agency conducts inspections at numerous ports of entry to prevent the entry and spread of potentially damaging pests and diseases. In addition, the agency deals with emergency outbreaks of plant pests and animal diseases, such as the gypsy moth outbreak in Oregon and the citrus canker infestation in Florida in 1985. The APHIS is also charged with overseeing the humane treatment of animals.

Grain Inspection, Packers, and Stockyards Administration

The Grain Inspection, Packers, and Stockyards Administration (GIPSA) facilitates the marketing of many agricultural products and promotes fair and competitive trade for U.S. commodities. Its Federal Grain Inspection Service was established in 1976 in response to international scandals in foreign grain sales. This agency establishes official standards for grain, conducts weighing and inspection activities, and inspects other agricultural products under the authority of the Grain Standard Act and the Agricultural Marketing Act of 1946. Inspection and weighing costs are recovered by user fees; they are required at all export ports and optional at interior locations. The GIPSA is also the regulatory agency that monitors livestock markets for potential unfair trade practices.

Farm and Foreign Agricultural Services

The Farm and Foreign Agricultural Services division oversees national efforts to keep the supply and demand of agricultural products in balance. During the 1960s, its primary efforts were directed toward reducing the supply of agricultural products, through massive storage programs and production control at the farm level. During the 1970s and 1980s, the major effort was directed toward increasing demand through the development of new foreign markets for agricultural products. However, attention is still being paid to supply control as well. This division currently includes the Farm Service Agency, the Foreign Agricultural Service, and the Risk Management Agency.

Farm Service Agency

The Farm Service Agency (FSA) is the result of the biggest change in the USDA structure in many years. The FSA is now a "one-stop shopping" area where the production agriculturalist can sign up for farm programs, obtain loans, and obtain assistance with soil conservation programs. The Farm Service Agency was created by the consolidation of the former Agriculture Stabilization and Conservation Service (ASCS) and the Farmers Home Administration (FmHA).

History. The FSA has the most direct contact with the modern production agriculturalist. During the 1950s, the old Agricultural Stabilization and Conservation Service received about 65 percent of the total USDA budget, but the FSA currently administers less than 50 percent; the percentage shifts are due primarily to major increases in the food assistance programs. The ASCS began in 1933 as the Agricultural Adjustment Administration (AAA), commonly known as the Triple A. In 1945, it became known as the Production and Marketing Administration, which survived until 1953. It was then divided into two agencies, known as the Agricultural Marketing Service and the Commodity Stabilization Service. The Agricultural Marketing Service (discussed earlier) continued its status as a separate agency, while the Commodity Stabilization Service became the ASCS until its merger into the Farm Service Agency.

The Farmers Home Administration (FmHA) was established in 1946 to provide financial assistance to farmers who were unable to obtain loans from private agencies, although it has taken on numerous other responsibilities as well. With these high-risk loans and housing loans, the borrower is expected to refinance the loan with a private lender as soon as possible. With the creation of rural development programs in the 1970s, FmHA made a higher percentage of farm loans because of the farm financial crisis. Rural development efforts are directed toward loans for rural home repairs, creation of rural water and waste disposal systems, and other development projects. Although the FmHA has been merged with FSA, loans for land are still made through the successor agency. However, loans for rural housing are made through the Rural Housing Service.

Responsibilities. The responsibilities of the FSA are to:

- stabilize the nation's agricultural economy through price support loans and purchases and crop and **income stabilization**
- preserve the nation's farm resources through a variety of conservation measures and land retirement programs
- protect the nation's food supply and food reserves through commodity storage programs and inventory management programs
- assist in the national defense through civil defense and defense mobilization activities
- assist with farm loans
- furnish information on commodities

Size. In addition to the national office in Washington, D.C., the FSA has 3 regional commodity offices, 50 state offices, and about 2,800 county offices that administer its various programs. It is officially directed at the local level by elected farmer committee members, and at the state level by committee members appointed by the Secretary of Agriculture. FSA employs more than 4,000 workers.

The FSA deals on a day-to-day basis with hundreds of thousands of producers, carriers, exporters, handlers, warehouse workers, and others. FSA responsibilities include the handling of millions of documents yearly.[28] Refer to Figure 15–6.

Commodity Credit Corporation

The Commodity Credit Corporation (CCC) was organized in 1933 to act as a purchasing agent for the federal government. The CCC loan program was the tool that acted as a price support mechanism. It was not until 1939 that the CCC was transferred to the Department of Agriculture.

28. Marcella Smith, Jean M. Underwood, and Mark Bultmann, *Careers in Agribusiness and Industry*, 4th ed. (Danville, IL: Interstate Publishers, 1991), p. 253.

Figure 15–6 This Farm Service Agency employee is advising a local production agriculturist with one of the agency's farm programs. (Courtesy of USDA)

The CCC was created to stabilize, support, and protect farm income and prices; to help maintain a balanced supply of agricultural commodities and products (foods, feeds, and fiber); and to assist in the orderly distribution of agricultural commodities and products. Basically, the CCC is the financial arm of the FSA, but it also finances other agricultural programs that are closely related, such as farm export activities of the Foreign Agricultural Service.

Foreign Agricultural Service

The Foreign Agricultural Service (FAS) has played a major role in making the United States the world's largest exporter of agricultural products. This agency is the export promotion and service agency for American agriculture, and **attachés** in 65 foreign countries maintain a constant flow of information on the international production and marketing of food and fiber. Import and export regulations are administered by the Foreign Agricultural Service in cooperation with numerous local, state, regional, national, and international trade groups. The Foreign Agricultural Service is assisted in its mission by the Office of International Cooperation and Development. Under Public Law No. 480, the FAS is responsible for maintaining an export credit sales program and financing the sale of agricultural commodities. The FAS employs about 75 people and operates with an annual budget of about $75 million.

The Office of International Cooperation and Development is responsible for cooperative international research, scientific and technical exchanges, and **liaison** with international agricultural organizations. It directs training and technical assistance programs in many developing countries.

Risk Management Agency

The Risk Management Agency is the new name of the former Federal Crop Insurance Corporation (FCIC). This agency provides farmers with insurance that reimburses crop production losses due to extreme weather, insect damage, diseases, and other uncontrollable natural conditions. The Risk Management Agency provides the only all-risk crop investment protection available to farmers in many major agricultural counties. An all-risk crop insurance bill was introduced in Congress in 1922 but was not enacted until 1938. Coverage was first available for wheat, but was soon expanded to cotton and other crops. Currently, insurance is available on 23 crops in 39 different states.

Since the enactment of the Federal Crop Insurance Act of 1980, substantial progress has been made in shifting the burden of crop insurance to the private sector. Currently, government insurance companies account for about 75 percent of crop insurance sales, and the remainder is handled by private agents through sales and service agreements.

Other parts of the mission of these agencies are to research, develop, and pilot new crop programs, insurance plans, and risk management strategies; to evaluate and make recommendations for improvement of existing crop programs, insurance plans, and risk management strategies; and to coordinate the development of and support for specialty crop programs.

Rural Development

Emphasis on rural development began to emerge in the late 1960s and early 1970s, when policy makers realized that a tremendous economic gap had developed between the urban and rural sectors of the U.S. economy. Rural development became part of the War on Poverty initiated during the Lyndon Johnson administration.

A new and sharper focus on rural development took shape with passage of the Department of Agriculture Reorganization Act of 1994. Rural development work was concentrated in three new organizations that report to the Undersecretary for Rural Development: the Rural Utilities Service, which offers telecommunications and electric programs along with water and sewer programs; the Rural Housing Service, which includes rural housing programs as well as rural community loan programs; and the Rural Business-Cooperative Service, which provides cooperative development and technical assistance, in addition to other business development programs and the Alternative Agricultural Research and Commercialization Center.

Rural Utilities Service

The Rural Utilities Service (RUS) was established as the Rural Electrification Administration (REA) in the early 1930s to help rural inhabitants obtain electric and telephone services. It did so primarily by making long-term loans to rural electric cooperatives and other power suppliers so that they could build and operate rural electric systems. This agency also makes loans, through the Rural Telephone Bank, to independent companies and cooperatives for extending and improving telephone service in rural areas. More than 1,000 cooperatives and other organizations have borrowed RUS funds to provide electric service. Loans are also made for water and waste disposal programs.

More than $2 billion has been loaned to provide electric service to more than 20 million rural inhabitants. In addition, more than 900 firms have borrowed more than $2 billion to provide modern telephone service to about 10 million rural people. These funds were loans, not **grants**, and must be repaid with interest. The repayment on these loans has totaled more than $1.5 billion since the REA (now RUS) was established as a part of President Franklin D. Roosevelt's "New Deal." This system, more than any other, has allowed rural people to enjoy the same modern conveniences as their urban neighbors.

Today, the mission of the Rural Utilities Service is to take the lead in improving the quality of life in rural America by administering its electrification, telecommunications, and water and waste disposal programs in a service-oriented, forward-looking, and financially responsible manner.

Rural Housing Service

The Rural Housing Service (RHS) operates credit programs for rural communities and rural housing. The biggest challenge for this agency is to make sure that loans go for housing in truly rural America, rather than in counties adjacent to metropolitan areas. Previously, rural housing loans were handled by the Farmers Home Administration.

Rural Business-Cooperative Service

This agency, originally called the Agricultural Cooperative Service, was first organized in the 1930s as part of the New Deal. The major functions of the Rural Business-Cooperative Service (RBS) are to provide technical assistance and conduct research on economic, financial, organizational, managerial, legal, social, and other issues that affect agricultural cooperatives.

Today, the mission of RBS is to enhance the quality of life for all rural Americans by providing leadership in establishing competitive businesses, including cooperatives that can build sustainable economic communities, compete successfully in the domestic arena, and develop emerging market opportunities in the international arena. RBS objectives are to invest its financial resources and technical assistance in businesses and communities and to create partnerships that leverage public and private resources to stimulate rural economic activity.

Natural Resources and Environment

The Natural Resources and Environment division is a relatively new organization composed of agencies moved from other divisions. The major objectives of these agencies are to conserve our natural resources and improve the environment in which we live and work. The financial support for this division has been reduced considerably, and user fees have thus been increased.

Forest Service

The U.S. Forest Service, created in 1905, is responsible for the conservation and management of the nation's forest and other wildlife resources. There are 154 national forests and 19 national grasslands, containing 187 million acres in 41 states and Puerto Rico. The major resources obtained from these forests and rangelands are wood, water, wildlife, forage, and recreation. The Forest Service conducts research in timber, range, water, recreation management, and forest product utilization, to ensure maximum benefits for the greatest number of people. The Forest Service also cooperates with state forestry agencies and private landowners to provide technical assistance for the protection and management of 480 million acres of state and private forest lands.

In addition to trees, the national forests and grasslands provide a home for more than 4 million big game animals and 79 species of threatened or endangered wildlife. The Forest Service activities, programs, and services benefit both production agriculturalists and the general public. Refer to Figure 15-7.

Natural Resources Conservation Service

The nation's leaders have been concerned about water and soil conservation since the 1800s, and numerous conservation acts have been passed. However, it was not until 1933 that a specific agency was created to deal with this problem. The Soil Erosion Service was first established within the Interior Department, but was transferred to the Department of Agriculture two years later and renamed the Soil Conservation Service (SCS). In 1994, it was renamed the Natural Resources Conservation Service. The NRCS administers programs to conserve soil and water resources, primarily by providing technical assistance to production agriculturalists and other landowners in soil and water conservation districts. It also provides technical assistance to communities and local governments through rural development and flood prevention programs. Soil classification, mapping, and surveys are also conducted by 2,950 local conservation districts that employ about 17,000 people.

Figure 15-7 The U.S. Forest Service is involved with recreation management as well as having responsibility for the conservation and management of the nation's forest and other wildlife resources. (Courtesy of Donald Waldrop)

Jobs. There are many interesting jobs in the NRCS:

- Soil conservationists work with production agriculturalists and other landowners on ways to conserve soil, build farm ponds, and reduce water pollution.
- Soil conservation technicians assist and supervise production agriculturalists and others with installation of conservation practices.
- Soil scientists map and classify soils; identify problems such as excessive wetness or erosion; identify soils using aerial photographs; and prepare soil descriptions and other information about soils.
- Range conservationists help plan grazing strategies and practices, suggest methods for brush control, and give advice on water management, as well as many other tasks.
- Engineers assist in erosion control, water management, structural design, construction, hydraulics, soil mechanics, and environmental protection.[29]

Research, Education, and Economics

The Research, Education, and Economics Division evolved from an agency created in 1978 that encompassed the activities of several research and education agencies. This division provides support, coordination, and joint planning relating to food

29. Ibid., pp. 253–255.

and agricultural research. It also provides extension and teaching efforts among federal departments, state extension services, experiment stations, universities, and other private and public institutions. The specific agencies in this division include the Agricultural Research Service (and National Agricultural Library); the Cooperative State Research, Education, and Extension Service; the Economic Research Service; and the National Agricultural Statistics Service.

Agricultural Research Service

The Agricultural Research Service (ARS) is the primary agency doing basic research on plants and animals.

Animal and Plant Research. The ARS investigates disease prevention and cure, as well as doing research aimed at improving the productivity of plants and animals. Current agricultural research priorities include improvements in animal reproductive efficiency; improved use and preservation of plant **germ plasm** (genes); removal of barriers to crop productivity; conservation of soil, water, and air; reduction of the effects of erosion on soil productivity; efficient energy use; and development of new pest control technology, among other initiatives.[30]

Marketing Research. The ARS also maintains marketing research programs related to development of new and improved products. Other goals are to expand domestic and foreign markets for agricultural products, to protect and maintain the quality of agricultural products in marketing channels, and to improve the efficiency of market facilities. Although much ARS work is done in cooperation with the states, the service also sends many researchers to foreign countries.

National Agricultural Library. The National Agricultural Library constitutes the most extensive source of agricultural information in the world, with its holdings of about 2 million books and more than 25,000 periodicals. The National Agricultural Library is located in Beltsville, Maryland, and its major users are USDA employees. Any individual can use NAL materials, but materials must be requested through a public or university library.

30. Ibid., pp. 256–257.

Cooperative State Research, Education, and Extension Service

The Cooperative State Research Service. This service administers federal grant funds for research in the production phase of agriculture, agricultural marketing, rural development, and forestry. Funds are provided to state agricultural experiment stations and other designated institutions in the 50 states, Puerto Rico, Guam, the Virgin Islands, and the District of Columbia. Cooperative State Research Service staff members review research proposals, conduct reviews of research in progress, and encourage cooperation between states.

Extension Service. The Extension Service (CSRES) is one of the major educational agencies of the Department of Agriculture. It is the federal arm of the Cooperative Extension Service, an entity run jointly by federal, state, and local governments. All three groups share in the financing, planning, and conducting of agricultural education programs. The Extension Service was created by the Smith-Lever Act of 1914, and its major purpose is to help the public learn about the latest technology developed through research by various agricultural institutions.

The Extension Service has educational faculty and facilities in land grant universities established in 1862 and 1890, and it staffs at least 3,150 county offices representing local governments. Extension personnel offer education programs to production agriculturalists, families, individuals, and communities.

Distance learning, teleconferencing, the Internet, and other communication technologies are used to deliver programs, research findings, science, technology, and other services. The Cooperative Extension Service identified eight initiatives for its work, in the following areas of concentration:

- competition and profitability of U.S. agriculture
- alternative agricultural opportunities
- water quality
- conservation and management of natural resources
- revitalizaton of rural America

- improvement of nutrition, diet, and health
- family and economic well-being
- development of human capital[31]

4-H Clubs. Depending on the size of the county or local district, the Agricultural Extension Service has one to five onsite agents. One of these is responsible for the adult programs and one is responsible for the youth program, the 4-H Club (Hands, Head, Heart, Health). Educators learned long ago that young people are more receptive to learning and adopting new ideas than are adults. Therefore, it was logical that the Agricultural Extension Service direct its attention to educating rural youth as well as adults, and the medium developed to reach young people was 4-H Clubs.

The purpose of the clubs is to give young people an opportunity to participate in problem-solving activities that affect themselves, their homes, and their communities. Members are encouraged to undertake projects that will improve their knowledge of agriculture, science, and/or home economics; in the process, they gain personal development, perform community service, and grow in many other areas.

More than 60 million Americans from all walks of life have been 4-H members. A few of the more famous 4-H alumni are Roy Rogers, John Denver, Dolly Parton, Alan Shepard, Don Meredith, Tom Wopat, Charley Pride, Orville Redenbacher, and Reba McEntire. Membership in 4-H Clubs in 2006 was estimated at about 6.5 million.[32] Members are active in all 50 states, Puerto Rico, the Virgin Islands, and numerous other countries. Youth organizations in more than 80 countries have adopted the 4-H system. Approximately $115 million is appropriated annually for 4-H activities from federal, state, and local governments, in addition to about $25 million from private sources. Thirty-eight states now have 4-H foundations, and more than 200 of the nation's major corporations contribute to 4-H each year. Most of this support is coordinated through the National 4-H Council. Refer to Figure 15–8.

31. Ibid., p. 263.
32. 4-H Club National Home Page (2007); available at http://www.4-H.org.

Figure 15–8 One of the objectives of 4-H is personal development. This member is participating in a 4-H style show. (Courtesy of Marlene Graves)

Economic Research Service

The Economic Research Service provides timely and reliable agricultural economic data, including historical research, forecasts based on major **economic indicators**, and **policy analysis** that serves as a basis for economic decisionmaking by production agriculturalists, consumers, extension workers, private analysts, marketing firms, input suppliers, and public officials. It employs the largest single group of agricultural economists in the world.

National Agricultural Statistics Service

The National Agricultural Statistics Service (NASS) has the unique responsibility of continuing the original mandate of the Department of Agriculture, which was to collect and distribute agricultural data. This agency provides agricultural data to fill in the gaps between agricultural census years. Crop and livestock estimates are made by a central office in Washington, D.C., and by state statistical offices, which are operated as joint state and federal services. The NASS also conducts and coordinates research on statistical techniques and data collected by satellite. Refer to Figure 15–9.

Figure 15–9 A Global Positioning System antenna on the top of the tractor operator's cab communicates with a satellite and a yield monitor to allow an onboard computer to plot corn yields about every six feet as the combine moves along. Data stored in the computer can later be used to produce color-coded yield maps for each field. (Courtesy of USDA)

Food Safety and Inspection Service

The Food Safety and Inspection Service is responsible for making sure that the nation's meat and poultry supplies are safe, wholesome, and properly labeled and packaged. This agency provides plant inspection of domestic meat- and poultry-processing operations, reviews inspection systems in foreign countries that export meat and poultry products to the United States, and inspects products at ports of entry. These services are funded by state and federal programs.

USDA Summary

In summary, the U.S. Department of Agriculture plays a very important role in modern agriculture. USDA activities affect not only those in the agricultural industry, but also the public at large. The USDA has expanded considerably since it was established in 1862, as the need for more and better services has increased. The USDA was established primarily as an agency to serve the needs of production agriculturalists, but has now expanded to encompass a group of agencies intended to serve the agricultural industry and the nation as a whole. Currently, more than 50 percent of the USDA's budget goes toward ensuring that the needy are provided at least a minimum level of food through the Food and Nutrition Service.

Currently, the USDA is funded at about $65 billion annually and maintains a payroll for about 100,000 people (including part-time employees). Most of the positions are under the jurisdiction of the U.S. Civil Service Commission. To qualify, you have to pass a civil service examination. There is stiff competition for the agricultural research federal civil service jobs. However, a job with this agency is desirable, as the federal government offers many outstanding benefits to its employees, such as vacations, sick leave, **flex time**, alternative work schedules, child care arrangements, credit unions, and recreation and fitness centers.

State Departments of Agriculture

Although most public support and regulation of agricultural programs are initiated and funded at the federal level, state governments also play a major role in modern agriculture. Presently, all 50 states have departments of agriculture or other, comparable agencies responsible for administering agricultural programs at the state level. State departments of agriculture are administered by secretaries, commissioners, or directors, most of whom are appointed by the state governor. However, some of these administrators are elected rather than appointed.

Mission

The state departments of agriculture are primarily regulatory agencies. However, numerous other activities are funded at the state level, including disease prevention and treatment programs, regulation of weights and measures, collection and dissemination of statistical data, supervision of agricultural fairs, enforcement of chemical fertilizer and seed quality laws, and agricultural promotion activities. The emphasis of the various activities varies by state.

National Association of State Departments of Agriculture

In an effort to influence the agricultural industry beyond the state level, the administrators of state departments of agriculture organized the National Association of Commissioners, Secretaries and Directors of Agriculture in 1915. In 1955, the name of this group was changed to the National Association of State Departments of Agriculture (NASDA). Since 1968, NASDA has maintained an office in Washington, D.C., headed by an executive secretary with a small staff, to gather and disseminate data to Congress and other federal agencies on agricultural matters.

NASDA staff members do not perform lobbying functions, but they do assist individual members of Congress and their staffs in preparing reports for congressional committees. Policy positions on current agricultural issues are often prepared by NASDA and presented to the proper federal agencies for consideration. NASDA has been, and continues to be, a very effective means of keeping the federal agencies abreast of grassroots problems in agriculture. "The interest of NASDA covers the full scope of agriculture from production through processing and transportation, consumer affairs, national regulatory programs and the fiscal policies of the Federal and State governments."[33]

Land Grant System

One of the most significant developments in the history of American agriculture was the establishment of what is known today as the land grant system. This system was designed to develop new technology through agricultural research and to make this technology available to the agricultural industry through a comprehensive educational system. It consists of the land grant universities, agricultural experiment stations, and the Agricultural Extension Service.

History of Land Grant Systems

The part of the system that developed first was the land grant universities or colleges. In 1853, Justin S. Morrill, a representative from Vermont, introduced a bill in the House requesting that public lands be donated "to the several states which may provide colleges for the benefit of Agriculture and Mechanical Arts."[34] Each state not in rebellion was entitled to 30,000 acres of public land for each senator and representative in Congress. The bill finally passed, by a narrow margin, in 1858, but was vetoed by President James Buchanan. Representative Morrill again introduced his bill in early 1862, but again it was rejected. However, the bill was later amended, passed by both houses, and signed by President Lincoln in July 1862.

The final content of the bill required that 30,000 acres would be allotted to each state for each senator and representative, with a maximum of 1 million acres in any one state. The proceeds from the sale of such land were to be used to support agricultural and mechanical programs at the college level. Each state was directed to designate one or more colleges as land grant colleges.

Colleges of 1890

In 1890, a second Morrill Act was passed, which provided annual appropriations to support teaching at the land grant colleges and added a "separate but equal" provision authorizing the creation of colleges for blacks, who, at that time, were not allowed to attend many of the existing universities. Sixteen southern states took advantage of this act and created colleges that are currently referred to as the "colleges of 1890."

Agricultural Experiment Stations

The first Morrill Act was followed by the Hatch Act in 1887, which authorized federal support for each state that established an agricultural experiment station in conjunction with the state's land grant college. An annual appropriation of $15,000 was approved for each state for setting up the stations and conducting experiments in the agricultural and mechanical arts. An important factor in the development of the land grant system was the establishment of the National Association of State Universities and Land-Grant Colleges in 1887. This organization became very powerful politically and had a major influence on the future expansion of research and education in agriculture.

Development of the Cooperative Extension Service

It soon became obvious that the new technology developed through agricultural experiment stations was not reaching people in rural communities. In response to this need, the Cooperative

33. National Association of State Departments of Agriculture, *NASDA—What's It All About?* (Washington, DC: Author, 1977).
34. Benjamin A. Hubbard, *A History of the Public Land Policies* (New York: Peter Smith, 1939), p. 348.

Extension Service was established in 1914 by the Smith-Lever Act. This act created an organizational structure through which the Cooperative Extension Service is funded cooperatively by federal, state, and local governments and administered by the land grant universities in each state. Refer to Figure 15–10 for a list of the land grant universities.

A List of the Land Grant Universities in All 50 States

State	Land-Grant Universities	State	Land-Grant Universities
Alabama	Auburn University	Missouri	University of Missouri
	Alabama A&M[a]		Lincoln University[a]
Alaska	University of Alaska	Montana	Montana State University
Arizona	University of Arizona	Nebraska	University of Nebraska
Arkansas	University of Arkansas	Nevada	University of Nevada
	Arkansas A&M[a]	New Hampshire	University of New Hampshire
California	University of California	New Jersey	Rutgers University
Colorado	Colorado State University	New Mexico	New Mexico State University
Connecticut	University of Connecticut	New York	Cornell University
Delaware	University of Delaware	North Carolina	North Carolina State University
	Delaware State University[a]		North Carolina A&M[a]
Florida	University of Florida	North Dakota	North Dakota State University
	Florida A&M[a]	Ohio	Ohio State University
Georgia	University of Georgia	Oklahoma	Oklahoma State University
	Fort Valley State College[a]		Langston University[a]
Hawaii	University of Hawaii	Oregon	Oregon State University
Idaho	University of Idaho	Pennsylvania	Pennsylvania State University
Illinois	University of Illinois	Puerto Rico	University of Puerto Rico
Indiana	Purdue University	Rhode Island	University of Rhode Island
Iowa	Iowa State University	South Carolina	Clemson University
Kansas	Kansas State University		South Carolina State University[a]
Kentucky	University of Kentucky	South Dakota	South Dakota State University
	Kentucky State University	Tennessee	University of Tennessee
Louisiana	Louisiana State University		Tennessee State University[a]
	Louisiana Tech Southern University	Texas	Texas A&M University
Maine	University of Maine		Stephen F. Austin
Maryland	University of Maryland, College Park		Prairie View A&M[a]
	University of Maryland, Eastern Shore[a]	Utah	Utah State University
Massachusetts	University of Massachusetts	Vermont	University of Vermont
Michigan	Michigan State University	Virginia	Virginia Polytech Institute
Minnesota	University of Minnesota		Virginia State University[a]
Mississippi	Mississippi State University	Washington	Washington State University
	Alcorn A&M[a]	West Virginia	University of West Virginia
		Wisconsin	University of Wisconsin
		Wyoming	University of Wyoming

[a]Colleges of 1890.

Figure 15–10 Land grant universities were created in 1862 and 1890.

Non-Land Grant Agricultural Programs

During the early years of higher education in the United States, the primary responsibility for teaching in agriculture was delegated to the land grant colleges and universities. However, it soon became evident that these institutions alone could not adequately serve all the young people who were interested in studying agriculture beyond the high school level. Even today, there are still some states that offer programs in agriculture only at the land grant universities. However, at least 34 states now have one or more agricultural programs in non-land grant institutions.

The non-land grant schools now play a major role in educating young people for agricultural occupations, especially at the undergraduate level. There are more programs available in non-land grant institutions, and they have about 50 percent of total agricultural enrollment. Many of the students in junior colleges transfer to land grant or non-land grant institutions, and many non-land grant graduates transfer to land grant institutions for graduate study.

Comparison of Land Grant and Non-Land Grant Universities

Land grant and non-land grant institutions differ in terms of financial support and objectives. Land grant institutions are supported by state and federal funds that are appropriated on an annual basis. The only question is how much they receive. Non-land grant institutions often receive state funds on a regular basis; they may receive federal funds for special projects, but not on a regular basis. Non-land grant schools are primarily, and in some cases solely, concerned with teaching, although most teachers are encouraged to do some research and public service. Land grant institution faculties have duties divided among teaching, research, and/or extension activities.

Sea Grant Program

Historical Development

In conjunction with increased efforts in the early 1960s to expand our knowledge about the availability of usable resources located in major bodies of water, Athelstan Spilhaus, chair of the Committee on Oceanography of the Natural Academy of Sciences, proposed the creation of sea grant universities, similar to the land grant universities established under the Morrill Act of 1862. The idea was further developed at a sea grant conference in 1965, which was sponsored by the University of Rhode Island. As a result, Senators Clairborne Pell of Rhode Island and Paul Rogers of Florida introduced the Pell-Rogers Sea Grant Bill. The National Sea Grant College and Program Act was signed into law by President Johnson in 1966, and the National Science Foundation was given administrative responsibility for the program. In 1970, the administrative responsibility for the sea grant program was transferred to the National Oceanic and Atmospheric Administration, an organizational unit of the U.S. Department of Commerce.

Mission

The national sea grant program is concerned with the exploration and development of the resources of the oceans, the Great Lakes, and U.S. coastal waters. It was established to promote research, education, and advisory services in marine resources, including conservation and socially and economically sound management.

Reason for the Sea Grant Name

The term *sea grant* was chosen to emphasize the parallel nature of this program with the land grant program, with which it shares a basic organizational pattern of research, teaching, and extension functions.

Funding

Through a matching fund program, the sea grant program provides grants to colleges and universities, as well as to private industry, state agencies, private foundations, consumer interest groups, and private individuals; the grants allow recipients to conduct research and educational projects in marine development. Usually, two-thirds of the funds come from federal appropriations and the remaining one-third is supplied by state, local, or private funds. Normally, 60 percent of sea grant funds is allocated to research, and the remaining 40 percent is used for education and advisory services programs. Grants awarded by the Office of Sea Grants are classified into four types: project awards, coherent project programs, sea grant institutional programs, and sea grant college programs.

Project Awards. Grants are awarded for specific projects in research, education, or advisory services that are appropriate to sea grant objectives. These awards are based on proposals submitted by competing individuals or institutions, with specific goals and completion schedules spelled out. Emphasis is placed on innovative projects designed to answer problems or meet needs in a specific area.

Coherent Project Programs. The purpose of the coherent project programs is to provide opportunities for both academic and nonacademic institutions to use their resources in solving marine-related problems. This program allows work on broad-based problems and permits more flexibility in methods and time of project completion than does the grant program. These projects are usually **interdisciplinary** in scope and may include research, education, and/or advisory services, depending on the resource capabilities of the sponsoring institutions.

Sea Grant Institutional Programs. Status as a sea grant institutional program is given to an institution, or **consortium** of institutions, with overall competence in marine affairs and a comprehensive marine program in research, education, and advisory services. An institution with coherent project status may advance its project to the classification of "Sea Grant Institutional Program" through the creation and development of a broad-based program over a period of time.

Sea Grant Colleges. The highest level of federal support is given to institutions of higher learning that have been designated as sea grant colleges. This status is conferred on institutions that have clearly demonstrated outstanding leadership in the quality, quantity, and productivity of performance in research, education, and advisory services for a period of at least three years. The designation of sea grant college is a highly coveted honor. It is sought by many but granted to only a limited number of universities. Only 29 sea grant college designations have been awarded. The location of these programs is shown in Figure 15–11.[35]

35. N. Omri Rawlins, *Introduction to Agribusiness* (Murfreesboro, TN: Middle Tennessee State University, 1999), p. 129.

International Mission of the Sea Grant Programs

In 1973, the Sea Grant Program Act was amended to provide a means for sharing the results of marine research with other nations through cooperative research and educational programs. The sea grant program has made a significant contribution, although it is only about three decades old. This program will probably play an even greater role in the future, as the provision of food and nonfood resources from the sea and major lakes increases in importance for the United States and the rest of the world.[36]

Agricultural Education

To complement agricultural research at the university level, agricultural education was developed to train young men and women at the high school level who want to pursue a vocation of production agriculture or agribusiness. Modern agricultural education is based on a cooperative federal, state, and local program set up under the Smith-Hughes Act of 1917. The act was named after Senator Hoke Smith and Representative Dudley Hughes, both of Georgia, who were the major sponsors. The program is administered by the U.S. Office of Education through state staff for agricultural education. Each state is responsible for presenting an agricultural education plan to the U.S. Office of Education. When this plan is approved, it becomes the responsibility of the state to implement its program through cooperation with local schools.

Agricultural education was originally designed to prepare high school students for farming, primarily through agricultural instruction planned around supervised, on-farm programs (Supervised Agricultural Experience Programs) for rural youth. Many current programs are still production-oriented, but others have expanded into numerous other agribusiness areas as well, depending on the particular needs of each community. Besides agricultural education teacher, there are many careers in the public agribusiness services. Refer to the Career Options section for an explanation of some of the careers in this area.

36. Ibid.

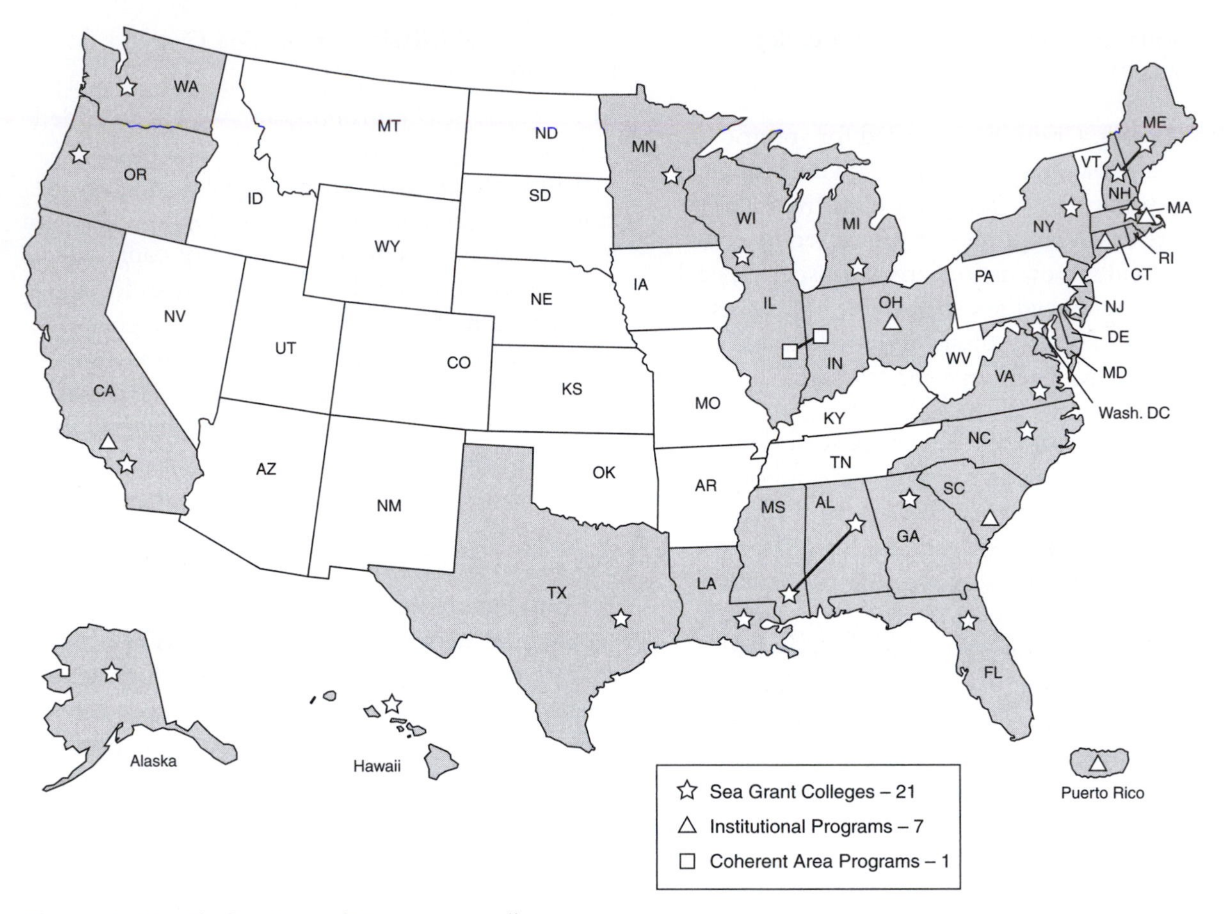

Figure 15–11 The locations of 21 sea grant colleges.

Future Farmers of America

The Future Farmers of America organization evolved from "agriculture clubs" that began forming shortly after the Smith-Hughes Act of 1917 was passed; that act provided federal support for vocational agriculture programs. However, it was not until 1928 that a national organization was formed. In that year, 33 delegates from 18 states met in Kansas City, Missouri, and formulated a constitution based on the organization of the Future Farmers of Virginia. Leslie Applegate of New Jersey was elected national president, Dr. C. H. Lane of Washington, D.C., became the first national advisor, and the first executive secretary-treasurer was Henry Groseclose of Virginia.[37]

Size. During the first 10 years of the FFA's existence, membership increased to 100,000, representing 4,000 chapters and 48 state associations. In 1939, the organization purchased 28½ acres of the George Washington estate in Alexandria, Virginia, which became the national headquarters and an FFA camp. In two years' time, FFA membership had increased to more than 245,000, with about 7,500 chapters. Currently, the national FFA organization serves almost 500,000 active members and has about 7,200 chapters located throughout the United States. In 2006, more than 54,000 FFA members, advisors, and guests attended the national convention.

The impact of FFA as a national organization is emphasized by the fact that numerous presidents, including Presidents Truman, Eisenhower, Johnson, Nixon, Ford, Carter, and Bush, have addressed national FFA conventions. The mission of the National FFA Organization is to make "a positive difference in the lives of students by developing their potential for premier leadership, personal growth and career success through agricultural education."[38] Refer to Figure 15–12.

37. Future Farmers of America, *Official FFA Manual* (Mt. Vernon, VA: Author, January 1997), p. 6.

38. http://www.ffa.org accessed August 27, 2007).

Career Options

Natural Resources Conservation Service, Air Quality Control, Federal and State Regulatory Agencies, Agricultural Extension and/or 4-H Leader, Agricultural Education Teacher

Soil and water conservation career options provide extensive opportunities for indoor and outdoor work. Work may be found in the Natural Resources Conservation Service. One may work in the field, served by such offices in most counties of the United States. Typical job titles are conservation technician, farm planner, soil scientist, and soil mapper.

The U.S. Department of Interior, state departments of natural resources, city and county governments, industry, and private agencies hire people with soil and water conservation expertise. Their employees manage water resources for recreation, conservation, and consumption. They may also work as consultants, law enforcement officers, technicians, administrators, or heavy equipment operators, among many other possibilities.

Air quality control employees monitor and help to improve air quality. Technicians collect, and chemists analyze, samples of air taken from various places in the atmosphere, buildings, and homes. Employees of environmental protection agencies and environmental advocacy groups strive to help maintain a healthful environment. Air quality specialists advise and assist industry in reducing harmful emissions from motor vehicles and industrial smokestacks.

Various federal and state governmental agencies provide services to aid production agriculturists in solving some of their farm problems. Employment is available, for example, in food inspection, disease prevention, natural resource protection, treatment program provision, weights and measures regulation, agricultural fair supervision, law enforcement, and agricultural promotion activities. These are just a few of the possible careers in government agriculture-related agencies.

Careers are also available as an agricultural extension agent/4-H leader or an agricultural education teacher/FFA advisor. Both these public agriservice jobs are very rewarding, as they involve helping young people grow and learn.

Agricultural extension agents and 4-H leaders have very rewarding jobs, including working with elementary school students. (Courtesy of USDA)

Jobs in Agricultural Education

There are more than 10,000 positions as agricultural education teachers throughout the United States. Most rural school systems and many urban systems have agricultural education programs. There is expected to be a shortage of agricultural education teachers in the future. Most programs require a candidate to have had four years of college to become a certified agricultural education teacher.

Conclusion

The federal and state governments have a major role in the agribusiness service sector of the agricultural industry. It would be difficult for

Figure 15–12 The mission statement of the National Future Farmers of America is to develop premier leadership, personal development, and career success. These FFA members are attending the FFA Washington Leadership Conference and meeting government officials from their states. (Courtesy of the National FFA Organization)

production agriculturalists to survive without governmental and public agribusiness services. Many government services are taken for granted, but it would be hard for agricultural industry members to accomplish anything without the various USDA agencies, state departments of agriculture, the land grant and non-land grant universities, and agricultural education programs. Most of these governmental services and programs exist because of agricultural policy decisions and mandates.

Summary

American leaders have long recognized the importance of a strong agricultural sector, possibly because many U.S. political leaders had their roots in farming. At least 23 presidents were involved in production agriculture, and numerous members of Congress were part- or full-time production agriculturalists before and during their terms of service. Government plays its major role in agriculture at the federal level. However, state and local governments also provide vital services in agricultural development and education.

The first federal aid to agriculture in the United States came in 1839, when Congress appropriated $1,000 to collect farm statistics and to collect and distribute seeds. In 1862, the Department of Agriculture was created, and in 1889 it was given Cabinet status. From this beginning, the USDA has expanded into a complex organization comprising several agencies, which is administered by a secretary, deputy secretary, assistant secretaries, and numerous administrative directors.

Agricultural policy is a course of action designed to achieve certain objectives in the food and fiber economy. Agricultural policies that embody the principles guiding government programs influence production, the resources utilized in production, domestic and international markets for commodities and food products, food consumption and nutrients, food safety, and the conditions under which people live in rural America.

Several forces may cause policy change: instability, globalization, technology, food safety, environmental issues, industrialization, politics, and unforeseen events. Many of these same forces directly affect agriculture and agricultural industry stakeholders, including production agriculturalists, agribusiness firms, food and fiber marketers, consumers, taxpayers, environmentalists, and others.

Government becomes involved in agriculture for several reasons and under several conditions. These include price and income instability, importance of the food supply, importance of food safety, poverty (economic justice), environmental externalities, and other public concerns and expectations.

When policies are developed, a need is defined and transformed into a goal by an individual, a legislator, or an interest group. The public is made aware of and educated about the merits of this goal. A proposed policy should be fully analyzed and revised on the basis of public comment and input. After analysis, the final draft policy can be written into a bill, introduced into the legislative process, and enacted.

Many groups and farm organizations influence the development of agricultural policies. These groups and organizations have special interests or agendas that lead to formation of new policies. These groups also often lead lobbying efforts to Congress regarding enactment, revision, or abandonment of policies.

It seems unlikely that total expenditures on agriculture programs, or agriculture's share of the federal budget, will increase or decline dramatically. Rather, the changes that do occur are likely to result from shifts in priorities, as legislators deal with issues that are of greatest immediate concern, such as globalization pressures to move toward freer markets, global economic and political pressures, resource scarcity, and environmental and food safety challenges.

Various governmental agencies and services are available to aid production agriculturalists in solving some of their farm problems. The government offers many services that we take for granted, such as inspecting food to ensure safety and quality, keeping foreign diseases from entering the country, and protecting our natural resources (among many others).

The U.S. Department of Agriculture plays an important role in the agricultural industry. Currently, the major divisions of the USDA are: Food, Nutrition, and Consumer Services; Marketing and Regulatory Programs; Farm and Foreign Agricultural Services; Rural Development; Natural Resources and Environment; Research, Education, and Economics; and Food Safety.

State governments also play a major role in modern agriculture. Presently, all 50 states have departments of agriculture or other, comparable agencies that are responsible for administering agricultural programs at the state level. The state departments of agriculture are primarily regulatory agencies. However, numerous other activities are funded at the state level, including disease prevention and treatment programs, regulation of weights and measures, collection and dissemination of statistical data, supervision of agricultural fairs, law enforcement, and sponsorship of agricultural promotion activities.

One of the most significant developments in the history of American agriculture was the establishment of what we know today as the land grant system. The system consists of land grant universities, agriculture experiment stations, and the Agricultural Extension Service. In 1862, the Morrill Act established land grant universities; in 1887, the Hatch Act provided funds for establishing agricultural experiment stations. In 1914, the Agricultural Extension Service was created by the Smith-Lever Act and added to the system. In addition to the land grant system, non-land grant schools play a major role in educating young people for agricultural occupations, especially at the undergraduate level. Approximately 50 percent of all undergraduate agricultural students are enrolled in non-land grant institutions. The major difference is that non-land grant schools are primarily concerned with teaching, whereas land grant schools are responsible for teaching, research, and extension activities.

In the 1960s, a sea grant program, similar to the land grant program, was developed to explore and develop the resources of the oceans, Great Lakes, and coastal waters. Through a matching fund, the sea grant program provides grants to colleges and universities, as well as to private industries, state agencies, private foundations, and others, to conduct research and educational projects in marine development.

The Smith-Hughes Act of 1917 provided funds for agricultural education programs at the high school level. These programs are designed primarily for young men and women who have selected some phase of agriculture as a vocation. The youth group associated with agricultural education is the Future Farmers of America, the mission of which is to develop premier leadership, personal development, and career success through agricultural education. The many 4-H Clubs are also organized and supervised by Agricultural Extension Service personnel.

End of Chapter Activities

REVIEW QUESTIONS

1. Define the Terms to Know.
2. Name eight American presidents who had close farm ties.
3. What are five agricultural policy areas that embody the principles guiding government programs?
4. What are five goals of U.S. agricultural policy?
5. Name eight forces that cause policy change.
6. What are six conditions and/or reasons leading to government involvement in agriculture?
7. Explain the process by which agricultural policy is developed before it reaches Congress.

8. What legislative steps do policies go through in Congress before they become law?
9. List two general farm organizations that influence agricultural policy development.
10. List five commodity organizations that influence agricultural policy development.
11. List eight agribusiness organizations that influence agricultural policy development.
12. What four agricultural agencies were created during the settlement period (1776–1929) of U.S. history?
13. What are seven future directional indicators of agricultural policy over the next few years?
14. List 24 services provided by the federal government.
15. What are the seven divisions within the USDA?
16. What programs are administered by the Food, Nutrition, and Consumer Service?
17. What are three agencies within the Marketing and Regulatory Services Division?
18. What are three agencies within the Farm and Foreign Agricultural Services Division?
19. What are three responsibilities of the Farm Service Agency?
20. What was the Risk Management Agency called before it acquired its new name?
21. Give three reasons why the Commodity Credit Corporation was created.
22. What is the mission of the Rural Utilities Service?
23. What is the mission of the Rural Business-Cooperative Service?
24. What are three agencies within the Rural Development division?
25. List and briefly describe five types of jobs with the Natural Resources Conservation Service.
26. What are the four agencies within the Research, Education, and Economics division of the USDA?
27. What are eight initiatives of the Cooperative Extension Service?
28. What is the purpose of 4-H Clubs?
29. What types of projects are 4-H members encouraged to develop?
30. Give seven examples of animal and plant research being conducted by the Agricultural Research Service.
31. List six activities conducted by state departments of agriculture.
32. What is the purpose and/or duties of the National Association of the State Departments of Agriculture (NASDA)?
33. What was the land grant system designed to do?
34. What three groups does the land grant system include at each designated university?
35. Many universities have *A&M* at the end of their name, such as Texas A&M. What does the A&M stand for?
36. What are the two major differences between land grant universities and non-land grant universities?
37. What is the mission of the national sea grant program?
38. Name and briefly explain the four types of grants awarded by the Office of Sea Grant.
39. What is the mission statement of the National Future Farmers of America Organization?

FILL IN THE BLANK

1. Agriculturalists want more research and educational services and less regulation, whereas consumer groups want more regulation and less funding spent on __________ and __________.
2. The first federal aid to agriculture in the United States came in __________, when Congress appropriated $1,000 to collect farm statistics and to collect and distribute seeds of various types.

3. On May 15, __________, President Lincoln signed the act creating the USDA.
4. __________ __________ was the first commissioner (now called secretary) of agriculture.
5. There are __________ national forests and __________ national grasslands, containing 187 million acres, in 41 states and Puerto Rico.
6. The national forests and grasslands provide a home for more than __________ million big game animals and __________ species of threatened or endangered wildlife.
7. Soil classification, soil mapping, and soil surveys are conducted by __________ local conservation districts employing about __________ people.
8. The __________ Act of 1862 established land grant universities.
9. The __________ Act of 1887 established Agricultural Experiment Stations.
10. The Cooperative Extension Service was created in 1914 by the __________ Act.
11. Agricultural education was established in 1917 by the __________ Act.

MATCHING

a. Farm Service Agency
b. Foreign Agricultural Service
c. Risk Management
d. Food, Nutrition, and Consumer Services
e. Food Safety and Inspection Service
f. Agricultural Marketing Service
g. Animal and Plant Health Inspection Service
h. Grain Inspection, Packers, and Stockyards Administration
i. Forest Service
j. National Resource Conservation Service
k. Agricultural Research Service
l. National Agricultural Library
m. Cooperative State Research, Education, and Extension Service
n. Economic Research Service
o. National Agricultural Statistics Service
p. Rural Utilities Service
q. Rural Housing Service
r. Rural Business-Cooperative Service
s. Commodity Credit Corporation

_____ 1. provides timely and reliable agricultural economic data
_____ 2. collects and distributes agricultural data
_____ 3. administers the Food Stamp Program
_____ 4. provides technical assistance and conducts research on issues that affect agricultural cooperatives
_____ 5. promotes the orderly and efficient marketing and distribution of agricultural commodities
_____ 6. protects plants and animals from economically dangerous diseases and pests
_____ 7. establishes official standards for grain, conducts weighing and inspection activities, and inspects other agricultural products
_____ 8. responsible for making sure that the nation's meat and poultry supplies are safe
_____ 9. maintains a good agricultural economy through price supports, makes loans and purchases, and conserves the nation's farm resources
_____ 10. loan program that acted as a price support mechanism
_____ 11. played a major role in making the United States the world's largest exporter of agricultural products
_____ 12. provides financial assistance to farmers who are unable to obtain loans from private agencies for farmland

_____ 13. provides farmers with insurance that reimburses crop production losses due to uncontrollable natural conditions

_____ 14. helps rural people obtain electric and telephone services by making long-term loans to rural cooperatives

_____ 15. responsible for the conservation and management of the nation's forest and other wildlife resources

_____ 16. administers programs to conserve soil and water resources through technical assistance to production agriculturalists and other landowners

_____ 17. helps the public learn about the latest technology developed through research by various agricultural institutions

_____ 18. the most extensive source of agricultural information in the world, with holdings of about 2 million books and more than 25,000 periodicals

_____ 19. major agency that does basic research on plants, animals, and marketing

_____ 20. administers federal grant funds for research in the production phase of agriculture, agricultural marketing, rural development, and forestry

_____ 21. makes loans for housing in rural areas

ACTIVITIES

1. Make a list of all the federal and state government jobs in your community that are related to the agricultural industry. Share your findings with the class. Get a total number of the agricultural-related federal and state jobs.
2. Suppose that Congress suddenly passed a law making taxes illegal. Explain if and how you would then pay for the agricultural-related federal and state services and jobs.
3. Figure 15–2 lists approximately 90 commodity, farm, and/or agribusiness organizations. Select the one that is of most interest to you, locate its Website, and give a report on this organization to your class.
4. Figure 15–3 lists approximately 30 legislative acts dealing with agricultural policy. Select one of these legislative acts, research it, and give a report to your class on its impact on the agricultural economy at the time it was passed.
5. The "Reasons for Government Services" section of this chapter lists more than 20 services of the federal government. Select the five that you feel are the most important and defend your answer.
6. Select five government services that you feel are the least important and explain why.
7. There are currently 19 agencies within the USDA. Prepare a brief report on one of these agencies and present it to the class. If your school library does not have enough information, you may wish to try the Internet, contact local agencies, or write to the agency for additional information.
8. List the fringe benefits of working for either the federal government or a state government.
9. Compare the purpose and/or mission of agricultural education and the FFA to the purpose and/or mission of the Extension Service and the 4-H organization.
10. Of all the public agribusiness business services discussed in this chapter, which is the most important to you personally? Prepare a brief paper on this service and its benefits and share it with the class.

Chapter 16

Private Agribusiness Services

Objectives

After completing this chapter, the student should be able to:

- Describe the general farm organizations.
- Discuss various agricultural commodity organizations.
- Explain the importance of agricultural research.
- Explain the importance of agricultural consultants.
- Describe the artificial insemination industry.
- Discuss the history and scope of veterinary medicine, and the duties of veterinarians.
- Explain the importance of agricultural communications.
- Describe the history and impact of agricultural cooperatives.
- Describe the Farm Credit System.
- Describe the Consolidated Farm Service Agency.
- Describe the role of commercial banks in agribusiness credit.
- Describe the role of life insurance in agribusiness credit.
- Describe the Commodity Credit Corporation.
- Discuss the role of individuals and others in agribusiness credit.
- Explain the career opportunities in agribusiness credit input services.

Terms to Know

agriservices
annuities
artificial insemination
collective bargaining
default
delinquencies
dividends
farm commodity
fraternal
glycerol
holding action
homogeneous
husbandry
inseminate
nonrecourse loans
open account
physiologist
policy statements
political power
prominent
regulatory medicine
resource specialization
rural sociology
securities markets
semen
sodium citrate
viable

Introduction

In the early colonial days of American agriculture, when farmers were basically self-sufficient in the production and distribution of food and fiber, there was very little need for specialized services. Each farm family produced and processed the food it needed, and shelters were built by the family members with trees from local forests. However, as the trend in farming shifted toward more specialized, commercial-production agriculture, in which the major production emphasis was to produce food and fiber for sale, the demand for specialized services began to develop. During the earliest phase of this development, special services were provided on an informal basis by trading services. When a major farm task had to be done, farmers would ask their neighbors to help, and each one was assigned the specific task for which he or she was best qualified. These services often served a social function as well as an economic one.

Resource Specialization

During the 1700s, Adam Smith, commonly recognized as the Father of Economics, pointed out the economic advantages of **resource specialization**. However, for resource specialization to be feasible, a minimum-size unit is required. The small family farm of early America was not very conducive to the specialization of labor. In recent years, however, the increasing complexity of modern farming and the trend toward large farm operations have caused the demand for specialized services offered by the private agribusiness services input sector of agriculture to virtually explode.

Private Agribusiness Services

It would be a challenge to discuss all the types of private services offered to production agriculturalists today. Nevertheless, an understanding of the major agencies that provide them is imperative to anyone seeking a minimum level of knowledge about modern agriculture, especially the agribusiness phase. This chapter is designed to provide a basic understanding of the major **agriservices** currently offered to production agriculturalists by private firms, the economic importance of agriservices to the total agricultural industry, and the major firms that provide these services.

Earlier, Chapter 13 discussed careers in the areas of supplies, machinery, and equipment. In Chapter 15, careers in agribusiness governmental services were examined. The agribusiness careers discussed in this chapter are in general farm organizations, agricultural commodity organizations, agricultural research, consulting, artificial insemination, veterinary services, agricultural communications, agricultural cooperatives, the Farm Credit System, the consolidated Farm Service Agency, commercial banks, life insurance, the Commodity Credit Corporation, and individual agribusinesses.

General Farm Organizations

Farmers learned a long time ago that some problems could be solved better by group action than by individual efforts. As a result, thousands of farm organizations have been created during the history of American agriculture. Numerous books have been written on the rise and fall of farm organizations and their impact on the society of their period. History shows that organizations that were formed in response to specific problems usually dissolved after the problems were solved, whereas organizations that sought to deal with a variety of issues and provide a variety of services (flexible depending on the immediate situation and needs) tended to last for longer periods of time.

Currently, five major general farm organizations are active in representing farmers' interests at the national level: the National Grange, the National Farmers Union, the American Farm Bureau Federation, the National Farmers Organization, and the American Agricultural Movement. There are also several other organizations that involve themselves more specifically in various agricultural issues, such as American Agri-Women (AAW), Women Involved in Farm Economics (WIFE), Concerned Farm Wives, Partners in Action for Agriculture, the National Catholic Rural Life Conference (NCRLC), and the Interreligious Task Force on U.S. Food Policy.

The National Grange

The National Grange is the oldest and second largest of the major general farm organizations. It was organized in Washington, D.C., during 1867 under the name the Patrons of **Husbandry**, but soon became known as the National Grange. It was organized as a **fraternal** organization, but its special emphasis was on strengthening farm families and rural communities. Its activities included a wide range of social, educational, recreational, and economic functions. Although Oliver Hudson Kelley, a former USDA employee, is usually called the founder of the Grange, several other people played important roles in setting up the organization. William Saunders developed the plan of organization, which goes from the local to the national order. He suggested the name of "Grange" for the meeting place and the title of "Patrons of Husbandry" for the order.[1] John R. Thompson was primarily responsible for the rituals of the Grange, and William Ireland developed the framework of its constitution and bylaws. Aaron Grosh, John Trimble, and Francis McDowell were also major contributors to early Grange development.

Accomplishments. The National Grange has played a major role in almost everything that has happened to improve agriculture and rural America during the past century. Some of the major accomplishments supported by the Grange include rural free delivery of mail, the founding of land grant colleges and experiment stations, the creation of the Agricultural Extension Service, vocational agriculture (now called agricultural education) and home economics education, antitrust legislation, the Farm Credit System, agricultural cooperatives, rural electrification, farm price supports, and many other initiatives. Today, Grange **policy statements** deal with issues that affect modern agriculture both directly and indirectly, including such areas as transportation, national welfare, conservation, foreign affairs, taxation, education, health, and safety.

Membership. Grange membership approaches 1 million in approximately 6,000 subordinate Grange units. Although about 35 states have Grange organizations, the membership is concentrated primarily in the north-central and northwestern states. The National Grange goals and objectives were summarized in its national policy manual as follows:

> The Grange is more than a farm organization. Its purpose is to serve the total interest of the rural community and the Nation. Thus policies and programs of the Grange pertain to a broad array of circumstances affecting the lives of rural Americans; they result from member action generated by total community and national interest—not agricultural interest alone.[2]

The current Grange mission and vision statements are available on its Website: http://www.nationalgrange.org.

National Farmers Union

The National Farmers Union (NFU) was first organized in the Southwest with an emphasis on improving the educational level of rural people and promoting agricultural cooperatives. In 1902, Newt Greshman, a farmer and newspaper editor, along with 10 other farmers in Paint, Texas, organized the Farmers Educational Cooperative Union, which later became the Farmers Union. By 1903 the organization had spread to Arkansas, Georgia, Louisiana, and Oklahoma, and a national organization was established in 1905.

Intended Purposes. Although one of the early purposes of the Farmers Union was to act as a fraternal organization, its primary purpose was to improve the financial condition of farmers. The union first tried to control prices by setting a uniform price and refusing to sell for less, but this tactic did not work. Next, it tried to limit output to improve prices, but this strategy also failed. As one NFU leader declared, "Whenever we tell the farmers to plant less cotton, they plant more."[3] Around 1920, the Farmers Union had more than 140,000 members, although bankers, merchants, and lawyers were barred.

Even though the NFU was not very successful in controlling farm prices, it was active in other efforts to support and improve the agricultural cooperative movement, and it also provided farm credit to members. Other farm-marketing programs were developed, especially for cotton, and the NFU was also very influential in programs for agricultural education.

1. W. L. Robinson, *The Grange, 1867–1967* (Washington, DC: The National Grange, 1966), p. 3.
2. The National Grange, *Legislative Policies, 1977* (Washington, DC: Author, 1977).
3. Gladys T. Edwards, *The Farmer's Union Triangle* (Denver: The Farmer's Union Education Service, 1941), p. 19.

American Farm Bureau Federation

The first county Farm Bureaus were organized between 1912 and 1914, primarily to act as sponsoring agencies for county agricultural agents. As the Farm Bureau movement expanded, local units organized into state groups and finally into a national organization. The first efforts to establish a national Farm Bureau occurred in February 1919, when representatives of 12 state groups met in Ithaca, New York, to discuss organizational plans. A second meeting was held in Chicago, in November 1919, to develop a constitution and bylaws. Adoption of the constitution and ratification of the bylaws occurred in March 1920, in Chicago, and the first annual meeting was held in Indianapolis in December 1920.

The American Farm Bureau Federation is the largest of the five major general farm organizations. Membership is based on the farm family, and only one membership fee per family is required. Currently, approximately 4.8 million families from 50 states and Puerto Rico are members of the national Farm Bureau.[4]

Insurance. Although the major thrust of Farm Bureau activities has been the representation of American agriculture in the political arena, the Farm Bureau offers numerous other services as well. A broad-based insurance company is operated for Farm Bureau members at the state level, and the American Farm Bureau Service Company provides group purchasing of farm supplies under the Safemark brand. Distribution is handled locally through about 3,100 outlets in 41 states and Puerto Rico. Since this service was initiated in 1965 (with 10 states participating), the volume of purchases has increased more than 700 percent.

Marketing. The American Agricultural Marketing Association (AAMA) is the marketing arm of the Farm Bureau. It was established in 1960 to provide cooperative marketing services for members. The AAMA is especially active in marketing livestock. Numerous other services are provided by state and local associations depending on the specific needs of the area.

Accomplishments. The Farm Bureau has given support to, and in many cases has led the battle for, a better life for American farmers. Some of the major areas supported by Farm Bureau include the establishment of rural electricity and telephone services, agricultural education, research and extension services, the federal farm gasoline tax refund, rural development, the establishment of the Farm Credit Administration, increases in the amount of inheritance allowed before taxes are imposed, and numerous others.

National Farmers Organization

The most controversial general farm organization is the National Farmers Organization (NFO), which was organized in Corning, Iowa, in 1955. Jay Loghry, who was a feed salesman, met with 35 farmers at Carl, Iowa, to discuss the idea of a new farm organization that would have some input into determining the prices received for farm products. A week later, a meeting was held in Corning, Iowa, to test the idea of "unionizing" farmers. About 1,200 farmers attended. However, the term *union* drew a negative response, so the group instead adopted the name National Farmers Organization.[5] Loghry was hired to assist in the development of the organization, and Dan Turner, a former Iowa governor, was selected as an unpaid advisor.

Strategy to Improve Farm Prices. The first national meeting of the National Farmers Organization was held at Corning, Iowa, in December 1955, with about 750 members attending. More than 55,000 members were on the roster at that time, and Oren Lee Staley was elected president. By April 1956, the organization had expanded to 140,000 members. The advisor to the NFO, former governor Dan Turner, recommended that the organization take action to improve farm prices through political means, and the NFO initially followed his advice with several visits to Washington and a major effort to elect NFO members to national office. However, by convention time in 1957, the members had decided that this technique was not proving successful, so they turned to **collective bargaining** to improve farm prices.

4. N. Omri Rawlins, *Introduction to Agribusiness* (Murfreesboro, TN: Middle Tennessee State University, 1999), p. 140.

5. Luther Tweeten, *Foundations of Farm Policy* (Lincoln: University of Nebraska Press, 1970).

Holding Actions. In 1959, the first **holding action** occurred to test market reaction and to give other farmers the opportunity to join the organization. This action involved livestock only and lasted just seven days because it was confined to a small area around St. Joseph, Missouri. Additional livestock holding actions were conducted in 1960, 1961, and 1962. Prices increased for a short period of time after each holding action, but very few, if any, long-term gains were realized. The major impact of the holding actions resulted from the violence that occurred during this period. Shots were fired at trucks hauling livestock to markets, tires were slashed, and nails were spread on highways.

In 1967, NFO members decided to apply their collective bargaining techniques to milk with a 25-state holding action. Thousands of gallons of milk were poured out on the ground, but the action had less effect on the market than anticipated. Again, numerous acts of violence were reported, generating ill feeling and bad publicity. Since that time, several other minor holding actions have been tried, regarding soybeans, grain, pork, and dairy products, but all showed similar results.

Price Improvement Challenge. Currently, the NFO is seeking new members and encouraging the participation of previously inactive members in NFO bargaining programs. the NFO estimates that participation of just 30 percent of the total production in any commodity would be sufficient to allow farmers to set their own prices, just as nonfarm businesses do. The NFO is attempting to improve farm prices by cooperative marketing programs, with a major emphasis on transporting farm commodities in bulk quantities from surplus areas to shortage areas. NFO officials contend that:

> When the volume of farm commodities moving through and committed to NFO programs reaches 30 percent, meetings will be called in no larger than ten-county areas and members will vote on their price based on cost of production plus a reasonable profit. These prices will be announced and contracts will be presented to the companies who handle and process farm production. If the companies won't agree to sign contracts to maintain prices based on cost of production plus a reasonable profit, NFO will advise members to keep their products on their farms until they do get their prices and their contracts.[6]

American Agriculture Movement

The American Agriculture Movement (AAM) began as a movement rather than an organization and operated several years without any formal structure. AAM members were primarily disenchanted with the NFO and National Farmers Union organizations, which primarily represented grain and cattle producers from Texas, Colorado, Kansas, North and South Dakota, and Georgia. The AAM members were generally young farmers who had expanded production during the early 1970s and then began experiencing difficulty because of low prices for their products. When the cost-price squeeze became severe in 1976 and 1977, these farmers turned to Washington for relief. The major technique for attracting attention was a "tractorcade" to Washington, during which they camped for several weeks on the Capitol Mall and disrupted traffic with their tractors. The goal of the AAM in Washington was to obtain price supports for farm products above the level guaranteed by the 1976 farm bill.

The early efforts of AAM drew considerable support from rural America, but their tactics soon lost favor with other farm organizations. In recent years, the AAM has officially organized as a national farm organization, opened an office in Washington, and begun to operate through more acceptable, traditional channels as do the other organizations. Refer to Figure 16–1 for the addresses of the five major organizations and six other associations providing farm general agricultural services.

Agricultural Commodity Organizations

Although most of the earlier farm organizations were general farm organizations, production agriculturalists soon learned that certain types of farmers had more in common than others. For

6. National Farmers Organization, *NFO: Historical and Background Information* (Corning, IA: NFO, Public Information Department, 1977).

General Farm and Service Organizations		
Agriculture Council of America (ACA) 1625 I Street, NW Washington, DC 20006	Farm and Land Institute 430 N. Michigan Avenue Chicago, IL 60611	National Farmers Organization 720 Davis Avenue Corning, IA 50841
Agricultural Trade Council 1028 Connecticut Avenue, NW Washington, DC 20036	National Agricultural Transportation League Box 960 Umatilla, FL 32784	National Farmers Union Box 2251 Denver, CO 80201
American Farm Bureau Federation 225 West Touhy Avenue Park Ridge, IL 60068	National Farm Workers of America La Paz Keene, CA 93531	National Grange 1616 H Street, NW Washington, DC 20006

Figure 16–1 Home office addresses of the major farm organizations and general agricultural service associations.

example, wheat farmers had more problems in common with each other than with livestock producers, who had different problems. As a result, production agriculture organizations began to develop around **farm commodity** lines. The reasoning was that a specialized commodity organization could represent special commodity interests better than a general farm organization. As a result, specialized farm commodity organizations sprang up rapidly during the early part of the 20th century.

Services Performed

Agricultural commodity organizations provide numerous services to their members. They assist with public relations, communications, promotions, publicity, and sales training. They offer auditing and recordkeeping services, provide marketing information, conduct research, and lobby for legislation. Personnel in agricultural commodity organizations monitor any legislation that may affect their specific sector. Often special reports and regular newsletters or periodicals are written and distributed to the particular industry group. Commodity groups might also engage in any or all of the following activities:

- make efforts to increase prices
- train producers on how to handle, store, produce, and market a commodity
- help wholesalers with problems of transportation, trucking, materials handling, and warehousing
- foster and improve trade relations among producers, distributors, and consumers
- promote development of new products
- keep members up-to-date on environmental issues
- render all services possible to group members
- organize research
- conduct market surveys
- aid in developing domestic and export markets
- demonstrate and secure adoption of special accounting systems
- study legislation and government policy concerning the commodity and commodity producers
- provide information on government regulations, and advise government when new regulations are proposed or considered [7]

Selected Types of Commodity Groups

Farm Commodity Groups. Today, there is one or more organizations operating at the state and/or national levels for almost every major farm commodity. Some of the larger ones include the National Cattle Beef Association, the American Dairy Association, the National Association of Wheat Growers, the National Cotton Council, and the National Potato Council, though there are

7. Marcella Smith, Jean M. Underwood, and Mark Bultmann, *Careers in Agribusiness and Industry*, 4th ed. (Danville, IL: Interstate Publishers, 1991), pp. 287–292.

many others. These organizations keep their members informed on matters relevant to the production and marketing of their products. In some cases, the **political power** of the commodity organization exceeds the power of some of the general farm organizations.

Agribusiness Commodity Groups. The organization of special commodity groups is not limited to production agriculture organizations. Some of the more **prominent** ones are the Farm and Industrial Equipment Institute, the Fertilizer Institute, National Independent Meat Packers, the National Canners Association, the National Association of Food Chains, and the Grocery Manufacturers of America (again, there are many others as well). These serve basically the same function as the special farm organizations. The trend toward the founding of special group organizations is expected to continue for production agriculture and agribusiness groups. Refer to Figures 16-2 through 16-10 for names and addresses of many of the selected commodity groups and breed associations.

Agricultural Research

Traditionally, farmers have not been involved to a large degree in basic agricultural research, primarily because they cannot afford to be involved. Farm profits are not sufficient to allow investment of money and time in research, partly because of the high degree of competition between production agriculturalists. Also, because the products produced by production agriculturalists are **homogeneous**, the benefits of new products cannot be controlled by those who invest the funds to develop them. As a result, production agriculturalists have depended on public funds for farm research, as the long-term benefits of research help primarily the agribusiness firms and the consumer.

Agribusiness firms have benefited to a great extent from public agricultural research, and they have also invested huge amounts in private research. Profits for agribusiness firms have been high enough to justify research expenditures, and patent laws allow them to restrict the use of the new products thus developed. Private agribusiness firms are continuously looking for new and better products, and many firms maintain extensive research facilities.

Areas of Agricultural Research

Research is an essential part of agribusiness. Research is conducted on all agricultural commodities and in many other areas of the agricultural industry, including conservation, machinery, equipment, forestry, management, marketing, processing, **rural sociology**, and soils. The amount spent by the private sector on research is comparable to that spent by the USDA.

Consulting

Similar to research is agricultural consulting; much of a consultant's job is research based. In recent years, private companies have emerged that offer technical services to production agriculturalists. These services are varied and may include research, laboratory, and consulting services. Firms that offer technical services vary by type. Some specialize in just one area, such as livestock consulting, while others may offer numerous services crossing several sectors.[8] Consultants screen and evaluate new products, evaluate systems, and work with production agriculturalists in many specific areas.

Crop Consulting

Many consulting jobs involve crops. Among the many services offered to production agriculturalists, consultants do soil sampling and make fertilizer recommendations; perform field checks and offer insect, weed, and disease control recommendations; and make equipment adjustments. Consultants may also conduct laboratory tests and analyses on soil, feed, water, or plant tissue.[9]

Livestock Consulting

Livestock consultants provide services that focus on improving livestock production. Their work may take the form of calculating feed ratios, automating silage feeding, testing feed supplies, and computerizing feed systems and farm budgets. Typically, a consultant will visit the farm or feedlot, make observations or conduct testing on the performance of livestock or the type of feed, analyze the results, and give recommendations for improvement.[10]

8. Ibid., p. 297.
9. Ibid.
10. Ibid.

General Crop Associations

American Corn Millers Federation
1030 Fifteenth Street, NW,
Suite 912
Washington, DC 20005

American Cotton Shippers Association
318 Cotton Exchange Building
P.O. Box 3366
Memphis, TN 38103

American Rice Growers Cooperative Association
211 Pioneer Building
Lake Charles, LA 70601

American Soybean Association
Box 158
Hudson, IA 50643

American Sugar Cane League of the USA, Inc.
416 Whitney Building
New Orleans, LA 70130

American Sugar Beet Growers Association
1776 K Street, NW, Suite 900
Washington, DC 20006

Burley Tobacco Growers Cooperative Association
620 South Broadway
Lexington, KY 40508

Durum Wheat Institute
710 North Rush Street
Chicago, IL 60611

Federated Pecan Growers Association of the U.S.
Drawer AX, University Station
Baton Rouge, LA 70803

Flue-Cured Tobacco Cooperative Stabilization Corporation
Box 12300
Raleigh, NC 27605

Grain Sorghum Producers Association
1708-A Fifteenth Street
Lubbock, TX 79401

Leaf Tobacco Exporters Association
P.O. Box 1288
Raleigh, NC 27602

National Association of Tobacco Distributors
58 East 79th Street
New York, NY 10021

National Association of Wheat Growers
1030 Fifteenth Street, NW,
Suite 1030
Washington, DC 20005

National Cigar Leaf Tobacco Association
1100 Seventeenth Street, NW
Washington, DC 20036

National Corn Growers Association
P.O. Box 358
Boone, IA 50036

National Cotton Council of America
1918 North Parkway, P.O. Box 12285
Memphis, TN 38112

National Cottonseed Products Association
P.O. Box 12023
Memphis, TN 38112

National Council of Commercial Plant Breeders
1030 Fifteenth Street, NW, Suite 964
Washington, DC 20005

National Grain Trade Council
725 Fifteenth Street, NW, Suite 604
Washington, DC 20005

National Hay Association, Inc.
P.O. Box 1059
Jackson, MS 49204

National Onion Association
5701 East Evans Avenue, Suite 26
Denver, CO 80222

National Pecan Shellers and Processors Association
111 East Wacker Drive, Room 600
Chicago, IL 60601

National Potato Council
Montbello Office Campus,
Suite 301
45th and Peoria
Denver, CO 80239

National Potato Promotion Board
1385 S. Colorado Boulevard
Denver, CO 80222

National Rice Growers Association
Box 683
Jennings, LA 70546

National Soybean Processors Association
1800 M Street, NW
Washington, DC 20036

National Sugar Beet Growers Federation
Greeley National Plaza, Suite 740
Greeley, CO 80631

North American Export Grain Association, Inc.
1800 M Street, NW, Suite 610 N
Washington, DC 20036

The Potato Association of America,
114 Deering Hall
University of Maine
Orono, ME 04473

Rice Council of America
Box 22802
Houston, TX 77027

The Shade Tobacco Growers Agricultural Association, Inc.
River Street, Box 38
Windsor, CT 06095

The Tobacco Institute
1776 K Street, NW
Washington, DC 20006

Tobacco Association of the U.S.
Box 1288
Raleigh, NC 27602

Tobacco Growers' Information Committee, Inc.
Cameron Village Station Box 12046
Raleigh, NC 27605

Wheat Flour Institute
1776 F Street, NW
Washington, DC 20006

Wild Rice Growers Association
Box 366 Highway 210 West
Aitkin, MN 56431

Figure 16–2 Home office addresses of general crop associations.

General Dairy Associations

American Butter Institute, Inc.
110 N. Franklin Street
Chicago, IL 60606

American Dairy Association
6300 N. River Road
Rosemont, IL 60018

American Dairy Science Association
113 N. Neil Street
Champaign, IL 61820

American Dry Milk Institute, Inc.
130 N. Franklin Street
Chicago, IL 60606

Milk Industry Foundation
1105 Barr Building
910 Seventeenth Street, NW
Washington, DC 20006

National Dairy Council
6300 N. River Road
Rosemont, IL 60018

National Independent Dairies Association
1730 K Street, NW, Suite 1302
Washington, DC 20006

National Milk Producers Federation
30 F Street, NW
Washington, DC 20001

Purebred Dairy Cattle Association, Inc.
Box 126
Peterborough, NH 03458

United Dairy Industry Association
6300 N. River Road
Rosemont, IL 60018

Whey Products Institute
130 N. Franklin Street
Chicago, IL 60606

Figure 16–3 Home office addresses of general dairy associations.

Dairy Breed Associations

American Guernsey Cattle Club
Peterborough, NH 03458

American Jersey Cattle Club
2105-J S. Hamilton Road
Columbus, OH 43227

American Milking Shorthorn Society
1722JJ S. Glenstone Avenue
Springfield, MO 65802

Ayrshire Breeders' Association
Brandon, VT 05733

Brown Swiss Cattle Breeders' Association
Box 1038
Beloit, WI 53511

Dutch Belted Cattle Association of America
Box 358
Venus, FL 33960

Holstein-Friesian Association of America
Box 809
Brattleboro, VT 05301

International Illawarra Association
313 S. Glenstone Avenue
Springfield, MO 65802

Figure 16–4 Home office addresses of dairy breed associations.

Artificial Insemination

One of the major special services offered to dairy and livestock producers today is **artificial insemination**. Although some production agriculturalists **inseminate** their own livestock, most animals are bred by local technicians who are trained specifically for this service. For dairy and livestock producers who prefer to do their own breeding, specialized equipment and **semen** are available.

Historical Developments

Although the use of artificial insemination was reported in several instances in the 18th century, it was first developed for practical application in livestock breeding in 1922 by a Russian **physiologist**. In the United States, large-scale use of artificial insemination began in the 1930s. Two of the major developments that stimulated the use of artificial insemination were the discoveries that egg yolk was valuable as a semen extender and that semen could be frozen without damage in a solution of **glycerol**, **sodium citrate**, and egg yolk. Artificial insemination is used predominantly with dairy cows, but beef and horse producers also use this service.

The Artificial Insemination Industry

The artificial insemination industry comprises semen collectors, distributors of semen and

Light Horse Breed Associations

American Albino Association
Box 79
Crabtree, OR 97335

American Andalusian Association
Box 1290
Silver City, NM 88061

American Association of Owners and Breeders of Peruvian Paso Horses
Box 2035
California City, CA 93505

American Bashkir Curly Registry
Box 453
Ely, NV 89301

American Buckskin Registry Association, Inc.
Box 1125
Anderson, CA 96007

American Hackney Horse Society
Box 174
Pittsfield, IL 62363

American Hanoverian Society
809 W 106th Street
Carmel, IN 46032

American Lipizzan Horse Registry
Box 415
Platteville, WI 53818

American Morgan Horse Association, Inc.
Box 1
Westmoreland, NY 13490

American Mustang Association, Inc.
Box 338
Yucaipa, CA 92399

American Paint Horse Association
Box 13486
Fort Worth, TX 76118

American Paso Fino Horse Association Inc.
Mellon Bank Building, Room 3018
525 William Penn Place
Pittsburgh, PA 15219

American Quarter Horse Association
Box 200
Amarillo, TX 79105

American Saddle Horse Breeders Association, Inc.
929 S. Fourth Street
Louisville, KY 40203

American Trakehner Association, Inc.
Box 268
Norman, OK 73069

Appaloosa Horse Club, Inc.
Box 8403
Moscow, ID 83843

Arabian Horse Registry of America, Inc.
3435 S. Yosemite
Denver, CO 80231

Chickasaw Horse Association, Inc.
Box 8
Love Valley, NC 28677

Cleveland Bay Association of America
Middleburg, VA 22117

Colorado Ranger Horse Association, Inc.
7023 Eden Mill Road
Woodbine, MD 21797

Galaceno Horse Breeders Association, Inc.
111 East Elm Street
Tyler, TX 75701

Hungarian Horse Association
Bitterroot Stock Farm
Hamilton, MT 59840

International Buckskin Horse Association, Inc.
Box 357
St. John, IN 46373

Missouri Fox Trotting Horse Breed Association, Inc.
Box 637
Ava, MO 65608

Morab Horse Registry of America
Box 413
Clovis, CA 93613

Morocco Spotted Horse Cooperative Association of America
Route 1
Ridott, IL 61067

National Chickasaw Horse Association
Route 2
Clarinda, IA 51232

National Trotting and Pacing Association, Inc.
575 Broadway
Hanover, PA 17331

Palomino Horse Association, Inc.
Box 324
Jefferson City, MO 65101

Palomino Horse Breeders of America
Box 249
Mineral Wells, TX 76067

Paso Fino Owners and Breeders Association, Inc.
Box 764
Columbus, NC 28722

(continued on the following page)

Figure 16-5 Home office addresses of light horse breed associations.

Light Horse Breed Associations *(concluded)*

Peruvian Paso Horse Registry of North America
Box 816
Guerneville, CA 95446

Pinto Horse Association of America, Inc.
Box 3984
San Diego, CA 92103

Royal International Lipizzaner Club of America
Route 7
Columbia, TN 38401

Spanish-Barb Breeders Association
Box 7479
Colorado Springs, CO 80907

Spanish Mustang Registry, Inc.
Route 2, Box 74
Marshall, TX 75670

United States Trotting Association
750 Michigan Avenue
Columbus, OH 43215

Standard Quarter Horse Association
4390 Fenton
Denver, CO 80212

Tennessee Walking Horse Breeders and Exhibitors Association of America
Box 286
Lewisburg, TN 37091

The Jockey Club
300 Park Avenue
New York, NY 10022

Ysabella Saddle Horse Association, Inc.
Prairie Edge Farm
Route 3
Williamsport, IN 47993

insemination supplies, and local service technicians. In the early years of the industry, most major artificial insemination firms were both collectors and distributors. However, because of mergers and affiliations, the industry now consists of a small number of semen collectors and a large number of distributors and technicians.

Artificial-Breeding Companies

The major collectors are Curtis Breeding Service, a division of Searle Agriculture, Inc.; Carnation Genetics, a division of Carnation Milk Company; American Breeding Service, a subsidiary of W. R. Grace Chemical Company; and Select Sires, a cooperative corporation owned and controlled by several regional and local service cooperatives. Each of these companies collects and sells semen from a wide variety of superior sires from all beef and dairy breeds. Many people believe artificial insemination to be one of this century's greatest technologies. Refer to Figure 16–11.

Veterinary Services

Although veterinary services have been in demand since the domestication of animals, the veterinary profession is not that old. The term *veterinarian* was first introduced in English writing in 1646 by Sir Thomas Browne, but the term was not generally accepted until 1846. Previously, veterinarians were called "cow doctors," "horse doctors," and similar names. One of the first indications of veterinary services in America was recorded by a medical historian named Blanton, who referred to a William Carter as an expert "cow doctor" who lived in James City, Virginia, in 1625. Richard Peters, a prominent patriot, lawyer, and country gentleman, was the first public figure to advocate the establishment of the veterinary profession in America.[11]

Father of Veterinary Medicine

John Busteed is often regarded as the "Father of Veterinary Medicine." He chartered the New York College of Veterinary Surgeons in 1857 and operated it until 1899. He was also the major supporter of the veterinary profession at that time. The profession gained some professional status when the American Veterinary Medical Association was organized in 1863. The first major goal of the association was to expand the number of veterinary colleges in the United States.

11. J. F. Smithcors, *The American Veterinary Profession* (Ames: Iowa State University, 1963).

Draft Horses, Ponies, and Jacks, Donkeys, and Mules

Draft Horses

American Shire Horse Association
6960 Northwest Drive
Ferndale, WA 98248

American Suffolk Horse Association, Inc.
672 Polk Boulevard
Des Moines, IA 50312

Belgian Draft Horse Corporation of America
Box 335
Wabash, IN 46992

Clydesdale Breeders Association of the U.S.
Route 3
Waverly, IA 50677

Percheron Horse Association of America
Route 1
Belmont, OH 43718

Ponies

American Connemara Pony Society
Route 1, Hoshiekon Farm
Goshen, CT 06756

American Gatland Horse Association
Route 2, Box 181
Elkland, MO 65644

American Shetland Pony Club
Box 435
Fowler, IN 47944

American Walking Pony Association
Route 5, Box 88
Upper River Road
Macon, GA 31201

National Appaloosa Pony, Inc.
Box 296 Gaston, IN 47342

Pony of the Americas Club
Box 1447
Mason City, IA 50401

Welsh Cob Society of America
Grazing Fields Farm
225 Head of the Bay Road
Buzzards Bay, MA 02532

Welsh Pony Society of America
Box A
White Post, VA 22663

Jacks, Donkeys, and Mules

American Donkey and Mule Society, Inc.
2410 Executive Drive
Indianapolis, IN 46241

Miniature Donkey Registry of the U.S., Inc.
1108 Jackson Street
Omaha, NB 68102

Standard Jack and Jennet Registry of America
300 Todds Road
Lexington, KY 40511

Figure 16-6 Home office addresses of draft horse, pony, and jack, donkey, and mule associations.

Beef Breed Associations

American Angus Association
3201 Frederick Boulevard
St. Joseph, MO 64501

American Beefalo Association
200 Semanin Building
4812 U.S. Highway 42
Louisville, KY 40222

American Blonde d'Aquitaine Association
217 Livestock Exchange Building
Denver, CO 80216

American Brahman Breeders Association
1313 LaConcha Lane
Houston, TX 77054

American Brangus Association
Box 1326
Austin, TX 78767

American Chianina Association
Box 11537
Kansas City, MO 64138

American Dexter Cattle Association
707 W.Water Street
Decorah, IA 52101

American Galloway Breeders Association
302 Livestock Exchange Building
Denver, CO 80216

American Gelbvieh Association
202 Livestock Exchange Building
Denver, CO 80216

American Hereford Association
715 Hereford Drive
Kansas City, MO 64105

American-International Charolais Association
1610 Old Spanish Trail
Houston, TX 77025

(continued on the following page)

Figure 16-7 Home office addresses of beef breed associations.

Beef Breed Associations *(concluded)*

American International Marchigiana Society
Route 2, Box 65
Lindale, TX 75771

American Maine-Anjou Association
564 Livestock Exchange Building
Kansas City, MO 64102

American Murray Grey Association
1222 N. 27th Street
Billings, MT 59102

American Normande Association
Box 350
Kearney, MO 64060

American Pinzgauer Association
1415 Main Street
Alamosa, CO 81101

American Polled Hereford Association
4700 East 63rd Street
Kansas City, MO 64130

American Polled Shorthorn Society
8288 Hascall Street
Omaha, NB 68124

American Romagnola Association
Box 8172
St. Paul, MN 55113

American Red Poll Association
3275 Holdrege Street
Lincoln, NB 68503

American Salers Association
Box 30
Weiser, ID 83672

American Scotch Highland Breeders Association
Box 240
Walsenburg, CO 81089

American Shorthorn Association
8288 Hascall Street
Omaha, NB 68124

American Simbrah Association
1 Simmental Way
Bozeman, MT 59715

American Simmental Association
1 Simmental Way
Bozeman, MT 59715

American South Devon Association
Box 248
Stillwater, MN 55082

American Tarentaise Association
Box 1844
Fort Collins, CO 80522

Amerispain Corporation
Box 4029
San Angelo, TX 76902

Barzolia Breeders Association of America
Box 142 1
Carefree, AZ 85331

Beefmaster Breeders Universal
G.P.M. South Tower, Suite 720
800 NW Loop 410
San Antonio, TX 78216

Beef Friesian Society
210 Livestock Exchange Building
Denver, CO 80216

Belted Galloway Society, Inc.
Box 5
Summitville, OH 43962

Devon Cattle Association, Inc.
Box 628
Walde, TX 78801

Foundation Beefmaster Association
Livestock Exchange Building,
Suite 200
4701 Marion Street
Denver, CO 80216

Galloway Cattle Society of America
Springville, IA 52336

International Ankina Association
Route 1, Box 323
Liberty, KY 42539

International Braford Association, Inc.
Box 1030
Fort Pierce, FL 33450

International Brangus Breeders Association, Inc.
9500 Tiaga Drive
San Antonio, TX 78230

International Illawarra Association
313 S Glenstone Avenue
Springfield, MO 65802

International South Devon Association
Lynnville, IA 50153

National Beefmaster Association
817 Sinclair Building
Fort Worth, TX 76102

National Buffalo Association
Custer State Park
Hermosa, SD 57744

North American Limousin Foundation
100 Livestock Exchange Building
Denver, CO 80216

North American Norwegian Red Association
Route 1, Box 346
Burns, TN 37029

Ranger Cattle Company
N. Pecos Station, Box 21300
Denver, CO 80221

Red Angus Association of America
Box 776
Denton, TX 76201

Red Poll Beef Breeders International
Dry Run Farm
Ross, OH 45061

Santa Gertrudis Breeders International
Box 1257
Kingsville, TX 78363

Sussex Cattle Association of America
Box AA
Refugio, IX 78377

Texas Longhorn Breeders Association of America
204 Alamo Plaza
San Antonio, TX 78205

United States Welsh Black Cattle Association
Route 1
Wahkon, MN 56386

Sheep Breed Associations

American and Delaine Merino Record Association
Aleppa, PA 15310

American Cheviot Sheep Society
Route 1
Carlisle, IA 50047

American Corriedale Association, Inc.
Box 29C
Seneca, IL 61360

American Cotswold Record Association
11903 Milwaukee Road
Britton, MI 49229

American Hampshire Sheep Association
Route 10, Box 199
Columbia, MO 65201

American North County Cheviot Sheep Association
Aitchner Hall, University of Maine
Orono, ME 04473

American Oxford Down Record Association
Box 401
Burlington, WI 53105

American Panama Registry Association
Route 2
Jerome, IN 83338

American Rambouillet Sheep Breeders Association
2709 Sherwood Way
San Angelo, TX 76901

American Romney Breeders Association
212 Withycombe Hall
Corvallis, OR 97331

American Shropshire Registry Association, Inc.
Box 1970
Monticello, IL 61856

American Southdown Breeders Association
Route 4, Box 14B
Bellefonte, PA 16283

American Suffolk Sheep Society
55 East 100 North
Logan, UT 84321

Black Top and National Delaine Merino Sheep Association
290 Beech Street
Muse, PA 15350

Columbia Sheep Breeders Association of America
Box 272
Upper Sandusky, OH 43351

Continental Dorset Club, Inc.
Box 577
Hudson, IA 50643

Debouillet Sheep Breeders Association
2662 University
San Angelo, TX 76901

Empire Karakul Registry
Route 1
Fabius, NY 13063

Finnsheep Breeders Association
Route 3
Pipestone, MN 56164

Montadale Sheep Breeders Association, Inc.
Box 44300
Indianapolis, IN 46244

National Lincoln Sheep Breeders Association
Route 6, Box 24
Decatur, IL 62521

National Suffolk Sheep Association
Box 324
Columbia, MO 65201

National Tunis Sheep Registry
1826 Nesbitt Road
Attica, NY 14011

Texas Delaine Sheep Association
Route 1
Burnet, TX 78611

U.S. Targhee Sheep Association
Box 40
Absarokee, MT 59001

Goat Breed Associations

American Angora Goat Breeders Association
P.O. Box 195
Rock Springs, TX 78880

American Dairy Goat Association*
P O Box 865
Spindale, NC 28160

International Nubian Breeders Association
4496 Main Road West
Emmaus, PA 18094

*Note: Information on the following breeds can be found here: Alpine, La Mancha, Oberhasli, Saanow, Toggenburg.

Figure 16-8 Home office addresses of sheep and goat breed associations.

Swine Breed Associations

American Berkshire Association
601 W Monroe Street
Springfield, IL 62704

American Landrace Association, Inc.
Box 111
Culver, IN 46511

American Yorkshire Club, Inc.
Box 2417
West Lafayette, IN 47906

Chester White Swine Record Association
Box 228
Rochester, IN 46975

Conner Prairie Swine, Inc.
14320 River Avenue
Noblesville, IN 46060

Hampshire Swine Registry
1111 Main Street
Peoria, IL 61606

Inbred Livestock Registry Association
14320 River Avenue
Noblesville, IN 46060

National Hereford Hog Record Association
Route 2
Lena, IL 61048

National Large Black Swine Breeders Association
Route 1, Box 44
Midland, NC 28107

National Spotted Swine Record, Inc.
110 W. Main Street
Bainbridge, IN 46105

OIC Swine Breeders Association
Route 2, Box 166
Cloverdale, IN 46129

Poland China Record Association
Box 71
Galesburg, IL 61401

Tamworth Swine Association
Route 2, Box 126A
Killsboro, OH 45133

United Duroc Swine Registry
1803 W. Detweiller Drive
Peoria, IL 61614

Wessex Saddleback Swine Association
4010 Clinton Avenue
Des Moines, IA 50310

Figure 16–9 Home offices addresses of swine breed associations.

Miscellaneous Agribusiness Service Organizations

Agricultural Research Institute, Inc.
2100 Pennsylvania Avenue, NW
Washington, DC 22037

American Beekeeping Federation
Route 1, Box 68
Cannon Falls, MN 55009

American Egg Board
205 Touhy Avenue
Park Ridge, IL 60068

American Mushroom Institute
Box 373
Kennett Square, PA 19348

American Poultry Association, Inc.
Box 70
Cushing, OK 74023

American Rabbit Breeders Association
1925 S Main Street, Box 426
Bloomington, IL 61701

American Seed Trade Association
1030 Fifteenth Street, NW
Washington, DC 20005

National Association of Animal Breeders
Box 1033
Columbia, MO 65201

National Association of Swine Records
112 1/2 N Main Street
Culver, IN 46511

National Livestock and Meat Board
444 N. Michigan Avenue
Chicago, IL 60611

National Livestock Exchange
Box 552
Memphis, TN 38101

National Livestock Producers Association
307 Livestock Exchange Building
Denver, CO 80216

National Pork Producers Council
4715 Grand Avenue
Des Moines, IA 50312

Poultry and Egg Institute of America
425 Thirteenth Street, NW
Washington, DC 20004

The Poultry Science Association, Inc.,
University of Missouri
Columbia, MO 65201

United Egg Producers
3951 Snapfinger Parkway, Suite 580
Decatur, GA 30035

Figure 16–10 Home office addresses of miscellaneous agribusiness service organizations.

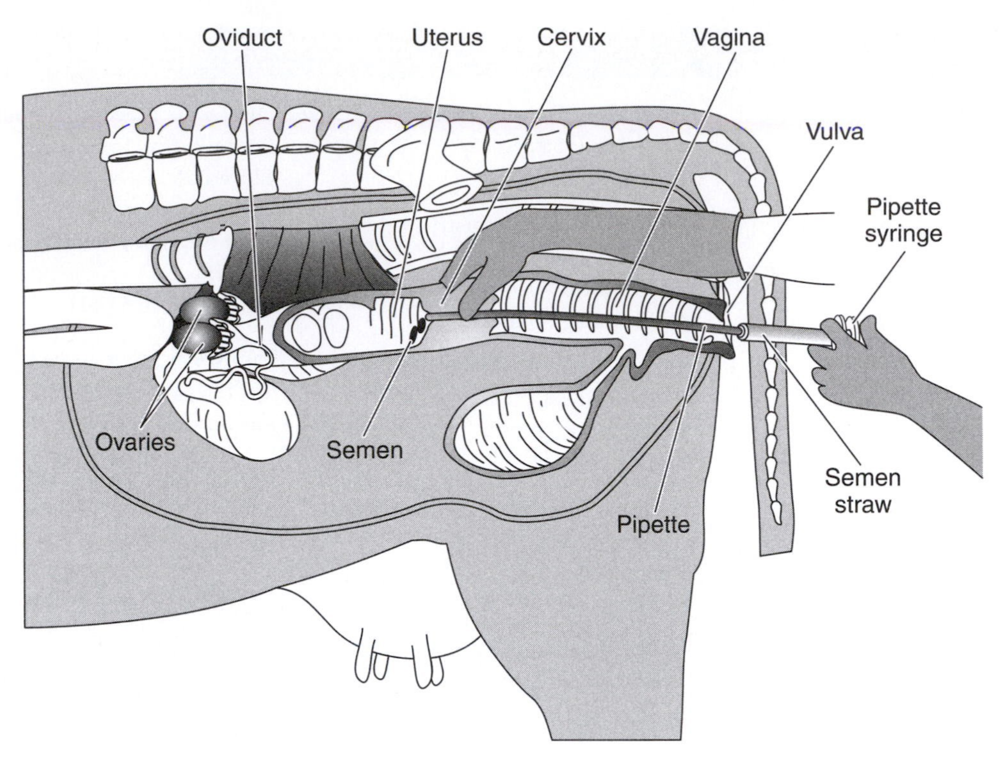

Figure 16–11 This diagram shows the artificial insemination process. Semen is deposited from the tube just inside the uterus.

First Veterinary Schools

The first veterinary schools were private ventures, and during the latter part of the 19th century the private schools flourished. However, they experienced numerous problems, and when pressures for veterinary research became strong, private schools could not support the extra cost; they were replaced fairly quickly by public schools. The last private veterinary school closed in 1927.

First College of Veterinary Medicine

The first public support for veterinary education took the form of special courses and programs, the first one being offered by Cornell University. The first college of veterinary medicine was established in 1879 by Iowa State University, and the second was formally established at the University of Pennsylvania in 1884. Thereafter, other veterinary schools were established rapidly throughout the United States.

Present-Day Veterinarians

There are currently about 86,000 veterinarians in the United States, providing a wide variety of services in private practice, teaching, research, public health, military service, private industry, and other special settings. The federal government employs about 1,200 veterinarians, primarily in the U.S. Department of Agriculture and the Public Health Service.[12] Veterinarians provide a valuable service to production agriculturalists. Among their many duties are diagnosing animals' medical problems, performing surgery, and prescribing and administering medicines and drugs. Veterinarians care for the nation's more than 60 million dogs, nearly 70 million cats, 10 million birds, more than 5 million pet horses, and millions of other companion animals. They also conduct medical research, work to prevent bio- and agroterrorism, serve in food safety positions, and contribute greatly to scientific knowledge and research worldwide.[13]

Other veterinarians engage in research, food safety inspection, or education. They also work in

12. Bureau of Labor Statistics, *Occupational Outlook Handbook, 2006-07 Edition: Veterinarians* (2006); available at http://www.bls.gov/oco/ocos076.htm; American Veterinary Medical Association (2007), available at http://www.avma.org/about_avma/history/history_avma.asp.
13. American Veterinary Medical Association (2007), available at http://www.avma.org/about_avma/history/history_avma.asp.

Figure 16–12 Most veterinarians operate small-animal clinics, but they are also employed in teaching, research, public health, private industry, and many other specialized service areas. (Courtesy of Cliff Ricketts)

regulatory medicine and public health in the areas of food inspection, do investigations of disease outbreaks, or conduct research in scientific laboratories. All states require veterinarians to be licensed. Applicants for licensing must have a Doctor of Veterinary Medicine degree and pass written and oral state board examinations.[14] Refer to Figure 16–12. In addition to careers in the field of veterinary medicine, many careers are available as breed association representatives and artificial insemination technicians. Refer to the Career Options section for a further explanation of these careers.

Agricultural Communications

Communications through the print media, computers, Internet, pictures, radio, television, videos, exhibits, and so forth are a part of all major segments of modern business and industry. Agribusiness is no exception. Many opportunities await those who have a good agricultural background and can combine it with the skills needed in communication. This part of the agribusiness private input sector includes advertising; photography; books, magazines, and newspapers; trade publications; computer/Internet-related services; and public relations, as well as many other areas. Few things offer more excitement, challenge, and opportunity to keep up-to-date with every segment of agriculture and all its related activities than agricultural communications. A person working in agricultural communications may interview production agriculturalists, scientists, and executives; attend conventions, demonstrations, and legislative sessions; report new developments, ideas, and trends; and report in general on anything that is of interest to the agricultural industry. Refer to Figure 16–13 for a list of the major publications in the agricultural industry.

Agricultural Cooperatives

A *cooperative* is a business that is financed, owned, and controlled by the people who use it. Cooperatives are made up of individuals who have similar interests, desires, and problems. By working together, they combine their investments and influence. This gives them financial strength, greater independence, and a stronger voice in their own business affairs.

History and Development of Cooperatives

During America's frontier days, friends and neighbors traveled from miles around to enjoy the fellowship of others while they helped a family "raise" a barn or build a house. There were also afternoons when women gathered at quilting bees, each sewing a patch to contribute to the project. Although these were not formal cooperative associations by any means, they were some of the nation's first cooperative efforts.

First Agricultural Cooperative. The first agricultural cooperative was established in Babylon around 2,000 B.C. Its objective was similar to the goals of modern agricultural cooperatives: namely, to provide greater flexibility and more independence for member farmers.

Forerunner of Modern-Day Cooperatives. The forerunner of modern-day cooperatives, the Rochdale Society of Equitable Pioneers, was founded in England in 1844. It started with 28 members, each of whom purchased one share of stock worth about $150 in today's economy. Its membership consisted of craftsmen such as weavers and shoemakers. Working together, they were able to sell their products under one roof and use a part of the earnings to purchase supplies in large quantities at economical prices. Another portion of the earnings

14. Smith, Underwood, and Bultmann, *Careers in Agribusiness and Industry,* p. 300.

General Farm Magazines

Publication	Address
Agribusiness Report	The Miller Publishing Company 2501 Wayzata Boulevard Minneapolis, MN 55440
Agri-Fieldman	Meister Publishing Company 37841 Euclid Avenue Willoughby, OH 44099
Agri-Finance	5520 Touhy Avenue, Suite G Skokie, IL 60076
Big Farmer	Big Farmer, Inc. 131 Lincoln Highway Frankfort, IL 60423
Doan's Agricultural Report	8900 Manchester Road St. Louis, MO 63144
Farm Journal	230 W. Washington Square Philadelphia, PA 19105
Progressive Farmer	P O Box 2581 Birmingham, AL 35202
Successful Farming	1716 Locust Des Moines, IA 50336
The Drovers Journal	1 Gateway Center Fifth and State Kansas City, KS 66101
Feedstuffs	Box 67 Minneapolis, MN 55440

Note: Numerous other agricultural magazines are offered at the regional level, especially in the livestock industry. In addition, almost all livestock breeder associations publish one or more breeder magazines which can be obtained from the various breeder associations.

Agricultural Communications Groups

American Agricultural Editors Association
2001 Spring Road
Oak Brook, IL 60521

American Association of Agricultural College Editors
23 Agricultural Administration Building
Ohio State University
2120 Fyffe Road
Columbus, OH 43210

National Association of Farm Broadcasters
Box 119
Topeka, KS 66601

Newspaper Farm of America
Editors
4200 Twelfth Street
Des Moines, IA 50313

Figure 16–13 Home office addresses of general farm magazines and agricultural communications groups.

was reinvested in the society so that the funds could continue to grow. The remainder was returned to the individual members in the form of refunds.

It was not until the Civil War and the coming of the Industrial Revolution that American agricultural cooperatives really took root. In the 1920s, the federal government and various state governments began to step in with laws designed to encourage the growth of agricultural cooperatives. Perhaps the most important new law was the Capper-Volstead Act of 1922, which allowed farmers to organize legally to buy and sell their products and is the law under which cooperatives operate today.

Present-Day Cooperatives. Today's cooperatives, whether in agriculture, utilities, banking, retailing, insurance, or wholesaling, are organizations in which people of similar needs and interests can work together to help themselves and each other. By doing so, they strengthen themselves economically. Agricultural cooperatives keep profits under their own control, and they stimulate free enterprise while helping to protect the family farm.[15] Therefore, cooperatives are organized to:

- improve bargaining power
- reduce costs
- obtain products or services that are otherwise unavailable
- expand new and existing market opportunities
- improve product or service quality
- increase income

Impact of Cooperatives

Today, the life of every American is touched in some way by cooperative enterprises. Approximately 45,000 separate cooperative organizations, with more than 90 million members, currently operate in this country. Cooperatives serve one out of every four citizens. Statistics show that of these 45,000 cooperatives, approximately 5,000 are farmer-owned, providing marketing, purchasing, and related services for farmers and producers. These agricultural cooperatives operate at a business value

15. Tennessee Council of Cooperatives, *What Is a Cooperative?* (Lavergne, TN: Author, n.d.).

Career Options

Breed Association Representative, Artificial Insemination Technician, Veterinarian, Animal Health Technician, Agricultural Communications

There are many career opportunities in agriculture-related areas. One may find employment as a field representative for a breed association, feed company, marketing cooperative, supply company, animal health products manufacturer or supplier, herd improvement association, or financial management firm. Also, many jobs reated with the huge improvements in animal production and performance that have come about as a result of selection and breeding; artificial insemination technician is only one example. With the increase in availability and use of various information and communications technologies, many individuals use their backgrounds in the agricultural industry by pursuing writing, publishing, and telecasting careers.

To sell what is produced in this country every year, it is imperative that the buying public be aware of the products available. Communication is the key to accomplishing this task, and specialists in the field of communications are in high demand, ranking 25th on the Bureau of Labor Statistics' National Industry–Occupation Employment Matrix (1995). In the agriculture industry, communications specialists find a wide variety of careers. An agriculture communications expert is knowledgeable about all aspects of production and can relay information in a positive manner. This position requires a bachelor's degree in public relations or agricultural communications and a good working understanding of agricultural production. Agricultural industry councils, such as those representing the beef, milk, or pork industries, employ many communications specialists as spokespersons, advertising salespeople, marketers, and advertising copywriters.

Veterinarians and their associates work constantly to treat animals with disorders and to improve animal health. The desire to become a veterinarian has been high among youth in recent years. This interest has permitted the supply of veterinarians to increase despite the rigor of the college curriculum and stiff competition to get into veterinarian schools. Colleges of animal sciences offer curriculum in other career areas in animal sciences, such as nutrition, breeding, education, and production. In urban and suburban areas, animal shelters, hospitals, kennels, and pet stores provide many career opportunities for people interested in becoming animal health technicians. In rural areas, typical positions in the animal health industry include large-animal veterinarian, veterinary assistant, laboratory veterinarian, and laboratory technician.

A career as a veterinarian is both challenging and rewarding for those willing to accept the rigor of the college curriculum and competition to get into veterinary schools. (Courtesy of USDA)

of more than $73 billion. Cooperatives market about 28 percent of all agricultural products and provide about 25 percent of all production supplies used on farms each year.

The cooperative business structure provides insurance, credit, health care, housing, telephone, electrical, transportation, child care, and utility services. Members use cooperatives to buy food, consumer goods, and business and production supplies. Farmers use cooperatives to market and process crops and livestock, purchase supplies and services, and provide credit for their operations.[16]

Types of Cooperatives

Chapter 5 briefly discussed three major types of agricultural cooperatives: purchasing or supply, marketing, and service cooperatives. A further explanation of each of these three types of cooperatives (co-ops) follows.

Purchasing or supply cooperatives are organized to purchase farm supplies for members in large volumes, which allows production agriculturalists to buy resources at a lower price and to have some control over their quality. In some cases these cooperatives produce their own products rather than buy them from someone else. Reports from the USDA *Agricultural Chartbook* (1997) indicate that about 25 percent of all farm supplies are purchased through co-ops. The major resources production agriculturalists purchase through their cooperatives are, in order of volume, petroleum products, feed, and fertilizer, which account for about 75 percent of total purchases from co-ops.

Marketing cooperatives provide one or more marketing services beyond the production process. Larger volume and consistency in quality give farmers more clout in negotiating higher prices for their products. Of course, some products are more adaptable to cooperative marketing than others. Grain, soybeans, dairy products, and livestock products account for about 85 percent of all farm products marketed by co-ops. More than 35 percent of all farm products are sold through more than 1,500 agricultural cooperatives, with annual gross sales of $47 billion. Cooperatives market 78 percent of milk, 33 percent of cotton, and 17 percent of fruits, nuts, and vegetables.[17]

Service cooperatives provide special services to production agriculturalists, such as credit, artificial breeding, irrigation, telephone, electricity, insurance, import-export, and numerous other functions. In addition to separate cooperatives that offer these services, many marketing and supply cooperatives provide special services as a part of their operations. Refer to Figure 16–14 for the addresses of several cooperative associations.

Service cooperatives account for about 2 percent of all agricultural cooperatives. The following briefly describes some service cooperatives.

- *Electric cooperatives.* Like other types of cooperatives, electric co-ops exist to serve members by providing a service that would otherwise be unavailable. Today, nearly 1,100 of these cooperatives provide electricity to 11 million consumers on farms and in many small towns and communities. Electric co-ops began to light the countryside in 1935, when an Executive Order by President Franklin Roosevelt created the Rural Electrification Administration (REA). Co-ops were needed because it was not economically feasible for existing power companies to build lines in the many sparsely populated areas of rural America. REA still provides loans on a continuing basis for rural electrification.
- *Telephone cooperatives.* Congress amended the Rural Electrification Administration Act in 1949 to include financing for telephone service as well as electric service in rural areas. Cooperatives were organized by citizens whose only interest was ensuring good, dependable telephone service in areas where there was little or no service and it was not economically feasible for commercial (for-profit) companies to do so. There are about 240 telephone cooperatives in the United States serving more than 2 million members.
- *Artificial breeders' cooperatives.* The purpose of artificial breeders' cooperatives is to offer

16. USDA, *What Is a Cooperative?* (Washington, DC: USDA, Rural Business and Cooperative Development Service, March 1995).

17. United States Department of Agriculture, *Farmer Cooperative Statistics: CS Service Report 43* (Washington, DC: U.S. Government Printing Office, 1993).

Cooperative Associations

American Institute of Cooperatives
1129 Twentieth Street, NW
Washington, DC 20036

American Rice Growers Cooperative Association
211 Pioneer Building
Lake Charles, LA 70601

Burley Tobacco Grower Cooperative Association
620 South Broadway
Lexington, KY 40508

Cooperative League of the USA
1828 L. Street, NW, Suite 1100
Washington, DC 20036

Farm Educational and Cooperative Union of America
12025 East 45th Avenue
Denver, CO 80251

Flue-Cured Tobacco Cooperative Stabilization Corporation
Box 12300
Raleigh, NC 27605

National Association Supply Cooperative
Box 303
New Philadelphia, OH 44663

National Council of Farmer Cooperatives
1129 Twentieth Street, NW
Washington, DC 20036

National Federation of Grain Cooperatives
1129 Twentieth Street, NW, Suite 512
Washington, DC 20036

National Livestock Producers Association
307 Livestock Exchange Building
Denver, CO 80216

National Milk Producers Federation
30 F Street, NW
Washington, DC 20001

National Rural Electric Cooperative Association
2000 Florida Avenue, NW
Washington, DC 20009

National Telephone Cooperative Association
2100 M Street, NW
Washington, DC 20037

Pacific Agricultural Cooperative for Export, Inc.
465 California Street, Suite 414
San Francisco, CA 94104

Figure 16–14 Home office addresses of cooperative associations.

members superior genetics in dairy and beef sires. They serve co-op members by conducting training seminars to teach the techniques of artificial insemination and the proper thawing and handling of semen; providing herd counseling in sire selection and equipment needs, and making deliveries of semen, nitrogen, and other supplies to farmer members at regular intervals.

- *Insurance.* Earlier we discussed the Farm Bureau as a general farm organization. The Farm Bureau, which is a cooperative, also provides insurance in most states. The various insurance service companies under the Farm Bureau umbrella offer home, automobile, and life insurance, and provide group health insurance through the Rural Health Program.
- *Credit.* Organizations such as the Farm Credit Service are farmer-owned cooperative businesses that exclusively serve rural communities. The primary goal of these banks and associations is to provide dependable and constructive credit and related services to production agriculturalists, agricultural cooperatives, and rural residents in their service areas.
- *Irrigation cooperatives.* Production agriculturalists, especially in the western states, formed mutual irrigation companies that now supply water to one-quarter of the land being irrigated. These co-ops acquire or develop sources of water supply and distribute the water to users. Irrigation makes it possible for production agriculturalists in dry lands to produce significantly greater yields.[18]
- *Dairy Herd Improvement Associations (DHIAs).* DHIAs are cooperative organizations of about 25 milk producers each. Each association employs a trained supervisor to keep monthly milk and butterfat production records on association herds. The milk producers then decide which cows in their herds are underproducing and should be removed.

18. Smith, Underwood, and Bultmann, *Careers in Agribusiness and Industry,* p. 310.

Export-Import Cooperatives		
Agway Inc. 333 Butternut Drive DeWitt, NY 13214	Ocean Spray Cranberries One Ocean Spray Drive Lakeville-Middleboro, MA 02349	Texas Citrus Exchange Expressway 83 and Mayberry Rd. P.O. Box 793 Mission, TX 78573-0793
California Almond Exchange 1802 C Street P.O. Box 1768 Sacramento, CA 95812	Sun Diamond Growers P.O. Box 1727 Stockton, CA 95201	Tree Top 220 East 2nd Avenue P.O. Box 248 Selah, WA 98942
Harvest States 1667 North Snelling Avenue P.O. Box 64594 St. Paul, MN 55164-0594	Sunkist Growers, Inc. P.O. Box 7888 Van Nuys, CA 91409-7888	

Figure 16–15 Home office addresses of export-import cooperatives.

- *Export-import cooperatives.* Export-import cooperatives play an important part in international trade of farm products. A large portion of the $54 billion-plus in annual exports involves companies that are cooperatives. The export business constitutes up to 50 percent of the total volume of some cooperatives.[19] Some cooperatives are importers as well as exporters. Refer to Figure 16–15 for some examples of export-import cooperatives.

In summary, agricultural cooperatives play a significant role in the private agriservices sector, as well as the input and marketing sector of agriculture. The economic importance of agricultural cooperatives is indicated by the fact that agricultural cooperatives are included in *Fortune* magazine's annual listing of the nation's 500 largest industrial companies, based on sales volume. In addition, several others are listed in the second 500 largest corporations. In 1999, the top agricultural cooperative by sales volume was Farmland Industries, with $9.2 billion in sales.[20]

Careers. Careers in cooperatives are as many and as varied as their diverse activities and enterprises. It is not possible to list here all the types of jobs and opportunities available with cooperatives. A few of the major categories of jobs include managers and assistant managers, public relations and communication specialists, laboratory and field technicians, product specialists, and providers of operations and business services, professional services, and finance, credit, and insurance services.

19. United States Department of Agriculture, *Agricultural Statistics, 1997* (Washington, DC: National Agricultural Statistics Service, 1997).
20. *Fortune* magazine (1999), available at http://www.pathfinder.com/fortune/global500/companies/499.html.

Farm Credit System

Many jobs are available through the agricultural credit system. These include jobs with Farm Credit Services, the Consolidated Farm Service Agency, commercial banks, life insurance companies, and the Commodity Credit Corporation.

Historical Beginning of Federal Land Banks

The Federal Land Banks were organized by the Federal Farm Loan Act of 1916, which established 12 Federal Land Banks across the United States. Federal Land Banks made long-term loans by first mortgages on real estate through more than 490 local Federal Land Bank Associations. The federal government provided the initial financial support to these banks (in 1916, $9 million, which was repaid in 1932, plus $189 million obtained in the following five years). The remaining money was then repaid to the federal government, and the

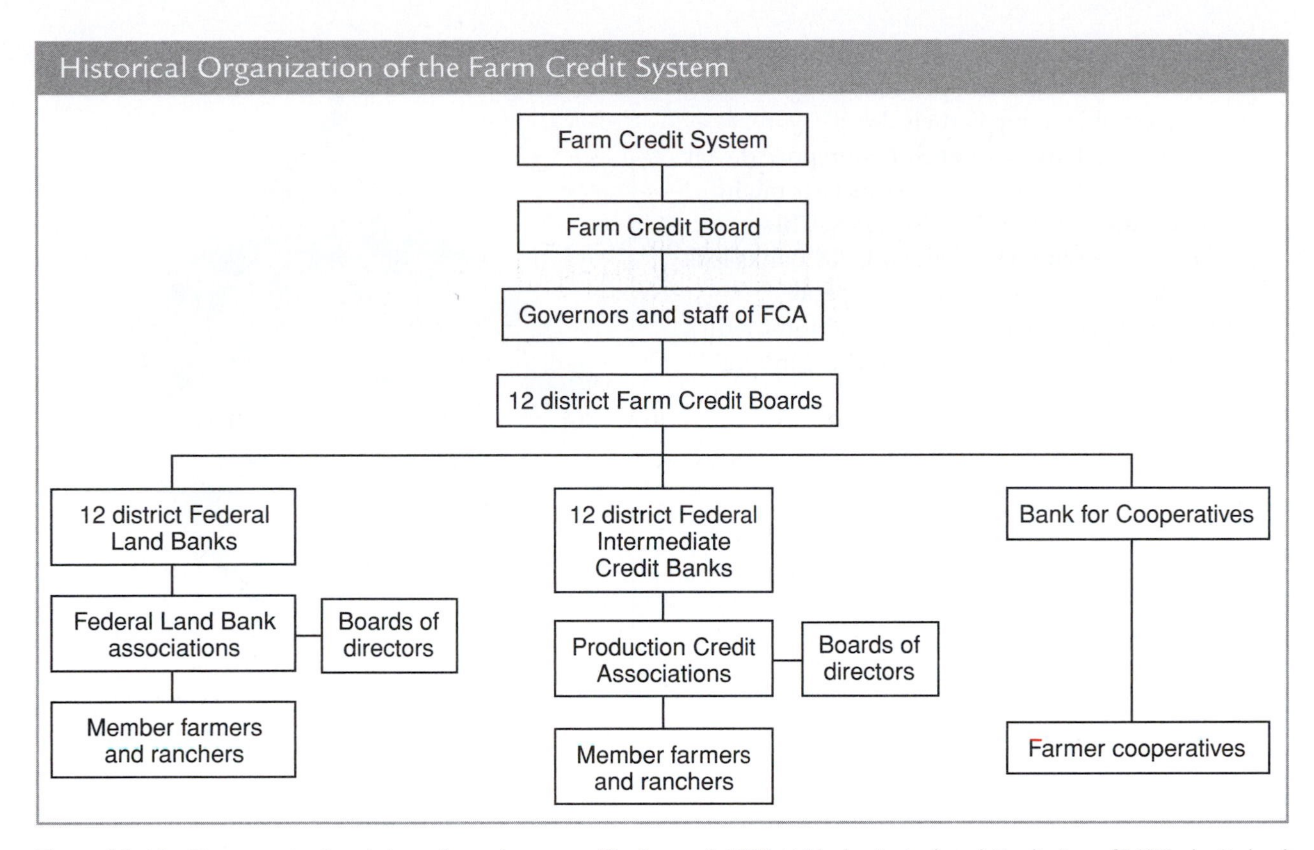

Figure 16–16 The organizational chart shown here was effective until 1986. With the Agricultural Credit Act of 1987, the Federal Land Bank Associations and Production Credit Associations merged into what is called Farm Credit Services in most districts.

Federal Land Banks became owned by their borrowers.[21]

Historical Beginning of Production Credit Associations

The production credit system was established under federal laws in 1923 and 1933 to provide short-term and intermediate-term loans to production agriculturalists and rural residents. The 12 Federal Intermediate Credit Banks were first authorized by the Farm Credit Act of 1923 to provide discounting services to banks for agricultural loans. In 1933, Congress authorized local Production Credit Associations to make direct short- and intermediate-term loans to individuals in the agricultural industry.[22] By 1986, there were 420 Production Credit Associations throughout the United States, with 1,500 full-time offices.

Historical Beginning of the Bank for Cooperatives

The Bank for Cooperatives was set up by the Farm Credit Act of 1933. It provides loan funds to farmer-owned cooperatives that are involved in marketing, supply, and service. Borrowing guidelines require that agricultural producers comprise at least 80 percent of the co-op's membership; in addition, at least 50 percent of the co-op's business must be conducted within its membership.[23] Refer to Figure 16–16 to see how the Farm Credit System was organized until 1986.

Restructured and Present-Day Farm Credit Systems

The nation's agricultural industry experienced rough times during much of the 1980s. After several years of rising land prices and increasing interest rates, plus a significant drop in the market for agricultural products, international competition, and

21. Gail L. Cramer, Clarence W. Jensen, and Douglas D. Southgate, Jr., *Agricultural Economics and Agribusiness,* 8th ed. (New York: John Wiley & Sons, 2001), p. 250.
22. *Agribusiness Management and Marketing* (8710-C) (College Station, TX: Institutional Materials Service, 1988), p. 1.
23. Cramer, Jensen, and Southgate, *Agricultural Economics and Agribusiness,* p. 253.

increased domestic production, farm product prices began to decline. As a result, production agriculturalists were less able to repay their debts.[24]

By 1985, the Farm Credit System reported that it held several billion dollars in loans that might go into **default** over the next few years. This turned out to be true. The Federal Land Banks lost about 66 percent of this total, the Federal Intermediate Credit Banks and Production Credit Associations about 33 percent, and the Bank for Cooperatives about 1 percent.[25]

Agricultural Credit Act of 1987. The financial problems of the Farm Credit System prompted Congress to pass the Agricultural Credit Act of 1987. This act merged the Federal Land Banks and Intermediate Credit Banks within the 12 credit districts. There is now a single institution, called the Farm Credit Bank, in each of the seven districts. These banks handle both long- and short-term credit needs of production agriculturalists.

In essence, the Federal Land Bank Associations and Production Credit Associations merged into what is called Farm Credit Services in most districts. The merger eliminated about half of the local offices of the Farm Credit System. In 1980, there were 915 associations, but by 1996 mergers and consolidations had reduced the number of associations to 228.[26] Refer to Figure 16–17.

Types of Farm Credit Service Loans. Besides real estate, farm, and land financing, the Farm Credit System loans money for the following: capital improvements, operating loans, family living expenses, farm equipment, livestock, specialty loans (such as for education), farm-related business loans, farm and rural home loans, and farm credit leasing. It also provides related services, including revolving lines of credit.[27]

Banks for Cooperatives. Ten Banks for Cooperatives merged with the Central Bank for Cooperatives to form a single institution, which was named CoBank and is now headquartered in Denver, Colorado.

24. Ibid., p. 256.
25. Ibid., p. 256.
26. Ibid., p. 257.
27. Ibid., pp. 257–258.

Figure 16–17 The Farm Credit Service Agencies make short-term, intermediate-term, and long-term loans. Loan purposes range from farm to educational loans. (Courtesy of Farm Credit Services)

Consolidated Farm Service Agency

Historical Beginning

Although the Farm Credit System was established by 1933, it did not fill all the farm credit gaps. Therefore, in 1935 the Farmers Home Administration was set up, originally as an independent government agency called the Resettlement Administration authorized under President Franklin D. Roosevelt's Executive Order 7027.[28] In 1937, the Resettlement Administration was taken over by the Farm Security Administration in an expansion of federal aid to agriculture. However, young production agriculturalists still found it extremely difficult to provide the collateral needed for a loan. Thus, the Farmers Home Administration Act of 1946 created the Farmers Home Administration (FmHA), now called the

28. Ibid., p. 253.

Figure 16–18 To save money at both the federal and state levels, many of the USDA agencies were consolidated. The old Agricultural Stabilization Conservation Service (ASCS) is now the Natural Resources Conservation Service. The Farm Service Agency (formerly FmHA) shares the facility shown here with other USDA and state agencies. (Courtesy of Marlene Graves)

Farm Service Agency (FSA). It was established within the U.S. Department of Agriculture.[29] Refer to Figure 16–18.

Purpose

The major purpose of the FmHA (FSA) was to provide supervised credit to production agriculturalists who were unable to get adequate credit from other sources at reasonable rates. Production agriculturalists also had to agree to refinance their loans with another agency as soon as it became economically feasible for them to do so. In other words, the FmHA was strictly a hardship agency for high-risk loans. The FmHA was also created to provide emergency credit to farmers affected by natural disasters such as extreme drought, hail, and flood. Fortunately, emergency loans have been only a minor portion of FmHA lending.

Change from FmHA to FSA

Being a lender of last resort for farm loans, the FmHA experienced a large number of loan **delinquencies** and defaults. Loan losses averaged about $1.7 billion per year from 1988 to 1995.[30] The Farmers Home Administration (FmHA) ceased operations upon passage of the Federal Crop Insurance Reform and Department of Agriculture Reorganization Act of 1994. Farm loans, including both direct and guaranteed loans, are currently handled by the Farm Service Agency (FSA). Loans made by the FSA are of the following general types:

- agriculture
- American Indian land acquisition
- business and industrial
- community facility
- farm emergency
- farm labor housing
- farm operating
- farm ownership
- grazing association
- home life development
- individual home ownership
- irrigation and drainage
- recreation enterprise
- rental and cooperative housing
- resource conservation and development
- self-help technical
- soil and water conservation
- watershed
- youth[31]

Nonfarm Portion of FmHA. In 1994, the nonfarm portions of the FmHA—the Rural Development Administration, Rural Electrification Administration, and Agricultural Cooperative Service—were merged into a new USDA organization called the Rural Housing Service, which is an agency within the Rural Development Division. (Note that before the restructuring of the FmHA, it had 1,700 field offices, normally in county seats.)

Commercial Banks

Since early colonial days, commercial banks have been the major source of short- and intermediate-term loans. Banks are still the major non–real estate lender.

29. USDA, *A Brief History of the Farmers Home Administration* (Washington, DC: U.S. Department of Agriculture, Farmer's Home Administration, February 1990).

30. Cramer, Jensen, and Southgate, *Agricultural Economics and Agribusiness*, p. 255.

31. Ibid., pp. 254–255.

There are thousands of commercial banks in the United States.[32] About 4,500 of these are classified as agricultural banks. To be designated as an agricultural bank, at least 25 percent of a bank's loans must be farm loans.[33]

Organization of Commercial Banks

Commercial banks are corporations created under federal or state law. They are owned by the stockholders who have invested in them. The stockholders elect a board of directors, which hires a staff to operate the bank. In reality, a commercial bank is a private business, although it is regulated by federal and state agencies to ensure the safety of depositors' funds.[34]

Agricultural Loans

Most commercial bank loans to agricultural producers are for short-term production expenses, such as fertilizer, cattle, feed, seed, fuel, labor, and chemicals. In the past decade or so, commercial banks have provided about 52 percent ($38 billion) in non–real estate loans to agriculture and about 28 percent ($22 billion) in real estate loans, representing a market share of 40 percent of total farm debt.[35]

Many banks have departments that specialize in agricultural credit needs. The banks employ skilled agricultural specialists, who can analyze farm business decisions and provide guidance in assisting production agriculturalists to use their capital wisely and meet their repayment schedules.[36]

Life Insurance

Life insurance companies play a large role in financing real estate loans. Most of the major life insurance companies have agents whose primary responsibility is to invest company funds, and a prime target for these funds is farm real estate. Insurance companies also finance real estate indirectly, through commercial banks. The banks negotiate the loan with farmers and then sell the note at a discount to an insurance company.

Life insurance companies finance about 6 to 7 percent of total farm debt and about 13 percent of the real estate loans (about $10 billion).[37] However, life insurance loans vary greatly among different regions of the country. For example, though they are seldom used in the Northeast, they are very popular in the Corn Belt and the Southwest.[38]

Size of Loans

Life insurance companies tend to focus on large farm businesses. Reportedly, at least one company will only make real estate loans of $250,000 or more. The average Farm Credit System loan in recent years has been only half the size of the average life insurance company loan. Typically, farm mortgages constitute 2 to 3 percent of a life insurance company's total investments.[39]

Source of Funds

Life insurance companies get most of their funds for loans from premium payments made by policyholders. Also, individual pension and retirement plans that provide **annuities** for their employees pay life insurance companies for administration and management of the annuities. Furthermore, life insurance companies receive **dividends** from the investments they have in the **securities markets**. These two sources also generate funds for insurance company loans.[40]

Generally, life insurance companies are a **viable** source of real estate financing as long as the interest rate is reasonable, the repayment period is satisfactory, and there is no penalty for early repayment.

Commodity Credit Corporation

The Commodity Credit Corporation (CCC) is the credit arm of the Farm Service Agency.

32. Ibid., p. 268.
33. Marcella Smith, Jean M. Underwood, and Mark Bultmann, *Careers in Agribusiness and Industry*, 4th ed. (Danville, IL: Interstate Publishers, 1991), p. 274.
34. Cramer, Jensen, and Southgate, *Agricultural Economics and Agribusiness*, p. 268.
35. Ibid.
36. Ibid.
37. Ibid.
38. John B. Penson, Rulon D. Pope, and Michael L. Cook, *Introduction to Agricultural Economics* (Englewood Cliffs, NJ: Prentice-Hall, 1986), p. 454.
39. Ibid.
40. *Agribusiness Management and Marketing* (8710-C), p. 4.

The purpose of the CCC is to finance the loan and price support programs of the FSA. The CCC was created through an executive order from President Roosevelt in 1933 and was set up as a government corporation, with $100 million in capital and authority to borrow additional funds if needed. In 1939, the CCC became a part of the Department of Agriculture, and in 1948 it received a permanent charter from Congress.

Nonrecourse Loans

One of the credit responsibilities of the CCC is to make **nonrecourse loans** to farmers through local FSA agencies. These loans are designed to prevent farmers from having to sell their commodities immediately upon harvest, when prices are usually at their lowest seasonal level. Farmers who qualify for these loans use their stored crops as collateral for a short-term period. If the crops are sold during the period of the loan, the loan is paid back with interest. However, if the farmer chooses not to sell the crop during this period, the crops become the property of the CCC and the loan is considered paid in full.

Limitations

The major limitation of CCC loans is that they are not available on all crops; also, only production agriculturalists who qualify for specific price support programs can obtain this type of loan. However, for those who can qualify, CCC loans are an excellent means of obtaining needed capital to pay current operating expenses while storing products until the price improves. In recent years, the CCC has become a major source of short-term funds for production agriculturalists. About $7.8 billion per year goes into direct loans to farmers.[41]

Individual and Other Loans

Individuals are a major source of credit for real estate loans. Most individual lenders are farmers who sell their land, receive a partial payment, and take a personal note on the remainder of the purchase price. Interest rates paid to individuals tend to be 1 to 2 percent lower than institutional rates, primarily because of lower overhead cost. Individuals play a very minor role in short- and intermediate-term lending, and their role in long-term loans is decreasing. The chief disadvantage for the borrower is that if he or she fails to make the payments, all payments previously made can be treated as rent, with the borrower losing any equity in the property and any money that has been paid so far.

Merchants

Merchants constitute a major part of the non-real estate loan market. They offer trade credit to managers of agribusinesses. The manager may establish an **open account** with the merchant to charge purchases. The balance on these purchases is normally due in 30 days.

Farm Implement Dealers

Farm implement dealers are only a small part of the non-real estate debt market. They provide financing on the machinery and equipment they sell, usually through credit institutions. The manufacturer sets up these institutions to provide credit at the dealerships. Dealer credit normally has longer terms but higher interest rates than bank or other loans.

Career Opportunities in Agribusiness Input Credit Services

There are many career opportunities in agribusiness credit or financial services for individuals with the needed qualifications. A knowledge of the economy in the agricultural industry, both locally and nationally, and the ability to understand and interpret the policies of agriculture are valuable assets for professionals in the field of finance.

Experience

Agribusiness financial institutions that make loans to production agriculturalists are often interested in hiring employees who have a basic knowledge of finance, accounting, risk management, and agricultural economics. A knowledge of production agriculture, land values, farm practices, production costs, and the marketing of agricultural products is also required. Lending institutions depend on their agricultural specialists to keep their farm credit activities sound.

41. Cramer, Jensen, and Southgate, *Agricultural Economics and Agribusiness*, p. 263.

Education

Of the people currently working in and for agribusiness financial institutions, a high percentage have college degrees. Others have had specialized vocational or technical training. More importantly, agribusiness financial institutions look for those who are willing to learn. For people who go to college, courses should include agricultural economics, accounting, money and banking, farm management, marketing, future markets and trading, crop and livestock production, farm finance, soil management, business law, insurance and risk management, management information systems, and computer applications.

Company training is highly desirable and is usually required by agribusiness financial institutions. These courses or training sessions provide understanding and knowledge of the methods, customs, and practices of the business. Such training programs provide practical applications for theories and methods learned during previous formal education.

Potential Available Jobs

The following are examples of jobs available with agribusiness financial institutions:

- credit officer for the Farm Credit System
- credit officer for the Farm Service Agency
- credit officer for a manufacturer of farm equipment or supplies
- credit officer for a wholesaler or for retail dealers
- credit officer for an agricultural cooperative
- collection officer for a manufacturer or supplier
- loan appraiser for a life insurance company
- appraiser of farmland and production agriculture enterprises
- adjuster for a farm crop insurance company
- loan representative for a federal or private lending agency
- salesperson for a farm crop insurance company
- staff member at any of the agribusiness financial institutions
- agricultural specialist for a rural bank[42]

42. Smith, Underwood, and Bultmann, *Careers in Agribusiness and Industry*, pp. 280–281.

Conclusion

There are many other jobs in the private agribusiness services. Those providing the most opportunities were discussed in this chapter. The following is a list of other jobs in the private agribusiness input sector that involve service to production agriculturalists: auctioneer, farm tax accountant, blacksmith, land appraiser, agricultural engineer, farm record service, livestock buyer, agricultural scientist, rural sociologist, and agricultural law. The many other possible jobs in the agribusiness input sector are discussed in other chapters.

Summary

The private agriservices sector of agribusiness is increasing dramatically, and this trend is expected to continue. The major reasons for this increase include the growing complexity of production agriculture and the expansion of farm units to the point that specialized services are economically feasible. Some of the major services offered by private agriservice groups include political representation, advertising and promotion by specialized commodity organizations, agricultural research, consulting, artificial insemination, veterinary services, communications, and purchasing, supplying, marketing, and servicing production agriculturalists through cooperatives.

General farm organizations play a major role in the private agriservices group. Although numerous organizations of this type are in operation, the four major ones are the National Grange, the American Farm Bureau Federation, the National Farmers Union, and the National Farmers Organization. Although the general goals of all these organizations are basically the same, they differ considerably in specific goals and the procedures for implementing them. Almost all production agriculturalists are members of at least one of these organizations, and many production agriculturalists are members of more than one.

In addition to general farm organizations, numerous specialized organizations have developed around specific farm commodities. The major purpose of these organizations is to promote the development of specific products through research, education, and market expansion efforts. There is an organization to assist in the development of almost every major farm commodity. Some examples of these specialized organizations include the National Cattle Beef Association, the American Dairy Association, the National Association of Wheat Growers, the National Cotton Council, and the National Potato Council, among many others.

Another major function of the private agriservices sector is to develop new technology through constant research on new ways of making agriculture more productive. Agribusiness firms provide a tremendous amount of funds for agricultural research in two major ways: providing grants to public research institutions and conducting applied research in their own research facilities. Similar to research is agricultural consulting, as much of a consultant's job is research based.

Another major service offered to dairy and livestock producers today is artificial insemination. Although some production agriculturalists inseminate their own livestock, most animals are bred by local technicians who are trained specifically to perform this service.

Veterinarians provide a wide variety of services in private practice, teaching, research, public health, military service, private industry, and other special settings. Veterinarians render valuable services to many production agriculturalists, particularly those whose business concerns animals and animal products.

Communications through print media, computers, the Internet, pictures, radio, television, videos, exhibits, and the like are a part of all major segments of modern business and industry. Agribusiness is no exception. This private agriservice offers many opportunities to those who have a good agricultural background and combine it with good communication skills.

Agricultural cooperatives also supply a wide variety of services and products for production agriculturalists. Today, the life of every American is touched in some way by cooperative enterprises. Approximately 45,000 separate cooperative organizations, with a business value of more than $73 billion and more than 90 million members, currently operate in this country. Cooperatives serve one out of every four citizens. Three types of cooperatives are purchasing and supply, marketing, and service.

There are several sources of credit for individuals in the agricultural industry. Production agriculturalists have access to all credit sources that other businesses have, plus some that have been set up strictly for their agribusinesses. The major sources of credit are the Farm Credit System, commercial banks, the Farm Service Agency, the Commodity Credit Corporation, life insurance companies, and individuals. All of these lenders also offer career opportunities.

There are many career opportunities in agribusiness credit or financial services for people with the needed qualifications. Experience, education, a knowledge of the economy in the agricultural industry both locally and nationally, and the ability to understand and interpret the policies of agriculture are valuable assets for a professional in the field of agribusiness finance.

There are many other jobs in the private agribusiness input sector. These include auctioneer, farm tax accountant, blacksmith, land appraiser, agricultural engineer, farm record service, livestock buyer, agricultural scientist, rural sociologist, and agricultural lawyer.

End of Chapter Activities

REVIEW QUESTIONS

1. Define the Terms to Know.
2. Why have private agribusiness services become more important in recent years?
3. What are the names of the five major general farm organizations?
4. What are 10 things in which the National Grange played a major role that served to improve agriculture and help rural America during the past century?
5. What is the Grange doing today regarding issues that affect modern agriculture?
6. What was the major purpose of the National Farmers Union when it was first organized?

7. What are four major efforts and/or services of the American Farm Bureau Federation?
8. What are seven accomplishments of the Farm Bureau that gave support to, or led the battle for, a better life for American farmers?
9. How does the National Farmers Organization attempt to improve farm prices in today's market?
10. Why did agricultural commodity organizations become popular?
11. List 15 services performed by agricultural commodity organizations.
12. Name five of the larger farm commodity groups.
13. Name six of the more prominent agribusiness commodity groups.
14. Name 10 subjects or areas in which agricultural research is conducted.
15. What are five jobs done by crop consultants?
16. What are five jobs done by livestock consultants?
17. What were two major developments that stimulated the widespread use of artificial insemination?
18. What are the three major parts of the artificial insemination industry?
19. Name six places where veterinarians might work.
20. What are eight duties of veterinarians?
21. What are 12 things (areas) that jobs in communications would include?
22. What are four major areas in the job of agricultural communications specialist?
23. What are six things that today's cooperatives are organized to do?
24. What are three major types of cooperatives?
25. What are seven examples of service cooperatives?
26. What four types of insurance does the Farm Bureau sell?
27. List nine other jobs from the private agribusiness input sector not previously mentioned.
28. When and why was the Federal Land Bank started?
29. When and why was the Production Credit Association started?
30. When and why was the Bank for Cooperatives started?
31. Why did the Farm Credit System have to be restructured?
32. List 10 types of loans made by the Farm Credit Service.
33. Briefly discuss the historical beginning of the Farmers Home Administration.
34. What were three purposes of the Farmers Home Administration?
35. Name 20 general-type loans made by the Farm Service Agency.
36. What are the names of three agencies of the nonfarm portion of the old FmHA, which merged in 1994 into a new USDA organization called the Rural Housing Service?
37. Explain the organization of commercial banks.
38. What percentage of agricultural non–real estate loans are made by commercial banks?
39. What percentage of agricultural real estate loans are made by commercial banks?
40. What size of loans for agricultural purchases are made by insurance companies?
41. What are the sources of funds for insurance companies for money to loan to agricultural borrowers?
42. What is the purpose of the Commodity Credit Corporation?
43. What are the limitations of CCC loans?

44. List one advantage and one disadvantage of borrowing from individuals to purchase land.
45. Explain how merchants and farm implement dealers make loans.
46. What experience is needed for individuals wanting to enter a career in agribusiness credit services?
47. Give 12 examples of potential jobs available with agribusiness financial institutions.

FILL IN THE BLANK

1. The NFO estimates that participation of __________ percent of the total production in any commodity would be sufficient to allow farmers to set their own prices.
2. In some cases, the political power of commodity organizations exceeds the power of some ________ ________ organizations.
3. Production agriculturalists have depended on public funds for farm research, as the long-term benefits of research help primarily __________ firms and the ________.
4. Artificial insemination was first developed for practical application in livestock breeding in 1922 by a ________ physiologist.
5. Artificial insemination is used predominantly with dairy cows, but beef and ________ producers also use this service.
6. ________ ________ is often regarded as the "Father of Veterinary Medicine."
7. The first college of veterinary medicine was established in 1879 by ________ ________ University.
8. Currently, there are about ________ veterinarians providing a wide variety of services.
9. A ________ is a business that is financed, owned, and controlled by the people who use it.
10. Approximately ________ separate cooperative organizations with more than _______ million members currently operate in the United States.
11. There are approximately _______ farmer-owned cooperatives providing marketing, purchasing, and related services at a business value of more than _____ billion dollars.
12. Cooperatives market about __________ percent of all agricultural products and provide about _____ ______ percent of all production supplies used on farms each year.
13. Cooperatives market __________ percent of milk, __________ percent of cotton, and __________ percent of fruits, nuts, and vegetables.
14. Today, nearly __________ cooperatives provide electricity to __________ million consumers.
15. For non-real estate loans, commercial banks supply __________ percent of financing/credit, the Farm Credit Service supplies __________ percent, the Farm Service Agency supplies __________ percent, the CCC supplies __________ percent, and individuals and others supply __________ percent.
16. For farm real estate loans, the Farm Credit System supplies __________ percent of financing/credit, life insurance companies __________ percent, commercial banks __________ percent, the Farm Service Agency __________ percent, and individuals __________ percent.

MATCHING

a. American Farm Bureau Federation
b. National Farmers Union
c. American Agriculture Movement
d. National Farmers Organization
e. The National Grange
f. Agricultural Credit Act of 1987
g. Federal Crop Insurance Reform and Department of Agriculture Reorganization Act of 1994
h. $7.8 billion
i. Consolidated Farm Service Agency
j. CoBank

_____ 1. oldest and second largest of the major general farm organizations

_____ 2. first organized in the Southwest in 1902

_____ 3. largest of the five general farm organizations

_____ 4. organized in Washington, D.C.

_____ 5. originally called the Farmers Educational Cooperative

_____ 6. first organized between 1912 and 1914 to sponsor county agricultural agents

_____ 7. has as members approximately 2.5 million families from 50 states and Puerto Rico

_____ 8. organized in Corning, Iowa, in 1955

_____ 9. originally called Patrons of Husbandry

_____ 10. most controversial of general farm organizations, because of holding actions regarding livestock and milk

_____ 11. originally, members were primarily disenchanted NFO and NFU members

_____ 12. primarily young farmers, who expanded the organization until prices dropped dramatically in the mid-1970s.

_____ 13. caused the Federal Land Banks and Production Credit Association to merge

_____ 14. new name following the merger of the Bank of Cooperatives and the Central Bank for Cooperatives

_____ 15. formerly called Farmers Home Administration

_____ 16. amount of CCC loans to production agriculturalists

_____ 17. legislation that terminated the FmHA

ACTIVITIES

1. Determine which general farm organization is the strongest in your community. Prepare a brief report on its mission and the things it has done recently to help improve the lives of production agriculturalists.
2. Determine which agricultural commodity organizations are strongest in your community. Make a list of the potential jobs available with these organizations.
3. Prepare a one- to two-page paper on a particular commodity and report on some research that has been done on the commodity. Share this report with your class. Indicate whether the research was conducted with private or public (government) funds.
4. Name three agricultural consultants in your community. If there are not three, give examples of three agricultural consultants.
5. Some believe that artificial insemination is one of the most important agricultural technological developments in this century. Give reasons why you agree or disagree with this statement.
6. Prepare a list of courses (classes) needed by a high school student who desires to become a veterinarian. Next, find out from your teacher, guidance counselor, or a college catalog what college courses are needed to qualify for veterinary school. Also, make a brief statement on the importance of grades.
7. Make a list of the potential jobs available within the agricultural communications profession in your community.
8. Make a list of the various cooperatives in your community. Identify for each whether it is a purchasing (supply), marketing, or service cooperative.
9. Select a job with an organization that makes loans to those in the agricultural industry. Prepare a one-page paper describing the experience, qualifications, and education needed for this particular job. Share this information with your class.

UNIT 4

The Agribusiness Output (Marketing) Sector

Chapter 17

Basic Principles of Agrimarketing

Objectives

After completing this chapter, the student should be able to:

- Explain agribusiness marketing.
- Discuss how marketing developed.
- Explain the importance of supply and demand.
- Describe the prerequisites of an efficient economic system.
- Examine the factors to consider in a consumer-driven market.
- Discuss farm commodity marketing.
- Discuss the marketing of agribusiness products.
- Describe commodity research and promotion boards.
- Explain how to conduct a market analysis.
- Discuss strategic planning and marketing.

Terms to Know

agrimarketing
allocation of resources
auctioneer
basis
blocked currency
check-off program
commission
cotton gins
demand
efficient economic system
emulsifier
free market economy
futures contract
futures options
hedging
institutional advertising
market analysis
marketing
marketing cooperatives
merchandising
niche
premium
prerequisite
pricing efficiency
product advertising
profitable
selling
spot price
supply
transformation
trichinae-safe pork
value adding
vertical integration
yardage fee

Introduction

Little is more important than marketing in the agricultural industry. In reality, production agriculturalists have become so efficient and effective in raising an abundance of food and fiber that many have produced themselves into bankruptcy. Production of food and fiber is not the problem; rather, it is marketing agricultural commodities to obtain a fair price that poses the challenge.

Agricultural marketing, or **agrimarketing**, is a large and important discipline in the agribusiness output sector. The cost of marketing U.S. domestic food amounted to $450 billion, while the cost of producing farm food products is $125 billion. Of the $544 billion spent by consumers for food products, less than one-fourth was returned to producers; more than three-fourths of that amount went for marketing costs.[1]

Marketing plays an integral role in all parts of the economy. Indeed, without effective marketing to drive an economy, a country's development will be greatly hindered. In agriculture, the role of marketing is to meet the consumer's needs while also ensuring that producers earn a sufficient profit. A well-designed marketing campaign is a key component in maximizing the profit of an agribusiness.[2]

The largest industry in the United States today is agribusiness. The largest sector within agribusiness is marketing. Consider these startling facts that indicate how important agribusiness marketing is to our country's economy:

- Four out of every five people employed in agribusiness work in the marketing field.
- Three-fourths of every dollar you spend on food goes toward the cost of marketing.
- Agribusiness marketing revenues account for more than 17 percent of the nation's annual gross domestic product.[3]

Marketing is one of the four major areas in which a production agriculturalist faces critical decisions, along with production, labor, and capital. Marketing shares a rank equal to these other functions. Managers make decisions on production, labor, and capital to keep the business running smoothly and efficiently. Refer to Figure 17–1. However, these decisions may not always be enough to keep the business **profitable** in a changing economic environment.[4] Without adequate markets offering reasonable prices, profit will not be attained.

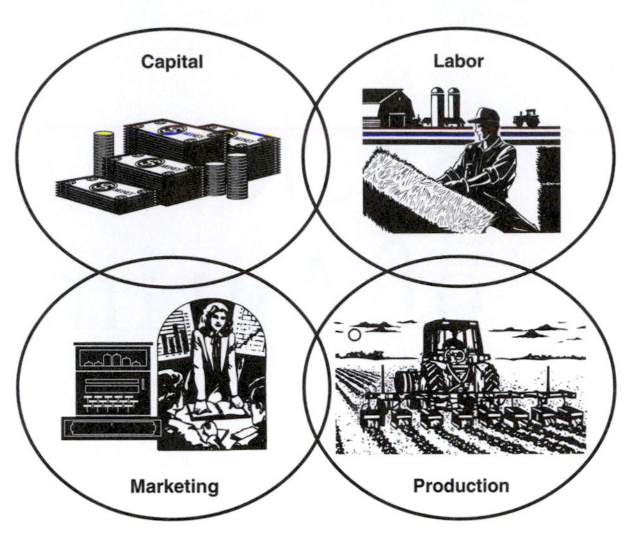

Figure 17–1 Marketing, capital, labor, and production are the four areas in which a production agriculturist faces major decisions. Which do you believe is the most important?

One factor that influences profits in a direct way is selling price. *Profit* is the difference between the selling price of a product and the total cost to produce that product. Therefore, the higher the selling price, the greater the profits.

What Is Agrimarketing?

Agrimarketing is the sum of the processes, functions, and services performed in connection with food and fiber from the farms on which they are produced until their delivery into the hands of the consumer. Agrimarketing includes many operations, such as buying, assembling, storing, packing, warehousing, communicating, advertising, financing, transporting, grading, sorting, processing, packaging, selling, **merchandising**, insuring, standardizing, regulating, inspecting, and gathering market information, among others.[5] Refer to Figure 17–2.

1. U.S. Department of Agriculture, *Agriculture Fact Book 1997* (Washington, DC: U.S. Government Printing Office, 1997); available at http://www.usda.gov/news/pubs/fbook97.
2. *Agribusiness Management and Marketing* (8720-B) (College Station, TX: Instructional Materials Service, 1988), p. 1.
3. USDA, *Agriculture Fact Book 1997.*
4. *Agribusiness Management and Marketing,* p. 1.
5. Ewell P. Roy, *Exploring Agribusiness* (Danville, IL: Interstate Printers & Publishers, 1980), p. 57.

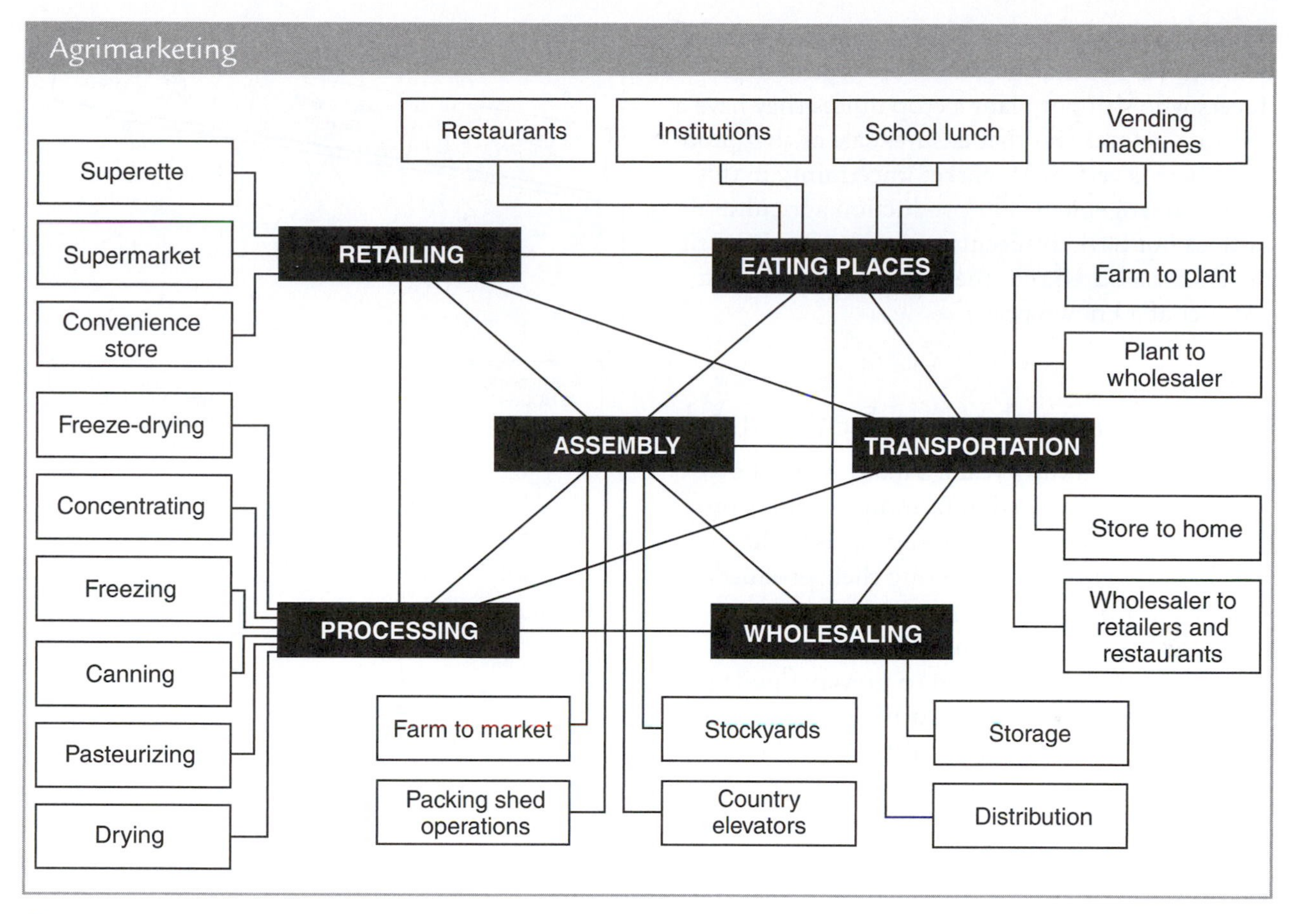

Figure 17–2 Agrimarketing is the name for the processes, functions, and services performed in connection with the movement of food and fiber from the farm on which they are produced until their delivery into the hands of the consumer.

The Difference between Selling and Marketing

Selling. Historically, production agriculturalists have taken any products not used on the farm and sold them locally for whatever available buyers would pay. If the price seemed too low, the production agriculturalists could try to bargain for a higher price, but buyers could usually get all they wanted at the posted price. Unable to bargain for higher prices, production agriculturalists' only choice was to take the wheat, hogs, eggs, or cream back to the farm. Usually they did not; they took the price offered.

This is not marketing; this is **selling**. When the businesses of individual production agriculturalists were small and the amount of excess commodities they wanted to sell was also small, the difference between selling and marketing did not matter very much. In reality, marketing was not an option in the early days. The production agriculturalists might feel cheated and go away disgusted, but there was always the hope of getting higher prices next year.[6]

Marketing. Many production agriculturalists now market their products instead of merely selling them. Good marketing begins by determining the demands of a given industry. There may be demand for a new kind of crop, a different variety of wheat, or leaner hogs. When production agriculturalists consider the market and are willing to produce a different crop or livestock animal than they did before, they maximize their potential for revenue and profit.[7]

Marketing contracts are becoming more popular in production agriculture, especially when

6. James G. Beierlein and Michael W. Woolverton, *Agribusiness Marketing: The Management Perspective* (Englewood Cliffs, NJ: Prentice-Hall, 1991), pp. 118–119.
7. Ibid., p. 119.

product quality or a specific food item must be guaranteed. For example, there are vegetable producers who will not plant a crop unless they have a contract beforehand that ensures sale of that product. There is very little market uncertainty in this type of arrangement. The production agriculturalist does not have to speculate about price or profit. Buyers are assured that they will have the desired product at a known price.[8]

How Agrimarketing Developed

In our country's early years, a majority of the workforce was involved in farming. Farmers produced foods and fiber for domestic use. They were also involved in processing their products for consumption: grain was milled into flour; fiber was woven into cloth; hides were turned into leather; meat was cured to prevent spoilage, and milk was processed for future use. Over the years, production agriculturalists learned that their land and their climate were most suitable for producing certain commodities and less successful for others.

As a result of the agricultural and industrial revolutions, a food-marketing industry developed in agriculture. Fewer people were growing their own food, so farmers were called upon to provide food for a greater number of people. Nonfarmers also wanted to buy food products in different ways than farmers were used to selling them. Marketing firms sprang up in an attempt to meet the new demands of the public. Today the largest sector of agriculture is dedicated to employing and funding marketing strategies.[9]

Supply and Demand

It is a basic fact of life: anything you will ever want to buy or sell has a price. That price is determined by how much of a product people are willing and able to sell and how much other people are willing and able to buy—in other words, supply and demand.

The potential profit from a product is determined by two critical factors: supply and demand.

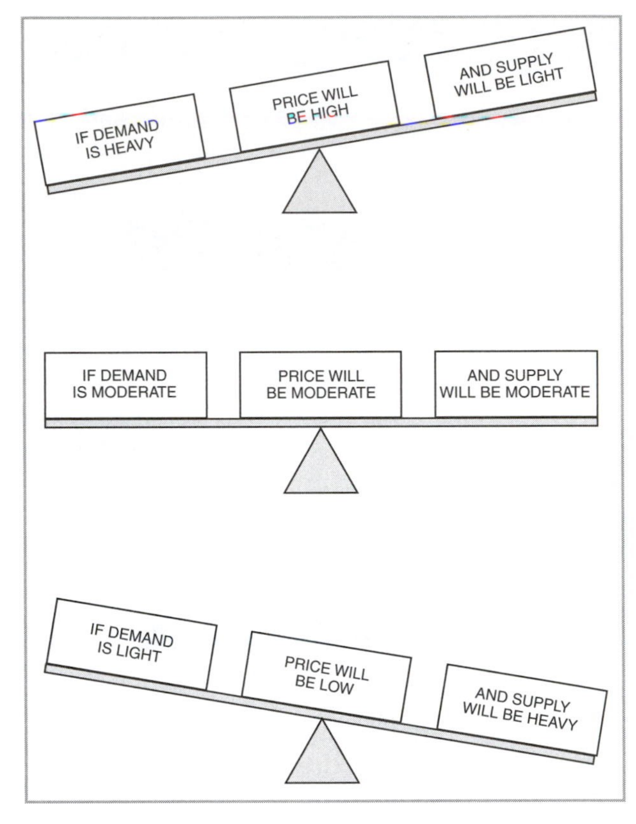

Figure 17–3 In a free market system, the relationship between supply and demand will determine price.

Supply is the quantity of a given product that is available at a specific time and price. The supply of a product is determined by several factors: the number of producers of the product there are in an area, the amount of the product that comes in from outside the area, and the historic profitability of the product within the area.

In contrast, **demand** is the quantity of a product that is needed at a specific time and price. It, too, is determined by several factors. Price is a primary determinant. Generally, the less expensive a product is, the more demand there will be for it. Competition from products of a similar nature tends to reduce demand. Some products are more in demand during certain seasons of the year. Finally, how much money consumers have to purchase the product strongly influences demand.[10] Refer to Figure 17–3.

8. Ibid., pp. 119–120.
9. Elmer L. Cooper, *Agriscience: Fundamentals and Applications,* 2nd ed. (Albany, NY: Delmar Publishers, 1997), pp. 653–654.

10. Ibid.

Prerequisites of an Efficient Economic System

When every **prerequisite** is met in the general economic system, the marketing system can lead, not only to efficient satisfaction of producer and consumer needs, but also to efficient allocation of limited societal resources. Three of the prerequisites are discussed in this section.

A Free Market Economy

A **free market economy** is a prerequisite. Consumers dictate what is produced, how much is produced, when it is produced, who produces it, and for whom it is produced. How well producers predict the consumer's dictates will determine their profitability and thus their success in the market.

Prices That Reflect the Full Value of Market Resources

Pricing efficiency is a second prerequisite. Pricing efficiency is determined by how closely an economic system comes to getting the full value for its resources. When the economy is achieving high degrees of pricing efficiency, its resources are being maximized. High pricing efficiency benefits both producers and consumers, in that:

- *Producers* use the most efficient technology available and the lowest-cost combination of inputs to make their products.
- *Consumers* purchase only products that maximize their total satisfaction.[11]

More Interaction between Consumers and Production Agriculturalists

When consumers and production agriculturalists interact directly, both parties gain knowledge of the variety and prices of goods available and therefore can make informed choices. The satisfaction of the following three prerequisites is desirable because it will lead to the development of an **efficient economic system** in which:

- consumers decide on the who, what, and where of the economic system
- a pricing system brings about efficient **allocation of resources**

Conflicting Needs of Producers and Consumers That Agrimarketing Seeks to Reconcile

Producers try to:	Consumers try to:
■ Maximize profit over the long term	■ Maximize the satisfaction they receive from the products they buy with their limited incomes
■ Sell large quantities of fewer products	■ Buy small quantities of many products
■ Obtain the highest prices	■ Obtain the lowest prices

Figure 17–4 Producers and consumers must work together to balance the needs in agrimarketing.

- consumer satisfaction is maximized
- producers' long-run profits are maximized[12]

Refer to Figure 17–4.

Factors to Consider in a Consumer-Driven Market

Consumers are ultimately the determinant of agrimarketing success. Commodities or services may be available to consumers, but if the consumer chooses not to buy and does not submit to persuasion, there will be no sale. In a democratic system, we have the right to produce and market goods and services for personal profit (within state and federal legal guidelines). Similarly, consumers have the basic right to refuse to buy goods and services. The agrimarketing system encompasses the interactions among producers and consumers, and also among all the people, processes, and jobs in between.[13]

Consumers are all those who use goods and services, including individuals, households, and businesses. Agrimarketing involves understanding what consumers need, how much they need, and what their preferences are. Agrimarketing also

11. Beierlein and Woolverton, *Agribusiness Marketing,* pp. 26–28.

12. Ibid., p. 28.

13. Cooper, *Agriscience,* p. 654.

provides a connection between the consumer and the producer. It helps the producer determine marketing strategies that will satisfy the most customers. The better producers can get to know the consumer, the better chance they have of offering superior and successful products.

Agrimarketing is the link between producing and consuming. It is crucial to understand the many kinds of consumers and how their choices affect the marketing process. The variety of consumers is why products must be marketed in a variety of ways.[14]

Meeting Modern Demands

Today, modern food stores and delicatessens and a huge array of restaurants (ranging from fast-food outlets to elite eateries) demonstrate the complicated, dynamic nature of agrimarketing. Markets now offer a wide choice of products, various systems of distribution, and many associated services, such as precooked meats and microwaveable meals. Much of the present market diversity results from food processors', manufacturers', and retailers' awareness that the market is consumer-driven. Consumers vote every day in the marketplace with their dollars, and the market hears what consumers are saying.

Lifestyles have been changing rapidly. Today, in many households both spouses work. With both men and women working outside the home, fewer people have time to prepare food. This has increased the demand for highly prepared foods and for restaurant meals. Product requirements of the prepared food and food service markets are different from the requirements of retail grocery stores. The high proportions of food sold through food service outlets or in highly prepared form are creating demands for new food characteristics.[15]

Predicting Consumer Demands Is Big Business

Finding out what consumers want and how they feel about various product characteristics has become big business. Management practices now involve **market analysis**, studying changes in consumer lifestyles and preferences and adjusting businesses to capitalize on those changes. The food-marketing industry spends billions of dollars each year on market research, new product development, and advertising. For example, the cost of advertising for 30 seconds in a Super Bowl is now more than $1.5 million.

Modern marketing involves product development, pricing strategies, efficient distribution, and product promotion. Marketing strategies integrate these components into an overall plan. Firms develop new products with desirable characteristics that were discovered through consumer research.

Once the characteristics and specifics of consumer demand have been evaluated, firms must develop the desired products and find appropriate marketing techniques. Successful firms have strategies for accomplishing these tasks. Despite huge investments, most new products fail in the marketplace within the first year after introduction. More than 10,000 new products or product variations are put on grocery store shelves yearly.[16]

The Consumer Population

The earth's current population is estimated at about 6.6 billion people. North America, including the United States, Canada, and Mexico, has about 439 million people. Within the United States, there are approximately 298.5 million people. The world's population increases at the rate of 2.8 people every second. This translates to a 10,080-person increase in population every hour. At the current rate of population growth, the world's population will double every 30 years. By the year 2050, there will be more than 9.3 billion people on earth.[17] They will need food and fiber. Agricultural production and marketing must keep pace if the needs of these people are to be met.

Farm Commodity Marketing

When agrimarketing is discussed, it is considered in two phases. The first phase is the marketing of commodities directly from the farm. After the

14. Jasper S. Lee, James G. Leising, and David E. Lawver, *Agrimarketing Technology: Selling and Distribution in the Agricultural Industry* (Danville, IL: Interstate Publishers, 1994), pp. 271–272.
15. Ewen M. Wilson, "Marketing Challenges in a Dynamic World," in *Marketing U.S. Agriculture: 1988 Yearbook of Agriculture,* ed. Deborah Takiff Smith (Washington, DC: U.S. Government Printing Office, 1988), pp. 2–4.
16. Ibid., p. 4.
17. http://www.infoplease.com/ipa/A0873845.html (accessed August 27, 2007).

product goes through the processes needed to get it ready for the consumer (**transformation**), it goes through another market phase, the agribusiness product-marketing phase. The phases are different yet interrelated. An agricultural product produced on the farm is sold, and marketed and consumed by people the way they like it. For example, both beef and swine now are leaner due to consumer demand.

Production agriculturalists are not always in a position to set the price they will receive for their products. Instead, they are accustomed to accepting the price that buyers of their products will pay, because production agriculturalists have traditionally had to take the current market price for each product. They have been "price takers" instead of "price makers." Under this approach, there is no way they can be assured that production of any particular agricultural product will be profitable. To counteract this market operation, production agriculturalists must regard marketing as an essential tool for determining the profitability of their commodities.[18]

In the agricultural industry, marketing is as important as the actual production of commodities. Thus, the time and effort devoted to production of animals and crops should be matched by the time and effort devoting to marketing them. Making good marketing decisions can be the difference between success and failure for an agribusiness.

Marketing Strategies for Farm Commodities

If the production agriculturalist is to make the most money from a farm commodity, successful marketing is essential. The following lists some strategies that can be used to market farm commodities more profitably:

- determine what types of markets are available to you
- determine the cost of various types of marketing
- determine transportation costs to each of the markets available to you, and sell where transportation costs are lowest
- determine the most profitable form in which to market your product (age, size, weight, degree of preparation)
- advertise to create markets where none existed before
- market seasonal products at the peak of demand
- shorten the marketing channels between you and your consumers[19]

Types of Livestock and Dairy Markets

Various types of agricultural markets are available to production agriculturalists. The one or more markets chosen by an individual producer are often a matter of what is available and what the producer prefers. Invest some time to carefully select the type of marketing you use. Intelligent marketing practices may well make the difference between profit and loss in a very competitive business.

Terminal Markets. Terminal markets hold and care for animals, often in stockyards, until they can be sold. The terminal market does not assume ownership of the animal, but instead markets it though a selling agent. The agent receives a fee, or **commission**, for selling the animal. The terminal market also receives a fee, called a **yardage fee**, for the care of the animal. These markets have been more popular in midwestern and western states than in other parts of the country. Today, terminal markets are used less often than in the past because owners are choosing alternate methods of selling their livestock.[20]

Auction Markets. In an auction market, members of the public bid on animals to purchase them. The highest bidder purchases the animal. The person who conducts the sales at this type of market is called an **auctioneer**.

Auction markets are now the most common and widespread method of marketing farm animals. These markets are popular across the country and can be found in most localities. They are most practical for small livestock producers.

The auction market receives a commission for the animals sold there. The commission varies depending on the type and size of the animal. Competition for animals at auction markets tends to be less than at a terminal market, and auctions are typically smaller.[21] In reality, animals are not marketed at an auction market—they are simply sold.

18. Cooper, *Agriscience,* pp. 657–658.

19. Ibid., p. 661.
20. Ibid., p. 661.
21. Ibid., p. 662.

Direct Sales. In direct sales, animal producers sell their livestock (or crops) directly to a processor. This type of marketing is especially popular in the beef cattle industry. It is beneficial because buyers do not charge commissions or yardage fees. Normally, the buyer comes to the farm to make purchases, so there are no transportation fees. Animals do not suffer the stresses and wear associated with being transported and penned with other animals.[22]

Cooperatives. **Marketing cooperatives** are formed by groups of producers who band together to market a particular commodity. Cooperatives are advantageous in several ways. They may process products themselves and sell the products directly to consumers. They may also choose to sell to processing plants. Some have the ability to transport product, set product quality standards, provide advertising, and even balance supply and demand. Marketing cooperatives of this type are utilized to market nearly 75 percent of the milk in this country. Refer to Figure 17-5.

Cooperatives are also used to market selected crops, although not to the same extent as with milk. An example is **cotton gins**. Some producers have formed ginning cooperatives to process their cotton at cost.

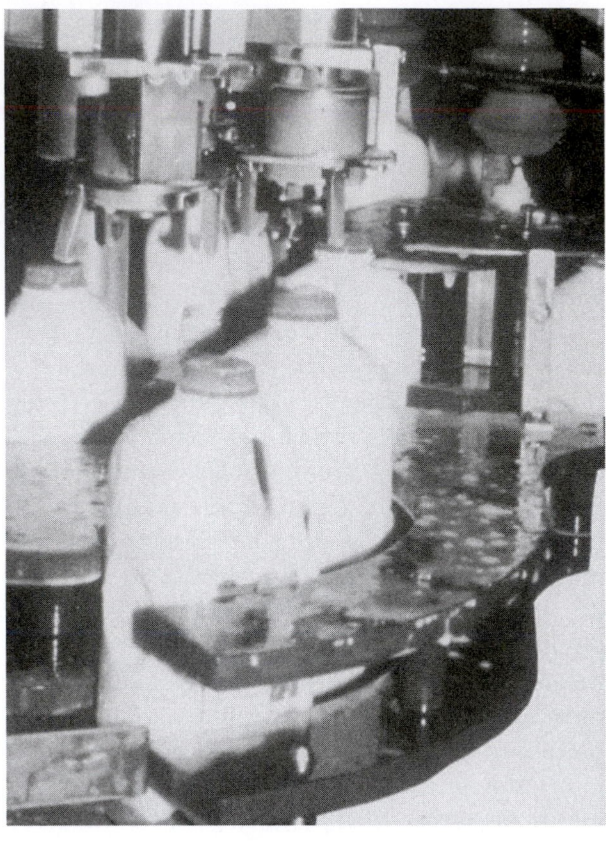

Figure 17-5 Approximately 75 percent of the milk marketed in the United States is sold through farmers' milk-marketing cooperatives. (Courtesy of USDA)

Government. The government purchase program also buys significant quantities of the milk produced in the United States. This milk goes either into the school lunch program or the program under Public Law No. 480. Public Law No. 480, also known as the Agricultural Trade Development and Assistance Act, was passed in 1954. This act allowed surpluses to be sold to developing nations for **blocked currency**. This means that money can be exchanged only within the nation making the purchase, to boost the economies of underdeveloped nations. The program also allowed donation of surpluses to nations that have experienced disasters.[23]

Vertical Integration. In the poultry market, almost 99 percent of the chickens produced for meat are grown using a system called **vertical integration** (in which several steps in the production, marketing, and processing of animals are combined). Refer to Figure 17-6. The use of vertical integration in the production and marketing of farm products allows for extremely large systems of production that can be very efficient. Only the number of animals that is anticipated to be needed by consumers is produced. There is less competition from other producers, and all phases of production can be controlled. Tyson Foods' poultry production system is an example of vertical integration. The swine industry is also fast adopting the vertical integration marketing system.

Futures Markets. **Hedging** is done by many livestock producers and feedlot operators. The Chicago Mercantile Exchange is the major exchange for livestock futures contracts. Refer to Chapter 18 for a comprehensive discussion of the **futures contract** and commodity (futures) marketing.

22. *Agribusiness Management and Marketing,* p. 6.
23. Ibid., pp. 6-7.

Figure 17–6 With vertically integrated production, marketing, processing, and even inspection are joined. USDA inspectors work onsite in the processing plants. (Courtesy of USDA)

Types of Grain Markets

Several types of markets are available to grain producers. These include futures contracts, forward contracting, harvest or spot pricing, and postharvest pricing. Futures contracts, which were mentioned previously as a type of marketing for livestock, are not addressed in this section because all of Chapter 18 is devoted to commodity (futures) marketing.

Forward Contracting. Cash-forward contracting is another way in which producers attempt to establish preharvest prices. As with futures contracts, a producer seeks to establish a fixed price to avoid a possible price decline. At the same time, however, "elevator" or forward contracting also fixes the amount of profit the producer can make.

The main purposes of forward contracting are to reduce the risk of price declines and to take advantage of pricing opportunities. Sale of a producer's harvest is guaranteed, and buyers are assured that they will have product. This is often attractive to lenders who are determining the creditworthiness of a firm or producer. In addition to setting a price, a forward contract specifies the quantity, quality, grading methods, and delivery times of the product. It may also outline the terms for settling disputes and for accounting for shrinkage and quality variations.

Forward contracting is not without risk. It is true that price risk is reduced with such contracts, but there is still the risk of product losses due to inclement weather, droughts, fires, and other types of crop damage. There are also legal requirements for forward contracting that must be met. You should consider all these things before entering into a forward contract.

Some advantages that forward contracts have over futures contracts are:

- buyer and seller of forward contracts can negotiate quantity, whereas futures contracts trade in units of 1,000 and 5,000
- forward contracts are not subject to brokerage fees and commissions; futures contracts are
- forward contracts do not require as much knowledge of the market as trading in futures contracts[24]

A drawback to using a forward contract is that the seller cannot withdraw from the contract. In contrast, if a producer hedges with a futures contract, he or she may withdraw at any time by lifting the hedge.

Harvest Pricing. Harvest pricing is the traditional way that producers set prices. They harvest the grain, ship it to the local elevator, and sell it for the going market price or **spot price**. This method has lost much of its popularity because price declines have occurred fairly frequently during grain harvest times in recent years.

Producers may decide to price grain during harvest time but wait to deliver and receive payment until later. They may do this through forward contracting or futures contracting. This method is best used when there is a wide **basis**. The advantages are that storage returns may be improved by holding the grain. There may also be tax benefits to deferring the income earned until later. One disadvantage is that the producer may lack the cash required to meet immediate needs.[25]

Postharvest Pricing. Another frequently used method of pricing grain and other crops is postharvest pricing. This strategy involves determining a price anytime from immediately following a harvest to 10 or more months later. A key component in effectively use of postharvest pricing is the producer's ability to store the grain for the desired waiting period.

24. Ibid.
25. Ibid.

Producers must realize that prices may rise or decline while the crop is stored. They may need to protect themselves against the risk they are incurring. This can be accomplished by purchasing **futures options**, a type of insurance that protects producers from price declines. A fee called a **premium** is paid to the insurer for a future option. Pricing risk may also be reduced through the use of forward contracts and futures markets.

Types of Fruit and Vegetable Markets

You must consider the product being produced, availability of markets, labor availability, and personal desires before you can decide which type of market to use. Roadside markets, farmer's markets, pick-your-own operations, local brokers, wholesaling firms, and direct sales to processors are all viable options.

Roadside Markets. Some producers choose to grow fruits and vegetables and market them only at roadside markets. Others purchase fruits and vegetables from fellow producers and sell them there as well. Although roadside markets are seasonal in many parts of the country, some have become major, year-round businesses. Consumers often believe that the "home-grown" products found at roadside markets are fresher. They may even be willing to pay a premium price for this perceived quality.[26]

Farmer's Markets. Farmer's markets have appeared in many areas, from small towns to large metropolitan areas, to cater to the demands of urban consumers. They give urban and suburban consumers access to fresh products directly from the producers. Farmer's markets give the producer access to markets that would seldom be available otherwise. The producer also has the opportunity to educate the consumer concerning the value of good farm products. Refer to Figure 17-7.

As with any business, there are costs involved with selling product at farmer's markets. Vehicles with cooling or heating mechanisms may be necessary to transport the product to market. Market organizers may impose participation charges, and costs are incurred in obtaining the appropriate shelving or displays for products. Often many vendors are selling the same product, so competition may be high. Finally, state and federal regulations may apply to the marketing of products in such places.[27]

Figure 17-7 An advantage of farmer's markets is that the producer has the opportunity to educate the consumer concerning the value of good farm products. (Courtesy of USDA)

Pick Your Own. "Pick your own" is a simple concept. The producer grows a large amount of a desirable garden crop, then allows customers to pick their own. Among the wide variety of fruits and vegetables that can be found at pick-your-own operations are strawberries, raspberries, peas, green beans, sweet corn, tomatoes, pecans, apples, and peaches.

People enjoy pick-your-own operations for several reasons. People like to eat fruits and vegetables that are as fresh as possible. Other people like to preserve fresh fruits and vegetables for later consumption. Also, some view picking or gathering their own food as a good family activity; some entertainment value is offered through pick-your-own operations.[28]

26. Lee, Leising, and Lawver, *Agrimarketing Technology*, pp. 189–190.

27. Cooper, *Agriscience*, p. 660.

28. Lee, Leising, and Lawver, *Agrimarketing Technology*, p. 189.

Food Brokers. Food brokers buy and sell fruits and vegetables and have them delivered to retail stores, restaurants, and institutions such as hospitals and public schools. The main characteristic of food brokers is that they may never actually receive or handle the product.

Wholesalers. Wholesalers buy a product and hold it at a warehouse. It is then graded, boxed, or packaged, and shipped to retail stores, restaurants, and institutions.

Processors. Processors often buy products directly from the producers, as discussed earlier with regard to livestock. The product goes directly to a processing plant, where the products are frozen, canned, or processed into some other form for the consumer.

Specialty Markets

Many types of markets are aimed at a particular **niche**. These markets work for some producers in unique situations. Although there are several types, we discuss only five here: fee fishing, organically grown crops, hormone-free beef, on-farm restaurants, and wineries.

Fee Fishing. Fish farms may offer "fee fishing," an arrangement in which consumers catch their own fish, for a price. Two types of fish commonly marketed this way are catfish and rainbow trout. Fish farms use different methods to determine their fees. Some charge a per-pound fee for any fish caught; others charge a base fee to fish plus a per-pound fee. Some include the cleaning of fish in the per-pound fee, whereas others charge extra for this service.[29]

Organically Grown Crops. An increasing number of people in this country prefer to eat only products that have been grown organically. Typically, they believe that foods grown without the aid of chemicals are more healthful and wholesome than those on which chemicals were used.[30]

Hormone-Free Beef. Another specialty is beef that is grown without the use of growth stimulants and other medications. This type of beef is referred to as *hormone-free beef.* Some consumers are willing to pay a higher price for this product because it is perceived to be healthier.[31] Much concern has been raised about the use of hormones to stimulate growth in beef animals. Some producers of hormone-free beef have established a market by selling beef directly from the farm. There is also a market for beef that is low in fat and cholesterol.

On-Farm Restaurants. Opening an on-farm restaurant is another strategy for shortening the gap between producer and consumer. By doing so, a producer of fish, for example, gains the advantage of removing several layers of intermediaries. The producer also has the advantage of marketing a value-added product. Patrons of the restaurant enjoy dining on fresh fish that has never been frozen.

Wineries. Wineries are another effective means of marketing a value-added product. Most wineries grow their own wine grapes, although some buy additional grapes from other producers. Grapes, in and of themselves, are not a particularly valuable crop, but when turned into wine the value soars.[32] To increase profits, wineries may also sell nonalcoholic juices. Products from wineries may be sold in onsite stores as well as in other markets.

Marketing Agribusiness Products

Once farm commodities are marketed, most of those commodities are processed and marketed again by agribusinesses. Agrimarketing firms and commodity groups spend more than $10 billion annually to advertise and promote food.[33] Five key aspects of marketing agribusiness products are advertising, price promotion, merchandising promotions, public relations activities, and coordination. Refer to the Career Options section for a further explanation of careers as salespersons and agrimarketing specialists, just two of the many careers in agrimarketing.

Advertising

Advertising distributes messages about a product by way of many different media, including television,

29. Ibid., p. 189.
30. Ibid., p. 191.
31. Ibid., p. 191.
32. Ibid., p. 192.
33. Wilson, "Marketing Challenges in a Dynamic World," p. 5.

Career Option

Salesperson/Agrimarketing Specialist

Agrimarketing firms and commodity groups spend more than $10 billion annually to advertise, market, and promote food. Effective salespersons and marketing specialists perform a variety of functions. These include creating an awareness of the product, motivating potential consumers, and reinforcing the value of purchases already made.

The educational level needed by a salesperson or marketing specialist depends on the company or the product with which the person works. Although educational level is important, equally important are the ability to communicate, a positive attitude, knowledge of the product, belief in the product, and excellent human relations skills. Also, salespeople must project a positive image of the company they are representing.

At this international food exposition in New Orleans, Louisiana, agrimarketing specialist Diana Tucker meets with industry representatives who have a potential interest in manufacturing products or developing technologies identified by the Agricultural Research Service. (Courtesy of USDA)

radio, billboards, newspapers, and magazines. The intent is to shape consumer perceptions and attitudes about the product or commodity and to establish a long-term market base.

Many products and services fail in the market, not because of their quality, packaging, or pricing, but because potential customers do not know they are available—or, if they are aware of the product, they do not know what it is or how to use it. If you are going to sell your product or service, you *must* promote it.[34]

What Is Advertising and What Will It Accomplish? Advertising is a way to send a sales message to possible consumers of a product. It is a critical part of any well-designed marketing plan. Advertising, at its best, will help keep existing customers, draw in new customers, and establish a recognizable presence in the marketplace for the business. However, it is not a cure-all. It cannot make a businessperson better or automatically fix a business that is not growing or sees few profits. Good (that is, effective) advertising is based on an analysis of the market that identifies and defines a specific target audience; such an analysis should be completed before any money is spent. You should view advertising as an investment in the future of your business rather than as just one more expense.[35]

Setting Advertising Goals. Advertising should be carefully planned and specifically used to accomplish certain goals. The business owner must decide what those goals are. However, for any business, a primary, overall goal is to get the right message to the right group of consumers at the right time. Another goal might be to establish an image of the business in the public eye. High-quality products, excellent customer service, and friendly, knowledgeable staff are all desirable images to convey. Yet another goal may be to make customers aware of the business's location, products, services, and prices. Advertising may make people walking by the business want to stop in. The possible goals are endless, and no single strategy will accomplish every goal. The idea is to find which method of advertising will work best to accomplish each goal.[36]

34. Gregory R. Passewitz and Nancy H. Bull, "Advertising: An Investment in Your Business's Future" (Ohio State University Fact Sheet CDFS-1276-95); available at http://ohioline.osu.edu/cd-fact/1276.html.

35. Ibid.

36. Ibid.

Advertising Budgets. The best way to budget for advertising is to spend no more than you need to accomplish what you want to get done. This requires a thorough analysis of the market and the types of advertising available. The budget should reflect yearly business goals and the cost to achieve each of the goals.

Many business owners make the mistake of spending more when sales are increasing and less when sales are declining. If advertising stimulates sales, though, it makes more sense to do just the reverse. Some agribusiness owners would say, "It's not what it costs, but what it retains that matters." As a rule of thumb, 10 percent of sales is considered fairly aggressive for an advertising budget. Of course, this figure depends on the nature of your agribusiness.

Types of Advertising. Market analysis provides a basis on which to choose what kinds of advertising you will do. When you know the types of customers you want to attract, you can choose from among various advertising media and types, selecting the ones that will most accurately and reliably reach your target audience.

Product advertising focuses on the product itself. Such advertising might stress the usefulness, durability, price, or value of a product, and/or the customer's need for the item. In contrast, **institutional advertising** is designed to create a favorable image of the firm or institution offering the products or services.

Price Promotions

Price promotions are used to stimulate quick sales. With "dealer price incentives," for example, retailers and wholesalers are allowed to purchase products from the manufacturers for limited periods of time at discounts. Retailers may pass all, part, or none of the incentive discounts on to consumers. Price incentives are often aimed directly at consumers through cents-off or percent-off coupons and rebates.

Merchandising Promotions

Merchandising promotions focus on the retail point of sale, such as a food store, restaurant, or fast-food outlet. Merchandising techniques include endcap displays, additional shelf facings, prominent signage, and in-store demonstrations. These are used to introduce new products and stimulate consumer responses in the form of immediate sales.

Public Relations Activities

Public relations activities draw attention to a product or manufacturer through support for local or national civic projects and sports events, public television and radio, educational programs and materials, and public-interest news releases. The goal is to stimulate sales by accomplishing one or more of three objectives:

- present the company in a favorable light
- present the product in a favorable light
- foster activities that call attention to the product[37]

Coordination

Another means of advertising effectively is to coordinate advertising with other marketing strategies, such as promotional activities, special pricing, public relations, unique selling techniques, merchandising plans, and product recognition efforts. Advertising should function to educate customers, gain brand recognition, prompt consumers to look for a specific product, and make consumers who have purchased the product glad they did.[38]

The Four Ps of Agribusiness Marketing and the Marketing Mix

As agribusiness marketing managers plan for success, they should include four controllable variables in the agrimarketing mix. These variables must be properly combined to meet the needs and preferences of the intended market. The agrimarketing mix is composed of four items, each beginning with the letter "P": *p*roduct, *p*rice, *p*lace, and *p*romotion.

- *Product*—the firm must develop the right product to give maximum satisfaction to members of the target market
- *Price*—the right product must carry the right price in light of market conditions
- *Place*—the right product, at the right price, must be in the right place to be purchased by members of the target market

37. Wilson, "Marketing Challenges in a Dynamic World," pp. 4–5.
38. Cooper, *Agriscience,* p. 655.

- *Promotion*—members of the target market must be told in the right way that the right product, at the right price, is available at the right location[39]

Agribusinesses and agrimarketing managers strive to develop the proper combination of product, price, place, and promotion that will adequately meet the needs of consumers. How well you manipulate the mix of these four Ps will ultimately determine the success of your agribusiness.[40]

Functions of Marketing

People are often unpleasantly surprised to learn that they will have to spend 80 percent of their time marketing their new business. That is the reality, however, and it is so because marketing encompasses many different parts of a business. Many people equate marketing with advertising. All advertising is a form of marketing, but marketing includes much more than just advertising.

As the owner of a small agribusiness, you perform many functions. You are the head of marketing and sales, the human resources director, the bookkeeper, and the director of production. You are probably even the janitor! Though it may not be obvious, all these tasks are part of marketing.[41]

The primary function of marketing is to attract people to your business so that they will buy a product or service for more money than it costs you to make that product or render that service. Many businesses fail to recognize this basic function; they think they are through with marketing after they put an advertisement in the newspaper or another local publication. This is actually just a first step in the right direction in an overall marketing plan.

The real marketing begins when the customers appear. Customer referrals, customer service, repeat business, and customer satisfaction are all marketing functions.[42] In all, there are eight main functions of marketing: buying, selling, transporting, storing, providing correct quantities and quality, financing, risk taking, and gathering marketing information.

Buying and Selling. These two functions are intimately related. With regard to selling, it takes more than just a good product for a business to succeed. A product may be superior, but also be obsolete; there is no consumer need and thus no demand for such a product. You must provide (sell) a product that fills a present and important need.

With regard to buying, you must purchase the best products available at the best possible price. You need to buy or make inputs and products that will fill the needs of your target market, and that you can sell at a profit.[43]

Transporting and Storing. The third function is transportation. If you purchase products for sale or resale, you must move them by some means so that they get to your customers. Even if you produce the product, you still have to make it easily available to consumers. As for storing, the fourth function, a seller may hold the product in inventory for you, or you may have to take delivery and incur storage/inventory costs yourself. This is inescapable, whether you buy the product or make it yourself.

Both transportation and storage can influence your marketing and advertising plan. If your product is stored elsewhere, shipping time may be an issue. If you store the product yourself, your delivery time will be quicker, but your inventory and handling costs may be higher.[44]

Providing the Right Quality and Quantity. Your inventory will include a certain amount of product and a certain quality of product. Your marketing strategy will influence how much attention you pay to quantity available and quality (or qualities) offered, and the decisions you make regarding each factor.[45]

39. William D. Perreault, Jr., and E. Jerome McCarthy, *Basic Marketing: A Global-Managerial Approach,* 12th ed. (Chicago: Irwin, 1996), pp. 50–53.
40. Beierlein and Woolverton, *Agribusiness Marketing,* p. 58.
41. Tom Egelhoff, "Small Business Marketing: How to Use the Eight Basic Marketing Functions"; available at http://smalltownmarketing.com/eightbasic.html.
42. Ibid.
43. Ibid.
44. Ibid.
45. Ibid.

Financing. The financial terms you are able to obtain from suppliers will influence your marketing. The financing arrangements may force you to raise prices or allow you to lower prices. Your advertising and marketing budgets will reflect prices that change because of your ability to obtain extended payment terms or volume discounts.[46]

Risk Taking. When you provide any good or service, you have three risks to consider. The first is the risk that consumers do not want or need what you are offering; the second is the risk that consumers want what you are offering but are not willing to buy at your price; the third is the risk that newer or better products or services render yours obsolete.[47]

Gathering Marketing Information. Information is a powerful tool in the marketplace. It is modified and exchanged rapidly, thanks to the Internet and other new technologies. You must stay current with the latest information related to your market. You must be willing and able to change your marketing strategies as new information becomes available. If your knowledge is not current or complete, your agribusiness may be in for trouble: outdated or poor-quality marketing information can seriously hamper your progress and reduce your profits.[48]

Other Marketing Functions. The eight marketing functions just discussed are the most important to any business. Though each will affect different businesses in different ways, each can have a huge positive or negative impact on your business.

Other marketing functions include: (1) awareness creation, (2) traffic building, (3) lead generation, (4) lead qualification, (5) direct selling, (6) service provision, (7) customer loyalty, (8) employee and vendor loyalty, (9) environmental analysis and marketing research, (10) marketing management, (11) scope broadening, (12) consumer analysis, (13) product planning, (14) price planning, (15) promotion planning, and (16) distribution planning.

46. Ibid.
47. Ibid.
48. Ibid.

Value Adding

Between the time the corn leaves the field and the cornflakes reach the consumer's table, many steps occur. With each step, value is added to the commodity. At each step, the value increases by more than the additional input—that is, **value adding** occurs. For example, after corn is ground into cornmeal, the value of the cornmeal is more than the value of the corn plus the cost of turning the corn into meal. Furthermore, when the cornmeal is made into cornflakes, the value of the cornflakes is more than that of the cornmeal and other ingredients that go into the flakes. By the time the cornflakes reach the consumer, the corn has been transported, ground, baked, packaged, marketed, displayed, and sold. Each of these processes adds value to the original commodity.[49]

Value Adding Can Occur at Any Point During Marketing. Value can be added by offering product in a better marketplace. It can be added by marketing the product in connection with a specified event. However, the most value is added when a part of the processing of the product is completed before the (new) product is sold.

Packaging. Packaging is a particularly important value-adding consideration in today's markets. Because many consumers make buying decisions at the point of sale (in the store), it is important to package products in eye-catching, attractive ways. Some of the ways in which packaging adds value to an agricultural product are:

- identifying the brand name
- advertising at the point of sale
- transporting the product with the least possible damage
- increasing shelf life
- increasing convenience[50]

Commodity Research and Promotion Boards

Food and fiber commodity research and promotion boards were established by congressional legislation; most such boards were created in

49. Lee, Leising, and Lawver, *Agrimarketing Technology,* p. 137.
50. Ibid., pp. 138–139.

the 1970s and 1980s. They are monitored by the Agricultural Marketing Service agency of the USDA, and they assist the food and fiber industry in developing new products through their research programs. Their success in increasing demand for their commodities depends on their ability to work with industry, universities, and government to develop new products and new uses for those commodities. By staying in touch with changes in consumer needs and disseminating their research to food technologists and manufacturers, these boards contribute significantly to the development of new food and fiber products.[51]

Check-Off Program

In a **check-off program**, sellers make available a pool of money to be used for advertising a specific commodity. The sellers take a small portion of the proceeds from each item they sell to fund the advertising pool; this is the "check-off" amount. Current estimates reveal that 9 out of 10 farmers in America participate in check-off programs, thereby financially supporting the 300-some state and federal programs that promote about 80 generic commodities. These programs spend up to half a billion dollars annually to fund commodity-specific research and promotion. Typical advertising media for agricultural products and services include newspapers, periodicals, radio, television, billboards and other signage, pamphlets, classified ads, telephone book ads, and catalogs.[52]

New Product Development by Commodity Boards

Examples of new product development that is financed by commodity research and promotion boards include a prototype new beef product, called *beef surimi*, that has a potential market of $70 million a year; a prototype all-cotton disposable diaper, which could be worth $240 million in additional annual sales to cotton growers; and an innovative dairy product with cultures that actually remove cholesterol from fermented dairy products such as sour cream.

51. J. Patrick Boyle, "Commodity Boards Help Develop New Products," in *Marketing U.S. Agriculture: 1988 Yearbook of Agriculture,* ed. Deborah Takiff Smith (Washington, DC: U.S. Government Printing Office, 1988), p. 160.

52. Cooper, *Agriscience,* pp. 655–656.

The following are examples of what the commodity boards are doing to monitor consumer trends and help industry introduce new products that can meet consumer demands and manufacturer needs.[53]

Potato Board. The Potato Board helps companies develop new food products by providing them with statistical information about potato production and consumption trends. It also furnishes companies with information about consumer attitudes and the nutritional benefits of the potato.

This board explores new ways in which potatoes can be used. For example, potato can be used in cereal and pasta as a binding agent. Potato products, such as starch and flour, may be used as an ice-crystal retardant in ice cream, as an **emulsifier** in salad dressings, as a fat extender in dairy products, and as a nutritional supplement in processed foods.

Cattlemen's Beef Promotion and Research Board. According to this board, a new 100 percent beef product, called *beef surimi*, is being developed and processed into high-protein foods in various forms. Beef surimi gets its name from a Japanese food process widely used for making fish sticks and patties. Beef surimi has high-bonding qualities and will contain only about 5 percent fat. Prototypes for a beef surimi hot dog and snack chip already have been developed.

National Honey Board. In a typical supermarket, you can find 40 to 50 products that contain honey. Honey is in demand because consumers increasingly prefer natural-tasting foods without added chemicals or preservatives. The National Honey Board has developed a list of new product applications, capitalizing on the known properties of this commodity. For example, honey attracts moisture and can reduce shrinkage in hams, cured meats, and baked goods. Honey also can clarify wine and is being tested as a seasoning for corn chips.

Cotton Board. Research funded by cotton growers has helped the textile manufacturing industry to develop new cotton fabrics and products.

53. Boyle, "Commodity Boards Help Develop New Products," p. 160.

Commodity Research and Promotion Boards	Congressional Act and Year Started
The American Egg Board 1460 Renaissance Drive, Suite 301 Park Ridge, IL 60068	Congressionally enacted by the Egg Research and Consumer Information Act of 1974
Cattlemen's Beef Promotion and Research Board PO Box 3316 Englewood, CO 80155	Congressionally enacted by the Beef Promotion and Research Act of 1985
The Cotton Board 5350 Poplar Avenue, Suite 210 Memphis, TN 38119	Congressionally enacted by the Cotton Research and Promotion Act of 1966
National Dairy Promotion and Research Board 2111 Wilson Boulevard, Suite 600 Arlington, VA 22201	Congressionally enacted by the Dairy and Tobacco Adjustment Act of 1983
National Honey Board 9595 Nelson Road, Box C Longmont, CO 80501	Congressionally enacted by the Honey Research, Promotion, and Consumer Information Act of 1984
National Pork Board PO Box 9114 Des Moines, IA 50306	Congressionally enacted by the Pork Promotion, Research, and Consumer Information Act of 1985
The Potato Board 1385 South Colorado Boulevard, Suite 512 Denver, CO 80222	Congressionally enacted by the Potato Research and Promotion Act of 1971

Figure 17–8 Food and fiber commodity research and promotion boards contribute significantly to the development of new food and fiber products, by offering funding and conducting research.

Since 1984, research supported by the Cotton Board, and conducted by its subsidiary, Cotton Incorporated, has developed new cotton products now used extensively: cotton-wool blends for year-round clothing; high-loft, resilient cotton for all-cotton pillows and mattress pads; and all-cotton stone-washed denim.

American Egg Board. Through its research programs, the American Egg Board has developed new products such as frozen egg-crust pizza, egg dip, egg sandwich loaf, and egg-filled burritos.

National Pork Board. The National Pork Board's most significant research was focused on creating **trichinae-safe pork**. The industry developed two distinctly different methods for safeguarding pork: irradiation and testing. Low-level irradiation of fresh pork was been approved by both the USDA and the Food and Drug Administration. While trichinae can be rendered harmless by freezing, curing, or cooking to a high internal temperature, irradiation and testing allow processors to sell trichinae-safe raw, fresh pork. The "other white meat" promotional campaign has also been very successful.

National Dairy Promotion and Research Board. The National Dairy Promotion and Research Board uses an annual product research budget of more than $4.4 million to support work at U.S. universities and laboratories. Twenty-five percent of the research budget goes toward new product research.

The board believes that biotechnology and genetic engineering can potentially revolutionize the dairy industry.[54] In the advertising arena, the board's "Got Milk?" promotion campaign, in which many celebrities are pictured with a milk mustache, has been very successful. Refer to Figure 17–8 for a list of the addresses of several major commodity research boards.

54. Ibid., pp. 160–163.

Market Analysis

Market analysis involves the collection and processing of information to determine if a product will sell. Careful attention to detail is required. To get accurate information, one must do *research,* a methodical study to gain knowledge about market potential. Conducting research for market analysis or the development of marketing plans typically involves six steps:

1. *Determine what information is needed.* Only collect information related to the marketing plan. The data needed will vary by product and target customer base.
2. *Develop a means for getting the needed information.* Information may be gathered through written surveys, structured interviews, and/or personal observations.
3. *Word questions carefully.* Questions should be structured so that you get the information you are seeking. For example, consider using rating scales if you want opinions and open-ended questions if you want more detailed answers.
4. *Gather the information.* Once you have completed the first three steps, you have to implement them. It may take a great deal of time to conduct the needed interviews or observations, and you may have to wait for survey responses to arrive in the mail.
5. *Tabulate and analyze the information.* Determine the best way to organize, categorize, and summarize the information you have gathered so that it can be analyzed. Once you have completed your analysis, decide what your results actually mean.
6. *Draw meaning and interpret the findings.* This step involves making judgments that lead to conclusions and recommendations. Try to avoid personal biases and desires; let the information speak for itself. Conclusions must always be based on the research findings.[55]

Strategic Planning and Marketing

Plan for Success

It takes time to build customer awareness. You must plan your marketing strategies based on your knowledge of the target market and your market research. Strategic planning considers a wide variety of issues, including advertising, employees, and customers.

Once you have a plan, implement it sufficiently. Never assume that a single advertisement, a single article, a business card, a small sign, or a Yellow Pages listing will be enough to bring your company into your customers' consciousness or to keep it there.[56]

Success Strategies. Some people work better with a list of points or considerations to follow as they develop strategies to build a successful agribusiness. Consider the following:

- How do you expect to do any business if no one knows that your business exists?
- Calling attention to your business promotes awareness of your business.
- A marketing (public relations) and promotion strategy is as important as your financial strategy.
- Your communications create the image of your company in the public eye and mind.
- Employees are part of your image. For example, customers who deal even once with an employee who has a negative attitude may take future business elsewhere. Customers want to be treated with courtesy by helpful, courteous employees who know their business. All employees in your business are part of your sales team, not just those who are called salespersons, because each will have contact with the public at some time. Any employee who is rude or incompetent can cost you business.
- Make sure your community sees your business as an asset, not a problem.
- Happy customers are valuable to your image. They are a good, credible source of positive advertising. In contrast, dissatisfied customers can be a marketing disaster, as their complaints will spread rapidly in your community—particularly among your targeted customer base.

55. Lee, Leising, and Lawver, *Agrimarketing Technology,* pp. 307-309.

56. Paul E. Adams, "Survival: Step Three: Pump Up Your Sales Efforts," (Khera Communications, 2002); available at http://www.morebusiness.com/running_your_business/management/d1013439481.brc.

- The image you want for your business is one of staying power, caring, and honesty.

If you are missing any of these points, change or expand your strategic plan to answer concerns in these areas.[57]

Characteristics of Strategic Planning

Planning means determining now what you will do later. It includes knowing what you will do, when you will do it, and how you are going to do it. *Strategic planning* determines a course of action over time that will match a firm's resources to its opportunities. The intent is to maximize opportunities and minimize threats to success.

Strategic planning is not an exact science. Rather, it is a creative process that uses intuition, but at the same time is analytical and systematic in nature. Strategic planning:

- **Considers the big picture.** Strategic planning encompasses your whole company. It attempts to both embody and implement the mission and vision of the organization.
- **Demands change.** Strategic planning has a broad scope. It must analyze past projections and try to predict the future. When you do so, you are likely to find that adjustments are needed—and some of those adjustments may require substantial changes in funding, product lines, cooperation with other companies, and so on.
- **Considers future environmental forces in the industry.** Your agribusiness will be affected by outside forces, such as the health of the economy, technological advances, new policies and regulations, buying trends, and additional competition, so you must include these factors in strategic planning.
- **Anticipates competitors' reactions.** A good strategic plan does not wait to see what the competition will do; it aims to influence what it will do. Quite simply, you want to be the leader in your business area.
- **Looks at a longer time horizon.** Strategic planning is usually long term. In the past, 10-year plans were common; however, given the speed of change and volatile economy in today's world, a five-year plan is more realistic. Long-term planning involves using past experience and current data to project future sales, costs, and technology.[58]

The Purpose of Strategic Planning

Strategic planning attempts to identify and isolate present actions and forecast how the results of those actions will influence the future. This kind of planning can help your firm gain a *competitive advantage*, if you do the following in the strategic planning process:

- Establish goals, objectives, priorities, and strategies to be completed within specified time periods. This sets a clear direction for management and employees to follow.
- Define, in *measurable terms*, what is most important for the firm.
- Establish a basis for evaluating the performance of management and key employees.
- Provide a management framework that facilitates timely responses to changed conditions, unplanned events, and deviations from plans.
- Anticipate problems and take steps to avoid or eliminate them.
- Allocate resources more efficiently to meet anticipated changes. Resources include labor, machinery and equipment, buildings, and capital.[59]

Financial Reasons for Strategic Planning.

Remember, it costs money to start a new agribusiness. Almost every business owner has to borrow money. Certainly, you develop a strategic plan and strategic marketing plan for your own benefit, but lending institutions need to see them too, to justify loaning you the money you are asking for. They have to be convinced that their loans are secure, and going to a business with a good chance of success.

---

57. Ibid.

58. Gerald B. White and Wen-fei L. Uva, *Developing a Strategic Marketing Plan for Horticultural Firms* (EB 200-01) (Ithaca, NY: Cornell University, Department of Agricultural, Resource, and Managerial Economics, College of Agriculture and Life Sciences, 2000), p. 2.

59. Ibid., p. 3.

Steps of the Strategic Planning Process

The contents of a strategic plan will vary considerably from business to business. After all, it is *your* plan, created specifically for *your* agribusiness. However, such a plan usually includes a mission statement, objectives, and alternative strategies. Consider the following steps:

- Define the mission of the agribusiness.
- Assess the external environment, noting influences from competition, technology, the economy, politics, laws and regulations, society, and culture.
- Identify major opportunities for and threats (such as recalls) to the agribusiness.
- Assess the strengths and weaknesses of the agribusiness.
- Establish objectives that are specific, measurable, attainable, rewarding, and timely.
- Develop and evaluate alternative strategies in areas such as product-market scope, growth directives, investment strategies, product lines, price and distribution, and assets.
- Select the best strategy or strategies, including but not limited to vision, mission, and objectives.
- Implement the strategic plan as an operating plan.
- Evaluate the results for the agribusiness and revise the plan as needed.

Developing the Strategic Marketing Plan

A marketing plan should be guided by the mission and objectives set out in the strategic plan for your agribusiness. The strategic marketing plan contains four basic elements: (1) an analysis of the agribusiness's situation, (2) a set of objectives, (3) a detailed strategy statement, and (4) a set of procedures for monitoring and controlling the plan, including a contingency plan.

Analysis of the Agribusiness's Situation

Your situation analysis includes consideration of the external environment of the marketing area, emphasizing factors and agents that affect the firm's marketing. The strengths and weaknesses of your agribusiness should be analyzed with marketing in mind. This section of the strategic marketing plan should also include a description of the products and services, target markets, and market potential of your agribusiness.

Objectives of the Agribusiness

You should develop one set of marketing objectives for the first two years and another set of marketing objectives for the next three to five years. You will also need to determine the sales and profit objectives for the same time periods.

Strategy Statements

Develop an overall strategy, a competitive strategy, and a pricing, distribution, and promotion strategy. Be sure to include marketing and advertising budgets. These strategies should all support accomplishment of the goals you have set for your agribusiness.

Procedures for Monitoring and Controlling the Strategic Plan

Be sure to include a financial impact evaluation in your plan. Also, include a section on potential problems in achieving your objectives for your agribusiness ventures and proposed solutions for identified problems. Develop timetables and benchmarks so that you can monitor the implementation of your tactical plan.[60]

Take some time to do an Internet search for some model or sample strategic marketing plans. You should be able to find free plans that will serve as excellent models. Refer to Figure 17-9 for a suggested outline of a strategic marketing plan.

Conclusion

Successful agrimarketing management is a matter of perspective. It is like the old story of two shoe salesmen who sail away to a distant land. Upon arrival, both note that people in this country do not wear shoes. The first salesman cables home, "Send return ticket immediately; no market here." The second one cables home, "Send warehouse plans; market appears

60. Ibid., pp. 4-6.

I. **Executive Summary**

II. **The Firm's Mission Statement**

III. **Situation Analysis**

A. Analysis of the external environment, emphasizing aspects that affect the firm's marketing plan

B. Analysis of the firm's strengths and weaknesses, with an emphasis on marketing

C. Description of products/services

D. Target markets

E. Market potential

IV. **Objectives**

A. Marketing objectives for next year and for the next three to five years

B. Sales and profit objectives for next year and for the next three to five years

V. **Strategies—How Will You Achieve Your Goals?**

A. Overall strategy

B. Competitive strategies

C. Pricing, place (or distribution), and promotion strategies

D. Marketing and advertising budgets

VI. **Financial Impact Evaluation**

VII. **Potential Problems in Achieving Objectives, with Proposed Solutions**

VIII. **Implementation or Tactical Plans (Timetable and Benchmarks)**

IX. **Monitoring and Control**

X. **Appendices** [include supporting documents]

Figure 17–9 Suggested outline for strategic marketing plan

unlimited." Agrimarket opportunities are in the eye of the beholder.

Summary

Little is more important than marketing in the agricultural industry. The cost of marketing U.S. domestic food amounts to $450 billion, whereas the cost of producing farm food products is about $125 billion. Agribusiness is the nation's largest industry, and marketing is its largest segment. More than 80 percent of those involved in agribusiness are employed in marketing.

Agribusiness marketing is those processes, functions, and services performed in connection with food and fiber, reaching from the farms on which they are produced until their delivery into the hands of the consumer. Selling is taking the farm product to the market and getting whatever the price is for that day. In contrast, marketing starts with analyzing the market to see what is needed before production begins.

In the formative years of this nation, most Americans were farmers producing food and fiber mainly for domestic consumption. As time passed, production agriculturalists found that it was more efficient to specialize. As the country became more industrialized, fewer people produced their own food; they became dependent on farmers to provide food for them. As a result, marketing firms began to develop the services that consumers were demanding.

Supply and demand are the factors that determine whether the production of a product is profitable. Supply is the amount of a product available at a specific time and price. Demand is the amount of a product wanted at a specific time and price.

When several prerequisites are met in the general economic system, the marketing system leads not only to efficient fulfillment of producer and consumer needs, but also to efficient allocation of society's limited resources. Three prerequisites of an efficient economic system are a free market economy, prices that reflect the full value of market resources, and a high degree of interaction between consumers and production agriculturalists. Consumers ultimately determine marketing success. Commodities or services may be on consumers' doorsteps, but if the consumer chooses not to buy and does not submit to persuasion, there will be no sale. Success in agrimarketing involves understanding the consumer and the role of consumption.

Production agriculturalists can do little to control the profitability of their products unless they approach marketing as a vital component of producing agricultural commodities. Various types of agricultural markets are available to production agriculturalists. The primary types of livestock and dairy markets are terminal markets, auction markets, direct sales, cooperatives, sales to government, vertical integration, and futures markets. The primary types of grain markets are forward contracting, futures contracts, harvest or spot pricing, and postharvest pricing. The primary types of fruit and vegetable markets are roadside markets, farmer's markets, pick-your-own operations, food brokers, wholesalers, and processors. Specialty markets are also important to producers that have found a special niche. These include fee fishing, organically grown crops, hormone-free beef, on-farm restaurants, and wineries.

Once farm commodities have been marketed, most of the commodities are processed and marketed again by agribusinesses. Agribusiness marketing firms and commodity groups spend more than $10 billion annually to advertise and promote food. Four key ingredients of marketing agribusiness products are advertising, price promotion, merchandising promotion, and public relations activities. The agrimarketing mix is composed of four items: product, price, place, and promotion.

Food and fiber commodity research and promotion boards exist because of congressional legislation. They have established check-off programs for their target commodities, in which sellers contribute a small amount for each unit sold to a money pool to be used to purchase advertising for the specific commodity. Commodity boards spent about $500 million annually for research and promotion. Some examples of commodity boards include the Potato Board, Cattlemen's Beef Promotion and Research Board, National Honey Board, Cotton Board, American Egg Board, National Pork Board, and National Dairy Promotion and Research Board.

Market analysis helps determine if a product will sell. Careful attention to detail is required. To get accurate information, research is needed. Conducting research for marketing plans typically involves six steps.

Strategic planning and marketing are crucial for a successful agribusiness. Planning considers a wide variety of issues, such as advertising, employees, and customers. The agribusiness owner needs a strategic marketing plan to analyze the agribusiness, set objectives, develop strategies, establish procedures, and monitor the business. Lending institutions also use strategic plans and marketing plans to decide whether to make loans to a business.

Successful agrimarketing management requires a positive attitude. A positive attitude can help you discover hidden opportunities.

End of Chapter Activities

REVIEW QUESTIONS

1. Define the Terms to Know.
2. Besides marketing, what are three other areas in which production agriculturalists face major decisions?
3. List 20 potential steps involved in agrimarketing.
4. Distinguish between selling and marketing.

5. What development or period in American history caused agrimarketing to develop at a faster rate?
6. Briefly explain how supply and demand offset prices.
7. What are three prerequisites to an efficient economic system?
8. What are four results of an efficient economic system?
9. List seven marketing strategies that can be used to market farm commodities more profitably.
10. List and briefly describe seven types of livestock and dairy markets.
11. List and briefly describe four types of grain markets.
12. What are four main objectives of forward contracting?
13. What are four potential risks associated with forward contracts?
14. List three advantages forward contracts have over futures contracts.
15. List and briefly discuss six types of fruit and vegetable markets.
16. List and briefly discuss five examples of specialty markets.
17. List and briefly discuss four key ingredients of marketing agribusiness products.
18. Name and define the four items ("Ps") that make up the agrimarketing mix.
19. What are five functions that packaging can perform in adding value to an agricultural product?
20. Who monitors the commodity research and promotion boards?
21. How are commodity research and promotion boards funded?
22. List seven commodity and research boards and give one example for each board of something it has produced or promoted.
23. List the six steps needed to conduct research for marketing plans.
24. What are eight points or considerations that will help you carry out your strategies and build a successful agribusiness?
25. What are five characteristics of strategic planning?
26. What are six purposes of strategic planning?
27. What are nine steps in the strategic planning process?
28. What are the four basic elements of a strategic marketing plan?

FILL IN THE BLANK

1. Little is more important than _________ in the agricultural industry.
2. Of the $511 billion spent by consumers for food products, less than _________ was returned to producers.
3. More than 80 percent of those involved in agribusiness are employed in _________.
4. _________ ultimately determine agrimarketing success.
5. _________ is the link between ___________ and ___________.
6. _________ vote every day in the marketplace with their dollars, and the market hears what they are saying.
7. The world's population increases at the rate of _________ every hour.
8. At the current rate of population growth, the world's population will double every _________ years.
9. Production agriculturalists have historically been ________ ________ rather than price makers.

10. Agrimarketing firms and commodity groups spend more than _________ annually to advertise and promote food.
11. Food and fiber commodity research and promotion boards exist because of ___________ ___________.
12. The commodity research and promotion boards collect approximately _________ dollars yearly.

MATCHING

a. 6.6 billion
b. terminal market
c. cooperatives
d. 10,000
e. vertical integration
f. 298.5 million
g. direct sales
h. auction market
i. 439 million
j. 450 billion
k. harvest pricing
l. postharvest pricing

_____ 1. cost (in dollars) of marketing U.S. domestic food in 1994
_____ 2. number of new products or product variations that are put on grocery store shelves yearly
_____ 3. earth's population
_____ 4. population of the United States, Canada, and Mexico
_____ 5. population of the United States
_____ 6. this market never actually owns the animals
_____ 7. place where animals are sold by public bidding on individual animals or groups of animals
_____ 8. producer sells animals (crops) directly to processors
_____ 9. group of producers who join together to market a commodity
_____ 10. arrangement in which several steps in the production, marketing, and processing of animals are combined
_____ 11. storage of grain in elevators is the important element in this type of grain market
_____ 12. traditional method of pricing used by producers

ACTIVITIES

1. Give an example of a farm product that you can sell and one that you can market. Write this in report form and present it to the class to see if they agree.
2. Give an example of an agricultural product and how supply and demand could affect its price. (Note: An example would be the mad cow disease scare.) Write this in report form and present it to the class.
3. List five examples of the way food is packaged (marketed) today that are the result of consumer demand and lifestyle. Share these with your class and compile other examples from your classmates.
4. Suppose you had 100 beef cows and sold more than 90 calves annually at weight of approximately 500 pounds each. Explain your marketing strategy in report form and present your report to the class.
5. Suppose you have 200 acres of either corn, wheat, or soybeans. Explain your marketing strategy in report form and present it to the class.
6. Suppose you had an acre of strawberries. Explain your marketing strategy in report form and present it to the class.
7. Many producers develop a specialty or niche market. Write a brief report about someone in your community (or elsewhere) who has developed a specialty or niche market. Present this report to your class.

8. Select a raw agricultural commodity; then select an end product of that commodity (for example, wheat and bread). List as many steps or processes as you can that add value to the product from the farm to the consumer.
9. A production agriculturalist must choose between selling corn at harvest or storing it for six months. The price at harvest is $3.10 per bushel. The elevator charges $0.05 per bushel for placing the corn in storage plus $0.015 per bushel per month storage fee. If the producer sells the corn at harvest, the proceeds will pay off a loan that has a 10 percent interest rate. What price must the corn producer receive to justify the storage costs and break even?
10. Production, labor, capital, and marketing are four areas in which a production agriculturalist faces major decisions. Write a brief report stating which you believe is the most important and why. Present this to your class.
11. A wheat producer's cost of production is $3.10 per bushel and the expected selling price is $3.65 per bushel. The producer expects to produce 30,000 bushels. What is the expected net income? Suppose the wheat producer could cut costs by 5 percent and increase selling price by 10 percent if the wheat is sold in another market. What is the new expected net profit?
12. A producer sells 82 steer calves for 74 cents per pound. The sales contract states that the calves are to be weighed at the ranch and that the pay weight or actual selling weight will be based on a 4 percent shrink. On the sale date, the 82 calves weighed a total of 34,400 pounds.
    a. What was the average pay weight per calf?
    b. How many total dollars did the producer receive from selling the calves?
    c. What did the calves cost in dollars per head?
13. On the computer, make a pie chart showing the different functions of marketing. Each part of the chart should represent a different function. For each part of the chart, list the function and give a short explanation of the function. You may add pictures or clip art that explain the function.

    For example: One part of the chart could represent distribution. Distribution will be listed on that part of the chart along with a short definition of distribution. Pictures of a ship, train, warehouse, airplane, and truck represent the different means of transportation of products.
14. Develop a strategic marketing plan for an agribusiness that you might consider opening some day. Go to the Internet to find models to assist you in preparation of the plan.

Chapter 18

Commodity (Futures) Marketing

Objectives

After completing this chapter, the student should be able to:

- Discuss the history of the commodity (futures) market.
- Describe commodity (futures) exchanges.
- Analyze the process of buying futures contracts.
- Describe hedgers and speculators in the futures market.
- Explain how the price of a commodity is determined.
- Explain how to make a futures trading transaction.
- Analyze the process of buying futures options.
- Discuss regulation of the commodity (futures) exchanges.

Terms to Know

basis
bear market
brokerage office
bull market
call
cash market
commodity
commodity (futures) exchanges
contract specifications
forward contracts
futures contracts
futures market
futures options
hedgers
initial margin
intrinsic value
long hedge
maintenance margin
margin call
margining system
minimum price fluctuations
offset
open outcry
performance bond margin
premium
price discovery
put
risk transfer
short hedge
speculators
ticks
time value
trading pit

Introduction

Since the early development of agricultural markets in the United States, production agriculturalists have attempted to protect themselves against falling commodity prices at harvest time. Many production agriculturalists ignored marketing techniques and sold their commodities at harvest regardless of the price. Today, product agriculturalists (whether livestock or crop producers) realize that a marketing strategy is equal in importance to production, capital, and labor strategies.

The development of **forward contracts** was a major step in allowing producers to reduce marketing risks. There are now many different strategies one can use in forward contracting. This chapter discusses use of the commodity (futures) market to reduce marketing risk.

History of Commodity (Futures) Markets

Commodity markets originated in Chicago in the 1800s to help producers reduce price risk. Chicago was the hub that joined all parts of the United States. Many of the country's main railroads, waterways, and roads passed through the city. Production agriculturalists brought their grain to sell by wagons, oxcarts, and river barges. Often the quantity so greatly exceeded the market demand that many producers ended up dumping their grain out on the streets of Chicago.[1]

No Early Standards

Standards, grades, and measures had not yet been developed. Therefore, some buyers and sellers attempted to cheat on their measurements, which caused heated disagreements.[2] Eventually, a uniform weight per bushel was set for selected grain commodities. For example, 56 pounds was set as the standard for a bushel of corn; previously, some people had sold bushels weighing as little as 46 pounds.

Few people understood how crucial agricultural futures were to maintaining the stability of markets. Without a price stability mechanism and standards, the price of grains would have continued to fluctuate drastically.

Formation of the Chicago Board of Trade

The price-fluctuation problems and increasing production created the need for a year-round, centralized marketplace. In 1848, 82 Chicago businessmen filled this need when they founded the Chicago Board of Trade (CBOT). This allowed buyers and sellers to get together in one place to exchange commodities. Both buyers and sellers soon began to see the advantages of making contracts to buy or sell commodities in the future. Forward pricing contracts were the result of this realization.[3]

As the market grew, the CBOT established rules and regulations to help maintain a truly competitive market. Members who held seats on the Chicago Board of Trade set up the rules. They then had to abide by those rules or be subjected to strict disciplinary action.

Forward Contracts Increase in Popularity

What made the Chicago Board of Trade increasingly popular as a centralized marketplace was the growing use of "to arrive" contracts. These contracts allowed buyers and sellers of agricultural commodities to specify delivery of a particular commodity at a predetermined price and date.

These early forward contracts were first used by river merchants who received corn from farmers in late fall and early winter but had to store the corn until it reached a low enough moisture content to ship and until the river and canal were free of ice. Seeking to reduce the price risk of storing corn through the winter, these river merchants would travel to Chicago, where they would enter into contracts with processors for the delivery of grain at an agreed-upon price in the spring. In this way, they assured themselves of both a buyer and a price for grain. The earliest recorded forward contract was made on March 13, 1851, for 3,000 bushels of corn to be delivered in June of that year.[4]

1. *Agribusiness Management and Marketing* (8720-B) (College Station, TX: Instructional Materials Service, 1988), pp. 2–3.
2. Ibid.
3. Ibid.
4. Chicago Board of Trade, *Action in the Marketplace* (Chicago: Chicago Board of Trade, Publications Department, 1994), p. 3.

Change from Forward Contracts to Futures Contracts

Cash-forward contracts did have drawbacks. They were not standardized in terms of quality or delivery time, and merchants and traders did not always fulfill their forward commitments. In 1865, the Chicago Board of Trade took a step to formalize grain trading by developing standardized agreements called **futures contracts**. Futures contracts, in contrast to forward contracts, were standardized as to quality, quantity, and time and location of delivery for the commodity being traded. The only variable was price, which was set through an auction-like process on the trading floor of an organized exchange.[5]

Standardized Agreements in Futures Contracts

Because futures contracts were standardized, sellers and buyers were able to exchange one contract for another and thereby offset their obligation to deliver the cash commodity underlying the futures contract. **Offset** in the futures market means taking a futures position opposite and equal to one's initial futures transaction; for example, buying a contract if a previous one is sold.

Because of the standardization of contract terms and the ability to offset contracts, use of the futures markets by agricultural firms increased rapidly. Grain merchandisers, processors, and other agricultural companies found that by trading futures contracts, they were able to protect themselves from erratic price movements in commodities they were planning to either sell or purchase.[6]

Initiation of the Margining System

In the same year the Chicago Board of Trade introduced futures contracts, it also initiated a **margining system** to eliminate the problems of buyers and sellers not fulfilling their contracts. A margining system requires traders to guarantee contract performance by depositing funds with the exchange or an exchange representative. Although early records were lost in the Great Chicago Fire of 1871, it has been quite accurately established that by 1865, most of the basic principles of futures trading as we know them today were in place.[7] Still, no one could have guessed how this infant industry would change and develop in the next century and beyond.

Growth in futures trading increased in the late 19th and early 20th centuries as more and more businesses made futures trading a part of their business plans. Nevertheless, most of the dramatic growth and successful contracts in the futures industry were yet to come. Refer to Figure 18–1 for historic highlights of the Chicago Board of Trade.

Commodity (Futures) Exchanges

Although most of our references so far have been to the Chicago Board of Trade, there are many other **commodity (futures) exchanges**. Refer to Figure 18–2. Each exchange tends to specialize in certain commodities, although there are overlaps and exceptions. For example, the Chicago Board of Trade specializes in grain commodities, whereas the Chicago Mercantile Exchange specializes in livestock commodities.

Exchanges Do Not Trade Anything Physical

The hardest concept for a beginner to understand is that neither an exchange nor the traders in it actually physically trade grain or livestock commodities. The exchange serves as a forum or meeting place for exchange members, buyers, and sellers of commodities. The traders are individual members and member firms who seek to trade agricultural commodities for their customers or themselves.

The questions become, "If grain is traded at an exchange, where is the physical commodity?" "Why are there no piles of grain outside of the Chicago Board of Trade?" Originally, farmers did bring their grain to the regional exchanges and markets at a certain time each year. However, they often found that the quantity immediately available far exceeded the market need; all the farmers were trying to sell their harvests at the same time, so competition was fierce. Grain buyers, knowing that there was such a large supply, had no reason to pay much. The result was ruinously low prices.

Today, the contracts that are traded are for grain to be delivered at some time in the future (say, September or December) at specified delivery points, if actual delivery is desired (this rarely

5. Ibid., p. 3.
6. Ibid., p. 3.
7. Ibid., p. 3.

The Chicago Board of Trade: A Chronological History

Year	Historic Event
1848	Chicago's strategic location at the base of the Great Lakes, close to the fertile farmlands of the Midwest, contributed to the city's rapid growth and development as a grain terminal. However, problems of supply and demand, transportation, and storage led to a chaotic marketing situation and the logical development of the futures market.
1851	The earliest "forward" contract (for 3,000 bushels of corn) is recorded. Forward contracts gain popularity among merchants and processors.
1854	The CBOT adopts standard weight of a bushel of wheat at 60 pounds, and oats at 32 pounds.
1865	Forward contracts create confusion for users and subsequent defaults. The CBOT formalizes grain trading by developing standardized agreements called *futures contracts*. The CBOT also begins requiring performance bonds, called *margin*, to be posted by buyers and sellers in its grain markets.
1866	First transatlantic cable completed. The time required to send a message to Europe from Chicago is reduced from three days to three hours.
1877	Futures trading becomes more formalized and speculators enter the picture. Growth in futures trading increases in late 19th and early 20th centuries as new exchanges are formed. Many types of commodities are traded on these exchanges, including cotton, butter, eggs, coffee, and cocoa.
1925	One of the CBOT's biggest years—26.9 billion bushels of grain traded. Western Union installs automatic ticker to replace slower Morse service for improved quotation system.
1967	New electronic price display boards are installed on the walls above the trading floors. Price reporting time is cut to seconds.
1977	The CBOT introduces the U.S. Treasury bond futures contract—today the most actively traded contract in the world.
1984	Trading in agricultural futures options begins as the CBOT launches options on soybean futures.
1990	The CBOT sets new world trading volume at 154 million contracts.
1995	Full membership trades at $710,000, a new record high.

Figure 18–1 Selected significant dates in the history of the Chicago Board of Trade.

happens). This keeps grain producers from having to accept terrible prices because of oversupply (quantity). In reality, at the time the contracts are traded, the grain is still in the fields, in grain elevators, or perhaps not even planted. This keeps too much grain from being delivered at the same time, thus stabilizing prices and keeping them at a reasonable level.

Other Commodities Are Also Traded

Although this chapter focuses on agricultural commodities (the futures contract process did start with agriculture), commodity (futures) exchanges trade many other commodities as well. Today, there are futures contracts for interest rates, insurance, stock indexes, manufactured and processed products, crude oil, nonstorable commodities, precious metals, and foreign currencies and foreign bonds, among other things.[8]

8. Ibid., p. 6.

Purpose of Commodity (Futures) Markets

Futures exchanges, no matter how they are organized and run, exist because they provide two vital economic functions for the marketplace: **risk transfer** and **price discovery**. Futures markets make it possible for those who want to manage price risk—**hedgers**—to transfer some or all of that risk to those who are willing to accept it—**speculators**. Discussion of these topics follows later in this chapter.

Buying Futures Contracts

When contracts are mentioned, you may think of pieces of paper with lots of fine print. Although there is substantial documentation and paperwork behind the existence of a futures contract, it is not a formalized contract written by an attorney. Futures contracts are legally binding agreements, made on the trading floor of a futures exchange, to buy or sell something in the future. That "something"

Major U.S. and Canadian Futures Exchanges		
Chicago Board of Trade Market Development Department 141 West Jackson Boulevard Chicago, IL 60604 312-435-3500 www.cbot.com	Minneapolis Grain Exchange 130 Grain Exchange Building 400 South Fourth Street Minneapolis, MN 55415 612-321-7101 www.mgex.com	THX Group PO Box 450 130 King Street West, 3d Floor Toronto, Ontario, Canada M5X 1J2 416-947-4670 www.tsx.com
Chicago Mercantile Exchange 20 South Wacker Drive Chicago, IL 60606 312-930-1000 www.cme.com	Montreal Exchange Stock Exchange Tower PO 61 800 Victoria Square Montreal, Quebec, Canada H4Z 1A9 514-871-2424 www.m-x.ca	Winnipeg Commodity Exchange, Inc. 400 Commodity Exchange Tower 360 Main Street Winnipeg, Manitoba, Canada R3C 3Z4 204-925-5000 www.wce.ca
ICE Futures U.S. One North End Avenue New York, NY, 10282-1101 212-748-4000 www.theice.com	New York Mercantile Exchange, Inc. World Financial Center One North End Avenue New York, NY 10282-1101 212-299-2000	
Kansas City Board of Trade 4800 Main Street Suite 303 Kansas City, MO 64112 816-753-7500 www.kcbt.com	Philadelphia Board of Trade 1900 Market Street Philadelphia, PA 19103 215-496-5000 www.phlx.com	
MidAmerica Commodity Exchange Market Development Department 141 West Jackson Boulevard Chicago, IL 60604 312-341-3000		

Figure 18–2 There are many commodity (futures) exchanges throughout the world. Those listed here are the exchanges in the United States and Canada.

could be corn, soybeans, gold, live beef futures, or some other commodity. In other words, a futures contract establishes a price today for a commodity that will be delivered later. Buyers and sellers in the futures markets look at current economic information (supply and demand) and anticipate how it may affect the price of a commodity.[9]

What Does a Futures Contract Look Like?

Although the futures contract is not a written contract, it is a valid and binding verbal agreement between a buyer and a seller (made on the floor of a futures exchange). Though not written, each futures contract specifies the time of delivery or payment, where the commodity should be delivered, and the quality and quantity of the item. This specificity is what makes futures contracts attractive to those who want to plan ahead and protect themselves from dangerous price swings and to investors wanting to profit from market fluctuations.[10]

The standard features (time of delivery, etc.) are called **contract specifications**. You will find most of them listed in the newspaper, along with prices from

9. Patrick Catania, ed., *Commodity Challenge* (Chicago: Chicago Board of Trade, Market Development Department, 1993), p. 19.

10. Chicago Board of Trade, *Agricultural Futures for the Beginner* (Chicago: Chicago Board of Trade, Market and Product Development Department, 1996), p. 4.

the previous trading day. The futures exchange where your commodity is traded also can provide contract specifications for the commodity you are studying. The *Wall Street Journal* is one of the best sources for commodity market information. Figure 18–3 gives a common example of how commodity prices may appear. In Figure 18–3, you will notice that several contract specifications are listed for each commodity. We will use corn as our example:

- name of commodity: corn
- where it is traded: Chicago Board of Trade
- contract months: Mar. '98, May '98, July '98, Sept. '98, Dec. '98, Mar. '99, July '99, Dec. '99
- contract size: 5,000 bushels
- price quote: price per bushel in 14-bushel increments

Open is the first price anyone paid for corn on January 5, 1998. *High* is the highest price anyone paid for corn on January 5, 1998. *Low* is the lowest price anyone paid for corn on January 5, 1998. *Settle,* or *settlement,* is the last price anyone paid for corn on January 5, 1998. *Change,* or *net change,* is the difference between the settlement price on January 5, 1998, and the previous trading day.

Practice Reading a Price Quote

Just for practice, read one line of price quotes for the July corn futures contract (following) and fill in the following blanks, on a separate sheet of paper. July corn opened at (1) __________ and closed or settled at (2) __________ on April 16. The lowest price anyone paid for July corn, on April 16, was (3) ________, and the highest price anyone paid for July corn on that day was (4) ____. The contract size is (5) ________, prices are quoted in (6) ________, and it is traded at the (7) ________ (answers follow).

This is just one example of how commodity futures prices can appear in a newspaper. Other papers may carry the same information or more, and they may publish it in a different format. Prices are listed in the paper according to the contract specifications. Thus, if corn is quoted in cents per bushel, then a price written as *232* means that corn is trading for *232* cents, or $2.32, per bushel.

Determining the Value of a Futures Contract

To figure out the value of a commodity, multiply the settlement price by the contract size: settlement price times contract size equals contract value. We will use an agricultural example (corn) to demonstrate how to determine the value of a futures contract.

Suppose the settlement price for December corn futures is 200 cents a bushel; that is, $2.00 a bushel. To calculate the dollar value of one corn contract, multiply the $2.00 settlement price by the contract size. In the case of CBOT corn futures, each contract equals 5,000 bushels of corn, so if 1 bushel of corn is worth $2.00, then a 5,000-bushel contract is worth $10,000: $2.00 per bushel times 5,000 bushels equals $10,000.

Futures contracts do not always trade in even numbers; sometimes they move in fractions. These fractions are the smallest price unit at which a futures contract trades and are called **minimum price fluctuations**. In futures lingo, minimum price fluctuations also are referred to as **ticks**. The tick size of a futures contract varies according to the commodity.

The minimum price fluctuation for a CBOT corn futures contract, for instance, is ¼ cent per

Friday, April 16

CORN (CBOT) 5,000 bu.; cents per bu.

	Open	High	Low	Settle	Change
May	231¼	231¾	227	227¾	–4
July	236½	237¼	232¼	233	–4¼
Sept	241¼	241¾	237½	237¾	–3¾
Dec	246	246¾	242¾	243½	–3½

Answers: (1) $2.36½, (2) $2.33, (3) $2.32¼, (4) $2.37¼, (5) 5,000 bushels, (6) cents per bushel, (7) Chicago Board of Trade

Monday, January 5, 1998: Open Interest Reflects Previous Trading Day

Grains and Oilseeds

CORN (CBT) 5,000 bu.; cents per bu.

	Open	High	Low	Settle	Change	Lifetime High	Lifetime Low	Open Interest
Mar	262½	266	261¾	265¾	+ 3	305	236	168,498
May	269	273	268¾	272¾	+ 3½	310	241¾	51,767
July	275	277¾	274	277½	+ 2¼	315½	245	60,236
Sept	275½	277¾	274½	277½	+ 2	301	244	7,593
Dec	280	281½	278½	281¼	+ 1½	299½	247	34,389
Mr99	285	287½	284½	287¼	+ 1½	305	283	1,356
July	293¼	294	293¼	294	+ 1¾	312	256¼	401
Dec	274½	275½	273¾	275½	+ 1½	291½	265	675

Est vol 40,000; vol Fri 8,115; open int 64,983, + 960.

OATS(CBT) 5,000 bu.; cents per bu.

	Open	High	Low	Settle	Change	Lifetime High	Lifetime Low	Open Interest
Mar	149¾	153¾	149¾	153¼	+ 3	180	148¼	8,269
May	153¼	157½	153¼	157	+ 2½	182½	151	2,184
July	157¾	160½	157½	160½	+ 2¾	184	153	750
Sept	161	162½	161	162½	+ 1¾	177	155	296
Dec	165	165	164	164	+ 1¼	177½	162	119

Est vol 1,000; vol Fri 56; open int 2,324, −8

SOYBEANS (CBT) 5,000 bu.; cents per bu.

	Open	High	Low	Settle	Change	Lifetime High	Lifetime Low	Open Interest
Jan	660½	669	660	667¼	+ 6¾	752	583	11,907
Mar	668	676	667	674	+ 6½	749½	593	56,345
May	675	682½	675	681½	+ 6¾	752	601	27,416
July	679½	687	679½	685½	+ 6¾	753	611½	26,720
Aug	682	684	681	683½	+ 6½	745	631	4,197
Sept	669	670½	667	669	+ 3¼	723	637	270
Nov	660½	663½	659½	662¾	+ 4½	717	597	10,711

Est vol 45,000; vol Fri 0-86; open int 27,566, +60

SOYBEAN MEAL (CBT) 100 tons; $ per ton.

	Open	High	Low	Settle	Change	Lifetime High	Lifetime Low	Open Interest
Jan	200.30	202.50	199.60	200.40	+ .80	239.50	185.50	13,621
Mar	197.40	200.00	197.00	198.20	+ 1.50	234.40	184.50	43,296
May	197.00	199.50	196.70	197.90	+ 1.50	231.00	185.50	25,507
July	199.50	201.00	198.20	199.80	+ 1.60	231.50	188.50	16,864
Aug	200.50	201.50	199.50	200.10	+ 1.60	231.50	189.00	5,340
Sept	201.00	201.80	199.50	200.40	+ .70	231.50	192.00	3,820
Oct	200.00	200.80	200.00	200.40	+ 1.70	226.00	190.00	903
Dec	202.00	202.00	200.50	201.20	+ 1.20	231.00	193.00	3,676

Est vol 16,000; vol Fri 24,832; open int 113,032, +2,042.

SOYBEAN OIL (CBT) 60,000 lbs.; cents per lb.

	Open	High	Low	Settle	Change	Lifetime High	Lifetime Low	Open Interest
Jan	24.85	24.90	24.70	24.79	− .05	27.45	21.98	4,534
Mar	25.15	25.25	25.05	25.12	− .03	27.50	22.20	55,929
May	25.53	25.55	25.33	25.41	− .02	27.55	22.35	18,641
July	25.70	25.75	25.55	25.58	− .06	27.40	22.40	12,104
Aug	25.55	25.55	25.50	25.50	− .04	26.70	22.72	3,716
Sept	25.40	25.45	25.31	25.31	− .10	26.15	22.90	1,379
Oct	25.15	25.20	25.05	25.05	− .05	26.20	22.80	679
Dec	24.98	25.10	24.90	24.95	− .08	26.30	23.00	2,184

Est vol 17,000; vol Fri 16,345; open int 99,183, −115.

WHEAT (CBT) 5,000 bu.; cents per bu.

	Open	High	Low	Settle	Change	Lifetime High	Lifetime Low	Open Interest
Mar	330¾	334	327	331	+ ¼	470	325	58,122
May	337¾	341	335	338¼	+ ½	439½	333	13,336
July	344½	347½	342	344¾	+ 1½	425	333	20,193
Sept	351¼	352½	350	350½	+ 1½	403	346½	1,069
Dec	361	363½	361	361	+ 1	417	358	2,907

Est vol 12,000; vol Fri 2,394; open int 19,135, +222.

WHEAT (KC) 5,000 bu., cents per bu.

	Open	High	Low	Settle	Change	Lifetime High	Lifetime Low	Open Interest
Mar	341½	344	337	338¾	− 2½	491	335	5,449
May	350½	351½	345½	345½	− 1½	450	343	1,477
July	357½	359	354½	354½	− 2¾	408½	333	1,913
Sept	359	359	359	359½	+ 2	410	335	124
Dec	369	370	367	367	+ 2	418½	363	103

Est vol 3,983; vol Fri 3,743; open int 45,449, +175.

WHEAT (MPLS) 5,000 bu.; cents per bu.

	Open	High	Low	Settle	Change	Lifetime High	Lifetime Low	Open Interest
Mar	366	370	364½	365½	− ¼	469	361	2,660
May	374½	376	371	372¾	− ½	440	363	788
July	380	382	378½	380¼	+ 1½	416	368	373
Sept	385	385	380½	383	+ 2	418	360	191

Est vol na; vol Fri 1,807; open int 20,179; +372.

CANOLA (WPG) 20 metric tons; Can. $ per ton

	Open	High	Low	Settle	Change	Lifetime High	Lifetime Low	Open Interest
Jan	380.70	381.60	380.10	381.60	+ 2.30	407.00	337.00	2,837
Mar	385.90	387.70	385.50	387.60	+ 2.20	412.30	339.00	27,090
May	390.50	392.50	390.50	392.50	+ 2.00	416.40	341.00	3,589
July	395.00	398.30	395.00	398.30	+ 2.50	417.40	366.00	975
Sept		....	....	376.10	....	391.00	363.50	220
Nov	376.50	377.80	376.00	377.60	− .60	384.00	365.00	514

Est vol 2,895; vol Fr 5,778; open int 35,305, +630.

WHEAT (WPG) 20 METRIC TONS; Can. $ per ton

	Open	High	Low	Settle	Change	Lifetime High	Lifetime Low	Open Interest
Mr98	163.00	164.50	162.60	164.50	+ 1.90	183.80	150.00	4,636
May		....	....	165.20	+ 1.40	184.40	152.00	1,289
July		....	....	166.20	+ 1.90	184.50	164.00	2,217
Oct		....	....	164.20	+ 1.70	177.00	164.60	1,161

Est vol 150; vol Fr 282; open int 9,303, −86.

BARLEY-WESTERN (WPG) 20 metric tons; Can. $ per ton

	Open	High	Low	Settle	Change	Lifetime High	Lifetime Low	Open Interest
Mr98	144.00	144.70	143.10	144.50	+ 1.60	161.00	126.30	6,911
May	146.80	147.90	146.70	147.90	+ 1.60	162.00	128.70	5,390
July	149.00	149.90	149.00	149.90	+ 1.40	162.60	145.00	1,140
Oct	148.00	148.90	148.00	148.90	+ .90	158.00	145.10	956

Est vol 175; vol Fr 330; open int 4,402, −37.

Livestock and Meat

CATTLE-FEEDER (CME) 50,000 lbs.; cents per lb.

	Open	High	Low	Settle	Change	Lifetime High	Lifetime Low	Open Interest
Jan	74.75	75.27	74.40	74.67	− .22	84.95	74.40	6,482
Mar	74.05	74.70	73.55	73.90	− .45	84.50	73.55	5,593
Apr	75.15	75.62	74.65	74.85	− .47	84.45	74.65	2,515
May	76.25	76.70	75.80	76.07	− .40	84.40	75.80	1,896
Aug	78.50	78.70	77.90	78.25	− .25	83.25	77.90	925
Sept	78.75	78.75	78.10	78.25	− .25	83.05	78.10	200
Oct	78.50	78.75	78.10	78.20	− .30	83.00	78.10	214

Est vol 4,298; vol Fr 3,477; open int 17,828, +151.

CATTLE-LIVE (CME) 40,000 lbs.; cents per lb.

	Open	High	Low	Settle	Change	Lifetime High	Lifetime Low	Open Interest
Feb	64.55	64.85	63.62	63.92	− 1.02	73.92	63.62	46,019
Apr	67.20	67.35	66.30	66.85	− .42	75.55	66.30	28,968
June	67.15	67.30	66.65	66.90	− .12	72.37	66.65	16,426
Aug	67.80	67.95	67.37	67.62	− .05	72.15	67.37	6,934
Oct	70.50	70.62	70.00	70.22	− .27	74.05	70.00	1,909
Dec	71.80	71.80	71.47	71.50	− .30	74.20	71.47	350

Est vol 22,842; vol Fr 18,360; open int 100,605, −859.

HOGS-LEAN (CME) 40,000 lbs.; cents per lb.

	Open	High	Low	Settle	Change	Lifetime High	Lifetime Low	Open Interest
Feb	57.60	57.60	57.07	57.20	− .32	71.90	57.05	21,937
Apr	56.52	56.70	56.22	56.62	+ .20	68.15	55.95	8,602
June	63.50	63.65	62.97	63.32	− .15	73.70	62.10	6,353
July	62.80	62.95	62.45	62.50	− .30	71.75	62.45	1,916
Aug	61.00	61.05	60.60	60.80	− .17	69.85	60.60	429
Oct	57.50	57.60	57.35	57.50	....	64.90	57.00	679
Dec	56.10	56.10	55.70	55.90	+ .15	58.50	55.25	197

Est vol 6,875; vol Fr 4,920; open int 40,169, +1,331.

PORK BELLIES (CME) 40,000 lbs.; cents per lb.

	Open	High	Low	Settle	Change	Lifetime High	Lifetime Low	Open Interest
Feb	50.45	51.20	49.75	50.17	− .90	81.02	49.50	5,585
Mar	50.50	51.10	49.35	49.62	− 1.22	79.40	49.35	1,525
May	51.25	51.80	50.60	51.30	− .47	79.00	50.50	1,167
July	50.25	50.85	49.50	50.20	− .65	79.00	49.50	678
Aug	46.80	47.50	46.80	47.05	+ .32	75.00	46.05	121

Est vol 2,028; vol Fr 1,373; open int 9,112, +170.

Food and Fiber

COCOA (CSCE)—10 METRIC TONS; $ PER TON.

	Open	High	Low	Settle	Change	Lifetime High	Lifetime Low	Open Interest
Mar	1,611	1,615	1,592	1,596	− 34	1,803	1,375	40,029
May	1,643	1,645	1,628	1,630	− 31	1,817	1,399	20,952
July	1,666	1,670	1,653	1,655	− 31	1,835	1,485	5,260
Sept	1,695	1,695	1,691	1,680	− 30	1,836	1,456	5,805
Dec	1,714	1,714	1,714	1,706	− 26	1,863	1,510	9,143
Mr99	1,741	1,741	1,738	1,735	− 26	1,901	1,634	8,579
May		....	....	1,753	− 26	1,911	1,688	2,165
July		....	....	1,770	− 26	1,759	1,705	601
Sept		....	....	1,784	− 26	1,778	1,773	416

Est vol 10,844; vol Wd 1,698; open int 92,950, −616.

COFFEE (CSCE)—37,500 lbs; cents per lb.

	Open	High	Low	Settle	Change	Lifetime High	Lifetime Low	Open Interest
Mar	161.90	171.00	161.90	170.70	+ 8.25	203.00	96.25	16,556
May	157.50	165.00	157.50	164.95	+ 7.20	195.00	101.00	5,673
July	151.50	159.00	151.50	159.00	+ 7.00	191.00	120.00	2,841
Sept	147.50	151.00	147.50	153.00	+ 6.75	186.00	122.50	1,354
Dec	142.25	145.00	142.25	148.50	+ 6.75	157.50	124.75	1,230
Mr99		....	....	145.50	+ 6.25	154.00	124.00	293

Est vol 7,528; vol Wd 4,254; open int 27,977, +76.

SUGAR-WORLD (CSCE)—112,000 lbs.; cents per lb.

	Open	High	Low	Settle	Change	Lifetime High	Lifetime Low	Open Interest
Mar	12.13	12.22	12.00	12.02	− .20	12.55	10.13	96,487
May	11.95	12.02	11.86	11.86	− .13	12.45	10.20	33,056
July	11.61	11.63	11.51	11.52	− .08	12.11	10.30	30,751
Oct	11.49	11.55	11.47	11.47	− .08	11.97	10.45	26,528
Mr99	11.38	11.43	11.38	11.38	− .06	11.87	10.41	7,362
May	11.33	11.33	11.33	11.33	− .07	11.68	10.75	1,220
July	11.30	11.30	11.28	11.28	− .06	11.68	11.28	1,399

EST VOL 26,357; VOL WD 9,808; OPEN INT 196,819, −457.

SUGAR-DOMESTIC (CSCE)—112,000 lbs.; cents per lb.

	Open	High	Low	Settle	Change	Lifetime High	Lifetime Low	Open Interest
Mar	21.89	21.89	21.89	21.89	− .03	22.51	21.81	2,585
May	22.10	22.10	22.10	22.10	− .03	22.56	22.06	3,038
July	22.40	22.40	22.40	22.40	....	22.67	22.33	2,713
Sept	22.50	22.50	22.50	22.50	− .03	22.68	22.35	2,555
Nov	22.43	22.43	22.43	22.43	....	22.51	22.00	1,048
Ja99		....	....	23.34	+ .02	22.44	22.24	301
Mar		....	....	22.37	+ .02	22.42	22.30	223

Est vol 722; vol Wd 448; open int 12,473, +121.

COTTON (CTN)—50,000 lbs.; cents per lb.

	Open	High	Low	Settle	Change	Lifetime High	Lifetime Low	Open Interest
Mar	66.98	67.85	66.90	67.54	+ .43	81.00	65.75	41,852
May	68.32	69.00	68.25	68.80	+ .38	81.00	67.40	14,985
July	69.70	70.25	69.55	69.94	+ .24	79.25	68.80	14,971
Oct	72.15	72.35	72.15	72.15	+ .30	78.00	70.80	1,444
Dec	72.50	73.35	72.50	73.14	+ .46	76.60	71.75	13,638
Mr99		....	....	74.14	+ .51	77.25	72.95	497
May		....	....	74.65	+ .40	75.30	73.40	313
July	74.95	75.00	74.95	75.15	+ .40	77.31	74.00	122

Est vol 15,000; vol Fr 4,147; open int 87,853, +213.

ORANGE JUICE (CTN)—15,000 lbs.; cents per lb.

	Open	High	Low	Settle	Change	Lifetime High	Lifetime Low	Open Interest
Jan	78.80	79.50	78.10	78.40	− 1.60	119.75	69.10	1,971
Mar	82.00	82.50	80.00	81.25	− 2.45	100.25	75.20	25,616
May	85.10	85.30	83.80	84.70	− 2.10	97.75	75.30	6,232
July	88.00	88.40	87.50	87.40	− 2.70	105.00	78.25	4,002
Sept	91.00	91.00	90.25	90.40	− 2.80	99.00	80.95	1,680
Nov	91.00	91.00	91.00	92.65	− 3.05	105.00	83.40	1,564
Ja99	97.00	97.00	97.00	94.90	− 3.30	107.50	86.05	444
Mar	95.70	95.70	95.70	97.15	− 3.55	109.50	90.80	495

Est vol 10,000; vol Wd 26; open int 42,075, −2,715.

Figure 18-3 Commodity contract specifications are listed in the format shown here. The *Wall Street Journal* is one of the best sources for commodity market information. (Courtesy of the *Wall Street Journal*)

bushel, or $12.50 per contract. Using fractions, you can figure out how we came up with $12.50. Because a 1-cent change in a 5,000-bushel contract equals $50 (5,000 × $0.01), a ¼-cent change equals $12.50 ($50 × ¼). A shorter calculation is: 5,000 bushels times $.0025 equals $12.50.

Here is a table to help in calculating the value of CBOT grain (corn) futures:

(cents/bu.; 5,000-bu./contract)

¼ cent per bu. = $12.50 per contract

½ cent per bu. = $25 per contract

¾ cent per bu. = $37.50 per contract
1 cent per bu. = $50 per contract

Keeping in mind that the minimum price fluctuation for CBOT corn futures is 14 cents per bushel, the next few higher prices above corn trading at 200 cents per bushel ($2.00/bu.) would be 200¼ cents, 200½ cents, 200¾, and 201 cents.

To calculate the value of a futures contract when fractions are involved, just use the same equation of settlement price times contract size. For example, if December corn futures are trading at 200¼ cents/bu. ($2.00¼), then the contract value is: $2.0025/bu. times 5,000 bushels equals $10,012.50.

Determining Profit or Loss on a Futures Contract

Suppose you read in the paper that soil moisture in the Midwest was below normal for the month of June and the forecast does not look promising for rain. Limited rainfall during the growing season could cause the production of corn to decrease, thus increasing the price. Anticipating higher corn prices, you buy one December corn futures contract at 250 cents/bushel ($2.50/bu). On July 1, if you were right and corn prices rise, you will make a profit. Remember, you want to satisfy the golden rule of trading: *Buy low, sell high.*

Throughout the month of July, there is no rain in the Corn Belt. The end result is higher corn prices, so you decide to offset your position on July 30 by selling one December futures contract at $2.55/bu.

Did you make a profit or a loss, and how much? First of all, ask yourself if you satisfied the golden rule of buy low, sell high. You did: you bought a corn futures contract at $2.50/bu. and sold it 5 cents higher at $2.55/bu. To figure out the actual dollar amount of your profit, use either one of the following calculations. (You know that the contract size is 5,000 bushels.)

Calculation #1

Jul. 1	BUY 1 Dec. corn futures at	$2.50/bu.
Jul. 30	SELL 1 Dec. corn futures at	$2.55/bu.
Profit		$.05/bu.

To calculate the profit per contract, multiply your profits by the contract size: $.05/bu. profit times 5,000 bushels equals $250 per contract. If you traded more than one contract, just multiply the number of contracts by $250 minus the profit per contract. Remember that brokerage fees are always subtracted from your profit.

Another way to figure out your profits is by calculating the values of the December corn futures contract on July 1 and July 30 and then subtracting the difference between the two.

Calculation #2

Jul. 1	BUY 1 Dec. corn futures valued at $12,500 ($2.50 × 5,000)
Jul. 30	SELL 1 Dec. corn futures valued at $12,750 ($2.55 × 5,000)
Profit	$250/contract
Total Profit	$250 × 1 contract traded = $250

As you can see, the results from both calculations are the same. Pick the method you feel most comfortable using to figure out trading profits and losses.

Hedgers and Speculators

There are two main categories of futures traders that utilize futures contracts. These are the hedger and the speculator. Hedgers either now own, or will at some time own, the commodity they are trading. Hedgers may be production agriculturalists, elevator owners, or any others in the agribusiness input or output sectors.

Speculators, in contrast, will likely never own or even see the physical commodity. In other words, they are not production agriculturalists or agribusiness owners. They just buy and sell futures contracts and hope to make a profit on their expectations and predictions of future price movements. Speculators always offset their positions by buying (selling) futures contracts they originally sold (bought). At planting time, for example, corn producers do not know what commodity prices will be at harvest. Hedging is a tool production agriculturalists and agribusinesses can use to shift some of the risk resulting from price uncertainty to speculators who trade in the futures market. Hedging involves taking a position in the futures market equal but opposite to what one has in the cash market. If prices fall, a producer who placed a hedge will be protected.[11] This is why hedgers willingly give up the opportunity to benefit from favorable price changes: to achieve protection against unfavorable changes.

Long (Buying) and Short (Selling) Hedgers

Two terms used to describe buying and selling are *long* and *short*. If you first buy a futures contract, this is called going long, or going **long hedge**. If you first sell a futures contract, this is called going short, or going **short hedge**.

Hedging in the futures market is a two-step process. Depending on your cash market situation, you will either buy or sell futures as your first position. For instance, if you are going to buy a commodity in the cash market at a later time, your first step is to buy a futures contract. In contrast, if you are going to sell a cash commodity at a later time, your first step in the hedging process is to sell futures contracts.

The second step in the process occurs when the cash market transaction takes place. At this time, the futures position is no longer needed for price protection, so it should therefore be offset (closed out). If your hedge was initially long, you would offset your position by selling the contract back. If your hedge was initially short, you would buy back the futures contract. Both the opening and closing positions must be for the same commodity, number of contracts, and delivery month.[12]

Example of a Long Hedge. A food processor is planning to buy 240,000 pounds of soybean oil over the next few months to cover production needs. Currently, soybean oil is quoted at 26 cents a pound, but the company management is concerned that prices will rise by the time it is ready to purchase and take delivery of the oil. To take advantage of current prices, the company decides to buy four CBOT September soybean oil futures contracts at 26 cents a pound. The standard contract size for CBOT soybean oil is 60,000 pounds.

The management's greatest fear comes true. The price of soybean oil jumps to 31 cents a pound by the time the food processor is ready to purchase it. The company offsets its futures position by selling four CBOT September soybean oil contracts for 31 cents a pound. The company's hedging activities result in the following:

Cash Market	Futures Market
June	**June**
Plans to purchase 240,000 lbs. soybean oil in the cash market at $.26 lb.	Buys 4 CBOT Sept. soybean oil futures contracts at $.26/lb.
August	**August**
Purchases 240,000 lbs. soybean oil in the cash market at $.31/lb.	Sells 4 CBOT Sept. soybean oil futures contracts at $.31/lb.
Purchase price of cash soybean oil	$.31/lb.
Less futures gain ($.31 − $.26)	$.05/lb.
Net purchase price	$.26/lb.

By using CBOT soybean oil futures, the food processor lowered its purchase price from 31 cents to 26 cents a pound. That was exactly what the company expected to pay. This type of hedge is a long hedge because the food processor initially purchased futures. If futures are initially sold, the hedge is a short hedge.[13]

Example of a Short Hedge. Rather than selling for whatever price the local elevator is willing to pay come harvest time, a producer decides to explore a variety of marketing alternatives, including futures. Her ultimate goal is to improve her bottom line. Suppose she figures that it costs her $2.00 to produce one bushel of corn. When corn prices enter a range where the producer can make a profit, she decides to hedge a portion of her crop by selling one CBOT December corn futures contract. (The standard contract size for one CBOT corn futures contract is 5,000 bushels.)

By early May, CBOT December corn futures hit $2.60 a bushel. To lock in a selling price of $2.60, the grain producer sells one CBOT December corn

11. Randall D. Little, *Economics: Applications to Agriculture and Agribusiness*, 4th ed. (Danville, IL: Interstate Publishers, 1997), p. 312.

12. Chicago Board of Trade, *Action in the Marketplace,* p. 9.

13. Chicago Board of Trade, *Agricultural Futures for the Beginner,* pp. 6–10.

futures contract. As it turns out, the Midwest experiences near-perfect growing conditions that year and corn yields are above normal, causing prices to drop. By harvest time, corn has fallen to $1.90 a bushel. The producer therefore offsets her futures position by purchasing one CBOT December corn futures contract. The corn producer's hedging activities result in the following:

Cash Market	Futures Market
May	**May**
Plans to sell 5,000 bu. corn in the cash market at $2.60/bu.	Sells one CBOT Dec. corn futures contract at $2.60/bu.
October	**October**
Sells 5,000 bu. corn in the cash market at $1.90/bu.	Buys one CBOT Dec. corn futures contract at $1.90/bu.
Sales price of cash corn	$1.90/bu.
Plus futures gain ($2.60 − $1.90)	$0.70/bu.
Net sale price	$2.60/bu.

By using CBOT corn futures, the producer increased her final sale price from $1.90 to $2.60 a bushel. That was exactly what she wanted to receive. Better yet, the final sale price was 60 cents per bushel higher than her production expenses.

Keep in mind that these examples are simplified to give you an overview of hedging. In both examples, the hedgers were holding two positions: a cash position and a futures position. By definition, a *hedge* consists of both a cash and a futures position, even when the cash position is anticipated, as shown in the first example.[14] As you will recall, the firm in the first example was planning to purchase 240,000 pounds of soybean oil in the cash market.

Transactions Can Be Unpredictable. Of course, the market does not always move as you predict or might anticipate—but a hedger's primary objective is to achieve price protection. This means that there could be times when you are not able to take advantage of a price move after you have initiated a futures hedge. However, experienced hedgers are willing to forfeit this opportunity so that they can gain price protection.

Relationships between the Cash and Futures Market. Understanding the relationships between the cash and futures markets is crucial to understanding the practice of hedging. A **cash market** is a market where actual commodities are bought and sold. A **futures market** is a market where traders buy and sell futures contracts. Many times the price of the underlying commodity and the related futures contract will move up and down together, but the price moves are not always equal.

Basis. In both of the preceding examples, the cash price of the commodity equaled the futures price, but in real life, this does not happen often. Typically, either the price one pays or receives from corn at a county elevator (cash market), for example, is either higher or lower than the futures price (Chicago Board of Trade). (There are many economic reasons why this happens, but this is beyond the scope of our discussion.) Anyway, the relationship between cash prices and futures prices is referred to as **basis**. The cash price minus the futures price of a commodity equals the basis. Successful hedgers understand how basis ultimately offsets their final results.[15]

Fertilizer Futures Now Traded

In 2004, the Chicago Mercantile Exchange (CME) added three fertilizer futures contracts to its list of agricultural commodities traded in the futures and options markets. The market price for these fertilizers has been extremely volatile in recent years, with annual price variations of up to 50 percent. This has been caused by volatility in the natural gas market, as natural gas is a significant input in the production of fertilizer. Such price variability can expose fertilizer merchandisers and producers to significant risk, and price swings can significantly alter per-acre cost of production.

Futures trading in fertilizer gives farmers and agribusinesses a tool with which to manage the risk associated with purchasing this important agricultural input. The size of the CME fertilizer futures contract is 100 tons, with delivery contract

14. Ibid.

15. Ibid.

months of March, May, July, September, and December.

Fertilizer Futures Contract Example. The fertilizer futures contract offers both farmers and fertilizer dealers/suppliers a way to manage the risk of changes in fertilizer prices. A farmer, for example, might be interested in locking in a favorable purchase price for a time in the future, whereas a fertilizer supplier may wish to cover cash-forward contracts it established with producers by locking in purchase prices using the futures market. In each of these cases, a long hedge would be used.

Assume, for example, that it is December 1 and that the current March futures price of fertilizer is $150/ton. The long hedger needs to secure 500 tons of fertilizer for use (either for application, in the case of a farmer, or to sell to farmers, in the case of a dealer). Because of volatility in the fertilizer and natural gas markets, the hedger decides to lock in the spring purchase at current futures prices by going long (buying) five March fertilizer futures contracts at $150/ton. Suppose also that the hedger expects that basis for March (difference between cash and futures price) will be $10/ton, and therefore anticipates a local cash price of $160/ton. Suppose that when March arrives, the March fertilizer futures price is $175/ton and basis is $10/ton as anticipated. The hedger would purchase the fertilizer in the spot market at $185/ton and simultaneously sell the futures contract (alternatively, the hedger could take physical delivery of the fertilizer specified in the futures contract). The futures position gain of $25/ton would make the realized cash purchase price $160/ton.

This hedge is illustrated in Figure 18-4. In this scenario, the hedger managed price risk, and would be able to obtain the fertilizer at a lower price than without the hedge. Note that although this fertilizer hedge is useful for mitigating price risk, it still leaves the user open to basis risk.

Speculators

Agricultural producers, commodity processors, exporters, food manufacturers, and others use the futures market to shift market risk (the risk of adverse price movements) to someone else. The party who assumes the risk is the *speculator*.

Gambler or Businessperson. A businessperson who is willing to take a price risk on a given product with the hope of making a profit is called a *commodity speculator*. The risk a speculator takes is not the same risk that a gambler takes in Las Vegas or in buying a lottery ticket. The risk in a casino is intentionally created; that is, risk of loss in games such as blackjack and roulette is built into the games. Those playing realize this, but hope to get lucky and make a profit even though the odds are not in their favor. This may be a source of entertainment for participants, but the losses are real, and society gains nothing.[16]

Speculators Are a Positive Influence in the Futures Market. In contrast to gambling, a commodity speculator assumes a naturally occurring risk rather than one that is deliberately created. In agribusiness, risk is inherent when one holds inventories of commodities. If speculators were not willing to assume some of this risk, a producer's costs would increase due to losses from adverse prices. Instead, agribusiness firms are able to produce items at lower costs because they have shifted some of their natural risk to commodity speculators. With this type of market, society benefits.[17]

Some have voiced concerns that commodity speculators could exercise unfair influence on, or have too much control over, futures market prices. This might have been true once, but it is far less likely now. Competition constrains those who try to unduly or artificially raise or lower marketplace prices. In addition, there are two tiers of oversight that prevent large-scale buyers and sellers from cornering a market or a group of speculators from manipulating prices.[18]

Speculators Are Not Interested in the Physical Commodity. In contrast to hedgers, speculators generally have no interest in buying or selling the actual commodity, whether it is corn, soybeans, or live cattle futures. Speculators buy and sell futures contracts expecting to make a profit based on their predictions of future price changes. The presence of speculators in the futures market increases the effectiveness of the market and provides hedgers with an opportunity to reduce their price risks.[19]

16. James G. Beierlein and Michael W. Woolverton, *Agribusiness Marketing: The Management Perspective* (Englewood Cliffs, NJ: Prentice-Hall, 1991), p. 296.
17. Ibid., p. 296.
18. Ibid., pp. 296–297.
19. Catania, *Commodity Challenge,* p. 22.

Fertilizer Futures Contract (Long hedge example)			
Date	**Cash**	**Futures**	**Basis**
12-1	No action	Buy 5 March CME UAN futures @ $150/ton	Expected 3-1 basis on Mar. 1 = $10/ton
3-1	Buy 500 tons UAN @ $185/ton	Sell 5 March CME UAN futures @ $175/ton	Actual basis on Mar. 1 = $10/ton
	Cash price paid = $185/ton	Net on futures = $25/ton	Difference between Actual & Expected basis on Mar. 1 = $0/ton

Figure 18–4 This is an example of the hedger managing price risk. Through the futures transaction, the hedger is able to obtain the fertilizer at a lower price than without the hedge.

The profit potential is proportional to the amount of risk that is assumed and the speculator's skill in forecasting price movement. Potential gains and losses are as great for the selling (short) speculator as for the buying (long) speculator. Whether long or short, speculators can offset their positions and never have to make or take delivery of the actual commodity.[20]

Trading Philosophy of Speculators. The golden rule of speculators is *buy low, sell high*. If a trader expects prices to rise, she will buy a futures contract. Then, at a later time, if prices rise, the trader will make an offsetting sale at a profit. The opposite also holds true: if a trader expects prices to fall, he will sell a futures contract. Again, at a later time, if prices fall, the trader will make an offsetting purchase at a profit. Experienced, knowledgeable speculators usually develop very sophisticated trading techniques.

Types of Speculators. Speculators can be classified by their trading methods: position trader, day trader, scalper, and spreader. A brief discussion of each follows:

- A *position trader* is a public or professional trader who initiates a futures or options position and holds it over a period of days, weeks, or months.
- A *day trader* holds market positions only during the course of a single trading session, and rarely carries a position overnight. Most day traders are exchange members who execute their transactions in the trading pits.
- *Scalpers* trades only for themselves in the pits. Scalpers trade in minimum fluctuations, taking small profits and losses on a heavy volume of trades. Like a day trader, a scalper rarely holds positions overnight.
- *Spreaders* trade on the shifting price relationships between two or more different futures contracts. Examples include: different delivery months for the same commodity, the prices of the same commodity on different exchanges, products and their by-products, and different but related commodities.[21]

Speculation Example. Let us say that, because of increasing world demand for corn, an investor anticipates that corn prices will rise, and so she buys one Chicago Board of Trade December corn futures contract at $2.75 a bushel. Within two weeks, the price rises to $3.00 a bushel. The investor offsets her position by selling one CBOT December corn futures contract at a profit of 25 cents a bushel ($3.00 selling minus $2.75 buying price). This equals a total profit of $1,250 ($.25 profit times 5,000 bushel contract size). Of course, if the price of the corn falls 25 cents per bushel, the investor loses $1,250 instead of profiting.

Determining the Price of a Commodity

Determining the price of a commodity is called *price discovery,* which also refers to the price of a futures contract agreed to by a buyer and a seller

20. Chicago Board of Trade, *Action in the Marketplace,* p. 20.

21. Ibid.

through a futures exchange. Futures exchanges themselves do not set price. They are free markets where the forces that influence prices are brought together in open auction. As this marketplace assimilates new information throughout the trading day, it translates the information into a single benchmark figure: a fair market price that is agreed on by both buyer and seller.[22]

Price Discovery Trading on the Trading Floor

The heart of a futures exchange is the **trading pit,** where buyers and sellers meet each day to conduct business. Each trader who enters these trading pits brings along specific market information, such as supply and demand figures, currency exchange rates, inflation rates, and weather forecasts. This information contributes to the ongoing price discovery function of the futures markets.[23]

For example, a grain exporter in the southern United States will be kept constantly informed of grain prices through the futures price quotes disseminated by the Chicago Board of Trade. This information will assist the export company in offering a purchase price for the grain it buys from American farmers and allow it to set an appropriate selling price for the grain it exports to other countries.

As trades between buyers and sellers are executed, the fair market value (price) of a given commodity is discovered and these prices are disseminated instantaneously by state-of-the-art telecommunications and computer throughout the world.[24] These prices are published in daily newspapers and carried by various online quotation services. If you inspect the newspaper listing of futures prices in Figure 18–5, you will notice the several contract specifications.

Forecasting Market Direction through Government Reports

Among the different factors market participants use to forecast market direction are USDA reports. Released throughout the year, the market anticipates the release of this data and uses the reported information in developing market opinion. Not surprisingly, attention to USDA reports is heightened during the growing season and climaxes at harvest. Some of the more closely watched reports are:

- *Crop Progress* (issued weekly during growing season) lists overall condition of selected crops in major producing states
- *Crop Production* (coincides with the growing season, issued around the 10th of each month) reports production estimates based on actual field surveys
- *Export Sales* (released weekly) summarizes U.S. export sales by importing country
- *Grain Stocks* (released quarterly) reports the number of stocks held by state
- *Supply and Demand* (issued around the 10th of each month) gives an overview of supply versus demand[25]

Making the Futures Trading Transaction

Opening an Account

Before you can do any futures trading, you must open an account with a **brokerage office** (commission house, wire house, futures commission merchant). Local brokerage offices are usually the most convenient outlet for futures transactions. These offices are located in most towns and cities throughout the United States. To be sure that you are dealing with a reputable broker, contact current and past customers and then check to see if the broker is registered with the National Futures Association (NFA) and that no complaints have been registered against that broker.[26] You can contact the NFA by calling 1-800-621-3570 (outside of Illinois) or 1-800-572-9400 (within Illinois). When you open an account, you should also inform the broker whether you will be using it for hedging or speculating purposes. Brokers charge customers a commission fee to buy or sell futures contracts. Commission rates will vary depending on the kind and amount of services offered, but rates are generally competitive. The customer also signs several

22. Chicago Board of Trade, *Agricultural Futures for the Beginner,* p. 12.
23. Ibid., pp. 10–11.
24. Ibid., pp. 10–11.
25. Ibid., p. 10.
26. Ibid., p. 22.

Sample Price Listing September 12

CORN (CBOT) 5,000 bu; cents per bu

	Open	High	Low	Settle	Change	Open Interest
Sep	293½	294½	290¼	292	+1¾	4,370
Dec	295	298¾	293¾	298½	+5¾	282,106
Mar 06	303	306¼	301	306	+5¾	78,564
May	305¼	308¾	304¼	308½	+5½	11,796
Jul	304¾	308¾	304	308½	+5½	30,727
Sep	284	287½	283¾	287¼	+4	2,292
Dec	269	273½	268½	272¼	+4	14,140
Dec 07	263	264	263	264	+2	201

Est vol 57,000; vol Mon 35,804; open int 424,916

Commodity: Chicago Board of Trade corn

Contract months: Sep '05, Dec '05, Mar '06, May '06, July '06, Sep '06, Dec '06, Dec '07

Contract size: 5,000 bu.

How the contract is quoted: cents per bu.

Figure 18–5 The price per bushel of corn shown here is not mandated by the futures exchanges. Instead, the price is set when a buyer and a seller agree on a price.

legal documents concerning both parties' responsibilities regarding the account.[27]

Types of Accounts. There are various types of futures accounts. The two most used by beginners are the individual and joint accounts. *Individual accounts* are used when trading decisions are made by the customer. *Joint accounts* are set up when both the broker and the customer are to have input into trading decisions. When an account is set up, money is deposited to get ready to trade.

Margins

When customers place orders, they are required to post a **performance bond margin**, which is a financial guarantee required from both buyers and sellers to ensure that they fulfill the obligation of the futures contract. Minimum margin requirements for futures contracts, which are set by futures exchanges, usually range between 5 and 18 percent of a contract's face value. However, brokerage firms may require a margin larger than the exchange minimum.[28]

The initial amount market participants must deposit into their accounts when they place an order is called **initial margin**. On a daily basis, the margin is debited or credited based on the close of that day's trading session (referred to as *marking to the market*) with respect to the customer's open position. A customer must maintain a set minimum margin known as **maintenance margin** in the account. If the debits from a market loss reduce the funds in the customer's account below a preset maintenance level, the broker will call the customer for additional funds to restore the account to the initial margin level. This request for additional money is termed a **margin call**. The customer may withdraw from the account any margin in excess of the required amount.[29]

Placing Orders

A vital function of the commission house is to properly relay customer orders. Orders must be written and entered without vagueness to ensure that they are properly handled on the trading floor. A mistake in this process can be very

27. Chicago Board of Trade, *Action in the Marketplace,* p. 12.
28. Ibid., p. 12.
29. Ibid., p. 12.

costly; therefore, both the customer and the commodity representative must communicate orders clearly and accurately. There are three major types of orders: the market order, limit order, and stop order. Keep in mind that each individual brokerage firm and exchange decides which kinds of orders it will accept. Be sure to talk to your broker before deciding which orders you want to use. A brief discussion of each type follows.

Market Order. A market order is the most common type of order. In a market order, customers state the number of contracts of a given delivery month that they wish to buy or sell. Customers do not specify the price; rather, they simply direct that the order be executed as soon as possible at the best possible price.

When a market order is filled, it is usually filled at a price close to the price that was trading at the time the order was placed. However, in a fast market, the actual price at which the order is filled could differ from what the price was when the order was entered.

Limit Order. A limit order states a price limit at which it must be executed. It can be executed only at that price or better. The advantage of a limit order is that customers know the worst price they will receive if the order is executed. The disadvantage is that the order may not be filled.

Stop Order. A stop order is normally used to liquidate earlier transactions. Assume that a customer bought a Chicago Board of Trade corn futures contract for $3.00 per bushel. To prevent a large loss in a **bear market** (in the event corn prices fall), the customer could place a stop order at $2.75 per bushel. The customer's position is liquidated if and when corn futures drop to $2.75. The stop order is executed when the stop price is reached, but not necessarily at the stop price.

A stop order also can be used to enter the market. Suppose a trader expected a **bull market** only if the market surpasses a specific price level. In that case, the trader could use a *buy-stop order* when and if the market reached this point.

One variation of a stop order is a *stop-limit order*. With a stop-limit order, the trade must be executed at the exact price (or better) or held until the stated price is reached again. If the market fails to return to the stop-limit level, the order is not executed.[30]

Order Route

Orders received from customers are sent as soon as possible by a commission house to the exchange on which the order is to be executed. Telephone stations are located on the outskirts of the trading floor to handle the many orders coming to the exchange from companies and individuals. All orders are time-stamped at various stages along the order route, as a check to ensure that the order is being handled in the most expeditious fashion. After the order is received by the exchange floor phone clerk, it is time-stamped and passed to a broker by a runner (messenger).

A pit broker may receive an order to "buy five [meaning 5,000 bushels] December wheat [at the market price, which is understood]." The pit broker looks for another trader who is selling. By voice and hand signals, the pit broker flashes the bid to buy, receives a "sold" signal from the seller, and the order is completed. The broker writes the price and the seller's identity on the order blank and endorses it, and a runner returns it to the floor phone clerk.

When the phone clerk receives the endorsed order indicating that the trade has been executed, he or she time-stamps the order and phones the report to the firm's wire room, where it is relayed to the initiating office. The commission house confirms the transaction with the customer.[31] Refer to Figure 18–6 for detailed, step-by-step procedures of the order flow.

Action on the Trading Floor

All market participants have indirect access to the trading floor through their brokers, but only exchange members have the privilege of trading on the floor. To the average person, the system appears to be totally chaotic, but in fact it is very well organized. At the Chicago Board of Trade, trading is done in pits (raised octagonal platforms), which permit all buyers and sellers to see each other. On each business day, trading is

30. Ibid., p. 13.
31. Ibid., p. 14.

Order Flow

1. After you place an order with your broker, he or she will call in the order to the floor of the Chicago Board of Trade.
2. The order is time-stamped.
3. A member firm employee, known as a runner, rushes the order to a floor broker in the trading pit. (Some orders are flashed by hand signals to brokers in the trading pit, and some fills are flashed back to the order station.)
4. The floor broker fills the order using open outcry and hand signals to communicate price and quantity.
5. Once the order is filled, the floor broker reports the price to the appropriate Chicago Board of Trade price reporter overlooking the trading pit.
6. The price reporter enters the price change into a laptop quotation computer.
7. The computer transmits the price information to electronic price boards facing the trading floor and around the world via information services.
8. The completed order is picked up by a runner and returned to the telephone desk.
9. The order is time-stamped again.
10. The customer is notified that his or her order has been filled.

Figure 18–6 Step-by-step procedure for the order flow of a commodity (futures) contract.

officially opened and closed by the clanging of a large gong. No trades can be made before or after this official signal. The rules and regulations of the Chicago Board of Trade require pit traders to use **open outcry** in buying and selling. In addition, they use standard hand signals to clarify their verbal bids and offers. Refer to Figure 18–7.

Because all trades must be recorded, traders wear colored badges indicating their trading privileges. Exchange rules dictate that personnel (both male and female) on the trading floor must wear jackets and ties. Everyone on the floor performs very specific and important job functions, and the colorful jackets help distinguish their jobs. So that brokerage firms can quickly and easily identify their staff on the trading floor, the staff members of each firm wear trading jackets of a certain color.

Futures Options

In 1982, another market innovation, *options on futures*, was instituted. In contrast to futures, options on futures allow investors and risk managers to define risk and limit it to the cost of a premium paid for the right to buy or sell a futures contract. At the same time, options can provide the buyer with unlimited profit potential.

Options contracts give production agriculturalists a choice; they do not "lock" the farmer into the market at a set price. With an option, Production agriculturalists have the right, but not the obligation, to buy or sell a specific commodity within a specific period of time at a specific price. Production agriculturists can use options contracts to establish a minimum selling price for their crops or livestock, while still retaining the right to any price increase. Livestock producers can utilize options contracts to establish a maximum cost for feed grains, while retaining the right to any price decrease. Both strategies involve buying an options contract.

Futures options are much more attractive to many hedgers and speculators than straight futures contracts. Futures options are similar to making a down payment on a house. Let us say you paid $1,500 to a real estate agent as a down payment on a $125,000 house that you intended to buy. While you were making arrangements to complete the contract, you discovered a $95,000 house with just as much floor space, which you liked as well as the $125,000 house. Therefore, you lose your $1,500 down payment that was used to hold the first house for you, but you avoid losing $30,000 by taking a better deal.

Here is another example of a simplified options contract. Consider a call option that conveys the

Figure 18–7 The action on the trading floor of an exchange appears to be chaotic, but it is very well organized.

right to purchase a used combine from your neighbor. You are debating whether to buy a used combine or to put up the capital for a new combine. You convince your neighbor to sell you an option to purchase the combine at any time before April 1. In turn, the neighbor gives you the right to buy the used combine for $10,000. For this right, you pay $2,000.

In options terms, the combine is the underlying commodity, and $10,000 is the *strike price*. April 1 is the expiration date, and the $2,000 you paid for the option is the **premium**. At any time before the expiration date, your option contract gives you the right to exercise your option and purchase the combine. However, you are not obligated to buy the combine. You may choose to not exercise your option—you can simply let your option expire. You may offset your current position by selling your option to someone else. Whatever measure you take, the *writer* (seller) of the option keeps the $2,000 premium.

With options, once you make a transaction, you can predict your maximum losses. For example, let us say you bought a futures options contract for December corn at $3.00 per bushel, anticipating that the price would go to $3.50 per bushel. The premium for the futures options contract cost $35. Unfortunately, because of overproduction, imports, embargoes, and other unexpected happenings, the price of corn drops to $2.00 per bushel. Since the market has gone against you, you can simply let your option expire. Your loss will be $35 plus a commission.

What if this transaction had been a futures contract? Remember, one CBOT futures contract for corn includes 5,000 bushels. Therefore, 5,000 bushels of corn times $3.00 per bushel equals $15,000, and 5,000 times $2.00 per bushel equals $10,000. ($15,000 minus $10,000 equals $5,000). As you can see, with the same drop in prices, a futures contract loses $5,000 plus commission, but a futures option contract loses only $35 plus commission. What if you had bought 10 futures contracts of corn?

Options allow the hedger and speculator to have their cake and eat it too. As you can see from the preceding examples, an option gives the buyer the right, but not the obligation, to buy or sell an agricultural futures contract at some specified time in the future for a price set at the time the option is purchased.

Career Option

Commodity Trader/Broker

Many of the agricultural crops grown in the United States are commodities valued around the world. Corn, cotton, wheat, sugar cane, sugar beets, live cattle, hogs, pork bellies, and orange juice are some examples of commodities.

A commodity broker buys and sells the predictions for crop success or failure on an open market. The Chicago Board of Trade is one of the best known commodity markets; there trade is conducted Monday through Friday every week of the year. Speculators in the commodity market are located all over the world. They buy and sell portions of these and other crops, or futures, through a broker or trader.

Experience and education are needed to be a commodity broker or trader. An understanding of agricultural production, crop trends over time, grains, animals, oils, and specialty crops is essential for success in this field. A bachelor's degree in agribusiness, agricultural economics, finance, or marketing is beneficial. A commodity broker or trader is licensed by the exchange or board to conduct business, and they all must adhere to a strict code of ethics.

Imports and exports of commodities are somewhat controlled by the futures market, and help the United States to maintain a favorable balance of payments with other countries. Commodity brokers generally work out of an office, taking buy and sell orders from speculators. They relay the orders to the commodity trader, who is present and active on the buying and selling floor of the market itself.

Related jobs that are important to the commodity market are farmer, hog or cattle producer, grain producer, orange grove owner, grain elevator operator, trucker, sugar processor, crop forecaster, and market reporter.

This is an action shot of commodity traders at the Chicago Board of Trade. (Reprinted with permission of the Chicago Board of Trade. From *Action in the Marketplace*. © 1994)

Calls (Purchases) and Puts (Sells)

A **call** is an option that gives the option buyer the right (without obligation) to purchase a futures contract at a certain price on or before the expiration date of the option, for a price called the premium which is determined in open-outcry trading in pits on the trading floor. A **put** is an option that gives the option buyer the right (without obligation) to sell a futures contract at a certain price on or before the expiration date of the option.

Premium Cost

The cost of a futures option is called the *premium*. In agricultural options traded at the Chicago Board of Trade, the premium is the only variable in the contract. The premium depends on market conditions, such as volatility, time until the option expires, and

Example of Newspaper Listing

CORN (CBOT) 5,000 bu; cents per bu

Strike		**Calls-Settle**			**Puts-Settle**	
Price	Dec	Mar	May	Dec	Mar	May
280	20¾	29¼	31½	2½	3⅛	4½
290	13¾	21¼	24½	5⅜	6½	8
300	8⅝	16	19	10⅝	10½	.....
310	5½	11½	14	17¾	16	.....
320	3⅜	8	11	25½	22	.....
330	1⅞	5¾	8	34	29½	.....

Est vol 20,000; vol Mon 9,120 calls; 4,021 puts

Open int Mon 209,380 calls; 120,729 puts

1. The first column lists several strike prices for Chicago Board of Trade corn options. Typically, the exchange lists corn options in 10-cent increments.

2–4. Columns 2–4 list option premiums for Dec, Mar, and May CBOT corn call options. CBOT corn options trade in 1/8 cent intervals and premiums are quoted in cents and eighths of a cent per bushel.

5–7. Columns 5–7 list option premiums for Dec, Mar, and May CBOT corn put options. Below the columns appear estimated trading volume figures for the day reported followed by actual trading volume and open interest for the preceding day divided between calls and puts.

Figure 18-8 Once you decide to purchase a futures option, locate a newspaper or other service to find the listing of the option's price. Notice that there are many choices of options, as illustrated here.

other economic variables. The price of an option is discovered through a trading process regulated by the exchange and the futures industry.

Different call and put options trade simultaneously. Trading months for options are the same as those for the underlying futures contracts, and strike prices are listed in predetermined multiples for each commodity. Initially, the strike prices are listed in a range around the current underlying futures price. As futures prices increase or decrease, additional higher and lower strike prices are listed. For each agricultural option traded, you will find several option prices listed. Each price listed represents the price for a given contract month and strike price.[32] You can find option prices in daily newspapers, through online quotation services, or at local grain elevators. Refer to Figure 18-8.

The premium is the only element of an option contract that is negotiated in the trading pit; all the other parts of an option contract—such as the strike price and expiration date—are predetermined or standardized by the Chicago Board of Trade or other mercantile exchange.[33]

The Most You Can Lose Is the Premium

Regardless of how much the market fluctuates, the most an option buyer can lose is the premium. You deposit the premium with your broker, and the money goes to the option seller. Because of this limited and known risk, buyers are not required to maintain margin accounts, as they must when entering into futures contracts. (*Margin* is a cash deposit used to secure a contract.) Option sellers, in contrast, face the same types of risks as participants in the futures market, and thus must post margin with their brokers. The amount of margin required for option sellers depends on

32. Chicago Board of Trade, *Agricultural Options for the Beginner* (Chicago: Chicago Board of Trade, Market and Product Development Department, 1996), p. 5.

33. Ibid., p. 5.

Determining Option Classifications		
	Call Option	**Put Option**
In-the-money	Futures price > Strike price	Futures price < Strike price
At-the-money	Futures price = Strike price	Futures price = Strike price
Out-of-the-money	Futures price < Strike price	Futures price > Strike price

Figure 18–9 Your decision as to whether you want to exercise, offset, or expire your futures options is based on the price of the commodity being in the money, at the money, or out of the money.

their overall position risk. At the Chicago Board of Trade, the exchange uses a margining system called SPAN®, which evaluates a customer's overall risk and determines the amount of required margin based on that risk. In addition to the premium, the broker you use to buy and sell an option will charge a commission fee.

Intrinsic Value

The option premium—the price of an option—equals the option's intrinsic value plus its time value. An option has **intrinsic value** if it would be profitable to *exercise* the option. Call options, for example, have intrinsic value when the strike price is below the futures price (meaning you can purchase the underlying futures contract at a price below the current market price). For example, when the December corn futures price is $2.50, a December corn call with a strike price of $2.20 has an intrinsic value of 30 cents a bushel. If the futures price increases to $2.60, the option's intrinsic value increases to 40 cents a bushel. If the futures price declines to $2.40, the intrinsic value declines to 20 cents a bushel.

Put options, in contrast, have intrinsic value when the strike price is above the futures price (meaning you can sell the underlying futures contract at a price above the current market price). For example, when the July corn futures price is $2.50, a July corn put with a strike price of $2.70 has an intrinsic value of 20 cents a bushel. If the futures price increases to $2.60, the option's intrinsic value declines to 10 cents a bushel.

Another way to say that an option has intrinsic value is to say that the option is *in the money*. Options also can be *at the money* (the option has no intrinsic value; strike price equals futures price), or *out of the money* (the option has no intrinsic value and exercise of the option would not be profitable). If an option has no intrinsic value, then the premium is equal to its **time value**.[34] Refer to Figure 18–9.

Time Value

Time value—sometimes called *extrinsic value*—reflects the amount of money that buyers are willing to pay hoping that an option will be worth exercising at or before expiration. For example, if July corn futures are at $2.16 and a July corn call with a strike price of $2 is selling for 18 cents, then the intrinsic value equals 16 cents (the difference between the strike price and futures price) and the time value equals 2 cents (difference between the total premium and the intrinsic value). Note that the time value of an option declines as the expiration date of the option approaches. The option will have no time value at expiration, and any remaining premium will consist entirely of intrinsic value. Major factors affecting time value include the following.

Relationship between the Underlying Futures Price and the Option Strike Price. Time value is typically greatest when an option is at the money. This is because at-the-money options have the greatest likelihood of moving into the money before expiration. In contrast, most of the time value in a deep in-the-money option is eliminated because it is fairly certain that the option will not move out of the money. Similarly, most of the time value of a deep out-of-the-money option is eliminated because it is unlikely to move into the money.

Time Remaining until Option Expiration. The greater the number of days remaining until expiration, the greater the time value of an option will be.

34. Ibid., p. 7.

This occurs because option sellers will demand a higher price because it is more likely that the option will eventually be worth exercising (in the money).

Market Volatility. Time value also increases as market volatility increases. Again, option sellers will demand a higher premium because the more volatile or variable a market is, the more likely it is that the option will be worth exercising (in the money).

Interest Rates. Although the effect is minimal, interest rates affect the time value of an option: as interest rates increase, time value decreases.[35]

Using Futures Options to Benefit Agribusiness

Food manufacturers, processors, millers, and other agribusinesses can use options to better manage their raw material positions so as to increase their profitability. Options are also a good way to protect the value of their on-hand inventories. *Elevator operators* can use agricultural options to offer their customers a wider variety of marketing alternatives, giving those elevator firms a competitive edge. One popular contract now possible with agricultural future options is a minimum-price contract, through which producers establish a minimum price for their crop but leave open the opportunity to increase the final sale price.

If you are a *production agriculturalist,* you can lock in a desirable selling price for your crops or obtain "price insurance" when you store your crop.

Feedlot managers can use options to protect themselves against rising feed costs by using agricultural futures options to set a maximum purchase price.

Investors have a number of option strategies, at a variety of risk levels, that can provide profit opportunities. For example, the purchase of agricultural futures options to profit from an expected price change can provide unlimited profit potential and limited risk.[36]

Ways to Exit a Futures Option Position

Once an option has been traded, there are three ways you can get out of a position: exercise the option, offset the option, or let the option expire.

Exercise. Only the option buyer can decide whether to exercise an option. When an option position is exercised, both the buyer and the seller of the option are assigned a futures position. The option buyer first notifies his or her broker, who then submits an exercise notice to the Board of Trade Clearing Corporation. The exercise is carried out that night. The Clearing Corporation creates a new futures position at the strike price for the option buyer. At the same time, it assigns an opposite futures position at the strike price to a randomly selected customer who sold the same option. The entire procedure is completed before trading opens on the following business day.

Offsetting. Offsetting is the most common method of closing out an option position. You do this by purchasing a put or call identical to the put or call you originally sold; or by selling a put or call identical to the one you originally bought.

Offsetting an option before expiration is the only way you can recover any remaining time value. Offsetting also precludes the risk of being assigned a futures position if you originally sold an option and want to avoid the possibility of being exercised against. Your net profit or loss, after a commission is deducted, is the difference between the premium you paid to buy (or received to sell) the option and the premium you receive (or pay) when you offset the option. Market participants always face the risk that there may not be an active market for their particular options at the time they choose to offset, especially if the option is out of the money or the expiration date is near.

Expiration. The other choice you have is to let the option expire by simply doing nothing. In fact, the right to hold the option up until the final day for exercise is one of the features that makes options attractive to investors. Therefore, if the change in price you anticipated does not occur, or if the price initially moves in the opposite direction, you have the assurance that the most you can lose is the premium you paid for the option. In contrast, option sellers have the advantage of keeping the entire premium they earned, provided the option does not move into the money by the expiration date.

Many options expire, worthless, on their last trading day. It is important to recognize that options on futures frequently follow a different

35. Ibid., pp. 8–9.
36. Ibid., p. 10.

Example #1

Situation: A farmer wants protection in case soybean prices fall by harvest

Strategy: Farmer buys a CBOT put option

If falling commodity prices are a threat to your profitability, buying put options allows you to establish a minimum selling price without giving up the opportunity to profit from higher prices.

Assume that you are a soybean producer and you have just finished planting your crop. You are worried that prices may fall before harvest and would like to establish a floor price for a portion of your expected soybean crop. You purchase a November soybean put with a strike price of $6.50—this gives you a minimum selling price of $6.50 a bushel (excluding basis, commissions, and the cost of the option).

If, just before harvest, the November futures price declines to $5.75 (your put option is in the money), you can exercise your option and receive a short futures contract at the $6.50 strike price. Buying back the futures contract at the lower market price of $5.75 gives you a 75-cent per bushel profit (the difference between the strike price and the futures price), which should roughly offset the decline in the price of soybeans. Or, rather than exercising the option, you might be able to earn more by selling back the option to someone else. This would allow you to profit from any remaining time value as well as the 75-cent intrinsic value, both of which would be reflected in the option premium.

If November soybean futures were trading at $5.75, the $6.50 soybean put would be worth at least 75 cents plus any remaining time value. Generally, when an option is in the money and has time value, it is common for someone to *offset*—sell back the option rather than exercising it. That is because exercising the option will yield only its intrinsic value. Any time value that remains will be forgone unless it is offset. Also, an extra brokerage commission may be incurred when exercising an option.

If the price of soybeans increases just before option expiration and is above the option strike price (your put option is out of the money), you can simply let the option expire, or sell it back prior to expiration to capture any remaining time value. In either case, your loss on the option position can be no larger than the premium paid, and you still will be able to sell your crop at the higher market price.

Figure 18–10 A grain producer who is worried about falling prices can buy a *put option*, as illustrated here.

trading schedule than the underlying futures contracts; typically they expire a month before the underlying futures contract does. For example, options on December CBOT corn futures stop trading in the third week of November, whereas the December CBOT corn futures contract does not stop trading until mid-December.

Options that have value are usually offset, rather than exercised, before they expire. However, if you are holding an in-the-money option at expiration, the Board of Trade Clearing Corporation will automatically exercise the option unless you give notice to the Clearing Corporation, before expiration, not to do so.[37] Refer to Figures 18–10 and 18–11 for two examples of futures options.

37. Ibid., pp. 11–13.

Regulation within the Commodity (Futures) Exchanges

Is Your Money Safe?

If you are seriously considering changing your current business practices to include futures, so that you have better protection against price fluctuations, one of your concerns is surely whether your money is safe. It is really in everyone's best interests to have the futures industry provide a safe, well-capitalized arena for trading futures contracts. Exchanges, market participants, and federal regulators work together to protect the integrity of this marketplace.

Exchanges are responsible for ensuring that their daily market operations are handled efficiently and in keeping with the highest standards of ethical business conduct. From an exchange

Example #2

Situation: A food processor wants protection against rising raw material costs

Strategy: Food processor buys a CBOT soybean oil call option

If rising commodity prices are a threat to your profitability, buying call options allows you to establish a maximum purchase price without giving up the opportunity to profit from falling prices.

For example, suppose that you are a food processor and you expect to buy soybean oil during the spring. You are worried that prices may rise over the next few months and would like to establish a ceiling price for your eventual purchase. So, you buy a May soyoil call with a 25-cent strike price: this gives you a maximum purchase price of 25 cents a pound (excluding basis commissions and the cost of the option).

If, in April, the May futures price increases to 30 cents (your call option is in the money), you can exercise your option and receive a long futures contract at the 25-cent strike price. Selling back the futures contract at the higher market price (30 cents per pound) gives you a 5-cent a pound profit, which should roughly offset the increase in the cost of soyoil. Or, rather than exercising the option, you might be able to earn an even larger profit by selling the option to someone else (offsetting the option position) at a higher premium. This would allow you to profit from any remaining time value as well as the increase in intrinsic value, both of which would be reflected in the option premium.

The 25-cent soyoil call would be worth at least its intrinsic value of 5 cents (the difference between the strike price and the futures price) plus any remaining time value. In most cases, when an option is in the money and has time value, it is common for someone to offset—in this case, sell back the option rather than exercising it. That is because exercising the option will yield only its intrinsic value. Any time value that remains will be forgone unless it is offset. Also, an extra brokerage commission may be incurred when exercising an option.

If the price of soyoil in April declines and is below the option strike price (your call option is out of the money), you can simply let the option expire, or sell it back prior to expiration. In either case, your loss on the option position can be no greater than the premium paid, and you will still be able to purchase your soyoil at the lower market price.

Figure 18–11 A food processor who is worried about rising prices can buy a *call option*, as illustrated here.

standpoint, one of the foundations of maintaining safe markets is the margining system.[38]

Protecting the Marketplace

As a regular part of its financial surveillance activities, the exchange monitors the positions of all traders with large orders within a firm. This continuous review of traders who hold large positions gives the exchange the ability to anticipate and counteract potential liquidity concerns created by open positions held by a member firm.

Protecting Investors in U.S. Markets

The Commodity Futures Trading Commission (CFTC) was created by Congress in 1974 to oversee the trading of contracts on U.S. futures exchanges. Empowered with jurisdiction over futures trading, the responsibilities of the CFTC include regulating exchange trading of futures contracts; approving the rules and regulations of all futures exchanges, approving all new types of contracts to be traded; and enforcing rules to protect customer funds.

In addition, the National Futures Association (NFA) was established in 1982 as a self-regulatory organization to ensure the highest standards of professional conduct among the individuals, firms, and organizations that make up the futures industry. Membership in the NFA is mandatory for futures commission merchants, commodity pool operators, introducing brokers, and commodity trading advisors. The overall objective of the NFA is to protect the futures-trading public. It registers companies and individuals that deal with public customers, enforces compliance with its own regulations and those of the CFTC, and provides a venue for resolving futures-related disputes.[39]

38. Chicago Board of Trade, *Agricultural Futures for the Beginner*, p. 13.

39. Ibid., pp. 15–18.

Conclusion

Rules and regulations of the exchanges are extensive and are designed to support competitive, efficient, liquid markets. These rules and regulations are scrutinized continuously and are periodically amended to reflect the needs of market users. The success of the system is obvious. Trading futures and futures options can be a very exciting and profitable learning experience. Before jumping into the trading market, though, do your homework, consider the risk, and evaluate your financial situation. The goal of trading futures is to improve your financial situation.

Summary

Since the early development of agricultural markets in the United States, production agriculturalists have attempted to protect themselves against falling commodity prices at harvest time. Production agriculturalists soon realized that a marketing strategy was equal in importance to production, capital, and labor strategies. Thus, forward contracting led to the development of commodity (futures) markets to reduce marketing risk.

Commodity (futures) markets originated in Chicago in the 1800s. Price fluctuations and increasing production created the need for a year-round, centralized marketplace. Eighty-two Chicago businessmen met this need when they founded the Chicago Board of Trade. In 1865, the Chicago Board of Trade developed standardized agreements called futures contracts.

Besides the Chicago Board of Trade, there are many other commodity (futures) exchanges. The hardest concept for a beginner to understand is that exchanges do not physically trade or handle grain, livestock, or other commodities. Trades are made for contracts of commodities, but delivery rarely occurs. Besides agricultural commodities, there are future contracts for interest rates, insurance, stock indexes, manufactured and processed products, nonstorable commodities, precious metals, and foreign currencies and bonds.

Futures contracts are legally binding agreements, made on the trading floor of a futures exchange, to buy or sell something in the future. Actually, there is no such thing as a written futures contract. It is a verbal—but binding—agreement between a buyer and a seller made on the floor of a futures exchange.

Two main types of futures traders use futures contracts: the hedger and the speculator. Hedgers either now own, or will at some time own, the commodity they are trading. Speculators will likely never own or even see the physical commodity. They just buy and sell futures contracts and hope to make a profit through correct predictions of future price movements.

Determining the price of a commodity is called price discovery. Futures exchanges do not set prices. They are free markets where the forces that influence prices are brought together in open action.

Before you can do any futures trading, you must open an account with a brokerage office. When you open an account, you should inform the broker whether you will be using your account for hedging or speculating. When customers place an order, they are required to post a performance bond margin, which is a financial guarantee required of both buyers and sellers to ensure that they fulfill the obligation of the futures contract.

A vital function of the commission house is to properly relay customer orders. There are three major types of orders: the market order, limit order, and stop order. Orders received from customers are sent, as soon as possible, by a commission house to the exchange on which the order is to be executed. After several steps on the trading floor are completed, the commission house confirms the transaction with the customer.

In 1982, futures options were instituted. In contrast to futures, options on futures allow investors and risk managers to define risk and limit it to the cost of a premium paid for the right to buy or sell a futures contract. Options can provide the buyer with unlimited profit potential.

It is in everyone's best interests to have the futures industry provide a safe, well-capitalized arena for trading futures contracts and options. Exchanges, market participants, and federal regulators work together to protect the integrity of the futures marketplace.

Trading futures and futures options can be a very exciting and profitable learning experience. Before you enter the trading market, do your homework, consider the risk, and evaluate your financial situation. Improving your financial situation is the goal of futures trading.

End of Chapter Activities

REVIEW QUESTIONS

1. Why was the Chicago Board of Trade started?
2. When was the Chicago Board of Trade started?
3. List the names of 12 futures exchanges other than the Chicago Board of Trade.
4. What is the hardest concept for a beginner to understand about futures exchanges?
5. Other than grain and livestock, list eight other futures contracts that are sold in futures exchanges.
6. Explain the difference between hedgers and speculators.
7. Explain how production agriculturalists use hedging to protect or enhance their financial situations.
8. Explain how speculators positively influence the futures market.
9. How are the prices of a commodity determined?
10. List five USDA reports that help to forecast market direction.
11. List 10 steps in the procedure for filling a futures contract order.
12. Compare futures contracts and futures options.
13. What are four major factors affecting the time value of a futures option?
14. Describe five ways in which using futures options can benefit agribusiness.
15. Name and briefly explain three ways to exit a futures option position.
16. List two organizations whose objectives are to protect commodity (futures) traders and investors.

FILL IN THE BLANK

1. A _________ strategy is equal in importance to production, capital, and labor strategies.
2. Standardization of contract terms and the ability to _________ contracts led to a rapid increase in the use of futures markets by agricultural firms.
3. By the year _________, most of the basic principles of futures trading as we know them today were in place.
4. Futures exchanges exist because they fulfill two vital economic functions for the marketplace: _________ and _________.
5. The primary objective of a hedger is to achieve _________.
6. _________ are not interested in the physical commodity.
7. Only ________ members have the privilege of trading on the exchange floor.
8. The rules and regulations of the Chicago Board of Trade require pit traders to use ________ ________ ________ in buying and selling.
9. On the trading floor, because all trades must be recorded, traders wear ________ __________, indicating that they have trading privileges.
10. _________ _________ are similar to making a down payment on a house.
11. A _________ is an option that gives the option buyer the right to purchase a futures contract.
12. A _________ is an option that gives the option buyer the right to sell a futures contract.
13. Another way to say that an option has _________ _________ is to say that the option is in the money.
14. _________ _________ reflects the amount of money that buyers are willing to pay in hopes that an option will be worth exercising at or before its expiration date.

MATCHING

a. spreader
b. premium
c. position trader
d. day trader
e. joint account
f. National Futures Association
g. individual account
h. brokerage office
i. stop order
j. maintenance margin
k. initial margin
l. limit order
m. market order
n. margin call
o. bank
p. scalper

_____ 1. a public or professional trader who initiates a futures or options position and holds it over a period of days, weeks, or months

_____ 2. holds market only during the course of a trading session and rarely carries a position overnight

_____ 3. trades in the trading pit only for himself or herself, using minimum fluctuations to make a profit (or suffer a loss)

_____ 4. trades the shifting price relationship between two or more different futures contracts

_____ 5. place you must go to open a futures account

_____ 6. place or organization to call to see if the brokerage office you selected is a reliable, quality business

_____ 7. used when trading decisions are made by customers

_____ 8. set up when both the broker and customer have input into trading decisions

_____ 9. amount of money market participants must deposit into their futures accounts when they place an order

_____ 10. request for additional money to be added to the futures account

_____ 11. customer states the number of contracts of a given delivery month without specifying the price at which to buy or sell

_____ 12. a trade that can be executed only at the desired price or better

_____ 13. a customer's order that is liquidated if a contract falls to a certain price

_____ 14. the most a futures option buyer can lose

ACTIVITIES

1. Write a short essay explaining how grain prices were stabilized by the formation of futures exchanges.
2. Complete the blanks in the following line of price quotes on soybeans, by referring to Figure 18-3: September soybeans opened at _____ and closed or settled at _____ on January 5. The lowest price anyone paid for September soybeans on January 5 was _____ and the highest price anyone paid for September soybeans on that day was _____. The contract is _____, prices are quoted in _____, and it is traded at the _____.
3. Suppose you buy two CBOT July wheat futures contracts on January 15 for \$3.30/bu. Then, on March 2, you offset the position by selling two CBOT July wheat futures contracts for \$3.50¼/bu. What is your profit or loss?
4. Give an example of a long hedge similar to the example in this chapter. Use a commodity with which you are the most familiar or would consider if you ever traded futures.

5. Give an example of a short hedge similar to the example in this chapter. Again, use a familiar commodity.
6. Some say that speculating is the same as gambling. Others say that speculating is a calculated business decision. Which view do you favor? Write a short essay defending your position. Present this to your class, along with other students, in the form of a discussion or debate.
7. Give an example of speculation similar to the one in this chapter.
8. If you decided to trade futures commodities as either a hedger or a speculator, would you trade futures contracts or futures options? Defend your choice.
9. Suppose you are raising a crop or livestock commodity that is traded at the Chicago Board of Trade and Chicago Mercantile Exchange. Give an example similar to the two examples in this chapter on the strategy you would use for futures options.
10. To gain experience as a commodity futures trader, complete the following activities. This assignment is designed for five weeks, but you can adjust it as your instructor directs. Samples are given to assist you with each activity. The four activities are tracking prices, charting prices, identifying economic factors, and taking trading positions.[40]

    Pick one contract month to trade and trade the same contract month throughout the trading exercise. Be sure to choose a contract month that is at least two months away from the time you begin your trading exercise. For example, if you begin your trading exercise on October 10, trade the contract month of December or later. If you begin the trading exercise on January 15, use the contract month of March or later.

    a. *Tracking Prices:* This activity is designed to get you used to monitoring the markets. Later you will use this information to create price charts.
        (1) **Every business day** during the five-week trading period, record the daily opening, high, low, and settlement prices, and net change for the contract month of the commodity you are trading. This information is published in the business sections of several newspapers, or you can contact a local commodity broker.
        (2) **Follow the same contract month** for the entire period. (You should be tracking prices of the same contract month and commodity you are trading.)
        (3) **Create a tracking table** like the one shown here, filling in the prices you collected during the five-week trading period. You should have 25 days of price history when you are done.
    b. *Charting Prices:* Although economic factors affect price, not every price change can be attributed to economic factors. In fact, some traders make buying and selling decisions based on what a particular chart looks like. Using charts to make trading decisions is known as *technical analysis*. One type of chart used by traders is a bar chart that shows the high, low, and settlement price for each trading day of a given time period.
        (1) Create a bar graph, using the prices you collected in the previous activity, "Tracking Prices." Your graph can be hand-drawn, computer-generated, black-and-white, color—whatever you like. Don't be afraid to be creative. Your bar chart should illustrate 25 consecutive business days of prices—high, low, and settlement—that you collected in "Tracking Prices."
        (2) On the vertical axis, list prices. Make sure to use price intervals appropriate to your commodity's current market prices. For example, if wheat prices ranged from $3.00 to $3.50, you would use 5-cent intervals.

---

40. These activities are adapted from the Commodity Challenge Program of the Chicago Board of Trade. This program was sponsored in conjunction with the National FFA Organization. Copies of the *Commodity Challenge* workbook can be purchased from Chicago Board of Trade, 141 West Jackson Blvd., Suite 2250, Chicago, IL 60604-2994.

SAMPLE: Tracking Prices (Prices obtained from quotes listed by CBOT below)					
Name	J. Trader		**Commodity**		Wheat
Trading Firm	Wheat Commodities		**Contract Month**		July
Date	**Open**	**High**	**Low**	**Settle**	**Change**
5/10	292½	294½	291	293¾	+¾
5/11	293¾	295	293¼	293½	−¼
5/12	295	298¼	295	297½	+4

Monday, May 10
WHEAT (CBOT) 5,000 bu.; cents per bu.

	Open	High	Low	Settle	Change
May	346	350	346	348	+1
Jul	292½	294½	291	293¾	+¾
Sep	295½	297¼	294¼	297	+1¼
Dec	307	308	305½	307¾	+1
Mar	311	313	310¾	313	+1½
May	311½	312	311½	312	+¾

Tuesday, May 11
WHEAT (CBOT) 5,000 bu.; cents per bu.

	Open	High	Low	Settle	Change
May	349	354	348	353¾	+5¾
Jul	293¾	295	293¼	293½	−¼
Sep	297	298¼	296¼	296½	−½
Dec	308	209	307	307¼	−½
Mar	313½	314¼	313	313	. . .
May	313½	313½	313	313	+1

Wednesday, May 12
WHEAT (CBOT) 5,000 bu.; cents per bu.

	Open	High	Low	Settle	Change
May	357	363½	356¾	363	+9¼
Jul	295	298¼	295	297½	+4
Sep	298½	300½	298	299¾	+3¼
Dec	309	311½	309	310¼	+3
Mar	314¾	316	314¾	315¾	+2¾
May	315	316	315	315½	+2½

(3) On the horizontal axis of your chart, list the 25 consecutive business dates of your five-week trading period.

(4) For every day of the trading period:

- Mark the day's high with a dot.
- Mark the day's low with a dot.
- Draw a vertical line connecting the two points. This line is referred to as the day's *trading range*.
- Draw a horizontal dash indicating the settlement price that intersects the vertical line.

(5) Draw all 25 daily bar charts on the same graph.

A sample bar chart is given here. The numbers used to create the chart come from the sample price table in section (a). Because only three days of prices are reported in the previous activity, the following sample includes charts for just those days. However, by the time J. Trader from Wheat Commodities is done, 25 bar charts will be recorded.

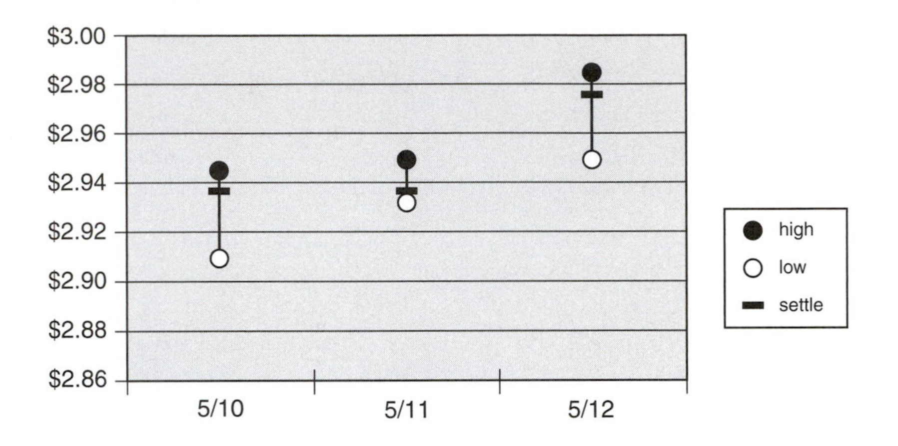

c. *Economic Factors:* Using economic information to make trading decisions is known as *fundamental analysis*. You make trading decisions based on your predictions of how economic factors and events may change the supply of and/or demand for your commodity and how prices may be affected. This combines your understanding of economic theory with real-world events.

(1) For the entire five-week trading period (25 business days), keep a daily diary of the economic factors/events that may have caused a change in the supply of and/or demand for your commodity and how price may have been affected. (A sample follows.) Daily newspapers, weekly magazines, radio/television broadcasts, and local market experts, such as commodity brokers, are excellent sources of information for your diary. Note that not every move in the market can be attributed to economic factors, but do include as many descriptions as possible. Your diary should include:

- date
- detailed summary of the economic event (can include local, domestic, and/or international news)
- description of how an event may affect supply of and/or demand for your commodity
- explanation of how an event may affect the price of your commodity—higher, lower, no change—and note of your trades

(2) In your diary, write down whether you made a trade for each of the 25 business days.

d. *Trading Positions:* Test your skills as a commodity trader using current economic news to make trading decisions. To begin trading in the futures market, you can either buy or sell one or more futures contracts. This is your opening position. At a later time, you can offset the position (or close out the position). To offset the position, you do just the opposite of your opening position. Suppose you bought one wheat futures contract on Monday. Then, on Friday, you decide to offset your position by selling one wheat contract. Likewise, if you sold one wheat futures contract on Monday, then on Tuesday, you decide to offset the position by buying one wheat contract.

Here is another trading tip: Each futures contract has specific trading months. If you buy a December corn futures contract, the only way you can offset it is if you sell a December corn futures contract. Contract months are included in your commodity's contract specifications.

(1) Trade the same contract month and commodity throughout the five-week trading period.

(2) Make at least two trades a week. The combination of an opening position and an offsetting position counts as one trade. Once you have completed the five-week trading exercise, you should have made

Sample: Economic Factors

Name	J. Trader	**Commodity**	Wheat
Trading Firm	Wheat Commodities	**Contract Month**	July

Date:	**Economic Factors & Effects on Supply, Demand, Price**
May 5	Dry conditions continue throughout the Midwest. Continued dryness could hurt the wheat crop. With a possible drop in the supply of wheat and no expected change in the demand, prices should increase. Bought wheat futures.
May 6	There's an announcement that Russia is in the market to buy wheat. Continued dryness in the Midwest. With a possible increase in the demand for wheat and continued concern over lower supply, wheat prices should increase. Bought wheat futures.
May 7	No important economic factors in the news. Market prices should remain stable. Did not trade.
May 8	Much needed rainstorms soak the Midwest. Reports of a big Canadian wheat harvest expected. Both of these factors would increase the supply of wheat. Without any expected change in demand, wheat prices should decrease. Sold wheat futures.

at least 10 trades. Note that you can make more than two trades per week. Decide if you want to open and offset them the same week or carry them over to another week. Example: Suppose you have made two trades in one week and then decide to open another position. You can either offset the third trade the same week you open it or offset it a week or two later.

(3) All open positions must be offset by the last day of the five-week trading period.

(4) Each trade should include 1 to 10 contracts. Example: If you buy three December wheat futures contracts on Tuesday, then on Thursday you sell three December wheat futures contracts.

(5) Calculate your profits and losses after you offset a position.

(6) Create a chart similar to the one following to record your trading positions, profits, and losses.

Sample: Trading Positions

Name	J. Trader	**Commodity**	Wheat
Trading Firm	Wheat Commodities	**Contract Month**	July

			#Contracts				**Total Profit/Loss**
						Unit Profit/ Loss	(contract size × unit profit/ loss × # of contracts)
Trade	**Date**	**Position**	**Buy**	**Sell**	**Price**		
1	5/5	Open	1		$3.50		
	5/6	Offset		1	$3.55	+.05	+$250
2	5/7	Open		1	$3.60		
	5/8	Offset		1	$3.70	−.10	−$500
3	5/9	Open	2		$3.68		
	5/13	Offset		2	$3.80	+.12	+$1,200
4	5/12	Open	1		$3.75		
	5/13	Offset		1	$3.80	+.05	+250

EXPLANATION OF THE TRADING POSITIONS

Trade 1: You start trading. You open a position on Monday, May 5, buying one July wheat futures contract at $3.50. On Tuesday, May 6, you offset the position by selling one July wheat futures contract at $3.55.

Trade 2: On Wednesday, May 7, you open another position—selling one July wheat futures contract at $3.60. On Thursday, May 8, you offset the position by buying one July wheat futures contract at $3.70.

Trade 3: On Friday, May 9, you buy two July wheat futures contracts at $3.68. You offset the position by selling two July wheat futures contracts at $3.80 on Tuesday, May 13.

Trade 4: On Monday, May 12, you buy another July wheat futures contract at $3.75. You offset that position on Tuesday, May 13, by selling one July wheat futures contract at $3.80. Note that because you bought two July wheat contracts on Friday, May 9, and one July wheat contract on Monday, May 12, you offset these positions by selling three July wheat futures contracts on Tuesday, May 13, at $3.80. In total, these positions account for Trade 3 and Trade 4.

Chapter 19

International Agriculture Marketing

Objectives

After completing this chapter, the student should be able to:

- Explain the importance of trade.
- Examine the basis for trade.
- Discuss the export of agricultural products.
- Discuss the import of agricultural products.
- Describe the agricultural trade balance.
- Differentiate between free trade and protectionism and describe the general economic importance of each.
- Explain export and import trade barriers.
- Describe international and regional trade agreements.
- Examine careers related to international agriculture.

Terms to Know

absurdities
advocate
commodity
common market
currencies
customs union
deficit
domestic
economic isolation
economy of scale
embargoes
European Union (EU)
exchange rates
exports
free trade
free trade area
humanitarian
import quotas
imports
law of comparative advantage
monetary policy
opportunity cost
protectionism
sanctions
subsidies
tariffs
trade barriers

Introduction

International trade has been a part of the agricultural industry since colonial times. Tobacco was the first of many agricultural **exports**. Early Americans quickly realized that an exchange of goods and services with other countries was best for both countries if the product exchanged for (imported) could be produced more cheaply abroad than at home. The colonists also quickly learned that international trade could increase the new country's standard of living.

On the surface, it might appear that international trade should be no different than trading within a nation. For example, it is easy for businesses in other states to trade with California or Florida for citrus products, Michigan for automobiles, Georgia for peaches, Maine for fish, or Texas for beef cattle. Factors of production (land, labor, capital, and management) differ substantially in different regions of the country. Because each region of the country has varying amounts of these production factors, each state has different advantages in producing products.

International trade is based on the same production factors. However, there are some differences that create challenges. Some of these challenges include: **currencies** and **exchange rates, monetary policy** or monetary systems, language barriers, technological status, national policies, national security, and politics. For example, former President Jimmy Carter canceled grain shipments to the USSR as punishment for its invasion of Afghanistan. Each country has its own **domestic** issues and problems and its own approach to addressing them. These challenges influence how each country views international trade.

Importance of Trade

Economic Benefits

Gains from international trade occur because of low prices of certain foreign products. With each country having varying levels of resources (land, labor, capital, and management), international trade allows a country to specialize in production of those products it produces best and most efficiently.[1]

A country is better off using its resources to produce those goods to which it is particularly suited and to buy from other countries the goods that those countries more efficiently produce. Trading and production specialization among nations permit higher living standards and real income than would otherwise be possible, and also permit more goods to be produced with a given set of resources. Both kinds of specialization allow a more efficient allocation of resources.

Quality of Life

We import many things that increase our quality of living and suit our eating preferences. For example, every time American consumers have a cup of coffee, a glass of iced tea, a banana split, or a chocolate candy bar; or add vanilla, cinnamon, or pepper to their food, they are benefiting from agricultural trade. The same is true when they wear silk shirts or put rubber tires on their vehicles. These and many other agricultural products consumed or used in the United States are produced elsewhere. Without **imports**, these products would be unavailable. Refer to Figure 19–1.

Jobs

Agricultural exports affect the economy in many ways, ranging from raising incomes to producing more jobs to creating positive trade balances. USDA data indicate that each dollar of agricultural exports results in $2.05 being added to our nation's economy. In addition, 75 cents of each dollar from food and fiber exports went to nonfarm enterprises. Finally, more than 1 million full-time jobs owe their existence to the agricultural export sector.[2]

U.S. exports provide increased income to production agriculturalists, as well as to those employed in agribusiness that supply purchased imports to production agriculturalists. U.S. exports also provide income to those that assemble, process, and

1. Larry B. Martin, ed., *U.S. Agriculture in a Global Economy* (Washington, DC: U.S. Government Printing Office, 1985), p. 21.
2. http://www.ers.usda.gov/Data/FATUS.

Figure 19–1 Besides common imported items such as coffee, tea, bananas, cocoa, vanilla, cinnamon, and pepper, foods from around the world (like those shown here) are finding a place on American tables. Can you identify the gooseberries, tomatillo, kiwi fruit, mountain soursop, papaya, and pomelo? (Courtesy of USDA)

transport agricultural exports to foreign countries. Although it varies from year to year, approximately 23 percent of farm income, or $1.00 out of every $4.26 received from the sale of farm products, is derived from exports.[3]

Impact on States. Because such a great variety of commodities is produced by production agriculturalists throughout the United States, all regions of the country benefit from agricultural exports. Agricultural exports account for a third to a half of total farm income in 16 states: Arkansas, Illinois, Indiana, Iowa, Kansas, Louisiana, Minnesota, Mississippi, Missouri, Montana, Nebraska, North Carolina, North Dakota, Ohio, Oklahoma, and South Carolina.[4]

If we abandoned the export market, the result would be a reduction in production of about 50 percent in wheat, rice, and soybeans, and 25 percent in feed grains. Refer to Figure 19–2. Loss of this market would be a disaster for U.S. production agriculturalists as well as for the related agribusiness input and output sectors. The agribusiness input sector could lose as many as 400,000 jobs, and the agribusiness output sector (export-related transportation, storage, merchandising, and port operations) could lose as many as 1 million jobs.[5]

Figure 19–2 If there were no exports, there would be a reduction in production of about 50 percent in wheat, rice, and soybeans, and of 25 percent in food grains. (Courtesy of USDA)

Consumer Benefit from International Trade

Without international trade, production agriculturalists would be burdened with large surpluses. These excess surpluses would be dumped into domestic markets or forced into government stocks. Prices would be severely depressed and farm income drastically lowered. More production agriculturalists would lose money and be forced out of business.

Initially, consumers would benefit from reduced food prices, but as production agriculturalists were forced to sell out because of depressed prices, and as excess stocks were depleted, food supplies would shrink and prices would rise again, to much higher levels than before. Without international trade, fewer production agriculturalists would be producing for a smaller domestic market. Profit would be substantially reduced, because of lack of **economy of scale**, and prices would be significantly higher than when the country engaged in international trade. Therefore, in the long run, consumers too would suffer from a lack of international trade.[6]

3. Gail L. Cramer, Clarence W. Jensen, and Douglas D. Southgate, Jr., *Agricultural Economics and Agribusiness,* 8th ed. (New York: John Wiley & Sons, 2001), pp. 37–38.
4. Gail L. Cramer, Clarence W. Jensen, and Douglas D. Southgate, Jr., *Agricultural Economics and Agribusiness,* 7th ed. (New York: John Wiley & Sons, 2001), pp. 465–467.
5. Ibid.
6. Martin, *U.S. Agriculture in a Global Economy,* p. 25.

Basis for Trade

At the time the Declaration of Independence was signed, Adam Smith stressed the **absurdities** of **economic isolation** and advocated free trade as a means for each nation to increase its wealth. In *The Wealth of Nations*, Smith stated, "It is the maxim of every prudent master of a family, never to attempt to make at home, what it will cost him more to make than buy."[7] This maxim reveals a basic concept of international trade. Relatively speaking, nations export the products that they produce most efficiently and import those that they produce least efficiently.[8]

Gains from trade occur when one nation is *relatively* more efficient than other nations in producing a certain commodity. This is referred to as the **law of comparative advantage**. Gain is not dependent on a nation being both *absolutely* more efficient (absolute advantage) than others in producing a certain commodity *and* absolutely less efficient at producing others.[9]

Absolute Advantage Example

	Soybeans (tons) Produced by 1 Man Day	Pepper (tons) Produced by 1 Man Day	Total
Before trade:			
Mexico	3	12	15
United States	15	5	20
	18	17	35

	Soybeans (tons) Produced by 1 Man Day	Pepper (tons) Produced by 1 Man Day	Total
After trade when countries specialize:			
Mexico	0	24	24
United States	30	0	30
	30	24	54

Figure 19–3 Mexico has an absolute advantage in pepper production, and the United States has an absolute advantage in the production of soybeans.

Absolute Advantage

An absolute advantage exists if one country can produce a **commodity** less expensively than another country can. Suppose that one resource unit (worker-day) in Mexico is needed to produce 3 tons of soybeans and another worker-day is needed to produce 12 tons of peppers. Further assume that one worker-day in the United States is used to produce 15 tons of soybeans and another worker-day is used to produce 5 tons of peppers. Figure 19–3 shows that when they do not trade, total production in the two countries together equals 18 tons of soybeans and 17 tons of peppers, or a total of 35 tons of raw food and fiber products. However, Mexico is more efficient in using a resource unit to produce peppers (one worker-day produces 12 tons of peppers compared to 5 in the United States) and less efficient in soybean production (3 tons compared to 15 in the United States per worker-day). Therefore, Mexico has an *absolute advantage* in pepper production and the United States has an absolute advantage in the production of soybeans.[10]

Comparative Advantage

The law of comparative advantage takes into account differences in **opportunity cost** to explain how trade can be mutually profitable between two countries even if one produces all commodities more efficiently than the other.

Assume that the efficiency of pepper production in the United States increases so that one worker-day can now produce either 15 tons of soybeans or 15 tons of peppers. Also assume that one worker-day in Mexico can still produce either 3 tons of soybeans or 12 tons of peppers. Refer to Figure 19–4. In this scenario, with two resource units available in each country and no specialization, world output would be 18 tons of soybeans and 27 tons of peppers. Notice that the United States has a greater productive capacity than Mexico for both commodities. The United States is 5 times (15 to 3) more efficient in producing soybeans and 1.25 times (15 to 12) more efficient in producing peppers.

7. Adam Smith (from *The Wealth of Nations*), quoted in Randall D. Little, *Economics: Applications to Agriculture and Agribusiness*, 4th ed. (Danville, IL: Interstate Publishers, 1997), pp. 497–498.
8. John B. Penson, Rulon D. Pope, and Michael L. Cook, *Introduction to Agricultural Economics* (Englewood Cliffs, NJ: Prentice-Hall, 1986), p. 475.
9. Ibid., p. 475.
10. Ibid., pp. 475–476.

Comparative Advantage Example

	Soybeans (tons) Produced by 1 Man Day	Pepper (tons) Produced by 1 Man Day	Total
Before trade:			
Mexico	3	12	15
United States	15	15	30
	18	27	45
	Soybeans (tons) Produced by 1 Man Day	**Pepper (tons) Produced by 1 Man Day**	**Total**
After trade when countries specialize:			
Mexico	0	24	24
United States	30	0	30
	30	24	54

Figure 19–4 The greatest comparative advantage of the United States would be in soybeans, and Mexico's smallest comparative disadvantage would be in pepper.

Assuming no transportation expenses, the law of comparative advantage indicates that it is better for each country to specialize in one product and then trade with each other. Although the United States produces both commodities more efficiently than Mexico, it has the *greatest comparative advantage* in producing soybeans. Mexico, in contrast, has the *smallest comparative disadvantage* in producing peppers.[11] Note this key concept: total production increases when both countries specialize.

Importance of Specialization

Again, refer to Figure 19–3. If each country specialized in producing what it has an absolute advantage in and traded with the other country, the United States could produce 30 tons of soybeans and Mexico could produce 24 tons of peppers. Thus, world output would jump from 35 tons to 54 tons, a gain of 19 tons of output.

Now, refer to Figure 19–4. With specialization on the basis shown in the figure, world output in our two-country example would still increase by 9 tons (54 tons minus 45 tons). *"The key point to remember is that even if a nation has an absolute advantage in both products, total world output can be increased if the opportunity costs differ among countries."*[12]

11. Ibid., pp. 477–478.

Exporting Agricultural Products

Because of excess production of many agricultural commodities in the United States, the export market has been extremely important for the agribusiness output sector. With demand for U.S. agricultural commodities increasing, it appears that those in the agribusiness output sector will have an opportunity to sell more of their commodities in foreign markets. Agricultural exports amount to approximately 8 percent of all merchandise exports from the United States.[13] In the 1990s, the United States made great strides in developing international markets for feed grains, soybeans, and wheat.

Major Commodities Exported

Figure 19–5 shows U.S. agricultural exports from 1970 to 1995, by individual product. Feed grains, wheat, and soybeans are the largest components of exports. Wheat and flour are next, followed by meats and meat products. Cotton and tobacco are still significant, but grains and preparations, oilseeds, fruits and vegetables, animals, and animal products are all important exports as well.

Currently, about 23 percent of U.S. farm cash receipts comes from the export market. The United States exports approximately 54 percent of its wheat crop, 35 percent of its soybean crop, 24 percent of its corn crop, 36 percent of its grain sorghum crop, and about 48 percent of its rice crop.[14] Refer to Figure 19–6.

Major Buyers of U.S. Exports

Asia and Western Europe are the largest markets for U.S. agricultural exports (Figure 19–7). Asian imports of rice, wheat, soybeans, and feed grains have steadily increased. Japan, the largest single importer of U.S. farm products, became our first $10-billion agricultural customer in 1999.[15]

12. Ibid, p. 478.
13. Cramer, Jensen, and Southgate, *Agricultural Economics and Agribusiness* (2001), p. 416.
14. Ibid., p. 419.
15. Ibid., pp. 419–420.

U.S. Agricultural Exports: Value by Commodity, Selected Calender Years

Commodity	1970	1980	1982	1986	1992	1995
			Million Dollars			
Animals and animal products						
Dairy products	127	175	347	438	726	711
Fats, oils, and greases	247	769	663	411	525	827
Hides and skins, excluding fur skins	187	694	1022	1521	1346	1621
Meats and meat products	132	890	9798	1113	3339	4522
Poultry products	56	603	515	496	1211	2345
Other	101	660	410	530	778	907
Total animals and animal products	850	3791	3935	4509	7925	10,933
Cotton, excluding linters	372	2864	1955	786	1999	3681
Fruits and preparations	334	1335	1376	2040	2732	3240
Grains and preparations						
Feed grains, excluding products	1064	9759	6444	4330	5737	8153
Rice, milled	314	1289	997	621	725	996
Wheat and flour	1111	6586	6927	3279	4675	5734
Other	107	357	273	398	3035	3654
Total grains and preparations	2596	17,991	14,641	8629	14,172	18,537
Oilseeds and products						
Cottonseed and soybean oil	244	915	692	343	432	786
Soybeans	1228	5880	6218	4321	4380	5400
Protein meal	358	1654	1447	1302	1398	1701
Other	91	944	784	493	980	1036
Total oilseeds and products	1921	9393	9141	6459	7190	8923
Tobacco, unmanufactured	517	1334	1547	1209	1851	1400
Vegetables and preparations	206	1188	1174	1024	2871	3889
Other	463	3360	2853	1349	4389	5211
Total exports	7259	41,256	36,622	26,064	42,929	55,814

Figure 19–5 Feed grains, soybeans, wheat and flour, and meats and meat products are among the largest U.S. agricultural exports. (Source: USDA, Economic Research Service, *Foreign Agricultural Trades of the United States*, January/February/March 1998)

Humanitarian Reasons for Exporting

Agricultural exports are also desirable for **humanitarian** reasons. Many parts of the world are incapable of producing enough food to feed their people. Others can produce the food, but only at a prohibitively high cost. Because the United States is a more efficient producer of food, it can help by donating or selling its products for very reduced prices. Countries with limited resources can buy food from the United States for substantially less than it would cost to produce that food themselves. The United States operates

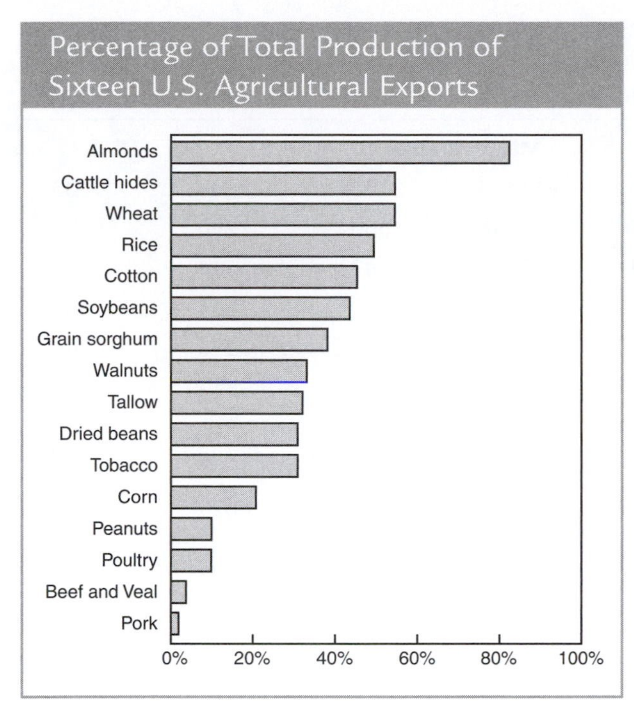

Figure 19–6 The percentages shown here represent the amount of total U.S. production of the commodity that is exported. (Source: USDA)

the world's largest food aid program, shipping to about 70 countries under the Food for Peace Program.[16]

Beneficial Side Effects of Exporting

Adequate nutrition and diets for a country's population improve the country's ability to develop and grow. As economic growth occurs in developing countries and incomes rise, these countries become more able to purchase the food they need, and also are more likely to want more and higher-quality food.

Because the United States is a major producer of agricultural products, it is in our best interest to promote world economic growth. Economic growth increases the size of our future markets for agricultural commodities. This growth in international demand is one of the underlying factors driving U.S. economic growth.[17]

Importing Agricultural Products

The United States is also one of the largest importers of agricultural products. Each year, the United States imports agricultural products worth more than $26 billion. Refer to Figure 19–8. Agricultural imports make up less than 5 percent of total U.S. imports.

Imports provide consumers with commodities that either are not produced domestically or are not available in sufficient quantities. Major imports not produced in the United States include bananas, some spices, coffee, tea, cocoa, and rubber. Domestic production of other goods, such as certain cheeses and tobaccos, is insufficient to satisfy domestic demand, so imports meet the excess demand.

Canada, Mexico, and Brazil are the top sources of agricultural products imported into the United States. However, if the **European Union (EU)** is considered as a single, collective entity, then it is the top exporter of agricultural goods to the United States.

Agricultural Trade Balance

When the outflow of dollars for purchase of imports exceeds the inflow of dollars from sale of exports, a country has a **deficit** in its international trade balance. International trade balance deficits can be corrected in the long run by reducing imports (purchases) and by increasing exports (sales). The monetary authorities in each nation track the trade balance so that they can adjust policy appropriately; another primary purpose of trade-balance tracking is to inform and aid banks, businesses, and individuals engaged in international trade in making business decisions.[18]

Although the media often portray the trade imbalance or deficit as having a major impact on the U.S. economy, it is not as bad as it may seem. For centuries, countries thought it was desirable to have an export trade surplus so that they could accumulate gold and other precious metals that were

16. Martin, *U.S. Agriculture in a Global Economy*, p. 27.
17. Ibid., p. 28.
18. Dominick Salvatore, *International Economics*, 5th ed. (New York: Macmillan, 1995), p. 362.

U.S. Agricultural Exports: Value by Commodity, Selected Calendar Years

Region	1973	1980	1982	1986	1992	1995
			Million Dollars			
Western Europe	5605	12,351	11414	7026	7804	9003
Enlarged EU-15	4526	9256	8398	6604	7290	8537
Other Western Europe	1079	3095	3016	423	514	466
Eastern Europe and FSU	1498	3285	2718	1091	2663	1639
FSU	920	1130	1871	658	2346	1346
Eastern Europe	577	2155	847	433	317	293
Asia	6509	15,037	13,675	10,537	17,923	27,939
West Asia	521	1418	1428	1334	1782	2478
South Asia	1564	734	800	450	629	1019
SE Asia, excluding Japan and PRC	1851	4281	4387	3586	1545	3012
Japan	2998	6331	5555	5106	8437	10,957
People's Republic of China (PRC)	575	2273	1505	61	545	2633
Latin America	1692	6176	4438	3641	6669	7926
Canada, excluding trans-shipments	1034	1905	1805	1533	4902	5738
Canada trans-shipments	677	12	15	14	0	0
Africa	583	2303	2287	1997	2570	3070
North Africa	307	1244	1223	1340	1461	2144
Other Africa	276	1059	1064	657	1109	926
Oceania	83	188	270	224	398	499
Total	17,680	41,256	36,622	26,064	42,929	55,814

Figure 19–7 Asia and Western Europe are the largest markets for U.S. agricultural exports. (Source: USDA, Economic Research Service, *Foreign Agricultural Trades of the United States*, January/February/March 1996)

considered measures of wealth. It was Adam Smith who pointed out (in *The Wealth of Nations*) that goods rather than gold are the true wealth of a nation.[19] However, the issue remains as to how much of a trade imbalance is too much. (This issue invariably arises during election years, when the economic track record of the majority political party is examined.)

Throughout most of its history, the United States has exported more than it has imported. After World War II, the U.S. balance of trade was positive because, on average, it exported $5 billion more than it imported. The balance of trade became negative starting in the 1970s. From 1977 to 1980, the United States averaged $30 billion more in imports than in exports. This negative balance peaked at nearly $160 billion in 1987. The balance of trade has improved since then, but in 1995, the U.S. still had a trade deficit of $140 billion.[20]

Agricultural exports have helped to reduce the trade deficit. Through 1995, there were 36 consecutive years of agricultural trade surpluses, in contrast to persistent trade deficits in nonagricultural trade. In 1995, the U.S. agricultural trade surplus was approximately $25 billion, the second largest trade surplus ever.[21]

Free Trade or Protectionism

Some people favor **free trade** and some favor **protectionism**. Which mechanism you favor usually

19. Cramer, Jensen, and Southgate, *Agricultural Economics and Agribusiness* (2001), p. 422.

20. Ibid.

21. Little, *Economics*, p. 514.

Value of Agricultural Imports, by Selected Major Commodity Group		
Commodity Group	**Value**	**Leading Countries of Origin**
	$1,000	
Competitive products		
Meat and products	2,657,548	Canada, Australia, New Zealand
Dairy products	963,376	New Zealand, Italy, Ireland
Grains and feeds	2,339,092	Canada, Italy, Thailand
Fruits, nuts, and vegetables	4,725,954	Mexico, Chile, Spain
Sugar and related products	1,128,712	Canada, Dominican Republic, Guatemala
Wine and malt beverages	2,084,650	France, The Netherlands, Italy
Oilseeds and products	1,562,955	Canada, Italy, Philippines
Tobacco, unmanufactured	613,182	Turkey, Brazil, Greece
Noncompetitive products		
Bananas and plantains	1,071,834	Costa Rica, Colombia, Ecuador
Coffee and products	2,485,433	Brazil, Colombia, Mexico
Cocoa and products	1,033,906	Canada, Ivory Coast, Indonesia
Rubber and gums	965,390	Indonesia, Thailand, Malaysia
Tea	185,515	China, Argentina, Germany
Spices	326,890	Indonesia, India, Madagascar
Total imports	26,818,015	Canada, Mexico, Brazil

Figure 19–8 More than $26 billion of agricultural exports are purchased yearly. Fruits, nuts, and vegetables are the largest imports. (Source: USDA)

depends on how you are personally affected. For example, suppose that you are a beef producer and you **advocate** free trade. However, you learn that three major processing companies have signed contracts for shipments of beef from neighboring countries. In the long run, the action taken by the beef-processing companies may help everyone except U.S. beef producers and ranchers. The prices they can now get could drop to below production costs. Now, are the beef producers and ranchers for or against free trade? This scenario could occur with any commodity.

Agricultural Trade Policy

Whether a country should adopt free trade policies or implement protectionism is not a clearcut issue; there are many gray areas. Because of these gray areas, trade policies are formulated in the "national interest." Effective trade policy takes into account the following objectives:

- maintaining certain industries, such as steel and aviation, in the interest of national security
- stabilizing economic activity
- positively affecting the balance of payments
- maintaining and safeguarding reasonable consumer costs
- increasing or maintaining employment
- promoting equality in income distribution[22]

General Economic Arguments for Free Trade

The leading arguments in favor of free trade center around the law of comparative advantage. Embedded in that law is the concept of specialization and the idea that society gains from specialization. It is argued that if markets are not obstructed by protectionist measures and private entrepreneurs are allowed to pursue their interests, resource allocation will be optimized and overall welfare will be maximized. In essence, the proponents of free trade say that resources will adjust across national boundaries as well as within a country.

22. *Advanced Agribusiness Management and Marketing* (8735-B) (College Station, TX: Instructional Materials Service, 1990), p. 235.

The French economist and statesman Frédéric Bastiat declared that free trade is always beneficial.[23] A nation must specialize in what it can produce best and trade with other countries to acquire goods at prices lower than the cost of purchasing those goods domestically. Again, countries thrive on free trade because of specialization. Specialization also increases output, making the product even more economical.

General Economic Arguments Supporting Protectionism

Those who favor protectionism present several economic arguments. These are briefly discussed here.

Infant Industry. This argument holds that industries in the startup phase (infancy stage) should be protected for a time until they are able to operate efficiently. This keeps an industry from being eliminated due to a lack of capital, lack of product demand, lack of volume, or higher unit costs—all of which are often associated with the infancy stage of existence.[24]

Balance of Payments Argument. This argument contends that free trade leads to excessive payment and resource imbalances during market adjustments. This line of thinking insists that economic policies be focused on growth and employment. Efficiency, the actual goal of a free market, is not emphasized.[25]

Unfair Competition Argument. This argument claims that unfair trade practices by certain countries harm industries in "unprotected" countries. This can occur when governments and banks provide such large **subsidies** to producers in certain countries that those countries gain a competitive advantage. Japan's "invasion" into the U.S. automobile market is cited by some as an example of unfair competition.[26]

Self-Sufficiency for National Security. This argument contends that trade restrictions should be used to fortify businesses that produce items necessary for survival in the event of military or economic war. The philosophy behind this argument is that during times of war, countries need to be self-sufficient. Food and military ordnance are examples of items that are essential for survival during such times.[27]

Diversification for Stability. This argument claims that restrictions on imports would cause countries to diversify their domestic markets and produce items for themselves rather than importing them. Greater economic independence and domestic stability are perceived benefits of this scenario.[28] Again, overall production efficiency is not a major consideration.

Protection of Wages and the Standard of Living. This argument contends that developed countries (like the United States) cannot compete with countries that pay workers low wages to do labor-intensive tasks. Mexican and Asian producers of shoes and clothing are often the targets of low-wage accusations.[29]

Which Is Best?

In the final analysis, the answer as to which system you will favor depends on your answer to the question, "How does it affect me?" Some say that protectionism results in lost jobs, higher prices, and higher taxes, and contributes to the debt crisis worldwide. These same people can back up their claims with statistics.[30] Those favoring free trade can also produce data supporting their views. Although academic economists generally support free trade, and can cite cutting-edge research as the basis for their opinions, overall trends do not help the individual production agriculturalist who suffers a sudden drop of 75 percent in sales of a product when another country decides to export that product and its cost of production is 80 percent less. You decide!

---

23. Vincent H. Miller and James R. Elwood, "Free Trade or Protectionism? The Case Against Trade Restrictions" (1988); available at http://www.isil.org/resources/lit/free-trade-protectionism.html.
24. Penson, Pope, and Cook, *Introduction to Agricultural Economics,* p. 499.
25. Ibid.
26. Ibid.
27. Little, *Economics,* pp. 504–505.
28. Ibid., pp. 505–506.
29. Little, *Economics,* p. 506.
30. Miller and Elwood, "Free Trade or Protectionism?"

Export and Import Trade Barriers

It stands to reason that if we are not going to allow free trade, there must be ways of protecting or restricting trade. These ways are called **trade barriers**. In fact, governments use many forms of protection with respect to international trade. The two major methods are those that influence imports and those that influence exports.

Import Controls

Tariffs are the taxes that governments charge when products cross national borders. Importers usually compensate for the extra tax by increasing the cost of their products. The desired outcome for the country imposing a tariff is that consumers will use more goods that are produced domestically. Tariffs are the most common means of restricting trade.

Import quotas set limits on how much of a certain product can be imported during a certain time. Imports of the specified product, regardless of price, cease once the quota has been met. Both tariffs and quotas are used to restrict trade; however, revenues from tariffs go to governments; revenues from quotas accrue to the importers of the protected item.[31]

Regulatory trade restrictions are other rules and policies established by governments to restrict trade. For instance, restrictions might be used to protect citizens from imported commodities that have been treated with undesirable pesticides.[32]

Export Controls

Export **embargoes** prohibit shipments of commodities to certain other countries. Nations use these embargoes or **sanctions** to maintain domestic supplies and to support foreign policy. For example, in 1973, a U.S. embargo on soybeans lowered soybean prices and increased American supplies of soybeans. An example of an embargo imposed for foreign policy purposes occurred in 1980, when a grain embargo was enacted against the USSR. This embargo was put into effect because of the USSR's incursion into Afghanistan.[33] More recently, sanctions against nations supporting terrorism have been instituted, both by individual countries and the United Nations.

However useful they may seem, economically or politically, export embargoes can result in global loss of confidence in a nation as a dependable supplier and result in the loss of trade. Also, during an embargo, the sanctioned country often attempts to increase domestic production of the prohibited commodities (or substitutes for those commodities), so that it is no longer dependent on imports.[34]

Voluntary export restraints are implemented when foreign governments agree to limit their exports of certain products to a certain amount. Fewer imports of the product allow domestic producers to produce and sell more of the item.[35]

International and Regional Trade Agreements

Trade agreements fall somewhere between the two extremes of free trade and protectionist trade barriers. Reducing trade barriers is an obvious way to increase trade. However, most governments are not willing to reduce these barriers without receiving something in return. Therefore, many governments enter into formal trade agreements with other nations. These agreements specify the terms for international trade.

The use of international and regional trade agreements has increased during the past 30-plus years. International agreements, such as the General Agreement on Tariffs and Trade and the World Trade Organization, have been ratified. Regional agreements have been created between nations that decide to liberalize trade amongst

31. Little, *Economics,* p. 500.
32. Ibid., p. 503.

33. Penson, Pope, and Cook, *Introduction to Agricultural Economics,* p. 504; *Introduction to World Agricultural Science and Technology* (8379B-2) (College Station, TX: Instructional Materials Service, 1991), p. 22.
34. Penson, Pope, and Cook, *Introduction to Agricultural Economics,* p. 504–505.
35. Little, *Economics,* p. 503.

themselves while maintaining standardized tariffs against others. The two most notable agreements of this type are the EU and NAFTA (discussed later in this section).[36]

International Approach to Trade Agreements

The General Agreement on Tariffs and Trade (GATT) was created in 1947 to secure a substantially free and competitive market, which was considered the most effective means of conducting international trade and serving the best interests of all nations. Since its creation, eight major negotiating conferences have been held under the auspices of the GATT. The last round, the Uruguay Round, was conducted between 1986 and 1994; 120 countries participated in these talks. The Uruguay Round resulted in an agreement to reform all the negotiating areas and to establish a new intergovernmental organization, the World Trade Organization. The five basic principles of the GATT are:

- Trade must be nondiscriminatory.
- Domestic industries receive protection mainly from tariffs.
- Agreed-upon tariff levels bind each signatory country.
- Consultations are provided to settle disputes.
- GATT procedures may be waived if other member countries agree and compensation is made to them.[37]

World Trade Organization

The World Trade Organization (WTO), which officially began functioning on January 1, 1995, essentially implements the GATT. Nations that ratified the GATT are members of the WTO. The structure of the WTO is intended to promote more effective decisionmaking and greater involvement in trade relations. The WTO has been given authority to oversee adherence to trade agreements on goods and services, as well as intellectual property. Five functions of the WTO are as follows:

- to facilitate the implementation, administration, and operation of the GATT terms and further the objectives established in the GATT
- to provide a forum for negotiation among member countries concerning multilateral trade relations, and then assist in implementing the results of those negotiations
- to oversee dispute settlement between member countries
- through the trade policy review mechanism, to improve adherence by member countries to rules, disciplines, and commitments made under their multilateral trade agreements
- to cooperate, as appropriate, with the International Monetary Fund and the International Bank for Reconstruction and Development (the World Bank), to achieve greater coherence in global economic policy making[38]

Mission. The World Trade Organization is the only international organization dealing with the global rules of trade among nations. Its main function is to ensure that trade flows as smoothly, predictably, and freely as possible.

The result is assurance. Consumers and producers know that they can access supplies and have wider choice of the finished products, components, raw materials, and services that they use. Producers and exporters know that foreign markets will remain open to them.

The result is also a more prosperous, peaceful, and accountable economic world. Almost all decisions in the WTO are made by consensus among all member countries, and they are ratified by members' legislative bodies. Problems regarding trade are handled by the WTO's dispute settlement process, where the focus is on interpreting agreements and commitments and deciding how to ensure that member countries' trade policies conform to those agreements. This system reduces the risk that disputes will spill over into political or military conflict. By lowering trade barriers, the WTO's system also breaks down other barriers between nations.[39]

---

36. Mordechai E. Kreinin, *International Economics: A Policy Approach,* 2d ed. (New York: Harcourt Brace Jovanovich, 1975), p. 307.
37. *Advanced Agribusiness Management and Marketing,* pp. 236–237.
38. Philip Raworth and Linda C. Reif, *The Practitioner's Deskbook Series: The Law of the WTO* (New York: Oceania Publications, 1995).
39. http://www.wto.org/english/thewto_e/whatis_e/inbrief_e/inbr00_e.htm.

Major Thrust. At the heart of the WTO system are the multilateral trading agreements, negotiated and signed by a majority of the world's trading nations, and ratified by their legislative/governing bodies. These agreements constitute the legal ground rules for international commerce. Essentially, they are contracts that guarantee member countries important trade rights. They are also binding on governments, which must keep their trade policies within agreed limits.

The agreements were negotiated and signed by governments, but their purpose is to help producers of goods and services, exporters, and importers conduct their businesses.[40]

Benefits. The goal is to improve the welfare of the peoples of the member countries. The major benefits of the WTO are that:

- the system helps promote peace
- disputes are handled constructively
- rules make life easier for all involved
- freer trade cuts the cost of living and raises incomes
- free trade provides more choice of products and qualities
- trade stimulates economic growth
- the basic principles make life and production more efficient
- governments are shielded from lobbying
- the system encourages good government[41]

The World Trade Organization is located in Geneva, Switzerland. It has 148 member countries. The budget in 2005 was 169 million Swiss francs (about $133 million in U.S. dollars). The staff consists of approximately 630 people.[42]

The World Bank

The World Bank is not a "bank" in the usual sense. It is an international organization owned by the 185 countries, both developed and developing, that are its members. The World Bank works to reduce poverty worldwide by promoting growth to create employment opportunities and helping poor people take advantage of those opportunities. The World Bank supports efforts by governments and member countries to invest in schools and health centers, provide water and electricity, fight disease, and protect the environment.

The World Bank was set up in 1944 as the International Bank for Reconstruction and Development. It first began operations in 1946, with 38 members. That number increased dramatically in the 1950s and 1960s, when many areas became independent nations and joined the organization. As membership grew and needs changed, the World Bank expanded; it is now made up of five different agencies (the International Bank for Reconstruction and Development, the International Development Association, the International Finance Corporation, the Multilateral Investment Guarantee Agency, and the International Centre for Settlement of Investment Disputes).

The World Bank is like a cooperative where members are shareholders. Through representatives on the Board of Executive Directors, the member countries set policy, oversee operations, and benefit from its work.

The World Bank is one of the world's largest sources of funding and knowledge for developing countries. Its main focus is on helping the poorest people and the poorest countries. It uses its financial resources, staff, and extensive experience to help developing countries reduce poverty, increase economic growth, and improve their citizens' quality of life.

The World Bank is headquartered in Washington, D.C., and has more than 100 offices around the world. It was established July 1, 1944, by a conference of 44 governments in Bretton Woods, New Hampshire. There are about 7,000 employees in Washington, D.C., and more than 3,000 in non-U.S. offices.[43]

International Monetary Fund

People sometimes confuse the World Bank with the International Monetary Fund (IMF), which was also set up at the Bretton Woods conference in 1944. Although the IMF's functions complement those of the World Bank, the IMF is a totally

40. Ibid.
41. http://www.wto.org/english/thewto_e/whatis_e/10ben_e/10b00_e.htm.
42. http://www.wto.org/english/thewto_e/whatis_e/inbrief_e/inbr02_e.htm.
43. World Bank Brochure; available at http://web.worldbank.org.

separate entity. Whereas the World Bank provides support to developing countries, the IMF aims to stabilize the international monetary system and monitor the world's currencies.[44] The 45 governments represented at the Bretton Woods conference sought to build a framework for economic cooperation that would avoid a repetition of the disastrous economic policies that contributed to the Great Depression of the 1930s. The main responsibilities of the International Monetary Fund are to:

- promote international monetary cooperation
- facilitate the expansion and balanced growth of international trade
- promote exchange stability
- assist in the establishment of a multilateral system of payments
- make its resources available to members that are experiencing difficulties with the balance of payments[45]

The International Monetary Fund is headquartered in Washington, D.C. It is governed by the governments of its 185 member countries. It has a staff of 2,716 people worldwide. The IMF has outstanding loans of $28 billion to 74 countries.[46]

Purpose. The IMF is responsible for ensuring the stability of the international monetary and financial system—the system of international payments and exchange rates among national currencies that enables trade to occur between nations. The IMF seeks to promote economic stability and prevent crisis; to help resolve problems when they do occur; and to promote growth and alleviate poverty. Its three main functions—surveillance, technical assistance, and financial assistance (lending)—are all intended to meet these objectives.

Surveillance is the regular dialogue and policy advice that the IMF offers to each of its members. On a regular basis, the fund conducts in-depth appraisals of each member country's economic situation. It discusses with the country's authorities the policies that will best promote stable exchange rates, growth, and prosperity.

Technical assistance and training are offered (mostly free of charge) to help enable member countries to design and implement effective policies. Technical assistance is provided in several areas, including fiscal policy, monetary and exchange rate policies, banking and financial system supervision and regulation, and statistics.

Financial assistance is available to give member countries the resources and time they need to solve balance-of-payment problems.[47]

Regional Approach to Trade Agreements

Three forms of regional integration are customs unions, common markets, and free trade areas. In a **customs union**, two or more countries agree to eliminate trade restrictions with each other, but establish common and uniform tariffs for other, nonparty countries. A **free trade area** is similar to a customs union except that there are no common and uniform tariffs on other countries' products. A **common market** is also a customs union, but it allows free mobility of factors of production.[48] The two best examples of regional trade agreements are the European Union and the North American Free Trade Agreement.

The European Union

The European Union (EU), a customs union, is an economic federation that evolved from the European Community (EC) and the European Economic Community (EEC). The EEC was organized in 1957 by the Treaty of Rome to promote economic integration of six western European countries. Presently, 15 countries belong to the EU: France, West Germany, Italy, Belgium, the Netherlands, Luxembourg, United Kingdom, Ireland, Denmark, Greece, Portugal, Spain, Sweden, Austria, and Finland. In the EU, free trade is expected among the member nations, much like the trade that is carried on between the states in the USA. The EU nations have agreed to a set of policies that govern agriculture, transportation, taxes, and international trade.[49]

44. Ibid.
45. "The IMF at a Glance" factsheet (April 2007); available at http://www.imf.org/external/np/exr/facts/glance.htm.
46. Ibid.
47. Ibid.
48. Cramer, Jensen, and Southgate, *Agricultural Economics and Agribusiness* (2001), p. 414.
49. Little, *Economics,* p. 512.

North American Free Trade Agreement

The North American Free Trade Agreement (NAFTA) was finalized in 1994. This agreement was created to establish a free trade area among the United States, Canada, and Mexico. Canada and Mexico are the second and third largest export markets for U.S. agricultural products (Japan is the largest). Under NAFTA, tariffs and other trade barriers among these countries were reduced throughout the 1990s. The outlook for increasing exports of corn, wheat, and oilseeds to Mexico is very good; meat exports have already expanded.[50]

NAFTA does not include any agreement to form common foreign policies, stabilize exchange rates, or coordinate welfare or immigration policies. The trade preferences apply only to goods made in NAFTA member countries, defined to be at least 62.5 percent domestic parts used to manufacture product.[51]

Benefits to U.S. Agriculture. In 2005, Canada and Mexico were, respectively, the first and second largest export markets for U.S. agricultural products. Exports to the two markets combined were greater than exports to the next six largest markets.

From 1992 to 2005, the value of U.S. agricultural exports worldwide increased by 46 percent. Over that same period, U.S. farm and food exports to the two NAFTA partners grew by 128 percent.[52]

Trade with Mexico. From 1999 to 2005, U.S. farm and food exports to Mexico grew to $9.4 billion—the highest level ever and the fourth record in 5 years under NAFTA. U.S. exports of soybean meal, red meats, dairy products, and poultry meat all set new records in 2005.

In the years preceding the adoption of NAFTA, U.S. agricultural products lost market share in Mexico as competition for that market increased. NAFTA reversed this trend. The United States supplied more than 71 percent of Mexico's total agricultural imports in 2005, in part because of the price advantage and preferential access that U.S. products now enjoy.

NAFTA kept Mexican markets open to U.S. farm and food products in 1995 during Mexico's terrible economic crisis. After the peso devaluation and the consequent economic problems, U.S. agricultural exports dropped by 23 percent, but came back to set new annual records. NAFTA softened the impact of the downturn and helped speed the recovery because of preferential access for U.S. products. In the midst of the crisis, rather than raising import barriers, Mexico honored its NAFTA commitments and continued to reduce tariffs.

In 2005, the 12th round of tariff cuts under NAFTA was made, further opening the subject market to U.S. products. U.S. commodities now eligible for duty-free access under Mexico's NAFTA tariff rate quotas include corn, dried beans, poultry, animal fats, barley, eggs, and potatoes. All tariffs are to be eliminated by 2008.[53]

Trade with Canada. The Canadian market for U.S. agricultural products has grown steadily under the U.S.-Canada Free Trade Agreement (CFTA), with U.S. farm and food exports reaching a record $10.6 billion in 2005, an increase of more than 81 percent since 1999. Fresh and processed fruits and vegetables, snack foods, and other consumer foods constitute close to 75 percent of U.S. sales to Canada.

U.S. exports of consumer-oriented products to Canada set records in 2005 in almost every category. Additionally, new value highs were recorded for vegetable oils, planting seeds, and sugars, sweeteners, and beverage bases. With a few exceptions, tariffs not already eliminated dropped to zero on January 1, 1998.

In 1996, the first NAFTA dispute settlement panel reviewed the higher tariffs that Canada had applied to its dairy, poultry, egg, barley, and margarine products, which were not subject to tariff barriers before the Uruguay Round agreements were implemented. The panel ruled that Canada's tariff rate quotas are consistent with its obligations under NAFTA, and thus need not be eliminated.[54]

Disputes. There are actually very few trade disputes between Mexico and the United States, especially

50. "Fact Sheet: North American Free Trade Agreement (NAFTA)" (March 2006); available at http://www.fas.usda.gov/info/factsheets/NAFTA.asp.
51. "North American Free Trade Agreement"; available at http://en.wikipedia.org/wiki/NAFTA.
52. "Fact Sheet: North American Free Trade Agreement (NAFTA)" (March 2006); available at http://www.fas.usda.gov/info/factsheets/NAFTA.asp.
53. Ibid.
54. Ibid.

given the large amount of trade conducted between the two countries. The few disputes that have occurred were minor and were resolved between the two countries alone or with the aid of NAFTA or WTO panels. Issues or commodities involved in disputes have been trucking, sugar, high-fructose corn syrup, and some agricultural products.

The United States and Canada have an ongoing dispute concerning the U.S. decision to charge a 27 percent duty on imports of Canadian softwood lumber. Canada requested dispute resolution from the NAFTA panel on several occasions. It won each time, most recently in March of 2006. Although the combined duties were reduced to about 11 percent in 2002, the United States still has not eliminated the duty completely. This has caused considerable debate in Canada, which has considered responding with duties of its own on American imports.[55]

Impact of Exchange Rates on International Trade

Demand for imports and exports can be affected by the values of various different currencies. For example, when the dollar is strong, the currencies of other countries appear inexpensive, and so do their products. Thus, demand for foreign products rises. So, as a general rule, when the U.S. dollar is strong, imports from other countries increase. U.S. importers benefit from this because they can buy products less expensively and pass their savings on to consumers by lowering prices. The lower the price, the higher the demand, and the higher the potential for profit. Electronics outlets, grocery stores, and gas stations tend to benefit when the dollar is strong.

A strong dollar also has negative effects. When the dollar is strong, buyers in other countries perceive our currency and our products as expensive. Therefore, the demand for U.S. exports decreases. U.S. exporters must lower prices to attract foreign buyers. This causes lower profits and may even put businesses at risk of failing. This tends to be true for farmers, computer companies, and automobile manufacturers. Generally speaking, the market effect of a strong U.S. dollar is to increase imports and decrease exports.

It All Evens Out. Currency exchange rates tend to correct themselves over time. When the U.S. dollar is strong, we tend to purchase more foreign currency to buy foreign products. Thus, the demand for the foreign currencies increases and causes them to appreciate in value. At the same time, there are more American dollars in the international market, and the demand for them falls. Eventually, the dollar depreciates in value. A weak dollar makes foreign imports less desirable in the United States because they cost more dollars. In contrast, U.S. exports are more attractive because of the stronger buying power of foreign currencies. In the United States, the market effect of a weak dollar is to decrease imports and increase exports. This cycle tends to repeat itself time and again.[56]

Careers Related to International Agriculture

Export-Import Services

The great volume and the technical nature of the export-import business require specialists who know the fundamentals of international trade and the commodities of agribusiness. International trade business needs efficient technical and commercial services: transportation, grading, storage, inspection, packaging, finance, marketing, and insurance. Export companies purchase commodities from U.S. production agriculturalists, store these products, arrange for their sale to foreign countries, and provide for shipment of the products.

Export companies need contracts and good relations with all foreign countries. They need people who can sell U.S. products, help expand markets, and develop new markets for those products. They need people who know how to listen as well as how to talk, and people who can communicate easily

55. "North American Free Trade Agreement," available at http://en.wikipedia.org/wiki/NAFTA; "Top Softwood Lumber Importing Countries," available at http://www.fas.usda.gov/ffpd/Newsroom/Softwood_Lumber_Importers.pdf.

56. "Exchange Rates and Exchange: How Money Affects Trade"; available at http://www.econedlink.org/lesson/index.cfm?lesson=EM342&page=teacher.

Career Option

International Agriculture Trader/Marketer

All nations strive to attain a higher standard of living for their people. This standard could not be maintained if the United States chose to isolate itself and not engage in a wide range of international trade activities. With each country having varying levels of resources (land, labor, capital, and management), international trade allows a country to specialize in production of those products that it produces best and most efficiently. As transportation and communication improve, we have become familiar with more places and people around the world, so more business is being conducted between and among people and nations of different continents. Without international trade, production agriculturalists would be burdened with large surpluses and forced to sell out because of low prices.

Careers in international trade are challenging and exciting. Some of the benefits include travel, working with people in different countries, and learning about the broad scope of world agriculture. Importers and exporters rely on professionals who can see the broad picture of production, processing, marketing, and trade. They employ those who can interpret complex policies and trade regulations, while also exhibiting competence in working with people of different cultures and speaking other languages. Production agriculturalists and agribusinesses in America will continue to depend on export markets for U.S. products, while consumers will continue to expect commodities and products from other countries to be available. Therefore, people who can sell U.S. products, help expand markets, and develop new markets for them will always be in demand.

Being an international trader requires the most up-to-date technological assistance, including, among other devices, fax machines, computers, and Internet access.

and effectively with citizens of foreign nations. Knowing more than one language is a distinct advantage.[57]

Foreign Agricultural Services

Global U.S. agricultural interests are represented by the Foreign Agricultural Service (FAS), an entity within the U.S. Department of Agriculture (USDA). It is headquartered in Washington, D.C., but has offices around the globe. More than 100 countries benefit from the expertise and assistance of the more than 70 offices that constitute the FAS. The mission of the FAS is to use its personnel and officers to promote trade of U.S. agricultural products, to improve trade conditions globally, and to track other nations' agricultural production. The FAS also provides information to the USDA that is used to formulate agricultural policy and develop programming.[58]

57. Marcella Smith, Jean M. Underwood, and Mark Bultmann, *Careers in Agribusiness and Industry*, 4th ed. (Danville, IL: Interstate Publishers, 1991), p. 340.

58. Ibid., p. 339.

International Agricultural Research

Agricultural research is vital. During the past 40 years, the amount of food produced has increased at a higher rate than the world population, in large part because of agricultural research. The major participants in agricultural research are public entities, universities, and agribusiness firms. The research data gathered is published or shared directly by professional associations and by others in the agribusiness sector. Research is applied to specific situations that scientists and production agriculturalists encounter.[59]

An important resource for both developed and developing countries is the Consultative Group on International Agricultural Research (CGIAR). This organization consists of a network of research centers around the world that specialize in agriculture. The following lists the individual centers:

- International Center for Tropical Agriculture (CIAT), Cali, Colombia
- International Center for the Improvement of Maize and Wheat (CIMMYT), El Batán, Mexico
- International Potato Center (CIP), Lima, Peru
- International Board for Planned Genetic Resources (BPGR), Rome, Italy
- International Center for Agricultural Research in the Dry Areas (ICARDA), Aleppo, Syria
- International Crops Research Center for the Semi-Arid Tropics (ICRISAT), Hyderabad, India
- The International Food Policy Research Institute (IFPRI), Washington, D.C.
- The International Institute of Tropical Agriculture (IIITA), Ibadan, Nigeria
- The International Livestock Center for Africa (ILCA), Addis Ababa, Ethiopia
- International Laboratory for Research on Animal Diseases (ILRAD), Nairobi, Kenya
- International Rice Research Institute (IRRI), Los Baños, Philippines
- International Service for National Agricultural Research (ISNAR), The Hague, Netherlands
- West Africa Rice Development Association (WARDA), Bouake, Côte d'Ivoire[60]

59. Ibid., p. 340.
60. Ibid., pp. 341–342.

The Peace Corps

The Peace Corps was started in 1961, based on an idea proposed by President John F. Kennedy. It was established by Congress to promote world peace and friendship. The Peace Corps Act contained three specific goals:

- to help the people of interested countries and areas meet their needs for trained workers and consultants
- to help promote a better understanding of Americans by the peoples served
- to help promote a better understanding of other peoples by Americans[61]

Volunteers in the Peace Corps take two-year assignments in other countries. They work with the people of those countries to help them become self-sufficient in the areas of education, health care, water sanitation, housing, and food production. Since its inception, the Peace Corps has sent more than 125,000 volunteers and staff to assist more than 94 countries. In the mid-1980s, the Corps was still active in 63 countries located in nearly every continent.[62]

Conclusion

The world of international trade is complex and ever-changing. International trade can be profitable, but it also carries significant risks. In the future, it is likely that an increasing percentage of world output will be traded internationally. The question for managers will no longer be, "Should we trade internationally?" but rather, "How can we trade more effectively?" Expertise in international trade will become an increasingly important skill for managers.

Careers in international trade are challenging and exciting. Travel, working with people in different countries, and learning the broad scope of world agriculture are all benefits of such jobs. Production agriculturalists and agribusinesses in America will continue to depend on strong export markets for U.S. products. Consumers will

61. Gerald T. Rice, *Peace Corps in the 80's* (Washington, DC: Office of the Peace Corps, Public Affairs, 1985), p. 6.
62. Smith, Underwood, and Bultmann, *Careers in Agribusiness and Industry*, pp. 342–343.

continue to expect that commodities and products from other countries will be available, safe and wholesome, and fairly priced.

To accomplish these tasks, importers and exporters will increasingly rely on professionals who understand international agriculture trade. They will need people who can see the broad picture of production, processing, marketing, and trade. They will need employees who can interpret complex policies and trade regulations. International companies with offices or production facilities in other countries seek—and promote to international positions—employees who are skilled in management and marketing. They also want these people to be competent and comfortable working with people of different cultures who speak languages other than English.

Summary

International trade has been a part of the agricultural industry since colonial times. Early Americans quickly discovered that exchange of goods and services with other countries was best for both countries if the imported product could be produced more cheaply abroad than at home. Gains from international trade occur because of low prices of selected foreign products. Because each country has varying levels of resources (land, labor, capital, and management), international trade allows a country to specialize in production of those products it produces best and most efficiently.

It is estimated that more than 1 million full-time jobs are related to agricultural exports. It is also estimated that for every dollar of agricultural exports sold by the United States, $2.05 is added to the U.S. economy. Additionally, it has been shown that for every dollar generated by food and fiber exports, 75 cents goes to nonfarm business sectors of the U.S. economy.

The key concept in understanding the basis of international trade is that countries export the commodities that they are relatively most efficient in producing and import commodities that they are relatively least efficient at producing. Gains from international trade do not depend on a nation being absolutely more efficient than other countries in producing some commodity; rather, gains from trade occur when a country is relatively more efficient in the production of some commodities than others.

Agricultural exports amount to approximately 8 percent of all merchandise exports of the United States. At present, about 23 percent of U.S. farm cash receipts are derived from the export market. The United States exports large percentages of its wheat, soybean, corn, grain sorghum, and rice crops. The United States is also one of the largest importers of agricultural products. Annually, the United States imports agricultural products worth more than $26 billion. However, agricultural imports make up less than 5 percent of total U.S. imports.

When the outflow of dollars from the purchase of imports exceeds the inflow of dollars from the sale of exports, a country has a deficit in its international trade balance. International trade deficits can be corrected in the long run by reducing imports (purchases) and by increasing exports (sales).

Some people favor free trade and some favor protectionism. The leading arguments in favor of free trade center around the law of comparative advantage. It is agreed that if markets are not obstructed by protectionist measures, and private entrepreneurs are allowed to pursue their own interests, resource allocation will be optimized and overall welfare will be maximized. Those that favor protectionism also pose several economic arguments supporting their view. These arguments concern infant industries, the balance of payments, unfair competition, self-sufficiency and national security, diversification for stability, and protection of wages and the standard of living.

If we allow free trade, we must have ways to protect or restrict trade; these ways are called trade barriers. Import trade barriers include tariffs, import quotas, and regulatory trade restrictions. Export trade barriers include export embargoes and voluntary export restraints. Between the two extremes of free trade and trade barriers lie trade agreements. Two major international trade agreements are the GATT and the WTO. Two major regional trade agreements are the EU and NAFTA.

There are many careers related to international agriculture. These include careers in export-import services, the Foreign Agricultural Service (an agency of the USDA), international agricultural research, and the Peace Corps.

End of Chapter Activities

REVIEW QUESTIONS

1. What are the four major factors of production?
2. Although production factors are the same internationally, what are eight challenges of international trade?
3. Name nine imports that increase our quality of living or suit our eating preferences.
4. List the 16 states in which a third to a half of total farm income comes from agricultural exports.
5. Explain how consumers benefit from international trade.
6. What is the key concept in understanding the basis for international trade?
7. Differentiate between absolute advantage and comparative advantage in relation to international trade.
8. Explain why it is important for a country to specialize even when it is able to raise an agricultural commodity.
9. List 10 agricultural commodities that are exported from the United States.
10. What are the humanitarian reasons for exporting?
11. Briefly explain the beneficial side effects of exporting.
12. Name six major imports not produced in the United States.
13. What is the main purpose of tracking the trade balance?
14. List six objectives taken into account by effective trade policy.
15. Explain the general economic arguments for free trade.
16. List and briefly explain the six general economic arguments for protectionism.
17. Briefly explain three trade barriers used to control imports.
18. Briefly explain two trade barriers used to control exports.
19. What are five basic principles of the GATT?
20. What are five functions of the WTO?
21. What are 10 benefits of the World Trade Organization?
22. What are five purposes of the World Bank?
23. What are five responsibilities of the International Monetary Fund?
24. What are two benefits of NAFTA to U.S. agriculture?
25. Name two disputes that have been subject to NAFTA resolution.
26. List the 15 countries that participate in the European Union.
27. List three general characteristics needed by people who are seeking careers in export-import businesses.
28. What are three goals of the Peace Corps?

FILL IN THE BLANK

1. ____________ was the first agricultural product to be exported from the United States.
2. International trade allows a country to ____________ in the production of those products it produces best and most efficiently.

3. ___________ and ___________ ___________ among nations permit higher living standards and real income than would otherwise be possible.
4. More than ___________ ___________ full-time jobs are related to agricultural exports.
5. For every dollar of agricultural exports sold by the United States, ___________ dollars were added to the U.S. economy.
6. The consequence of abandoning the export market would be a reduction in production of about ___________ percent of wheat, rice, and soybeans and ___________ percent of feed grains.
7. The United States exports approximately ___________ percent of its wheat crop, ___________ percent of its soybean crop, ___________ percent of its corn crop, and about ___________ percent of its rice crop.
8. ___________, ___________, and ___________ are the top countries (sources) of agricultural products imported into the United States.
9. International trade balance deficits can be corrected in the long run by ___________ imports and ___________ exports.
10. Through 1995, there were ___________ consecutive years of U.S. agricultural trade surpluses.
11. The ___________ helps promote trade of U.S. agricultural products, works to improve trade conditions through the world, and monitors agricultural production in other countries.
12. The Peace Corps idea was proposed by ___________.
13. The World Trade Organization headquartered in ________________.
14. The World Bank has its headquarters in ________________.
15. The International Monetary Fund is headquartered in ________________.

MATCHING

a. 25 million
b. Peace Corps
c. Germany
d. 8 percent
e. 23 percent
f. 15 percent
g. 26 billion
h. 600,000
i. USSR
j. 5 percent
k. 1 billion
l. free trade area
m. customs union
n. common market
o. NAFTA
p. GATT
q. CGIAR
r. Japan

______ 1. percent of U.S. farm income derived from exports
______ 2. without exports, number of jobs lost in U.S. agribusiness input sector
______ 3. without exports, number of jobs lost in U.S. agribusiness output sector
______ 4. percent of total exports that were agricultural
______ 5. became the United States' first $10 billion agricultural customer
______ 6. dollar value of agricultural products imported annually to the United States
______ 7. percentage of total U.S. imports that are agricultural products
______ 8. approximate dollar value of annual U.S. agricultural trade surplus
______ 9. agreement created to establish free trade among the United States, Canada, and Mexico

_______ 10. an economic and political organization between two or more countries abolishing trade restrictions amongst themselves and establishing a common and uniform tariff for outsiders

_______ 11. worldwide group of research centers that support agriculture

_______ 12. volunteers spend two years working with people in other countries

ACTIVITIES

1. Prepare a report (research paper) on a major international trade agreement that influences the marketing of U.S. agricultural products.
2. Read about and study both free trade and protectionism. Choose the stance you favor and defend your decision by presenting data and arguments supporting it to your class.
3. Write a one-page essay supporting or criticizing the opinion that U.S. agriculture should feed a hungry world. Present this paper to your class.
4. Is a negative U.S. trade balance detrimental to the country? Explain.
5. Prepare a paper stating whether you are for or against NAFTA. Read articles from various agriculture magazines and other sources. Present your paper to the class or debate the topic.
6. Go to your local grocery store and identify as many products as you can that are imported. If possible, list the country of origin for each.
7. Explain comparative advantage. In what commodities does the United States have a comparative advantage? Can the United States have a comparative advantage in the production of every commodity? Explain.
8. Prepare a one-page essay on how the arguments for and against free trade relate to the law of comparative advantage.
9. Explain the impact of the exchange rate on international trade.
10. To calculate exchange rates, begin with the equation of one currency equal to another. For example, the table below indicates that the value of a U.S. dollar in terms of British pounds is 0.70 pounds; therefore, US$1 = £0.70, and £1 = US$1.43. You can then use that information to answer other questions. Use the following exchange rate information (from April 2002) to answer questions a through f.

1 British pound = $1.43	$1 = 0.70 British pounds
1 Canadian dollar = $0.63	$1 = 1.59 Canadian dollars
1 Egyptian pound = $0.22	$1 = 4.61 Egyptian pounds
1 Russian ruble = $0.03	$1 = 31.21 Russian rubles
1 South African rand = $0.09	$1 = 11.14 South African rands
1 Swiss franc = $0.60	$1 = 1.66 Swiss francs

   a. How much does it cost in U.S. dollars to buy one South African rand?
   b. How much does it cost in Russian rubles to buy one U.S. dollar?
   c. Suppose you wanted to buy a sculpture that costs 2,000 Egyptian pounds. How much would the sculpture cost in U.S. dollars?
   d. Suppose you wanted to buy a watch that costs 50 Swiss francs. How much would the watch cost in U.S. dollars?
   e. If you planned a vacation to the United Kingdom and wanted to exchange your money before you left, how many British pounds could you get for US$1,000?
   f. Why do American business owners try to avoid accepting Canadian coins?

Chapter 20

Agrimarketing Channels

Objectives

After completing this chapter, the student should be able to:

- Describe the historical evolution of agrimarketing channels.
- Discuss assemblers of agricultural products.
- Discuss agricultural commodity processors and manufacturers.
- Explain agribusiness wholesaling.
- Explain agribusiness retailing.
- Analyze the food service industry.
- Explain the services related and contributory to the marketing channels.
- Analyze careers in agrimarketing.

Terms to Know

agricultural commodities
agricultural products
agrimarketing channels
assemblers
blanching
broilers
chain store
commercial food establishment
convenience stores
grading
kilns
merchandising
noncommercial food establishment
perishable
specialized food stores
standardization
superettes
supermarket
superstores
warehouse/limited-assortment supermarket
wholesalers

Introduction

Former Commissioner of Agriculture for the State of Tennessee William Wayne Walker III claimed that while he was in office, 50 percent of the workforce was employed because of the agricultural industry. Although the statistical reporting service does not exactly substantiate this claim, former Commissioner Walker based his claims on the number of people employed in the agribusiness input sector, production agriculture, and especially the agribusiness output sector.

The agribusiness output sector, which includes the marketing channels and those jobs and careers that provide support to the agrimarketing channels, employ millions of workers. Few outside the agricultural industry realize just how many jobs exist because of agriculture. The primary objective of this chapter is to explore the jobs and career opportunities that are available because of the marketing channels of **agricultural commodities** and **agricultural products**.

Agrimarketing channels are the path that an agricultural commodity follows from the "farmer's gate to the consumer's plate."[1] The length of this path depends on the product. For example, pick-your-own strawberries have a very short path: from producer to consumer. However, the market channel for most food products is much longer: producer–assembler–food processor–food wholesaler–food retailer–consumer.[2] Jobs and careers that provide support to these marketing channels include storing, warehousing, financing, insuring, risk taking, transporting, promoting, advertising, communicating, inventorying, packaging, labeling, merchandising, selling, distributing, inspecting, and regulating. As you can see, Commissioner Walker could substantiate his claims if all the facts were known. Refer to Figure 20-1.

1. Kevin J. Bacon and Robert Birkenholz, *Careers I Unit for Agricultural Sciences I Core Curriculum* (Columbia, MS: University of Missouri, Instructional Materials Laboratory, 1988), p. 1.
2. James G. Beierlein and Michael W. Woolverton, *Agribusiness Marketing: The Management Perspective* (Englewood Cliffs, NJ: Prentice-Hall, 1991), pp. 129–130.

Historical Evolution of Agrimarketing Channels

In the early years of this nation, most Americans were farmers, producing food and fiber primarily for personal use and home consumption. They did more than just produce it: they grew or made most of the inputs needed to produce these products. They also processed farm products into consumable forms: grain into flour, flour into bread, fibers into cloth, and hides into leather. They packaged their produce as required by consumers. They cured meat and processed milk so these products could be stored for future consumption. They took their produce to town or to docks for export. They found buyers and took care of all the financial matters and procedures involved in transferring ownership of their produce. All of these activities—those beyond growing the basic raw farm commodities—added value to the raw farm materials.[3]

Evolution of Specialists

As new technologies were developed and applied by both farmers and nonfarmers, and as people began to claim areas of expertise, it became physically impossible for farmers to perform some of these functions, and economically infeasible for them to perform others.

Specialists evolved to provide building supplies, machinery, and tools; to process and package foods; to transport goods; and to buy and sell farm produce. They were not only more efficient at these tasks, they could also reap the benefits of economies of scale with large operations. Farmers also became better at producing raw materials, as they no longer had to divide their attention and skills between farming and a myriad of other activities.[4]

The Farmer's Share of the Consumer's Dollar Falls

Early farmers received almost all of the consumer's food expenditures, but the food was expensive because producers were not particularly skilled or

3. Milton C. Hallberg, "U.S. Food Marketing—A Specialized System," in *Marketing U.S. Agriculture: 1988 Yearbook of Agriculture,* ed. Deborah Takiff Smith (Washington, DC: U.S. Government Printing Office, 1988), p. 12.
4. Ibid., p. 12.

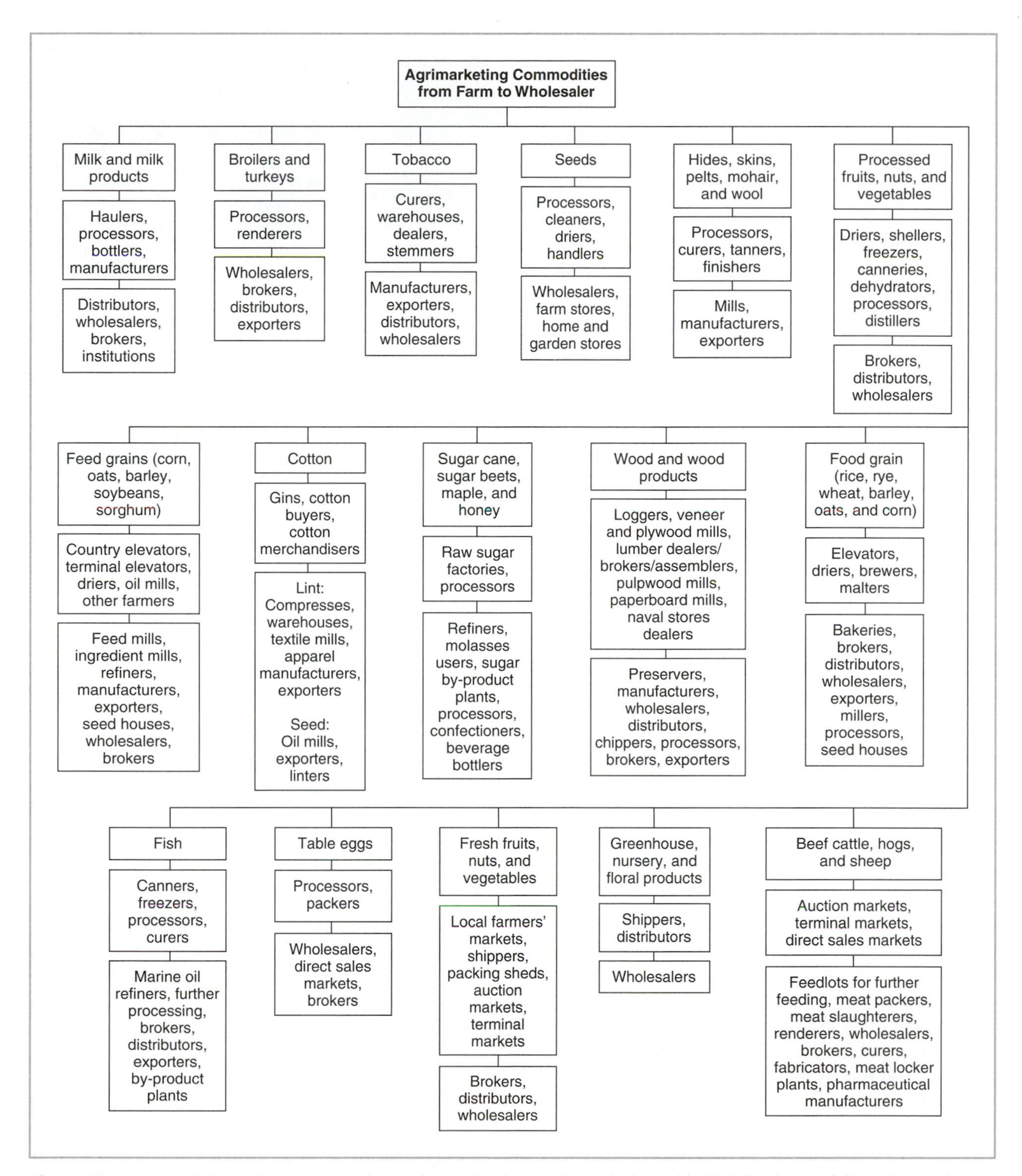

Figure 20–1 Many jobs and processes make up the agrimarketing channels through which foods travel from the producer to the consumer. This chart stops at the wholesaler. Can you complete the chart as the food continues to the consumer?

efficient at providing marketing functions in addition to producing raw materials. Today, production agriculturalists capture less than 30 cents of the consumer's food dollar, but as a whole, the price of food with all of the services that the marketing sector adds is less than it would otherwise be.[5]

Both Producer and Consumer Gain

In a system with specialized production and marketing functions, everyone gains. Consumers have access to a larger quantity and variety of products at a lower unit cost; production agriculturalists can sell more produce at a higher unit price; and the agrimarketing sector can employ more people, because the increased demand for food and fiber (and the related services required to process products) requires more workers.[6]

Assemblers of Agricultural Products

More than 10,000 **assemblers** may be interposed between the production agriculturalist and the processor.[7] Assemblers do not usually change the form of a product. Instead, they put small lots of a commodity together to provide a larger, more economical unit. This permits processors to capture more of the economies of scale possible in shipping and plant operation. For grain, this function is performed by assemblers such as the local elevators. Refer to Figure 20–2. For livestock, this function is performed by a commission buyer who, for a fee, buys animals from individual producers on behalf of a processor. Other examples are vegetable packing sheds; livestock sale barns; transporters that go from one farm to another, such as bulk milk trucks that pick up raw milk on the farm; and egg-sorting plants. A further explanation of assemblers follows.

Resident buyers are people who own or operate an area business that purchases products. These people may represent themselves or another company.

Order buyers purchase products for wholesale or chain stores. They usually specialize in livestock.

Figure 20–2 Grain elevators are an example of food assemblers in the agrimarketing channel. (Courtesy of USDA)

Typically, they meet consumer wants by purchasing items of a specific type and quality. Order buyers earn a salary from a company or are paid a commission.

Traveling buyers move from area to area during crop season. They purchase large quantities of a product directly from the production agriculturalist. They may represent themselves or terminal buyers.

Auction markets are places where farmers take products so that bidders can purchase the products. Commodities such as tobacco, fruits, vegetables, and animals may all be sold at such markets. A great deal of product is sold in this manner; in fact, more than 90 percent of tobacco is sold by this method.

Having consumers as assemblers is a form of direct marketing that ties production agriculturalists to their consumers. In this form of marketing, producers sell their products directly to consumers. Consumers like this customer-friendly marketing and tend to view the commodities they receive as fresher, more wholesome, and more economical than store-bought products. More than 13,000 production agriculturists operate farmer's markets, pick-your-own operations, and roadside stands, all of which put consumers in the role of assemblers.[8]

As you can see, the role of the assembler is to increase the efficiency of the marketing system by getting "the right quality of products, at the

5. Ibid., p. 12.
6. Ibid., pp. 12–13.
7. Ewell P. Roy, *Exploring Agribusiness* (Danville, IL: Interstate Printers & Publishers, 1980), p. 69.
8. Ibid., pp. 72–73.

right time, at the right price, at the right place, and in the right quantity to meet the needs of the processor."[9]

Agricultural Commodity Processors and Food Manufacturers

Although we have used the words *commodity* and *product* interchangeably through this book, food manufacturers differentiate the terms. A *commodity* is an agricultural crop, livestock, or item (such as milk) that comes off the farm in its raw state. Once the raw commodity has been altered by a processor so that it has been changed from its initial form, it becomes a food *product*.[10] For example, milk is a commodity, whereas cheese is a product.

Technology in Commodity Processing and Food Manufacturing

Advances in technology have allowed **perishable** food items to last longer without spoilage. The advances have created an entire manufacturing industry that is able to process commodities at lower costs and in large quantities.[11]

Size, Volume, and Structure

Size. There are approximately 22,000 firms in the commodity processing and food manufacturing sector. They add more than $202 billion of value to the items they handle. About half of the sector's output comes from the 100 largest firms.[12] They include such well-known companies as General Foods, General Mills, Kelloggs, Pillsbury, Sunkist Growers, Land O' Lakes, and a host of others.

Volume. The food-processing sector in the United States was very profitable in the early 1990s. Processed food shipments, which total about $395 billion annually, account for about 13 percent of all U.S. manufacturing activity and represent the largest single sector in the economy.[13]

American consumers can choose from among more than 230,000 packaged food products. Including new size introductions, more than 16,000 new grocery products are introduced in some years. However, industry estimates put the failure rate of new food products at 90 to 99 percent.[14]

Structure. The structure of the food-processing industries has changed significantly. Many smaller processors have sold out or closed because they were unable to find capital and make the needed operational changes. More than 1,250 businesses were sold between 1982 and 1992. To survive in some form, others merged with stronger companies: almost 2,900 mergers took place during that same decade. The 50 largest food processors control 47 percent of sales.[15]

Competition. Despite the increased size of the larger firms, competition among the 16,000 firms in the 49 food-processing industries is fierce. Even in bad economic times, food processors use price and nonprice competitive strategies to gain both consumer acceptance and retail shelf space in the $160 billion brand-name retail food market and in the $240 billion food-processing market.[16]

Processors

The processor is normally the first one in the marketing system to alter the form of a raw agricultural commodity. Processing involves many different activities. It includes all the changes that occur as a product is prepared for consumption. There are 11 principal types of food processors: canned food, frozen food, sugar, beverage, meat, dairy, bakery, grain mill, confectionery, paper, and cotton. These food processors conduct their businesses using more than 20,000 plants.

---

9. Beierlein and Woolverton, *Agribusiness Marketing*, p. 130.
10. Ibid., p. 131.
11. Ibid., p. 128.
12. U.S. Department of Agriculture, *Agricultural Statistics, 1998* (Washington, DC: U.S. Government Printing Office, 1998).
13. Randall D. Little, *Economics: Applications to Agriculture and Agribusiness*, 4th ed. (Danville, IL: Interstate Publishers, 1997), p. 325.
14. Ibid., p. 326.
15. Ibid., p. 325.
16. Ibid.

Figure 20–3 Processing is part of the agrimarketing channel. The workers shown here are making beef into hamburger patties. (Courtesy of Oklahoma Department of Agriculture)

Livestock processing is the manner in which animals (such as cattle) are slaughtered and prepared in various ways for sale. Hamburger, steak, hot dogs, chops, and ribs are examples of the different forms that processed meat may take. Additional frozen and canned food items may be produced from some or all parts of the animal carcasses. Refer to Figure 20–3.

Potato processing takes many forms. Some potatoes are sold whole, in bags, at supermarkets. Potato chips and instant mashed potatoes are potato items that require significant processing.

Cotton processing starts with separating seed from lint or fiber. Once separated, the fiber is woven into cloth. The cloth is made into clothing and other products.

Wheat processing involves milling grain into flour and baking a flour product. Processing high-quality, healthful wheat products involves a multitude of detailed steps.

Milk is *pasteurized* (heated to a specified temperature to kill bacteria) and *homogenized* (treated to disperse fat particles throughout the milk to keep the cream from rising to the top). It is packaged in cartons of convenient sizes for customer use. Milk must be carefully stored in refrigerated areas at all phases of processing and marketing to maintain its quality.

Poultry processing involves slaughtering the animal, removing the feathers and internal organs, preparing the desired cuts, and packaging the final product for sale. Prior to use, fresh poultry must always be refrigerated or frozen.

Canning and freezing fruits and vegetables involves cleaning, peeling, cutting, shelling, breaking (beans), **blanching**, cooking, canning, freezing, and other processing depending on the commodity. Warehousing is a critical factor for canning and freezing plants, as they cannot immediately sell all of what they have processed.

Paper processing takes raw wood and makes it into a variety of products. It is estimated that each person uses the equivalent of 500 pounds of paper in the course of a year. This equates to about 3/4 of an acre of commercial wood growth for each American annually. Business use is much higher: for example, producing the paper for the Sunday edition of one large New York newspaper alone requires the yearly growth from 6,000 acres of commercial forest land.[17]

Some of the large and better-known processors include Cargill (grain and meat), IBP (meat), Central Soya (oilseeds), AMPI (dairy), and Monfort (beef).

Food Manufacturers

After processing, food manufacturers continue to add value to livestock and crops. The goal is to improve the quality of the product, preserve it better, and make it more convenient for the consumer to use. Food manufacturers produce baked goods, breakfast cereals, ice cream, and a huge range of other products. In each of these various industry segments, the food manufacturer takes a raw agricultural commodity (milk, for instance) that has been changed in some way by a processor (pasteurized milk) and alters it into a food product that consumers desire (ice cream).

Through this manufacturing process, an agricultural commodity loses the identity it started with and becomes part of a whole new food product. In essence, it becomes one of many ingredients in a manufactured and branded food product—in our earlier example, ice cream. In the agribusiness marketing sector, the conversion of a raw product into a food product is a significant transition or separation point.[18]

17. Roy, *Exploring Agribusiness,* p. 78.
18. Beierlein and Woolverton, *Agribusiness Marketing,* p. 131.

Agribusiness Wholesaling

The transfer of food and fiber from processing and manufacturing plants to wholesalers and then to retailers and other establishments is an important step in agrimarketing channels. **Wholesalers** are operators of agribusinesses engaged in the purchase, assembly, transportation, storage, and distribution of groceries and food products for sale to retailers, institutions, and business, industrial, and commercial users.

Food wholesalers purchase large quantities of food products. They transport the product (or have it delivered) to their warehouses, where it is separated into cases or pallets that can be sold to individual retailers in the food service industry.[19] Refer to Figure 20–4.

Figure 20–4 Wholesaling is part of the agrimarketing channel. The grain is being stored in the unit shown here. It can then be divided into smaller units for shipment to individual food retailers and food service establishments. (Courtesy of Oklahoma Department of Agriculture)

Size and Volume of Sales

There are more than 476,000 wholesale business establishments with annual sales totaling more than $1 trillion. More than 6.3 million persons are employed in this agribusiness subsector, with an annual payroll of more than $181 billion. Wholesaling of grocery and farm-related raw materials employs more than 880,000 people, through 54,700-plus firms with $500 billion in sales annually.[20]

Trend Toward Wholesale-Retail Affiliation

At one time, the wholesaler occupied the major economic position in the food marketing channel. The large number of small, independent retailers were dependent on wholesalers to stock the shelves, provide credit for retail inventories, and fulfill numerous other functions. On the other end, processors looked to wholesalers as the major outlet for their products. However, the economic power of wholesalers began to decrease with the development of the food retail chain stores.

Chain stores began to bypass independent wholesalers and deal directly with processors. The large volume handled prompted retailers to establish their own warehousing facilities. Processors also began to increase in size and, essentially, do their own wholesaling. To survive, independent wholesalers aligned with small, independent retailers who were also being squeezed by the development of retail chain stores. Some marketing specialists contend that the trend toward wholesale-retail affiliation is the most significant development in food wholesaling in many years.

Reasons Why Wholesalers Exist

Food wholesalers exist in the agrimarketing food channel because they provide food retailers with a range of services, including credit, savings, and variety.

Credit. Wholesalers make a wide assortment of products available to retailers, and in some cases provide short-term credit.

Savings. By purchasing large quantities of single products, wholesalers obtain price savings in the form of volume discounts that retailers would be unable to obtain for themselves.

Variety. Wholesalers maintain a variety of products and inventory levels that retailers could not sustain.[21]

19. Ibid., pp. 131–132.
20. Little, *Economics,* p. 328.
21. Beierlein and Woolverton, *Agribusiness Marketing,* pp. 132, 144.

Types of Wholesalers

Wholesalers are classified in many ways, depending on the interest of the classifier. The most common methods of classification are based on economic structure, the number of functions performed, the variety of goods handled, and affiliation with retailers. It should be obvious that any given wholesaler may fit several of these classifications at one time. Based on economic structure, wholesalers are normally classified as (1) merchant wholesalers, (2) manufacturer's sales branches, or (3) agents and brokers.

Merchant Wholesalers. Merchant wholesalers take title to the goods they handle. This is the largest group of wholesalers, numbering more than 50,000 in the food industry with total sales of more than $253 billion annually. These agribusiness wholesalers constitute about 30 percent of all merchant wholesalers in the United States and handle about 50 percent of all sales volume. These statistics suggest that food wholesalers are larger than wholesalers of most other products. Merchant wholesalers can be divided into several types:

- *General-line wholesaler merchants* handle a wide range of dry groceries, health and beauty aids, and household products.
- *Limited-line wholesaler merchants* handle a smaller range of dry groceries; most of their stock tends to be canned foods, coffee, spices, bread, and soft drinks.
- *Specialty wholesale merchants* handle perishables, such as frozen foods, dairy products, poultry, meat, fish, fruit, and vegetables.
- *Wholesale clubs* are hybrid wholesale-retail establishments selling food, appliances, hardware, office supplies, and similar products to their individual and small-business members at prices slightly above wholesale.[22]

During the late 1980s and early 1990s, the number of wholesale clubs grew rapidly. As a result, intense competition developed among the different wholesale clubs and between the wholesale clubs and supermarkets. Existing supermarkets began offering services that the wholesale clubs did not. Examples include in-store specials, bakery departments, and restaurant (deli) style foods.[23]

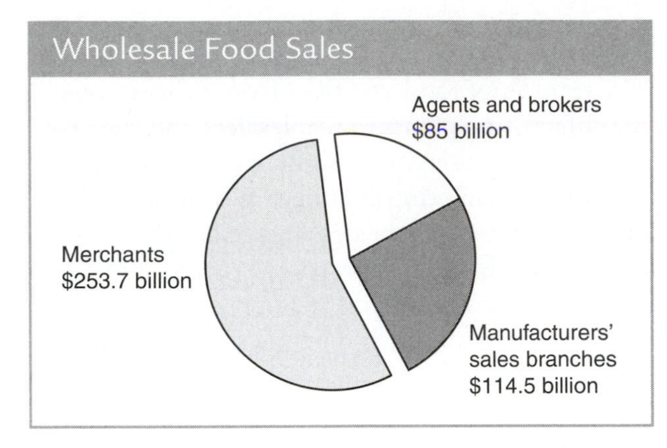

Figure 20–5 Food wholesalers are usually classified as either merchant wholesalers, manufacturer's sales brokers, or agents and brokers.

Manufacturers' Sales Branches. Manufacturers' sales branches are extensions of food-processing firms' marketing activity at the wholesale level. They are owned and operated by food manufacturers, and they provide a wide variety of services. They are the second largest group by number, and they usually deal in larger volume than merchants or brokers. A major portion of all food manufacturers' sales branches are controlled by companies that have more than 25 units in different locations.

Agents and Brokers. Agents and brokers are wholesalers that do not take title to the goods they handle. The ownership of the goods is retained by the clients. Therefore, agents and brokers have no price risks in their business. This type of wholesaler tends to be smaller than other types, and many are highly specialized, although some do handle a wide variety of goods. Also, agents and brokers deal more with other wholesalers than with retailers. Refer to Figure 20–5.

Agribusiness Retailing

Agribusiness retailers include businesses selling groceries, prepared foods, soft drinks, floral products, clothing, shoes, furniture, home furnishings (made from agriculturally derived products), and other products. There are more than 588,000 retail food outlets in the United States.[24]

22. Little, *Economics,* pp. 328–329.
23. Ibid., p. 329.
24. Ibid., p. 330.

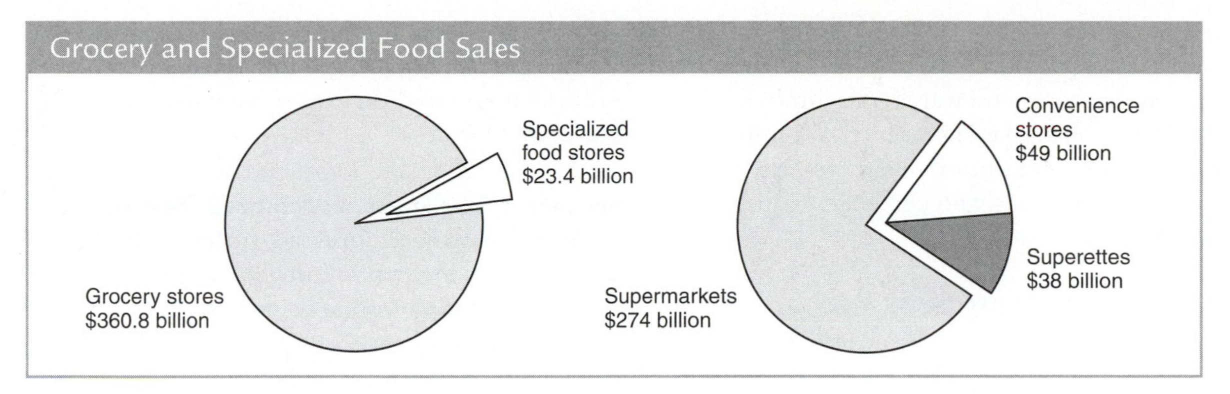

Figure 20-6 Grocery stores are usually classified as either supermarkets, convenience stores, or superettes. Specialized food stores, which account for about 6.5 percent of food sales, include candy and nut stores, dairy stores, produce markets, meat and fish markets, retail bakeries, and miscellaneous food stores.

The history of food retailing in this country is marked by three major developments. The first was the development of the retail food chain store. The second was the development of the supermarket concept.[25] A third development is the advent of the superstore and **warehouse/limited-assortment supermarket**. Chain stores are discussed next. Superstores and warehouse limited-assortment establishments are discussed as types of grocery stores.

Chain Stores

A **chain store** is defined as the operation of 11 or more stores under a single owner. Grocery chain stores now account for about 65 percent of food store sales, up from 37 percent in 1948 and 25 percent in the 1920s. The chain store's share of total grocery store sales was consistent through the mid-1980s.[26]

The chain stores radically changed the way food was sold at retail. Their lower prices and wider selection overwhelmed many of the family-run, single-outlet food stores that are sometimes called "Ma and Pa" stores. Some of the larger retail food chains include Safeway, Kroger, and A&P. The largest chains operate as many as 5,000 stores. In addition to these, there are many other chains that operate fewer stores on a local or regional basis.[27]

Types of Grocery Stores

The retail grocery store is near the end of the marketing channel, where agricultural products, as well as other products, are purchased by the consumer. There are about 165,000 grocery stores with annual sales of more than $361 billion, employing about 3.2 million people.[28] A retail grocery store is usually classified as one of three types: supermarket, convenience store, or superette. Refer to Figure 20-6.

Supermarkets. **Supermarkets** offer the customer a full line of 10,000 to 15,000 items, in addition to many nonfood items such as cleaning products, health and beauty aids, and so forth. Surprisingly, the supermarket concept was introduced by independent stores (the "Mom and Pop" stores) rather than by food chain stores. When the concept became popular and successful, chain stores reacted by adopting the idea. Thus, by the mid-1950, the supermarket was a well-established food retailing method.[29]

Although just under 10 percent of all retail food stores are made by supermarkets, these businesses account for almost 72 percent of total grocery sales.[30] Two key trends affecting supermarkets are the growth of superstores and the appearance of warehouse/limited-assortment supermarkets.

Superstores offer a greater variety of products than conventional supermarkets. About 25 percent of all supermarkets are superstores, which do 33 percent of supermarket sales volume. *Warehouse/limited-assortment supermarkets* offer

25. Beierlein and Woolverton, *Agribusiness Marketing,* p. 141.
26. Little, *Economics,* p. 333.
27. Beierlein and Woolverton, *Agribusiness Marketing,* p. 142.
28. Little, *Economics,* p. 331.
29. Beierlein and Woolverton, *Agribusiness Marketing,* pp. 142–143.
30. Little, *Economics,* p. 331.

larger sizes of fewer items at lower prices than the typical supermarket. These nonconventional formats (superstores and warehouse/limited-assortment supermarkets) presently account for more than 50 percent of all supermarkets and make approximately 66 percent of supermarket sales.[31]

Convenience Stores. **Convenience stores** first appeared in the late 1950s in the South and West. In many growing suburbs, they played the same role that the "Ma and Pa" grocery stores did in older communities. A number of convenience stores started as dairy stores, with milk accounting for as much as 50 percent of sales. These helped to fill the declining niche of home milk delivery. When gasoline prices soared in the 1970s and 1980s, many convenience stores added self-service gasoline pumps. Now carry-out foods, including hot sandwiches, and in-store eating have become staples of convenience stores.[32] Few consumers do all their weekly grocery shopping at convenience stores; however, most do buy a few items there on occasion. One can now find chain convenience stores throughout the country. The largest of these chains is 7-Eleven.[33]

Convenience stores are responsible for 12.6 percent of all grocery sales. They constitute about 20 percent of all retail stores. Fairly recently, gasoline stations have begun to add convenience stores, and have been able to effectively compete with the traditional convenience store. Because food sales are typically less than half of the overall sales in a gasoline station, these integrated convenience stores are not considered food stores.[34]

Superettes. Smaller grocery stores, or "Ma and Pa" stores, are referred to as **superettes.** Thirty-seven percent of retail food outlets are superettes. They are responsible for about 10 percent of sales. They normally stock both food and nonfood grocery items. Superettes are found in areas underserved by supermarkets—typically in sparsely populated regions or very densely populated locales.[35]

Specialized Food Stores. **Specialized food stores** account for 6 percent of retail grocery sales. They usually sell a single food category. Examples include retail bakeries; meat and fish markets; candy and nut stores; natural and health food stores; coffee, tea, and spice stores; and ice cream stores.[36] Figure 20-7 shows the percentage of sales of various products sold in grocery stores.

Other Agribusiness Retailing

Besides food sales, agribusiness retailing also includes sales of alcoholic beverages, tobacco products, floral products, clothing and shoes, and furniture and home furnishings. These products are mainly derived from grain, tobacco, cotton, wool, mohair, hides, skins, feathers, wood, nursery, and flowering and ornamental plants.[37]

Food Service Industry

The food service industry in hotels, restaurants, and institutions is one of the fastest-growing sectors in agribusiness. This industry is represented by the people and companies that serve food in hotels, restaurants, fast-food chain stores, schools, military bases, hospitals, and prisons. It is estimated that currently approximately 33 percent of retail food sales are made by these outlets—but the figure continues to rise.[38]

Growth and Volume of the Food Service Industry

In 1954, before the federal interstate highway system paved the way to suburbia, and before Ray Kroc opened his first McDonald's, the U.S. food service

31. Ibid., p. 333.
32. Alden C. Manchester, "Food Marketing Industry Responds to Social Forces," in *Marketing U.S. Agriculture: 1988 Yearbook of Agriculture,* ed. Deborah Takiff Smith (Washington, DC: U.S. Government Printing Office, 1988), p. 10.
33. http://www.7-11.com/about/history.asp (accessed August 27, 2007).
34. Little, *Economics,* p. 332.
35. Ibid., p. 332.
36. Ibid., p. 333.
37. Roy, *Exploring Agribusiness,* p. 113.
38. George J. Seperich, Michael W. Woolverton, and James G. Beierlein, *Introduction to Agribusiness Marketing* (Englewood Cliffs, NJ: Prentice-Hall Career & Technology, 1994), p. 140.

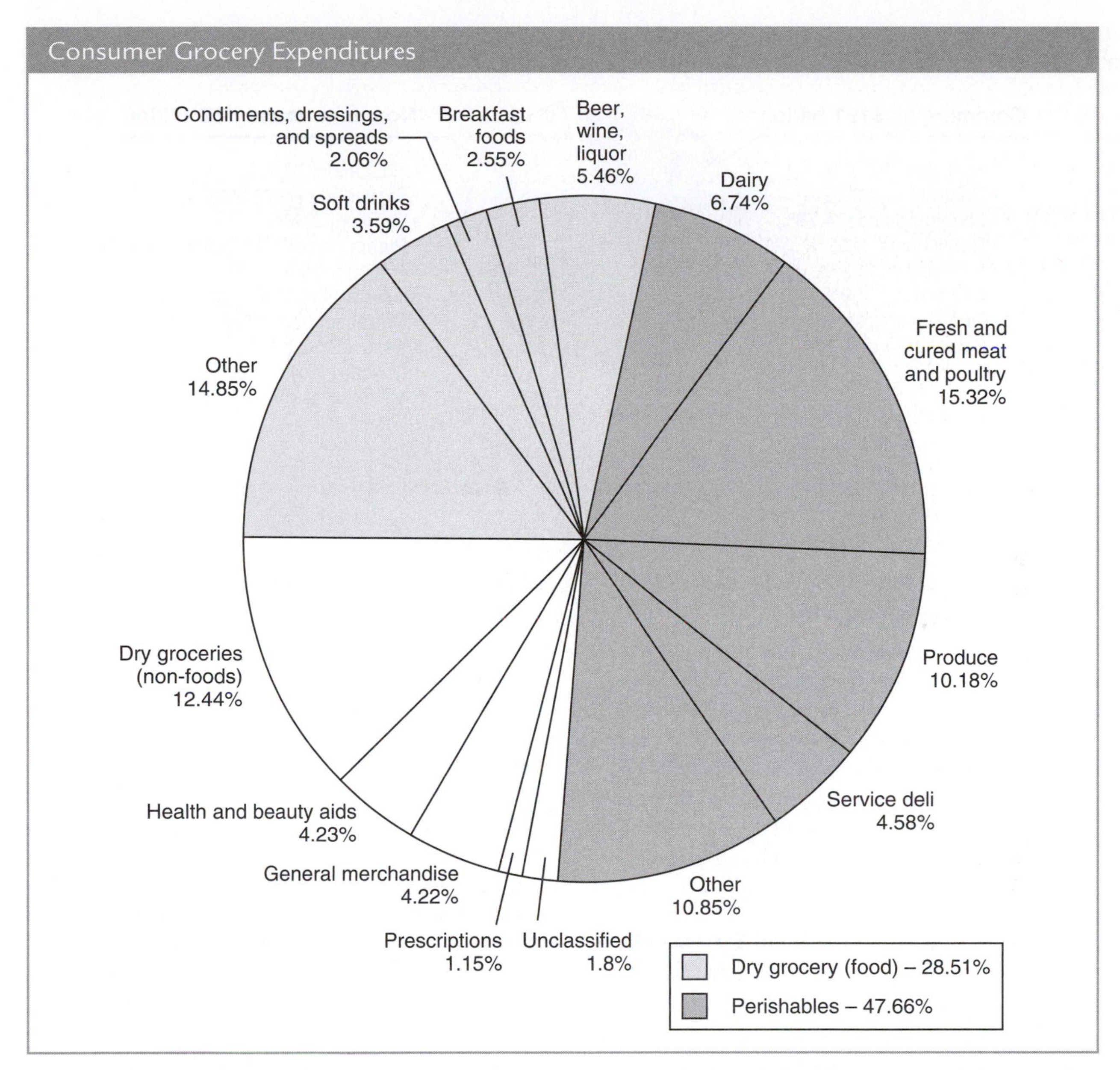

Figure 20–7 Do you ever wonder what percentage of sales of various products are sold in grocery stores? If so, this graph should help.

industry was a $15-billion-a-year business. Except for the rich, Americans generally ate out only when they were truly away from home, on business trips or on vacation, or for a very special occasion. Only 195,000 restaurants existed in the country at that time, and food service accounted for only 25 percent of total consumer spending for food.

Today, eating out is an integral part of our lives, and food eaten away from the home accounts for more than 50 percent of total food expenditures. The food service industry has increased its market share over its traditional rivals, the supermarkets and retail grocery stores. It has expanded to become a $260-billion-plus industry, with more than 600,000 outlets all across America. It employs approximately 10 million people.[39]

39. Steven D. Mayer, "U.S. Food Service Industry: Responsive and Growing," in *Marketing U.S. Agriculture: 1988 Yearbook of Agriculture,* ed. Deborah Takiff Smith (Washington, DC: U.S. Government Printing Office, 1988), p. 86.

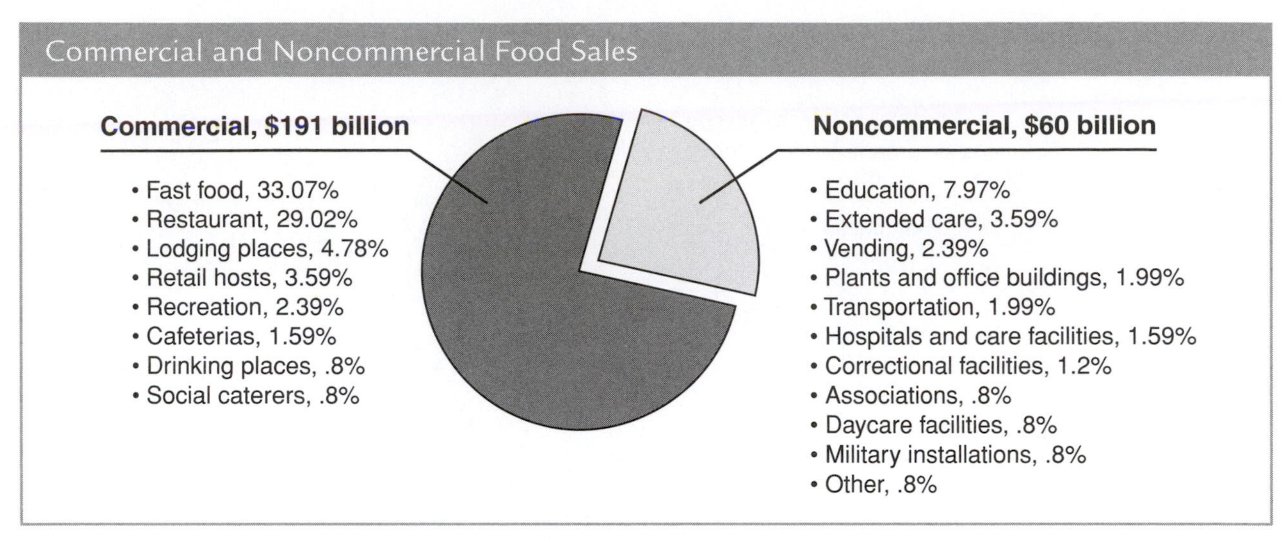

Figure 20-8 Commercial food sales account for approximately 76 percent of all food sales, and noncommercial food sales make up the remainder.

Percentage of Selected Foods Consumed in the Food Service Industry

Every day, approximately 100 million Americans (42 percent of the total U.S. adult population) eat out at least once. More than 78 billion meals and snacks are served by the food service industry each year. This industry uses more than 40 percent of all the meat, 55 percent of all the lettuce, 60 percent of all the butter, 65 percent of all the potatoes, and 70 percent of all the fish produced in the United States.[40]

Reasons for Growth in the Food Service Industry

Growth in the food service industry reflects the changes in Americans' lifestyles. In many households, both the husband and wife work; single parents also must work outside the home. Simply put, there is not enough time to shop, prepare meals, and clean up afterward. People are willing to pay to eat away from home to gain more time. In addition to the time factor, Americans are much more mobile now; they tend to be away from home at mealtime. These factors have made eating in restaurants more a necessity than a luxury for many.[41]

Types of Food Service Establishments

Food service establishments are divided into two major categories: commercial and noncommercial.

Commercial. **Commercial food establishments** make sales and profits from preparing and serving meals and snacks for consumers. They include full-service restaurants, lodging places, retail hosts, recreation places, cafeterias, drinking places, and social caterers. Approximately 76 percent of food expenditures (more than $191 billion) is made annually in commercial food establishments.[42] Refer to Figure 20-8.

Fast foods are familiar to everyone. Three of the larger fast-food establishments are Pizza Hut, KFC, and McDonald's. Agribusiness corporations are owners of many fast-food establishments. Examples are Ralston Purina's former operation of the Jack-in-the-Box fast-food chain; Pillsbury's ownership of Burger King and Steak-n-Ale; and Pepsico's holdings of Pizza Hut, Taco Bell, and KFC.[43]

Full-service restaurants are places to eat a sit-down meal when people are away from home. At these establishments, a wait staff serves full meals (breakfast, lunch, and dinner) to customers. The success of these restaurants depends on quality food, good service, pleasant surroundings, and reasonable prices.

Lodging places are locations that offer temporary accommodations and food service to consumers. Hotels, motels, and tourist courts are examples of lodging places.

40. Ibid.
41. Beierlein and Woolverton, *Agribusiness Marketing,* p. 150.
42. Little, *Economics,* p. 335.
43. Beierlein and Woolverton, *Agribusiness Marketing,* p. 151.

Retail hosts are businesses that offer food service within another retail outlet. They operate as a separate entity or as part of the department store, drugstore, or other retail outlet in which they are located.

Recreation places are entertainment locales such as theaters, billiard halls, bowling alleys, and similar establishments that contain a food service operation. Food service businesses of this type may also be found in country clubs, at golf courses, and in other amusement-oriented operations.[44]

Cafeterias are similar to restaurants except that most serve via a walk-through line where customers select their own salads, bread, vegetables, meats, desserts, and beverages from a display. As a rule, cafeterias tend to be less formal than restaurants.

Drinking places are operations that serve alcoholic beverages to consumers on site and that may also offer food services. (Often local law requires that food be available at such establishments.) Common examples of drinking places are bars, taverns, nightclubs, and saloons. Beer sales account for 57.5 percent of the alcohol sales in drinking places; distilled spirits constitute 30.9 percent, and wine sales generate 11.6 percent.[45]

Social caterers bring food on site for special occasions such as weddings, graduation parties, company picnics, and so forth.

Noncommercial Establishments. Approximately $60 billion worth, or 24 percent, of all food service sales occur in **noncommercial food establishments**. These are places of business that offer food service as a secondary service rather than as their main function. For example, a hospital may operate a cafeteria, but the primary function of the hospital is still to offer health care. Food service is offered as a convenience or support service. Other places that may have noncommercial food establishments within them are plants, office buildings, hospitals, military bases, and day care facilities.[46]

Taste Trends and New Operating Styles

Does the food service industry really adapt to meet existing consumer needs, or does it often anticipate consumer needs and wants and create new markets where none existed before? Salad bars, the away-from-home breakfast market, the frozen pizza market, ethnic foods, and, more recently, gourmet cookies, soups, and pasta are all cases in point. Was the industry merely responding to consumer demand, or was it, through its marketing know-how and skill, creating new ways to expand the market and make eating away from home more inviting and enjoyable to consumers?[47]

Future of the Food Service Industry

Like the food industry overall, the food service industry is relatively recession-proof. It is more affected by employment, consumer confidence, and disposable income than by deficits, interest rates, and gross domestic product. Despite economic uncertainties, the food service industry has shown that it is able to adapt and respond to a changing environment. Staying attuned to consumers always has been, and will continue to be, the strength of the food service industry.[48]

Related and Contributory Services

Many functions relate or contribute to the agrimarketing channels. These services are essential in the movement of food and fiber from the farm to the consumer. Some of these related and contributory services include standardization and grading, packaging, storage, order processing, inventory control, transportation, market communications, financing, product development, market research, advertising/marketing, and regulatory services.

Standardization and Grading

Standardization means the establishment of standards for matters such as quality, size, weight, and color. Once standards are set, **grading** is possible using the criteria established in the standards. In food markets, for example, meat can be graded prime, choice, or good. The use of standardization and grading make for easier contractual and exchange agreements between marketers.

The USDA is extensively involved in grading agricultural products. For example, in one year, the

44. Little, *Economics,* p. 336.
45. Ibid., p. 335.
46. Ibid., p. 336.
47. Mayer, "U.S. Food Service Industry," p. 88.
48. Ibid., p. 90.

USDA graded approximately 37 percent of eggs, 95 percent of butter, 55 percent of frozen fruits and vegetables, 81 percent of beef, 80 percent of turkeys, 56 percent of **broilers** and other poultry, 97 percent of tobacco, and 97 percent of cotton.[49]

Packaging

Packaging serves two primary purposes: to preserve contents and to merchandise and advertise the product. With the availability of self-service retailing, the packaging of a product does the majority of the selling. Many times the color, shape, label, and other characteristics of a package entice consumers into purchasing a product—exactly as the sellers intended and hoped. Packaging is a multimillion-dollar industry.

Storage

Timing is essential in the marketing process. Making goods available at the time consumers need them has a beneficial effect on prices of the commodity as well, as it allows a commodity to be marketed throughout the year rather than flooding the market at harvest time.

Unprocessed, as well as processed, agricultural products may be stored. Warehouses, bins, sheds, coolers, elevators, tanks, freezer lockers, and **kilns** are but a few of the types of storage available.

Order Processing

The specific procedure a firm follows when it receives an order is called *order processing*. The process usually involves choosing the desired items from inventory, preparing them for shipment, preparing a correctly totaled invoice, delivering the product, and collecting payment.[50]

Inventory Control

Inventory control refers to stocking sufficient amounts of product to meet consumer needs. It also involves producing adequate amounts of product and keeping accurate records of receipts and shipments. This function is computerized in many agribusinesses today.[51]

49. Little, *Economics,* p. 303.
50. Jasper S. Lee, James G. Leising, and David E. Lawver, *Agrimarketing Technology: Selling and Distribution in the Agricultural Industry* (Danville, IL: Interstate Publishers, 1994), pp. 390–392.
51. Ibid., p. 392.

Transportation

Products have to be moved from where they are produced to where they are needed. This involves several loading and unloading steps, including farm to local market, local market to processor, and processor to retailer. The four most common methods of transporting commodities are by truck, railroad, boat, and airplane.

Truck Transportation. The main benefits of truck transportation are rapid movement of perishable foods, flexible schedules, lower costs, dependability, speed, and less loss in transit. Truck transport is much more common than other ways of hauling raw and processed foods and fibers. Grain, livestock, milk, vegetables, and fruits account for about 85 percent of the farm produce hauled by truck in the United States.

Railroads. Railroads provide a means of moving large volumes of bulk commodities safely and economically over long distances. Rail service is the oldest form of general transportation. Approximately one-fifth of the railroads' profits come from transporting farm commodities. The primary commodities carried by rail are grains, animal feeds, canned goods, lumber, and wood products.

Boat Transport. Boat transport on both rivers and oceans has increased. Water transport is best adapted to transport of grains or large-volume bulk commodities for which speed is not critical.

Air Transport. Commodities that have a high value per unit are well suited to air transport. The speed and dependability of air transport make it the preferred method for shipping very perishable commodities such as flowers, nursery items, seafood, and specialty crops.[52]

Market Communications

In distributing farm supplies, agribusinesses use newspaper advertising, news articles and press releases, telephone, radio, television, direct mail, faxes, the Internet, and satellites. Production agriculturalists utilize newspapers to obtain farm stories, advertising, and general farm news. Use of the telephone

52. Little, *Economics,* pp. 307–308.

and fax machines is varied and increasingly important. Radio is relied upon for market, advertising, and farm news reports.

Daily and weekly marketing reports are available for nearly all commodities. These reports are issued through more than 170 USDA market news offices across the country.[53] Trade associations represent and educate their members, as well as inform the general public concerning trade activities, problems, and responsibilities. Refer to Figure 20-9. The communication industry is massive and hires thousands of people.

Financing

Lags between the time of production and of sale, and between the time of sale and receipt of payment, are reasons for financing. Financing must be provided if plants are to operate and perform the functions necessary to market a commodity.

Product Development

Product development refers to research on and development of new food products and the modification of existing products. These efforts are conducted to gain market share from competitors, to develop new products or new, more attractive forms of existing products, and to innovate and create new technology.

Market Research

Producers and marketers conduct marketing research through the use of customer interviews, taste-test panels, store experiments, special displays, and a multitude of other methods. The desired result is a thorough understanding of the consumer so that marketing and production efforts can be maximized.[54]

Merchandising

Merchandising concerns all phases of marketing. That is, it refers to all the characteristics associated with the distribution, display, and sale of a product. The goal of merchandizing is to successfully market a commodity that is appealing and satisfies a consumer desire or need.

53. Ibid., p. 309.
54. Ibid., p. 315.

Advertising

Agrimarketing firms spend about $11 million annually on advertising. Consumers are the primary target of advertisers. Twenty-two percent of advertising is done on television, 7 percent is heard on radio, and 5 percent is found in magazines. Food and food product advertising account for nearly 16 percent of all advertising done on television and nearly 7 percent of all advertising done in magazines.[55]

Regulation

The USDA carries out many regulatory functions, such as inspection and grading, as mentioned earlier. The quality of agricultural commodities is a very important consideration in agrimarketing channels. The Grade Standards Program of the USDA's Agricultural Marketing Service helps to ensure that purchasers of agricultural commodities receive the quality they want and for which they pay.

Careers in Agrimarketing

The number of careers in the field of agrimarketing is massive. Many jobs have been mentioned throughout this chapter, including assemblers, food processors, food wholesalers, food retailers, and the food service industry. Careers are also abundant in the many related areas that support and contribute to the agrimarketing channels. These areas include careers in standardization and grading, packaging, storage, order processing, inventory control, transportation (truck, rail, water, and air transport), market communication, financing, product development, market research, merchandising, advertising, and regulation.

Space does not permit a discussion of the volume and size of each of these areas. However, we will briefly discuss the opportunities in the meat and livestock, feed, fruit and vegetable, ornamental horticulture, cotton, and dairy industries. The Career Options section illustrates the magnitude of the career areas in agrimarketing.

Meat and Livestock Industry

There are more than 1,400 federally inspected slaughter facilities in the United States, plus many

55. Ibid., pp. 309–310.

Trade Associations

American Association of Nurserymen
1250 Eye Street, NW, Suite 500
Washington, DC 20005

American Bakers Association
1111 14th Street, NW, Suite 300
Washington, DC 20005

American Cotton Shippers Association
1725 K Street, NW, Suite 1210
Washington, DC 20006

American Dairy Products Institute
130 North Franklin
Chicago, IL 60606

American Farm Bureau Federation
600 Maryland Avenue, SW, Suite 800
Washington, DC 20024

American Feed Industries Association
1701 North Fort Myer Drive
Arlington, VA 22209

American Frozen Food Institute
1764 Old Meadow Road, Suite 350
McLean, VA 22102

American Goat Society
Route. 2, Box 112
De Leon, TX 76444

American Meat Institute
1700 North Moore Street
Arlington, VA 22209

American Plywood Association
7011 South 19th, P.O. Box 11700
Tacoma, WA 98411

American Quarter Horse Association
Amarillo, TX 79160

American Rabbit Breeders Association
1925 South Main Street, Box 426
Bloomington, IL 61701

American Seed Trade Association
1030 15th Street, NW
Washington, DC 20005

American Sheep Producers Council
200 Clayton Street
Denver, CO 80206

American Soybean Association
600 Maryland Avenue, SW
Washington, DC 20024

Burley and Dark Leaf Tobacco Export Association
1100 17th Street, NW, Suite 902
Washington, DC 20036

California Avocado Commission
17620 Fitch, 2d Floor
Irvine, CA 92714

California Cling Peach Advisory Board
P.O. Box 7111
San Francisco, CA 94120

California Pistachio Commission
5114 East Clinton Way Suite 113
Fresno, CA 93727

California Raisin Advisory Board
P.O. Box 5335
Fresno, CA 93755

California Table Grape Commission
P.O. Box 5498
Fresno, CA 93755

Corn Refiners Association
1001 Connecticut Avenue
Washington, DC 20036

Cotton Council International
1030 15th Street, NW
Suite 700, Executive Building
Washington, DC 20005

Farm and Industrial Equipment Institute
410 N. Michigan Avenue, Suite 680
Chicago, IL 60611

The Fertilizer Institute
1015 18th Street, NW, Suite 11
Washington, DC 20036

Florida Department of Citrus
1115 East Memorial Boulevard
P.O. Box 148
Lakeland, FL

Flue-Cured Tobacco Cooperative Stabilization Corporation
Box 12600
Raleigh, NC 27605

Food Marketing Institute
1750 K Street, NW, Suite 700
Washington, DC 20006

Grocery Manufacturers of America
1010 Wisconsin Avenue, NW
Washington, DC 20007

Independent Bakers Association
Box 3731
Washington, DC 20007

International Apple Institute
6707 Dominion Drive, Box 1137
McLean, VA 22101

International Ice Cream Association
888 16th Street, NW
Washington, DC 20006

Millers National Federation
600 Maryland Avenue, SW, Suite 305
Washington, DC 20024

Minnesota Grain Exchange
Minneapolis, MN

National Agricultural Chemicals Association
1155 15th Street, NW, Suite 900
Washington, DC 20005

National Association of Conservation Districts
1025 Vermont Avenue, NW, Suite 730
Washington, DC 20005

National Association of Meat Purveyors
8365-B Greensboro Drive
McLean, VA 22101

National Association of Wheat Growers
415 Second Street, NE, Suite 300
Washington, DC 20002

National Broiler Council
1155 15th Street, NW, Suite 614
Washington, DC 20004

National Cheese Institute
699 Prince Street, Box 20047
Alexandria, VA 22320

National Corn Growers Association
201 Massachusetts Avenue, NE,
Suite C4
Washington, DC 2002

National Cotton Council of America
1030 15th Street, NW, Suite 700
2111 Wilson Boulevard, Suite 600
Arlington, VA 22201

National Food Processors
1401 New York Avenue, NW
Washington, DC 20005

National Forest Products Association
1250 Connecticut, NW, Suite 70
Washington, DC 20005

National Hay Association, Inc.
P.O. Box 99
Ellensburg, WA 98926

National Milk Producers Federation
1840 Wilson Boulevard.
Arlington, VA 22201

Northwest Cherry Growers
1005 Tieton Drive
Yakima, WA 98902

Northwest Horticultural Council
P.O. Box 570
Yakima, WA 98907

Nursery Marketing Council
1250 Eye Street, NW, Suite 500

National Pecan Marketing Council
741 Piedmont Ave., NE
Atlanta, GA 30308

National Potato Promotion Board
1385 South Colorado Boulevard, #512
Denver, CO 80222

National Pork Producers Council
1015 15th Street, NW, Suite 200
Washington, DC 20005

National Turkey Federation
11319 Sunset Hills Road
Reston, VA 22090

North American Blueberry Council
P.O. Box 166
Mamora, NJ 08223

North American Export Grain Association
1747 Pennsylvania Ave, NW
Suite 1175
Washington, DC 20006

Northern Hardwood and Pine Manufacturers Association
P.O. Box 1124
Green Bay, WI 54305
Washington, DC 20005

Oregon-Washington-California Pear Bureau
Woodlark Building
Portland, OR 97205

Papaya Administrative Committee
First Insurance Building
1100 Ward Avenue, Room 860
Honolulu, HI 96814

Protein Grain Products International
6707 Old Dominion Drive
Suite 240
McLean, VA 22101

Rice Council for Market Development
P.O. Box 740123
Houston, TX 77274

Rice Millers Association
1235 Jefferson Davis Highway, Suite 302
Arlington, VA 22202

Southern Forest Products Association
P.O. Box 52468
New Orleans, LA 70152

Sugar Association
1511 K Street, NW
Washington, DC 20005

Tobacco Associates
1101 17th Street, NW
Washington, DC 20036

United Egg Producers
3951 Snapfinger Parkway, Suite 580
Decatur, GA 30035

United Fresh Fruit & Vegetable Association
727 North Washington street
Alexandria, VA 22314

United States Beet Sugar Association
1156 15th Street, NW
Washington, DC 20005

U.S. Brewers Association
1750 K Street, NW
Washington, DC 20006

U.S. Chamber of Commerce
1615 11 Street, NW
Washington, DC 20062

U.S. Feed Grains Council
1400 K Street, NW, Suite 1200
Washington, DC 20005

U.S. Meat Export Federation
3333 Quebec Street, Suite 7200
Stapleton Plaza
Denver, CO 80207-2391

U.S. Wheat Associates, Inc.
1620 I Street, NW, Suite 801
Washington, DC 20006

USA Dry Pea and Lentil Council, Inc.
Stateline Office
P.O. Box 8566
Moscow, ID 83843

Washington State Apple Commission
P.O. Box 18
Wenatchee, WA 98801

Western Wood Products Association
1500 Yeon Building
Portland, OR 97204

Wine Institute
165 Post Street
San Francisco, CA 94109

Figure 20–9 Trade associations communicate to consumers about their members' products. Many jobs and careers are created by trade associations.

smaller plants providing local custom slaughter and meat processing under state and local inspection. There are several thousand wholesalers of meat and meat products and about 261,000 retail markets that sell meat directly to consumers. More than 15,000 of these stores specialize in meat and seafood. Some 350,000 restaurants market more than 40 percent of the meat products sold in the United States.[56]

Feed Industry

The feed industry is one of the largest in the United States, with more than 7,000 animal-feed processing firms generating more than 110 million tons of feed annually. Proceeds from feed products come to approximately $18 billion each year. Farmers and feeders devote 14 percent of their total costs to formula and concentrated feeds. Farmers themselves may provide feed grains that leave the farm, are processed, and return as commercial feed. For example, corn may return to the farm as antibiotics, enzymes, pharmaceuticals, amino acids, molasses, or corn gluten feed or meal.

The feed industry employs a great number of people. There are as many as 100,000 feed mill workers and custom grinders and mixers. Another 2,000 are wholesale feed dealers, and more than 20,000 work in retail sales of feed products (including several thousand hatcheries).[57]

Fruits and Vegetables

Approximately $375 billion worth of canned, frozen, dried, or specialty fruit and vegetable products are manufactured annually in more than 22,000 processing plants that employ nearly 220,000 people.[58]

In recent years, the value of all fruit and tree nuts sold has exceeded $8 billion, and 11 percent of all agricultural crops grown in this country are in this category. The average annual per capita consumption of fruits is nearly 217 pounds. The annual consumption of fresh fruit is currently at 98 pounds per person, representing an increase of 17 pounds over a 17-year period.

The value of all vegetables sold to producers has topped $13 billion in annual sales. The average American eats 325 pounds of vegetables each year. Statistics show that Americans are eating about 130 pounds of potatoes, sweet potatoes, and dried peas; 100 pounds of canned vegetables; 50 pounds of frozen vegetables; and 70 pounds of fresh vegetables annually.[59]

Ornamental Horticulture

Ornamental horticulture agribusinesses produce cut flowers, bedding plants, cultivated forest greens, and flowering and foliage plants. There are approximately 23,500 establishments of this type, and their total annual sales exceed $22 billion. Another 16,000 agribusinesses sell nursery stock and lawn and garden supplies. Annual sales in these operations are more than $3.4 billion.[60]

Cotton Industry

The cotton industry receives cotton from producers and through a variety of processes transfers it to users in the United States and around the world. As a result, textile mills are able to employ nearly 730,000 people. The apparel sector provides jobs for more than 1 million others.[61]

Dairy Industry

The dairy industry employs more than 145,000 workers, plus supervisory staff and other specialists who work in thousands of plants, distribution centers, and retail outlets. The industry purchases more than 17.9 billion gallons or 134 billion pounds of milk annually. The average American consumes 224 pounds of milk and dairy products each year. This represents sales totaling nearly $66 billion, or 12 percent of all consumer food purchases. Yearly farm income from dairy products is approximately $22 billion.[62]

---

56. Marcella Smith, Jean M. Underwood, and Mark Bultmann, *Careers in Agribusiness and Industry*, 4th ed. (Danville, IL: Interstate Publishers, 1991), pp. 137–138.
57. Ibid., p. 119.
58. Seperich, Woolverton, and Beierlein, *Introduction to Agribusiness Marketing*, p. 121.
59. U.S. Department of Agriculture, *Agricultural Statistics, 1998* (Washington, DC: U.S. Government Printing Office, 1998).
60. Ibid.
61. Ibid.
62. Ibid.

Career Options

Inspector, Quality Controller, Seller, Processor, Trucker, Wholesaler, Distributor, Produce Manager, Meat Cutter, Packer, Grader, Auctioneer, Co-op Manager

The food industry is massive and includes both plant and animal products. It includes producers, processors, distributors, wholesalers, retailers, fast-food establishments, and restaurants. Careers in food science, store management, produce management, meat cutting, laboratory testing, field supervision, research, diet and nutrition, health and fitness, and promotion are all possibilities in the food industry.

The processing-and-packaging industry is enormous and employs a large number of individuals in the United States. Field supervisors and coordinators direct the work of crews to harvest crops at their peak of quality and transport them to processing or packing plants. In many cases, huge packing and/or processing machines are used right in the field or orchard or on boats. Quality-control personnel collect food specimens, label them, test them, and maintain records to ensure quality control on each batch of food coming out of the plant.

Excellent jobs are available in supermarkets for produce stockers and foremen. Products include fruits, vegetables, and, possibly, ornamental plants. Tasks may include inventorying, ordering, handling, stocking, and displaying produce to keep it fresh and attractive. In large cities as well as small towns, street vendors are frequently seen selling premium fruit and vegetables. Roadside stands provide opportunities for younger members of families to develop business skills while earning money for present and future needs.

Corn, wheat and other small grains, sugar cane, soybeans, sugar beets, and other specialty crops are grown on large acreages in the United States, Canada, and many other nations. In the United States, huge amounts of grain, oil, and specialty crops are exported, and do much to help maintain our balance of payments in foreign trade. Grain brokers, futures brokers, market reporters, grain elevator operators, and others owe their jobs to these crop enterprises.

Along with crop enterprises are jobs in building and storage construction, systems engineering, machinery sales and service, welding, irrigation, custom spraying, hardware sales, agricultural finance, chemical sales, and seed distribution.

After agribusiness (food and fiber) products go through the agrimarketing channels, they end up with the consumer. Satisfaction, quality, and food safety are all goals of the individuals in agrimarketing. (Courtesy of USDA)

Career Options

The marketing of agricultural products occurs in all segments of our society on a continuing basis. This is evidenced by the overflowing, year-round displays of produce, dairy, meat, seafood, bread, frozen foods, canned goods, deli foods, and ready-to-eat foods in the supermarkets.

Marketing careers may be launched through programs in technical schools, colleges, or universities in the plant and animal sciences, food science, agricultural economics, marketing, or communications. Additionally, studies in the biological sciences are appropriate for laboratory work associated with agribusiness products and marketing.

Conclusion

Agribusiness is big business, especially in the agribusiness output sector or agrimarketing. Marketing channels define the path that a food product follows from the "farmer's gate to the consumer's plate." The future belongs to those who prepare for it. Many careers are available in the agrimarketing sector for those who are willing to apply themselves. Refer to Figure 20-10 for a list of careers and employment projections.

Summary

Marketing channels are the paths that an agricultural product follows from the farm to the consumer. The length of the marketing channel depends on the product. Pick-your-own produce has a short path: from producer to consumer. However, the market channel for most food products is producer to assembler to food processor to food wholesaler to food retailer to consumer.

Early farmers took the food from production to retailing. As farmers and nonfarmers alike developed and applied new technologies, specialization became more possible and popular. Farmers became better at producing the raw materials, as they no longer had to divide their attention and efforts between farming and a host of other activities. In the early days, farmers captured almost all of the consumer's food dollar. Today, production agriculturalists capture less than 30 cents of the consumer's food dollar, because of specialization and agrimarketing.

More than 10,000 assemblers can be identified between the production agriculturalist and the processor. Assemblers do not usually change the form of a product. They put small lots of a commodity together to provide a larger, more economical volume.

There are approximately 22,000 firms in the commodity processing and food manufacturing sector. They add more than $215 billion of value to the items they handle. The processor is normally the first party in the marketing system to alter the form of a raw agricultural commodity. Food manufacturers continue the job started by the processor, adding value to crops and livestock by increasing the level of preservation, convenience, and quality.

The transfer of food and fiber from processing and manufacturing plants to wholesalers and then to retailers and other establishments is an important step in agrimarketing channels. There are more than 476,000 wholesale business establishments with annual sales totaling more than $1 trillion. The wholesaling of grocery and related products and farm-related raw materials involves more than 880,000 employees in some 54,700 firms making $500 billion in sales annually.

Agribusiness retailers include businesses that sell groceries, prepared foods, soft drinks, floral products, clothing, shoes, furniture, home furnishings, and other products. There are more than 588,000 retail outlets in the United States. The history of food retailing in this country is marked by three major developments: of retail food chain stores, of supermarkets, and of superstores and warehouse/limited-assortment supermarkets.

One of the fastest growing segments of agribusiness is the food service industry. This industry segment includes all those people and firms involved in serving food at hotels, restaurants, fast-food outlets, schools, hospitals, prisons, military installations, and other away-from-home eating locations. The food service industry has increased its share of the market relative to retail groceries and supermarkets. It has become a $260-billion-plus industry with 600,000 outlets across the country employing more than 10 million people.

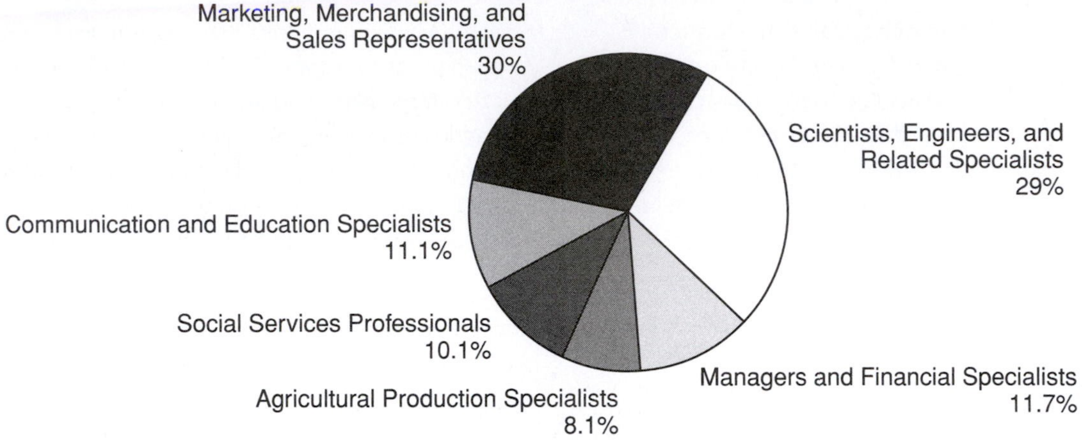

Marketing, Merchandising, and Sales Representatives

Account Executive
Advertising Manager
Commodity Broker
Consumer Information Manager
Export Sales Manager
Food Broker
Forest Products Merchandiser
Grain Merchandiser
Insurance Agent
Landscape Contractor
Market Analyst
Marketing Manager
Purchasing Manager
Real Estate Broker
Sales Representative
Technical Service Representative

Managers and Financial Specialists

Accountant
Appraiser
Auditor
Banker
Business Manager
Consultant
Contract Manager
Credit Analyst
Customer Service Manager
Economist
Financial Analyst
Food Service Manager
Government Program Manager
Grants Manager
Human Resource Development Manager
Insurance Agency Manager
Insurance Risk Manager
Landscape Manager
Policy Analyst
Research and Development Manager
Retail Manager
Wholesale Manager

Communication and Education Specialists

College Teacher
Computer Software Designer
Computer Systems Analyst
Conference Manager
Cooperative Extension Agent
Editor
Educational Specialist
High School Teacher
Illustrator
Information Specialist
Information Systems Analyst
Journalist
Personnel Development Specialist
Public Relations Representative
Radio/Television Broadcaster
Training Manager

Scientists, Engineers, and Related Specialists

Agricultural Engineer
Animal Scientist
Biochemist
Cell Biologist
Entomologist
Environmental Scientist
Food Engineer
Forest Scientist
Geneticist
Landscape Architect
Microbiologist
Natural Resources Scientist
Nutritionist
Pathologist
Physiologist
Plant Scientist
Quality Assurance Specialist
Rangeland Scientist
Research Technician
Resource Economist
Soil Scientist
Statistician
Toxicologist
Veteranarian
Waste Management Specialist
Water Quality Specialist
Weed Scientist

Social Services Professionals

Career Counselor
Caseworker
Community Development Specialist
Conservation Officer
Consumer Counselor
Dietitian
Food Inspector
Labor Relations Specialist
Naturalist
Nutrition Counselor
Oudoor Recreation Specialist
Park Manager
Peace Corps Representative
Regional Planner
Regulatory Agent
Rural Sociologist
Youth Program Director

Agricultural Production Specialists

Aquaculturalist
Farmer
Feedlot Manager
Forest Resources Manager
Fruit and Vegetable Grower
Greenhouse Manager
Nursery Products Grower
Farm Manager
Rancher
Turf Producer
Viticulturist
Wildlife Manager

Figure 20–10 Many careers are available in the agrimarketing channels and related services. (Courtesy of USDA)

Besides the obvious agrimarketing channels, many other functions relate or contribute to the agrimarketing channels. Some of these related and contributory services include standardization and grading, packaging, storage, order processing, inventory control, transportation, market communications, financing, product development, market research, and regulatory services. Thousands of jobs are available in these areas.

There are millions of careers in the field of agrimarketing. Many jobs and the number of people employed in various segments of the agrimarketing channels have already been presented. Other jobs in various areas include:

- *Meat and livestock industry*—261,000 retail markets sell meat, 15,000 stores specialize in meat and seafood, and 350,000 restaurants sell meat products
- *Feed industry*—employs more than 100,000 workers and sells more than $18 billion worth of feed each year
- *Fruit- and vegetable-processing industries*—provide more than 219,000 jobs in 2,000 processing establishments producing $30 billion worth of products
- *Ornamental horticulture*—includes 23,500 establishments producing $1 billion annually, and 16,000 establishments selling lawn and supplies and nursery stock making a total $3.1 billion annually
- *Cotton industry*—employs 729,000 persons in textile mills and 1 million in the apparel industry
- *Dairy industry*—employs more than 145,000 and produces $54 billion worth of dairy products.

End of Chapter Activities

REVIEW QUESTIONS

1. Define the Terms to Know.
2. Briefly explain the historical evolution of agrimarketing channels.
3. Give five examples of assemblers, and briefly describe each.
4. Give three examples of consumers as assemblers (direct-marketing outlets).
5. Differentiate between an agricultural commodity and an agricultural product.
6. Give a brief explanation of eight examples of processors.
7. Explain the role of food manufacturers.
8. List five functions that wholesalers perform.
9. Briefly discuss the trend toward wholesale-retail affiliation.
10. Name and briefly discuss three reasons wholesalers exist.
11. Name and briefly discuss three types of wholesalers.
12. Who started supermarkets? Why?
13. Explain the difference between traditional supermarkets, superstores, and warehouse/limited-assortment supermarkets.
14. List seven other agribusiness retail items besides food.
15. List seven segments of the food service industry.
16. What percent of each of the following food items are served in the food service industry: (a) meat, (b) lettuce, (c) butter, (d) potatoes, (e) fish?
17. List and briefly discuss eight types of commercial food establishments.
18. List 10 noncommercial food establishments.

19. Production agriculturalists sell corn, but they also buy it. Name seven products that production agriculturalists buy that include corn.
20. How many pounds of each of the following products do Americans consume annually: (a) potatoes, sweet potatoes, and dry peas; (b) fresh vegetables; (c) canned vegetables; (d) frozen vegetables; (e) milk and dairy products?

FILL IN THE BLANK

1. Production agriculturalists capture less than ___________ cents of the consumer's food dollar.
2. More than ___________ assemblers can be identified between the product agriculturalist and the processor.
3. The commodity processing and food manufacturing industries owe their existence to advances in ___________.
4. There are approximately ___________ firms in the commodity processing and food manufacturing sector.
5. American consumers have more than ___________ packaged food items from which to choose.
6. Industry estimates put the failure rate of new food products at ___________ to ___________ percent.
7. There are more than ___________ wholesale businesses, with annual sales totaling more than ___________ trillion and employment exceeding ___________.
8. The wholesaling of grocery and related products and farm-related materials involves more than ___________ employees in some 54,7000 firms with $500 billion of sales annually.
9. There are more than ___________ retail food outlets in the United States.
10. Chain stores account for ___________ percent of food store sales.
11. There are about 165,000 grocery stores, which make annual sales of about $360 billion and employ about ___________ million people.
12. Superstore and warehouse supermarkets presently account for more than 50 percent of all supermarkets and make approximately ___________ percent of supermarket sales.
13. Convenience stores make up about 20 percent of retail food stores, and account for ___________ percent of grocery sales.
14. Superettes represent about 37 percent of all retail food stores and account for ___________ percent of sales.
15. Specialized food stores make up ___________ percent of retail food sales.
16. Eating out accounts for ___________ percent of total food expenditures.
17. The food service industry has become a ___________-dollar industry employing ___________ ___________ people.
18. Each day, approximately ___________ percent of Americans eat out at least one meal.
19. Commercial food service sales account for ___________ percent, whereas noncommercial accounts for ___________ percent of sales.
20. Newspapers receive about ___________ percent of all advertising expenditures.

MATCHING

a. order processing
b. inventory control
c. merchandising
d. packaging
e. transportation
f. product development
g. storage
h. regulatory agency
i. financing
j. advertising
k. market communication
l. standardization and grading

_____ 1. warehouses, bins, sheds, elevators, tanks, freezers, lockers, and kilns

_____ 2. needed often because of the time of sale and receipt of payment

_____ 3. ensures that purchasers of agricultural commodities receive the quality they want and for which they pay

_____ 4. consumers are the primary targets

_____ 5. makes for easier contractual and exchange agreements between marketers

_____ 6. newspaper advertising, news articles and press releases, telephone, radio, television, faxes, the Internet, and satellite

_____ 7. maintaining an adequate stock of products

_____ 8. involves picking what the customer wants from inventory and preparing it for shipment

_____ 9. involves researching new food products

_____ 10. distribution, display, and sale of a product

_____ 11. moving products from where they are produced to where they are needed

_____ 12. does the majority of the selling

ACTIVITIES

1. Identify the jobs of your six closest relatives or guardians. Determine if their jobs are a part of the agrimarketing channel and/or related services (example: accountant at grocery store). Along with the others in your class, compile a list of the total workers and determine the percentage of those employed because of agrimarketing.
2. Give five examples of assemblers of farm commodities in your community or adjoining communities.
3. List the agricultural commodity processors and/or food manufacturers in your community and adjoining communities.
4. List the agribusiness wholesalers in your community that sell or deliver products to restaurants, grocery stores, and/or institutions.
5. Compile a list of 10 of your friends or schoolmates. Determine if and where they work part-time. What percent of them are employed in segments of the agricultural industry?
6. List the commercial and noncommercial food service establishments in your community and surrounding areas.
7. Have the teacher divide the class into two groups and debate the following topic: Does the food service industry adapt to meet consumer needs, or does it anticipate consumer needs and wants where none existed before?

8. List as many services related and contributing to the agrimarketing channels as you can that exist in your community and/or surrounding areas. Compare your list with lists produced by others in your class. How many jobs could you identify?
9. Numbers in the several thousands and, in some cases, millions were given for jobs in several different areas of agrimarketing channels and their related and contributory services. List the job areas and the employment statistics for each area. What is your total? Remember, this is only an estimate.
10. Beyond this chapter, look at other areas in the agribusiness input sector and list job areas and numbers. This information can be found in this book and other areas.
11. Refer to Figure 20-1. Complete the chart, being sure to include food retailers, the food service industry, and related and contributory services.

Appendix

Laws of Supply and Demand

Once you have decided to take advantage of the opportunity to operate an agribusiness, the next decision is what to grow. Many considerations will be analyzed in the decisionmaking process. As you decide which crop(s) to raise and market, another economic principle to consider is the law of supply and demand. Selection of a crop that has already saturated the market would not be economically sound. There would be little *demand* for your product. In contrast, selection of a crop that would not be competing with a strong market would be an economically sound decision. The low supply of a crop with a high demand should lead to strong revenue.

Factors Affecting Price

Three factors determine the price of almost everything bought and sold: supply, demand, and the general price level. A single factor cannot explain price change. For example, a rose grower sold roses for $12.00 a dozen. A year later, she sold a dozen roses for $8.00 per dozen. There may be several reasons for the difference in price between the two dozen roses. Perhaps the number (supply) of roses available increased. Maybe consumer preference (demand) for roses decreased. Perhaps the general price level of goods caused the change in price. Forces that have little to do with rose production directly can change the general price level. Wars, depressions, recessions, periods of recovery, and other things affect the general price level.

Law of Supply

Supply is the amount of a product available at a given time and place at a specific price. It is a direct price and quantity relationship that states what producers are willing to supply at a given price. The law of supply is as follows: As the price of a product increases, producers will supply more of that product. As the price falls, the quantity supplied also falls.

Factors That Can Increase or Decrease the Supply of Products

The following are some of the important factors that can increase or decrease the supply of agricultural products.

Total Units in Production. Reporting services provide statistics such as numbers (supply) of cattle on feed or total acres of corn planted. This type of data is a good indicator of future supply.

Weather. A sudden freeze or continued drought limits production and reduces supply in the affected area, causing the supply curve to shift to the left. A freeze or drought must be very severe and widespread to have a major influence on total supply. Nevertheless, the price of crops grown in limited areas, such as Georgia peaches, will increase substantially in response to a spring freeze after the trees have blossomed.

Land. In the past, production agriculturalists increased agricultural production by cultivating more land. From 1950 to 1980, 30 percent of the increase in world agricultural production resulted from cultivating new land. In the future, less new land will be available for cultivation.

Water. The amount of water available (rainfall or irrigation water) can exert a major influence on the supply of agricultural products. Irrigation can greatly increase the productivity of land.

Changes in Technology. Since 1950, technology has accounted for two-thirds of the increase in world agricultural production. Tractors, fertilizers, pesticides, biotechnology, and hybrid seed are examples of agricultural technologies that have made farmers more productive and efficient.

Amount in Storage. Wholesalers, producers, and government agencies store many commodities. Storage is a normal part of distributing products to consumers. However, when storage facilities fill to capacity, average production levels may result in average supply.

Government Action. Federal policy often affects supply. When a price support program requires producers to limit production, supply decreases. When a government places sanctions on the export of a product, domestic supply increases.

Time Lags. When a price is unusually high, producers typically decide to increase production. However, it takes time to bring idle resources into full production. *Time lag* is the period of time it takes for such decisions to be made and production to be ramped up so that it affects the supply and thus the market. Time lags usually correlate to a growing season or animals' reproductive cycle (for beef, 27 to 30 months).

Credit. Abundant, low-interest credit permits faster expansion of production and increases supply. "Tight" money and high interest rates limit expansion and thus supply.

Input Price Changes. Changes in the prices of inputs used in the production process also affect the supply. Higher input prices would tend to decrease production, whereas lower input prices would tend to increase production.

Prices of Other Outputs. The prices of other products that can be produced using the same resources also influence the supply of a given product. Producers will try to make the products that brings the greatest profit; thus, supply of the less profitable products tends to decrease.

Anticipated Prices. If production agriculturalists expect the price of a commodity—corn, for example—to increase, then they will plant more corn, thereby increasing the supply.

Number of Sellers. If existing producers are generating profits, new producers have an incentive to produce the same item so to capture the profits, which results in increased supplies. In contrast, producers who suffer continued losses will exit the industry, resulting in decreased supplies.

Law of Demand

Similar to supply, the economic concept of demand is based on a price and quantity relationship. How much of a product is demanded depends on the amount that consumers are willing and able to purchase at a particular price. The law of demand states that as the price of a product rises, consumers will demand less of that product.

Factors That Can Increase or Decrease the Demand for Products

As with the supply curve, there are important factors relating to demand.

Income Level. When people have more money, they buy more goods and services. Thus, demand for certain commodities increases.

Population. As the number of people trying to buy the same item increases, the demand for that item also increases.

Individual Taste. Some consumers will buy a product because it is needed or because it is a small luxury. Others will buy a product because it fits their lifestyle; it may be perceived as the "in" thing to have or to serve. This motivation is harder to measure exactly and may change relatively fast. Still, it can be a useful tool for predicting the success of a new product or determining the value of a proven product.

Competing Products. If the price of a competing product is much lower than what is usually paid for an original (older) product, eventually someone will try it. If consumers stay with the new product, their shift will lower the demand for the old product. Manufacturers of newer products use this fact as a tool to build demand for their products. They offer coupons, taste improvements, and better results to gain customers.

Weather. Extremes in temperature can affect the demand for some commodities. For example, people eat more pork in the winter and drink more lemonade in the summer.

Seasonal Demand. Consumers demand more turkey during Thanksgiving and more ham at Christmas. Saint Valentine's Day and Mother's Day cause an increase in demand for flowers.

Advertising and Promotions. Advertising and promotions are ways of trying to increase consumers' awareness and stimulate their desire to have a certain product—that is, to increase demand. Some methods used are television and radio commercials, magazine ads, and billboards. Promotions include such things as free trial samples and giveaways.

Glossary

absurdities–Things that are meaningless, lacking order or value.

accelerated cost recovery system (ACRS)–Method under which an asset is depreciated more in the early years of its useful life.

accounting cycle–The sequences of accounting procedures used to record, classify, and summarize accounting information.

accounts payable–The liability arising from the purchase of goods and services on credit.

accounts receivable–A balance due from a debtor on a current account.

accrual adjustments–Adjustments to income.

accrued–Accumulated.

accumulated depreciation–Total amount of depreciation claimed (to date) during the ownership period of an asset.

actuarial interest rate–Interest rate equal to the actual rate charged on a loan.

A.D.–*Anno Domini* ("in the year of the Lord"), indicating a year since the birth of Christ.

add-on interest–Interest that is added to the loan by taking the total interest paid on a loan and adding it to the loan amount. This allows the lender to divide by the number of periods to get the installment payment amount.

advocate–To plead in favor of.

aggregate–Total.

agribusiness–The manufacture and distribution of farm supplies to the production agriculturalist, and the storage, processing, marketing, transporting, and distribution of agricultural materials and consumer products that were produced by production agriculturalists.

agribusiness input sector–All resources that go into production agriculture to produce farm commodities.

agribusiness output sector–Any phase of the agribusiness sector after the commodity leaves the farm, such as sales or marketing.

agricultural commodities–Agricultural animals, crops, and vegetables in their original, unaltered state.

agricultural economics–An applied social science dealing with how humans choose to use technical knowledge and scarce productive resources such as land, labor, capital, and management to produce food and fiber and to distribute agricultural commodities and products for consumption to various members of society over time.

agricultural policy–Outlines the steps that will be taken to reach certain goals in the food and fiber economy. Typically such policies affect the resources, production, and markets related to agricultural products and services.

agricultural products–Agricultural commodities that have been altered from their original state; for example, butter made from milk.

agricultural statistics–Facts and figures relating to the agricultural industry.

agrimarketing–Those processes, functions, and services performed in connection with food and fiber from the farms on which they are produced until their delivery into the hands of the consumer.

agrimarketing channels–The paths that an agricultural product follows from "the farmer's gate to the consumer's plate."

agriscience–All jobs relating in some way to plants, animals, and renewable natural resources. Also, the application of scientific principles and new technologies to agriculture.

agriservices–Services offered to production agriculturalists by private firms, such as agricultural research, consulting, artificial insemination, veterinary, communications, and supply businesses.

agriservices sector—The portion of the agriculture industry concerned with researching new and better ways to produce and market food, to protect food producers and consumers, and to provide special, custom services to all the other phases of agriculture.

agronomic—Any part of agriculture having to deal with field crop production and soil management.

alien—Person who is not a citizen of the United States.

allocation—The distribution or apportioning of something for a specific purpose.

allocation of resources—Land, labor, and equipment that are equally distributed to all segments of the economy.

amortize—To set up equal installment payments.

annual percentage rate (APR)—Common name for actuarial interest rate; the rate charged for interest on a loan.

annuities—Sum of money payable yearly or at other regular intervals.

applied science—The application of basic scientific theories to specific problems.

appreciate—To increase in value.

APR. See *annual percentage rate.*

artificial insemination—The deposition of spermatozoa into the female genitalia by artificial rather than natural means.

assemblers—Person who puts small lots of a commodity together to provide a larger, more economical unit.

assets—Economic resources that are owned by a business and are expected to benefit future operations. Some examples include cash, receivables, inventory, investments, equipment, buildings, and prepaid accounts.

astute—Crafty, shrewd, keen in judgment.

attaché—A person on the official staff of an ambassador or minister to a foreign country.

auctioneer—The person who is conducting the sale at an auction market.

augmentation—The act or process of making greater, more numerous, larger, or more intense.

average fixed costs (AFC)—Total fixed costs divided by the quantity.

average product (AP)—Amount of output produced divided by the number of units of input; total product expressed relative to some level of input.

average total cost (ATC)—Total fixed costs plus the total variable costs divided by the quantity.

average variable cost (AVC)—Total variable costs divided by the quantity.

balance sheet—Reports the business's financial condition on a specific date.

balancing a checkbook—Making sure your checkbook register records are the same as the bank's records.

bank money—The bank credit that banks extend to their depositors.

barter—The exchange of goods or services for other goods and services, in which no money changes hands.

base mixes—A feed mixture produced by a feed manufacturer to which home-grown feed, such as corn, is added.

base year—Typical years of consumer prices used to help compile the consumer price index.

basic science—Research done to investigate general relationships between two or more variables.

basis—The difference between the price of a commodity and the price of a related futures contract; for example, cash price minus futures price equals basis.

B.C.—Before Christ; used to identify years.

bear market—A period of declining market prices.

biopesticides—Effective pesticides that will not be harmful to water, air, soil, wildlife, humans, or the food supply.

biotechnology—Technology concerning the application of biological and engineering techniques to microorganisms, plants, and animals; for example, gene splicing, cloning, and DNA mapping.

blanching—Scalding in water or steam to remove the skin, whiten, or stop enzymatic action (as done to food before freezing).

blocked currency–The idea that money could be exchanged only within the nation making the purchase when surpluses were sold to developing nations; used to support or enhance the economies of underdeveloped nations.

board of directors–A group of individuals chosen to make decisions for a company.

bovine somatotropin (BST)–A growth hormone found in cattle. Injections of extra BST are used to increase the production of meat and milk.

break-even point–When total receipts equal total costs (or when your expenditures equal your income).

broach–A cutting tool for removing material from metal or plastic to shape an outside surface or a hole.

broilers–Chickens from 8 to 12 weeks old, weighing 2½ or more pounds, sufficiently tender to be broiled.

brokerage–Firm that buys and sells stock or other products for a commission.

brokerage office–Most convenient outlet for futures transactions, such as commission houses, wire houses, and futures commissions merchants. Such offices are located in most towns and cities throughout the United States.

budget–Financial plan for an agribusiness.

budgeting–The orderly fashion in which owners of an agribusiness put their financial plans on paper.

bulk feed–Feed sold in large quantities.

bull market–A period of rising market prices.

business cycle–The repeated rise and fall of economic activity over time.

business plan–A written description of a new business venture that describes all aspects of the proposed agribusiness. It helps you focus on exactly what you want to do, how you will do it, and what you expect to accomplish.

business survey–A survey that a business conducts to determine the market potential for a selected business.

buyer's fever–When a buyer acts on quick, thoughtless decisions; buying without thinking.

call–An option that gives the option buyer the right (without obligation) to purchase a futures contract at a certain price on or before the expiration date of the option for a price called the premium; determined in open-outcry trading in pits on the trading floor.

callus–Plant tissue that is not differentiated into leaf, root, stem, or other specialized tissue.

capital–Goods used to produce other goods and services; the investment that the owner has put in the business.

capital expenditures budget–Budget that shows how money projected for capital expenditures is to be allocated among various divisions or activities within the agribusiness; lists projects (equipment, etc.) that management believes to be worthwhile, together with the essential cost of each.

capital-intensive–Having a high capital cost per unit of output; requiring greater expenditure in the form of capital than of labor.

capitalism–An economic system in which individuals own resources and have the right to use their time and resources however they choose.

cash-flow budget–Budget that summarizes the amount and timing of income that will flow in and out of the business during the year.

cash-flow statement–Summarizes all of the cash receipts and cash expenditures throughout the year.

cash market–Where actual commodities are bought and sold.

casting–An impression taken from an object with a liquid or plastic substance; used to make a mold.

centralization–The placing of a complete line of farm production supplies and services in a central location.

chain store–The operation of 11 or more stores under a single owner.

check-off program–A program in which the seller contributes a small amount for each unit sold to a money pool to be used to purchase advertising for the specific commodity.

chronological–Arranged in, or according to, the order of time.

classical economic theory–Classification of economics that contends that an economic system is self-sufficient in itself and that any outside interference by government does more harm than good.

cloning—A process through which genetically identical organisms are produced.

collateral—Something of value deposited with a lender as a pledge to secure repayment of a loan.

collective bargaining—Negotiation between an employer and a labor union, usually concerning wages, hours, and working conditions.

commercial food establishment—Category of food service establishment that is open daily to the general public. Includes fast-food outlets, restaurants, and lodging places.

commission—A fee the selling agent receives for work in selling (such as animals or futures contracts or options).

commodity—A transportable resource product with commercial value.

commodity (futures) exchanges—Central marketplaces with established rules and regulations where buyers and sellers meet to trade futures and options on futures contracts.

commodity money—A good the value of which serves as the value of money.

common market—Same as a customs union, except that it includes the free mobility of factors of production.

common stock—Partial ownership of a business; gives the owner the right to vote on business matters.

communism—Economic system in which the government has total control of economic matters and private individuals have none.

compatible—Agreeing with or capable of existing together in harmony.

complete budget—Also called *whole-business budgeting* or *operating budget;* the most comprehensive of all budgets; a financial plan for all segments of the business.

complete feed—Feed given to animals without any additional preparation; accounts for 80 percent of feed produced.

complete fertilizer—One that contains nitrogen, phosphoric acid, and potash.

compound interest—Interest computed on the accumulated unpaid interest as well as the original loan principal.

compression—The process of squeezing the fuel mixture in a cylinder of an internal combustion engine.

concentrates—Any feed high in energy (usually grain); stock feed low in fiber content and high in digestible nutrients.

consolidation—The process of bringing a complete line of farm production supplies and services together in a location.

consortium—An agreement, combination, or group (as of companies) formed to undertake an enterprise beyond the resources of any one member.

consumer price index (CPI)—General measure of retail prices for goods and services usually bought by urban wage earners and clerical workers. It includes the prices of about 400 items, such as food, clothing, housing, medical care, and transportation.

contract specifications—The specifics of an agreement, such as time and amount of delivery.

contraction—A noticeable drop in the level of business activity, which indicates a slowdown in the growth of the economy.

contractual interest rate—Interest rate not always equal to the actual rate charged on the loan.

convenience stores—Type of retail grocery store that was intended to replace the "Ma and Pa" grocery stores in older communities. These stores usually sell necessities, including bread, milk, and eggs.

cooperative—An organization in which the profits and losses are shared by all members.

Cooperative Extension Service—Organization that is responsible for programs in four major areas: agricultural and natural resources, home economics, community development, and 4-H youth development. The Cooperative Extension Service of the land grant colleges and universities was created under federal legislation (the Smith-Lever Act of 1914). The Cooperative Extension Service philosophy is to help people identify their own problems and opportunities, and then to provide practical research-oriented information that will help them solve their problems and take advantage of their opportunities.

corporate charter–A license to operate in a specified state granted when the articles of incorporation of a business are in agreement with state law.

corporation–A legal classification of an entity with authority to act and have liability separate from its owners.

cost–Amount paid or charged for something.

cost-effective–Economical in terms of tangible benefits produced by money spent.

cost function–Formula providing valuable information that an agribusiness manager needs to determine the level of output for profit maximization.

cost of sales–The expense and labor that are directly associated with the production or acquisition of products held for sale.

cost-price squeeze–Condition in which a production agriculturalist is pressured by increasing cost of resources on the one hand and lower prices received from marketing firms on the other.

cotton gin–A machine used to separate the cotton seed from the lint; invented by Eli Whitney in 1793.

crawler-type–A diesel-powered farm tractor developed by Caterpillar Tractor Company in 1931.

credit–An entry on the right-hand side of a ledger constituting an addition to a revenue, net worth, or liability account.

creditor–The person or company to whom an account payable is owed.

crop rotation–Growing annual plants in a different location in a systematic sequence. This helps control insects and diseases, improves the soil structure and productivity, and decreases erosion.

cultivate–To loosen the soil and remove weeds from among desirable plants; cultivation is the planting, tending, harvesting, and improving of plants.

currencies–Something (coins, treasury notes, bank notes, etc.) that is in circulation as a medium of exchange.

current ratio–Number obtained by dividing total current assets by total current liabilities.

custom grinding–Grinding and mixing of feed for a certain person and use that produces a specified feed content mixture. It is mixed according to the farmer's needs.

customs union–An economic and political organization between two or more countries abolishing trade restrictions among themselves and establishing a common uniform tariff for outsiders.

cylinder–The piston chamber in an engine.

Dairy Herd Improvement Association (DHIA)–Group that, among other things, periodically weighs the milk of cows to provide official milk records for dairy cattle.

debit–An entry on the left-hand side of a ledger constituting an addition to an expense or asset account or a deduction from a revenue, net worth, or liability account.

debt-equity ratio–The ratio of debt to equity allowed in the business. For example, if the ratio is 2, one can have twice as much debt as equity.

debt-to-asset ratio–Number equal to total liabilities divided by total assets.

default–Failure to pay financial debts.

deficits–An excess of expenditure over revenue.

deflation–A prolonged decline in the general price level.

defoliant–A type of chemical that, when applied to a plant, causes the foliage to drop off.

delinquencies–Debts on which payment is overdue.

demand–The amount of a product wanted at a specific time and price.

demand curve–A graphical illustration of the points within a demand relationship.

demand schedule–A table outlining the various price and quantity combinations that exist within a specific demand relationship.

deoxyribonucleic acid (DNA)–The molecule that carries the genetic information for most living systems. The DNA molecule consists of four bases (adenine, cytosine, guanine, and thymine) and a sugar-phosphate backbone, arranged in two connected, twisted strands that form a double helix. DNA contains the genetic material for every cell and is found in every cell.

depreciable–Losing value over time.

depreciable asset–An asset that is important to the farm and loses value over time.

depreciation–The systematic write-off of the value of a tangible asset over its estimated useful life.

depression–A severe drop in income and prices for an area or a nation.

desiccant–A type of pesticide that attempts to destroy pests by drying out or reducing water or moisture content.

diligence–Steady application; constancy in an effort to accomplish something.

discount (prime) rate–Rate set by the Federal Reserve System. These rates are influenced by the supply and demand factors for money.

discouraged workers–Workers who are not actively searching for jobs.

distance learning–Communication technology in which services are delivered to people via satellite or Internet from another location.

distillation–The process of separating the components of a mixture by differences in boiling point; the evaporation and subsequent condensation of a liquid, as when water is boiled in a retort and the steam is condensed in a cool receiver.

dividends–Money paid to a shareholder.

domestic–Of, relating to, or originating within a country.

domesticate–To bring wild animals under the control of humans over a long period of time for the purpose of providing useful products and services. This process involves careful handling, breeding, and care.

double-entry bookkeeping–Practice of writing every financial transaction in two places.

double taxation–System under which regular corporations must first pay taxes on the profits of the firm and then the income is taxed again as individual income when stockholders receive their dividends.

draft animals–Animal used for work stock; in some countries, these animals are still used for plowing and pulling heavy loads.

drought–A period of insufficient rainfall for normal plant growth which begins when soil moisture is so diminished that vegetation roots cannot absorb enough water to replace that lost by transpiration.

E-85–fuel blend that is 85 percent ethanol and 15 percent gasoline (formerly called gasohol).

ecologist–One who is concerned with the interrelationship of organisms and their environment.

economic depreciation–Calculated by following a declining-balance schedule for the remaining value of a machine or piece of equipment where the amount of depreciation remains constant for the life of the machine.

economic fluctuations–The ups and downs in economic activity.

economic indicators–Important data or statistics that measure economic activity and business cycles.

economic isolation–Description of countries that do not trade with other countries.

economic policy–A course of action that is intended to influence or control the behavior of the economy.

economic recovery–A rise in business activity.

economics–Many and varied definitions; most often defined as the science of allocating scarce resources (land, labor, capital, and management) among different and competing choices and utilizing those resources to best satisfy human wants and needs.

economists–People who specialize in the study and application of economics and concern themselves chiefly with description and analysis of the production, distribution, and consumption of goods and services.

economy of scale–The most efficient production size.

efficiency–Ratio of output per unit of input.

efficient economic system–System in which there is decisionmaking by consumers as to the who, what, and where of the economic system, and a pricing system that brings about an efficient allocation of resources; the maximization of consumer satisfaction; and the maximization of producers' long-run profits.

elasticity—The price responsiveness of an item.

embargoes—Also called sanctions; bans that prohibit shipments of commodities to certain other countries.

embryo splitting—A form of cloning that is accomplished by dividing a growing embryo into equal parts using a surgical procedure performed with the aid of a microscope.

embryo transfer—Procedure for placing living embryos obtained from a donor animal into the reproductive tract of a recipient female animal.

emulsifier—Substances that aid in the uniform dispersion of oil in water.

enterprise budget—An inventory of all the estimated income and expenses associated with a specific enterprise.

enterprises—Projects on a farm, such as the production of any crop or livestock.

entrepreneur—Person who accepts all the risks pertaining to forming and operating a small business.

entrepreneurship—The art or process of organizing, managing, and perhaps even owning a business. Also entails performing all business functions associated with a product or service and includes social responsibility and legal requirements.

environmentalist—Any person who advocates or works to protect the air, water, and other natural resources from pollution.

equal product curve—Graphic illustration of the concept whereby different resources can be substituted for one another in the production of a given product. There are numerous combinations of land, labor, and capital that will produce the same amount of a product.

equilibrium price—The price in a market at which quantity supplied and quantity demanded are equal. See also *point of equilibrium.*

equity—Money value of a property or of an interest in a property in excess of claims or liens against it.

equity capital—Venture capital; capital invested or available for investment in the ownership element of a new or fresh enterprise.

eradicate—To do away with completely.

erroneous—Mistaken, wrong; containing or characterized by error.

ethanol—The alcohol product of fermentation that is used in alcoholic beverages and for industrial purposes.

European Union (EU)—Customs market comprising many European countries; top exporter of agricultural goods into the United States.

exchange rate—The ratio at which the principal unit of one currency may be traded for another currency.

exports—Shipments of commodities or products, such as agricultural commodities, to foreign countries.

externalities—Factors beyond the control of an individual, group, or business.

extrapolate—To infer from known data.

fallow—Cropland left idle to restore productivity through accumulation of moisture.

farm asset—Something that is of value or a complement to the farm, whether real estate or non-real estate.

farm commodity—Agricultural product, such as wheat, corn, or cattle, that is raised on the farm.

farm contracting—Services performed by individuals who use (and usually own) specialized equipment; services include clearing land, installing drainage systems, and aerial spraying.

fascism—Economic system in which productive property, though owned by individuals, is used to produce goods that reflect government or state preferences.

Federal Land Banks—Several banks that are a part of the Farm Credit Administration of the United States; through local national farm loan associations, the Federal Land Banks make long-term (5–40 years), amortized, first-mortgage, farm real estate loans.

Federal Reserve System—A central banking system in the United States, created by the Federal

Reserve Act of 1913 and designed to assist the nation in attaining its economic and financial goals.

fermentation–The processing of food by means of yeasts, molds, or bacteria.

fiat money–Money that has value because a government order (fiat) has established it as acceptable for payment of debts.

fidelity shares–Low-cost shares of money invested.

financial assets–Funds or capital that a person, corporation, or other party has in its possession.

financial efficiency ratios–Financial figures that illustrate the percent of gross farm revenue that went to pay interest, operating expenses, and depreciation, and how much was left for net farm income.

financial institution–Any business that lends money.

financial resources–Any place where a consumer or business person can acquire money.

financial security–Having no financial strain or monetary problems; being able to pay monthly bills without stress.

fiscal policy–Economic strategy designed to influence economic activity; involves using government spending and taxation to influence the economy.

fixed expenses (costs)–Items that can be used over and over for a long period of time and results in the same cost (expense) each year; bills that are due every month, such as house payments.

flaming–The process of setting a field on fire to destroy all organic matter.

flex time–A system that allows employees to choose their own times for starting and finishing work within a broad range of available hours.

foreclosure–A procedure whereby a lender takes steps to obtain ownership of property given as security for a loan (collateral).

forge shop–Place where heavy hammers pound red-hot steel billets into shape.

formulation–The product of a mixture, such as feed.

forward contracts–A cash contract in which a seller agrees to deliver a specific cash commodity to a buyer sometime in the future. These are privately negotiated and are not standardized.

foundry–Place where molten iron is poured into molds to make castings.

franchise–A contract in which a franchisor sells to another business the right to use the franchisor's name and sell its products.

franchisee–Person or firm that purchases a franchise.

franchisor–Person or firm that grants a franchise.

fraternal–Of, relating to, or involving brothers.

free enterprise–Economic system that allows individuals to organize and conduct business with a minimum of government control; individuals privately own what they produce.

free market economy–Exists where consumers provide the answers to the questions of what to produce, how much to produce, when to produce, who should produce, and for whom goods should be produced.

free markets–The movement of goods and services among nations without political or economic obstruction.

free trade–Trade between businesses in different countries without restrictions from any government.

free trade area–Group of nations that abolishes trade restrictions among themselves without also imposing common tariffs on other nations.

full-line companies–Produce and sell tractors as well as a wide variety of other equipment.

fumigant–A type of pesticide that produces gas, vapor, fume, or smoke intended to destroy insects and other pests.

fungicide–A type of pesticide used to control fungi and molds that attack plants and animals.

futures–Legally binding agreements made on the trading floor of a futures exchange; trading by means of buying or selling agricultural commodities that are to be produced or are in the process of being produced.

futures contract—A legally binding agreement, made on the trading floor of a futures exchange, to buy or sell something in the future. A futures contract specifies the quantity, quality, time and form of delivery, and a negotiated price.

futures market—Where traders buy and sell futures contracts.

futures options—Type of insurance to protect against price decline.

gasohol—Registered trade name for a blend of 90 percent unleaded gasoline with 10 percent fermentation ethanol.

gender selection—Managing the reproductive process to produce animal offspring of the desired sex.

gene mapping—The process of finding and recording the locations of genes on a chromosome.

gene splicing—The process of removing a gene from its location on a chromosome and replacing it with another gene.

general partnership—Association of two or more people who, as owners, manage a business together.

genetic engineering—The practice of modifying the heredity of an organism by inserting new genes from other organisms into the recipient organism's chromosome structure.

germ plasm—The heredity material of the germ cells. It is the chromosomes in germ cells that transmit heredity characteristics to offspring.

glycerol—A substance obtained from fats used as a solvent and plasticizer; used as a semen extender.

goal-setting theory—Management theory holding that the setting of specific, attainable goals creates high levels of motivation and performance if the goals are accepted by managers and accompanied by feedback.

goods—Things that have economic utility or satisfy an economic want.

grading—The arranging of something, such as food, according to established standards and criteria for quality.

grants—An award given for something deserved or merited after careful weighing of pertinent factors; usually public funds given to a school or individual for a specific project.

gross domestic product (GDP)—Total value of goods and services produced in a country in a given year.

gross pay—Total amount earned before deductions are taken from a paycheck.

hatcheries—Companies where eggs are incubated; usually a commercial establishment where newly hatched chicks or fish are sold.

Hawthorne effect—Refers to people's tendency to behave differently when they know they are being studied.

hedgers—Those who use the futures market to establish either a buying or selling price for a commodity they own or are expecting to own; those who manage price risk. Hedgers own or will own the actual cash commodity.

hedging—Buying or selling futures contracts to protect one's profit margin against a possible price change of a cash commodity that he or she plans to buy or sell.

herbicide—A type of pesticide used to control undesirable plants (weeds).

hog cholera—An acute, contagious, viral disease of swine characterized by sudden onset, fever, high morbidity, and high mortality.

holding action—A strategy used to try to improve farm prices. Prices usually increase for a short time after each action, but few long-term gains have been realized thereby.

homogenous—Of the same kind or a similar kind or nature.

hormone—A substance produced in the body and carried by body fluids to tissues where it causes specific body functions to occur.

host—A living plant or animal in which a parasite lives.

human resource management—Management of people and employees; includes finding the right people, motivating them by providing incentives in an ideal environment, and evaluating them.

humanitarian—Philanthropic; having human attributes or qualities.

humus–Organic matter in the soil that has reached an advanced stage of decomposition and has become colloidal in nature. It is usually characterized by a dark color, a considerable nitrogen content, and chemical properties such as high cation-exchange capacity.

husbandry–The care of a household or scientific control and management of a branch of farming.

hybrid–The result from a cross between parents that are genetically unlike; hybrid plant seed was developed for better-quality, higher-producing crops.

hydraulic lifts–Machines that provide a means of raising or lifting by pressure transmitted when a quantity of liquid is forced through a small orifice, such as a tube or hose.

hydroponics–A plant production system in which plant nutrients are provided in a water solution, and plants are grown without soil.

hydrostatic transmission–A transmission in which fluids tend to be at rest.

hygiene factors–Have to do mostly with the job environment, and may cause dissatisfaction if they are missing, but would not necessarily increase motivation if they were improved or increased.

hypothesis–A tentative assumption made so that its logical consequences can be identified or tested.

impairments–Real or perceived handicaps or disabilities.

implant–A device placed under the skin or in the female reproductive tract to slowly release hormones into the bloodstream of the animal.

implements–Tools that aid a person by making work and effort more productive and effective.

implicit GDP price deflator–Price index that removes the effect of inflation from GDP so that the overall economy in one year can be compared to another year.

import quotas–Impose a limit on the amount of a good that may be imported in a given period.

imports–Shipments of commodities and products, such as agricultural commodities, from foreign countries for sale, use, or donation.

incidental expenses–Any expenses that occur rarely but are business related, including unexpected emergencies due to accidents or weather damage.

income stabilization–Steps taken by the Farm Service Agency to stabilize the nation's agricultural economy through price support loans and purchases.

income statement–Reports revenues, costs, expenses, and profits (or losses) for a specific period of time and shows the results of business operations during that period.

incomplete fertilizer–Fertilizer containing only two of the three necessary materials (nitrogen, phosphoric acid, and potash).

indigo–A leguminous plant that produces a blue dye used for various purposes.

individual retirement account (IRA)–Account in which the government allows the owner to invest $2,000 yearly, with taxes deferred until retirement; contributions are subject to income restrictions.

Industrial Revolution–Time when advances in technology allowed many firms to become more efficient and productive, and enhanced communications and transportation allowed businesses to expand sales into larger areas, thereby increasing sales and the need for employees.

industrialization–The process of becoming mechanized or industrial.

inflation–An increase in the volume of money and credit relative to available goods and services, resulting in a continued rise in the general price level.

inherent–Built in or inbred; innate.

initial margin–The initial amount a market participant must deposit into his or her account when he or she places an order.

initiative–Individual willingness to take risks and make a profit.

initiatives–Programs and actions instituted to implement policy.

input–Supplies and services farmers use to produce crops, livestock, and other items; any resource used in production.

insecticide–Type of pesticide used to control various insects.

inseminate—To place semen in the vagina of a female animal.

institutional advertising—Designed to create a favorable image of the firm or institution offering the product or service.

integrated pest management (IPM)—The use of natural insect enemies and limited chemical applications to control harmful insects while providing protection for useful insects.

interdisciplinary—Involving two or more academic, scientific, or artistic disciplines.

interest—The charge for or cost of borrowed money.

interest rates—The price charged by lending institutions for use of money.

intrinsic value—The amount by which an option is in the money.

inventory—A physical count of all assets in a business, with their estimated worth.

investment portfolio—A record of all investments; the goal is diversification to get greater returns.

investors—One who provides money to another, usually a business, for later income or profit.

irrigate—To furnish water to the soil for plant growth in place of, or in addition to, natural precipitation, by surface flooding or sprinkling and subirrigation methods, using surface water or water from underground sources.

isocost line—Illustrates the different combinations of two inputs that can be purchased with a specific amount of money; also shows the amount of one input that would cost the same if another unit of the other input was purchased.

isoquant curve—A graphical illustration of the principle of resource substitution. It shows the set of all pairs of inputs (X_1 and X_2) that can be used to produce a specific output (Y). Additionally, it indicates the amount of one input that can be replaced by another input, while sustaining the same level of output.

job description—Lists the duties and responsibilities of each job as well as the educational level, skills, and special characteristics required for that post.

job enlargement—Strategy in which several tasks are combined into one job to make that job more challenging, interesting, and motivating.

job rotation—System in which new employees move from job to job until they learn the various tasks.

journals—Books where accounting data are first entered.

Keynesian—Classification of economics contending that economic systems are not always self-sufficient and sometimes need outside help.

kiln—An oven, furnace, or large heated room for the curing of lumber, tile, or bricks.

laissez-faire—Noninterference by government; leaving coordination of an individual's wants to be controlled by the market.

lathes—Machines in which work is rotated around a horizontal axis and shaped by a fixed tool.

law of comparative advantage—Principle stating that producers tend to produce the product or products for which they have the highest economic advantage or the least disadvantage.

law of diminishing returns—States that as you apply (input) additional units of a variable resource to the production of a particular product, there will come a point when the next unit of input will not increase total output as much as the previous one did.

law of increasing cost—States that to produce equal extra amounts of one product or service, an increasing amount of another product or service must be given up or foregone.

least-cost combination of inputs—Provides the decision rule for producers to decide what combination of inputs is most efficient in a particular production process.

ledger—An accounting system that includes a separate record for each item. It consists of three parts: a title (name of the particular asset, liability, or owner's equity), the debit (left) side, and the credit (right) side.

legal classification. See *legal structure.*

legal entity—Status under law whereby a corporation is separate from the people who own it or work for it.

legal structure—The way a business is set up for legal or tax purposes, such as a corporation, cooperative, or sole proprietorship.

liabilities—Amounts that are owed; debt. Examples include accounts payable, notes payable, and mortgages.

liaison—A close bond or connection; a go-between.

lien—The lender's right to take possession of the asset in collateral if the borrower fails to repay the loan.

limited liability—Status in which limited partners are not responsible for the business's debts beyond the amount of their investment; their personal assets are not at risk.

limited partner—Person who risks whatever investment he or she makes in the firm, but has limited liability and cannot legally help manage the company.

limited partnership—Special type of partnership in which some partners are not completely liable for other partners' debts. This partnership is for investment purposes only.

liquid income—Money that is available quickly, such as cash.

liquidity—The ability to pay debts as they come due.

locked into—Committed; condition in which someone cannot get out of what he or she is doing because debts exceed assets.

long hedge—Buying of a futures contract.

long-line companies—Firms that produce and sell a wide variety of general farm equipment, including self-propelled combines, but no tractors.

long-term credit—The extension of credit by the decision of the lender. One source includes Farm Credit Services.

machine shop—Place containing lathes, milling machines, cutters, planes, broaches, and automatically controlled machine tools.

macroeconomics—Concerned with the study of the economy on a large scale, or nationally.

maintenance margin—The set, minimum margin in a customer's account.

management—Accomplishing tasks through people.

managers—Individuals who make sure that things get done.

margin—In economics, an additional or incremental unit of something; in futures trading, the amount market participants must deposit into or have in their accounts when they place orders.

margin call—A request for additional money.

marginal analysis—Examination of the consequences of adding to or subtracting from the current state of affairs.

marginal benefits—The value of the additional goods or services that could be produced.

marginal cost (MC)—The change in total cost associated with each additional unit of output.

marginal factor cost (MFC)—The additional cost of adding one more unit of a variable input to the production process.

marginal product (MP)—The change in total product associated with each additional unit of input.

marginal rate of substitution (MRS)—An equation to assist in making decisions regarding resource substitution; the amount of one input that is replaced by an additional unit of another input.

marginal revenue–marginal cost method—Based on the premise that for each unit sold, marginal profit equals marginal revenue minus marginal cost.

marginal value product (MVP)—Demonstrates the change in total returns received through adding one more unit of input.

margining system—A system initiated to eliminate problems of buyers and sellers not fulfilling their contracts and obligations. It requires traders to deposit funds with the exchange or an exchange representative to guarantee contract performance.

market—The interaction between potential buyers and potential sellers of a good or service.

market analysis—Collecting information to determine if a product will sell.

market basket—About 400 goods, which are sold in about 21,000 outlets, including food, housing, transportation, clothing, entertainment, medical care, and personal care; used to determine the consumer price index.

marketing—Process that starts with analyzing the market to see what is needed before beginning production, not just selling of the product.

marketing cooperatives—Organizations that assist production agriculturalists in marketing agricultural products by finding buyers who will pay the highest price.

mechanical power—Power originating from mechanized sources or objects, such as the steam engine.

Medicare—The U.S. national health insurance program that was established under the Federal Insurance Contributions Act.

merchandising—Sales promotion as a comprehensive function, including market research, development of new products, coordination of manufacturing and marketing, and effective advertising and selling.

microeconomics—Concerned with the study of the economy on a small scale.

milling machines—A machine tool on which work (usually of metal) secured to a carriage is shaped by rotating cutters.

minimum price fluctuation—The smallest price unit at which a futures contract trades. See also *tick*.

modified accelerated cost recovery system (MACRS)—Method under which assets are depreciated more at the beginning of their lives; the years of an asset's useful life are mandated by the 1986 Tax Reform Act, so MACRS applies to any property placed into service after 1987.

monetary policy—Management of the money supply and interest rates; government economic policy, designed to influence economic activity, that involves controlling the supply of money and credit to influence the economy.

motivators—Factors that provide satisfaction and stimulate people to work.

mutual funds—Investments made solely by a company on behalf of others. By law, a mutual fund must invest in at least 30 different stocks or other investments; most invest in many more.

needs—Things that are really crucial to daily living.

net farm income—Overall income earned from the farm manager's capital and labor; calculated by adding capital gains to or subtracting losses from the net farm income from operations.

net farm income from operations—Figure arrived at by subtracting gross farm expenses from gross farm revenue (net income generated from the production and marketing activities of the operation).

net pay—Amount taken home in the paycheck after deductions.

net profit—Revenue minus the cost of sales; also called profit margin.

net worth—The difference between total assets and total liabilities.

net worth statement—Also called *balance sheet*; a listing of property owned (assets) and debts owed (liabilities).

niche—A place or position in the community or industry suitable for a person or thing.

noncommercial food establishment—Category of food service establishment in which meals and snacks are provided as a supportive service rather than as the primary service. These include educational facilities, extended care facilities, vending areas, plants, and office buildings.

nondepreciable asset—Thing that does not lose value (depreciate) or is not kept for more than one year, such as land, breeding livestock, and anything raised on the farm.

nonexempt employees—Employees who must be paid overtime if they work more than 40 hours in a standard work week.

nonrecourse loans—Loans designed to prevent farmers from having to sell their commodities immediately upon harvest, when prices are usually at their lowest seasonal level.

nonruminant—An animal, such as a pig, without a functional rumen.

normative economics—Subjective statements about economic issues based on opinion only, often without a basis in fact or theory; value-based, emotional statements that focus on "what ought to be."

note payable—A formal written promise to pay a certain amount of money, plus interest, at a definite future time.

offset—Most common method of closing out an option position; done by purchasing a put or call identical to the put or call originally sold or by selling a put or call identical to the one originally bought.

open account—Trade credit account through which managers of agribusinesses charge purchases with a merchant. The balance on these purchases is normally due in 30 days.

open outcry—Method of public auction for making verbal bids and offers in the trading pits or rings of futures exchanges.

operating budget—Summarizes the expected sales or production activities and related cost for the year; an estimate of sales and income plus the fixed and variable expenses that the agribusiness should experience during the year.

operating costs—Expenses such as repair and maintenance, fuel and lubrication, labor costs, and any other costs associated with the operation of machinery on a farm.

operating expenses—Any regular expenses associated with the operation of the business, including rent, salaries, utilities, insurance, and depreciation.

opportunity cost—The value of a service or product that must be given up or foregone to obtain another good or service.

organic fertilizer—Usually refers only to natural, organic, proteinaceous materials of plant and animal origin; excludes synthetic, organic, nonproteinaceous materials such as urea.

output—A marketable product of a farming operation, such as cash crops, livestock, and so forth; the result of the production process.

outstanding loans—Loans that have not yet been paid.

owner equity—Money the owner invests in the agribusiness.

ownership costs—Depreciation, interest, taxes, and outlays for housing/storage and maintenance facilities.

parent company—The franchisor; firm that prepackages all the business planning, management training, and assistance with advertising, selling, and day-to-day operations.

partial budget—Measures the change in net farm income that can be anticipated from a projected change in a farming operation.

partnership—An organization whereby two or more people legally agree to become co-owners of a business.

part-time (avocational) enterprises—Supplemental pursuits that provide extra income in addition to a full-time occupation.

pathogens—Disease-causing organisms.

Payment in Kind (PIK)—A government program established to help farmers by giving them products as payment for reducing acres planted of certain crops, for example, giving soybeans instead of cash.

peak—Highest level of economic activity in a business cycle; indicates prosperity and means the economy is expanding rapidly.

perfect complements—Exist when inputs cannot be substituted for each other; adding more of one input (X_1) will not change or replace the amount of the other input (X_2) used.

perfect substitutes—Exist when one input (X_1) always replaces a consistent amount of another input (X_2); one input (X_1) always replaces a consistent amount of another input (X_2).

performance appraisals—Evaluations of an employee's performance and achievements.

performance bond margin—A financial guarantee required of both buyers and sellers to ensure that they fulfill the obligation of a futures contract.

perishable—Liable to spoil or decay, such as fruit, vegetables, butter, or eggs.

perseverance—Steady persistence in a course of action.

pessary—A sponge or other material to which hormones have been added and that is implanted under the skin or inserted in the female reproductive tract.

pesticide—Chemical used to control weeds, insects, and diseases that affect crops, livestock, or people. Pesticides include herbicides, insecticides, fungicides, nematocides, and rodenticides.

pheromone—A chemical substance emitted by animals and insects, used to attract mates through the sense of smell.

physiologically—Characteristic of, or appropriate to, an organism's healthy or normal functioning.

physiologist—Person who is concerned with the functions and activities of life or of living matter.

plastic prosperity disease—The impression that no money was spent when a credit card was used, because no cash changed hands.

pneumatic—Adapted for holding or inflated with compressed air.

point of equilibrium—Graphical point at which, given a price and a quantity, the amount supplied equals the amount demanded; the point at which the demand curve intersects the supply curve for the same product. See also *equilibrium price.*

policy—Set of specific rationales that will govern action, and will guide present and future decision-making in achieving agreed-upon goals and objectives in a given environment.

policy analysis—Study from which forecasts are made that serve as a basis for economic decision-making by production agriculturalists, consumers, and others.

policy statements—Pronouncements or publications dealing with issues that affect modern agriculture both directly and indirectly, including such areas as transportation, national welfare, conservation, foreign affairs, taxation, education, health and safety, and numerous other issues.

political power—Power relating to governmental procedures and administration of government policies.

porcine somatotropin (PST)—Type of growth-enhancing hormone used on pigs.

positive economics—Statements based on economic theory rather than emotion or social philosophy; objective statements dealing with matters of fact and questions about how things actually are.

posting—The process of transferring debits and credits from the general journal to the ledger account.

power take-off (PTO)—A supplementary mechanism allowing the operator to control mounted and drawn equipment with the tractor's engine.

precision farming—A crop management system that adjusts applications of fertilizers and other crop inputs on the basis of production capacity differences within a field.

preferred stock—Gives a person the opportunity to invest in a business, and, it is hoped, to receive a reasonable return on investment.

premium—The cost of providing insurance for a stored crop in the futures market; also, the cost of a futures option.

premix—Feed containing only vitamins and minerals. Premixes are added to feed grains and a protein source, such as cottonseed or soybean meal, to provide a complete ration; used at the rate of less than 100 pounds per ton.

prepackages—Business assistance provided by a franchisor or parent company; generally includes management training and assistance with advertising, selling, and day-to-day operations.

prerequisite—Something that is necessary to achieve an end or to fulfill a function.

price discovery—The generation of information about future cash market prices through the futures markets.

price support—The price for a unit of a farm commodity that the government will support. These are determined by law and set by the Secretary of Agriculture.

pricing efficiency—How close an economic system comes to achieving the point when prices reflect the full value of resources and resources are allocated to their highest and best use.

principal—A capital placed at interest, due as a debt, or used as a fund.

principle of diminishing returns–input basis—Concerned with varying the amount of one input while keeping all other inputs constant within the production process; takes into consideration the relationship among product curves, and assists producers in determining the most effective production level for their operations.

principle of resource substitution—Addresses two or more variable inputs within the production equation; determines the combination of two or more variable inputs (that can be substituted for each other in varying amounts) that will produce a specific amount of a given product with the least cost of production.

private agriservices—Any services not included in federal, state, or local government programs; for example, the Farm Credit System.

producer price index—An index that shows the cost of resources needed to produce manufactured goods during the previous month.

product advertising—A way of advertising that focuses on the product itself, such as its usefulness, durability, price, value, and customer need for the item.

production agriculturalists—Farmers; those who actually produce food and fiber. Their output is taken by agribusiness companies that process, market, and distribute agricultural commodities and products.

production controls—Government policies that control production of agricultural commodities.

production efficiency—Receiving optimum output from a reasonable input.

production function—The relationship between inputs and outputs.

production possibilities—A type of economic analysis in which the possibilities for the producer are determined by the type and amount of resources available in combination with a specific level of technology.

profit—A valuable return.

profitability—The difference between an operation's income and its expenses.

profitable—Yielding advantageous returns or results.

prominent—Standing out or readily noticeable.

prospectus—Document explaining how mutual funds work and where the money is invested.

prosperity—The condition of being successful.

protectionism—The idea of establishing government economic protection for domestic producers through restrictions on foreign competitors.

public agriservices—Groups at the federal, state, and local levels that provide special services to production agriculturalists and others. The major areas of emphasis include research, education, communication, and regulation.

public policy—The steps the government will follow to achieve specific objectives or to solve pressing problems that affect the general public.

put—An option that gives the option buyer the right (without obligation) to sell a futures contract at a certain price on or before the expiration date of the option.

quantity supplied—The amount of a good or service a supplier is willing to provide at a particular price.

rapport—A harmonious or sympathetic relationship.

raw material—Crude or processed material that can be converted by manufacture, processing, or combination into a new and useful product.

reaper—Used to cut or harvest, as a crop of grain; Cyrus McCormick invented the mechanical reaper to reduce hand labor in the harvest of grain.

receivables—A type of asset, which is owned by a business and is expected to benefit future operations.

recession—A period of reduced economic activity.

reconcile—To compare, as in balancing a checkbook, to see if whether one record matches another.

reconciling—Checking a financial account against another record or account to ensure accuracy.

regulatory medicine—Government law and policy applicable to drugs and medicines.

repayment capacity—The adequacy of cash generated by the farm to pay the principal and interest on loans as they come due.

resource specialization—The economic theory underlying production of a single agricultural commodity or resource rather than diversifying into two or more agricultural commodities.

resources—The means available for production; include land, labor, capital, fertilizer, chemicals, machinery, transportation, marketing, and the like.

restriction enzyme—A specific enzyme capable of removing a particular gene from its location on a chromosome.

résumé—Concise listing that provides detailed information about work history, personal traits, and skills. The main purpose of a résumé is to highlight a person's qualifications and abilities.

revenue—The value of what is received from goods sold, services rendered, and other financial sources; money coming into the agribusiness; income.

risk management—Making investments to get the highest returns with the least possible risk.

risk transfer—A concept used in futures trading to transfer the risk of the crop (hedger) to a speculator.

roughages—A feed high in fiber and low in digestible nutrients, such as straw, hay, haylage, and silage.

ruminant—Any of a class of animals, including sheep, goats, and cows, that have multiple stomachs.

rural sociology—Area of study dealing with the social relationships of country life and its people.

salvage value—The value of a piece of equipment once it has been fully depreciated.

sanctions—Embargoes that prohibit shipments of commodities to certain other countries.

scarce—Deficient in quantity or number in relation to demand.

scarcity—Economic term describing a situation in which there are not enough resources available to satisfy people's needs or wants.

scenario—A synopsis of a possible course of events or actions.

secured—Relieved from duty; made certain or guaranteed.

securities markets—Investments backed by some tangible asset that is pledged to the originator (inceptor) if the principal is not paid back (e.g., secured bonds).

selective breeding—The breeding of selected plants or animals chosen because of certain desirable qualities or fitness, as contrasted with random or chance breeding.

self-actualization—A person's need to accomplish personal goals and develop to his or her full potential.

self-esteem—A person's (positive) perception of his or her own worth.

selling—Taking the farm product to the market and getting whatever the price is for that day.

semen—A fluid substance produced by the male reproductive system containing spermatozoa suspended in secretions of the accessory glands.

services—Nontangible products that cannot be held, stored, or touched. They can include benefits, activities, or satisfactions that are offered for sale.

short hedge—Selling of a futures contract.

short-line companies—Produce highly specialized equipment such as planters and cultivators, forage equipment, and milking equipment.

sickle—A sharp, curved metal blade fitted with a short handle; used for cutting weeds or grasses; one of the earliest hand implements used for harvesting small grain.

simple interest—Interest that applies to loans with a single payment. Because no borrowing fees or down payment are required, the APR is the same as the rate charged.

single-entry bookkeeping system—The simplest bookkeeping system, in which entries are made only in a checkbook or the checkbook and a journal.

single (sole) proprietorship—An organization that is owned, and usually managed, by one person.

small business—A business that is independently operated, is not dominant in its field, and meets certain size standards in terms of number of employees and annual receipts.

Small Business Administration (SBA)—A division of the federal government that oversees small businesses.

social maturity—Ability to get along with and work with many types of people even if a particular person is disliked.

socialism—Economic system in which there is public ownership of all productive resources. The government, or the state, directs all decisionmaking.

sodium citrate—A crystalline salt used chiefly as a buffering agent, emulsifier, or alkalizer; used as a semen extender.

Soil Conservation Service (SCS)—A bureau of the USDA established by the Soil Conservation Act of 1935. Its basic purpose is to aid in bringing about physical adjustments in land use and treatment that will conserve natural resources, establish a permanent and balanced agriculture, and reduce the hazards of floods and sedimentation.

solarization—A cultural method that inhibits the accumulation of starch in leaves in the presence of intense illumination.

solvency—Condition in which there is more than enough cash to cover all liabilities if all assets were converted to cash.

specialized food stores—Establishments that usually sell a single food category. Examples include retail bakeries; meat and fish markets; candy and nut stores; natural health and food stores; coffee, tea, and spice stores; and ice cream stores.

speculative investments—Assumption of unusual business risk in hopes of obtaining commensurate gain.

speculators—Those who are willing to accept price risk. They buy and sell futures contracts and hope to make a profit by correctly predicting future price movements.

spot price—Another name for the prevailing cash price.

standard of living—How well people are living in a particular nation; depends on the amount of goods and services available to the people of that nation.

standardization—The establishment of set criteria (standards) for matters such as quality, size, weight, and/or color.

startup expenses—Costs incurred before the business begins operations, such as attorney's fees, incorporation expenses, and cost for site development.

statement of cash flows—Reports cash receipts and disbursements, related to the major activities of a business: operations, investments, and financing.

statutes—Laws; bills that have been formally enacted by a legislature and approved by the executive branch of government.

stock—Shares in the ownership of a corporation.

stock price—Cost of a share of ownership in the corporation that issued the stock.

stockholders—The owners of corporate stock, who pay a set price for their shares; each stockholder has one vote in the major decisions made by the corporation for each share of stock purchased.

straight fertilizer—Fertilizers containing only one of the three primary materials (nitrogen, phosphoric acid, and potash).

Subchapter C—Regular corporation; company that sells stocks to investors.

Subchapter S—Small business or family corporation.

Subchapter T—Type of corporation for cooperatives.

subsectors—Parts of the economy: household sector (all consumers), business sector (all firms), and government sector (all government agencies).

subsidies—Government grants of money to aid or encourage a private enterprise that serves to benefit the public.

superette—A type of retail grocery store; usually a smaller grocery store that offers a variety of food and nonfood grocery products, and usually located in areas not adequately served by supermarkets.

supermarket—Type of retail grocery store that offers a full line of 10,000 to 15,000 food items, plus many nonfood items.

superstores—Offer a greater variety of products than conventional supermarkets; constitute 25 percent of all supermarkets and result in 33 percent of supermarket sales.

supplements—Formula feeds requiring the addition of grain to make a complete ration.

supply—Amount of a product available at a specific time and price.

surveying—The practice of measuring a tract of land for size, shape, or the position of boundaries.

synthesis—The combination of parts or ideas into a whole.

synthetic chemicals—Chemicals produced artificially; used to control weeds, insects, and diseases.

target customers—The intended client or consumer; for example, the full-time and part-time production agriculturalists whom farmer supply chain stores desire to serve.

tariffs—Government taxes imposed on goods when those goods cross national boundaries.

tax shelter account (TSA 403b)—An investment allowed by the government for teachers and hospital employees that allows annual investment of 20 percent of income and on which taxes are deferred until retirement.

taxable income—The difference between revenue and expenses; gain or loss by an agribusiness.

tax-deferred—Postponement of taxes until retirement.

tax-deferred savings—Savings on which taxes are not due until the savings owner retires.

teleconferencing—Communication technology in which several people to talk to each other on the same telephone line, as in a conference.

term life insurance—Insurance for which a yearly premium is paid, but that never accumulates a cash value.

terminal—A station for delivery or receipt of produce; a depot or warehouse needed to store items for later distribution in smaller quantities to retail food stores or to restaurants.

tertiary—Third-rank; occurring third in a series of steps or operations.

therapeutic activity—Something done as a stress reliever.

threshing machine—Separates the grain from the waste of the plant.

tick—The smallest price unit at which a futures contract trades. See also *minimum price fluctuation.*

time value—Also called *extrinsic value*; reflects the amount of money that buyers are willing to pay in hopes that an option will be worth exercising at or before its expiration date.

tissue culture—The development of roots, stems, and leaves from callus tissue using a solution containing nutrients and hormones.

torque amplification—A feature that expands or increases the force that tends to produce rotation.

total cost (TC)—Sum of the fixed cost and the variable cost.

total fixed costs (TFC)—Costs that do not change as level of production changes.

total product (TP)—Represents the total amount of a product (corn, cotton, soybeans, etc.) produced in a given period of time with a given amount or group of resources.

total revenue–total cost method—Based on the fact that total revenue minus total cost equals profit: $TR - TC = Profit$. Data plots effectively illustrate the relationships between revenue, cost, and profit.

total variable costs (TVC)—Costs that change as production levels or amounts of use change.

trade barriers—Ways to protect or restrict trade.

trading pit—The heart of the futures exchange, where buyers and sellers meet each day to conduct business.

transformation—The act of changing one configuration or expression into another.

transgenic—An animal that has developed from a genetically modified cell to which a gene from another living organism has been transferred.

trial balance—The proof of the equality of debit and credit balances.

trichinae-safe pork—Pork that is safeguarded from trichinosis by means of irradiation and testing.

tricycle-type—A tractor introduced by International Harvester in 1924, which was very popular for cultivating as well as plowing.

trough—The lowest level of business activity in a particular business cycle.

turbocharger—A centrifugal blower driven by exhaust gas turbines and used to supercharge an engine.

turpentine—A substance used as a fuel in early internal combustion engines.

undercapitalization—Not enough capital (assets) to get a company started; may lead to business failure.

universal life insurance—Insurance for which a yearly premium is paid and that accumulates a cash value in addition to providing a death benefit.

unlimited liability—Legal status in which creditors can make claims on the owner's personal assets, as well as the business assets, for payment of business debts.

utilization of resources—Using whatever resources you already have to limit expenses, or purchases made for greater economic efficiency.

vaccines—A substance that contains live, modified, or dead organisms or their products that is injected into an animal in an attempt to protect the host from a disease caused by that particular organism; discovered by Edward Jenner.

value-added—Condition of having had an operation performed that increases worth and price, such as being packaged or processed.

value adding—Increasing the value of a product through processing, packaging, or other improvement after the product leaves the site of production.

value judgments—Used in normative economics to assess the performance of the economy and economic policies.

variable expenses—Expenses that change regularly, such as recreation and entertainment, in which one has control over the amount spent.

variables—Quantities that may assume any one of a set of values.

vertical integration—Operation in which several steps in production, marketing, and processing of animals are combined or joined together.

vertically integrated—Status in which a single firm controls two or more stages in the chain of production, processing, and distribution.

viable—Capable of working, functioning, or developing adequately.

wants—Things that are not crucial to daily living.

warehouse/limited-assortment supermarket—Offers larger sizes of fewer items at lower prices than the typical supermarket.

whole-farm planning—Also called *comprehensive farm planning*; a process used by the farm family to balance important life aspects, such as quality of life, with important business aspects, such as the farm's resources, the need for production and profitability, and long-term stewardship.

wholegoods—Machinery or equipment in the completely assembled state.

wholesalers—Intermediary or middle person between producer and retailer; sells in large quantities to volume buyers and retail stores.

working capital—The difference between current assets and current liabilities. This figure indicates the amount of cash available for meeting daily operating costs.

write-off—To depreciate or reduce the estimated book value of.

xenotransplantation—Transplantation of organs and tissues between species.

yardage fee—The fee a terminal market charges the seller for caring for animals until they are sold.

Index